A World of WEATHER

Fundamentals of Meteorology

Sixth Edition
REVISED PRINTING

Jon M. Nese | Lee M. Grenci

Kendall Hunt
publishing company

CAUTION: In-class activities and pages contained in this textbook are protected by copyright law. Photocopying of these pages is in violation of copyright law.

Kendall Hunt
p u b l i s h i n g   c o m p a n y

www.kendallhunt.com
Send all inquiries to:
4050 Westmark Drive
Dubuque, IA 52004-1840

FOR

Kathie and Mack

—Lee M. Grenci

FOR

Dad, who dedicated his life to education

—Jon M. Nese

■ BRIEF CONTENTS

◼ CONTENTS

CHAPTER 14 MID-LATITUDE III: SPAWNING SEVERE WEATHER 395

CHAPTER 15 A CLOSER LOOK AT TORNADOES 423

CHAPTER 16 MID-LATITUDE IV: WINTER WEATHER 449

■ NEW TO THIS EDITION

The Sixth Edition of *A World of Weather: Fundamentals of Meteorology* has many updates and features that continue our long tradition of educating students in order to make them better-informed weather consumers.

There are several new additions to Chapter 16 which explore the topic of winter weather. After roughly a decade of severe drought, California's winter of 2016–2017 was dramatically wet, with waves of heavy rain and heavy mountain snow emphatically putting a profound dent in the state's long-term water deficits. In this context, students will learn about atmospheric rivers of fast-moving, moisture-rich air from the tropics and subtropics and their role in fueling substantial precipitation along the West Coast of the United States.

A continent away on the East Coast, a nor'easter produced an historic snowfall on January 22–24, 2016, across parts of the Middle Atlantic and Northeast States. We provide students with a fresh, in-depth account of the meteorology that dissects this memorable snowstorm (and others like it). In concert with this historic blizzard, we also introduce the Regional Snowfall Index (RSI), a relatively new technique to rank snowstorms regionally according to their societal impacts.

We also updated Chapter 10 (Tropical Weather) with a discussion of the 2015–2016 El Niño, one of the strongest on record. We recount the dire predictions for excessive rain along and near the West Coast during the 2015–2016 cold season, forecasts that were issued in response to the expected barrage of fierce Pacific storms that sometimes accompany a strong El Niño. Such dire predictions began in August 2015, when an oceanographer at NASA's Jet Propulsion Laboratory made national news by christening the burgeoning El Niño in the tropical Pacific, "Godzilla." As it turned out, much of California received less than 60% of its seasonal precipitation that winter. In this new lesson, we teach students that not all strong El Niños have the same impacts on the West Coast, emphasizing the importance of other atmospheric and oceanic players in determining seasonal weather patterns.

We updated Chapter 11 (Hurricanes) with new material that focuses on the steering layer of a tropical cyclone. Research has shown that this layer is determined largely by the storm's minimum pressure. The idea is that a lower minimum pressure translates to a deeper vortex, thereby requiring the storm's steering influences to extend higher in the atmosphere. We study this new concept using Hurricane Harvey (2017), which stalled for several days after landfall, producing historic rainfall in parts of Texas and Louisiana. The 2017 hurricane season also included Hurricanes Irma and Maria, which caused utter destruction, especially in the Caribbean. An example using Irma solidifies the claim of lightning around the core of powerful hurricanes by introducing the Global Lightning Mapper on GOES-16, one of NOAA's new generation geostationary satellites.

We did not forget about Hurricane Sandy (2012). There are references to this infamous hurricane in several places of the Sixth Edition, including Chapter 5 (Satellite and Radar Imagery). Indeed, we updated our thorough discussion of satellite imagery in the context of Sandy.

Also in Chapter 5, we introduce students to dual polarization radar, a relatively new tool in a weather forecaster's toolbox. Dual-pol radar transmits pulses of microwave radiation in both the horizontal and vertical dimensions, allowing forecasters to gain a sense of the shape and size of radar targets and, thereby, make important decisions about precipitation type. Moreover, in tandem with other Doppler radar capabilities, dual-pol helps forecasters to decide, in real time, whether a tornado is present.

In contemporary times, climate change is at the forefront of public discussion. In response, we put considerable effort into updating Chapter 18 (The Human Impact on Weather and Climate), including the most current scientific consensus from the fifth report published by the Intergovernmental Panel on Climate Change (IPCC). We trace the history of the IPCC's consensus about its confidence (and uncertainties) regarding human impacts on climate due to increasing greenhouse gases. Chapter 18 also includes a new focus on the Earth's cryosphere (derived from the Greek "kryos," for "icy cold") by taking a "hard" look at the melting of Arctic sea ice, the Greenland ice sheet, and smaller glaciers in the Northern Hemisphere. We then discuss this melting in the context of sea-level rise (both past and future, the latter in the setting of the IPCC's fifth report). We also discuss the state of Antarctica's ice shelves and its voluminous ice sheets as well as the risk that global warming poses to this icy continent. In addition, we dispel the misinformed belief that climate model projections decades to centuries into the future "must be incorrect" simply because the models that predict weather lose their skill in a matter of days. Students will learn that such spurious arguments used against the scientific consensus of global warming are unequivocally false; there are indeed important differences in the models that predict weather and those that predict climate.

The entire list of updates in the Sixth Edition is too long to reveal here. Nonetheless, we feel compelled to mention our online assessment and our online learning and teaching resources for students and instructors. We created more than 100 new laboratory and interactive exercises. The online format affords us the opportunity to take advantage of hyperlinks that help students to have relevant data and resources at their fingertips.

Of all the editions that we've written over almost a quarter of a century, the Sixth Edition of *A World of Weather: Fundamentals of Meteorology* has excited us the most; it is current, thorough, conversational, and, most importantly, scientifically sound.

Louis Agassiz (1807–1873), a Swiss-American naturalist and geologist whose research included the study of glaciers, once said: "A smattering of everything is worth little. It is a fallacy to suppose that an encyclopaedic knowledge is desirable. The mind is made strong, not through much learning, but by the thorough possession of something."

We agree. We have long held that memorizing a litany of facts about meteorology, as one might do by reading an encyclopedia, amounts to fleeting knowledge because facts tend to slip away as time goes by. Our goal in the Sixth Edition is for our students to gain a "thorough possession" of the underlying scientific principles of meteorology and to be able to think critically about them. That's the formula, we believe, for students to become lifelong learners.

Instead of presenting meteorology as an encyclopedia of facts to memorize (and eventually forget), we encourage our students to delve deeper into the underpinning science. For example, in many introductory textbooks on weather, authors will claim that "warm air holds more water vapor than cold air" (a statement supposedly used to explain the net condensation of water vapor). Never mind that we can immediately present a counterexample to show that such universal claims are false. Consider air at a temperature of 20°C (68°F) and a relative humidity of 80%, compared to air at a temperature of 30°C (86°F) and a relative humidity of 20%. Guess what? The colder air "holds" more water vapor, actually about twice as much.

The encyclopedic mantra that "warm air holds more water vapor than cold air," or even another variant, "warm air *can* hold more water vapor than cold air" lends the impression that "air is like a hotel," to quote Dr. Craig Bohren, a professor emeritus in the Department of Meteorology and Atmospheric Science at Penn State. This comparison suggests that "the hotel can be full" (if the air is cold) or "there can be vacancies" (if the air is warm). But the hotel metaphor quickly breaks down. You see, if we assume that the diameter of an air molecule is about one-tenth of the separation between air molecules, we arrive at the inevitable conclusion that approximately one part in 1000 of "air" is occupied by something that has mass (an air molecule). Yes, air is mostly empty space! In other words, there's ample room, regardless of the air's temperature.

In the Sixth Edition, students will learn basic principles that will allow them to understand the process of net condensation without invoking some incorrect mantra put forth as gospel in many other sources. In other words, we avoid such trappings by approaching the net condensation of water vapor in a scientifically sound way. The bottom line is that we don't want our students to simply recite some oversimplified and erroneous rule that they memorized. Rather, we want students to be able to critically think about basic scientific principles.

Fundamentally, we want our students to become educated weather consumers. That's because they will be bombarded by all types of weather information throughout their lives. Some of it will be scientifically accurate, while some will be fuzzy and misleading (such as "warm air holds more water vapor than cold air"). By immersing themselves in *A World of Weather*, students will develop the ability to think more critically about scientific concepts. As a payoff, students will sharpen their ability to discriminate between accurate and misleading information that they hear on television or glean from the Internet. In this way, cultivating critical thinking in our students paves the way for lifelong learning.

As weather enthusiasts and instructors, we also aspire to cultivate a passion for meteorology in our students. By studying historic weather events such as the devastating flooding caused by Hurricane Harvey in Texas in August 2017, we want our students to vicariously "rub shoulders" with professional weather forecasters at the National Hurricane Center and to work with the same tools that they used to assess the threat of prodigious rains from a stalling tropical cyclone. By the same token, we want our students to walk in the shoes of forecasters at the National Weather Service who sometimes must issue tornado warnings based on radar reflectivity, Doppler velocities, and correlation coefficients derived from dual-pol radar.

There is an expansive online component of *A World of Weather* that houses online laboratory and interactive exercises, flash cards, and other resources to help you during your weather apprenticeship. The exercises are replete with real-life examples and non-traditional problems that use weather data from previous storms. In this way, we hope to enhance your learning via this hands-on approach.

The online exercises and resources make our book a valuable teaching tool for introductory courses at both large and small universities, including colleges that do not have a meteorology program and offer only one course in atmospheric science. Though the text is streamlined for college students with a non-meteorology major, we firmly believe that *A World of Weather* will also serve as a well-rounded foundation for students intending to major in meteorology.

In the final analysis, we believe that the Sixth Edition of *A World of Weather* will give students a "thorough possession" of the basic principles of meteorology (to again quote Louis Agassiz). Rather than memorizing a litany of encyclopedic facts that will eventually be forgotten, we believe that students who attain a "thorough possession" by studying *A World of Weather* will become lifelong learners.

Dear Student of Meteorology:

Welcome to the Sixth Edition of *A World of Weather*! To prepare you for your apprenticeship in meteorology and weather forecasting, we want to give you a heads-up on what to look for as you explore *A World of Weather*.

APPROACH OF THIS BOOK

The first thing you will discover is that this book is written in a nontraditional style. We purposely abandoned the "encyclopedia" approach to meteorology that some other textbooks systematically follow. Rather than presenting encyclopedic facts and flooding the pages with idealistic examples and schematics, *A World of Weather* shows you how fundamental concepts fit together to form a conceptual pyramid that explains how the atmosphere really works. More importantly, *A World of Weather* routinely shows you how to apply the concepts you learn by offering you examples based on real weather data. We believe that, by frequently exposing you to real weather data, *A World of Weather* provides you with a unique apprenticeship in meteorology.

LIFELONG LEARNING DESIGN

You see, we want you to discover meteorology for what it truly is—a pyramid of scientific concepts. *A World of Weather* will provide you with the design, tools, and building blocks so that you can construct your own pyramid of learning. Along the way, we hope you realize that *A World of Weather* paves the way for lifelong learning.

Just so you know exactly what we mean by lifelong learning, consider that the topic of computer weather models is increasingly in the mainstream news, particularly when significant storms such as hurricanes or East Coast snowstorms are involved. *A World of Weather* teaches you how to interpret and apply these computer forecasts that professional meteorologists use every day. This might sound overly ambitious to you, but we have students majoring in business, history, and communications (to name a few) who can go to the Internet, access computer forecasts, and decide on what they need to pack for a trip to Europe. Don't get nervous. *A World of Weather* will be with you every step of the way. For example, we will show you where you can find computer forecasts on the Internet so that you can access these tools and create your very own weather forecast. In this way, your apprenticeship prepares you for lifelong learning about weather forecasting.

CONVERSATIONAL STYLE

You will quickly discover that the text is written in a very conversational style. As teachers of meteorology, our goal is not only to impart knowledge and understanding, but to nurture an enthusiasm for the science that we love and respect. We believe that our conversational style is an effective way to encourage you to engage the material. We hope that our passions are contagious!

HANDS-ON EXERCISES

Keep in mind that there is an expansive online component to *A World of Weather* where you will find laboratory and interactive exercises that will take advantage of the online format. These online laboratory and interactive exercises will afford you the opportunity to apply the concepts you learn in the textbook. More importantly, the online component of the Sixth Edition will give you "hands-on" experience with real weather data. In some cases, you will analyze some of the greatest storms in history.

OTHER TOOLS TO USE IN YOUR APPRENTICESHIP

Book Layout

Weather Folklore and Commentary. These appear in many chapters. Weather folklores are homespun forecasts based on observations of nature. Weather commentaries are consumer reports on products dealing with the dissemination of weather information.

Focus on Optics. These essays on optical phenomena are scattered throughout the text and expose you to the colorful wonders of the sky.

Electronic Resources. Due to the fluid nature of Internet resources, this list is monitored for availability and is posted on our website at:

www.kendallhunt.com/aworldofweather6e.html.

Glossary. Serves as a ready reference for vocabulary terms.

Index. Provides for quick page reference for important terms from the text.

☐ ACKNOWLEDGMENTS

We gratefully acknowledge our many friends and colleagues who offered suggestions, guidance, and data during this project: Richard Alley, Scott Bachmeier, Craig Bohren, Eugene Clothiaux, Steve Corfidi, Michel Davison, Dale Durran, Jenni Evans, Greg Forbes, Mike Fritsch, Fred Gadomski, Jose Galvez, Rani C. Gran, Christopher Juckins, Mark Klein, Rebecca Lindsey, Scott Lindstrom, Paul Markowski, Tim Martin, Ray Najjar, Tim Olander, Art Person, Jon Racy, Dave Santek, Trent Schindler, Tim Schmit, Nels Shirer, Cathy Smith, Mark Thornton, Chris Velden, Nathaniel Winstead, and George Young.

We also want to acknowledge the professionalism, hard work, and infinite patience of those at Kendall Hunt Publishing Company, especially Lynne Rogers.

ABOUT THE AUTHORS

Jon M. Nese is Associate Head for Undergraduate Programs in the Department of Meteorology and Atmospheric Science at Penn State where he teaches a variety of undergraduate courses, including *Introduction to Weather Analysis* and *Synoptic Meteorology Laboratory*. He also hosts the Emmy Award-winning series WxYz for the department's long-running weather magazine show *Weather World*. Prior to joining the Penn State faculty, Dr. Nese was an on-air Storm Analyst for The Weather Channel and Chief Meteorologist at The Franklin Institute Science Museum in Philadelphia. While there, he co-authored *The Philadelphia Area Weather Book*, which received the 2005 Louis J. Battan Author's Award from the American Meteorological Society.

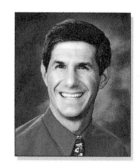

Lee M. Grenci retired from the Department of Meteorology and Atmospheric Science at Penn State in 2012 after 28 years of teaching. From 1986 to 2001, he was also a member of the Penn State Weather Communications Group, whose responsibilities included the daily preparation of the weather page for the *New York Times*. Among many honors, he received the Wilson Award for Outstanding Teaching in Penn State's College of Earth and Mineral Sciences in 1993, the College's 2003 Mitchell Award for Innovative Teaching, and the College's 2008 Faculty Mentoring Award. In retirement, Lee is active in raising money to help people afflicted by multiple sclerosis. He has raised over $32,000 for research to find a cure for MS. He is also an avid cyclist.

David M. Babb prepared much of the artwork within the text as well as the digital media used in many of the exercises. Dr. Babb received his B.S. in meteorology from the University of Kansas, during which time he worked summers at the National Severe Storms Laboratory in Norman, OK. In 1996, he earned his Ph.D. from Penn State studying cloud-drop sizes using millimeter-wave radar. Currently, he teaches several online meteorology courses and resides as a fellow in the John A. Dutton e-Education Institute at Penn State, specializing in instructional media design and assessment. His use of media and interactive learning tools for a nationally acclaimed online forecasting course at Penn State has been honored by the University Continuing Education Association.

Weather Analysis: The Tools of the Trade

1

LEARNING OBJECTIVES

After reading this chapter, students will:

- Be able to describe the basic composition of the atmosphere
- Gain a qualitative sense of the vertical scale of the atmosphere compared to the size of the earth
- Be able to interpret the system of longitude and latitude on weather charts
- Understand the various distortions associated with the map projections used in meteorology
- Be able to convert between Coordinated Universal Time and local time

- Be able to convert readings from any temperature scale to the other two temperature scales
- Understand the meaning of common statistics such as probability of precipitation and "normal" high temperatures
- Be able to isopleth charts of surface weather data and locate areas with large (or small) gradients, and interpret their meaning
- Understand the basic weather variables routinely reported by weather observers and automated observing systems
- Be able to decode surface station models
- Be able to decode meteograms

Upon the canvas of our lives, the atmosphere often remains in the background, appealing subtly to our senses, from the tepid touch of a tropical breeze to the gentle blues of a cloudless day. Indeed, this sky-blue tint stamped onto the thin envelope of air that hugs our planet makes a first-class postcard when viewed from space (see Figure 1.1).

Formally, the **atmosphere** collectively includes all the gases, clouds, and other particles that make up this envelope of air. Without any postage due, the building blocks of earth's present-day atmosphere of oxygen, nitrogen, water vapor, carbon dioxide, and other gases originally arrived from the cosmos via special delivery (bulk rate, of course) as our planetary system formed in a congealing swirl of gases more than four billion years ago. The concentrations of these gases have changed over the millenia. Table 1.1 shows their current levels in the atmosphere. These invisible gases not only make life possible on this "third rock from the sun" but they also help to regulate global temperatures, preventing our health-conscious planet from running a high fever or catching a big chill.

In essence, we live at the bottom of an ocean of air. Like earth's seas of water, the atmospheric ocean is teeming with waves of all sizes. Some waves are large, rolling swells of air in the high-altitude jet stream, which is a fast, relatively narrow current of air over the mid-latitudes at altitudes near 9000 m (30,000 ft). Large waves traveling in the jet stream transport warm air poleward and cold air towards the equator, helping to keep north-south temperature contrasts from getting too large. On the other end of the wave spectrum, small acoustic waves allow us to hear a tree falling in the forest. Even what we see and smell are affected by atmospheric conditions.

TABLE 1.1	Principal Gases of the Present-Day Atmosphere Near the Earth's Surface

PERMANENT GASES
(concentrations essentially constant)

Gas	Symbol	Percent (by volume in dry air)
Nitrogen	N_2	78.08
Oxygen	O_2	20.95
Argon	Ar	0.93
Neon	Ne	0.0018

VARIABLE GASES
(concentrations vary from place to place and/or over short time scales)

Gas	Symbol	Percent (by volume)
Water Vapor	H_2O	0 to 4
Carbon Dioxide	CO_2	0.041
Methane	CH_4	0.00018
Nitrous Oxide	N_2O	0.00003

Although the atmosphere is a jack-of-all-trades (its vast repertoire also includes driving ocean currents and eroding the landscape), it is not a formidable physical feature. This unassuming veil of air around our planet is precariously thin (revisit Figure 1.1)—indeed, the distance an average person could walk in a few hours, if directed upward, would terminate at an altitude where the air is too thin to sustain life. For many, this very thin veneer of air is only fully appreciated during dramatic weather events such as hurricanes, blizzards, or thunderstorms.

Our first real glimpse of **meteorology**, the study of atmospheric science, begins with the ancient Greeks,

FIGURE 1.1 When a portion of the satellite image inside the rectangle shown on the inset is enlarged, the relatively thin atmosphere hugging our planet becomes visible. This thin protective covering can be likened to the skin on an apple (courtesy of NASA).

who christened the study of the atmosphere and the heavens "meteorologica." The Greeks collectively referred to raindrops, hailstones, and snowflakes as "meteorons," which translates to "things in the air" (not coincidentally, these precipitation types are examples of what we now call "hydrometeors"). Today, one of the jobs of meteorologists (atmospheric scientists) is to prepare weather forecasts so the public can make informed decisions about their daily schedules.

Whether the atmosphere is in the background or at the forefront of our daily routines, it continuously offers unparalleled beauty and physical truths that are worthy of attention. In order to appreciate these opportunities, you should carry along a few tools so that your understanding can be forged from raw observation. Just as a doctor uses a variety of instruments and medicines to diagnose and treat illnesses, a meteorologist's diagnosis and forecast of a pending storm requires a familiarity with some basic

concepts involving time and space scales of weather systems, geography, and even a slight sprinkling of mathematics. Let's investigate these tools of the trade!

MEASURING THE FUNDAMENTALS: THE WHERE, THE WHEN, AND THE WARMTH

We commonly refer to locations and times using phrases such as "10 miles west of here" or "20 minutes ago." Or we might describe today's temperature by saying that it's "15 degrees lower than yesterday's." These descriptions are *relative*: They describe a place or time or temperature (or some other quantity) in terms of human perception or experience. Relative descriptions are useful, but not always sufficient. Sometimes you must know location, time, temperature, or some other quantity in *absolute* terms, using a frame of reference and a scale of measurement.

Spatial Scales: Going the Distance

In 1919, a young Lieutenant Colonel participated in a convoy of military vehicles that traveled from Washington, DC to San Francisco. The trip took 62 days and covered muddy, winding, locally designed roads. Mindful of this trip (and after seeing the German Autobahn), President Dwight D. Eisenhower signed legislation in 1956 authorizing the creation of a national system of interstate highways across the United States. This network of roads represents a standardization of space that allows the efficient transport of people and goods. Similarly, maps describing locations on the surface of the earth are standardized according to two main categories: physical (showing natural features such as mountains, oceans, and continents) and political (showing boundaries and borders set by governments).

To a meteorologist, the physical structure of the earth's surface is important because it strongly influences the flow and characteristics of the air above—from a swirl of leaves around the corner of a house to a fast-flowing jet stream several miles above the ground. Political descriptions of locations allow the convenient communication of where weather events occur, from the detection of dangerously shifting winds at the west end of a major airport to the report of a large high-pressure system over Utah. Figure 1.2 shows North America with many important physical and political features labeled. Because patterns of weather and climate can be dramatically affected by mountainous areas and large bodies of water, it is essential to know the locations of the Rockies, the Appalachian Mountains, the Atlantic and Pacific Oceans, the Great Lakes and the Gulf of Mexico.

To specify location anywhere on earth, we have long relied on a quantitative system of latitude and longitude. This system became important during the seventeenth century when navigators sailed with increasing frequency and confidence into the New World. As shown in Figure 1.3, **latitude lines** are circles that connect points of equal distance from the **equator**, the special line of latitude that equally divides the earth into Northern and Southern Hemispheres. Latitude varies from a minimum value of 0° at the equator to maximum values of 90° at the poles. To be complete, a location's latitude must include either an "N" (for north) or "S" (for south), the exception being the equator. The **tropics** are loosely defined to lie between latitudes 23.5°N (called the **Tropic of Cancer**) and 23.5°S (known as the **Tropic of Capricorn**). Adjacent to the tropics, extending to about 35° in each hemisphere, are the **subtropics**. The **polar regions** are generally defined to stretch from latitude 66.5°N (the Arctic Circle) to the North Pole and from latitude 66.5°S (the Antarctic Circle) to the South Pole. The latitude bands between the subtropics and the polar regions in each hemisphere are commonly referred to as the **mid-latitudes** (some definitions of the mid-latitudes include the subtropics).

Longitude lines, also known as **meridians**, run from pole to pole along the earth's surface (see Figure 1.3). Lines of longitude meet at the poles and diverge to a maximum separation at the equator: They are everywhere perpendicular to lines of latitude. Since 1884, the line of longitude passing through the British Royal Observatory in the village of Greenwich, England (a borough of London) has been recognized internationally as the 0° line of longitude (in recognition of Britain's dominance of the colonial-era seas aided by a storm that wiped out the Spanish Armada in 1588). This special line of longitude is called the **prime meridian**. Longitude lines extend both east and west of the prime meridian to the opposite side of the globe, where the maximum longitude of 180° (the approximate location of the International Date Line) is found. A precise longitude specification requires either "E" (for east) or "W" (for west), with the exceptions being at the prime meridian and at 180°. Check Figure 1.2 to get a sense of North America's location in terms of latitude (Northern Hemisphere) and longitude (Western Hemisphere).

When mapmakers attempt to represent the spherical earth (or a portion thereof) on a flat chart, strange things can happen to longitude and latitude lines. Indeed, they do not always intersect at right angles. For this reason, weather forecasters routinely take into account the layout of longitude and latitude lines on weather maps.

Map Projections: Going Flat Out

Thank goodness weather maps are flat! Imagine how inconvenient it would be for forecasters to plot weather data on a real globe and then carry the "Earth" around with them all day!

But there's a trade-off in using two-dimensional charts. To see what we mean, imagine taking off the skin of an orange in one piece and then trying to make the peel lie flat without breaking or distorting it. Good luck to you! Without reservation, you'll run into the same kinds of problems if you tried to make a flat map of the spherical earth. No matter what you do, there would always be distortions.

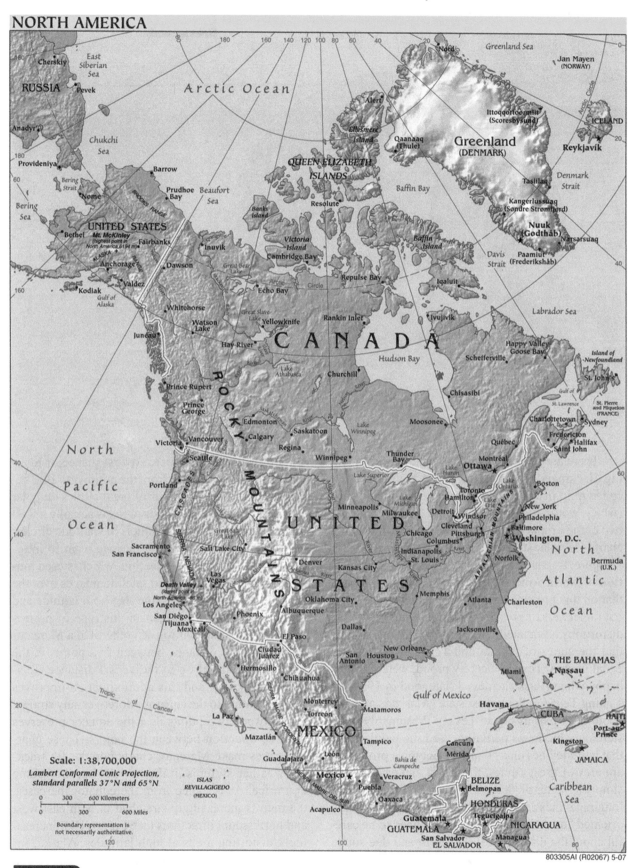

FIGURE 1.2 A map of North America showing many of the most important physical and political features (courtesy of the Central Intelligence Agency).

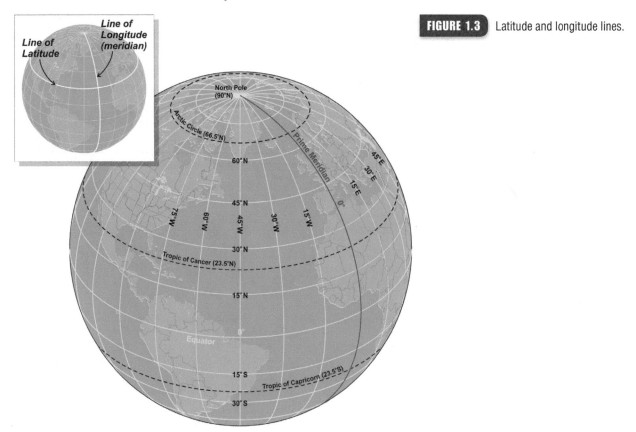

FIGURE 1.3   Latitude and longitude lines.

There are several imperfect techniques for making flat maps of the earth. These techniques are called **map projections**. Without getting into details (which would require delving into solid geometry), two of the primary map projections used for weather charts are polar stereographic and mercator.

Figure 1.4 shows a **polar stereographic projection** of North America and adjacent oceans. Notice that on this projection, latitude lines are not straight but rather arcs of great circles. As an example of the distortions associated with map projections, notice that the states of Alaska and Texas appear nearly equal in area. Texans, fiercely proud, were not very happy when Alaska achieved statehood in 1959, replacing Texas as the largest state (Alaska is actually about 2.5 times as big as Texas). Presumably polar stereographic maps partially eased the pain! Given that latitude lines on a polar stereographic projection are arcs of great circles, then west and east are not simply a matter of "left and right." And even though longitude lines are, in fact, lines, they are not simply oriented "up and down," and so you need to be careful about the directions of due north and due south. So what does all this have to do with weather?

Here's an example. In Figure 1.4, suppose the arrow off the Pacific Northwest coast depicts the wind direction in that area. At first glance, it looks like this wind is blowing from the northwest. But, because latitude lines are arcs of great circles on polar stereographic projections, the wind is actually a west wind because the arrow parallels the latitude circles.

The other map projection that we want to introduce is the **Mercator projection**, which is used routinely on weather maps and satellite images over the tropics (see Figure 1.5). Note that both latitude and longitude lines are straight on this type of projection and perpendicular to each other. On a Mercator projection, the distance between two points in the tropics is pretty close to the *actual* distance (relatively speaking), and this approximation improves with proximity to the equator. Moreover, any straight line between two points near the equator preserves the true direction between the points. These characteristics make Mercator charts popular for tracking weather systems in the tropics, but they come at a price. As distance from the equator increases, the map is increasingly stretched in both east-west and north-south directions (see Figure 1.5), increasingly distorting areas at high latitudes. Alaska, for example, dwarfs Texas (a distortion Alaskans can probably live with). The Mercator projection is essentially unusable at latitudes greater than about 70°.

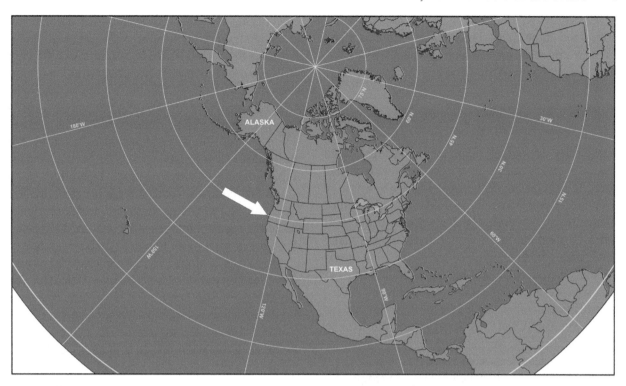

FIGURE 1.4 A polar stereographic projection.

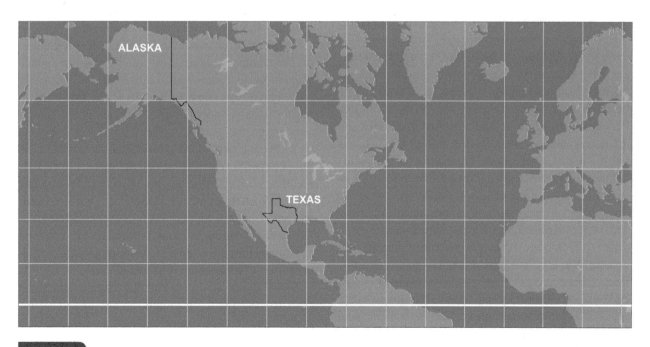

FIGURE 1.5 An example of a Mercator map, a popular choice of map projection for tracking storms in the tropics (the equator is marked by a thick white line).

Besides spatial scales, weather forecasters also are concerned about time—but not just their local time. In the words of the music group Chicago, "Does Anybody Really Know What Time It Is"?

Time Scales: Synchronized Sands through the Hourglass

In the early days of television news, the wall behind the anchor desk commonly was covered with clocks, each dutifully providing the local time in a foreign city such as London, Moscow, or Tokyo. In the world of meteorology, however, it makes sense to use a standardized time because weather systems span many time zones, and tagging simultaneous weather observations taken in different time zones with their local times would lead to great confusion. Enter Coordinated Universal Time, or UTC (formerly Greenwich Mean Time, or GMT), the local time at the British Royal Observatory, which lies on the prime meridian. For years, meteorologists have used this 24-hour clock system, also frequently called Z-Time, to temporally standardize all weather data. Whenever and wherever a weather observation is taken, the time of the observation is recorded in UTC, not in the observer's local time. This avoids the confusion associated with local times, and allows simultaneous weather observations all over the world to be effectively grouped together.

But because nearly everyone uses local time in their daily lives, adjustments between UTC and local time are often necessary. These conversions are made by adding or subtracting a given number of hours based upon location with respect to the prime meridian. Figure 1.6 shows some international time zones and the corresponding local times for 1500 UTC. In the United States, the Eastern Time Zone is five hours behind UTC on Standard Time. Once the switch to Daylight Saving Time occurs, the difference becomes four hours (although some locations, such as Hawaii and most of Arizona, stay on Standard Time all year). Table 1.2 on page 10 can be used to convert your local time to UTC (or vice versa). Given the time difference between UTC and some local times, it is possible for a U.S. weather map to have tomorrow's date (in UTC), but still represents today's data (in local time). For example, 00 UTC on January 1, 2016 is 7 P.M. EST on December 31, 2015.

Temperature Scales: A Matter of Degrees

Everyone has an intuitive and qualitative notion of temperature in terms of sensing hot and cold.

Quantifying changes in temperature is fundamental to the study of the atmosphere. Meteorologists, and scientists in general, define temperature as a function of the motions of molecules (it can't get any more basic than that).

Molecules of all substances, even liquids and solids, are in constant motion, vibrating, wiggling, and colliding with one another. Metaphorically, atoms and molecules continuously buck, like horses refusing to be broken. But some are tamer than others. Molecules that compose solids are the tamest. Although they wriggle and vibrate, these molecules are tightly packed in regular arrangements. In other words, they don't move from place to place (they are penned up in "rigid stalls"). Molecules that compose liquids have more freedom than their solid counterparts. Although they are close together, molecules in liquids vibrate, move about, and slide past each other in less confining "corrals." Gas molecules are the wildest, in constant motion colliding with each other at high speeds (there are no "fences" to confine roaming gas molecules). **Kinetic energy** describes the motion of molecules, and the **temperature** of a substance is a measure of the average kinetic energy of its molecules. Greater average molecular speeds correspond to higher temperatures. And air molecules aren't slowpokes: Room temperature corresponds to an average molecular speed of around 450 m/s (about 1000 mph)!

Based on this molecular definition, it is logical to define a temperature scale on which zero corresponds to no molecular motion. Such a scale, used primarily by scientists, is called the **Kelvin scale**. There are no negative temperatures on the Kelvin scale, making it convenient for scientific calculations: Zero Kelvins (0 K) is called **absolute zero** and corresponds to the temperature at which all molecular motion hypothetically ceases (note that the unit of temperature on this scale is written simply as K, not °K). Absolute zero has never been achieved, even in a controlled laboratory environment, although scientists have slowed molecules enough to produce a corresponding temperature of just a fraction of one Kelvin!

The **Celsius scale** (called the Centigrade scale prior to 1948) is the standard temperature scale for popular use in most countries that use the metric system. The United States, like a professional athlete seeking a contract renegotiation, has held out. The scale is named in honor of the eighteenth-century Swedish astronomer Anders Celsius, who designed a scale based on dividing the interval between the freezing and boiling points of pure water into 100

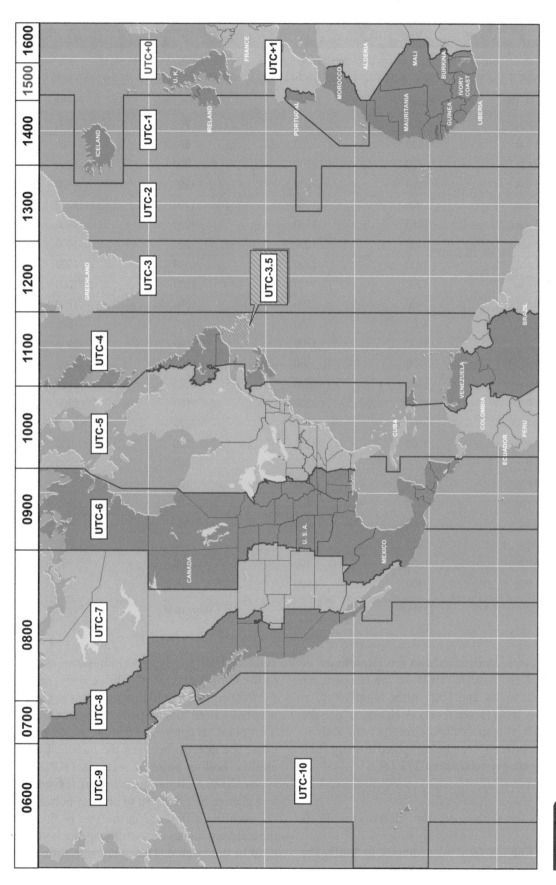

FIGURE 1.6 Some time zones of the world. The time across the top is the local Standard Time (on a 24-hour clock) corresponding to 1500 UTC. On Daylight Saving Time, add one hour to the local times.

TABLE 1.2

To convert your local Standard Time to UTC, find the intersection of the row corresponding to your local time with the column corresponding to your time zone. For Daylight Saving Time, subtract one hour from the designated UTC.

LOCAL TIME	Eastern	Central	Mountain	Pacific
Midnight	500	600	700	800
1 A.M.	600	700	800	900
2 A.M.	700	800	900	1000
3 A.M.	800	900	1000	1100
4 A.M.	900	1000	1100	1200
5 A.M.	1000	1100	1200	1300
6 A.M.	1100	1200	1300	1400
7 A.M.	1200	1300	1400	1500
8 A.M.	1300	1400	1500	1600
9 A.M.	1400	1500	1600	1700
10 A.M.	1500	1600	1700	1800
11 A.M.	1600	1700	1800	1900
noon	1700	1800	1900	2000
1 P.M.	1800	1900	2000	2100
2 P.M.	1900	2000	2100	2200
3 P.M.	2000	2100	2200	2300
4 P.M.	2100	2200	2300	000
5 P.M.	2200	2300	000	100
6 P.M.	2300	000	100	200
7 P.M.	000	100	200	300
8 P.M.	100	200	300	400
9 P.M.	200	300	400	500
10 P.M.	300	400	500	600
11 P.M.	400	500	600	700
LOCAL	**Eastern**	**Central**	**Mountain**	**Pacific**

equal intervals. In deference to the importance of water, the Celsius scale ties 0° to the temperature at which ice melts, and 100° to the temperature at which water boils (at sea level). A change of one degree Celsius corresponds to a change of one Kelvin. Degrees Celsius (°C) can be computed directly from Kelvins by simply subtracting 273.15:

$$°C = K - 273.15.$$

Thus, absolute zero on the Celsius scale is –273.15°C.

In a few English-speaking countries, including the United States, temperature is still officially reported on the **Fahrenheit scale**. G. Daniel Fahrenheit, an eighteenth-century German physicist, assigned 32° to the temperature at which ice melts and 212° to the temperature at which water boils at sea level (he originally used human body temperature as a reference point, assigning it 96°—not a bad guess). Because the interval between the melting point of ice and the boiling point of water is 180° Fahrenheit (versus 100° Celsius), a change of 1°C corresponds to a change of 1.8°F. So to convert between °C and °F, both the difference in the size of the units and the different values of the melting and boiling points must be considered. The conversion formulae are

$$°F = (9/5) \ °C + 32 \qquad °C = 5/9 \ (°F - 32)$$

MATHEMATICAL TOOLS: HOW TO CRUNCH NUMBERS AND LIKE IT

Sometimes numbers are just absolutely necessary. Imagine trying to qualitatively compare baseball legends. You might say, "Babe Ruth was a great hitter, but wow, Hank Aaron was really amazing . . ." or "Ted Williams was the best hitter, but Barry Bonds was the most feared." Obviously, there are more great hitters in the annals of baseball history than there are adjectives to uniquely describe each of them. So we describe these players quantitatively in terms of statistics: Aaron's 755 lifetime home runs, Williams's .406 average in 1941, Ruth's career 2204 runs batted in, and Bonds's record 73 home runs in 2001. Similarly, numbers are essential for describing the weather: Observations of variables such as temperature, wind, pressure, and relative humidity produce a continuous avalanche of data that meteorologists need to understand the present weather situation and forecast future weather conditions. Making sense of these numbers requires a basic familiarity with a few mathematical concepts, including the International System, unit conversion techniques, and some simple statistics.

Converting Units: Just for Good Measure

On September 23, 1999, NASA lost the Mars Climate Orbiter spacecraft because engineering teams failed to convert from English to metric units (the spacecraft likely burned and broke into pieces in the Martian atmosphere). The failure to properly convert units cost American taxpayers $125 million. Meteorology uses both English and metric units, so it is important that you have the ability to switch from one system to the other, although the consequences of an improper conversion likely won't cost you that much.

The scientific system of standardized units is the International System (SI), which has its roots in the metric system. The basic SI units are meters (m) for length, seconds (s) for time, kilograms (kg) for mass, and Kelvins (K) for temperature. Because the United States uses the English System of measurement, with basic units the inch (in), second (s), pound (lb), and degree Fahrenheit (°F), conversion from one unit of measure to another is often necessary in meteorological applications. To convert from one system to another, you can use conversion factors such as those listed in Table 1.3. In most cases, we use the International System of units in the text, the exceptions being

---

EXAMPLE 1

In the Northwest Pacific Ocean, hurricanes are called typhoons. A typhoon with maximum sustained winds of at least 130 knots is known as a super-typhoon. On November 7, 2013, Supertyphoon Haiyan bore down on the Philippine Islands in the western Pacific with a maximum sustained wind speed of 170 knots. Convert this wind speed to miles per hour (round to the nearest mile per hour).

SOLUTION

$$170 \text{ knots} \times \frac{1.15 \text{ mph}}{1 \text{ knot}} = 195.5 \text{ mph} \sim 196 \text{ mph}$$

EXAMPLE 2

Between July 1, 1998 and June 30, 1999, Mount Baker Ski Area in Washington recorded 1,140 inches of snow, setting a United States record for seasonal snowfall. Convert this amount to meters (round to the nearest meter).

SOLUTION

$$1140 \text{ in} \times \frac{2.54 \text{ cm}}{\text{in}} \times \frac{1 \text{ m}}{100 \text{ cm}} = 28.956 \text{ m} \sim 29 \text{ m}$$

TABLE 1.3	Common Unit Conversions
Length:	1 kilometer (km) = 1000 meters (m) = 3281 feet (ft) = 0.62 miles (mi)
	1 mile (mi) = 5280 feet (ft) = 1.61 kilometers (km) = 0.87 nautical miles (nm)
	1 centimeter (cm) = 0.39 inch (in)
	1 inch (in) = 2.54 centimeters (cm)
	1 foot (ft) = 12 inches (in)
Time:	1 hour (hr) = 60 minutes (min) = 3600 seconds (s)
Mass:	1 kilogram (kg) = 2.2 pounds (lb)
Speed:	1 knot (kt) = 1 nautical mile per hour = 1.15 miles per hour (mph)
	1 mile per hour (mph) = 0.45 meters per second (m/s) = 1.61 kilometers per hour (km/hr)

when certain quantities are traditionally measured in units from the English System.

The two examples on this page illustrate how to use some of the conversion relationships in Table 1.3. Common to both is the following idea: Just as we reduce a fraction by factoring the same number from both the numerator and denominator and then canceling that number out (think about reducing 8/12 to 2/3 by factoring out a 4), you can cancel units whenever they appear in both the numerator and denominator during a unit conversion operation.

Spatial and Temporal Scales: The Big One That Got Away

Perhaps you know a fisherman who talks about "the big one that got away." Such fish tales usually bring a smile because most folks suspect that there's probably some exaggeration in the story.

When warnings for tornadoes urgently scroll across television screens, the impression often conveyed to viewers is that everyone in the warned area is in immediate and dire danger. You will learn later that most tornadoes have diameters of only a few hundred meters or less and last only about five to perhaps twenty minutes. So warnings for entire counties are, for the most part, exaggerations. Indeed, most folks in a warned area will not be affected by a tornado.

In order to become better weather consumers, you must know the spatial and temporal scales on which specific types of weather occur. To see what we mean, consider Figure 1.7, which is a photograph

FIGURE 1.7 A well-placed penny gives an observer a basis for judging the true size of these small Rocky Mountain flowers.

of flowers growing in the Rockies near Denver, CO. Without the penny to give you the basis on which to judge the spatial scale of the flowers, you might think that they are much bigger than they really are. That's our two cents on this issue.

Figure 1.8, the weather version of the flower-penny photograph, illustrates how atmospheric motions span a broad range of temporal and spatial scales. For clarity, meteorologists often refer to atmospheric phenomena by their scales. For example, turbulence commonly encountered during aircraft flight is considered a **microscale** phenomenon because it usually lasts only seconds and has horizontal dimensions of only tens of meters. At the other extreme, the high- and low-pressure areas on the typical mid-latitude weather map, whose influence can span many days and reach over hundreds of thousands of square kilometers, are considered **synoptic-scale** phenomena. That the atmosphere is composed of a hierachy of waves and whirls of air of many sizes, all superimposed on one another, makes air motions and weather patterns extremely complex. This complexity limits our ability to obtain a complete description of the present state of the atmosphere, and thus helps to explain why the accuracy of forecasts (which are descriptions of the atmosphere's future state) decreases considerably beyond a few days.

In light of these large differences in magnitudes of the spatial and temporal characteristics of atmospheric phenomena, it is no surprise that meteorologists and their weather prediction models occasionally hit a sour note based upon data that are incorrectly interpreted. Consider, for example, a surprise storm that struck England in October 1987, downing trees and power lines as hurricane-force winds swept over London. The computer forecast called for the storm to form in the North Sea, 325 km (about 200 mi) to the east. However, a single weather observation was incorrectly entered into a computer model used by forecasters at the time. The error was caused by a shift in the decimal place of the number, causing the value to be off by several powers of ten, or **orders of magnitude**. The result was an incorrect shift in the storm's predicted location.

Fortunately, numbers that are gigantically or minutely unruly can be reformed using a numeric shorthand called **scientific notation**. The scientific notation form of a number, shown in the far right column of Table 1.4, is extremely valuable for keeping track of very large and very small numbers. Numbers

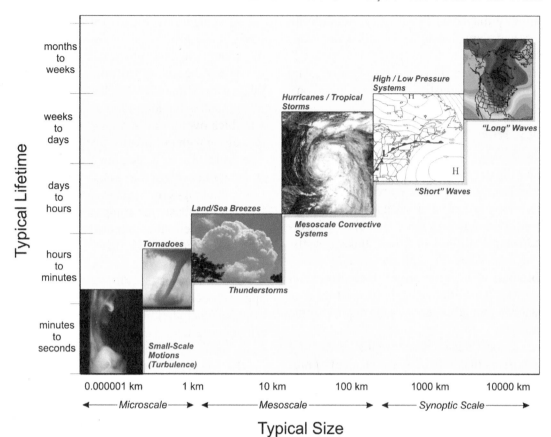

FIGURE 1.8 The typical sizes and typical lifetimes of various weather phenomena.

TABLE 1.4	Some Metric (SI) Orders of Magnitude			
Prefix	**Abbreviation**	**Value**	**Regular Notation**	**Scientific Notation**
exa	E	one quintillion	1 000 000 000 000 000 000	$= 10^{18}$
peta	P	one quadrillion	1 000 000 000 000 000	$= 10^{15}$
tera	T	one trillion	1 000 000 000 000	$= 10^{12}$
giga	G	one billion	1 000 000 000	$= 10^{9}$
mega	M	one million	1 000 000	$= 10^{6}$
kilo	k	one thousand	1 000	$= 10^{3}$
hecto	h	one hundred	100	$= 10^{2}$
deka	da	ten	10	$= 10^{1}$
———	—	one	1	$= 10^{0}$
deci	d	one tenth	0.1	$= 10^{-1}$
centi	c	one hundredth	0.01	$= 10^{-2}$
milli	m	one thousandth	0.001	$= 10^{-3}$
micro	μ	one millionth	0.000 001	$= 10^{-6}$
nano	n	one billionth	0.000 000 001	$= 10^{-9}$
pico	p	one trillionth	0.000 000 000 001	$= 10^{-12}$
femto	f	one quadrillionth	0.000 000 000 000 001	$= 10^{-15}$

are commonly displayed in scientific notation on calculators and computers to better define variations in orders of magnitude.

So there are powers in scientific notation. And there is power in the statistics of weather. Read on.

Statistics: From the Mean to the Extreme

Mark Twain once credited nineteenth-century British Prime Minister Benjamin Disraeli with the now much-celebrated quote: "There are three kinds of lies: lies, damned lies, and statistics." Although Twain's attribution to Disraeli was probably incorrect, the message still hits the mark. Indeed, statistics can be misleading, but nonetheless they are an invaluable part of many analyses, meteorological and otherwise. Thus, the analysis of weather data spanning any temporal or spatial scale requires some basic statistics.

One of the most misinterpreted statistics used by meteorologists is the probability (or chance) of precipitation (POP for short), which is included in most one- and two-day forecasts issued by the National Weather Service. For example, a 60% probability of rain in tonight's forecast means that there's a six in ten chance that a point in the forecast area (such as your rooftop) will receive at least 0.01 inches of liquid precipitation by morning. Or, considering ten such forecasts, it should rain on your house six of the ten times. A 60% POP does not mean that it will rain 60% of the time tonight in the forecast area, nor does the POP imply anything about how hard the rain will fall.

Probably the most frequently used statistical measure in meteorology (and in general) is the mean (or average), which is simply the sum of all observed values divided by the number of observations. For example, weather observers usually arrive at an official snow-depth measurement by taking the average of several nearby measurements. In this way, they smooth out fluctuations in depth caused by wind and other local effects. So, if an observer measured snow depths of 2.5, 3.2, 3.0, and 2.7 inches, the officially reported snow depth would be (2.5 + 3.2 + 3.0 +2.7) ÷ 4 = 2.9 inches. When calculating such a mean, an observer should scrutinize the measurements for values considerably different from the rest. For example, had an observer made a fifth measurement of 8.3 inches in a snow-covered ditch, the average would increase to 3.9 inches (a value not truly representative of the actual snow depth)!

Averaging is one way to turn weather observations into climate information. Weather refers to day-to-day changes in atmospheric conditions—essentially, the state of the atmosphere now and in the near future. Observations quantify weather conditions. **Climate**, in contrast, is the long-term average of atmospheric variables over months, years, or even longer periods, along with information about weather extremes. The collection of a location's climate information is sometimes referred to as its **climatology**—essentially, the long-term averages and extremes that can be used as a basis for comparison. In the words of Robert Heinlein, the 20th century American novelist and science fiction writer, "Climate is what you expect, weather is what you get."

The backbone of U.S. climate statistics is a network approximately 10,000 volunteers who make observations of precipitation and maximum and minimum temperature once per day. It is very possible that someone in your hometown participates in this **cooperative observer network**. Let's apply some basic statistical analysis to a month's worth of data recorded at one of these locations. The daily maximum (high) and minimum (low) temperatures and daily precipitation for State College, PA, are shown in Table 1.5 for February 2014. This month was relatively cold and snowy, and it is interesting to calculate how conditions for the month compared to average. The mean daily maximum temperature and the mean daily minimum temperature for the month, shown at the bottom of those columns in the table, were determined by summing the respective daily readings and dividing by 28, the number of days in February. These two values are then averaged to give the overall monthly mean temperature:

$$(32.0°F + 15.5°F) \div 2 = 23.8°F.$$

In order to compare this monthly mean temperature to climatology, meteorologists usually define "average" by basing it on thirty years of data. In meteorology, such averages are sometimes called "normals." The National Weather Service updates normals at the beginning of every decade. For example, weather observations taken from 1981 to 2010 have governed the normals since 2011. Based on this thirty-year period, the normal monthly temperature for February at State College is 29.6°F (computed by averaging the mean maximum of 37.5°F and the mean minimum of 21.7°F). So February 2014 was 29.6°F – 23.8°F = 5.8°F below the 30-year average.

TABLE 1.5

Daily maximum and minimum temperatures, daily solid precipitation (in inches), and daily liquid precipitation (rain plus melted snow and sleet, in inches), for State College, PA in February 2014. The monthly means of the maximum and minimum temperatures appear below their respective columns. The total monthly solid precipitation and the total monthly liquid precipitation appear below their respective columns. The notation "T" in the precipitation columns stands for "Trace," meaning that precipitation was observed, but measured less than 0.1 inches of solid precipitation or 0.01 inches of liquid precipitation.

Date	Max Temp °F	Min Temp °F	Liquid Precipitation (inches)	Solid Precipitation (inches)
1	37	22	T	T
2	44	31	T	0
3	46	29	0.15	1.3
4	30	13	0.20	1.7
5	28	13	0.81	4.5
6	35	16	0.15	T
7	25	16	T	T
8	25	8	T	T
9	23	8	T	T
10	23	13	0.18	2.0
11	23	5	T	T
12	21	−1	T	T
13	23	−1	0.13	1.5
14	31	15	0.54	8.0
15	34	25	0.19	2.7
16	31	9	0.08	1.2
17	24	6	T	T
18	28	5	0.10	1.5
19	34	19	0.12	2.0
20	47	24	0.20	T
21	45	25	0.02	0
22	46	33	0.13	0
23	50	29	0	0
24	46	22	0	0
25	28	19	0	0
26	26	20	0.02	0.3
27	21	12	T	T
28	22	−1	T	T
Mean	32.0	15.5		
Totals			3.02	26.7

Now focus your attention on the columns for precipitation in Table 1.5. These values are totaled not averaged, primarily because precipitation does not occur every day (in contrast to routine daily maximum and minimum temperatures). The normal liquid precipitation for February at State College is 2.53 inches, while the normal solid precipitation (mostly snowfall, but also including sleet) is 11.0 inches. Thus, February 2014 was very wintry by snowfall standards and thus, not surprisingly, also well above the thirty-year average for liquid precipitation.

Glancing over the February 2014 temperature data again, we can identify the extreme values for the month, which are the highest daily maximum temperature and the lowest daily minimum temperature. These values, 50°F and −1°F, define the **range** of temperatures observed at State College during February 2014. Extremes can also be determined for each calendar day and are often reported as the daily "record high" and "record low" temperatures. These represent the highest and lowest temperatures observed on a particular day over the entire length of time that an observing station has been operational (called the **period of record**).

It's important to note that many meteorologists prefer to say "30-year average" or "climatology" instead of "normal" because normal, by definition, implies a regular or usual mode of behavior for the weather. In fact, it would be unusual (in most places) for temperatures (or precipitation) for a day, month, season, or year to exactly match the "normal" for that time period (one noteworthy exception is the equable tropics, where high and low temperatures vary little from day to day and daily averages nearly match seasonal averages). As an analogy, think of your heart rate: An average value might be 70 beats per minute, but much of the time your rate will be considerably higher (during exercise, for instance) or lower (during sleep, for example). A heart rate of 75 beats per minute measured during your last doctor's visit would not be considered abnormal, but rather within a range of normality.

Similarly, in the mid-latitudes, it would be "normal" for a particular day's high (or low) temperature to fall within a range of temperatures centered on the 30-year average high (or low) for that day, but not necessarily hit the "normal" exactly. As an example, we'll again consider data observed at State College, PA, a typical mid-latitude location, on an early spring date, April 9 (see Figure 1.9). The "normal" maximum temperature for this date in State College, based on the period 1981–2010, is 57°F. Historically, about 25% of all highs on this date have fallen within 5°F of this value—that is, from 52°F to 62°F—and more than 50% have fallen within 10°F. However, as Figure 1.9 shows, a high temperature of exactly 57°F has never been observed in State College on April 9 during the period of record! Indeed, at most mid-latitude locations, an observed daily high or low temperature that exactly matches the "normal" happens a lot less frequently than you might think, especially outside of the warm season.

Nonetheless, climatology serves as a general guide for forecasters. In fact, the routine of preparing a forecast for the high (or low) temperature at a given city on a given day begins with a check of the date's averages so that forecasters have a ballpark figure in mind. Indeed, you should give pause before you forecast a high (or low) temperature that's, say, 25°F above (or below) the average high for the date. Not that such large departures from climatology are impossible. They're just not a common occurrence.

Starting the forecasting process for temperature with climatology tends to work better in summer than in winter. That's because during summer, the relatively high angle of the sun and the infrequency of dramatic changes in air masses allow warm-season temperatures to be less variable, on balance, than their winter counterparts. Only on summer days when clouds and rain have the upper hand do high temperatures stray far below climatology. Figure 1.10 shows that July 15, 2013 was such a day over much of Texas as a thick canopy of clouds and occasional rain suppressed daytime heating and set the stage for dozens of locations across the Lone Star State to set records for the lowest daily maximum temperature. Readings only reached the 70–75°F range at many places from Midland to Dallas, well below the daily average high which is near 95°F.

WEATHER ANALYSIS: SPATIAL TOOLS

Before weather forecasters can predict future conditions, they strive to thoroughly understand the present and recent past states of the atmosphere. Back in the 1950s, it would have taken decades for a meteorologist to be exposed to as much weather data as a modern-day forecaster looks at in just one eight-hour shift. Analyzing maps of weather data and culling from them the essential features of the atmosphere is the trademark of the forecaster. Before the age of computers, this tedious analysis was done

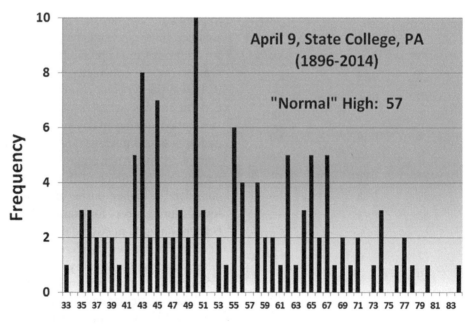

FIGURE 1.9 This histogram, or frequency distribution, shows the number of times that a particular daily maximum (high) temperature has been observed on April 9 in State College, PA, during the period 1896 to 2014. The "normal," or average high, on this date of 57°F has not been observed on April 9 during the period of record.

FIGURE 1.10 A satellite image on the afternoon of July 15, 2013, with some high temperatures on that day superimposed. Beneath the shield of clouds that covered much of Texas, high temperatures remained well below average, setting dozens of records for lowest daily maximum temperature.

exclusively by hand. Now, most of the processing and plotting of the raw meteorological data is left to the computer, but the interpretation of the data remains very much with the meteorologist.

Without certain techniques of analysis to help them make sense of weather data, forecasters would have their hands full. That's because, to modernize an old English proverb, sometimes you just can't see the forest through the trees. This pearl of wisdom, which counsels that you can lose sight of life's big picture if you pay too close attention to the details, also applies to meteorology. To get an inkling of how meteorology can mimic life, see Figure 1.11a, which shows a forest of surface temperature observations taken across a broad area of the nation on a day in early March. At first glance, the thicket of numbers in this observational forest might seem highly disorganized to you, just like the impression you might get from the photograph of a forest of pine trees in Figure 1.12a. But, if you change your perspective a little, an organized pattern of trees emerges (see Figure 1.12b).

Similarly, patterns typically emerge when weather data is analyzed on a map. In order to change how they view a set of weather data (like the thicket of temperatures shown in Figure 1.11a), forecasters see the meteorological forest (weather pattern) by using a technique called **isoplething**. Like trail-blazing hikers marking a path through a virgin forest (in an environmentally sound way, of course), forecasters draw **isopleths** ("iso" meaning "equal," "pleth" meaning "value") which connect points on a weather map that have the same value of some quantity of

interest (such as temperature, humidity, or pressure). Figure 1.11b shows isopleths of temperature, known as **isotherms**, for the map of data in Figure 1.11a. By drawing isotherms, you can plainly see patterns of warmer and colder air where, before, all you probably saw were a bunch of numbers. The patterns in the isotherms help us visualize where colder air is plunging southward (in this case, the Central Plains) and where warmer air is bulging northward (in the lower Ohio River Valley).

But we are not "out of the woods" yet. Isoplething will be a recurrent theme throughout this text, so you need to become proficient at the mechanics of isoplething and particularly the interpretation of isopleths. Then, you will be able to spot forests of important weather systems from maps with lots of trees of data. So think of yourself as a novice hiker who is about to learn how to blaze through a wilderness of weather data.

Topographic Maps: Elevating the Discussion on Isoplething

Scouts are trail blazers, ably learning to follow maps that can guide them through wilderness. Interpreting topographic maps, which are simply isoplethed charts of elevation above sea level, is one of the skills that scouts learn during their outdoor apprenticeship. If we follow their learning lead, we will be one step closer to understanding the art of isoplething and thereby be poised to organize and make sense of the avalanches of weather data that come careening down on forecasters every day.

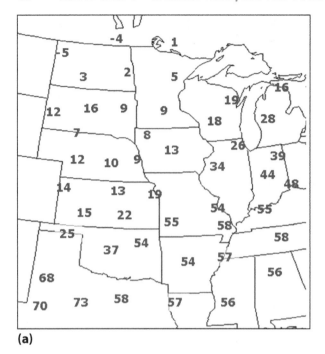

(a)

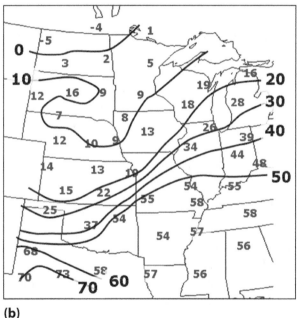

(b)

FIGURE 1.11 (a) A forest of temperature observations from 00 UTC on March 5, 2003. (b) The data in part (a) with lines of equal temperature, called isotherms, drawn.

(a)

(b)

FIGURE 1.12 (a) Given the camera angle, this forest of pines looks like a haphazard collection of trees; (b) From a different perspective, the forest of pines shows a highly organized pattern of rows. In similar fashion, isoplething allows meteorologists to see important patterns in seemingly disorganized forests of weather data.

Hiking on trails through the wilderness can be fun, but it's always a good idea to take along a map, just in case. If the map is of good repute, it provides spatial information, conveying the horizontal location of waterfalls, mountains, white-water rapids in rivers, and other natural features. A flat map shows these features in two dimensions in terms of east-west (the x-axis) and north-south (the y-axis) directions. However, a description of a third dimension (the vertical, or z-axis) is often needed as well. In Figure 1.13a, the axes representing these three dimensions are included with a photograph of Mt. Marcy, the highest peak in New York State at 1629 m (5344 ft) above sea level. In Figure 1.13b, the trail from the point where the photograph in Figure 1.13a was taken (looking toward the summit of Mt. Marcy) appears in two dimensions, relative to the x- and y-axes. Without the **contours** of constant elevation on this topographic map, the nearly straight trail looks like a leisurely jaunt in the horizontal rather than an arduous, steep climb to the summit—if you look at a cross section of the trail in Figure 1.13c, you get an inkling of the grade of the remaining ascent to the top. Instead of climbers marking every foot of the trail with elevation signs (which would be a real eyesore for Mother Nature), information about the change in elevation is traditionally shown on a two-dimensional map (a topographic map).

To better visualize contours, let's take matters into our own hands! Set the palm of your hand flat on a table, then make a fist. Grab a pen with your other hand, and carefully draw a circle around your highest knuckle, parallel to the surface of the table. Now draw another circle parallel to and slightly below the first circle, and continue the process, dropping the same distance each time you begin a new curve. The result of such hand-painting is shown in Figure 1.14. If you imagine that your knuckles are a series of four mountains, Figure 1.14a shows a side, or **cross-sectional view** (cross section) of this "mountain range": This two-dimensional view displays the vertical dimension and one horizontal dimension. Figure 1.14b shows a

top-down, or **plan view** of the knuckles, showing the two horizontal dimensions: This view demonstrates that contours provide information about elevations in a two-dimensional setting.

Note in the plan view that the highest "mountain," Mount Index Finger Knuckle, has the greatest number of contours below its "peak"—these contours, by necessity, must be closer together (on average) in the vicinity of this highest "peak" than elsewhere. This relationship between the packing of isopleths and the amount of change in the isoplethed variable will serve us well as we now delve into the interpretation of isoplethed weather maps.

Isoplething Weather Maps: Getting Down and Dirty with Data

There are some important differences between drawing contours on your hand and drawing isopleths on a

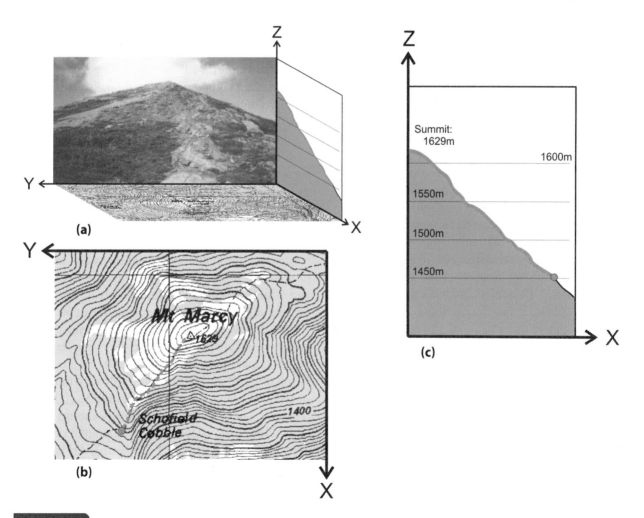

FIGURE 1.13   (a) Photograph of Mount Marcy, NY, with three-dimensional coordinate system included; (b) Two-dimensional (*x-y* plane) topographic map showing the trail from the point at which the photograph in Figure 1.13a was taken to the summit of Mount Marcy. Elevation is in meters; (c) A cross section (*x-z* plane) of the trail on Mount Marcy.

(a)

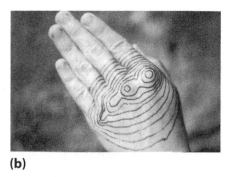

(b)

FIGURE 1.14 (a) Side or cross-sectional view of a contoured hand; (b) Top-down or plan view. From *Boy Scout Handbook*, 10th Edition by R. Birkby. Copyright 1990 by Boy Scouts of America. Reprinted by permission.

weather map. First, the contours in Figure 1.14b loop around and close on themselves. In doing so they help pinpoint the "mountain peaks." When you isopleth weather data, this won't always happen. Yes, sometimes an isopleth will close on itself and thus identify a "pocket" of higher or lower value—this is particularly true of air pressure. But more often than not, an isopleth on a weather map will simply stop at the edge of the map, where the data end.

Second, while the three-dimensional surface of the hand provides elevation information at all points, weather data are typically observed only at selected locations. As a result, it is usually necessary to estimate the values between observation sites using interpolation. This technique often requires a little detective work, using values from several nearby locations. Isoplething is based on the principle that the parameter being isoplethed is continuous in space: In other words, if a parameter has a value of 10 at one location and a value of 20 at another, the parameter must assume every value between 10 and 20 at least once between the two locations. Thus, an 11-isopleth, a 12-isopleth, a 13-isopleth, and so on, would "pass between" the two points. Furthermore, we might reasonably estimate that a value of 15 occurs at a point halfway between the sites with values 10 and 20.

To visualize how to isopleth a field of weather data (let's use temperature as an example), imagine a rollerblader who has to race through a course while successfully maneuvering through a series of flags (see Figure 1.15a). With this image in mind, imagine your pencil as the rollerblader and temperature observations on a weather map as the flags on the course. For the sake of argument, let's suppose you want to draw the 70°F isotherm, so think of your pencil as a rollerblader who's wearing a "70" on his or her T-shirt. In order to draw the 70°F isotherm (and thereby complete the course), your pencil can only "skate" between a

flag marked by a number greater than 70 and another flag marked with a number less than 70 (see Figure 1.15a). You can follow the rollerblading approach to isoplething in Figure 1.16a. Note how your pencil should "skate" between "temperature flags," on one side lower and on the other side higher than 70; you should "hit" temperature flags that are marked by a 70. If you're not careful where your pencil "skates" and you happen to pass between values of 63 and 68 (for example), you'll draw an illegal isopleth (in rollerblading competition, you'd be disqualified).

After completing the 70°F isotherm, suppose you now wish to isopleth temperatures for every multiple-of-ten that appears on this particular weather map. No problem—we just need more rollerbladers! So a skater wearing number 60 might logically be next to go. A racing rollerblader would have to skate through flags marked by numbers slightly greater and slightly less than 60 (see Figure 1.15b). Similarly, successfully drawing the 60°F isotherm requires you to guide your pencil between temperature flags marked by numbers slightly greater than and slightly less than 60 (see Figure 1.16b), while passing directly over any temperatures of 60.

Next, follow a similar strategy for drawing the 80°F isotherm, except you should pass between flags with values greater than and less than 80 (Figure 1.15b). If you draw all possible isotherms that are multiples-of-ten, you can now see the forest in spite of the trees (Figure 1.16c). Note that all of the isotherms started and ended at the edges of the temperature data (except, in this case, the 90°F isotherm). An isotherm that loops around and closes on itself marks a local maximum (or minimum) of temperature—essentially a "pocket" of relatively warm or relatively cold air.

To make your isoplething as accurate as possible, there's one other guideline to keep in mind. As an example, let's consider drawing a 40 isopleth between

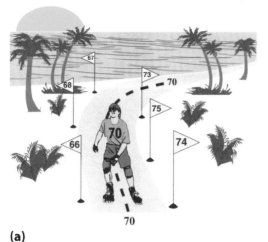

(a)

FIGURE 1.15 (a) Isoplething can be thought of as rollerblading between "flags" of data. The rollerblader represents a specific value of the parameter to be isoplethed (in this case, 70), and must maneuver between flags so that values greater than 70 remain to one side of his path, while values less than 70 remain to the other. A flag marked with 70 would be crossed directly (painful for the rollerblader, but essential for the isoplether); (b) Here, the 60 and 80 isopleths are added.

(b)

data points of 39 and 44. In doing so, the 40 isopleth should pass closer to the 39 than the 44 simply because 40 is closer to 39 than 44. So the general rule is this: The "best-fit" location where an isopleth passes between two points should reflect the relative closeness of the given isopleth to the values of the points it's passing between.

Of course, temperature is not the only weather variable that meteorologists routinely isopleth—others include air pressure, wind speed, and relative humidity, just to name a few. Even snowfall from a winter storm can be isoplethed. Indeed, Figure 1.17a is an isoplethed map of total snowfall from Hurricane Sandy. Surprised? Although Hurricane Sandy is infamous for the destruction that it inflicted on parts of the Mid-Atlantic Coast, the storm produced a few feet of snow in parts of the central Appalachians (the trademark of snowstorms in the eastern United States is that the heaviest snows often fall at high elevations, in part because mountains force air to rise and rising air is associated with clouds and precipitation).

Note that in this area, isopleths close on themselves, a tell-tale sign of a maximum (or minimum).

As a matter of practice, meteorologists typically use a constant interval between isopleths (for example, 5°F and 10°F intervals are common for isotherms). However, some fields of data require other isopleths in order to provide a more complete analysis. Indeed, note that Figure 1.17a contains the 1-inch and 3-inch isopleths, which give you a better sense for the spatial extent and overall pattern of Sandy's lighter snowfall. Also note that the 30-inch contour was not drawn, even though parts of West Virginia and North Carolina received snowfalls above 30 inches. This isopleth was omitted in order to reduce the clutter on the map. Indeed, interpreting the map would get pretty difficult if this value had been drawn, especially in the mountains of North Carolina where isopleths are already very close together.

An alternative, more colorful way to present the snowfall information in Figure 1.17a is shown in Figure 1.17b. Here, different colors represent various

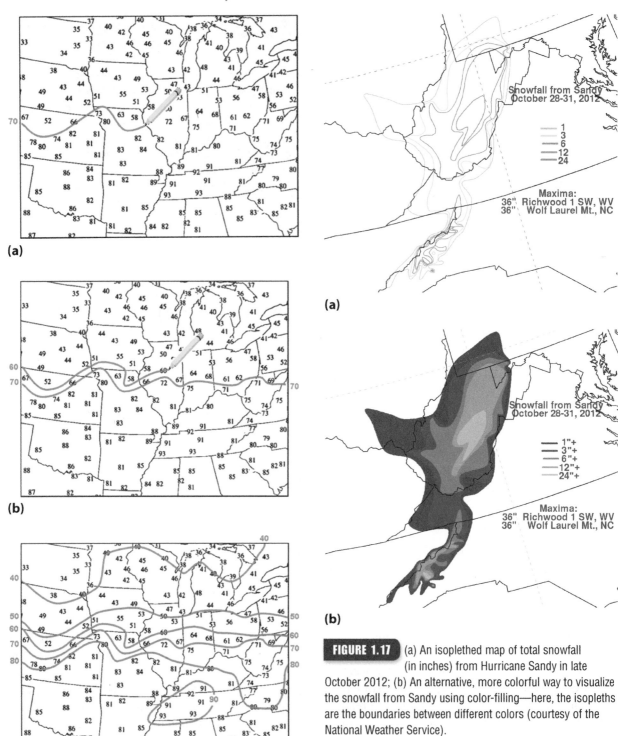

(a)

(b)

(c)

FIGURE 1.16 (a) Starting to draw the 70°F isotherm. The pencil must "skate" between observations that are greater than 70°F and those that are less than 70°F; (b) Once the 70°F isotherm is completed, work can begin on the 60°F isotherm; (c) Once all multiples-of-ten are isoplethed, patterns in the data emerge, including the pocket of relative warmth centered in Tennessee.

(a)

(b)

FIGURE 1.17 (a) An isoplethed map of total snowfall (in inches) from Hurricane Sandy in late October 2012; (b) An alternative, more colorful way to visualize the snowfall from Sandy using color-filling—here, the isopleths are the boundaries between different colors (courtesy of the National Weather Service).

ranges of snowfall totals. For example, the lightest blue shading represents regions that recorded snowfalls greater than 24 inches. Note the relationship between Figure 1.17a and 1.17b: The isopleths in Figure 1.17a are essentially the boundaries between the filled colors in Figure 1.17b. Sometimes, instead of an array of colors, different gray shades or patterns of hatching are used on such maps.

As a matter of practice, the following general rules should be used as guidelines when isoplething a map of data:

- The analysis should be neat and smooth, and all isopleths should be labeled at least once.
- Only the area on the map where data are available should be isoplethed.
- An isopleth should either close on itself or end at the edge of the data field. Isopleths should never branch or fork.
- Isopleths should be drawn at equal intervals (although for some data, it may be informative to plot a special isopleth, such as the 32°F isopleth for temperature).
- Once an isopleth is drawn, it can be used in combination with the available data to place additional isopleths.

The technique of isoplething is indispensable for detecting patterns in meteorological data and estimating values of the data at locations where observations are not taken. Throughout this text, you will get plenty of practice at interpreting isoplethed maps.

Caution! Gradient Ahead: Weather Conditions May Change Rapidly

Figure 1.16 shows large variations in temperature from northeastern Kansas into Nebraska, while Figure 1.17 shows dramatic changes in snowfall amounts from north to south in northeastern Pennsylvania and southern New England. These large changes over relatively small distances can be diagnosed by the tight packing, or closeness, of the isopleths in these areas (or, in the case of Figure 1.17b, by a rapid change in color). The change in a variable over a given distance is known as the **gradient** of that quantity. The word gradient is related to "grade," a term often used to describe the steepness of a slope of a mountain or hill.

Figure 1.18 is a topographic map of a portion of the Vail Ski Resort in Colorado, with contours drawn every 200 ft. The peak of Vail Mountain lies at the center of the closed contour marked by an elevation of 11,250 ft in the lower right corner. Assume the thick line between dots represents a ski trail from the top (point A) to the bottom (point G) of Vail Mountain. This ski trail is broken into six segments, each a half-mile long: AB, BC, CD, DE, EF, and FG. Which of these segments is the steepest, on average? In other words, on which segment does elevation change most

rapidly with horizontal distance? Qualitatively, we're interested in the largest gradient. Visually, we're looking for the part of the ski trail on which the contours are packed closest together, on average. Mathematically, the gradient of any portion of the ski trail is computed as follows:

$$\text{Gradient between two points} = \frac{\text{difference in elevation between the points}}{\text{distance between the points}}$$

The elevation of each point, in feet, is shown on Figure 1.18 (if these values weren't given, they could easily be estimated from the contours). Then, because each segment is a half-mile long:

gradient of segment AB = (11250 ft − 10550 ft) / 0.5 mi
= 1400 ft/mi

gradient of segment BC = (10550 ft − 10480 ft) / 0.5 mi
= 140 ft/mi

gradient of segment CD = (10480 ft − 9680 ft) / 0.5 mi
= 1600 ft/mi

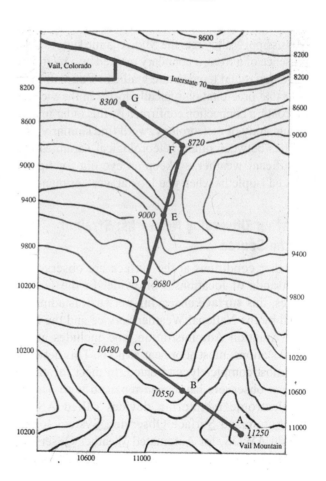

FIGURE 1.18 Contour map of a portion of the Vail Ski Resort. Contours are drawn every 200 feet. The thick line between dots represents a hypothetical ski trail from the top to the bottom of Vail Mountain.

gradient of segment DE = (9680 ft – 9000 ft) / 0.5 mi
= 1360 ft/mi

gradient of segment EF = (9000 ft – 8720 ft) / 0.5 mi
= 560 ft/mi

gradient of segment FG = (8720 ft – 8300 ft) / 0.5 mi
= 840 ft/mi

Because segment CD has the largest gradient, it is the steepest, on average. On the other hand, segment BC would be most suitable for novice skiers because it is the least steep.

Just as skiers carefully analyze the "grade" of their paths, meteorologists often look for large gradients of temperature, pressure, and other weather variables as they define areas of active or rapidly changing weather. For an example, check out Figure 1.19 which is a radar image (like ones you've probably seen on television or the web) from 12 UTC on March 3, 2014 during one of the many storms that affected the eastern United States during that long winter. Surface isotherms are overlaid on the radar image. The relatively tight packing of the isotherms from the lower Mississippi Valley into the Mid-Atlantic suggests the presence of a front, a boundary marked by a relatively large horizontal temperature gradient. You will learn later that precipitation typically forms in the vicinity of fronts, a connection confirmed by this radar image. In our study of meteorology, we'll find similarly useful ties between other meteorological variables and significant weather, so keep an eye out for tightly packed isopleths when you look at weather maps!

Weather Observing Networks: Straight to the Source

Weather conditions at the surface are observed at thousands of locations worldwide. In the United States, this **surface observing network** is administered by the National Weather Service and the Federal Aviation Administration, and includes about 1500 stations, most at or near airports. This network is almost completely automated. In most locations, human observers have been replaced with a suite of electronic monitoring devices called ASOS, for A̲utomated S̲urface O̲bserving S̲ystem (or its cousin AWOS, for A̲utomated W̲eather O̲bserving S̲ystem). An ASOS is shown in Figure 1.20. Typically, an ASOS measures temperature, humidity, air pressure, wind speed and direction, cloud cover and cloud height, horizontal and vertical visibility, and weather (such as rain, snow, or fog). The automation of the observing network allows more frequent and

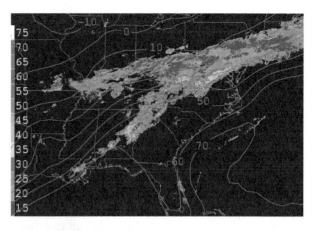

FIGURE 1.19 A radar image at 12 UTC on March 3, 2014, with isotherms (red, in °F) overlaid. The blues, greens, and yellows indicate a swath of precipitation associated with a storm and its associated fronts moving through the eastern United States. Note the general correlation between the relatively large gradient in temperature and the location of the precipitation (courtesy of the National Weather Service).

more remote weather observations, and the new system has performed well for measuring basic weather elements such as temperature, pressure and wind. But ASOS cannot measure snowfall depth (among other things), so a human observer still augments the reports at some ASOS sites. If the truth be told, some meteorologists still long for the days when humans took all the observations.

Increasingly, other organizations are setting up their own surface observing systems to supplement (on a smaller spatial scale) the federal government's network. For example, the Iowa Environmental Mesonet (for M̲eso̲scale Net̲work) is a partnership between the University of Iowa, Iowa State University, the Iowa Department of Transportation, and the National Weather Service. Figure 1.21 was produced using data gathered from this Mesonet on January 4, 2014, as Arctic air poured into the Midwest behind a powerful cold front that eventually produced temperatures around −20°F. Many other states and private organizations have similar systems, though the quality of the data varies from network to network.

To supplement surface weather observations on land, a network of buoys collects hourly weather data over U.S. coastal waters and the Great Lakes (see Figure 1.22). In addition to standard weather variables such as temperature, wind, these buoys typically measure water temperature, wave height, and wave frequency.

FIGURE 1.20 An Automated Surface Observing System, or ASOS. The weather instruments in this array typically measure temperature, humidity, air pressure, wind speed and direction, cloud cover and cloud height, horizontal and vertical visibility, current weather (such as rain, snow, or fog) and precipitation amount (courtesy of the National Weather Service).

FIGURE 1.22 Two technicians from the National Data Buoy Center make repairs to a NOAA weather buoy moored about 40 miles south of Jonesport, Maine. The instruments aboard the buoy had been damaged in a storm a few months earlier. The Coast Guard cutter *Thunder Bay* stands by (courtesy of NOAA and the United States Coast Guard).

Systematic observations of pressure, temperature and moisture above the surface are made at regular intervals using **radiosondes**, packages of weather instrumentation carried aloft by balloons. Since the late 1930s, radiosondes have been the meteorologist's principal tool for routinely observing weather conditions in the lowest 30 km (19 mi) or so of the atmosphere. Most radiosondes are tracked so that wind direction and speed can also be computed as the balloon ascends. In con-

trast to surface observations, these systematic upper-air measurements are made at fewer locations (about 100 in North America) and at less frequent intervals (twice a day, at 00 UTC and 12 UTC). Sites in this **upper-air observing network** are shown in Figure 1.23.

Increasingly, radiosonde data are supplemented by observations from aircraft and from satellites. These additional sources of upper-air data are vitally important, primarily because the wide spacing between radiosonde observing sites and the infrequency at which such observations are made limit our ability to accurately describe the state of the atmosphere aloft (to a lesser extent, this is true of surface observations as well). One reason that these limitations are important is that the accuracy of computer model forecasts (formally called numerical weather prediction) is sensitive to the quality of the observations fed into the models to begin the simulation—that is, the more accurately a computer model can virtually replicate the initial state of the atmosphere, the better the computer forecast should be (if you put "garbage in," you'll get "garbage out").

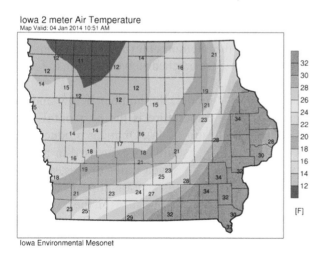

Iowa 2 meter Air Temperature
Map Valid: 04 Jan 2014 10:51 AM

Iowa Environmental Mesonet

FIGURE 1.21 Temperature (in °F) is color-coded on this map of Iowa from the morning of January 4, 2014, produced by the Iowa Environmental Mesonet. Each temperature marks the location of an observing site (courtesy of Iowa Environmental Mesonet).

The Station Model: Templates of Local Observations

On weather maps, meteorologists use a standard template to display the data observed at a particular location at a given time. In this template, appropriately called a **station model**, certain types of data (such

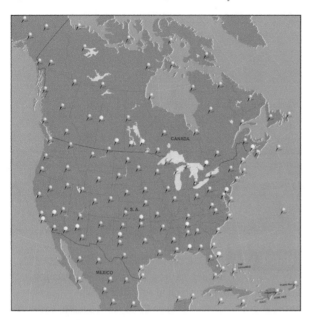

FIGURE 1.23 Red dots mark the North American locations where radiosondes are routinely launched each day at 00 UTC and 12 UTC, while yellow dots indicate sites where launches are made irregularly.

as temperature, wind, and pressure) must appear in designated places. An abridged version of the station model for data observed at the surface is given in Figure 1.24a. In this simplified surface station model, nine weather variables can be simultaneously displayed. We will paint most of the nine weather variables onto the surface station model canvas here: cloud cover (F), temperature (T) and dew point (T$_d$), wind speed (S) and wind direction (D), visibility (V), and obstructions to visibility (W).

Representing cloud cover in a station model is fairly straightforward. The fraction of the sky covered by clouds over a station determines how the inside of the circle is filled in, using the guidelines shown in Figure 1.24b. Informally, "Few" would be equivalent to mostly sunny or mostly clear, "Scattered" corresponding to partly cloudy or partly sunny, "Broken" akin to mostly cloudy, and "Overcast" similar to cloudy. "Sky Obscured" means that the observer (or observing equipment) cannot determine what percentage of the sky is covered by clouds because of obstructions to visibility (heavy snow, for example).

Temperature ("T" in the template) appears in the ten o'clock position relative to the cloud-cover circle. The units of temperature depend on the country of origin of the map: For example, weather maps produced at the National Centers for Environmental

Prediction in Washington, DC, use °F, while maps drawn by meteorologists at the European Center for Medium Range Forecasting use °C.

The number that appears to the lower left of the station circle in the station model is the dew-point temperature, or **dew point** for short (T$_d$ in the template). By way of introduction, the dew point is the approximate temperature to which the air must cool (at constant pressure) in order for water vapor (a gas) to condense into liquid water (see Figure 1.25). We'll delve deeper into the concept of dew point in Chapter 4. In general, the higher the dew point, the more water vapor there is in the air. To "wet" your whistle here, check out Table 1.6 which shows how the value of the dew point relates to your general level of "comfort."

The line representing wind direction in the station model ("D" in Figure 1.24a) points in the direction *from which* the wind is blowing. The **wind vane**, the standard device used to measure wind direction, also points into the wind. The standard compass angles allow us to get our directional bearings, with 0° corresponding to north, 90° to east, 180° to south, and 270° to west (see Figure 1.26).

The wind speed in the station model is given in units of **knots** (1 kt = 1.15 mph). This unit of measure owes its existence to the regularly spaced knots tied in ropes that were cast overboard from mercantile ships in bygone days to determine the vessel's speed. As Figure 1.24c shows, long and short "feathers" (representing 10 kt and 5 kt, respectively) are attached from the end of the wind-direction line to indicate the wind speed. The standard meteorological instrument for measuring wind speed is the **anemometer** (both an anemometer and a wind vane are perched atop the pole in the center of Figure 1.20).

In the station model, the number below and to the left of the temperature is the horizontal visibility ("V" in Figure 1.24a). On maps produced in the United States, visibility is expressed in miles. If visibility lowers to seven miles or less, a symbol for the offending obstruction appears beneath the temperature and to the right of the visibility ("W" in the station model template). There is one exception to this rule: Precipitation, no matter how light or how little it affects visibility, is always shown on station models. Symbols for some of the most frequently observed obstructions to visibility are shown in Figure 1.24d.

The Surface Station Model

(a)

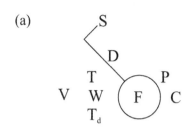

F	Fraction of cloud cover
S	Wind speed (in knots)
D	Direction from which the wind is blowing
T	Air temperature (in °F)
V	Visibility (in miles and fractions)
W	Obstruction to visibility (weather)
T_d	Dew point temperature (in °F)
P	Sea-level pressure
C	Change in sea-level pressure over the last 3 hours

(b) Fraction of Cloud Cover
(classified by fraction of sky which is cloudy)

Clear (0/8 coverage)

Few (1/8 - 2/8 coverage)

Scattered (3/8 - 4/8 coverage)

Broken (5/8 - 7/8 coverage)

Overcast (8/8 coverage)

Sky Obscured

(c) Wind Speed and Direction

S	MPH	Knots
◎	Calm	Calm
	1-2	1-2
	3-8	3-7
	9-14	8-12
	15-20	13-17
	21-25	18-22
	26-31	23-27
	32-37	28-32
	38-43	33-37
	44-49	38-42
	50-54	43-47
	55-60	48-52
	61-66	53-57
	67-71	58-62
	72-77	63-67
	78-83	68-72
	84-89	73-77

(d) Obstructions to Visibility

Light snow

Moderate snow

Heavy snow

Snow shower

Light rain

Moderate rain

Heavy rain

Rain shower

Light drizzle

Thunderstorm with rain

Severe Thunderstorm

Sleet

Freezing rain

Fog

Haze

FIGURE 1.24 (a) Template for a simplified station model, with short descriptions of entries; (b) Guidelines for entering cloud cover into the station model; (c) Guidelines for entering wind direction and wind speed into the station model; (d) Symbols used in the station model for some common obstructions to visibility.

FIGURE 1.25 As the air temperature fell to the dew point on a clear, cool morning, the stage was set for water vapor to condense into drops of dew (released into the public domain by the photographer via Wikipedia.).

TABLE 1.6

Values of the dew point and descriptors of the general level of human comfort associated with those values.

Dew Point	General Level of Comfort
60°F	For most people, the air starts to feel a bit humid
65°F	The air feels moderately humid
70°F	The air feels sultry and tropical
75°F	The air feels oppressive and stifling

Figure 1.27 is an example of a station model that depicts some very interesting weather. When decoded, here's how this information reads: The temperature is 78°F, the dew point is 75°F, and the sky is overcast (cloudy). Winds are from the east at

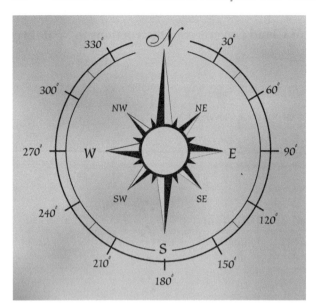

FIGURE 1.26 A compass shows the standard wind directions used in a station model. Remember, the wind is reported in terms of the direction *from which* it blows. So a west wind (from a 270° direction) blows *from* the west.

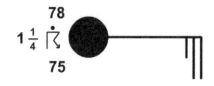

FIGURE 1.27 An example of a station model.

25 kt (the actual wind could be anywhere from 23 to 27 kt). The visibility is one and a quarter miles, and a thunderstorm (with rain) is occurring.

Figure 1.28 is an analyzed map of station models at 00 UTC on April 28, 2011, a few hours after a swarm of violent tornadoes tore through parts of Mississippi and Alabama, including the city of Tuscaloosa. **Isobars** (isopleths of air pressure—more on this topic in Chapter 6) are shown, enabling the placement of centers of high and low pressure and several fronts (the lines with triangles and circles on them). All the symbols, lines, and numbers on each station model are like branches and foliage on individual trees—drawing isobars is one way to help you see the forest through the trees. In the forthcoming chapters, you will learn more about all the weather variables in the station model and their isopleths. For now, note the southerly winds to the east of the cold front (the blue line)—this warm and moist flow helped to transport relatively high values of dew point (in the 65–75°F range) northward, paving the way for the severe thunderstorms that spawned the deadly tornadoes that day.

Data collected by ocean buoys are also organized into station models, although buoy station models include wave heights and other measures of the state of the sea. Ocean buoys can provide crucial information to meteorologists, particularly to tropical forecasters,

because buoys and ships provide the only surface observations in and around hurricanes while these monstrous storms are totally over water. Figure 1.29 shows the increase in significant wave height at a buoy off Cape Hatteras, NC, as Hurricane Sandy approached, peaking above 30 feet on October 28, 2012. For the record, "significant" indicates that only the average height of the highest one-third of waves is being considered. It's worth noting that National Hurricane Center forecasters often reference buoy data in their discussions as providing key information in determining the intensity of tropical weather systems, not surprising given that such data may be the "only game in town" when it comes to surface observations over the ocean.

Maps of station models give forecasters a broad perspective of weather conditions at a single time. Forecasters also like to look at how weather conditions evolve with time at specific locations. To graphically organize local weather data with time, forecasters use a template of graphs called a meteogram. Think of them as messages in time.

METEOGRAMS: MESSAGES IN TIME

Meteorologists often find it helpful to display weather observations or forecasts at a single location as a function of time—typically, information is plotted every hour over a 24-hour period. Such a "time display" of weather data is called a **meteorogram**. In some circles, "meteogram" is preferred instead, probably because it's easier to say (try saying "meteorogram" five times fast). Although the glossary of the American Meteorological Society lists only meteorogram, we will use "meteogram" in this text.

Figure 1.30 is a meteogram for Fargo, North Dakota, from 11 UTC on January 19, 2013, to 12 UTC on January 20—the heading reveals the weather station identifier for Fargo's Hector International Airport, KFAR. For quick reference, the start and end times appear along the bottom of the meteogram (we have rounded 1053 UTC on January 19 to 11 UTC, and 1153 UTC on January 20 to 12 UTC). At first

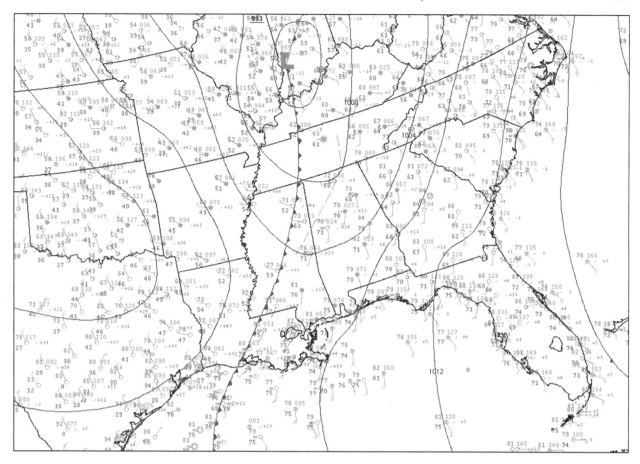

FIGURE 1.28 Station models on an isoplethed map of air pressure, from 00 UTC on April 28, 2011. The data in the station models enable the analysis of the map, including isobars, fronts, and highs and lows of pressure (courtesy of the National Weather Service).

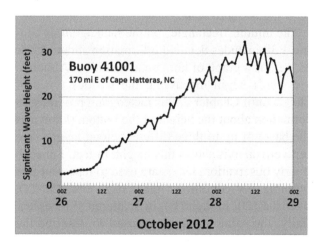

FIGURE 1.29 The significant wave height measured at Buoy 41001, moored approximately 170 mi (275 km) east of Cape Hatteras, NC, from 00 UTC on October 26, 2012 to 00 UTC on October 29, as Hurricane Sandy approached and then passed the buoy (courtesy of National Data Buoy Center).

glance, interpreting such a diagram might seem like a daunting task, but it's really rather straightforward once you get your bearings—we will take you through the basics, but keep in mind that many variations of this standard meteogram exist.

Let's start with the time period. For reference, short, vertical tic marks staggered along the bottom of the meteogram indicate the date and time of the observations—here, observations are plotted hourly (which is typical). Focus your attention on the top-most "rectangle" of the meteogram which includes three graphs that show the time variations of air temperature (in purple), dew point (green), and relative humidity (blue). The scale for temperature and dew point (in °F) lies on the left, while the scale for relative humidity (in percent) appears on the right. Note that the hourly air temperatures are listed below the graph (the row labeled "TMPF" on the left), so you don't have to interpolate a specific hourly

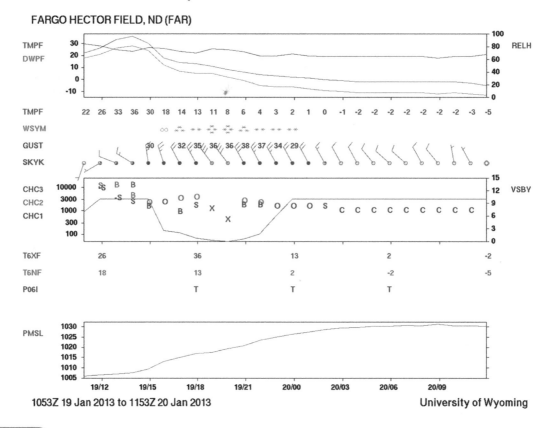

FARGO HECTOR FIELD, ND (FAR)

1053Z 19 Jan 2013 to 1153Z 20 Jan 2013   University of Wyoming

FIGURE 1.30   A standard meteogram, in this case documenting weather observations in Fargo, North Dakota, during the 25-hour period beginning at 11 UTC on January 19, 2013 (courtesy of Department of Atmospheric Sciences, University of Wyoming).

temperature from the temperature trace. So, for example, the temperature at 20 UTC on January 19 was 8°F. In contrast, we must estimate the dew point and relative humidity from their graphs—certainly more challenging—roughly, we might guess 5°F for the dew point and a bit below 80% for the relative humidity at 20 UTC (by the way, we'll have plenty more to say about relative humidity in Chapter 4).

Below the line of hourly air temperatures is the current weather ("WSYM"). In this row, precipitation and obstructions to visibility are represented using the same conventions used in the surface station model. At 20 UTC on January 19, for example, heavy snow was falling at Hector Field. As in the station model, precipitation is always reported if it is occurring, but non-precipitating obstructions to visibility (such as fog and haze) are only reported when the horizontal visibility is seven miles or less.

Peak wind gusts ("GUST"), expressed in knots, appear on the next line, followed (on the line below) by sky conditions ("SKYK"), wind direction, and wind speed. Again, the same conventions for cloud

cover and winds that you learned for surface station models also apply here. So, for example, at 20 UTC on January 19, the sky was obscured, sustained winds were 30 kt (in reality, the range would be from 28 to 32 kt), gusting to 36 kt.

The middle "rectangle" on the meteogram in Figure 1.30 includes detailed information about clouds as well as a graph of horizontal visibility. Although we won't begin discussing the various types of clouds until Chapter 2, this meteogram provides information about the height of the bottom (known as the base) of up to three different cloud layers—the left vertical axis shows this height, in feet. For each hourly observation, letters are used to represent five possible sky conditions: clear (C), obscured (X), overcast (O), broken (B), or scattered (S)—essentially, the same terminology used to describe the fraction of cloud cover in the surface station model (refer back to Figure 1.24). As an example, at 17 UTC on January 19, 2013, low clouds were broken (B), with the associated cloud base near an altitude of approximately 1000 feet. The sky was also overcast

(O) at an altitude slightly above 3000 feet (but below 10,000 feet). We note that you'll sometimes see coding such as "-S" and "-O" on meteograms—the "-" stands for "thin," so "-S" means "thinly scattered" or, more succinctly, "a few clouds." Similarly, "-O" indicates a thin overcast.

The middle "rectangle" also includes a graph of horizontal visibility (the thin brown line), in miles (the visibility scale is on the right). In this case, visibility tells an interesting story—at 12 UTC on January 19, the horizontal visibility was 10 miles, but, eight hours later at 20 UTC, it had dropped to nearly zero with heavy snow. With the wind howling at 30 kt at that time, gusting to 36 kt, the makings of a blizzard were in progress at Fargo's Hector International Airport. We say "makings of" because, officially, a blizzard requires three consecutive hours of sustained winds (or frequent gusts) of 30 kt (35 mph) or higher with visibility reduced to less than a quarter mile by falling and/or blowing snow. Truth is, this extreme combination is rarely achieved for three straight hours—and it wasn't in this case—but blizzard or not, conditions were clearly very harsh.

Continuing on, note the three rows of data below the cloud layer and visibility information, labeled T6XF, T6NF, and P06I. The row marked T6XF shows the maximum temperature measured in a six-hour period ending at 00 UTC, 06 UTC, 12 UTC, and 18 UTC—these are known as the *standard* synoptic times (03 UTC, 09 UTC, 15 UTC, and 21 UTC are *intermediate* synoptic times). So, for example, the maximum temperature at Fargo's Hector Field in the six-hour period from 18 UTC on January 19 to 00 UTC on January 20 was 13°F. Similarly, T6NF is the minimum temperature measured in any six-hour period between successive standard synoptic times. Thus, the minimum temperature in the six-hour period ending at 00 UTC on January 20 was 2°F.

The row of data labeled P06I represents the total amount of precipitation (rain or the liquid equivalent of snow or sleet) that was measured in the six-hour period between successive standard synoptic times. The precipitation amount is expressed in inches, and if rain or liquid equivalent is less than 0.01 inches, the official measurement is registered as a trace (T)—that's what fell, for example, during the six-hour period ending at 00 UTC on January 20. But wait a minute! Wasn't heavy snow reported during this time? What's going on? Well, here's an example of a weakness of the automated observing system. In the days prior to January 19, several episodes of light snow dropped an inch or two in the region around Fargo. Then, as winds really increased on January 19 (revisit the meteogram), the snow covering the ground was whipped into the air—that is, there was lots of blowing snow. The ASOS at Hector Field failed to properly discriminate between blowing snow and falling snow and erroneously reported heavy snow. We point out that some very light snow showers did occur at Fargo that day, but almost all of the snow in the air that afternoon came from the ground, not the clouds.

Finally, the bottom graph on the meteogram is a time trace of **sea-level pressure**. Instead of expressing air pressure in bicycle-tire units of "pounds per square inch," weather forecasters typically use **millibars** (abbreviated "mb") as units of air pressure. We'll save the mathematical basis for these units for Chapter 6, but here we'll give you some meteorological markers so that you can put the values you read off meteograms into better context.

For starters, the average sea-level pressure is approximately 1011 mb. Assuming that weather associated with this value is often rather "average," let's entertain the notion of a big snowstorm moving northward along the East Coast during winter. Such a formidable low-pressure system (an "L" on a weather map) can have a sea-level pressure in the neighborhood of 960 to 990 mb.

On the flip side, a representative value for the sea-level pressure of a strong high-pressure system (an "H" on a weather map) that comes southward from Canada during winter and numbs the northern United States with an Arctic chill usually lies in the neighborhood of 1030 to 1040 mb. Such a high pressure system was generally in control during the time period represented by Figure 1.30.

We close this discussion with a friendly heads-up. The format and style of the meteograms, station models, and isoplethed charts that you'll find on the Web vary from site to site. They don't always contain the same weather data or plot it in the same way that we outlined in this chapter. Regardless of differences in content or style of these diagrams, rest assured that you now have the fundamentals necessary to begin to diagnose changes in the weather, both in time and in space. We'll have more to say about weather systems such as highs, lows and fronts in Chapter 3. But first, in Chapter 2, you'll learn how radiation from the Sun drives Earth's weather machine. We'll also reveal that the Sun is not the only energy source that shapes our weather.

Weather Folklore and Commentary

One of our goals in writing this book is to make students better weather consumers. To this end, we have reserved space in many chapters for a weather commentary or a weather folklore.

A weather folklore is sort of a homespun forecast that is based upon observations of nature. When European settlers came to the New World, there weren't any meteorologists to forecast the weather. For a developing agrarian society in an unfamiliar land, knowing as much as possible about weather and climate was vital for successful farming. To get inklings about changes in weather, farmers carefully observed their surroundings, noting nature's subtle signals and recording subsequent weather patterns. From these humble observations came folklores, some with solid footing on the ground of scientific truth. Others, shall we say, got stuck in the mud of unreliability.

Weather lores are sprinkled throughout this book. Learning about signs from the atmosphere and what they mean about weather will enrich your understanding of meteorology and make you better weather consumers. For example, a ring of light around the sun or moon, together with lowering air pressure, turns out to be a useful predictor of the arrival of precipitation within 36 hours. We will discuss this lore and more in the coming chapters.

In the world of fast-paced television, we are continually bombarded with short sound bites on the evening weather report. For example, you may have heard that the average number of tornadoes per year has more than doubled in the last 50 years. At first, this sounds alarming, but should it be? Furthermore, the age of "weatherspeak" has sometimes introduced canned explanations and expressions about the weather, which are, at the worst, incorrect, and at best, misleading. In some upcoming chapters, we will offer commentaries that address a few of these issues. Think of these weather commentaries as consumer reports on products dealing with the dissemination of weather information.

Focus on Optics

Sky Lights

While driving through bucolic central Pennsylvania, you sometimes see bumper stickers on cars that read: "If God isn't a Penn State fan, why is the sky blue and white?" (Our apologies to North Carolina Tarheel fans.) Though this light-hearted explanation of atmospheric colors may evoke a chuckle, the scientific bases for blue skies and white clouds are at the heart of a branch of meteorology called atmospheric **optics**. Rainbows, mirages, and red sunsets are a few additional phenomena that fall under the heading of atmospheric optics (see Figure 1.31 for a couple others). In short, atmospheric optics is the study of "sky lights." In many of the chapters that follow, we will delve into a fundamental topic in optics that relates to the chapter material, hoping to shed scientific light on the kaleidoscope of colors that paint our sky.

FIGURE 1.31 Crepuscular rays over Lake Balaton in Hungary in July 2012. These rays of sunlight, which stream through breaks in clouds, appear to radiate from the area of the sky where the Sun is located. Their name comes from the Latin "crepusculum," meaning twilight, in reference to their common occurrence early and late in the day (courtesy of Zsofia Biro and Jim Foster).

The Global Ledger of Heat Energy

LEARNING OBJECTIVES

After reading this chapter, students will:

- Be able to describe the spectrum of electromagnetic radiation based on the wavelength of the radiation
- Be able to describe the emission spectra for the Sun and the Earth in terms of the wavelength associated with each body's peak emission
- Understand the roles of absorption and emission in determining the temperature of any object

- Be able to describe the roles of atmospheric absorption, transmission and scattering in determining the amount of solar radiation that reaches the ground
- Be able to explain why the bases of some clouds appear dark while others appear white
- Gain an appreciation for the albedo of Earth's surface and the albedo of various kinds of clouds
- Be able to scientifically argue the shortcomings of the popular notion that the sky is blue because air molecules preferentially scatter blue light
- Understand the roles of conduction and convection in transferring heat energy from the earth's surface
- Be able to argue that clouds do not "act like a blanket" to keep the ground and the overlying air warmer at night
- Understand that the natural greenhouse effect is necessary for life to exist on Earth and that global warming is an enhancement of this natural effect
- Gain an appreciation for the buoyancy of an air parcel and its dependence on the density of the parcel versus the density of its environment

Light from the Sun travels through space in a blur, covering the 150,000,000 km to Earth (about 93,000,000 mi) in a cosmic blink of about eight minutes. But the sunshine that whizzes through space is more than just visible light. It is an ensemble of visible and invisible energy that we call **solar radiation** (see Figure 2.1). Though the word *radiation* sometimes has negative connotations, the heating provided by solar radiation maintains life on this planet and drives the Earth's weather machine.

GOES-15 SXI   04-Jun-2010 17:24:14 UTC
TinMesh   0.500s   0x1C320200
NOAA/NASA/Lockheed Martin

FIGURE 2.1   An "X-ray" of the sun. Not only does the sun give off X-rays, it emits an ensemble of visible and other invisible radiation (courtesy of NASA).

The Sun is not the only source of energy for the Earth's surface. Our planet also absorbs radiation emitted by the atmosphere (yes, clouds and gases also radiate). In fact, the Earth's surface receives more energy from the Earth's atmosphere than it does directly from the Sun! In addition, the atmosphere absorbs radiation emitted by the Earth's surface. This mutual exchange of **terrestrial radiation** ("terrestrial" means "of the Earth") between the Earth and the atmosphere, in concert with incoming solar radiation, governs how our planet warms and cools. No wonder that climate scientists are concerned that human activity (such as the burning of fossil fuels) is changing the composition of the atmosphere, thereby affecting Earth's delicate energy balance.

Lest we leave you with the impression that the Earth and its atmosphere are a closed system, we note that some terrestrial radiation is directed to space. Otherwise, the heat would really be on!

If you've ever walked barefoot on hot sand at the beach on a sunny day, you've already felt first-hand that the Earth's surface absorbs incoming solar energy and heats up. The hot sand would be even hotter if it didn't radiate energy away like crazy! But heat energy is also transferred from the hot sand to the overlying air by other processes such as conduction and convection that will be discussed later in this chapter.

The bottom line here is that the Sun is the engine that fuels our weather, and the Earth is the transmission that drives it. To get a better idea of what we mean, hop in, fasten your seat belt, and let's take a spin around the block to see how solar and terrestrial radiation, and the processes that help distribute that energy, conspire to generate weather.

RADIATION: A DOSE OF REALITY

The mere mention of the word *radiation* often elicits images of disasters and mishaps from the pages of history. Chernobyl (1986) and Fukushima (2011) are two examples of meltdowns of nuclear reactors where radiation showed its ugly side, feeding the nuclear-age notion that all radiation is inherently dangerous and figuratively speaking, causes "the sky to fall." If you are one of these believers, rid your mind of this conviction and consider this enlightening fact: *Everything emits radiation at all times*. And there are no exceptions. Everything means everything: Trees, dogs, snow, the oceans, the atmosphere, your body, and even this page, all radiate. Most of this radiation is invisible to humans and harmless in the doses present near the earth's surface. Moreover, radiation emitted by invisible gases in our atmosphere is extremely beneficial and even essential for life. In reality, we should thank our lucky stars for radiation.

The Electromagnetic Spectrum: More Than Meets the Eye

There has been much debate about the possible link between brain cancer and exposure to radiation from cellphones. According to the Food and Drug Administration, "the weight of scientific evidence does not show an association between exposure to radio frequency from cellphones and adverse health outcomes. Still, there is consensus that additional research is warranted." Research on this

connection focused on the effects of the electromagnetic fields (EMFs) which form anywhere electrical charges are in motion. To a much lesser and safer degree, we all are exposed to EMFs all the time. Indeed, every atom in our bodies has its own EMF. How are these electromagnetic fields generated?

Within each atom, negatively charged electrons orbit a nucleus of neutrons and positively charged protons. Every charged particle has an electrical field surrounding it, and moving charges generate magnetic fields (think of the magnetic field surrounding an ordinary wire carrying electricity—electric current is simply charge in motion). When an electron oscillates, it distorts its electrical and magnetic fields. Like moving your hand rapidly back and forth in a swimming pool to test the water, oscillating electrons send out ripples of energy that have magnetic and electrical properties. We call such pulses **electromagnetic radiation** (see Figure 2.2).

Like the free exchange of currency, radiation spent (emitted) by one body often gets deposited in the energy wallet (absorbed) by other bodies (which are also radiating). Whether a body emits more radiation than it absorbs or absorbs more than it emits will dictate whether its temperature will increase or decrease (or remain the same if the body has a balanced energy budget). But more on this later.

For now, the $64,000 question is: How do these energy transactions take place? A useful way to think about how radiation is transmitted is to visualize the motion of swells of water in the ocean. Much like waves of water, radiation travels as waves of energy of varying **wavelength**. As illustrated in Figure 2.2, wavelength is defined as the distance between the crest of one wave and the crest of the next wave. Considered as a whole, all wavelengths of radiation form the **electromagnetic spectrum**, shown in Figure 2.3. You are undoubtedly familiar with much

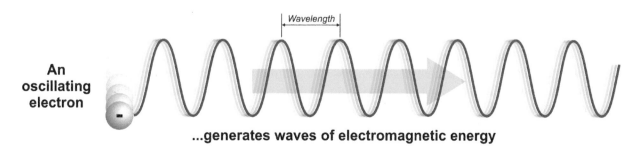

FIGURE 2.2 Oscillating electrons produce waves of electromagnetic radiation.

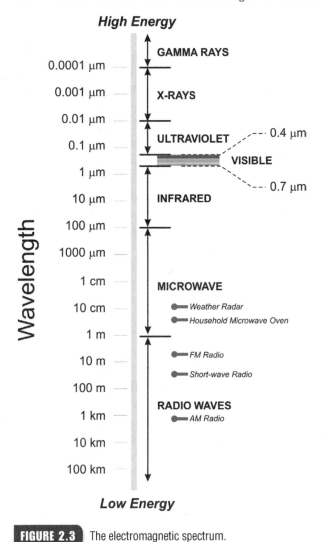

FIGURE 2.3 The electromagnetic spectrum.

and wavelength might seem like a neat and tidy way to present how nature works, but it robs us of the whole truth: A given color can typically be produced in more than one way.

Let us explain. Perhaps you remember being told that an orange looks orange (or an apple looks red) because it absorbs all the colors of the visible spectrum except orange (or red in the case of the apple), which it reflects. And because popular science equates orange with a wavelength around 0.6 μm, it follows that oranges only reflect wavelengths of (or near) 0.6 μm. Well, that's simply not true. Actually, an orange "reflects" *all* the wavelengths of visible light. Yes, all of them, but longer wavelengths more than shorter wavelengths (see Figure 2.4). This broad, lopsided spectrum of visible light reflected by the orange in all directions enters our eyes and strikes our retinas, which, in turn, send messages to our brains. Our brains then interpret the orange's spectrum of reflected light as a single color, which we learned as children to call "orange." To drive home our point about colors being produced in our brains and not our eyes, consider that we sometimes dream in color despite the absence of light on our retinas.

We will elaborate on the traps of equating color with wavelength later in the chapter when we answer "Why is the sky blue?" For now, just rid yourself of the notion that specific colors in nature are associated exclusively with a single wavelength.

Although we are literally bathed in electromagnetic radiation, and blind to most of it, the effects of this invisible radiation are all around. For example, because microwave radiation is very efficient in heating water, and most foods contain some water, microwave ovens are popular for cooking food. The signal from your favorite radio station really is carried over the "airwaves"— the sounds are transmitted as electromagnetic radiation, which is decoded by a special receiver within your radio. And the navigation systems of so-called "heat-seeking missiles" rely on sensing differences between the infrared radiation emitted by the target and infrared radiation emitted by the surrounding air or water. More fundamentally, infrared radiation emitted by the earth's surface and the atmosphere plays a central role in keeping our planet habitable, a role that will be explored in detail later in this chapter. On the down side, large doses of ultraviolet radiation (the primary source of such radiation is the sun) can cause sunburns. Overexposure to such short wavelength radiation has been linked to increased occurrences of skin cancer and cataracts.

of the terminology used here to describe various types of radiation—for example, infrared goggles can be used to "see" at night, while ultraviolet wavelengths are used in tanning beds.

Although nature imposes no boundaries between types of radiation, the human eye does (with the brain as interpreter). Our two "electromagnetic receptors" are sensitive only to the narrow range of wavelengths between approximately 0.4 and 0.7 micrometers (or microns, abbreviated μm: 1 μm = 0.000001 m). This range of wavelengths is called **visible radiation**. You may note in Figure 2.3 that we matched specific wavelengths in the visible spectrum to specific colors, but we do so with some reservations. Why, you ask? Consider that a wavelength of 0.4 microns is routinely associated with the color violet. Indeed, if radiation composed of only that wavelength were observed, you would perceive violet. Such a specific match between color

Spectrum of an Orange

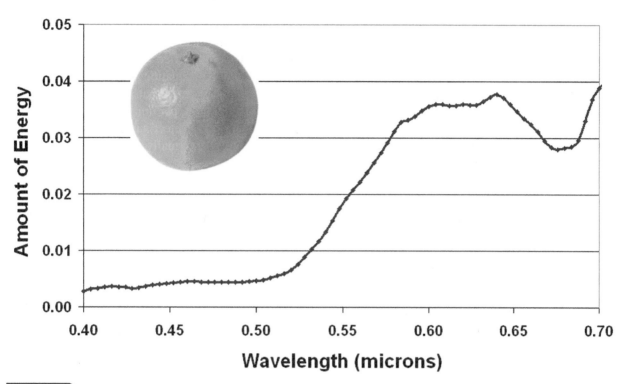

FIGURE 2.4 The spectrum of the intensity of reflected light from an orange as a function of wavelength (in microns). An instrument called a spectrophotometer collected these data (courtesy of Eugene Clothiaux).

However, even radiation that is harmful in large doses can be helpful in moderation. For example, minute doses of X-ray radiation are commonly used to diagnose bone injuries and tooth damage. Other forms of radiation, in extremely small doses, can be useful to meteorologists. For example, gamma radiation is lethal in large doses, but naturally occuring uranium in the soil emits minuscule amounts. Meteorologists take advantage of these harmless gamma-ray emissions to estimate the water equivalent of snowpacks covering some areas of the United States, particularly the mountains of the West. Aircraft flying over remote areas before the first snows of winter are equipped with instruments that detect gamma radiation emitted by the surface. In spring, aircraft fly over the same areas (see Figure 2.5), again taking measurements of the intensity of gamma radiation emitted from below. Because snow absorbs gamma radiation fairly efficiently, the intensity of gamma rays reaching the aircraft decreases with increasing water content in the snowpack. A quick comparison of the "before and after" measurements then allows meteorologists to estimate the water equivalent of the snowpack.

Knowing this water equivalent helps **hydrologists** (scientists who study the movement and distribution of water) gauge the potential for flooding in the spring when rain and melting snow can combine to produce rapid runoff. The Red River of the North, which forms the border between North Dakota and Minnesota, commonly experiences such flooding. In

FIGURE 2.5 As part of the *Airborne Snow Survey Program*, a Rockwell Aero Commander measures gamma radiation over a remote snowpack (courtesy of NOAA).

early April 2013, a 20–40 inch snowpack covered much of the drainage basin of this northward-flowing river. Figure 2.6a shows the estimated amount of water stored in that snowpack (called snow-water equivalent)—many areas draining into the Red River held 4 to 8 inches of liquid. Before and after pictures of the river taken near Fargo, ND on April 18 and April 30, 2013 (see Figure 2.6b) vividly illustrate the impact that snowmelt combined with rain can have in this river basin.

It's now time to establish an important connection between an object's temperature and the intensity and wavelength of the radiation that it emits.

Temperature and Radiation: The Movers and the Shakers

Atoms and molecules do not all move at the same speed. There are some tortoises and some hares. The effect is that every object emits *some* radiation at *all* wavelengths. However, every object has a wavelength at which it emits *peak* (the most) radiation. To shed light on this assertion, consider an ordinary light bulb. Though an incandescent bulb "brings all good things to light," the spectrum of radiation it emits does not peak in the visible range. Rather, the peak emission of an incandescent light bulb lies in the infrared. Where the wavelength of peak emission of an object lies in the electromagnetic spectrum depends solely on the temperature of the object.

To visualize the relationship between temperature and wavelength, attach one end of a rope to a post and hold the other end. Shake the free end of the rope quickly to represent high temperatures. Numerous short-wavelength ripples travel along the rope. Shaking the rope less vigorously, to represent radiation emitted by objects of lower temperatures, produces ripples of longer wavelength. Recall from the definition of temperature that the hotter an object, the faster its atoms and molecules vibrate, on average. The conclusion: *the hotter an object, the shorter the wavelength at which its peak emission occurs.*

Next let's investigate how the intensity of the peak radiation emitted by an object varies with temperature. Take a look at Figure 2.7 which shows the intensity of the radiation emitted as a function of wavelength for the Sun (average surface temperature near 6000 K, or 10340°F) and the Earth (average surface temperature 288 K, or 59°F). Note that the intensity of the energy emitted (per unit area) by the much hotter

Sun (scale on left) dwarfs the peak intensity emitted by the cooler Earth (scale on right). This difference in energy emission suggests another useful (and not surprising) relationship: *the higher the temperature of an object, the more energy it emits.* This relationship is quantified in the **Stefan-Boltzmann Law** (warning: an equation is approaching—do not panic). This law states that the amount of energy emitted by an object, per unit area (let's call it E), is proportional to the object's temperature T raised to the fourth power. Formally:

$$E = \sigma T^4 \qquad (2.1)$$

Here, the Greek letter σ (sigma) is called the Stefan-Boltzmann constant, and temperature T must be expressed in Kelvins. If $\sigma = .0000000567$ (units of Watts per square meter per Kelvin to the fourth power, or W/m^2K^4), the units of energy E are Watts per square meter (W/m^2). The use of Kelvins as the unit of temperature is essential to convey the notion that all objects emit radiation, as long as their temperature is above absolute zero (0 K).

Figure 2.7 also reinforces the previous relationship between temperature and wavelength. The Sun radiates primarily at much shorter wavelengths than does the Earth. The Sun emits about 45% of its radiation at visible wavelengths, and about 90% of its energy at wavelengths less than 1.5 μm. The wavelength of peak emission for the Sun is approximately 0.5 μm: to the brain, radiation of this wavelength produces a blue-green sensation. The cooler Earth emits its maximum radiation at infrared wavelengths, with the peak emission near 10 μm. In fact, terrestrial objects emit nearly all of their radiation at infrared wavelengths.

So far, we have focused on the sources of radiation (that is, the objects doing the emitting). Now it's time to talk about what happens to radiation on the receiving end.

On the Receiving End of Emitted Radiation: Distance and Angle

The intensity of radiation that reaches an object depends on the distance from the source of the radiation and the angle at which the radiation ultimately impinges upon the object. Let's deal with distance first, using the Sun as the source of the radiation in our example.

Start by imagining a hollow sphere surrounding the Sun. Because the Sun radiates energy in all directions,

Modeled Snow Water Equivalent for 2013 April 10, 0:00 Z

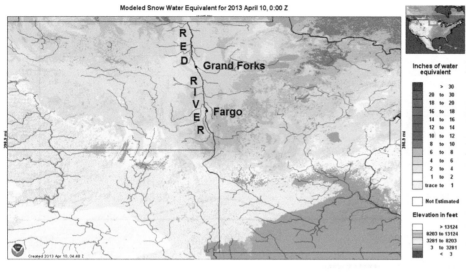

(a)

before

(b) after

FIGURE 2.6 (a) An estimate of the liquid equivalent in the snowpack in the upper Middle West on April 10, 2013. Between 4 and 8 inches of water were held in the snowpack (which was as deep as 40 inches in places) in much of the drainage basin of the Red River of the North (courtesy of NOAA). (b) Before and after photographs of the Red River of the North near Fargo, ND, on April 18 and April 30, respectively. Spring snowmelt and several bouts of rain combined to produce this flooding which was well-predicted several weeks in advance, in part using aircraft measurements of water equivalent in the snowpack (Minnesota Public Radio News. © 2013 Minnesota Public Radio®. Used with permission. All rights reserved. "http://www.mpr.org" www.mpr.org.).

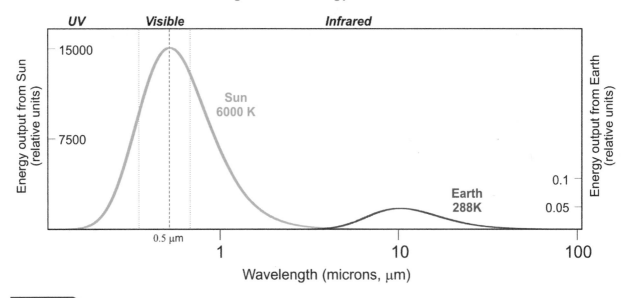

FIGURE 2.7 Comparison of the emission spectra of the Sun and the Earth. Note the huge disparity in the amount of energy emitted by the Sun (left-hand scale) and the Earth (right-hand scale). Also note that the Sun emits ultraviolet, visible, and infrared radiation, while the cooler Earth emits almost exclusively at longer wavelengths.

we can assume that sunlight penetrates the surface of the sphere uniformly. Now imagine that the sphere expands, and you are standing on its surface. As the sphere gets bigger and bigger and you retreat from the Sun, you occupy an increasingly smaller percentage of the sphere's surface. Because the total amount of radiation reaching the sphere is fixed, you receive less and less radiation as the sphere expands. Using geometry, it's possible to show that the amount of radiation that you receive decreases with the square of your distance from the Sun—this means, for example, that if your distance doubles, the amount of radiation that you receive decreases to one-fourth the original amount, and if you triple your distance from the Sun, you receive one-ninth of the original amount. And so on. By the time the sphere expands to reach the Earth, the intensity of the solar radiation has been greatly reduced (yet it still packs enough wallop to heat our planet—more on that in the next section).

Before we delve into exactly what happens when radiation interacts with an object, we must first deal with the angle at which the radiation strikes the object. To motivate this discussion, consider a boxer striking a punching bag—the more directly (straight on) that the boxer hits the bag, the more energy he transfers into the bag. But if the boxing gloves make contact at an angle, most of the boxer's energy glances off the bag.

Now, leave the gym and consider the Sun's energy incident upon the earth's surface. Figure 2.8a shows solar radiation striking the surface around local noon in early July at a location near 40°N latitude (such as Harrisburg, PA, Champaign, IL, or Boulder, CO). Figure 2.8b shows the same situation, but at local noon in early January. Note that the larger the Sun's **altitude**, or angle above the horizon, the more concentrated the radiation is when it strikes the surface. The same amount of radiation (represented schematically by segment AB) must warm a larger portion of the surface in winter (segment EF) when the Sun is lower in the sky, meaning the radiation is less concentrated at any given point on the surface. On the flip side, radiation is more intense, and consequently more efficient in heating, when it strikes a surface at the more direct angle (segment CD).

Now that we know how the intensity of the radiation impinging on an object depends on the angle at which it strikes, we can move on and discuss what can happen to that radiation as it interacts with the object.

The Three Fates of Emitted Radiation: Absorption, Transmission, and Scattering

When electromagnetic energy strikes an object, three things can happen (see Figure 2.9). Some (or all) of the incident radiation can be absorbed,

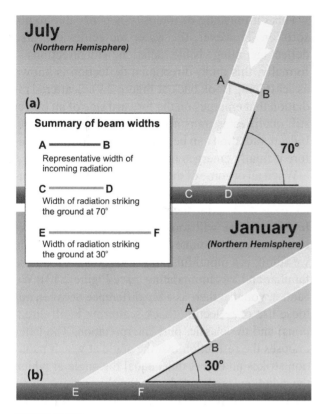

FIGURE 2.8 Comparison of noon sun angle at a location near 40°N in (a) early July and (b) early January. The lower the altitude of the Sun, the more the radiation spreads out on the surface and the less concentrated the radiation at any location on that surface.

altering the object's energy budget and likely raising its temperature. Some (or all) of the radiation can be transmitted, passing directly through the object without interacting with any of its molecules and atoms. Finally, the radiation can be scattered, which means deflected in any direction. Typically, all three processes occur at once, and the exact breakdown depends (in part) on the wavelength of the incident radiation and the physical characteristics of the object. For our purposes, the most important "objects" are the atmosphere and the Earth's surface, but the concepts apply to any substance.

The physical characteristics of the object on the receiving end play a major role in the fraction of incoming radiation that is absorbed, scattered, or transmitted. In the visible range of wavelengths, the fraction of energy absorbed by an object (appropriately called its **absorptivity**) depends, in part, on its color.

Darker-colored surfaces absorb more visible radiation than do lighter-colored surfaces: Thus, don't wear black on a sunny summer day if you hope to stay cool. As another example, snow back-scatters a large percentage of incident visible light (leaving little for absorption), a fact well known to skiers who don protective eyewear on sunny days to prevent temporary "snow blindness" (see Figure 2.10).

Absorptivity for "invisible" wavelengths of radiation is selective as well, depending on factors such as

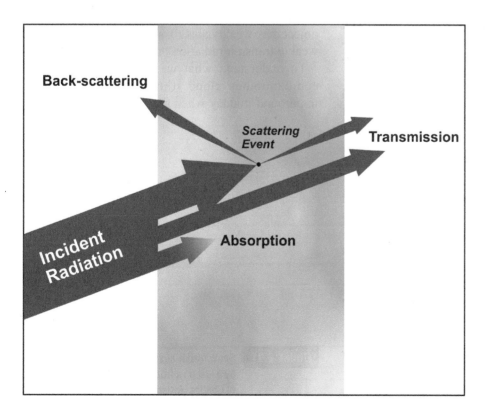

FIGURE 2.9 The possible fates for radiation when it strikes an object. Radiation can be absorbed or it can pass directly through the object without any interaction (pure transmission). The object can also scatter radiation in all directions. Scattered radiation emerging on the forward side of the object still qualifies as transmission. Scattered radiation can also be redirected back toward the source, a process appropriately called back-scattering.

FIGURE 2.10   Snow back-scatters a high percentage of the visible light that strikes it. On sunny days, sunglasses are almost always a necessity with snow on the ground (courtesy of Larry Nittler).

molecular structure. We continue the saga of snow by noting that its distaste for absorbing visible light is countered by its huge appetite for infrared radiation. Occasionally, there are subtle signs of snow's ability to absorb infrared radiation. Consider Figure 2.11, which shows a single leaf sitting on a frozen pond covered by a thin powder of snow. Incoming solar radiation is mostly scattered backwards by snow, but the darker leaf absorbs much more sunlight. Accordingly, its temperature rises and the amount of infrared radiation it emits increases. The surrounding snow absorbs infrared radiation from the leaf and starts to melt, a subtle sign of snow's huge appetite for the infrared.

Granted, the amount of infrared radiation emitted by the leaf was nothing to write home about, but it was enough that the energy absorbed by the snow exceeded the energy emitted by the snow. This surplus of energy resulted in a warming of the snow surrounding the leaf, setting the stage for it to melt.

Of course, an object may not absorb all radiation that is incident upon it. Some (or all) of the energy can pass through the object without interacting with its molecules (see Figure 2.12 where enough visible radiation has penetrated below the ocean surface to allow you to see the dolphins). Formally, in the purest sense, **transmission** of radiation means that there is no diminishing of the intensity as the radiation passes through the object. Clearly, however, in a body of water, the intensity of visible radiation diminishes as you go deeper—after all, below a certain depth, you won't be able to see unless you bring your own light source.

This diminishing of intensity is partly due to absorption. However, the water molecules can also deflect incoming visible radiation in all directions—formally, this multi-directional deflection is known as **scattering** (look back at Figure 2.9). Scattered radiation that emerges on the forward side of an object still qualifies as transmission. It's simply not pure. Indeed, radiation can be scattered multiple times before it finally emerges from the object.

When all or some of the incident radiation gets scattered back in the general direction of the source, meteorologists describe the process as **back-scattering**. In this case, we will sometimes use the word **reflection** as a loose substitute for "back-scattering" (which, we admit, is a mouthful, plus, reflection is a lot more familiar and less intimidating – see Figure 2.13). We caution you that there is a big difference between our loose use of reflection (back-scattering in all directions) and the classic, pure interpretation. The latter imposes the restriction that the angle at which radiation strikes an object must equal the angle at which the radiation is redirected from the object (think about how a billiard ball bounces off a bumper on a pool table). Just keep in mind that our sometimes loose substitution of "reflection" for "back-scattering" allows for radiation to bounce back at many different angles.

You already know from experience that there are often gray areas in life. And so it is with radiation. To varying degrees, absorption, reflection, and transmission can all happen at the same time. For example, an object may weakly absorb, strongly back-scatter and weakly transmit radiation that strikes it.

To understand what we mean, picture a thick cumulonimbus cloud (thunderstorm) approaching around midday when the Sun is high in the sky

FIGURE 2.11   Snow readily absorbs infrared radiation emitted by a leaf on a frozen lake, and melts (courtesy of Charles Hosler).

FIGURE 2.12 Water transmits some of the sunlight that strikes the surface. Thus, to some depth, you can see under water.

FIGURE 2.13 Water also reflects some of the solar radiation incident upon it.

(see Figure 2.14). Water drops and ice crystals in the tall cloud are actually weak absorbers of sunlight, but because there are so many of them, sunlight is strongly back-scattered by a cumulonimbus. Thus,

relatively little sunlight gets transmitted below the base of the cloud and the cloud bottom appears dark. Meanwhile, the top of a tall thunderstorm (and the side facing the Sun) will often appear bright—that's scattered sunlight from the cloud reaching your eyes.

Air molecules also back-scatter incoming sunlight (for more details on the back-scattering abilities of air molecules, see *Why Is the Sky Blue?* later in this chapter). Thus, the longer the path that solar radiation must travel through the atmosphere, the greater the chance the intensity at the surface will be diminished. Figure 2.15 compares the paths that solar radiation must travel in order to reach the surface at locations near the equator and at locations near the poles. Given the Earth's curvature, solar radiation impinging upon polar regions must traverse more atmosphere to reach the surface than radiation striking equatorial regions.

In fact, Figure 2.15 greatly underestimates the difference between the atmospheric path lengths for solar radiation penetrating the atmosphere. For example, radiation striking the Earth at an angle of 10° (points near the poles in Figure 2.15) must pass through nearly six times as much atmosphere to reach the surface as radiation arriving at an angle of 90° (near the equator in Figure 2.15). The longer path through the atmosphere increases the chance that the radiation will never reach the surface. This same argument helps to explain why on a

FIGURE 2.14 The base of a tall supercell thunderstorm in Weld County, CO, on June 6, 1995, was dark because little sunlight was transmitted through the thick cloud.

clear day the midday sun is almost impossible (and dangerous) to view directly, while glancing at the rising or setting sun causes less squinting.

Now that you understand what can happen to radiation when it strikes an object, let's see what effect the radiation has on the object.

Balance of Radiation: The Ins and Outs of Energy

All objects emit radiation. We hope we've convinced you. As it turns out, most objects absorb at least some of the radiation that strikes them. It's time to look at

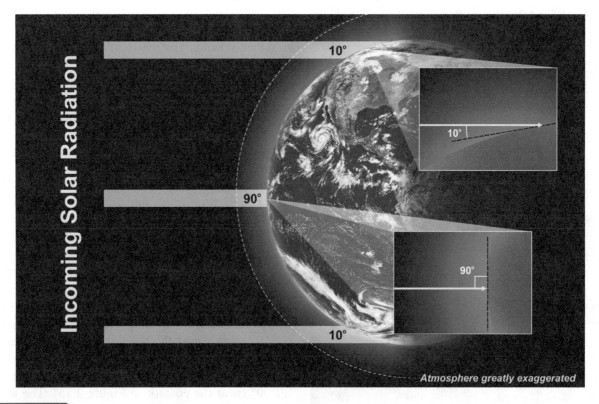

FIGURE 2.15 Solar radiation striking the earth at high latitudes must travel through more atmosphere to reach the surface than radiation striking equatorial regions. This longer path length through the atmosphere depletes the intensity of the radiation (atmosphere not to scale).

what happens to the object when both absorption (energy gain) and emission (energy loss) are taken into account. After all, that's how Nature works.

Objects that absorb more energy than they emit usually undergo a *net* warming: They warm more from the energy gained by absorption than they cool from the energy lost from emission. Objects that radiate more energy than they absorb usually undergo a *net* cooling. Objects that absorb and radiate equal amounts of energy usually experience no change in temperature (we say *usually* because the presence of water complicates matters—we will consider the consequences in detail later).

So it's fair to say that the temperature of an object is determined by its energy budget. For example, the ground absorbs radiation from the Sun (during the daytime only) and the atmosphere, representing an energy gain (radiational warming). On the other hand, the ground emits infrared radiation, which is an energy loss (radiational cooling). The temperature of the ground increases when radiational warming exceeds radiational cooling (we would call that *net* radiational warming). On the flip side, the temperature of the ground decreases when radiational cooling exceeds radiational warming (*net* radiational cooling).

The concept of a net change in temperature (or any variable) can be better understood by considering your bank account. Suppose that you deposit a $300 check while simultaneously withdrawing $100 cash. You have made a net deposit of $200. However, if you deposited that same $300 check but withdrew $500 cash, then you would have made a net deposit of –$200 (that is, a withdrawal). Now, change dollars into units of energy, like the electric company does with your money each month. Treat energy absorption as deposits and energy emission as withdrawals. On average, at the First National Bank of the Earth, as much radiation is emitted as is absorbed. However, some latitudes run in the black (more absorption than emission) while others run in the red (more emission than absorption). Figure 2.16 indicates which latitudes are in energy debt and which have an energy surplus. Regions within about 35° of the equator receive more radiation than they emit, and thus undergo a net warming. Latitudes closer to the poles emit more radiation than they receive, and thus experience a net cooling.

Despite these imbalances, the earth's surface and atmosphere do not warm continuously at low latitudes nor do they cool continuously at polar latitudes. The difference in net radiation between polar and equatorial regions sets in motion a pattern of

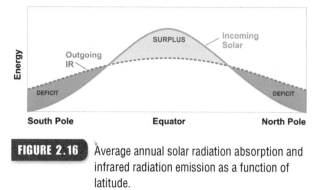

FIGURE 2.16 Average annual solar radiation absorption and infrared radiation emission as a function of latitude.

winds in the atmosphere and currents in the oceans that helps to partially mitigate the energy imbalance. In essence, equatorial regions make interest-free loans of energy to their higher-latitude counterparts. We'll learn more about this **general circulation** of the atmosphere in Chapter 10. In the meantime, don't leave the bank without realizing that the concept of net change is important in understanding your own finances as well as in understanding meteorology.

HEATING THE ATMOSPHERE: FROM THE GROUND UP

The earth receives less than one billionth of the total energy emitted by the sun. However, the planet does just fine with what it gets: At the top of the atmosphere, a plane perpendicular to the direct rays of the Sun receives about 1361 Watts per square meter (W/m^2) of solar radiation, an amount known as the **total solar irradiance**. This variable is also commonly referred to as the **solar constant**, although this term is misleading because the value is not constant. Regardless of nomenclature, to visualize its meaning, consider a completely efficient solar panel of area 1 m^2 at the top of the atmosphere, always pointing directly at the Sun—the energy collected by this panel could power thirteen 100-watt light bulbs and one 60-watt bulb, with a little left over. That doesn't sound like much energy, but this minuscule portion of the Sun's total output makes life on our planet possible and powers the Earth's weather machine.

The Solar Energy Budget: Savings and Loans

To open and maintain any bank account, an initial deposit is required. The focus here will be on the deposits of solar radiation made to the earth on a daily basis. For simplicity, let's assume that 100 units of solar energy are deposited daily to the top of the Earth's atmosphere. We'll use Figure 2.17 to help balance the books. Much like a service charge is deducted from an account, 30 of these 100 units of energy are reflected back to space, predominantly by air molecules and clouds (because the Earth reflects 30% of the solar radiation incident upon it, we say that the planet has an **albedo** of 0.30). Of the remaining 70 units, 22 are absorbed in the atmosphere by clouds and by air molecules, while the other 48 units are absorbed at the surface. Thus, on average, only 22% of the solar radiation that reaches the top of the atmosphere is used to heat the air directly, while 48% of this energy is used for heating the surface.

One consequence of this partitioning of the Sun's energy is that temperature, on average, decreases with height in the lowest 10 km (6 mi) or so of the atmosphere, a layer called the **troposphere**. This word derives from the Greek "tropos" which means "turn," suggesting that the combination of upward and downward motions of air (that is, overturning) plays an important role in the structure and behavior of weather in this layer. Why does temperature in the troposphere typically decrease with increasing altitude? The answer lies in Figure 2.17. Notice that the atmosphere does not absorb as much solar radiation (22 units) as the Earth's surface does (48 units), and so more of the Sun's energy is directly available to heat the surface than the air. The surface, warmed by this absorption,

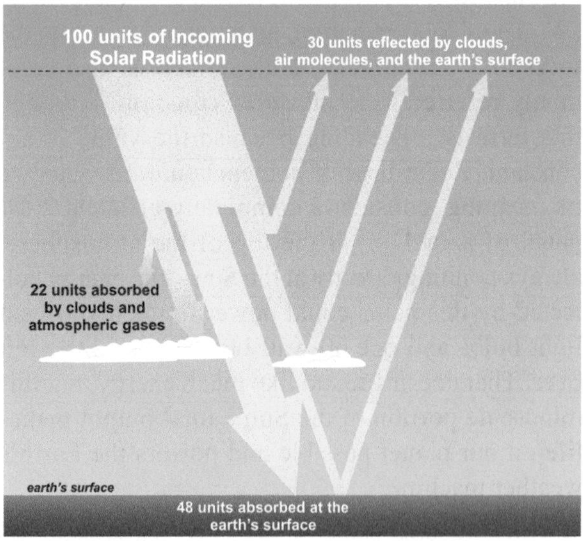

in turn heats the air. Thus, although the Sun provides the raw power that ultimately warms the atmosphere, the Earth's surface acts as a "middleman," processing and distributing energy to the atmosphere. Indeed, one of the basic tenets of meteorology is that *the Sun warms the ground, and the ground warms the air*.

But remember, the Earth's surface receives energy from another source—the atmosphere—a source that cannot be overlooked. In fact, life as we know it depends on it.

Clouds and Greenhouse Gases: Bonus Energy for the Surface

During the daytime, clouds (especially the thick ones) can dramatically back-scatter incoming solar radiation, thereby reducing the overall heating at the ground. At night, clouds (especially the low and relatively warm ones) act like space heaters because their downward emissions of infrared radiation partially offset the radiation losses at the surface. Clouds radiate in all directions, but the downward emission of infrared energy—formally called **downwelling radiation**—is what really makes a difference at the surface.

To see what we mean, consider Figure 2.18a, which is a plot of the variation of downwelling infrared radiation with time at Penn State University for a 12-hour period beginning at 8 P.M. on a January evening. Time runs along the horizontal axis, while the intensity of infrared radiation, expressed in units of energy (W/m^2), runs along the vertical. From 8 P.M. to 10 P.M, the sky over State College was clear. Note that downwelling radiation from gases in the mostly clear atmosphere was relatively small. However, around 10 P.M., clouds dramatically increased and a low overcast built over central Pennsylvania. In response, the intensity of downwelling radiation dramatically increased.

The effects on surface temperature were striking, as shown in Figure 2.18b. Notice the increase in the temperature after 11 P.M., largely a result of downwelling infrared radiation from the low clouds.

We mentioned earlier that gases in the clear atmosphere also emit infrared energy to the surface (remember, folks, everything emits radiation). To understand the impact of these emissions, consider Figure 2.19. The bottom row corresponds to the absorptivity of the atmosphere as a whole, while the other rows correspond to the absorptivity of the specifically labeled gases. The vertical axis for each row varies from 0 (no absorption) to 1 (total absorption).

FIGURE 2.17 Average distribution of 100 units of solar radiation arriving at the top of the Earth's atmosphere, based on satellite and surface data from the period 2000–2010. Some uncertainty still exists with these values, particularly the distribution between absorption at the surface and absorption in the atmosphere.

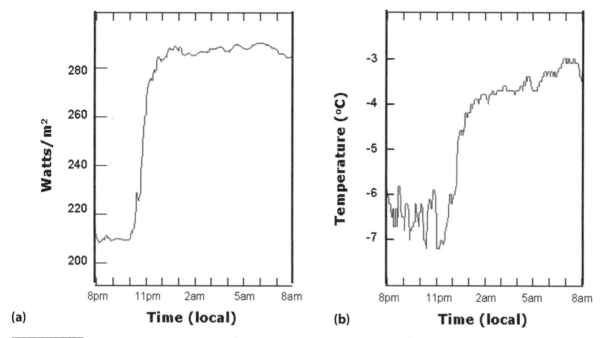

FIGURE 2.18 (a) The intensity of downwelling radiation over Penn State University on a January night dramatically increased around 10 P.M. as low clouds arrived overhead. (b) As a result, temperatures began to climb after 11 P.M.

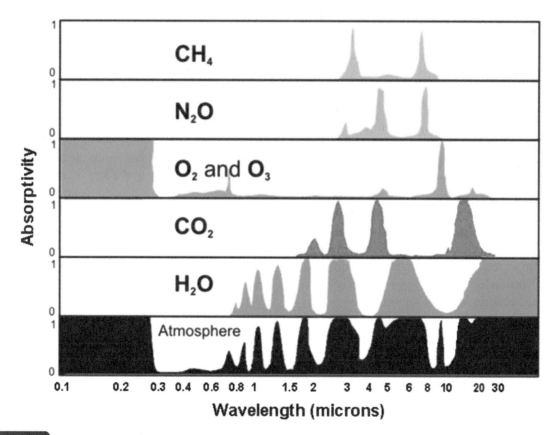

FIGURE 2.19 The absorptivity of various gases of the atmosphere and the atmosphere as a whole as a function of the wavelength of radiation. An absorptivity of zero means no absorption of radiation at the corresponding wavelength, while a value of one means complete absorption. The dominant absorbers of infrared radiation are water vapor (H_2O) and carbon dioxide (CO_2). Oxygen (O_2) and ozone (O_3) absorb much of the Sun's harmful ultraviolet radiation.

The horizontal axis represents the wavelengths of the electromagnetic spectrum, expressed in microns.

As an example of how to interpret this graph, focus your attention on water vapor (second from the bottom). At a wavelength of approximately 6 μm, the absorptivity of water vapor is one, meaning that water vapor absorbs all infrared radiation at this wavelength. In contrast, water vapor absorbs only a small fraction of radiation at a wavelength of 10 μm.

Why is this "selective absorptivity" important? A tenet of physics called **Kirchoff's Law** states that a body that is an efficient absorber of radiation of a particular wavelength is an efficient emitter of radiation of that same wavelength. Thus, because water vapor has a high absorptivity at several infrared wavelengths, it also has a high **emissivity** at those same wavelengths. As a result, water vapor accounts for some of the downwelling infrared radiation absorbed by the earth's surface. Carbon dioxide (CO_2), whose absorptivity appears third from the bottom in Figure 2.19, accounts for much of the rest. Note that these gases absorb little (if any) visible radiation (wavelengths between 0.4 and 0.7 μm), meaning that much of the solar radiation entering the atmosphere makes it all the way to the surface.

Both water vapor and carbon dioxide qualify as **greenhouse gases** because they, along with a few other gases such as methane and nitrous oxide, produce a planetary heating known as the **greenhouse effect**. This term was dubbed in the early 1800s, when it was thought that greenhouses stayed warm because the panes of glass allowed solar radiation in, but prevented radiation emitted by the contents of the greenhouse from escaping. We now know that this factor only partly explains the warmth of a greenhouse (a greenhouse also stays warm simply because the air inside is unable to mix with cooler air outside). Nonetheless, the term *greenhouse effect* has stuck, to the chagrin of many scientists.

Greenhouse Effect and Global Warming: The Same, but Different

It was during the record hot summer of 1988 in the United States that news of the greenhouse effect emerged from research institutions and academic seminar rooms into the mainstream media, often accompanied by apocalyptic warnings of the impending warming of our planet. Since then, it seems that the greenhouse effect has developed a bad reputation.

In reality, the greenhouse effect is essential to life on this planet. Furthermore, the greenhouse effect is far from new: For as long as greenhouse gases such as water vapor and carbon dioxide have been in the atmosphere (several billion years), the greenhouse effect has warmed our world.

Astronomers believe that about four billion years ago, when our Sun was a sputtering new star, its energy output was only about 70 percent of its present yield. With the Sun running at this reduced power, the Earth should have been a frozen ball of ice at that time. But ancient sedimentary rocks found near Isua, West Greenland (and elsewhere), suggest that liquid water existed on the surface 3.8 billion years ago (because such rocks form as layer upon layer of sediment settles in a body of water). This apparent inconsistency is known as the "Faint Young Sun Paradox." One popular explanation for this paradox is that excessive amounts of carbon dioxide were present in Earth's early atmosphere, enhancing the greenhouse effect. Today, the absorption and emission of infrared radiation by greenhouse gases keeps the average surface temperature of our planet about 60°F (33°C) higher than it would be without these gases. An earth with no greenhouse effect would be inhospitable, to say the least. Thus, some greenhouse effect is not only tolerable, but desirable.

If the greenhouse effect is so beneficial, why has it developed such a bad reputation? The answer is simple: Human beings, by burning carbon-rich fuels such as coal, natural gas, and oil are adding carbon dioxide to the atmosphere. Since the beginning of the Industrial Revolution in the late eighteenth century, atmospheric carbon dioxide levels have increased from about 280 CO_2 molecules per million air molecules (written 280 ppm) to around 400 ppm (see Figure 2.20). Because carbon dioxide is second in importance (to water vapor) as a greenhouse gas, there is legitimate concern that adding carbon dioxide to the air will cause the planet's atmosphere to warm. Simply put, human activities are enhancing the greenhouse effect, and thus are likely contributing to a "global warming." Indeed, it might be possible to have too much of a good thing. The possibility of a greenhouse effect enhanced by human activities, along with other potential human impacts on weather and climate, will be explored in detail in Chapter 18.

First, however, we still have an outstanding issue to address regarding the energy exchange between the earth and the atmosphere. Remember that the sun really doesn't heat the atmosphere directly. Instead,

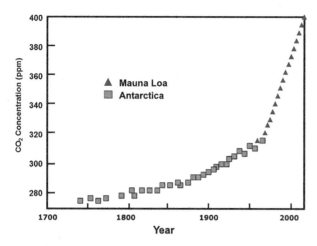

FIGURE 2.20 Observations of atmospheric carbon dioxide (in ppm) derived from air trapped in Antarctic ice and from direct observations taken on Mauna Loa in Hawaii. Carbon dioxide concentrations have increased since the beginning of the Industrial Revolution. By far, the greatest increases have occurred in the last century.

the surface warms the atmosphere much more efficiently than the Sun does. Emitting radiation to the atmosphere is one way to transfer energy. But that's only part of the story.

Conduction: A Touchy Subject

Besides radiation, there are other modes for transferring energy from the surface to the atmosphere. Let's start with **conduction**, which, as you will soon see, is a "touchy subject."

That's because conduction, in the meteorological sense, usually refers to the transfer of energy between the surface and the air directly in contact with it. To understand why conduction relies on touch, let's review some basic principles. Recall that the molecules and atoms in warm objects have relatively high kinetic energy, on average. The kinetic energy of molecules in cold objects is much more low key. When warm and cold objects come into contact, fast-moving molecules collide with slower ones, transferring kinetic energy during the collisions.

To understand what we mean, think of wildly gyrating teenagers (molecules in warm objects) dancing to rock 'n' roll. On the same dance floor, retired couples dance cheek-to-cheek to a slow ballad. To no one's surprise, the teenagers bump into the older folks. In turn, slow dancers gain unwanted kinetic energy and lurch awkwardly across the floor. Meanwhile, ruffian teenagers lose some of their kinetic energy during collisions.

Let's leave the dance floor and talk some science. Suppose a relatively warm object comes in contact with a colder one. In light of the dance-floor metaphor, the warmer object gets colder as its molecules lose kinetic energy and the colder object gets warmer as its molecules gain kinetic energy. For example, your hand feels cool when you grasp a metal object in your apartment or house. That's because metal has a high **thermal conductivity** (a measure of a material's ability to conduct heat energy). In other words, metal conducts kinetic energy rapidly away from the fast-vibrating molecules in your skin. As a result, your hand feels cooler.

Unlike most metals, air has low thermal conductivity. Alternatively, air is highly insulating. That's why porous materials such as wool are effective thermal insulators (porous means that there are small pockets for air to occupy). Given air's insulating character, it shouldn't come as a surprise that conduction between the surface and the overlying air proceeds at a relatively slow pace.

To get a sense of what we mean, suppose you were to press a slab of wood a few inches thick against a hot electric burner on your kitchen stove. In a short time, the temperature of the wood in contact with the oven burner is pretty close to that of the burner (your kitchen's probably smoking up by now). But the side of the wood slab not in contact with the stove can be touched without any pain (at least for a little while). Why? It takes time for the entire block of wood to heat up because the thermal conductivity of wood is sufficiently low so that the transfer of molecular kinetic energy is slow.

In contrast, imagine placing a similarly thick slab of aluminum on the hot burner instead of a block of wood. If you're inclined to touch the slab, watch out, you'll burn your hand! Why? The thermal conductivity of aluminum is hundreds of times greater than that of wood!

The wood slab on your kitchen stove (and its low thermal conductivity) is akin to a slab of air overlying the hot ground on a sunny summer day. After sunrise, the surface typically warms rapidly as it absorbs solar radiation. Incoming solar energy concentrates in the first few inches of the ground (indeed, on a sunny, hot day, you don't have to dig very far down to reach cool soil). As a result, a very thin layer of air in contact with the hot ground warms dramatically by conduction, albeit gradually. Over paved roads, temperatures in this thin layer of air can reach as high as 140°F (60°C) (the living conditions of an insect can be extreme indeed). At the same time,

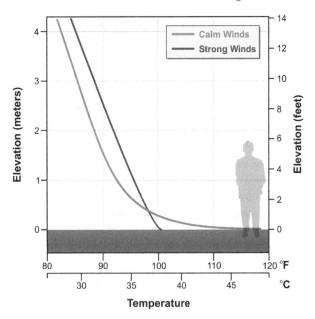

FIGURE 2.21 Large vertical temperature gradients sometimes exist very close to the ground, especially on days with little wind. Winds tend to mix the lower atmosphere vertically, reducing the gradient.

however, temperatures at nose level might be only 90°F (32°C) or so, marking a rapid drop-off with height. Hot bare feet (ouch!) but tolerable nose-level temperatures prevail, in part, because of the air's low thermal conductivity.

Figure 2.21 shows representative vertical variations of temperature during calm and windy conditions on a typical summer day. Note that vertical gradients of temperature are largest when the wind is calm—the red curve shows this large drop-off over a short vertical distance. Large vertical temperature gradients near the surface help create mirages, such as the apparent "puddle of water" that sometimes shimmers in the distance over a hot roadway. On breezy days, however, the wind mixes the lower atmosphere (more on this "stirring" topic in the next section), allowing ground-heated air to mix more freely with cooler air above. As a result, the vertical temperature gradient near the surface lessens (blue line on Figure 2.21).

Conduction also plays a role on clear, calm nights in creating relatively large vertical temperature gradients near the surface—but in this case, the coolest air resides near the ground. Once the sun sets, the surface emits more radiation than it gains from the atmosphere and starts to cool. In turn, a thin layer of air in contact with the surface cools as downward heat transfer via conduction gets underway. Gradually,

if the wind is relatively light, the chilling layer in contact with the cooling ground thickens. That's because the layer of air directly above the layer in contact with the cooling surface is initially warmer. But once there is a temperature difference between the bottommost layer and the overlying layer, the overlying layer starts to cool via conduction. Then the same process occurs between the second-from-the-bottom layer and the layer of air directly above it. And so on, and so forth.

Keep in mind that this is a relatively slow process because of the air's low thermal conductivity. Nonetheless, temperatures soon start to decrease several feet above the surface and even at modest altitudes higher up. Meanwhile, warmer air resides above the slowly deepening layer of nocturnal chill. Here, the downward heat transfer is painstakingly slow because of the increasing thickness of the insulating layer of air below it. The bottom line here is this: On a clear night with light winds, the delay in cooling the air not in direct contact with the surface results in the air temperature increasing with height. Meteorologists refer to this atypical vertical temperature profile as an **inversion**—specifically, in this case, it's called a **nocturnal inversion** because it forms at night.

Often, you'll hear this nocturnal low-level chill explained by invoking "radiational cooling," suggesting that such cooling is a process that only occurs at night. Do air molecules, like many streetlights, have special photo-detectors that allow them to sense sunset so that they can begin radiating? Nonsense. Actually, the surface radiates away more energy during the day than at night because of its higher daytime temperature. But since the surface receives a surplus of solar energy during the day, temperatures increase despite great losses of infrared energy. On a clear, calm night, without any energy input from the Sun, the surface undergoes *net* radiational cooling. In other words, it emits more infrared energy than it receives from the cloudless atmosphere. Thus, the surface conducts energy from the air in contact with it, cooling the lowest levels of the atmosphere and creating a chilly layer of air near the ground.

To demonstrate that the coldest air gathers next to the surface before dawn after a clear, calm night, consider Figure 2.22. It is a photograph of a clematis plant on a light post after several consecutive autumn nights with clear, calm, and frosty conditions. Note that the lower half of the plant is brown and dormant, while the top of the plant is green and vibrant,

FIGURE 2.22   A clematis plant on a light post after several consecutive frosty autumn nights. The lower half of the plant has browned after suffering frost damage, while the top of the plant remains green.

seemingly unaware of the freeze damage below. We submit Figure 2.22 into evidence as Exhibit A. It removes any reasonable doubt.

The chill of the night often lingers near the surface beyond sunrise, until the rays of the morning sun warm the ground sufficiently to initiate another important mode of vertical energy transport.

Convection: Up, Up, and Away in My Beautiful Hot-Air Balloon

On a sunny, summer day, temperatures a few feet above the ground would go ballistic if it weren't for a process called **convection**, the transfer of heat

energy vertically by the movement of air. Just like a hot-air balloon lifting off, blobs of surface-warmed air (called **air parcels** by meteorologists) rise, transporting heat energy skyward. As a result, convection limits the thickening of the layer of very hot air in contact with the surface on a sunny, hot day.

To understand the nuts and bolts of convection, we start with the simple idea of buoyancy. Suppose that, while taking a swim, you submerge your favorite beach ball (a rubber duck, if you prefer) and then let it go. In a heartbeat, the ball (or the duck) will bob to the surface of the water. Next, submerge a rock and then release it. In short order, it falls to the bottom of the pool. In formal scientific terms, we say that the beach ball (or rubber duck) has **positive buoyancy**, while the rock has **negative buoyancy** (an observation we're sure you'll never "take for granite").

The differences in buoyancy relative to water result from differences in **density**, which is simply the mass (weight) of an object divided by its volume. The rock has a high density compared to water because its weight is relatively large for such a small volume. The beach ball's density is much less than that of water because its weight is small for such a relatively large volume (after all, the ball is filled with air).

Near the Earth's surface, temperature is the primary controller of air density. To understand the link between air temperature and air density, let's hit the dance floor again. After a slow song ends and a fast one begins, dancers naturally spread out and occupy a greater space on the dance floor. And so it is with air molecules. When the air temperature is low, air molecules "dance" relatively slowly. As the temperature increases, molecules "jitterbug" (move faster) and naturally occupy a greater space (volume). Assuming that the number of air molecules selected for our dance experiment doesn't change (in other words, the mass remains constant), the density of warmer air is less than the density of cooler air (because density equals mass divided by volume, and the volume is larger in the warmer air).

We arrive at the following important result: *Increasing the temperature of an air parcel lowers its density* (and vice versa). In turn, the buoyancy of the parcel increases and, as a result, the parcel shows a tendency to rise when it's "submersed" in cooler air with higher density (analogous to a beach ball that's forcibly submerged in a swimming pool).

Now let's connect this discussion about density and buoyancy to convection. On a summer day, the Sun heats the surface which, in turn, heats a thin layer of air in contact with it. But the surface heats the overlying air unevenly, so there are spots that are hotter than others. For example, think about a sunny summer day and the torridly hot air in contact with an asphalt parking lot. Now think about the cooler air that overlies the surrounding grassy area. The air over the parking lot is less dense than the surrounding air and therefore more positively buoyant. In turn, air parcels rise more readily from the parking lot, transferring heat energy upward. This transfer is, of course, a consequence of convection. Manifestations of convection vary from very tall cumulonimbus clouds (recall Figure 2.14) that produce lightning and heavy rain to the invisible **thermals** that hawks and hang gliders routinely ride (see Figure 2.23).

As you will learn in subsequent chapters, rising air will produce clouds if enough water vapor in the ascending air condenses into water. In fact, convection is so important as a mechanism for cloud formation that when meteorologists use the term, they generally are referring to the process of clouds forming as water vapor condenses in buoyant parcels of rising air, rather than simply the process of transporting energy skyward. Furthermore, as water vapor condenses, energy stored in that vapor (known as **latent heat**) is released, providing yet another mechanism for transporting energy vertically (albeit a sneaky one—latent means "hidden").

We emphasize that thermals are invisible to the naked eye. But, through an ingenious process called Schlieren photography, we can "see" invisible convection. Figure 2.24 is a Schlieren photograph of one of the authors taken at the Gas Dynamics Laboratory in the Department of Mechanical and Nuclear

Engineering at Penn State. The turbulent aura above his head and shoulders is the Schlieren image of the "invisible" convection initiated by his warm skin. Now imagine this happening on a grand scale with the Earth's surface instead of a human body. With the uneven heating of land and water, field and forest, pavement and bare ground, and the huge variety of surface coverings, is it any wonder that convection on various spatial scales plays a significant role in the vertical transport of energy?

Eddies: The Essence of Convection

There's a method to our madness here. We showed you this Schlieren photograph so you could see that **eddies**, which are turbulent swirls of air, are the essence of convection. Figure 2.25 shows an idealized representation of an eddy. Essentially, this turbulent swirl circulates warm air upward and cooler air downward, relaxing large temperature gradients near the surface and, for all practical purposes, mixing the air.

As a final note on convection, we point out that eddies can develop simply when the wind blows over the earth's rough surface, creating a different form of convection in which the catalyst is not the warm ground.

To see why we're attempting to draw a distinction between thermally-driven and wind-driven convection, consider Figure 2.26. The relatively short wind arrow closest to the surface suggests that the rough ground markedly reduces wind speed there. A little higher up, the wind speed is greater (as indicated by the longer wind arrow). Here, the retarding force of the surface (friction) is weaker. With faster winds blowing over slower winds, eddies develop (think of

FIGURE 2.23 Hang gliders depend on rising pockets of relatively warm air to stay aloft (copyright Peter Dyrynda).

FIGURE 2.24 Schlieren photography captures invisible convection above one of the authors, proving, once and for all, that he's a warm, Italian guy (courtesy of Dr. Gary Settles, Director of the Gas Dynamics Lab, Penn State University).

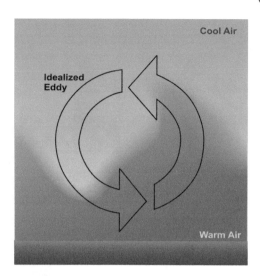

FIGURE 2.25 An idealized sketch of an eddy associated with convection from the warm surface. Note that the eddy circulates warm air upward and cooler air downward.

placing a pencil on your palm and then moving your other hand forward over the pencil—it rolls!). Given the somewhat mechanical way the wind generated these eddies, meteorologists refer to the resulting convection as **mechanical convection**.

Like their thermal counterparts, mechanical eddies mix the air. On a windy night, the downward circulation of relatively warmer air toward the ground by mechanical eddies prevents a big nocturnal chill from forming near the surface. That's why windy nights tend to be warmer than calm nights (all other factors being equal). And on a windy, sunny day, mechanical eddies help to lessen the vertical temperature gradient near the surface by circulating warm air rapidly away from the ground, just like eddies driven by thermal convection.

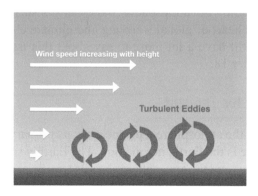

FIGURE 2.26 Mechanical eddies develop near the ground because friction with the earth's surface slows the air closest to the ground more than air higher up.

Needless to say, convection is important—so important that we must include the process in the overall energy budget for the Earth-atmosphere system. And that's how we close the chapter.

A Complete Energy Budget: A Full Accounting

Now that we've tracked what becomes of incoming solar radiation (recall Figure 2.17) and considered the sources and sinks of terrestrial radiation, it's time to combine the two into a complete Earth-atmosphere budget. And that's what we do in Figure 2.27, which includes the individual allocations of energy for the surface and the atmosphere. This budget is based on satellite and ground measurements for the period 2000–2010, and like Figure 2.17, for simplicity sake assumes that 100 units of solar radiation reach the top of the atmosphere.

On the plus side of the ledger, the Earth-atmosphere system retains a total of 48 + 22 = 70 units of energy from the Sun, once we remove the 30 units which are back-scattered to space. On the flip side, the atmosphere radiates 64 units of infrared energy to space, while 6 units of infrared energy emitted by the surface also escape to space (the suite of wavelengths at which the atmosphere allows infrared radiation to pass through is known as the **atmospheric window**). So, on the minus side of the ledger, a total of 70 units of energy are lost to space—the message here is that the Earth-atmosphere system has a balanced energy budget. Looking good so far.

If we focus on the individual energy exchanges between the surface and the atmosphere, the balance sheets get a little more complicated. As we already mentioned, 48 units of solar energy are absorbed by the surface. In addition, the surface receives 102 units of energy (infrared) from the atmosphere. So, on the plus side of the energy budget, the surface receives a total of 150 units of energy. On the minus side of the ledger, the surface emits 117 units of terrestrial radiation. Moreover, 7 units of energy are transported to the atmosphere via the processes of conduction and convection.

Last but not least, 26 units of energy are transferred to the atmosphere in the form of latent heat. We will explore this concept in more detail in Chapters 4 and 8, but for now simply recognize that some of the energy absorbed by the Earth's surface is used to evaporate water. In doing so, that energy is stored in the water vapor molecules, essentially hidden for later use (thus, the word "latent"). When the evaporation process is reversed and water vapor

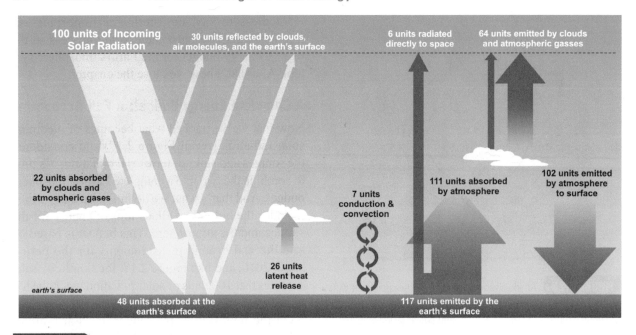

FIGURE 2.27 The energy budget of the Earth-atmosphere system takes into account not only the incoming solar energy but also terrestrial radiation as well as the many processes by which energy is transferred from the surface to the atmosphere.

condenses into water droplets in the atmosphere to form clouds, this energy is released as the latent heat of condensation—and voila, an energy transfer from surface to atmosphere is completed!

Now accounting for all the energy losses at the surface, we get $117 + 7 + 26 = 150$ units of energy, which balances the 150 units of energy that the surface receives from the Sun and the atmosphere. So far, our audit of energy for the Earth-atmosphere system is right on the mark!

All that's left is the energy budget of the atmosphere. Once again, we turn our attention to Figure 2.27 and first consider the plus side of the atmosphere's energy ledger. As we've already established, the atmosphere absorbs 22 units of energy from the Sun. Of the 117 units of energy emitted by the surface, 111 units are absorbed by the atmosphere (the other 6 units of energy radiated by the surface escape to space through the atmospheric window). And, as we've already suggested, the atmosphere also receives 7 units of energy via conduction and convection from the surface as well as 26 units of energy via the release of latent heat of condensation.

So, on the plus side of the energy budget, the atmosphere receives $22 + 111 + 7 + 26 = 166$ units of energy. Let's see if the minus side of the energy ledger for the atmosphere agrees. The atmosphere radiates 64 units of energy to space and 102 units of

energy to the Earth's surface, totaling 166 units of energy. So our energy audit passes muster. No need to call in the IRS.

The overall message you should take from this discussion is that all the pluses and minuses ostensibly balance. We caution, however, that there is uncertainty associated with satellite and ground measurements, so the numbers you see in Figure 2.27 aren't set in stone. Moreover, as human activities such as the burning of fossil fuels continues to add greenhouse gases to the atmosphere, more of the energy emitted by the surface will be absorbed by the atmosphere, which, in turn, increases the amount of energy emitted by the atmosphere to the surface (and to space). This increasing amount of energy emitted by the atmosphere to the Earth's surface is at the heart of global warming and climate change. We will have a lot more to say about this issue in Chapter 18.

Food for Thought

From the atmosphere's perspective, its primary heat source is the Earth, not the Sun. The Earth's surface radiates and the atmosphere absorbs terrestrial radiation at a variety of infrared wavelengths. Conduction and convection occurring at or near the earth's surface also transfer heat energy to the air, but they play second fiddle to radiation.

Thus, somewhat curiously, the atmosphere also plays a major role in its own heating. Given that the atmosphere is a selective absorber of terrestrial radiation, it has a higher radiating temperature than it otherwise would have. Indeed, without greenhouse gases and clouds, the atmosphere's temperature would be dramatically lower and life, as we know it, would not exist.

Quantitatively, the Earth's surface receives nearly double the energy from our atmosphere than it does from the Sun! That's because, although the Sun is much hotter, it occupies much less of the sky than our atmosphere. As a result, radiation coming from the direction of the Sun pales in comparison to the sum total of radiation emitted by each portion of the atmosphere that comprises the entire sky.

The bottom line here is that the atmosphere is warmer because the Earth's surface is warmer. And the surface is warmer because it receives a great deal of energy from our atmosphere.

Focus on Optics

Why Is the Sky Blue?

Air molecules scatter all wavelengths of visible sunlight. Figure 2.28 shows the spectrum of the light emanating from a blue sky as a function of wavelength. An instrument called a spectrophotometer collected these data. It was pointed upward away from the direct rays of the sun.

Notice that the scattering is not uniform across the visible spectrum. Why not? First, consider the input—remember from earlier in the chapter that the Sun's emission spectrum peaks at a wavelength near 0.5 microns, so incoming sunlight is already predisposed toward wavelengths in the blue-green part of the spectrum. But wait, there's more! Air molecules, because of their small size, disproportionately scatter shorter wavelengths more readily than longer wavelengths. These two factors weigh heavily in producing the spectrum of skylight in Figure 2.28, which clearly shows that an assortment of wavelengths of scattered visible light enters our eyes from a clear sky— this assortment strongly favors shorter wavelengths. But now physiology comes into play—the human eye is less sensitive to violet than to blue. So when the brain interprets the assortment of visible wavelengths captured by our eyes, we perceive this brain-integrated ensemble of light as blue (see Figure 2.28).

Make sure you understand what we're saying here. Technically, our eyes do not see the sky as "blue." They merely act as collectors of skylight. Only our brains really "see" the blue sky. As evidence, consider that we may dream in color despite the absence of light on our retinas.

This explanation of the intimate relationship between color and the human brain probably goes against the grain of what you have previously learned. Indeed, most explanations rely on assigning a specific color to a certain wavelength of visible light. The explanation then proceeds by asserting that air molecules scatter blue light, thereby giving the sky its distinctive color. But to reduce color to a specific wavelength of visible light essentially removes our brains from the entire process. And that's simply not correct.

In the final analysis, light scattered by air molecules is not blue in any objective sense of the word. Indeed, the light scattered by air molecules is composed of all visible wavelengths (with a peak in the shorter wavelengths). Only our brains can integrate this complicated assortment into a single color that we learned, as children, to call "sky blue."

Why Is the Martian Sky Butterscotch?

The color of the Martian sky is typically butterscotch (a yellow-brown hue), as seen in Figure 2.29, a true-color mosaic from the Curiosity robotic rover in September 2012. Fine dust particles suspended in the Martian atmosphere contain an iron oxide called magnetite. This mineral preferentially absorbs shorter wavelengths of sunlight. Meanwhile, larger dust particles uniformly scatter all wavelengths of sunlight. The absorption of sunlight by magnetite, combined with the uniform scattering of all the wavelengths of light by large dust particles, account for the butterscotch sky color on Mars.

(Continued)

(Continued)

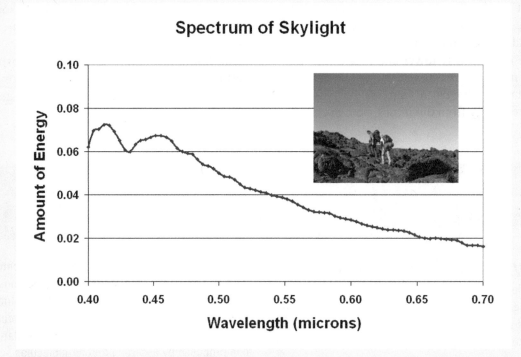

The spectrum of skylight (light from the sky), measured by pointing a spectrophotometer upward but away from the sun (courtesy of Eugene Clothiaux).

This raw color mosaic of dozens of images was captured on September 20, 2012 from the Mast Camera on NASA's Mars Curiosity Rover. The butterscotch-colored sky of Mars is visible behind Mount Sharp, a layered mound rising more than 5 km (3 mi) above the floor of Mars' Gale Crater, where Curiosity landed in August 2012 (courtesy of NASA/JPL-Caltech/MSSS).

Global and Local Controllers of Temperature

3

LEARNING OBJECTIVES

After reading this chapter, students will:

- Be able to scientifically explain why the Earth has seasons and argue against the popular notion that the change in seasons is determined by the Earth's distance from the Sun
- Gain an appreciation for the differences between astronomical and meteorological seasons
- Gain an appreciation for the seasonal lag in temperature following the summer and winter solstices

- Gain an appreciation for the salient differences in climate between locations near a large body of water and those in the heart of a continent
- Be able to scientifically argue why temperature decreases with increasing altitude (on average) in the troposphere
- Understand the latitudinal variation in altitude of the tropopause and its dependence on convection
- Gain an appreciation for the role that ocean currents play in weather and climate
- Be able to interpret surface weather analyses by identifying warm, cold and stationary fronts
- Understand why the maximum temperature on a given day often occurs a few hours after the time corresponding to the most direct solar radiation
- Gain an appreciation for why the nighttime minimum temperature typically occurs just before sunrise
- Understand the atmospheric and surface conditions that control daytime and nighttime temperatures
- Gain an appreciation for the various instruments that measure temperature

In the *Wizard of Oz*, Dorothy clicked her heels and was quickly whisked from the distant Land of Oz to Kansas. It all seemed so magical back in the years after the film was released in 1939.

Who would have ever thought that real life would one day take on a Wizard-of-Oz quality? Think about it. By clicking a computer mouse or launching an app on a smartphone (akin to Dorothy's heel taps), you can be instantly whisked away to faraway places to vicariously experience wondrous sights. And the modern notion of a closely connected world doesn't end with the Internet. In a blink of a camera's eye, news networks beam television signals to communication satellites orbiting the earth, dispersing breaking news to the world.

From this technological perspective, we live in a very small world indeed. Yet, from other perspectives, continents seem worlds apart. Consider that tourists visiting the Australian outback can be sweltering in broiling heat while back home in Buffalo, NY, their neighbors dig out from two feet of snow. It is an irrefutable fact of life on this planet that while the Northern Hemisphere shivers in winter, the Southern Hemisphere simultaneously basks in summer, and vice versa (see Figure 3.1). There are just some things that technology can't shorten or shrink.

Seasonally speaking, why are the hemispheres so far apart? In addition to seasonal variations, what processes control the average pattern—in other words, the climatology—of global temperature? What smaller-scale processes, both in space and in time, contribute to local, day-to-day variations in temperature? To gain insights into the answers to these and other fundamental questions about this most basic of weather variables, we'll start with the annual cycle.

SEASONAL TEMPERATURE VARIATIONS: LIKE PINBALL, THEY DEPEND ON TILT

The change of seasons has more to do with the Earth than the Sun. The Earth makes one complete revolution around the Sun every 365.25 days (the extra quarter of a day is accounted for every four years in leap year). This motion would not be important in the changing of the seasons were it not for the angle of 23.5° between the plane of the Earth's equator and the plane of the Earth's orbit around the Sun—this is often shortened to "the Earth is tilted 23.5° on its axis" (see Figure 3.2). Because the axis points in the same direction in space at all times during its orbit, the orientation of the Earth with respect to the Sun's rays is continuously changing over the course of a year. Without the inclination of the Earth's axis, there would be no major seasonal temperature variations. Presently, the axis of rotation points almost directly toward Polaris, the North Star. Figure 3.3 is a time-lapse photograph of the sky above Hawaii's Mauna Kea on a December night. The rotation of the Earth made distant stars, which are fixed in space, appear to trace out circles (appropriately named star trails). The center of these concentric star trails is the

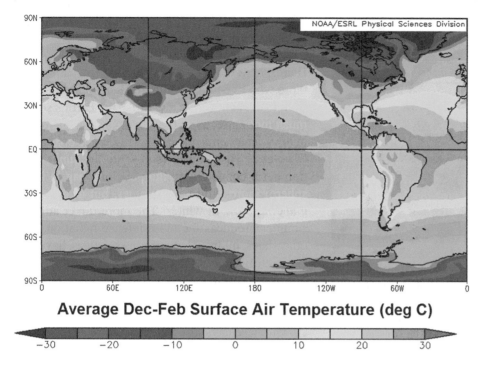

Average Dec-Feb Surface Air Temperature (deg C)

FIGURE 3.1 Average temperatures (in °C) for the three-month period from December 1 to February 28. The lowest temperatures are shaded in purple, the highest in orange and red. It is winter in the Northern Hemisphere and summer in the Southern Hemisphere (courtesy of ESRL).

direction to which the Earth's axis of rotation points. The bright segment very close to the center of the concentric circles is Polaris.

Figure 3.2 shows the position of the Earth at four different times during its annual journey around the Sun. At one point each year, when the Earth reaches position A, the Sun's rays directly shine upon latitude 23.5°N, the Tropic of Cancer. For those who live in the Northern Hemisphere, this astronomical milestone, which occurs around June 21, is the **summer solstice**, an instant in time that marks the "official" start to summer. At no other time of the year are the direct rays of the Sun as far north. At all locations below the Arctic Circle in the Northern Hemisphere, the number of daylight hours reaches its annual maximum on this day.

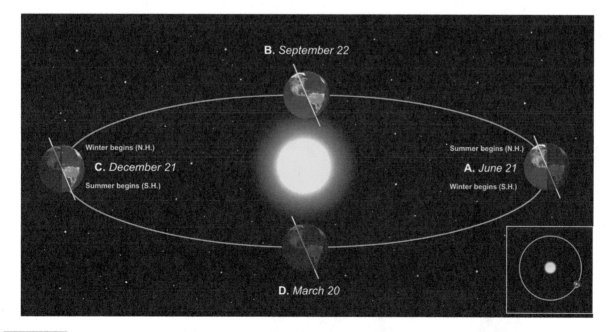

FIGURE 3.2 The position of the Earth with respect to the Sun on the first day of the four seasons. The elliptical shape of Earth's orbit is greatly exaggerated in this side view. In reality, if we looked down on the orbit from above (see inset), it would appear nearly circular.

A time-lapse photograph of the sky above Hawaii's Mauna Kea on a December night shows concentric, circular star trails. The center of these concentric star trails is the direction to which the Earth's axis of rotation points. The bright segment very close to the center of the concentric circles is Polaris (courtesy of Gemini/AURA.)

From the Arctic Circle to the North Pole, the sun never sets for a period of a few days to six months around the time of the summer solstice. To visualize this "endless daylight," imagine the Earth at Point A rotating on its axis. Folks living at high latitudes such as Barrow, AK, will never lose sight of the Sun, as Figure 3.4 strikingly illustrates. A forecast of "Partly sunny tonight" is not far-fetched in this "Land of the Midnight Sun." In contrast, south of the Antarctic Circle, the Sun never peeks above the horizon on this day. In the Southern Hemisphere, this astronomical moment is the **winter solstice**, the first day of astronomical winter.

Six months later, at position C, the situation is reversed: The Sun's direct rays reach their southernmost latitude, directly striking 23.5°S, the Tropic of Capricorn. The Sun remains below the horizon all day at all latitudes north of 66.5°N. December 21 or 22 marks the winter solstice for the Northern Hemisphere and the start of astronomical summer for those who live in the Southern Hemisphere.

Notice that the Sun's rays strike the Earth's surface much more directly in the summer hemisphere than in the winter hemisphere. Remember the boxer from Chapter 2? His punches (substitute the Sun's radiation here) connect with the punching bag (a hemisphere) more solidly in summer, but at more of a glancing blow in winter. The more direct angle in summer produces the more intense radiation. Two other dates, midway between the solstices, have special significance: In the Northern Hemisphere, the **autumnal equinox**, the astronomical first day of fall, occurs around September 22, while the first day of spring, the **vernal equinox**, falls on or near March 20. On these days, the Sun's rays directly strike the equator: Neither hemisphere has an edge with regard to incoming solar radiation.

Speaking of the equator, recall from Chapter 2 that the sun gets close to the **zenith** every day in the tropics (the zenith is the point that lies directly overhead). Thus, seasonal variations are small by mid-latitude standards. Any noticeable seasonal variations in temperatures at low latitudes often result from seasonal changes in clouds and precipitation. For example, Figure 3.5 shows the average monthly maximum temperature and average monthly rainfall at Bhopal, India (latitude 23°N). Average highs peak above 100°F (38°C) in May, but fall to almost 80°F (27°C) in August (a month that you might expect to be really, really hot). The explanation lies with clouds and recurrent rains during the summer monsoon in Southeast Asia, evident by the peak in rainfall from

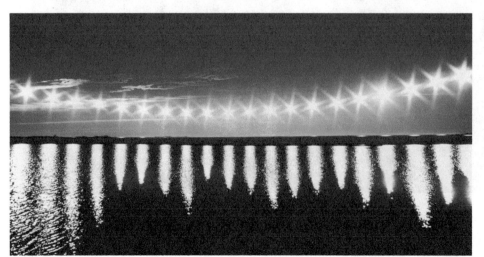

This time-lapse photograph on a mid-summer day shows that the sun never sets at high latitudes during this time of year. As a result, these latitudes are sometimes called the "Land of the Midnight Sun" (© Tony Stone/ Getty Images).

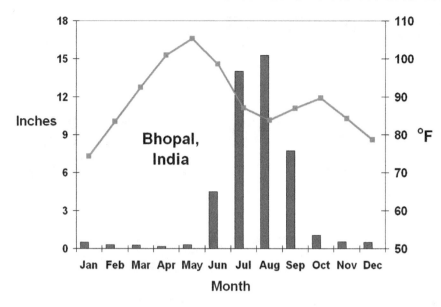

June to September (more on this topic in Chapter 10). Essentially, clouds and precipitation act as a thermostat to limit daytime temperatures.

In defining the seasons in terms of Earth-Sun relationships, we have carefully used the term *astronomical*. However, experience suggests that nature often does not abide by the "official" beginning (or end) of a season. Many locations in the mid-latitudes of the Northern Hemisphere have experienced searing heat waves in May or early June, while bone-chilling cold and accumulating snow often arrive well before astronomical winter begins. Thus, meteorologists have slightly different definitions of the seasons, based simply on average temperature.

The three coldest calendar months in the Northern Hemisphere are December, January, and February, on average, and weather forecasters consider this period to be "meteorological winter" in this hemisphere. Similarly, June 1 through August 31, which encompasses the three warmest months (on average), qualifies as "meteorological summer." The transitional periods between these two seasons are "meteorological spring" and "meteorological autumn." Of course, these "meteorological seasons" are reversed in the Southern Hemisphere. This definition of the seasons is commonly used in compiling climate statistics simply because using full months makes averaging easier.

Surface temperature, averaged over months, seasons, years, or even longer periods, is one of the hallmarks of climate. Let's now explore the factors that shape the long-term average patterns of global and regional temperatures.

AVERAGE PATTERNS OF GLOBAL AND REGIONAL TEMPERATURE: LOCATION, LOCATION, LOCATION

Ask a real estate agent to name the three most important factors to consider when purchasing a home, and inevitably the answer is "location, location, location." In this context, "location" may refer to certain neighborhoods, easy access to major thoroughfares, or proximity to good schools or convenient shopping.

The pattern of average surface air temperature across the earth's surface is also highly location dependent: In this context, "location, location, location" equates to latitude, proximity to large bodies of water, and elevation.

Latitudinal Gradients: Another Boxing Lesson

Figure 3.6 shows average temperature data from January 1979. The top and middle panels show average daytime and average nighttime temperatures, respectively. Maroon marks the warmest spots, with progressively lower temperatures indicated in red, yellow, light blue and dark blue (which marks the coldest spots). Clearly, it was summer in the Southern Hemisphere (note that South Africa, southern South America, and Australia are the warmest places on earth) and winter in the Northern Hemisphere (note the blue over much of North America and northern Eurasia).

We can apply our boxing lesson to explain the large-scale temperature variations in Figure 3.6. No matter which hemisphere you choose in the top two panels, the colorful horizontal bands of temperature are

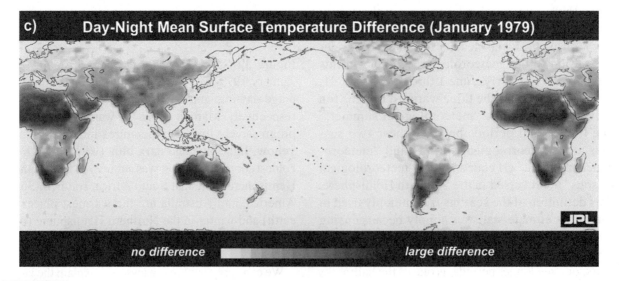

FIGURE 3.6 The top panel shows average daytime temperatures in January 1979. The middle panel depicts average temperatures at night during that month. The bottom panel shows the difference between average daytime and average nighttime temperatures (courtesy Moustafa Chahine, JPL).

generally aligned with latitude, with high latitudes colder than the tropics (the poles are much, much colder than the tropics in the winter hemisphere compared to the summer hemisphere). Regardless of which hemisphere more directly faces the sun, the curvature of the earth dictates that solar radiation will always strike our planet more directly over the tropics than over higher latitudes.

We head to Barrow, AK to confirm that the angle at which solar radiation strikes the earth plays a pivotal role in surface air temperatures. At latitude 71.3°N, Barrow is the northernmost primary observing location maintained by the National Weather Service (see Figure 3.7). Residents of Barrow experience the long periods of continuous darkness and continuous daylight typical of high latitude locations. We plotted the number of hours of daylight at Barrow as a function of calendar day in Figure 3.8: At Barrow, the sun stays above the horizon from around May 10 to early August of each year. However, during this period of more than 2000 hours of continuous daylight, the sun spends much of its time low in the sky, never rising more than about 42° above the horizon (revisit Figure 3.4). This low sun angle is one reason why the average surface air temperature at Barrow is only 36°F (2.2°C) in June and 41°F (5.0°C) in July, even though the sun is always above the horizon these months. Another primary factor is the proximity of the chilly Arctic Ocean, which still has ice in early summer (see Figure 3.9).

The temperature contrast between the poles and tropics, which is a consequence of the Sun's uneven heating of the Earth, is a fundamental reason why there's changeable weather, especially across the mid-latitudes. Yes indeed, north-to-south differences in temperature set the stage for the wind to blow and clouds and precipitation to form. Winds from the sub-tropics move warm air poleward, while winds from polar regions transport cold air toward the equator in a never-ending attempt by the atmosphere to erase temperature contrasts between the tropics and poles. In this futile process, there are inevitable confrontations between cold and warm air (especially in the mid-latitudes), often resulting in clouds and precipitation. We will elaborate on the showdowns between warm and cold air later in this chapter. Until then, let's continue with our discussion of temperature controllers.

More Uneven Heating: Land versus Water

We have not yet discussed the bottom image in Figure 3.6. This panel shows the difference between average daytime temperatures (Figure 3.6a) and average nighttime temperatures (Figure 3.6b) during January 1979. In Figure 3.6c, large day-to-night differences are depicted in red and maroon, and small differences in white and blue. One of the most striking features on the bottom panel is the large area of white over the oceans, indicating virtually no change, on average, between daytime and nighttime temperatures during the month. In contrast, average daily temperature swings over land were very large.

Water is particularly slow to warm (and cool): In fact, a volume of water must absorb about three times as much energy as a similarly sized volume of land in

FIGURE 3.7 Given its Arctic location, Barrow also serves as a site for global climate research. The Barrow Observatory, which is located just outside the town, is maintained by the Earth System Research Laboratory (ESRL). This research branch of the National Oceanic and Atmospheric Administration (NOAA) monitors atmospheric gases that are capable of changing earth's climate and air quality (courtesy of NOAA).

FIGURE 3.8 Number of hours that the sun is above the horizon at Barrow, AK, on the Arctic Ocean at latitude 71.3°N, as a function of calendar day.

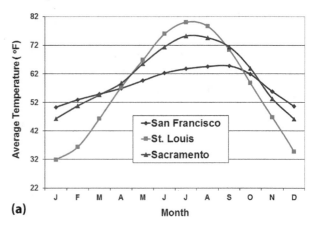

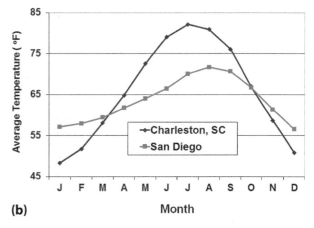

FIGURE 3.9   Fragmented edge of the fast ice (ice attached to shore) just north of Barrow, AK, in early summer. Pressure ridges cross-cut the floating ice, with upended slabs of ice as much as 15 feet tall. Northerly winds often push drifting ice against the shore around Barrow even in the summer (courtesy of Andy Mahoney).

FIGURE 3.10   (a) Average monthly temperatures at three cities at approximately the same latitude—San Francisco, CA, St. Louis, MO, and Sacramento, CA—demonstrating the effect of proximity to a large body of water on the annual variability of temperature; (b) Comparison of average monthly temperatures at two coastal cities that are at approximately the same latitude: Charleston, SC, on the East Coast, and San Diego, CA on the West Coast. There is much less seasonality at the West Coast location.

order to achieve the same temperature increase. Not surprisingly, a volume of water must lose about three times as much energy as a similarly sized volume of land to achieve the same temperature decrease. Formally, we say that water's **heat capacity** is three times that of land. As a result, given that the air is primarily heated (or cooled) by the earth's surface, changes in surface air temperature tend to be less over water than over land.

Another important factor that helps to reduce the change in air temperature over the oceans is that winds frequently stir and mix the topmost layers, causing water from greater depths to rise to the surface. So, for example, during the daytime in the summer, breezes might allow cooler water from below to mix to the ocean surface, limiting temperature increases in the overlying air.

As a result of water's sluggishness to warm (and cool), locations downwind of a large body of water tend to show relatively small swings in temperature between the summer and winter (in other words, relatively small **seasonality**) compared to land-locked areas. Figure 3.10a compares average monthly temperatures at three U.S. locations that lie at nearly the same latitude but differ in their proximity to a large body of water. The curve with the least variation from warmest to coldest month represents San Francisco, which lies on the central California Coast downwind of prevailing winds from the Pacific Ocean. The most variation occurs at St. Louis, deep in the interior of the lower forty-eight states. Figure 3.10b delves further into the issue of seasonality, comparing the

average monthly temperatures at Charleston, SC and San Diego, CA, two coastal cities at about the same latitude but on different sides of the continent. Note that the seasonality is much smaller in San Diego, where prevailing westerly winds blow off the ocean. In Charleston, the moderating effect of the nearby water is not as pronounced because prevailing winds are predominantly westerly and have largely traversed the continent before reaching the city.

The tendency of water to warm and cool slowly also helps to explain why maximum average air temperatures, which occur in late July or early August in the Northern Hemisphere, lag the peak in the intensity of solar radiation, which occurs on the first day of summer (on or about June 21 in the Northern Hemisphere).

To understand this lag, think about going to your refrigerator to retrieve a piece of leftover pizza. You place the cold pizza into the oven, already preheated to 300°F (about 150°C). After a few moments, you surrender to the tempting aroma filling your kitchen and remove the slice. In mouth-watering expectation, you take a bite, but are instantly annoyed because the inside of the pizza is still cold. Lesson learned: It takes time for cold, leftover food to heat up.

After more heating, you finally are able to enjoy hot pizza. But still hungry, you decide to have dessert. You reach into the cupboard and pull out a box of banana pudding (it has lots of "a-peel"). You pour the pudding mix into a pan of milk, stir, bring the pudding to a boil, and place it in your refrigerator to congeal. But after only five minutes, your impatient taste buds again get the best of you, and you remove the treat. As you slurp your first runny spoonful, you realize that the pudding is still very warm. Again, you learn another important cooking lesson: It takes time for hot pudding to cool.

These kitchen lessons can be used to explain seasonal temperature lag. In late June, the input of solar energy to the Northern Hemisphere is set on maximum. But, like the cold pizza in the oven, it takes time for the surface to heat up in response (and since the atmosphere is primarily heated by the surface, it takes time for the overlying air to warm). Moreover, considering that 70 percent of the surface is covered by relatively slow-warming oceans, it is no surprise that maximum average air temperatures are not realized until well after the summer solstice. On the other hand, even though the sun's energy hits rock bottom in late December (in the Northern Hemisphere), it takes time for continents and oceans (and the air above them) to chill, delaying the bottoming out of average temperatures until a month or so after the winter solstice. Again, the oceans play the primary role in the delay because they cool much slower than land.

Speaking of land, note in the bottom panel of Figure 3.6 the large changes in temperature between day and night over Australia and the Sahara Desert in Africa (depicted in maroon). Though these regions are among the hottest places in the world by day in the summer, they cool off rapidly at night. Desert air typically contains little water vapor, a greenhouse gas that preferentially absorbs and emits infrared radiation. The lack of water vapor (and clouds) allows much of the radiation emitted by the sandy soil to escape to space. The result is a rapid cooling of the desert floor and overlying air after sunset. We will discuss more about predicting nighttime temperatures later in this chapter. Until then, we will whet your whistle with one general temperature forecasting rule: Higher elevation usually means lower temperature.

Temperature and Altitude: Snow-Capped Tropical Mountains

The top and middle panels of Figure 3.6 are chock full of information—one subtle feature we'd like to point out is the chill atop tall mountains. For example, the narrow yellow strip over western South America (most noticeable in the middle panel) coincides with the mighty Andes, one of the tallest mountain chains of the world with an average height of about 3660 m (12,000 ft). The peak of the volcano Aconcagua lies at an altitude just over 7000 m (about 23,000 ft), making it the highest mountain in the Western Hemisphere. Halfway around the world, just to the north of the Bay of Bengal near India, pockets of dark blue mark the deep chill of the Himalayas, including Mt. Everest, the world's tallest peak at 8853 m (29,035 ft).

On average, temperature decreases about 6.5°C per kilometer of ascent (3.6°F/1000 ft) in the troposphere, which roughly includes the lowest 10 km (6 mi) or so of the atmosphere. This rate of temperature decrease in the troposphere is called the **average environmental lapse rate** (the term *lapse rate* formally means "rate of decrease in temperature with increasing height"). As discussed in Chapter 2, the reason temperature generally decreases with increasing altitude is that the troposphere is heated primarily by the surface, which absorbs solar radiation and then doles it out from the bottom up via the warming processes of conduction, convection, and radiation.

They say a picture is worth a thousand words, and Figure 3.11 says a lot. That's the snow-capped peak of Mauna Kea on the big island of Hawaii (at about 20°N latitude, in the tropics). At an altitude of 4205 meters (13,796 ft), Mauna Kea (which means "white mountain") is often snow covered during the winter months, while down at sea level, palm trees sway in the warm breezes. Blizzard conditions sometimes buffet the summit and skiers frequent the slopes. Even in the balmy paradise of Hawaii, the decrease of temperature with altitude is inescapable.

FIGURE 3.11 A snow-capped Mauna Kea, on the island of Hawaii, is a reminder of the high-altitude chill that stands in dramatic contrast to palm trees flourishing at warmer low elevations (courtesy of Steven Parente).

For now, let's get back to the issue of how temperature varies in the vertical. We already mentioned that temperature decreases, on average, about 6.5°C for every kilometer above the surface. We emphasize "average" in reference to the environmental lapse rate because much of the atmosphere is dynamic, frequently changing its vertical temperature structure in time.

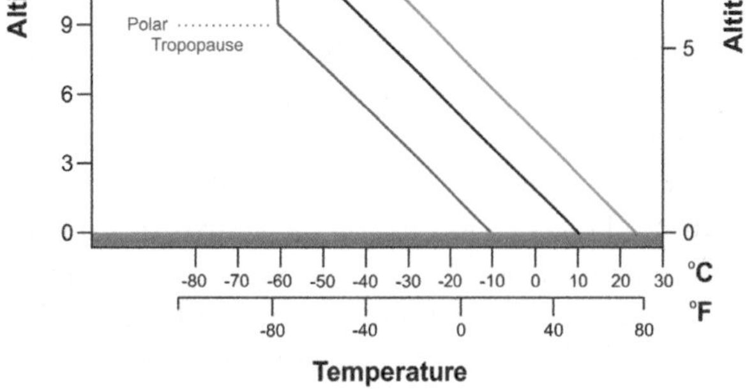

But not only does the atmosphere's vertical temperature structure vary with time, it also varies with latitude. Figure 3.12 shows average lapse rates over the poles, tropics, and mid-latitudes. Though the slopes of the three lines are nearly identical in the lower part of the diagram, the height at which temperature stops decreasing with altitude varies significantly. Indeed, the altitude of the **tropopause**, the transition zone between the troposphere and the next layer up, the **stratosphere** (where temperature remains constant and then increases with increasing altitude), varies from more than 16 km (10 mi) over the tropics to about 9 km (5.6 mi) over the poles.

The lofty tropopause over the tropics results from the towering convection that occurs over these very warm and humid regions. Imagine blowing bubbles over a hot plate set on "high." The bubbles will accelerate toward the ceiling as they are carried by rising currents of hot air. And so it is over the tropics. In contrast, the solar "hot plate" is set on "low" over the poles—bubbles of air don't rise nearly as high, so the polar tropopause is much lower. Figure 3.13 shows the average temperature of the tropopause worldwide. Note that the top of the deep tropical troposphere is colder than the top of the shallower troposphere over higher latitudes, testimony to the high altitude to which tropical parcels can rise.

FIGURE 3.12 The rate of decrease in temperature between the surface and the tropopause varies with time and latitude, but, on average, is fairly close to 6°C per kilometer. The height of the tropopause, however, which largely depends on the vertical extent of convective air currents, varies dramatically from the tropics to polar regions.

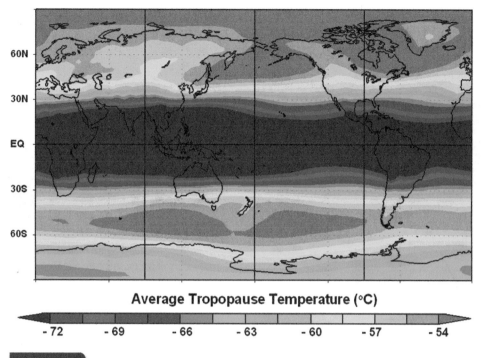

Average Tropopause Temperature (°C)

-72   -69   -66   -63   -60   -57   -54

FIGURE 3.13 A global view of the average annual temperature at the tropopause level. Purple represents the lowest tropopause temperatures, while red indicates the highest. Notice that the tropopause is colder over tropical latitudes because it lies at a much higher altitude (courtesy of the Earth System Research Laboratory).

In contrast to the troposphere, vertical motions in the stratosphere are dramatically weaker. Meteorologists characterize any layer of air, no matter where it resides in the atmosphere, as "stable" if it resists strong vertical motion. Stability is a concept you will learn more about in Chapter 8. For now, it suffices to say that the way in which temperature varies vertically in any layer of air determines the degree of vertical motion in that layer. So it should come as no surprise that the average (typical) vertical variation of temperature helps determine the primary layers of the atmosphere. Figure 3.14 shows the vertical variation of temperature (in red) and pressure (in blue) for the **standard atmosphere** over the mid-latitudes. By international agreement, the standard atmosphere is representative of typical vertical profiles of temperature and pressure.

As the vertical variations in temperature shown in Figure 3.14 suggest, the atmosphere can be partitioned into four distinct layers: the troposphere, the stratosphere, the mesosphere, and the thermosphere. We will focus almost all our attention on the troposphere and the stratosphere (where the absorption of ultraviolet radiation by ozone drives the increase in temperature with height), but some interesting observations can be made about the other layers. For example, very thin, cold clouds called

noctilucent clouds (see Figure 3.15) can form near the top of the mesosphere at altitudes of 76–85 km (47–53 mi). Also known as polar mesospheric clouds, noctilucent clouds can sometimes be observed around twilight, when these rarefied clouds are still bathed in sunlight.

The bottom line of this discussion is that elevation (which, of course, is just "location" in the vertical) plays a key role in the global distribution of temperature. At the Earth's surface, we have already established that proximity to large bodies of water is also very important—this is especially true when there's an ocean current offshore.

Ocean Currents: Giant Gyres That Team with the Wind

To gain additional insight into global temperature patterns, we return to the ocean. This time, we consider the key role played by the "ocean in motion" in regulating global temperatures. We're talking about ocean currents.

Arguably, the most famous of all ocean currents (at least to folks living in the Western Hemisphere) is the **Gulf Stream**. At one time mapped by Benjamin Franklin, this well-defined current of

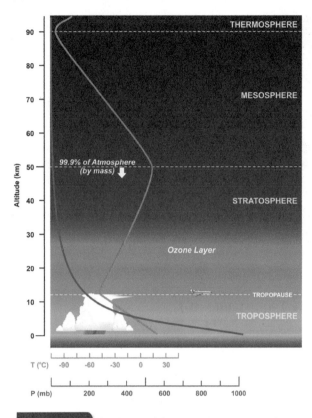

FIGURE 3.14 The layers of the atmosphere are strongly determined by temperature variation with height. On average, temperature decreases with height in the troposphere and mesosphere, and increases with height in the stratosphere.

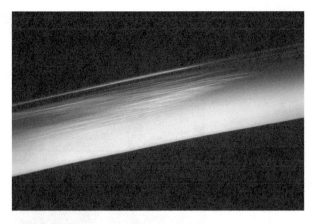

FIGURE 3.15 Noctilucent clouds, viewed from the International Space Station. These clouds form near the boundary of the mesosphere and thermosphere when water molecules gather and crystallize onto specks of debris, possibly from disintegrating meteors (courtesy of ISS/NASA).

relatively warm water originates near the Straits of Florida and flows northeastward along the East Coast of the United States. Figure 3.16 shows sea-surface temperatures on October 28, 2012, as Hurricane Sandy was making its way northward off the coast (Sandy's position at various times is marked by the standard symbols for tropical storm and hurricane). The Gulf Stream stands out as the plume of red and orange that parallels the Southeast Coast and then heads northeastward into the North Atlantic Ocean. Sandy actually intensified as it passed over the Gulf Stream on its way to an evening landfall on October 29 along the southern New Jersey coast.

The Gulf Stream is a member of a rather exclusive club of warm and cold ocean currents that circulate within the oceans of the world (refer to Figure 3.17). Figure 3.18 is a visualization of the speeds associated with the surface ocean currents in the Atlantic Ocean. Shades of green represent the fastest-moving water (the Gulf Stream typically

moves at an average speed of three or four knots, with a maximum speed approaching five knots). As the Gulf Stream flows into the North Atlantic Ocean, it slows and becomes more diffuse. Nonetheless, a weaker extension of the Gulf Stream, the **North Atlantic Drift**, continues toward Iceland, the British Isles, and Scandinavia, helping to moderate the winter climate at high latitudes in extreme western Europe (the degree of moderation is still the subject of some academic debate).

Arguably one of the Gulf Stream's most dramatic effects on weather occurs along the East Coast of the United States during winter. Then, with cold air dominating over the continent, the relatively warm Gulf Stream (which warms the air above) increases the temperature gradient in the lower troposphere along the coast. In turn, as you will learn later, low-pressure systems feed off this enhanced temperature gradient, frequently leading to explosive development of winter storms near the East Coast.

Driven by systems of prevailing winds over the oceans, the giant gyres shown in Figure 3.17 aid in trying to erase the large temperature contrasts between the poles and tropics. For example, one study showed that poleward-flowing currents account for nearly 40 percent of the transfer of heat energy out of the tropics toward latitude 30°N. Though the wind and ocean currents are a formidable team, the large temperature contrasts between the poles and tropics will never be erased—as long as the sun continues to heat our planet unevenly.

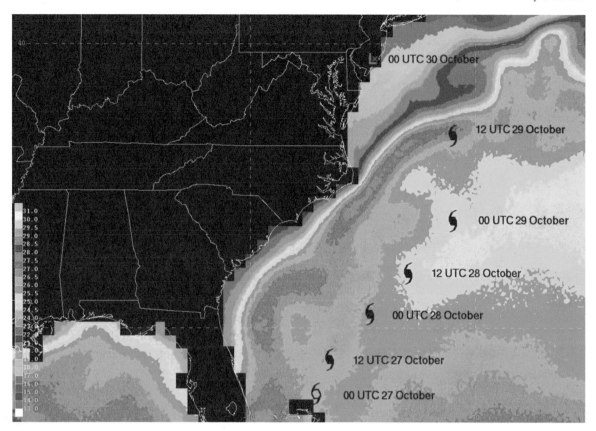

FIGURE 3.16   Sea-surface temperatures (SSTs, in °C) on October 28, 2012 unveil the Gulf Stream as the plume of red and orange flowing northeastward off the East Coast of the United States. SSTs were derived from a blend of satellite and computer model data. The location of Hurricane Sandy at various times is shown by the standard tropical storm and hurricane symbols. Prior to landfall in New Jersey, Sandy intensified a bit as it passed over the Gulf Stream (courtesy of the National Weather Service).

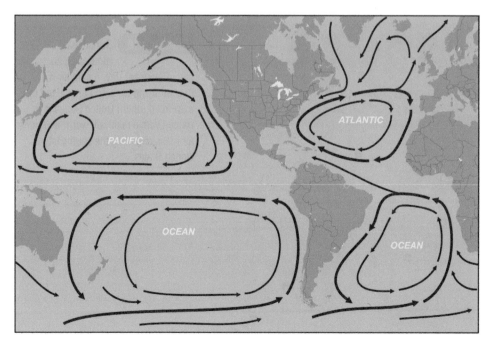

FIGURE 3.17   Principal surface currents of the Atlantic and Pacific Oceans.

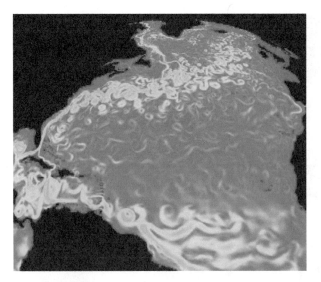

FIGURE 3.18 A visualization of the speeds associated with the surface ocean currents in the Atlantic Ocean. Brighter colors represent the fastest-moving water (courtesy of NOAA's Science on a Sphere).

All of the controllers of global temperature that have been discussed so far shape the average temperature patterns on this planet, a key element of climate. But to the mid-latitude meteorologist charged with predicting the weather over the next few days, climatology merely provides a general context for the forecast. To track day-to-day temperature variations, forecasters must follow the movement of large masses of warm and cold air and the boundaries that separate them. Here, we take a first look at those boundaries, the common features on weather maps called fronts.

FRONTS: THE BATTLE BETWEEN WARM AND COLD AIR

Large blobs of air that loiter over a region of the earth tend to acquire temperature and moisture properties consistent with the underlying surface. For example, air lingering over the tepid waters of the Gulf of Mexico in summer becomes warm and very humid, while lethargic air hovering over Siberia in winter is frigid and bone-dry.

Formally, an **air mass** is a large volume of air with fairly uniform properties of temperature and moisture at any horizontal level. On average, air masses have a horizontal breadth on the order of several thousand kilometers. In Figure 3.19, shading marks two different air masses influencing North America on an early December day. The warm and relatively humid air mass affecting the Southeast States originated over subtropical seas and is therefore called a maritime tropical air mass (abbreviated mT). *Maritime* refers to its oceanic origin (relatively moist), while *tropical* is self-explanatory—the air mass originated over a relatively warm part of the world.

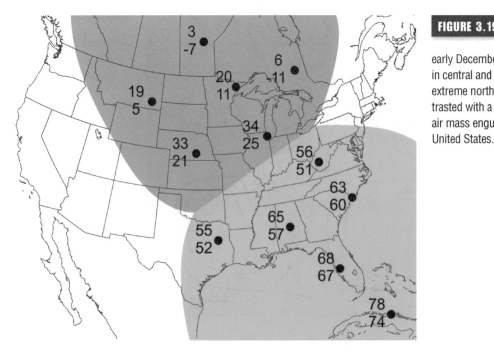

FIGURE 3.19 Surface station models on an early December day. A cold air mass in central and western Canada and the extreme northern United States is contrasted with a relatively warm and moist air mass engulfing the southeastern United States.

The cold air mass over Canada and the northern U.S. acquired its characteristics over frigid high latitudes and is therefore called a continental polar air mass (abbreviated cP). *Continental* refers to the air mass' land-locked source region (relatively dry), while *polar* refers to the air mass's origin over a relatively cold part of the world.

Mixing and matching, the four basic classifications of air masses are cP, cT, mP, and mT, with a special classification of cA (continental Arctic) reserved for those truly frigid air masses that originate over hyperborean regions such as Siberia during winter. Figure 3.20 shows the birthplaces, or **source regions**, of air masses that affect the contiguous United States.

If the tropics and poles did not regularly liquidate their hot and cold inventories, north-to-south temperature contrasts across the globe would go sky high. But the atmosphere runs a sensible business, continually sending out large shipments of cold and warm air to the mid-latitudes. "UPS" is the operative word for cold and warm air in transit because wherever and whenever they meet, relatively light, warm air goes up and over relatively heavy, cold air, often forming clouds and precipitation. The boundaries between colliding shipments of cold air from high latitudes and warm air from lower latitudes are called **fronts**. If cold air *advances* at

the front, the boundary between warm and cold air is a **cold front**. By convention, cold fronts on surface weather maps are designated by a chain of blue triangles (see Figure 3.21), with the tips of the triangles pointing toward the warm air. If cold air *retreats* at the front, the boundary is a **warm front**, which is designated on surface weather maps by a chain of red semicircles, with each semicircle directed toward the cold air (see Figure 3.21). When cold air moves very slowly or not at all, the boundary is a **stationary front**, designated on surface weather maps by a chain of alternating blue triangles and red semicircles.

Note in the definitions of the various types of "front" that the colder air is always "the boss"—its movement relative to the front determines the type of front. To understand why colder air is always in control, recall from Chapter 2 that colder air tends to be denser than warmer air (assuming you make the comparison at the same level in the atmosphere). So, in essence, fronts are boundaries between air masses of different densities, and the less dense, warmer air cannot advance at the surface unless the more dense, colder air retreats. Cold air's stubbornness leads to all kinds of weather action at fronts.

Because fronts mark the boundaries between air masses of differing temperature and moisture

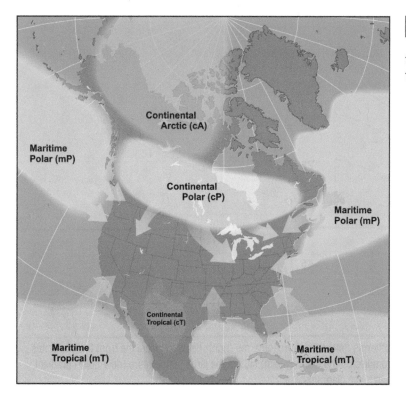

FIGURE 3.20 Source regions for air masses affecting North America, and the common paths those air masses follow into the contiguous United States.

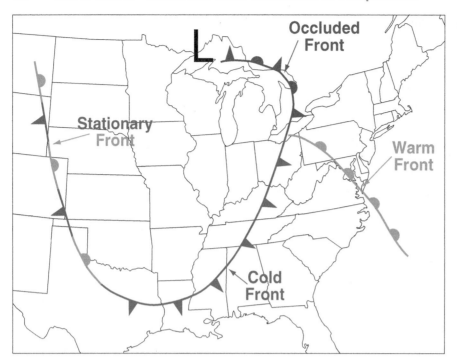

FIGURE 3.21 The symbols that meteorologists use to indicate fronts on surface weather analyses. Cold fronts are blue lines with blue triangles on the warm side of the line (think of the triangles as pointing in the direction of movement of the front). Warm fronts are red lines with red semi-circles on the cold side of the front (the humps "point" in the direction of movement of the front). The stationary front symbol is essentially an alternation of the warm and cold front symbols. An occluded front is purple, with the triangles and semi-circles alternating on the same side of the front, "pointing" in the direction of movement of the front. Occluded fronts will be discussed in detail in Chapter 13.

characteristics, we should expect temperature and dew-point contrasts across fronts to be relatively large. That expectation is borne out in Figure 3.22, which shows isotherms (left) and **isodrosotherms** (isopleths of dew point, right) in the vicinity of the boundary between the air masses in Figure 3.19. Note the relatively tight packing of the isotherms and isodrosotherms from the lower Great Lakes into

Kansas and Oklahoma. By convention, a front is placed just on the warm side of the relatively large temperature gradient (alternatively, just on the moist side of the relatively large dew-point gradient). Or think of it this way: a front is placed at the leading edge of the cold (or dry) air mass. These relatively large temperature and dew-point gradients are classic signatures of fronts.

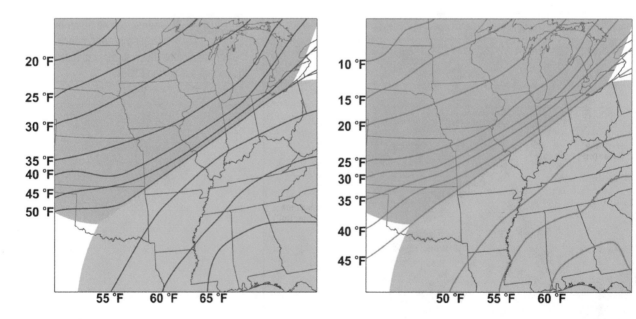

FIGURE 3.22 The relatively tight packing of the isotherms (left) and isodrosotherms (right) helps to locate the front that separates the air masses in Figure 3.19. The front lies on the warm (moist) side of the large temperature (dew-point) gradient, approximately where blue meets red in these diagrams.

The passage of a front can be meteorologically dramatic. The winds might change direction sharply, and the temperature and dew point might decrease (or increase) rapidly. Figure 3.23a documents a cold-frontal passage through Dallas, TX between 22 and 23 UTC on January 12, 2007. Notice that, after remaining relatively steady during the afternoon hours, both the temperature and dew point fell more than 20°F (11°C) in two hours in the wake of the front, and the wind changed direction by almost 180 degrees. The colder, drier air continued barreling southward, reaching San Antonio the next afternoon (see Figure 3.23b). Meteorologists attending the Annual Meeting of the American Meteorological Society in San Antonio experienced a temperature drop from 70°F (21°C) to 52°F (11°C) in two hours as winds shifted from southerly to northeasterly (interesting weather always seems to follow the AMS meeting—later in the week, freezing rain would fall in San Antonio).

Clearly, the analysis of fronts is critical to weather forecasting in the mid-latitudes. And because fronts represent showdowns between huge masses of air of differing characteristics, a poor prediction of a front can result in large errors in forecasted temperatures for locations close to the boundary. Let's investigate this idea a little more closely, linking the wind to the changes in temperature that take place in concert with the approach and passage of fronts.

A Short Primer on Temperature Forecasting: When Advection Is Important and When It's Not

Let's consider a cold front that extended from Lake Superior to southern Texas on an early March afternoon (see Figure 3.24a). Ahead of the front, milder southerly breezes transported warmer air northward. Instead of the rather generic "transported," meteorologists say that the southerly wind **advected** warm air northward. They describe the process as **warm advection**. Behind the front, winds from the north and northwest advected cold air southeastward (**cold advection** occurred). In the grand scheme of things, these temperature advections are the atmosphere's way of trying to even out temperature contrasts between the tropics and polar regions. In the short term, forecasters weigh these advections when they predict temperature. Isopleths of temperature advection at this time are shown in Figure 3.24b, in units of degrees Fahrenheit per hour. Ahead of the cold front, values of warm air

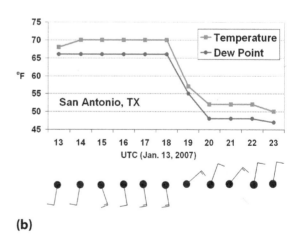

(a)

(b)

FIGURE 3.23 A meteogram of temperature, dew point, cloud cover, and wind direction and speed documents the passage of a southward-moving cold front through (a) Dallas, TX around 22 UTC on January 12, 2007 and (b) San Antonio, TX just after 18 UTC the following day. After holding fairly steady in the hours before the frontal passage, the temperature and dew point decreased dramatically as colder, drier air arrived behind the cold front. Winds shifted from a predominantly southerly direction to a northerly direction as the cold front passed.

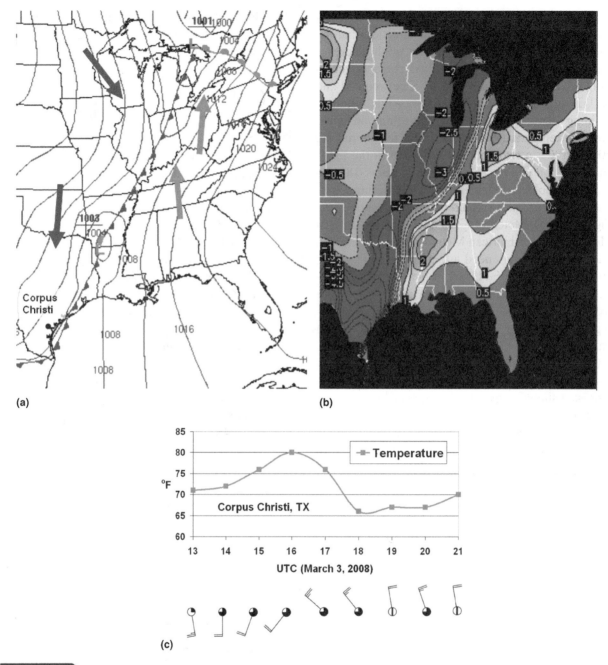

FIGURE 3.24 (a) Surface weather map at 18 UTC on March 3, 2008, shows a cold front stretching from Lake Huron to Texas (courtesy of NOAA); (b) Temperature advection at this time, isoplethed in units of °F/hr. Purples indicate the largest cold advection, while orange and red represent the strongest warm advection (courtesy of Bob Hart); (c) The cloud and wind portions of the station model, and the trace of temperature, at Corpus Christi, TX, from 13 UTC to 21 UTC on March 3, 2008.

advection exceeded 2°F per hour in parts of the lower Mississippi River valley and northern Ohio. In other words, temperatures would have increased at these rates if warm advection were the only influence on temperature at this time.

Now notice the values of advection west of the front: in central Illinois, –3°F per hour; in parts of south Texas, –4.5°F per hour or less. In other words,

temperatures would have decreased at these rates if advection were the only influence on temperature at this time. In reality, however, cold advection had some competition. In some locations, the sun came out after the cold front passed. Figure 3.24c shows the temperature and the cloud and wind portion of the station model from Corpus Christi (in south Texas) in the hours before and after the front passed. Cold advection alone indicated that temperatures should decrease on the order of 4 to 5°F per hour. Indeed, temperatures did drop significantly in the two hours after the front passed. But solar heating eventually overcame the cold advection and temperatures actually increased a few degrees after 18 UTC.

Of course, March is the start of meteorological spring, and in the Deep South during this time of year, the sun is pretty strong. But in winter, advections of cold and warm air largely control the daytime high temperature over the mid-latitudes simply because the sun tends to remain low in the sky, weakening its heating power. So temperatures can fall steadily in the wake of strong cold fronts, even in full sunshine, because cold advection overwhelms weak solar heating.

Similarly, on winter nights, strong warm advection can cause temperatures to reach their maximum in the wee hours before dawn. Forecasters must weigh these issues carefully when predicting temperature in the vicinity of fronts.

On the other hand, if there isn't an air mass change at a given location from one day to the next (in other words, if no front goes through, which is the case more often than not), some simple forecasting rules may apply. For example, forecasters might take note of one day's high temperature and use it again for tomorrow's predicted maximum. Such an approach is an example of **persistence forecasting**.

Typically, however, forecasters will use a related method called **modified persistence** if they expect the forecast area to stay within the same air mass from one day to the next. In such cases, forecasters will use one day's high temperature as the starting point for the next day's predicted high. For example, suppose the high temperature within an mT air mass on a sunny, late-spring day in Atlanta is 85°F (29°C). Tomorrow, the forecaster anticipates skies to be cloudy much of the day, though still in the warm, humid air. Taking this into account, the forecaster might predict a high of 80°F (27°C) for tomorrow, thinking that the clouds will suppress temperatures several degrees.

Or suppose a thin veil of cirrus clouds dimmed sunshine over Washington, DC, during a cold January afternoon when the high reached 27°F (−3°C). Assuming the city stays within the same unchanging air mass the next day and that the clouds will move east of the region, a forecaster might predict a high of 30°F (−1°C), thinking that full sunshine (though relatively weak) will help to boost temperatures a few degrees above the previous day's maximum.

The use of persistence and modified persistence in temperature forecasting isn't restricted to just high temperatures. Assuming there aren't any changes in the air mass at a given city, weather forecasters can also apply these methods to predict the low temperature at night.

The difference between the highest and lowest temperature on a particular day is called the **diurnal range**. In many ways, daily temperature variations mimic those of a calendar year, and some of the same factors that control seasonality also influence the magnitude of the diurnal range. But the high and low temperatures on any particular day depend on so much more. Read on.

DAILY VARIATIONS OF TEMPERATURE: OFTEN LAGGING BEHIND

Experience tells us that, on most days, the air near the ground reaches its maximum temperature during the mid to late afternoon. On closer inspection, this timing is actually somewhat puzzling considering that the sun is highest in the sky around local noon and thus solar radiation is most intense at that time. Equally intriguing is the timing of the overnight low temperature, which typically occurs not in the middle of the night, but rather around sunrise.

We call on our earlier cooking lessons (with pizza and pudding) to kick-off our explanation. Just as cold pizza placed into a preheated oven takes time to warm up to the desired eating temperature, the ground also takes time to heat up during the day. Additional time is then required for the ground to transfer its energy to the air above, via conduction, convection, and radiation. Similarly, just as hot pudding placed in a refrigerator takes time to cool, the ground steadily cools throughout the night, typically starting shortly before sunset. The chilling ground then conducts energy away from the air above, setting the stage for the lowest air temperature to typically occur around sunrise.

Figure 3.25 generically illustrates this idea for a typical day in the mid-latitudes, around the time of the equinoxes, when the number of hours of daylight and darkness are approximately equal. Incoming solar radiation (blue curve) increases during the morning, peaks at local noon, then decreases, but the highest air temperature (red curve) occurs several hours after the maximum in solar radiation. Looked at from the perspective of an energy budget, temperatures keep rising as long as incoming solar

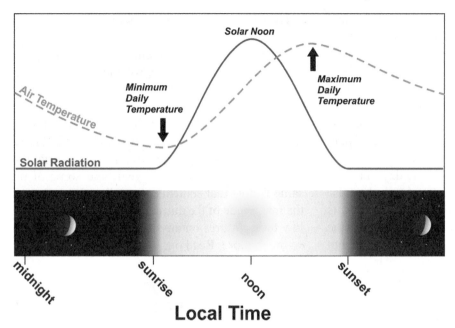

FIGURE 3.25 A comparison of incoming solar radiation (solid blue curve) and air temperature (dashed red curve) during a typical 24-hour period at a mid-latitude location around the equinoxes. The daily high temperature typically lags the maximum in solar input by a few hours.

radiation exceeds outgoing infrared radiation. On the other hand, temperatures fall overnight because outgoing infrared radiation exceeds any other energy input, reaching a minimum approximately around sunrise (at night, solar radiation is zero but the ground does receive some infrared radiation from the atmosphere).

The difference between the daily high and low temperature can vary quite a bit, and once again location plays a key role. We'll now investigate geographical influences on the diurnal range as well as factors that are tied more directly to day-to-day changes in the weather.

Variations in The Diurnal Range: More Location, Location, Location

Like the annual range of temperature, the average diurnal range depends on several geographical factors:

- PROXIMITY TO WATER BODIES: The daily variation of water temperature in oceans and large lakes rarely exceeds a few degrees Fahrenheit. Thus, coastal locations, particularly those downwind of a body of water, experience smaller diurnal ranges, on average, than more continental locations. For example, the average diurnal range in Monterey, CA, on the Pacific Coast, is about 17°F (9°C). Meanwhile, 180 km (112 mi) inland from the Pacific, the average diurnal range at Fresno, CA, is more than 26°F (14°C).

If we consider just the effects of proximity to a water body on maximum temperatures, easterly winds blowing off the cool Atlantic Ocean reduce daytime readings at Boston, MA, particularly in spring. In a similar vein, northwest winds during an April afternoon blowing across the chilly waters of Lake Erie keep daytime temperatures at Erie, PA, and Cleveland, OH, close to surface water temperatures, even if the sun is out. On the other hand, these locations typically experience their first freeze in autumn a little later than locations farther from (or upwind of) the lake, benefiting from the moderating influence of the relatively mild water during the fall. Many northern vineyards are located along the shores of Lake Erie (and other Great Lakes) because of the lengthened growing season.

- TOPOGRAPHY: In general, temperature decreases with elevation (and this would, of course, include the daily high and daily low temperature). But forecasters must consider another effect of topography: at night, especially under clear, calm conditions, the coldest (densest) air slowly drains down from higher elevations and collects in the valleys. On such nights, valley floors are typically colder than the surrounding hillsides. For this reason, vineyards in Northern States are often situated on mountainsides (see Figure 3.26) because the latest freeze in spring and the earliest freeze of autumn typically occur in the valleys, making the growing season longer on nearby hillsides. Vineyards in Northern States also tend to be located on southeast-facing slopes, which receive slightly more sunshine than the valleys. Mountain slopes also tend to be breezier than valleys, helping grapes to dry out more quickly after summer rains (too much moisture can cause rot and fungus).

- URBANIZATION: In urban areas, buildings and asphalt streets absorb vast amounts of solar energy by day and then emit the reservoir of heat energy steadily at night. As a result, cities are generally warmer than surrounding less-developed areas, and the larger the city the greater the difference tends to be, especially on clear, calm nights in winter. As an example, Figure 3.27 shows minimum temperatures on a February morning in 2014, in the vicinity of Philadelphia, Pennsylvania. Urbanized areas are colorized purple in this NASA image—Philadelphia County, the most urbanized area, lies approximately in the middle of the image, a relatively warm "urban heat island" surrounded by a sea of nocturnal chill.

Research has shown that the impact of urbanization tends to be larger at night—note in this example that the difference between the minimum temperature in the most urbanized areas of Philadelphia county and some of the surrounding areas exceeded 15°F (8°C). As a result, in general, the diurnal range within a city tends to be larger than the diurnal range in less-developed suburban and rural areas. We'll have more to say

about the effects of urbanization on weather and climate in Chapter 18.

Variations in the Diurnal Range: Weather or Not

On a day-to-day basis, local weather conditions, such as the amount of cloudiness and the strength of temperature advections, largely control the day's highest and lowest temperatures, and thus by default, the diurnal range. Here's a rundown of the most important factors:

- WIND: Earlier in the chapter when we discussed fronts, we emphasized that temperature advection (which, of course, requires wind) can play a major role in disrupting typical daily temperature variations, especially during the cold season. As a result, temperature advection can strongly impact the diurnal range.

But even wind that produces zero temperature advection—that is, transports air of roughly equal temperature—can affect the diurnal range. Recall from Chapter 2 that inversions often form at night near the surface, especially under clear, calm conditions, with the coolest air residing near the ground. On a windy night, however, mechanical eddies prevent the formation of a

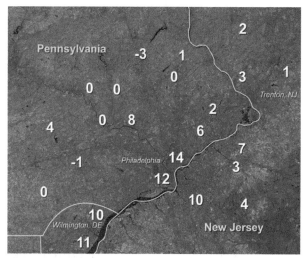

FIGURE 3.27 Low temperatures in and around Philadelphia on the morning of February 12, 2014. The background is colorized by land use—forested areas are green, agricultural areas are pink, and urbanized areas are purple. Note that the highest overnight temperatures tend to occur in the most urbanized areas (courtesy of NASA).

FIGURE 3.26 Mount Nittany Winery, outside State College, PA, is situated on a hillside, not in the valley, delaying the first freeze and thus prolonging the growing season.

deep layer of chilly air next to the ground by mixing down slightly warmer air from higher up. Thus, all other factors being equal, surface air temperatures on windy nights with little or no temperature advection are higher than on nights with light or calm winds, thereby reducing the diurnal range.

- CLOUDS AND WATER VAPOR: Clouds back-scatter sunlight, but because they are composed of water and/or ice, they preferentially absorb infrared radiation. During the day, the amount of solar radiation back-scattered by clouds exceeds the amount of infrared radiation that the clouds absorb. Thus, clouds suppress daytime temperatures, especially during the warm season when solar radiation rules. At night, with no incoming solar radiation, the absorption and emission of infrared radiation by clouds keeps the lower troposphere warmer than it would be on a clear night (recall the dramatic effect of downwelling infrared radiation in Figure 2.23). Thus, all other factors being the same, cloudy days are usually cooler than clear days, but cloudy nights are usually warmer than clear nights (particularly those with a low overcast and no precipitation).

Even if clouds are not present, invisible water vapor can play a big role in orchestrating nighttime lows. Water vapor, the most potent greenhouse gas, is not distributed evenly around the globe, so the amount of infrared radiation lost to space varies greatly as well. The effect of this nonuniformity in terrestrial radiation loss is most noticeable at night. For instance, the humid atmosphere of St. Croix, in the Virgin Islands, keeps overnight minimum temperatures above 70°F (21°C) most of the summer, after daytime highs typically in the 80–90°F range (27–32°C). In contrast, mountains typically shield the west Texas town of El Paso from the advance of moist low-level air from the Pacific Ocean and the Gulf of Mexico (see Figure 3.28). In El Paso, days with high temperatures in the 80° to 90°F range are often followed by nighttime low temperatures near 50°F (10°C). Taking the notion of dry air to an extreme, some of the largest diurnal ranges occur in arid desert regions, where cloud-free skies allow maximum solar heating during the day, while dry air permits large losses of infrared energy at night (check back to Figure 3.6c).

- PRECIPITATION: Once clouds begin to precipitate, surface air temperatures typically decrease. That's because some of the smaller raindrops (or snowflakes) evaporate in the air before reaching the ground. Evaporating water requires energy, which comes from the surrounding air. This **evaporational cooling** lowers the air temperature. The same process gives you goose bumps when you get out of the pool on a sunny, breezy day: Water evaporating from your skin steals energy from your body, giving you a chill.

- SNOW COVER: As discussed in Chapter 2, nighttime temperatures on a clear, calm night are lower when there are at least a few inches of snow on the ground. How can we explain this observation? It turns out that snow is an efficient insulator, effectively sealing off the heat energy stored in the ground. When the ground is not covered by snow, heat energy stored below ground helps to limit the drop in nighttime air temperatures by keeping the topsoil warmer. With a snowpack sealing off this heat energy, the temperature of the air in contact with the rapidly chilling snow can "drop like a rock." All other meteorological factors being equal, nighttime low temperatures in snow-covered rural areas can be 10°F (6°C) or more lower than had the ground been bare. Such large downward spirals in nighttime temperatures over snow-covered areas are often set into motion during the previous day by the highly reflective snowpack. Because of snow's relatively high albedo, temperatures at sunset are likely lower than they otherwise would be, so nocturnal cooling has a head start.

TEMPERATURE MEASUREMENT: DON'T TRUST BANK THERMOMETERS OR SPORTSCASTERS

We'll end this chapter on temperature with a discussion of how we measure this most fundamental of atmospheric variables. Getting an accurate gauge of the air's temperature isn't quite as simple as it might seem.

Consider what sometimes happens during the televising of sporting events during the warm season. The sportscasters in the broadcast booth often request a report about weather conditions on the

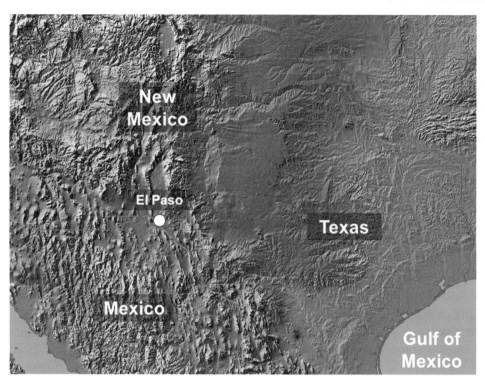

playing surface. In response, a reporter stationed on the field checks a thermometer that reveals an extraordinarily high reading, perhaps 120°F (49°C) or above. The temperature displays on some bank signs (and perhaps on your own outdoor thermometer at home) often register similarly high temperatures during the summer. Yet the meteorologist on the evening news that night might report an "official" high temperature that day of only 90°F (32°C) or so. What's going on here?

Thermometers: All Designed with One Goal in Mind

Designing thermometers to measure air temperature starts with one basic principle: Most substances contract slightly when cooled and expand slightly when heated. With that in mind, let's explore the different designs of thermometers.

Liquid-in-glass thermometers are reasonably accurate and inexpensive. Their design has remained basically unchanged for more than 200 years (see Figure 3.29). A liquid, usually mercury or red-colored alcohol, is free to move within a thin opening inside a glass enclosure. The opening is so narrow that even small temperature changes cause

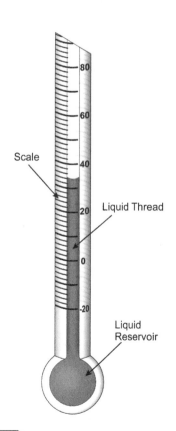

FIGURE 3.29 Schematic of a common liquid-in-glass thermometer.

a relatively large change in the length of the liquid. The glass also expands and contracts as temperature changes, but its expansion and contraction is negligible compared to that of the liquid. Modified liquid-in-glass thermometers have been designed specifically for measuring daily maximum and minimum temperatures. In these special thermometers, the length of the liquid thread at either its maximum expansion or maximum contraction is preserved until the thermometer is reset.

Circular thermometers primarily used outdoors or in refrigeration units are examples of **bimetallic thermometers** (see Figure 3.30). These devices contain no liquid: Instead, two different pieces of metal are welded together to form a single strip that is usually arranged in a spiral. The two metals, often iron and brass, must differ significantly in the amount they expand and contract as temperature varies. Temperature changes cause deformation in the strip as it expands and contracts unevenly. The deformation is magnified by a system of levers that controls a pointer on a calibrated dial. In a **thermograph**, a pen is connected to the pointer, allowing a continuous record of temperature to be traced on a clock-driven rotating drum (see Figure 3.31).

Electrical thermometers rely not on the expansion or contraction of a liquid or metal but rather on the relationship between temperature and electrical resistance or electrical current flow. For example, a **thermistor** is an electrical thermometer composed of a solid material, usually ceramic, whose resistance to the flow of electricity changes at a constant rate as temperature changes. Electrical thermometers are commonly used in the ASOS network (recall Chapter 1) as well as in radiosondes, the packages of weather instrumentation carried aloft by balloons to measure weather conditions above the Earth's surface.

To ensure accurate measurement of outdoor air temperature, thermometers are housed in an enclosed shelter which protects the instruments from precipitation and direct sunlight. Historically, cooperative weather observers (recall Chapter 1) used liquid-in-glass thermometers placed inside a "cotton-region shelter" (see Figure 3.32a), sometimes called a Stevenson Screen because the father of writer Robert Louis Stevenson was interested in meteorology and designed such a thermometer-shielding enclosure in 1864. To a large extent, these older shelters have been replaced by a newer design which looks like a beehive (see Figure 3.32b) and which houses a thermistor called a Maximum/Minimum Temperature Sensor, or MMTS. In both types of shelters, the sides have open vents to allow outside air into contact with the instruments. Shelters are painted white to maximize the reflection of sunlight, and are usually mounted at a standard height of five to six feet above ground level to minimize the effect that the underlying surface might have on the official observed temperature.

FIGURE 3.30 A bimetallic thermometer. The bimetallic strip (not visible) is usually arranged in a spiral on the back of the thermometer.

FIGURE 3.31 A thermograph, a device for measuring and recording temperature.

(a)

(b)

FIGURE 3.32 (a) A cotton-region instrument shelter, or Stevenson Screen, used for housing liquid-in-glass thermometers; (b) A Maximum/Minimum Temperature Sensor, or MMTS, is a thermistor housed in a shelter which resembles a beehive.

Weather Folklore and Commentary

Do Clouds Act Like a Blanket?

You may have heard the saying that "Cloudy nights tend to be warmer than clear nights." This popular observation is accurate (all else being equal), but the explanation that's frequently given—that clouds "act like a blanket"—is misleading. We can use Schlieren photography to provide a firmer understanding of what's really going on.

Consider the Schlieren photograph on the left in Figure 3.33, and note the convection emanating from the bare arm of one of the authors as his skin continuously warms a thin layer of nearby air. On the right photograph of Figure 3.33, his arm has been covered with a cotton blanket, which is relatively porous. Nonetheless, the blanket markedly reduces convection, affording us insight into one impor-

tant way that blankets keep us warm at night: They suppress convection of body-warmed air, thereby reducing the transfer of heat energy away from the body.

So, do nighttime clouds behave like blankets in this way, by suppressing convection originating at the surface? Not really. Remember that after sunset, the ground typically cools as it loses more energy via infrared radiation than it gains from the atmosphere. In turn, the lowermost layer of air cools. But recall that convection requires that the air in contact with the surface be heated by that surface. Thus, there simply won't be any convection for the clouds to suppress at night! And as a result, clouds at night really don't act like blankets, at least not in this "convective" sense. In fact, clouds act more like space heaters, emitting much more infrared radiation (some of it toward the surface) than a clear sky emits. So,

(Continued)

(Continued)

on cloudy nights, especially those with an overcast of relatively warm, low clouds, the surface simply receives and absorbs more infrared radiation from

clouds than it would have had the sky been clear. As a result, surface air temperatures tend to be higher on cloudy nights.

 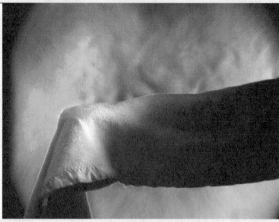

FIGURE 3.33 Schlieren photographs of a bare arm (left) and the same arm with a blanket (right) (courtesy of Dr. Gary Settles, Director of the Gas Dynamics Lab, Penn State University).

So what's the problem with measuring air temperature using a thermometer placed in direct sunlight? Recall that air molecules are in perpetual motion, constantly vibrating and colliding with one another, and air temperature is a measure of the average speed of these molecules. Let's apply this definition to a liquid-in-glass thermometer—it actually measures the temperature of the bulb which is assumed to be the same as the temperature of the air. But compared to air, the bulb of a thermometer is a more efficient absorber of solar radiation, so when the bulb is exposed to direct sunlight, its temperature would be higher than that of the ambient air. So direct sunlight is a game-changer—to be truly representative of the *air* temperature, thermometers must be shaded from the Sun.

This argument suggests one possible explanation for air temperature observations that seem too high—the thermometer may have been exposed to direct sunlight. Having said this, we should point out that those high temperature readings "down on the field" might not be completely inaccurate, even if the thermometer is protected from direct sunlight. Recall from Chapter 2 that large vertical temperature gradients typically exist close to the ground on sunny

days. Thus, a thermometer placed directly above an artificial playing field or the asphalt parking lot of a bank will read much higher than a thermometer at eye level at those locations, which is approximately the height at which official air temperature readings are made.

To get a sense for the range of officially observed surface air temperatures on Earth, note that the lowest temperature ever recorded at the surface is –129°F (–89°C), at Vostok, Antarctica (a Russian research station near 80°S latitude), on July 21, 1983. The lowest temperature on record in the United States is –79.8°F (–62.1°C) at Prospect Creek, AK, on January 23, 1971. On the warm side, the highest officially observed surface temperature in the world, 134°F (57°C), was recorded at Death Valley, CA, on July 10, 1913. It's worth noting, however, that until 2012, the all-time heat record was claimed by Al Azizia, Libya, a temperature of 136°F (58°C) on September 13, 1922. However, this record was invalidated in 2012 by the World Meteorological Organization which concluded that an error was made in measuring the temperature. Temperature extremes for each continent are summarized in Figure 3.34.

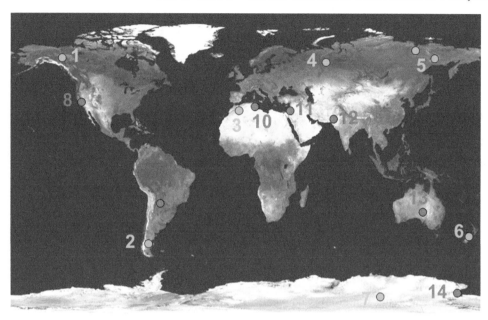

Low Temperature Extremes				
	Continental Area	**Lowest Temp (°F)**	**Place**	**Date**
1.	North America	−81.4	Snag, Yukon, Canada	Feb. 3, 1947
2.	South America	−27	Sarimento, Argentina	Jun. 1, 1907
3.	Africa	−11	Ifrane, Morocco	Feb. 11, 1935
4.	Europe	−72.6	Ust'Schugor, Russia	Dec. 31, 1978
5.	Asia	−90	Oimekon, Russia	Feb. 6, 1933
	Asia	−90	Verkhoyansk, Russia	Feb. 7, 1892
6.	Australia	−14	Eweburn, New Zealand	Jul. 17, 1903
7.	Antarctica	−129	Vostok	Jul. 21, 1983

*exact date unknown

High Temperature Extremes				
	Continental Area	**Highest Temp (°F)**	**Place**	**Date**
8.	North America	134	Death Valley, CA	Jul. 10, 1913
9.	South America	120	Rivadavia, Argentina	Dec. 11, 1905
10.	Africa	131	Kebili, Tunisia	Jul. 7, 1931
11.	Europe	129	Tirat Tsvi, Israel	Jun. 21, 1942
12.	Asia	128	Mohenjo-daro, Pakistan	May 26, 2010
13.	Australia	123	Oodnadatta, Australia	Jan. 2, 1960
14.	Antarctica	59	Vanda Station, Scott Coast	Jan. 5, 1974

FIGURE 3.34 Global temperature extremes, by continental area, based on regions defined by the World Meteorological Organization, or WMO. The high temperature record for Asia is still under investigation.

Focus on Optics

Subsuns and Sun Pillars: Magical Mirrors

Magician Harry Houdini, best known for his escape from a packing case that had been locked and then secured with rope and steel tape and dropped into New York harbor, attributed all of his feats of magic to natural, physical effects.

The sun is also a magician that relies on natural, physical effects. But unlike Houdini, the sun is not an escape artist. Rather, with a simple electromagnetic wave of its radiation wand, the sun creates magical displays of atmospheric optics such as rainbows and halos. Although halos are popularly regarded as circles of light around the sun or moon, the scientific definition includes any rings, arcs, pillars or spots of light in the sky that are caused by light reflecting off or bending through ice crystals.

The sun's faithful assistant in these optical sleights-of-hand is the atmosphere, whose props include water drops, ice crystals, air molecules, and other airborne particles. These props act like lenses, mirrors, and prisms, which focus, bend, reflect, and scatter the sun's rays into a kaleidoscope of sky lights. Because this chapter is about temperature, we'll restrict this discussion to some of the optical effects performed by ice crystals, which require the temperature to be below the melting point of ice (32°F, or 0°C).

Ice crystals come in many shapes and sizes (just think of the myriad of appearances of snowflakes to convince yourself of this assertion). Of all the different kinds of ice crystals, just two types are responsible for creating halos: plates and columns (see Figure 3.35). The aerodynamic properties of plates and columns (in other words, how they "fly" toward earth) depend on their shapes and sizes. For example, if the crystals are very tiny, with diameters less than 15 to 20 microns, they can be jostled around by air currents. In this instance, flying formations of squadrons of tiny crystals might resemble peanuts in a bag—their arrangements are quite jumbled.

Large plates assume a more structured flying formation, falling with their flat sides facing up and down, looking more or less like dishes sitting on a dining room

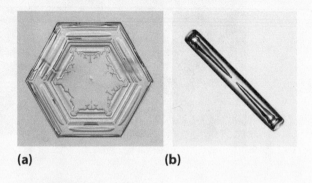

(a)   **(b)**

FIGURE 3.35 Photographs of ice crystals in the form of (a) plates, and (b) columns. Of the myriad shapes of ice crystals found in the atmosphere, only two—hexagonal plates and hexagonal columns—account for almost all of the ice-crystal optical phenomena observed in the sky (courtesy of Kenneth G. Libbrecht).

table. Large columns, on the other hand, fall with their long, horizontal axis parallel to the ground. Their horizontal axes are free to rotate or swivel in any direction, so their flying formation resembles pencils that are scattered randomly on the floor.

It's probably not surprising to you that different ice crystals in different flying formations interact with sunlight differently and therefore produce different types of halos. Since some magicians have been accused of using mirrors to perform their feats, we will restrict this discussion to ice crystals that act as mirrors.

When squadrons of large plates assume flying-dish formations, rays of sunlight that strike their large, flat faces are reflected back at the same angle that they intercepted the plates. Thus, in order to see this reflected light, you must be higher than the plates: You might be, for example, atop a mountain, in an airplane, or at some other lofty position. If you are in such a high place and the clouds below you contain large plate crystals (for this to occur, the cloud tops need to be sufficiently cold), you will see a halo like the one shown in Figure 3.36. It is called a **subsun** because it is an image of the sun that lies below the observer.

(Continued)

(Continued)

FIGURE 3.36 A subsun appears beneath an airplane as large ice-crystal plates in cold cloud tops reflect incoming sunlight (courtesy of Mila Zinkova).

In tune with the law of optics that states that the angle of incidence (the angle at which light strikes a surface) equals the angle of reflection (the angle at which the light is reflected back), a subsun appears as far below the horizon as the sun appears above the horizon. Also, it should be noted that large plates do not fall in a perfectly level formation—they get a bit tipsy from being jostled by up-and-down air currents. As a result, the span of angles taken by sunlight reflecting off tipped faces is slightly wider than the span of angles of rays bouncing off perfectly flat plates. Thus, the subsun often appears a bit elongated, resembling a slightly bulging ellipse as opposed to a perfect circle.

Ice crystals in the shape of large columns can also act as atmospheric mirrors. Such squadrons, while flying in scattered-pencil formations, produce a pillar of

FIGURE 3.37 A sun pillar forms above the sun as large ice-crystal columns reflect light to the observer (courtesy of Rick Stankiewicz/EPOD).

light like the one shown in Figure 3.37. These **sun pillars** often appear before or just after sunrise or sunset. More often than not, sun pillars appear in winter when ice crystals are in large supply. However, sun pillars can also be observed in summer, when large columns in high, icy cirrus clouds hovering on the horizon go into their mirror act and reflect sunlight. With an ice-crystal "Abracadabra!," a magical sun pillar then appears out of thin air.

We think that Harry Houdini would approve.

Weather Folklore and Commentary

The Dog Days of Summer

By the time most people think the "Dog Days of Summer" are just getting started, they are actually ending. This expression originates to some extent with the ancient Egyptians who believed that exceptionally hot weather was directly related to the appearance of the Dog Star Sirius in the constellation Canis Major. Sirius is visible in the Egyptian sky between approximately July 3 and August 11. Thus, technically, the Dog Days as they relate astronomically to the appearance of Sirius are over by the middle of August. A more modern interpretation of the Dog Days has been fashioned to portray the seemingly endless heat that often characterizes the month of August.

The Egyptians believed that energy from Sirius combined with energy from the Sun to produce heat waves. Sirius is about 23 times brighter than our Sun, but the Dog Star is 546,000 times farther from Earth than the Sun.

Thus, the intensity of the radiation that reaches Earth from Sirius is minuscule compared to the energy Earth receives from our Sun. Those ancient Egyptians who held the notion that the Dog Star could actually bolster heat waves on Earth were "barking up the wrong tree."

We have seen in this chapter that astronomy and meteorology are often out of synch. For example, the astronomical beginnings of the seasons (on the equinoxes and the solstices) often do not coincide with what is happening meteorologically. Astronomy dictates that the snowstorms and cold that sometimes occur over Northern States in November or the first weeks of December are part of autumn, not winter. With regard to the Dog Days, experience with August heat waves dictates that their astronomical end on August 11 is premature. It seems that late-August weather just doesn't want to heel at the urgings of subtle astronomical cues.

The Role of Water in Weather

4

LEARNING OBJECTIVES

After reading this chapter, students will:

- Gain an overall appreciation for the role that the hydrologic cycle plays in weather and climate
- Be able to argue how evaporation alters temperature and dew point
- Understand the subtle but important distinction between condensation and net condensation (similarly, between evaporation and net evaporation)
- Be able to distill the definition of relative humidity in terms of vapor pressure and equilibrium vapor pressure into an explanation based on temperature and the amount of water vapor in the air

- Gain an appreciation for the diurnal variation of relative humidity that's dictated solely by the diurnal variation in temperature
- Understand that haze is an ostensible sign of poor air quality and, as such, signals a risk to public health
- Be able to argue against the popular and scientifically erroneous notion that "warm air holds more water vapor" (alternatively, "cold air holds less water vapor")
- Be able to explain the way that clouds form as a result of forcing air to rise
- Be able to describe how ground fog and mixing clouds form
- Gain an appreciation for how forecasters use dew point to help predict the weather

You may have heard Earth referred to as the "Water Planet" because oceans cover so much of its surface. To get a sense for just how much water there is on Earth, consider Figure 4.1, where the large blue sphere represents all the water stored in the oceans, ice caps, lakes, rivers, groundwater, and atmosphere. The medium-sized sphere depicts only Earth's freshwater, while the very tiny sphere (shown in this image approximately in the state of

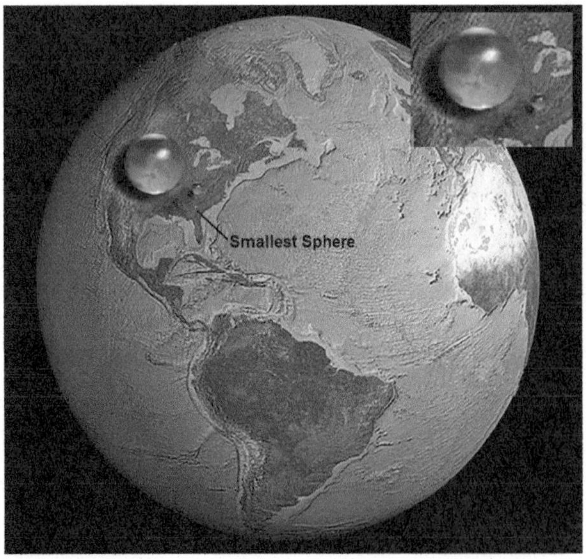

FIGURE 4.1   The largest blue sphere, which is 860 mi (1385 km) in diameter, represents all of Earth's water, including that in the oceans, liquid freshwater, and water in lakes and rivers. The medium-sized sphere, which is 170 mi (274 km) in diameter, represents only freshwater, while the smallest sphere, just 35 mi (56 km) in diameter, represents freshwater in lakes and rivers (credit: Howard Perlman, USGS; Jack Cook, Woods Hole Oceanographic Institution; Adam Nieman).

Georgia) represents only the freshwater stored in rivers and lakes.

Without reservation, the oceans account for the lion's share of the largest sphere's volume, containing well over a billion cubic kilometers—that's about 97% of all the planet's water. Given the small volume of this sphere in relation to the Earth, you might have an underwhelming impression about exactly how much water that is. So here's another way to visualize the amount of water stored in our oceans. Think of a perfectly smooth, spherical planet equal in size to the Earth. If all of the water in Earth's oceans were spread uniformly over this hypothetically smooth planet, the water would be about 2.5 km (1.7 mi) deep! Yes, that's a lot of water!

Now let's focus on water in the atmosphere. Take a moment to look back at Table 1.1, which lists the most common permanent and variable gases in the atmosphere and their concentrations. The high variability of water vapor compared to the other variable gases sticks out like a sore thumb. Even over warm tropical oceans, concentrations of water vapor seldom exceed four percent of the total atmospheric composition. In fact, the amount of water in our atmosphere accounts for only 1/3000 of Earth's total water—in Figure 4.1, this would be one-seventh the volume of the tiniest sphere. Yet, this seemingly paltry amount is enough to fuel ferocious hurricanes such as Sandy in late October 2012 (Figure 4.2) or the supercell thunderstorm that produced the tornado that devastated Moore, Oklahoma on May 20, 2013.

As it turns out, water is rather quirky. That's because all three phases of water—liquid, solid

FIGURE 4.2 Satellite image of Hurricane Sandy on October 28, 2012 (courtesy of NOAA).

(ice), and gas (water vapor)—can exist in the same environment at the same time. Later, in Chapter 8, we will explore how this coexistence leads to the production of precipitation. Of course, you usually can't get precipitation without clouds, so we'll start there.

Formally, a **cloud** is a vast collection of tiny water drops and/or ice crystals (if the cloud is sufficiently cold). Clouds form when water vapor (which is invisible) condenses onto tiny airborne particles called **condensation nuclei**, or, in the case of very cold clouds, when water vapor "deposits" onto **ice nuclei**,

which are small airborne particles that encourage the production of ice crystals. The simultaneous existence of the three phases of water in the atmosphere sets the stage for the Bergeron-Findeisen process, an in-cloud, give-and-take between water, ice, and water vapor that creates much of the precious precipitation that falls to earth (more on this process in Chapter 8).

In Figure 4.2, much of the water vapor that condensed to make cloud droplets (or deposited to make ice crystals) probably evaporated from the oceans. The tracing of the journeys of water, ice, and water vapor as they change phase and shuttle back and forth between the earth and the atmosphere is called the **hydrologic cycle**—a maze of possible paths that water molecules can take on their many varied voyages (see Figure 4.3). Let's go for a loop.

HYDROLOGIC CYCLE: THE TORTOISE AND THE HARE

We've already mentioned that nearly all the water on or close to the Earth's surface resides in the oceans. Of what's left, only about 1/3000 resides in the atmosphere, and most of that is in the form of water vapor. And what little water vapor the atmosphere has at any given moment, it doesn't have for long.

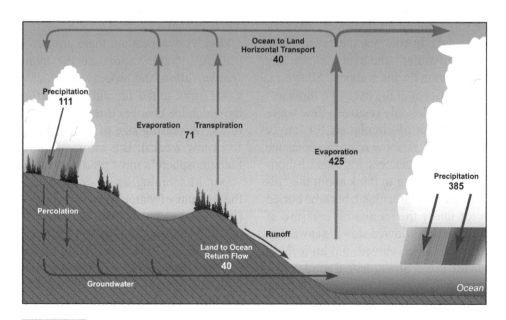

FIGURE 4.3 The hydrologic cycle. Numbers are expressed in units of thousands of cubic kilometers of water transferred per year. Note that, in terms of the volume of water transferred, evaporation and precipitation are the dominant processes.

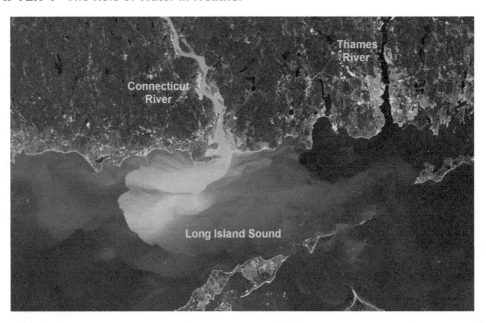

FIGURE 4.4   This view from space on September 2, 2011, shows muddy runoff in the Connecticut River flowing into Long Island Sound, the result of 6 to 10 inches of rain from Hurricane Irene about a week earlier. Farther east, the Thames River water is relatively sediment-free because its drainage basin received much less rain.

For water to evaporate into the air, move with the wind for a distance, ascend in a rising current of air, condense to help form a cloud droplet, and precipitate back to earth—only eleven days elapse, on average. Indeed, once in the atmosphere as vapor, water speeds through the hydrologic cycle like a hare. Returning to the surface, water may quickly recycle into the atmosphere or it may have a short layover as it makes its way through groundwater, lakes, or rivers (see Figure 4.4). In contrast, the oceans have quite a hold on their water—the average time any water molecule resides in the sea is about 2,800 years, a lengthy stay reminiscent of the pace of a tortoise.

In the hydrologic cycle, only relatively few water molecules get taken out of circulation for longer periods—those that fall as snow over the polar ice sheets can get buried in glaciers for tens of thousands of years. It is sobering to think about the "ill-fated" snow that fell over Greenland, became buried and turned to ice under the increasing pressure of more accumulating snow, flowed slowly seaward in a tortoise-like glacial river, calved into an iceberg and then toured the Davis Strait, drifted south in the Labrador Current, and then sank the mighty Titanic in 1912, taking perhaps hundreds of thousands of years to complete its sinister mission.

The hydrologic cycle is called a "cycle" for another good reason. Consider that mean annual precipitation, averaged over the globe, is about one meter (39 in). Because the amount of water that resides on or near the earth's surface is essentially constant over relatively short time scales, average global precipitation must be, for all practical purposes, balanced by average global evaporation. In other words, over the course of a year, water vapor that gets into the air must be nearly balanced by water that falls out of the air. Now that's a cycle!

Besides evaporation, there are two other ways to introduce water vapor into the hydrologic cycle: **transpiration**, the process by which plants release water vapor to the air, and **sublimation**, the process by which ice changes directly to water vapor (think of the decreasing size of ice cubes left in a freezer for a few weeks). But evaporation remains by far the atmosphere's most prodigious supplier of water vapor. Meteorologists generally define **humidity** as the amount of water vapor in the air at any instant (later in this chapter, you will learn that we differentiate between humidity and relative humidity). As you probably know through experience, humidity can vary greatly in time and space, from oppressively high humidity along the Gulf Coast in summer to low, lip-chapping humidity in northern Minnesota during winter.

As we already mentioned, most of earth's water is held in detention in the ocean. Relatively speaking, only a few select water molecules get paroled to take a whirlwind leave into the free atmosphere at any given time. Over the next few sections, we will follow water on its temporary leave from evaporation to precipitation.

EVAPORATION: WATER MOLECULES CAN CHECK OUT, BUT THEY CAN NEVER LEAVE . . . WELL ALMOST NEVER

We've already used the term *evaporation,* appealing to the experience you already have with this process. For the record, **evaporation** is the process by which water molecules break their bonds with their neighbors and escape to the air as a gas (water vapor).

At very high altitudes, intense solar radiation can **dissociate** water vapor. By "dissociate," we mean that the Sun's energy breaks water molecules apart into constituent hydrogen and oxygen. Once freed from oxygen, hydrogen is sufficiently light to escape the earth's gravitational pull and move out into space. Losing hydrogen to space is tantamount to losing water because hydrogen is a building block for water molecules.

Scientists estimate that earth loses the equivalent of a small lake to space each year. Not to worry. The total loss of water over geologic time probably amounts to less than 0.2% of the water contained in all of the oceans, an infinitesimal loss at the time scales at which earth's present-day hydrologic cycle operates. Temperature, however, changes on much shorter time scales, and the hydrologic cycle responds accordingly.

Temperature: The Warden of Evaporation Rates

Water molecules are like captives in minimum security detention—the bonds that bind them to the liquid phase are relatively lax. Thus, water molecules can get time off for kinetic behavior whenever their molecular motions break these bonds, allowing them to take a furlough in the free atmosphere in the form of water vapor. Even less energetic molecules imprisoned in the rigid lattice of ice—they're doing "hard time"—also can be paroled to the vapor state.

How are these liquid bonds broken? Every now and then, a water molecule gets enough of a boost in energy from its neighbors that it can overcome their attraction and escape into the air above the water. Think of water molecules in a puddle of water as a crowded dance floor with couples rockin' to a fast song. Given the fast tempo, it's inevitable that dancers bump into each other, and, occasionally, a dancer gets knocked off the dance floor. And so it is with water molecules that evaporate.

Like air molecules, water molecules oscillate faster as the water temperature increases. In turn, this boost in kinetic energy allows more water molecules to break free from their liquid bonds. In terms of our dance metaphor, the beat quickens and couples bump into each other more frequently (and more "energetically"). Thus, the rate of evaporation quickens as water temperatures increase.

The bottom line here is that water temperature is a primary controller of the evaporation rate. Qualitatively, a puddle of water evaporating from a street once the sun returns after a brief summer shower is an example of a relatively high evaporation rate. In contrast, water sitting in a vase of flowers on the kitchen table evaporates at a slower rate.

Whatever its rate, evaporation is a cooling process. That's because water molecules with the greatest kinetic energy are more likely to wriggle free and break their molecular bonds. The loss of these highly energetic molecules lowers the average kinetic energy of the water, which, in turn, translates to a lower water temperature. As an apprentice weather forecaster, you will need to be concerned about the effects of evaporation on air temperatures. In fact, some of the most difficult temperature forecasts arise when precipitation is slated to occur.

You may have noticed that surface air temperatures often decrease once it starts to rain. That's because some of the smallest raindrops evaporate on descent toward the ground. The heat energy required for this evaporation comes from the air surrounding the drops. As a result, the air usually cools when it rains. In fact, whenever water changes its phase (for example, by evaporating or freezing), energy is either required or released. These energy exchanges are summarized in a diagram that we like to call the "energy staircase," shown in Figure 4.5. Here we document all the phase changes involving water and the energy transactions that go with those phase changes.

By its position on this "staircase," ice is the lowest energy phase of water. It takes energy to melt 1 g of ice, 80 calories to be precise (a calorie is the amount of energy required to increase the temperature of 1 g of water by 1°C). Continuing up the energy staircase, we observe that it takes 600 calories to evaporate 1 g of water (at 0°C and typical air pressures). Considering the definition of one calorie, that's a lot of

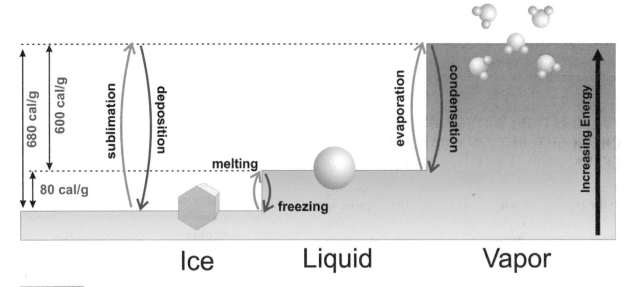

FIGURE 4.5 The "energy staircase" for water. Ice is the lowest energy phase of water. It takes 80 calories to melt one gram of ice (a calorie is the amount of energy required to increase the temperature of one gram of water by 1°C). On the flip side, when one gram of water freezes, 80 calories are released as latent heat of fusion. It takes 600 calories to evaporate one gram of water. Clearly, water vapor is the highest energy phase of water. When one gram of water vapor condenses, 600 calories are released as latent heat of condensation.

energy! So, without reservation, water vapor is the highest energy phase of water. None of this energy is lost, of course. For example, when water vapor condenses back to water, energy is released to the tune of 600 calories per gram. That **latent heat of condensation** has implications to weather forecasting, and we'll provide an example later in the chapter.

For now, let's look at evaporational cooling in action. Consider Figure 4.6, a "before-rain" and "during rain" example of the temperature and dew-point profile of the atmosphere between the surface and the base of a precipitating cloud. In Figure 4.6a (before rain begins), the temperature and dew point are far apart below the cloud base. Figure 4.6b shows the situation once rain reaches the ground. The temperature between the cloud base and the ground decreases in response to evaporational cooling, while the dew point increases as some raindrops evaporate, increasing the water vapor content of the air.

All weather forecasters worth their salt weigh the effects of evaporational cooling on predicted air temperatures when they expect precipitation to fall. If forecasters anticipate high evaporation rates, they know that the effects on temperature can be dramatic. When precipitation is underway during winter, for example, relatively high rates of evaporation at altitudes near 1500 m (about 5000 ft) can sometimes cool the air enough to tip the scales toward frozen precipitation instead of rain. We'll discuss how weather forecasters make such calls in Chapter 16, which deals with winter weather.

During summer in particular, evaporational cooling is important to human comfort. That's because evaporation, no matter where it occurs, requires energy. When you sweat, some perspiration evaporates from your skin, and much of the energy required for that evaporation comes from your body. This cooling process helps keep you from overheating. On hot, humid days, however, this evaporational cooling is typically not as efficient. Conclusion: There are other controllers of evaporation rates besides temperature. Let's investigate.

Net Evaporation: Evaporation Edges Condensation

Not all hot days are created equal. For example, a 95°F-day in June in New Orleans, LA, is typically accompanied by high humidity. It can be very difficult to get comfortable on such a day in the "Big Easy." On the other hand, a 95°F-day in June in Phoenix, though still uncomfortably warm by some people's standards, will not feel nearly as oppressive. The difference in the "feel" of the air in the two cities is the result of differences in the concentration of atmospheric water vapor—the overall evaporation rate (of perspiration) and, thus, the associated evaporational cooling in New Orleans is simply less

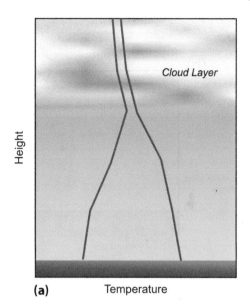 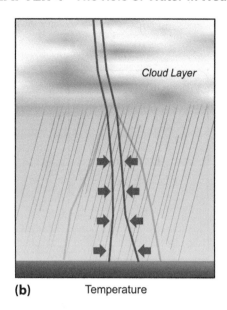

(a) Temperature   **(b)** Temperature

FIGURE 4.6 Sample temperature (red) and dew point (green) profiles of the atmosphere between the ground and the base of a cloud (a) before rain begins; and (b) after some rain has reached the ground. Note that the dew point increases while the temperature decreases (the two profiles get closer together).

than in Phoenix. That begs the question: Why would greater concentrations of atmospheric water vapor mean lower evaporation rates?

The key lies in recognizing that the reverse of evaporation, condensation, is occurring simultaneously. Formally, **condensation** is the process by which water vapor returns to the liquid state (you could say it's the "reverse" of evaporation). Even so, it turns out that evaporation and condensation usually take place at the same time. Figure 4.7 is a photograph taken by

FIGURE 4.7 A photograph inside the eye of Hurricane Katrina during aircraft reconnaissance in late August 2005 (courtesy of the Hurricane Research Division, NOAA).

aircraft reconnaissance from inside the eye of Hurricane Katrina in August 2005. Note the relatively low overcast in the eye of Hurricane Katrina (the shadowed swirl of clouds; yes, the eye of a hurricane can be rather cloudy at times). Here, condensation rates exceeded evaporation rates, so there was **net condensation**. In other words, clouds formed. Above this layer of clouds inside the eye, the sky was clear. Indeed, in this clear part of the eye, net evaporation ruled because evaporation rates exceeded condensation rates.

It's often said that a picture is worth a thousand words. But, in the case of Hurricane Katrina's eye, it's not obvious that both evaporation and condensation were in progress at the same time. Let's conduct a simple experiment to illustrate why this was the case.

Consider a closed container like the one shown in Figure 4.8a. Although there are no such containers suspended in the real atmosphere, we'll use this idealized example to make our point. Assume that, initially, the closed system contains only dry air (no water vapor). But because we plan to add some water to the container, we'll need to make provisions for keeping track of it.

For sake of argument, let's suppose that we are able to count water molecules as they move back and forth between the liquid and vapor states. Because vapor molecules will bounce off the walls of the container, each will exert a pressure; collectively, the

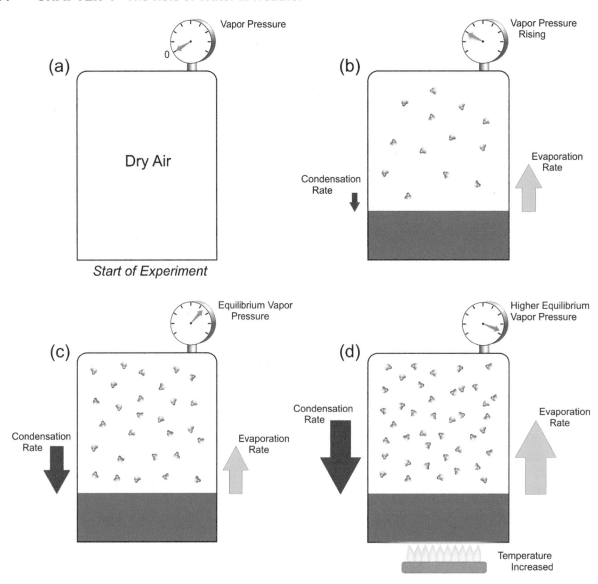

FIGURE 4.8 At the start of the experiment, with just dry air inside the container, the "vapor pressure meter" reads zero; (b) When water is added to the container, evaporation dominates condensation at first, and the vapor pressure of the air increases;(c) Eventually, an equilibrium is reached when condensation and evaporation are equal; at this equilibrium, we say the air is saturated; (d) If the temperature of the water is increased, a new equilibrium state is achieved with greater rates of evaporation and condensation, and thus a greater equilibrium vapor pressure.

pressure they exert is called, appropriately enough, **vapor pressure**. To understand this concept, we first need to briefly introduce you to the concept of air pressure (a more thorough treatment awaits in Chapter 6). We've already established that all air molecules (be they oxygen, nitrogen, argon, water vapor, and so on) have kinetic energy. Because air molecules are in perpetual motion, they collectively exert a force when they strike objects. Air pressure is simply the force exerted per unit area. And vapor pressure is that part of the total air pressure exerted by water vapor.

Now let's add water at a temperature of, say 60°F (16°C), to the container (see Figure 4.8b). Within a short time, some water molecules break their bonds with their neighbors and evaporate. As they do, these molecules pass through the "water-air interface," a very thin partition between the two mediums that theoretically has the properties of both water and air. In this interface, we hypothetically monitor evaporation rates by counting molecules on their way from the liquid to gaseous state. We also monitor condensation rates, as a few vapor molecules inevitably also reenter the interface on their way back to the liquid phase. Initially, the evaporation rate exceeds the condensation rate (there is net evaporation), and the vapor pressure increases.

But as the number of water vapor molecules in the air increases, the chances that vapor molecules will reenter the water-air interface and condense increases. Eventually, an equilibrium between rates of evaporation and condensation will be established (see Figure 4.8c). We define the **equilibrium vapor pressure** to be the vapor pressure associated with this balanced state. Moreover, we refer to this balanced state, when the evaporation rate equals the rate of condensation, as **saturation** (the equilibrium vapor pressure is sometimes referred to as the saturation vapor pressure).

Now let's increase the water temperature to, say, 80°F (27°C). At this higher temperature, water molecules have greater kinetic energy and thus are more likely to break their bonds with their neighbors and evaporate. So, in quick fashion, the evaporation rate increases (and, in turn, so does the vapor pressure). In response to increasing water vapor, the comeback-kid condensation rate rises to the task, and the frequency of water vapor molecules entering the water-air interface increases. Eventually, there's a new balance between evaporation and condensation rates, only this time at a higher equilibrium vapor pressure (Figure 4.8d).

Carrying out the experiment at a myriad of different temperatures yields the curve in Figure 4.9, which shows the equilibrium vapor pressure as a function of temperature. For any given temperature, points that lie below the curve represent vapor pressures that fall short of equilibrium. In other words, below the curve, the rate of condensation can't keep up with the rate of evaporation—there is **net evaporation**. In contrast, points that lie above the curve represent vapor pressures that exceed equilibrium—there is net condensation and, ladies and gentlemen, a cloud is born!

Clearly, water temperature is a primary player in determining evaporation rates. Thus, the water vapor content of the air over and near warm bodies of water such as the Gulf of Mexico is, on average, relatively high. On the other hand, vapor pressures over and near relatively cool bodies of water, such as the offshore waters of the Pacific Northwest, are low, on average.

But as our experiment shows, the amount of water vapor already in the air also matters because some water vapor condenses into water and essentially reduces net evaporation (that is, saturation is reached at a lower evaporation rate because there are already water vapor molecules in the air). Now the question becomes: How do meteorologists gauge the degree to which condensation will affect the evaporation rate?

Vapor Pressure Gradient: A Proxy for the Net Evaporation Rate

Right after we first introduced water into the experiment with the closed container, evaporation had the upper hand (big time) because there was no water vapor in the air that could condense into water and offset evaporation. Had there initially been water vapor in the air, equilibrium would have been reached sooner and, consequently, the amount of water lost from its initial supply would have been less. Thus, when humidity is relatively high, net evaporation is reduced. Your experience supports this observation: On a humid summer day, evaporation of sweat from your skin (a process that helps to cool your body) is limited. As a result, most people feel uncomfortably warm.

Another way to look at this discussion is through the lens of the **vapor pressure gradient**, which is the difference in vapor pressure between the air and the water-air interface. When there's little water vapor in the air, the vapor pressure gradient is large because the vapor pressure of the air is low and the vapor pressure of the moist water-air interface

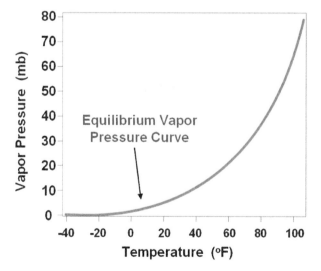

FIGURE 4.9 Recall from Figures 4.8c and 4.8d that the equilibrium vapor pressure (the vapor pressure when the rate of condensation equals the rate of evaporation) increases as water temperature increases. This curve, called the equilibrium vapor pressure curve, formally quantifies the relationship between temperature and equilibrium vapor pressure.

is high. So net evaporation is large. In contrast, when humidity is high, the vapor pressure gradient is small because the vapor pressure of the air is high and the vapor pressure of the water-air interface is also high. So net evaporation is dramatically reduced.

Wind also affects the vapor pressure gradient. If wind blows over the water (imagine screen walls in the closed-container experiment), it would whisk water vapor away, keeping the vapor pressure of the air relatively low. Thus, the vapor pressure gradient would remain relatively large, setting the stage for large net evaporation. This observation is also consistent with your experience: Wet swimsuits hung outside typically dry more quickly when the wind blows, and the faster the wind, the faster the clothes dry.

Now that you have a sense of what makes evaporation and condensation tick, let's look at the next intermediate step in the hydrologic cycle—clouds.

NET CONDENSATION: A CLOUD IS BORN

At any given moment, nearly half of the earth is covered by clouds. Some are short-lived, evaporating in seconds, like the fleeting cloud from your breath on a cold winter's day. Others are long-lived—the low, precipitation-barren clouds left in the wake of deep low-pressure systems of late autumn and winter can last for a week or more. In terms of the recycling mission of the hydrologic cycle, both of these cloud types fail, dissipating without ever having produced a drop of precipitation. Other clouds, such as tall,

regal cumulonimbus (see Figure 4.10), can produce flooding summer rains in a matter of tens of minutes. Such prodigious rainmakers are important cogs in the hydrologic cycle.

Clouds and Relative Humidity: There's Nothing Magical about 100%

One winter day a few years ago, a hot-water pipe burst above the second floor of the meteorology building at Penn State University. After flooding several rooms, a fog formed above the standing warm water. One of the authors quickly snapped a picture (see Figure 4.11) that he knew would illustrate the foggy consequences of high evaporation rates (let's face it, folks, some meteorologists are simply "weather weenies"—they eat, sleep, and drink the science).

This indoor flood of warm water affords us an opportunity to introduce the concept of relative humidity (you've probably already heard this term many times). To define it, let's return to the experiment we conducted in the closed container. **Relative humidity** is the ratio of the actual vapor pressure and the equilibrium vapor pressure (at the observed temperature), converted to a percent. In mathematical form,

$$\frac{\text{Relative}}{\text{Humidity}} = \frac{\text{Vapor Pressure}}{\text{Equilibrium Vapor Pressure}} \times 100\%$$

If we apply this definition to the state of saturation in the closed container, the relative humidity is

FIGURE 4.10   A towering cumulonimbus cloud (background) drifts over Swifts Creek in Victoria, Australia. Note the rising, turbulent turrets associated with developing cumulus clouds in the foreground (courtesy of Henry Firus, Flagstaffotos, http://www.flagstaffotos.com.au/gallery2/main.php).

FIGURE 4.11 Fog forms in a classroom at Penn State University after a hot-water pipe burst during the winter.

100%. For the record, the relative humidity in the foggy, flooded classroom was 100% as well.

There's one important stipulation underlying this definition of relative humidity that sometimes gets overlooked. This definition applies to the air space above a *flat* surface of *pure* water. Why do we make these distinctions? If the truth be told, cloud droplets are neither flat nor pure. As a result, it is actually possible to have relative humidities in the atmosphere that exceed 100%. In fact, the relative humidity inside a cloud is, on average, a few tenths of a percent above 100% (that is, net condensation rules)! So the common reference that clouds form when the relative humidity reaches 100% (in other words, the air reaches saturation) is generally not true in the strictest sense. What's up with all of this?

Weather Folklore and Commentary

Human Hair and Relative Humidity

In the popular vernacular of the 1990s, a humid, damp day is a "bad-hair day." Indeed, human hair is very sensitive to changes in atmospheric moisture. When humidity rises, for example, human hair (as well as many other organic substances such as wood) absorbs a proportional amount of moisture and its length increases. For people with curly hair, the extra length means that there will be extra curl. In the 1970s, such hair was referred to as the "frizzies." Added length from increasing moisture can also cause hair to "lose its body" and look limp.

There is an instrument that measures relative humidity by taking advantage of hair's response to changes in atmospheric moisture. It is the hair hygrometer and its design is relatively simple (**hygrometer** is the general term for an instrument that measures the amount of moisture in the air). A bundle of hair is attached mechanically to a pointer. As relative humidity ebbs and flows, hair length decreases and increases, causing the pointer to pivot over an indicator calibrated between 0% and 100% (see Figure 4.12). Typically, hair length varies 2 to 3% as relative humidity varies from 0% to 100%.

Blond is the color of choice because it responds more quickly to changes in relative humidity. Even so, one of the major drawbacks of hair hygrometers is that they do not respond quickly enough, particularly

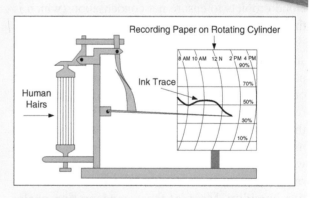

FIGURE 4.12 A hair hygrometer. As relative humidity changes, the strands of hair contract and expand in length. Using a series of levers, the changes in hair length are amplified so that changes in relative humidity move the pen and trace a time series.

at low temperatures. Hair hygrometers have other shortcomings, too. They are not as accurate as other types of hygrometers such as hygristors (used in radiosondes) which depend on changes in electrical resistance as humidity varies. Plus, hair hygrometers are in constant need of calibration.

So if you desire carefree measurements of relative humidity and really don't want to spend money on a more sophisticated instrument, relax. The limpness of your own hair is a reliable indicator of high relative humidity and the possible arrival of rain.

To answer this question, you'll need a little background regarding the size of atmospheric particles. Cloud droplets are spherical with typical diameters on the order of 10 μm (recall that 1 μm = 0.000001 m). Figure 4.13 helps to put this size in perspective. The average diameter of the finest drizzle or the tiny drops that make up fog is about 100 μm. Small raindrops are about 1000 μm in diameter, while large raindrops may reach 5000 μm.

Now let's consider the fate of the tiniest, embryonic cloud droplets, much smaller than even those shown in Figure 4.13. If no other processes were at work except for evaporation and condensation, the relative humidity of their environment would have to be very high (actually about 300%) for them to grow into mature cloud droplets and not rapidly evaporate. Why would the relative humidity have to be so high? It turns out that the rate of evaporation from a spherical drop is much higher than the evaporation rate from a flat surface. Moreover, the evaporation rate increases as the drops get smaller. So there would have to be a lot of water vapor (relatively speaking) around embryonic cloud droplets to ensure net condensation (which is the situation necessary for cloud droplets to grow).

However, relative humidities on the order of 300% simply do not exist in the real atmosphere. Fortunately for life on earth (which ultimately depends on the precipitation produced by clouds), such high relative humidities aren't necessary. That's because the natural presence of condensation nuclei, those microscopic particles of dust and dirt onto which water vapor readily condenses, allows net condensation to occur at much less inflated values of relative humidity. Many of these condensation nuclei

are **hygroscopic**, which means that they tend to attract water vapor (salt is hygroscopic, for example). As a result of this propensity for water vapor, hygroscopic condensation nuclei generally reduce the relative humidity required for clouds to form to just a few tenths above 100%.

That's not to say that net condensation doesn't occur at lower relative humidities. For example, when relative humidity in the lower troposphere exceeds 70% or so, fine particles of soil, smoke, sea salt, and pollution begin to swell by net condensation, forming **wet haze** (see Figure 4.14). This hallmark of summertime in the eastern United States reduces visibility and gives the sky a milky-white appearance. Although some people associate hazy skies with the good times of summer, the appearance of wet haze often translates to poor air quality, which poses a risk to public health. The connection between

(a)

(b)

FIGURE 4.14   (a) The view of the mountains surrounding State College, PA, is sullied as humid southwesterly winds imported pollution from the Ohio Valley, setting the stage for wet haze to develop; (b) The same view on a day with a fresh air mass from Canada (those are clouds over the mountains, not haze).

fine drizzle
fog droplets
(100 μm)

small raindrop
(1000 μm)

cloud droplets
(10 μm)

human hair
(50-100 μm)

FIGURE 4.13   The relative sizes of cloud droplets, drizzle and fog droplets, and small raindrops, compared to the diameter of a typical human hair.

pollution and hazy skies is most evident near and downwind of industrial areas in the Ohio Valley and Mid-Atlantic, for example.

The notion of relative humidity exceeding 100% inside a cloud may not be very intuitive for you. So here's a more straightforward approach: For a cloud to form, the rate of condensation must simply exceed the rate of evaporation (or, more succinctly, there must be net condensation). Recall that the relative humidity equals 100% at saturation, when the rate of evaporation equals the rate of condensation. Thus, when net condensation occurs, the relative humidity must be slightly greater than 100%.

It's now time to apply what you've learned to the real atmosphere.

A Tried-and-True Recipe for Clouds: A Pinch (of Nuclei), a Dash (of Vapor), Then Chill

So, how is net condensation achieved in the atmosphere? The most common way is simply by cooling the air.

As air temperature decreases, water vapor molecules slow down. As they become increasingly sluggish, more and more of them huddle closer to condensation nuclei (of course, the same is true of other air molecules, such as oxygen and nitrogen, but when it comes to cloud formation, we don't care about them). Once there's a "quorum" of water vapor molecules, the stage is set for net condensation. It's actually pretty straightforward.

Along these lines, we often hear the explanation that the reason clouds form as air cools is that there's "not enough room for as much water vapor in cold air" or that "cold air can't hold as much water vapor as warm air." The implications are that the air (which is predominantly composed of nitrogen and oxygen) takes some sort of active role in determining how much water vapor can occupy any given space. That's simply not true. Cold air is not like a hotel that hangs out a "No Vacancy" sign once all the rooms for water vapor are filled. To the contrary, net condensation occurs when cooling causes a sufficient number of sluggish water vapor molecules to huddle close around condensation nuclei. This may sound like an argument over semantics, but the idea that there is "limited space" in the air is simply unscientific.

When concentrations of water vapor are relatively small, cooling must be substantial to reach net condensation (especially when the air is warm and water vapor molecules are very energetic). On the other hand, when concentrations of water vapor are high, less cooling is needed to achieve net condensation.

Now the question becomes: What are the mechanisms for cooling the air?

Lifting the Air: Caution! Air Molecules at Work

When parcels of air rise, they expand and cool (see Figure 4.15). To understand this basic tenet of meteorology, we first observe that air pressure decreases with increasing altitude. Let's explore this idea.

Figure 4.16 shows climber Tom Rippel wearing an oxygen mask atop the summit of Mount Everest on May 21, 2008. At an elevation of 29,035 ft (8849 m), the amount of available oxygen can barely support life. Not surprisingly, the density of air (the number of air molecules per unit volume) at this altitude is low, providing evidence that air density decreases with increasing height. Air pressure follows suit, mirroring the vertical profile of air density; in a nutshell, fewer air molecules in a given volume translates to lower air pressure, so air pressure decreases with height.

Let's look more carefully at an air parcel as it rises. Initially at rest near the ground, the parcel is in balance with its environment—the air pressure inside the parcel equals the air pressure outside the parcel. As the parcel rises and moves through air of lower pressure, the balance in air pressure is lost because

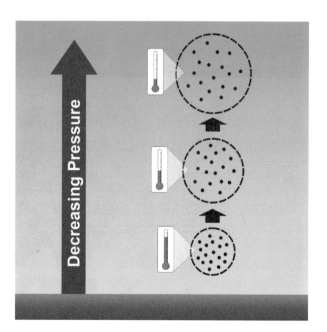

FIGURE 4.15 When a parcel of air rises, it expands as it moves through air at lower pressure. As the parcel expands, it cools.

FIGURE 4.16 Tom Rippel at the summit of Mount Everest on May 21, 2008 (courtesy of Peak Freaks Expeditions, Inc.).

the pressure inside our idealized parcel is greater than the pressure outside. To try to achieve a new balance, air molecules inside the parcel push out the sides—in other words, the parcel expands. In pushing out the sides of the parcel, air molecules lose some of their kinetic energy (the work required to push out the sides of the parcel comes at an expense—there are no free passes in the atmosphere). As a result, the temperature decreases inside the parcel. This cooling proceeds as long as the air parcel continues to rise and expand. If the parcel contains water vapor and it is lifted high enough to produce sufficient cooling, net condensation occurs and a cloud is born.

Of course, we can't ride along on a rising parcel of air to witness net condensation, but we can do the next best thing—simulate what happens. So it's time for an experiment, which we'll document in Figure 4.17. We'll use a glass bottle with about an

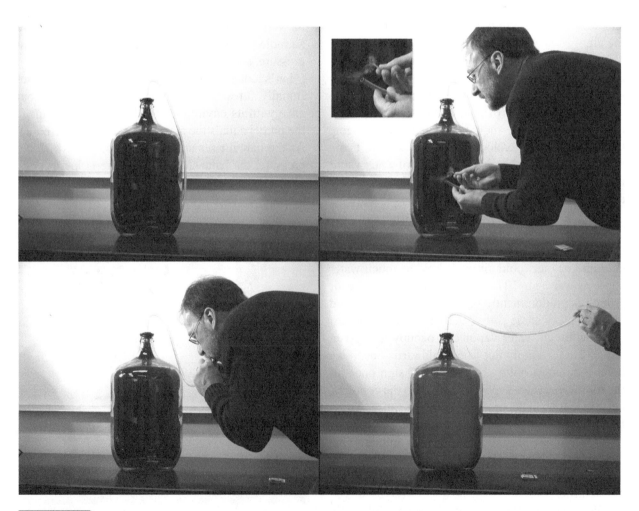

FIGURE 4.17 Like a genie, one of the authors makes a cloud in a bottle. The bottle has a layer of water in the bottom, and condensation nuclei are added in the upper right photograph. The air is cooled (bottom left), and a cloud forms (bottom right).

inch of water on the bottom (upper-left photograph). A rubber stopper seals the top of the bottle, although plastic tubing inserted through the center of the stopper allows us to blow air into or suck air out.

Like any good chef who follows a recipe, we first added some condensation nuclei to the air inside the bottle. We did this by sucking air out of the bottle (through the plastic tubing), pinching off the tube (to prevent air from rushing back in), holding the extinguished end of a match close to the open end of the tube, and then quickly releasing the pinch (upper-right photograph). Because we had first sucked air out, the air pressure inside the bottle was less than the air pressure in the room, so air rushed into the container to regain balance. In the process, some smoke particles were drawn into the bottle as well.

However, simply adding condensation nuclei to the air in the bottle was not enough to make a cloud because the rate of condensation on these newly introduced particles did not exceed the rate of evaporation. Somehow, in order to make the cloud, we had to tip the scales toward net condensation.

Next, we blew hard into the tube (lower-left photograph), thereby increasing the air pressure (by adding more air molecules). Then we pinched off the plastic tubing. Once the pinch was released, air rushed out of the container like air escaping from a bicycle tire that's been punctured. In the process, the escaping air expanded. As you just learned, air cools when it expands. As a result, the rate of evaporation, which strongly depends on temperature, decreased dramatically.

How about the rate of condensation? Though some water vapor was lost as air rushed out of the bottle, some water vapor was also added from the author's hot, moist breath. Nonetheless, the rate of condensation also likely decreased. But the real loser was the rate of evaporation, which is tied so strongly to temperature. The now-lowered temperature of the air inside the bottle (from expansion) meant that the rate of condensation was able to exceed the markedly reduced rate of evaporation. Translation: There was net condensation, and a cloud was born (lower-right photograph).

To drive home our point about the consequences of cooling the air by expansion, carefully observe a bottle of soda whenever you open it. The air in the neck of the bottle above the soda, which has a greater air pressure than the air outside the bottle, rapidly expands when you remove the cap (listen for the "whoosh"), cooling the air and paving the way for a fleeting cloud to form in the neck of the bottle.

We confess that the cloud that forms in the neck of a soda bottle is a bit different than the one we made in our experiment because there aren't any condensation nuclei in the air above the soda. But, after you open the soda bottle, the cooling of the expanding air is so dramatic that water vapor molecules become really sluggish. By chance, sluggish water vapor molecules congregate and initiate net condensation without the benefit of condensation nuclei. Such clouds do not form in the atmosphere because there are always foreign particles suspended in the air.

Hopefully we've convinced you that air cooling by expansion during ascent can lead to clouds. Now let's briefly discuss a few common lifting mechanisms. Low-pressure systems and their attendant fronts regularly lift the air to produce clouds and precipitation (there's more to come on lows, fronts, and lifting throughout the book). As you learned in Chapter 2, convection is a form of lift, with uneven heating of the ground caused by differences in surface covering (for example) leading to the positive buoyancy of some air parcels (see Figure 4.18). And mountains,

FIGURE 4.18 In this satellite image from July 2014, convective clouds formed over the land in the vicinity of Lake Michigan, while convection was suppressed over the cooler water and a narrow strip of land along the eastern shore, where cooler lake air had advanced inland (courtesy of NOAA).

which act as barriers to the wind, are "heavy" lifters, forcing air moving horizontally to abruptly ascend the sloping terrain.

Formally, the forced lifting of air by the terrain is called **orographic lifting**. In mountainous areas, this process is the dominant cause of precipitation. In fact, some of the snowiest and rainiest places in the world are located on the **windward** side of mountains—the side that faces into the prevailing winds. The opposite side of the mountain is known as the **leeward** (or lee) side. Originally, windward and leeward were nautical terms, as sailors referred (and still refer) to the side of the sail catching the wind as the windward side and the side sheltered from the wind as the lee side (see Figure 4.19).

Figure 4.20a shows the average annual liquid precipitation in northern California, while Figure 4.20b is a topographic map of the same area. For the record, the prevailing winds blow from the west over California. Note how, in general, precipitation amounts increase as the terrain slopes upward and decrease as the terrain slopes downward. In other words, relatively heavy precipitation generally falls at the highest elevations or where the terrain slopes upward. The Central Valley of California, where air tends to downslope off the coastal mountains

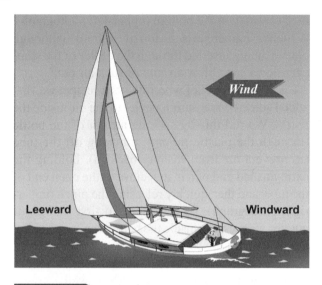

FIGURE 4.19 The windward side of a sail is the side that catches the wind. The lee (or leeward) side is sheltered from the wind.

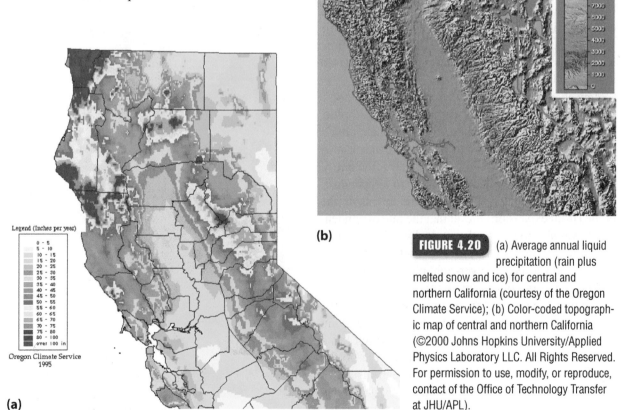

(b)

(a)

FIGURE 4.20 (a) Average annual liquid precipitation (rain plus melted snow and ice) for central and northern California (courtesy of the Oregon Climate Service); (b) Color-coded topographic map of central and northern California (©2000 Johns Hopkins University/Applied Physics Laboratory LLC. All Rights Reserved. For permission to use, modify, or reproduce, contact of the Office of Technology Transfer at JHU/APL).

to the west, is a relative minimum in precipitation. When a precipitation minimum occurs on the lee side of a mountain range, the zone is called a **rain shadow**. To the east of the Central Valley, precipitation increases again as the prevailing westerly wind flow encounters the Sierra Nevada Mountains. There's an indication that extreme eastern California is the gateway to the vast rain shadow that is the Great Basin, which lies to the east of the Sierras.

There are a couple other ways to produce clouds without forcing the air to rise. Let's investigate.

Clouds without Lifting: Special Recipes

In Chapter 2, you learned that on a clear night with light winds, the ground can impart a big chill to the overlying layer of air (primarily by conduction). When the nocturnal chill is spread throughout a sufficiently deep layer by light winds (only a few miles per hour), net condensation can occur and **ground fog** forms by dawn. A typical ground fog (sometimes called **valley fog**) is shown in Figure 4.21. We'll discuss the details of how fog evaporates in Chapter 8. However, to get you thinking about this process, consider that the fog in the room shown in Figure 4.11 dissipated after the window had been opened for several minutes.

Fog can also form over lakes, rivers, and streams when water is relatively warm and nights are chilly, conditions particularly common in early autumn. Such fog can happen at other times of the year as well—for an example, see Figure 4.22. Here, the water in the relatively shallow pond was warmer than the overlying air, so a completely different mechanism must have caused this **steam fog** to form. Indeed, steam fog is a form of **mixing cloud**. As its name suggests, a mixing cloud forms when warm, moist air mixes with cooler drier air. In the case of Figure 4.22, the mixing occurred between warm moist air in the water-air interface over the pond and cooler drier air above.

The most common example of a mixing cloud is the fleeting cloud that forms from the exhalation of your breath on a cold winter day. Conditions must be optimal because, on other days or indoors, you won't see "your breath." To gain insight, we must refer to the equilibrium vapor pressure curve in Figure 4.23, which is sort of an operator's manual for determining whether a cloud will form or not.

In Figure 4.23a, let point A represent the temperature and vapor pressure of your breath (approximately body temperature and very moist) while point B represents the temperature and vapor pressure of the air on a winter day (cold and, in this case, at or near equilibrium). The points on the line connecting A and B represent all possible states that might result from the mixing process. Note that all the points lie above the curve within the jurisdiction of net condensation. Voila! You see your breath as a cloud. As another example, consider Figure 4.23b, where point A still represents your breath but point C represents the conditions inside your house (taken to be 70°F (21°C) and not close to saturation). Now note that all points on the line connecting A and C lie below the curve, where net evaporation dominates. Sorry, you won't see your breath.

FIGURE 4.21 Valley fog lingers well into the morning in the Conyngham Valley of northeastern Pennsylvania, after forming on a clear night with light winds.

FIGURE 4.22 Steam fog over a lake in North Carolina during the early morning hours. The colorful leaves on the trees indicate that the season was autumn, when the early morning air can be chilly and the water still relatively warm, conditions that are optimal for steam fog to form (courtesy of Tim Martin).

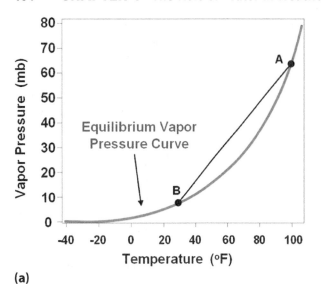

(a)

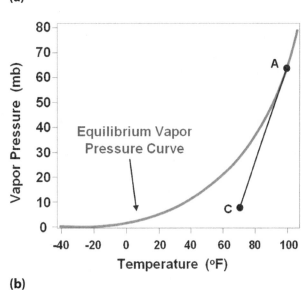

(b)

FIGURE 4.23 Whether a mixing cloud forms depends on the temperature and vapor pressure of the air involved in the mixing. (a) Mixing of warm, moist air from your breath (represented by point A) and the outside air on a cold winter day (represented by point B). All points on the line connecting A and B lie above the equilibrium curve, indicating net condensation. Thus, a cloud forms; (b) Mixing of warm, moist air from your breath (point A) and the warm dry air from inside your house (point C). The line connecting points A and C lies below the equilibrium curve, so no cloud forms.

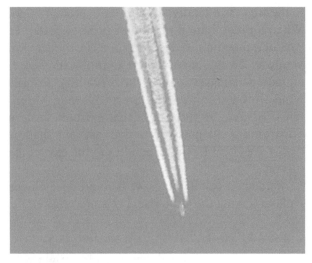

FIGURE 4.24 Contrails form as the hot, moist exhaust from a jet engine mixes with colder, drier air. Note the clear gaps between the jet engines and the contrails, indicating the zone where the mixing occurred.

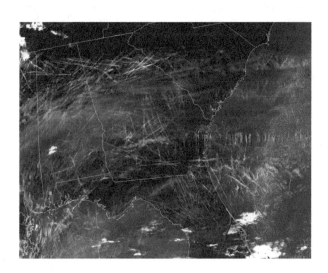

FIGURE 4.25 This view from a weather satellite shows persistent contrails over the southeastern United States (courtesy of NOAA).

Like your breath on a cold day, the hot, moist exhaust from jet aircraft mixes with cold, relatively dry air, forming mixing clouds called **contrails** (short for *con*densation *trails*). In Figure 4.24, note the clear gaps between the jet engines and the contrails, indicating the zone where the mixing occurred. Sometimes, contrails will rapidly evaporate when the air is very dry. At other times, they spread out horizontally into a long ribbon of high, wispy clouds. Ultimately, the life span of a contrail depends on the relative humidity of the air at cruising altitudes. If the air is close to saturation, for example, contrails may persist for hours (although winds may spread them out or break them up). Around busy airline hubs such as Atlanta, GA, long-lived contrails can dramatically increase high cloud cover (see Figure 4.25).

If you read the last paragraph closely, you probably noticed that we initially stated that contrails can form in air that's relatively dry. Then we said that contrails persist when the air is close to saturation. Are these two statements contradictory? Not at all, when you consider that measures of moisture can be relative (as in relative humidity) or absolute. Read on.

ASSESSING MOISTURE: APPLICATIONS TO WEATHER FORECASTING

Even when the water vapor content of the air is low, the relative humidity can be very high. Before you think that the authors have lost their moisture marbles, let us explain.

Relative Humidity versus Dew Point: To Each His Own

Recall the definition of relative humidity that we introduced earlier in this chapter: vapor pressure divided by equilibrium vapor pressure, multiplied by 100%. Also recall that the denominator in this quotient depends on temperature. What this means is that we have a (so-called) measure of atmospheric moisture that is not exclusively a function of the amount of moisture in the air. So relative humidity, despite its popularity, is not an *absolute* measure of moisture in the air—it is a *relative* measure because it also depends on temperature.

With this caveat in mind, let's return to the brain teaser we posed at the start of this section. On a chilly December morning in New England, under the control of a dry Canadian high-pressure system, for example, the relative humidity may be quite high (see Figure 4.26). Yet the air feels crisp and far from

humid. The apparent paradox results from the low morning temperature, which dictates that the equilibrium vapor pressure is also low. So the actual vapor pressure, which is low (remember, this is a dry Canadian air mass), divided by the low equilibrium vapor pressure, yields a relative humidity that's pretty high.

There's another rub to relative humidity. Consider a day on which the actual amount of water vapor in the air stays nearly constant. On such a day, the vapor pressure would remain nearly constant as well. Nonetheless, the relative humidity will change on this day as the temperature changes—as the air warms during the day, the equilibrium vapor pressure increases, so the denominator in the relative humidity formula increases. As a result, relative humidity decreases. At night, when temperatures typically decrease, the equilibrium vapor pressure decreases, so relative humidity increases. This rollercoaster behavior of relative humidity is depicted schematically in Figure 4.27.

Before we move from the relative to the absolute, let us say something positive about relative humidity. A relative humidity near 100% gives us a clue that the air is close to saturation and that additional cooling will produce net condensation. Thus, relative humidity is a tool that forecasters use to predict the formation of fog and clouds. For example, Figure 4.28 is a meteogram that includes the time trace of relative humidity (blue line, in percent) at Red Bluff, CA, beginning in the pre-dawn hours on a December day. Note that fog formed as the relative humidity reached 100%. Still, relative humidity alone cannot tell us how much water vapor is in the air, information that has profound implications for precipitation forecasting as well as human comfort.

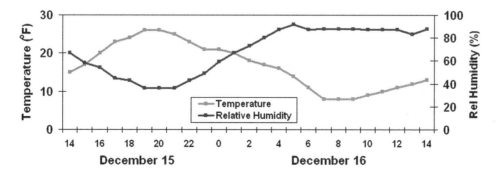

FIGURE 4.26 The time traces of temperature (red line, left axis in °F) and relative humidity (blue line, right axis in percent) at Orange, MA, from 14 UTC on a December day to 14 UTC the next day. Note that the relative humidity in the pre-dawn hours was close to 90%. Yet the temperature was near 10°F, and the frigid morning air was far from feeling humid.

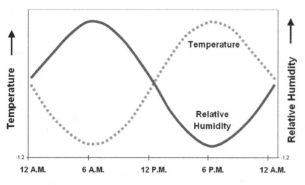

FIGURE 4.27 Suppose, for sake of argument, that the actual amount of water vapor in the air remained nearly constant on a particular day. Even so, the relative humidity would still change throughout a typical day, but only because temperature varied. Early in the morning on such a day, the relative humidity would reach a maximum because the temperature is lowest. On the other hand, relative humidity would be lowest when the temperature is highest—in the late afternoon.

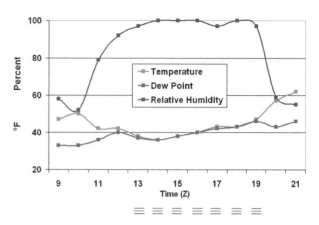

FIGURE 4.28 Meteogram showing temperature (red line, in °F), relative humidity (blue line, in percent), and dew point (green line, in °F) at Red Bluff, CA, from 09 UTC to 21 UTC on a December day. All three variables share the numerical axis labels on the left. Obstructions to visibility are shown along the bottom. Note that fog formed at 13 UTC as the relative humidity neared 100% and persisted as long as the relative humidity remained that high.

So what measure do forecasters routinely use to represent the absolute amount of water vapor in the air? Answer: the dew point, which is the temperature to which air must be cooled (at constant pressure) in order to reach saturation. To understand this concept, think of a clear, calm autumn morning with dew on the grass. How did the dew get there? The previous night, air in contact with the chilling ground cooled to the temperature at which saturation was reached, and then a tad lower so that net condensation began and dew formed. The temperature at which dew formed on the grass is the dew point. Now look back to Figure 4.28 and note that, as the temperature (red line) fell toward and eventually reached the dew point (green line), fog formed as net condensation occurred in a layer of air just above the ground.

When the dew point is low [say 40°F (4°C)] and the temperature is much higher [say 80°F (27°C)], a great deal of cooling is required for the air to reach saturation. The amount of water vapor in the air (consistent with the definition of dew point) is therefore low and the air feels dry. Suppose that the temperature is the same but the dew point is now high [say, 70°F (21°C)]. In this case, not as much cooling is required to reach saturation—there is ample water vapor in the air and it feels very humid. As we introduced in Chapter 1, the air starts to feel a bit humid when the dew point reaches 60°F (16°C). By the time the dew point reaches 65°F (18°C), most folks would agree that the air is humid. A dew point of 70°F (21°C) is downright tropical while at dew points of 75°F (24°C) or higher, the air feels oppressive and

stifling. The dew point seldom exceeds 80°F (27°C) in the United States—the highest average dew points in the world are found in coastal areas of the Persian Gulf, Gulf of Aden, and Red Sea (see Figure 4.29). These water bodies are relatively shallow, and water temperatures can soar in summer above 90°F (32°C), promoting extremely high evaporation rates and thus extremely high dew points.

Before we close this section on dew point, we note that when the air temperature is 32°F (0°C) or less, the **frost point** becomes the absolute measure of moisture in the air, although many meteorologists continue to loosely use "dew point" at such low temperatures.

You already know that isopleths of temperatures are called isotherms. What about isopleths of dew point? Well, the Greek word for "dew" is "drosos," and dew point is, after all, a type of temperature. So it's not a big surprise that an isopleth of dew point is called an **isodrosotherm**. Make sense? Just as patterns of surface isotherms are constantly changing, particularly across the mid-latitudes, so are patterns of surface isodrosotherms. So it behooves weather forecasters to understand the processes that control the patterns of surface dew points.

How to Change Dew Points: From Up, Down, and All Around

One of the most obvious ways to change the dew point is to simply evaporate water into the air. For

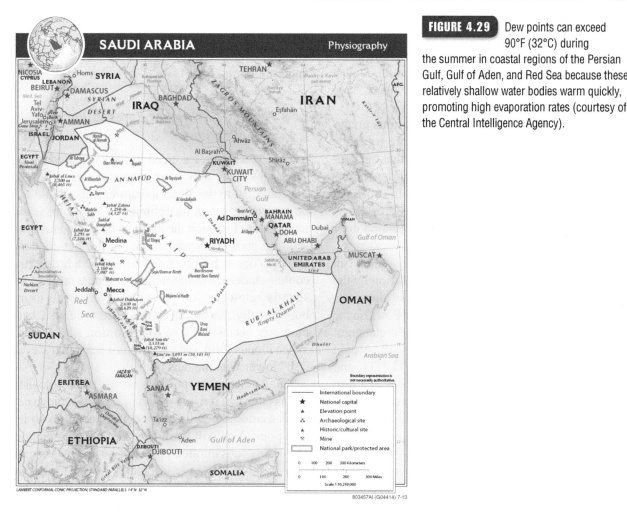

803457AI (G04414) 7-13

FIGURE 4.29 Dew points can exceed 90°F (32°C) during the summer in coastal regions of the Persian Gulf, Gulf of Aden, and Red Sea because these relatively shallow water bodies warm quickly, promoting high evaporation rates (courtesy of the Central Intelligence Agency).

example, chilly, dry Canadian air moving over the relatively warm Great Lakes in autumn and early winter creates large vapor pressure gradients just above the water surface. As a result, evaporation rates are large and the Canadian air gets moistened as it crosses the Lakes. This moistening sets the stage for "lake-effect snow" (see Figure 4.30), which we will discuss in more detail in Chapter 16.

Recall that dew points typically increase below the base of clouds after the onset of precipitation because some raindrops evaporate on the way to the surface. For folks driving or hiking in the Rockies or Appalachians, there are implications. Indeed, when it rains or snows in mountainous regions, the cloud base (ceiling) often lowers below the peaks as increasing dew points and decreasing temperatures bring the air closer to saturation. As a result, ridge-top fog often shrouds mountain peaks during wet weather.

Another way for surface dew points to change is for eddies to mix drier air downward (this is more noticeable during summer). In general, water vapor

FIGURE 4.30 Outside the National Weather Service office in Buffalo, NY, in the aftermath of lake-effect snow that brought in excess of 80 inches to parts of Erie county in late December 2001 (courtesy of Tom Niziol, National Weather Service, Buffalo, NY).

concentrations decrease with increasing altitude. Thus, on a sunny, hot and humid summer day, dew points can decrease a bit in the afternoon when eddies generated by solar heating mix drier air from higher up (typically about a kilometer or so) toward the ground.

Figure 4.31 shows this process in action. It's a portion of a meteogram for Dallas, TX, on an August day dominated by a hot and humid air mass. Note that the dew point was a very muggy 75°F (24°C) at 13 UTC. As the day grew hotter, dew points gradually decreased, bottoming out at a more tolerable 64°F (18°C) at 21 UTC [the same time that the temperature peaked at 98°F (37°C)]. Lower dew points could be traced to heights around 1800 m (6000 ft), about the highest altitude from which eddies generated by major-league solar heating circulated drier air toward the ground.

Of course, air of a different moisture content can also be transported *horizontally* into a region by the wind, causing dew points to increase or decrease. For example, **dry advection** typically follows on the heels of a cold front, as the portion of the meteogram for Atlanta, GA, illustrates in Figure 4.32. A cold front moved through Atlanta between 11 UTC and 12 UTC on this December day. In response, the dew point (and temperature) started to decrease beginning at 12 UTC in response to the west-northwesterly winds in the wake of the cold front. Also note that modest **moist advection** took place prior to the frontal passage as dew points increased in concert with southerly winds from the Gulf of Mexico.

The overall message that you should take from this discussion is that dew points are an absolute measure of the amount of moisture in the air. In addition to giving a sense of how humid or dry the air feels, dew points also play a pivotal

role in other facets of weather forecasting. We'll finish the chapter by discussing a few of these applications to further enrich your forecasting apprenticeship.

Dew Points and Weather Forecasting: A Few Applications

In case you haven't noticed in the examples we've given so far, the dew point always seems to be less than or equal to the air temperature. This should not be surprising, given its definition: Dew point is the temperature to which the air must be cooled (at constant pressure) to reach saturation. For net condensation to occur, however, there must be a little extra cooling. Indeed, the air temperature must fall ever so slightly below the dew point whenever clouds form, but we really can't measure this difference on a thermometer because it's so tiny. So don't be fooled. When clouds form, the temperature is a "gnat's eyelash" lower than the dew point.

For all practical purposes, however, meteorologists can assume that a rising parcel of air will achieve net condensation when its temperature decreases to its dew point. You will learn later that the temperature of a rising, unsaturated parcel of air decreases at a rate of about 10°C per kilometer of ascent (about 5.5°F per 1000 feet). As it turns out, the dew point of an ascending parcel also decreases, but at a much

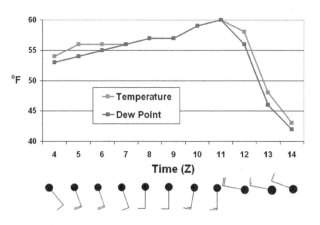

FIGURE 4.32 The temperature (red line), dew point (green line), cloud cover, and wind portion of the meteogram from Atlanta, GA, from 04 UTC to 14 UTC on a December day. A cold front moved through around 11 UTC. Ahead of the front, moist advection occurred on southerly breezes, and the dew point increased. Behind the front, westerly and northwesterly winds imported drier air, and the dew point decreased. Note that from 07 UTC to 11 UTC, the air was saturated so the temperature and dew point curves overlap.

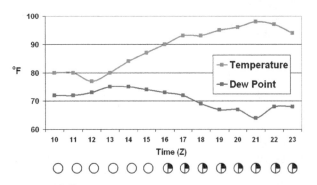

FIGURE 4.31 An August day started off warm and very humid in Dallas, TX, with a temperature of 80°F and a dew point of 75°F at 13 UTC. But as the mercury rose toward 100°F (red line) on a mainly sunny day (see cloud portion of meteogram), convective eddies mixed drier air toward the ground, causing dew points (green line) to decrease to a more comfortable 64°F by 21 UTC.

smaller rate. The exact numbers aren't important here. What's important is that the temperature of a rising air parcel decreases faster than its dew point. The altitude at which the temperature of a rising parcel falls imperceptibly below its dew point (they are essentially equal) is called the **lifting condensation level** (LCL for short). In Figure 4.33, you can get a sense of the level of the LCL by eyeing the flat bottom of the clouds.

For a given surface temperature, the LCL will be closer to the ground when surface dew points are relatively high. That's because higher dew points translate to more water vapor in the air, which means that less cooling is required to reach net condensation; in other words, air parcels do not have to rise as high for clouds to form.

Conversely, where surface dew points are often very low (the interior of the Western United States, for example), the LCL tends to lie at a higher altitude, especially during late spring and early summer when temperatures are high and dew points are relatively low (see Figure 4.34). When high-based thunderstorms erupt in this environment, raindrops often evaporate before reaching the ground. That's because they have to fall through a thick layer of relatively dry air below the cloud base. Thunderstorms with lightning but no rain can spark wildfires, so it behooves weather forecasters to alert the public when high-based thunderstorms are likely. Dew points play a pivotal role in making these kinds of forecasts.

The forecasting utility of dew points doesn't end there. On clear, calm evenings, the dew point often

FIGURE 4.33 The flat bottom of this layered cloud in the mountains of North Carolina is found at an altitude called the lifting condensation level (courtesy of Tim Martin).

FIGURE 4.34 A small, high-based afternoon shower over the Cascade Mountains of Washington in August 2007 likely produced rain that evaporated before reaching the ground (courtesy of Krzysztof Z. Gajos).

serves as a lower bound for the upcoming night's low temperature, provided weather conditions stay pretty much the same throughout the night. For example, Figure 4.35 shows the temperature, dew point, relative humidity, wind, and cloud cover components of the meteogram for State College, PA for the 24-hour period beginning at 15 UTC on September 7, 2014. Note that the dew point around sunset on September 7 (00 UTC on September 8) is between 50°F and 55°F. On nights with light winds and clear skies (such as the night in question), the dew point typically doesn't change much because there's little or no advection. So, after sunset, the temperature falls toward the dew point—in this case, by 07 UTC, they were within a degree of each other, and foggy patches had likely formed in the area (though fog was not officially reported until 10 UTC). Note that from 07 to 13 UTC, the temperature remained very close to the dew point, but never fell below it. That's because the net condensation associated with the fog released latent heat, which was just enough to keep the temperature from falling measurably below the dew point. In this way, on clear, calm nights, the dew point around sunset often serves as a reasonable first guess for the overnight low.

As a final example, analyses of dew points also help forecasters pinpoint areas at risk for severe thunderstorms, which produce damaging winds, large hail, and tornadoes. Consider Figure 4.36a

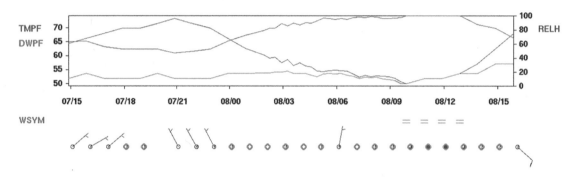

FIGURE 4.35 Meteogram showing graphs of the temperature (purple), dew point (green), and relative humidity (blue), as well as the cloud cover and wind portions of the station model, at State College, PA, from 15 UTC on September 7, 2014 until 16 UTC the next day. The evening dew point served as a reasonable estimate of the overnight low on this clear, relatively windless night, and fog formed late at night as the temperature fell to the dew point.

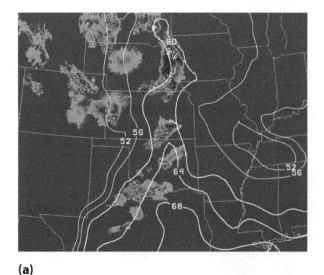

(a)

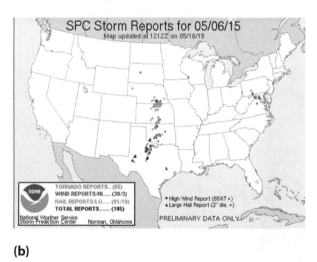

(b)

FIGURE 4.36 (a) An analysis of isodrosotherms at 03 UTC on May 7, 2015, shows a tongue of high dew points surging northward into Oklahoma, Kansas, and Nebraska. The moist air helped fuel strong thunderstorms indicated by oranges and reds on the underlying radar image. (b) The powerful thunderstorms produced a swarm of tornadoes from Nebraska to Texas, indicated by the red dots in this storm report from the Storm Prediction Center in Norman, OK (courtesy of NOAA).

which shows isodrosotherms superimposed on a radar image at 03 UTC on May 7, 2015. Note the surge of higher dew points northward into Oklahoma, Kansas, and Nebraska, indicative of moist advection associated with southerly winds. These relatively high dew points indicated plenty of moisture that served as fuel for powerful thunderstorms, which appear as the red blobs on the radar image. These thunderstorms produced a swarm of tornadoes from southern Nebraska to north-central Texas, indicated by the red dots in Figure 4.36b. You will learn more about forecasting severe thunderstorms later in the book.

Of course, relatively high dew points in the lower troposphere also can signal the potential for flooding, particularly during summer when slow-moving thunderstorms can produce prodigious rainfalls in relatively short periods of time. We'll delve into flooding from thunderstorms in Chapter 9. Until then, we'll end this chapter with a brief overview of the instruments that meteorologists use to measure precipitation.

RAIN GAUGES: PUTTING THE PRECIPITATION PUZZLE TOGETHER

Beginning on the evening of October 30, 2013, heavy thunderstorms tracking over the same area brought as much as 12–14 inches of rain in a narrow swath southwest of Austin, TX, with many surrounding areas receiving at least 6 inches of rain. Most of the heaviest rain fell in rural areas; however, based primarily on measurements taken from the cooperative observing network, the National Weather Service was able to put the pieces together and get a good sense of the pattern of rainfall (see Figure 4.37).

Among cooperative weather observers, the most common instrument used to measure rainfall is the standard National Weather Service rain gauge, shown in Figure 4.38. The outer shell of this standard rain gauge is an eight-inch diameter cylinder with a funnel on top. Inside the outer shell lies a second cylinder (just over 2.5 inches in diameter) into which rain funnels. Given that the cross-sectional area of the outer cylinder is ten times that of

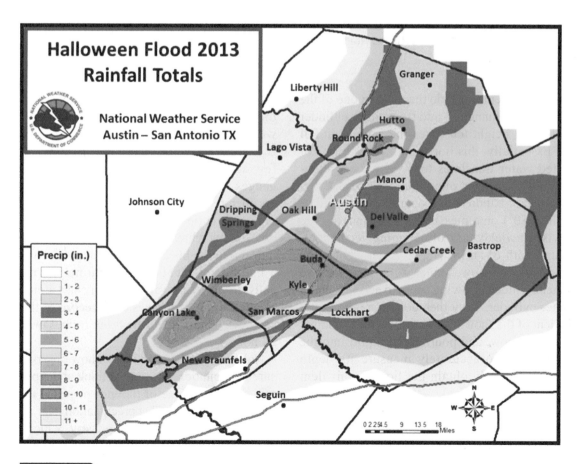

FIGURE 4.37 Map showing rainfall in and around Austin, in central Texas, on October 31, 2013.

FIGURE 4.38 The standard National Weather Service rain gauge (courtesy of the National Weather Service).

FIGURE 4.39 A tipping-bucket rain gauge (courtesy of the National Weather Service).

the inner cylinder (you'll do the math in one of the exercises), rain funneling into the smaller cylinder will rise to a height ten times the actual rainfall depth. Essentially, this horizontal "squeezing" of collected water translates to a vertical "stretching." Thus, a meager one-hundredth of an inch (0.01″) of rain balloons to a depth of one-tenth of an inch (0.10″) on a measuring stick, making the otherwise unwieldy task of obtaining accuracy to one-hundredth of an inch a lot easier. When the forecast calls for snow, weather observers remove the inner cylinder, allowing snow to accumulate in the outer cylinder. They then determine the liquid equivalent of the snow by taking the gauge indoors, melting the snow, and pouring the liquid into the small cylinder to accurately measure its depth (observers similarly obtain the liquid equivalent of hail and sleet this way).

Another instrument for measuring rainfall is the tipping-bucket rain gauge (see Figure 4.39). This gauge works by collecting rain and funneling it into a two-compartment bucket; 0.01″ of rain will

fill one compartment and overbalance the bucket so that it tips, emptying the water into a reservoir and moving the second compartment into place beneath the funnel. As the bucket tips, it completes an electrical circuit which then records the rainfall. The National Weather Service's Automated Surface Observing System (ASOS) network (recall Chapter 1) includes tipping-bucket rain gauges that are heated to prevent freeze-up during cold weather.

There are several other kinds of rain gauges. A few are very sophisticated, such as optical rain gauges, which measure rainfall using a laser. Others operate in a simpler way. For example, a weighing gauge collects rain and funnels the water into a hole, below which is a catch bucket. As water accumulates in the bucket, its weight gets converted to a liquid-equivalent depth. Routinely, a mechanically driven pen scrolls across a chart attached to a rotating drum and traces out a record of this depth as a function of time (the drum rotates once every 24 hours).

Now that you have an idea of how observers measure precipitation, it's time to think about predicting rain and snow. One of the indispensable tools that forecasters use to predict precipitation over relatively short time periods is radar imagery, which will be one of the fascinating topics we'll explore in the next chapter.

Focus on Optics

The Color of Clouds: Of *Star Trek* and Milk

In the 1960s, the popular television series *Star Trek* (the original version) always prided itself on the quality of its scientific content. But, on a few episodes, writers and producers committed "Bad Science." In the near-vacuum of space, visible light from phasers fired by the Starship Enterprise should not have been able to be seen by any neutral observer simply because there aren't enough molecules in space to scatter light to the observer's eyes. Only if an observer assumed a position in the line of fire, looking directly at the incoming phaser energy, would the observer be able to see the light. And taking such action would not be a prudent way to prove a scientific point.

Back on Earth, the scattering of visible light plays a crucial role in determining the color of clouds perceived by sky gazers. Some clouds are white while others are not, and we'll try to shed some light on the differences.

Let's start with a tall glass of milk (homogenized, of course). Milk contains lots of fat globules, which are more efficient at scattering the shorter wavelengths of visible light than the longer wavelengths. Thus, when

"white light" (which, like sunlight, contains the wavelengths of the visible spectrum) impinges on milk, these fat globules scatter shorter wavelengths efficiently and longer wavelengths not so efficiently. But, because there are so many globules, longer wavelengths, after rattling around a bit, eventually get their "scattering due" (as do the other wavelengths of visible light), and white light emerges from the milk. Thus, the glass of milk looks white!

And so it is with clouds. Trillions of tiny water drops (and some ice crystals) act like milk globules to scatter white light to our eyes. Think of a smattering of puffy, fair-weather cumulus clouds (see Figure 4.40) as just glasses of milk in the sky.

From above, clouds always look white, but the view from below can sometimes be as different as day and night. The bases of cumulonimbus and cumulus congestus clouds (see Figures 4.10 and 4.41), which are tall and thick, typically appear dark to a nearby observer on the ground.

We again turn to milk for our explanation. Check out Figure 4.42, which is a sequence of four photographs of an experiment we conducted to convince you that thick clouds transmit very little sunlight. The upper-left photograph shows a plain bottle of water. We then

FIGURE 4.40   Fair-weather cumulus clouds.

FIGURE 4.41   A thick cumulonimbus cloud (rain shaft on the right) has a very dark base.

(Continued)

(Continued)

FIGURE 4.42 The upper-left photograph shows a bottle filled with plain water. From the upper-right to the lower-left and then finally to the lower-right photograph, an increasing amount of milk was added. In this order, the albedo of the mixture increased. As a result, light from behind the bottle was increasingly back-scattered by more and more milk globules, and less light was transmitted to the camera. Thus, you cannot see through the bottle in the lower-right photograph because the albedo of the mixture is so high.

FIGURE 4.43 A waterspout looks dark against the bright background (courtesy of the U.S. Navy).

began the experiment by mixing in just a little milk (upper-right photograph). Although the bottle looks a bit "milky," you can still see right through the mixture because visible light from the window behind the bottle was largely transmitted to our camera. In other words, the relatively few milk globules back-scattered a little visible light, but not enough to prevent some light from the window blinds from getting through. The bottom line is that you can't see through the bottle as well as you could before we added milk.

Next, we added more and more milk (lower-left and then lower-right photographs). In each case, the greater number of milk globules back-scattered an increasing amount of light, to the point that you cannot see through the bottle. In effect, the mixture with the greatest concentration of milk globules extinguished light from the window behind the bottle.

In turn, the mixture in the lower-right photograph looks relatively dark.

Now think of the mixture of milk and water in the lower-right photograph as a thick cumulonimbus cloud. If we liken the camera to the eye of a nearby observer on the ground looking at the base of the cloud, you'll understand why its base looks dark to the observer: The "optically thick" cloud extinguished (back-scattered) a lot of sunlight.

This dark perception is enhanced when the observer views these clouds against a brighter background. Indeed, we cannot say enough about the importance of background in determining the perceived shade and color of clouds. One of the authors just went outside with his "white" handkerchief to prove a point. The handkerchief looked very white against the dark green siding of a house, but appeared almost pink against the bright white backdrop of freshly fallen snow.

And so it is with clouds. A single cloud that might appear gray or dark against a surrounding brighter sky might look almost white if it were viewed against the backdrop of an approaching dark thunderstorm. Yes, background can make a world of difference in how clouds appear! Figure 4.43 shows the swirling cloud of a waterspout that looks dark against a bright background. Meanwhile, Figure 4.44 shows a tornado whose funnel cloud looks bright when viewed against the dark backdrop of the optically thick parent thunderstorm.

Writers and producers of *Star Trek*—don't get your hopes up! Even though the backdrop of space is pitch black (save for some tiny, distant stars), you still won't be able to see the phasers of the Enterprise unless you're directly in the line of fire. And that view won't last very long.

FIGURE 4.44   A white tornado (copyright Ian Wittmeyer).

Satellite and Radar Imagery: Remote Sensing of the Atmosphere

5

LEARNING OBJECTIVES

After reading this chapter, students will:

- Gain an appreciation for the orbital characteristics of geostationary and polar-orbiting satellites, and be able to describe the salient advantages and disadvantages associated with each kind of weather satellite
- Be able to relate cloud depth and the general number and size of constituent water drops (and/or ice crystals) to the albedo and appearance of clouds on visible satellite imagery
- Be able to distinguish snow cover from clouds on visible satellite imagery
- Be able to identify low and high clouds on infrared satellite imagery

- Be able to argue why fog may be indistinguishable from the ground on infrared satellite imagery, particularly at night
- Understand why clouds with high tops routinely "contaminate" water-vapor imagery
- Be able to argue against the popular and scientifically erroneous notion that dry air depicted on water-vapor imagery always translates to dry air near the earth's surface
- Be able to specifically explain how radar qualifies as an active remote sensor
- Gain an appreciation for the deceptively low values of radar reflectivity associated with snow falling relatively far from the radar site
- Understand the underlying scientific principles for interpreting radar images that display Doppler velocities

If you ever watch professional golfers on a breezy day, they'll often toss blades of grass into the air to get a better sense of the prevailing wind direction. Using grass to determine the wind direction on a golf course is an example of an **in-situ observation**. For the record, an instrument takes an in-situ observation whenever it is in direct contact with the medium it "senses" (in the case of a golf pro using turf to assess wind direction, the blades of grass were in direct contact with the moving air). Other common examples of in-situ weather instruments are liquid-in-glass thermometers, rain gauges, and wind vanes.

In contrast, an instrument that takes a **remotely sensed observation** is not in direct contact with the medium it "senses." There are two kinds of remote sensing: active and passive. Active remote sensors typically emit electromagnetic radiation that is then "collected" by the sensor after it strikes "targets." For example, X-ray machines used by doctors and dentists as a tool for diagnosis qualify as active remote sensors because they emit low doses of X-ray radiation into the body. X-rays passing through the body strike a plate coated with special film, causing a chemical reaction that produces an image. The process works because bones absorb X-rays more readily than "soft tissues" such as muscle, skin, and blood vessels. This difference in absorptivity is translated into different shades on an X-ray: Bones look white, while soft tissues appear gray.

Passive remote sensors simply collect electromagnetic energy emitted or scattered by objects. For example, our eyes are passive remote sensors because they process visible light that is scattered and emitted by objects. Figure 5.1 is an image of the Sun in ultraviolet radiation. This image was created using passive sensors on the Earth-orbiting Solar Dynamics Observatory,

a sun-pointing spacecraft that takes nearly continuous observations of the Sun. Here, the ultraviolet emissions come mainly from atoms in the solar corona at temperatures near 1,000,000°F (about 555,540°C).

In this chapter, you will learn about weather radar, which is an active remote sensor used, in part, to detect the location and intensity of precipitation. You will also learn about weather satellites that passively collect visible and infrared radiation backscattered and emitted (respectively) by the earth and the atmosphere. Data collected by these satellites (see Figure 5.2) are beamed back to Earth, where computers process and electronically massage the data into cloud images and other forecasting products.

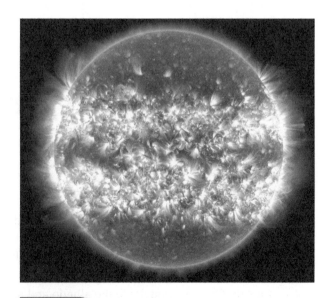

FIGURE 5.1   A false-color composite image of the Sun in ultraviolet radiation. The hottest areas, which emit the greatest intensity of ultraviolet radiation, are show in bright white (courtesy of NASA Solar Dynamics Observatory).

FIGURE 5.2 An artist's rendition of a polar-orbiting satellite, which scans the Earth in relatively narrow swaths (courtesy of NASA).

WEATHER SATELLITES: DIFFERENT VIEWS, DIFFERENT ORBITS, DIFFERENT MISSIONS

North American forecasters rely heavily on two high-flying **geostationary satellites** called GOES-East and GOES-West, which orbit the Earth at altitudes near 36,000 km (about 22,500 mi). To kick off our discussion, we point out that GOES is an acronym for Geostationary Operational Environmental Satellite.

As it implies, "Geostationary" means that the satellite doesn't move with respect to the Earth ("geo" is synonymous with "Earth"). Indeed, GOES-East and GOES-West orbit over fixed points on the equator at a speed matching the Earth's rotation rate. For the record, GOES-East and GOES-West are parked above the equator at 75°W and 135°W longitude, respectively (see Figure 5.3). Continuing with our deciphering of the GOES acronym, we note that "Operational" distinguishes GOES from experimental satellites, while "Environmental" refers to the satellites' capability to measure temperature, moisture, and winds at various altitudes in the atmosphere. Among other duties, GOES also relays distress signals from people, aircraft, and ships as well as data from ocean buoys and the radio collars on roving bears.

GOES-East and GOES-West have fixed, full-disk views that, together, cover the Western Hemisphere, allowing weather forecasters to monitor the development and evolution of storms (see Figure 5.4). Note how the curvature of the Earth limits the view of high latitudes in both the Northern and Southern Hemispheres, a drawback of all geostationary satellites. Of course, you'll seldom see full-disk images on television news programs. Indeed, computer software routinely crops full-disk images so that weather forecasters, research scientists, and broadcast meteorologists can narrow their focus to smaller regions (the contiguous states, for example).

Given that GOES routinely provides images every 15 minutes (in "super rapid scan" mode, images can be collected every minute or even less), animations of GOES images afford meteorologists the capability to assess the movement (speed and direction) of evolving weather systems. As you will learn later, these "satellite loops" also enable meteorologists to assess

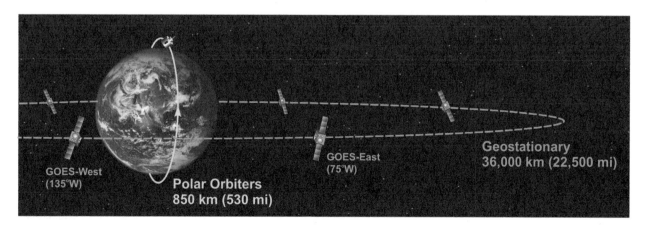

FIGURE 5.3 The orbits of geostationary and polar-orbiting satellites. A geostationary satellite sits above the equator and orbits the earth at the same rate that the planet rotates, so such a satellite is essentially "parked" over a fixed spot. Polar-orbiting satellites can image the entire surface, but only in swaths.

(a)

(b)

FIGURE 5.4 Full-disk satellite images from (a) GOES-East and (b) GOES-West, on January 9, 2015. Note how the curvature of the Earth limits the view of high latitudes (courtesy of NOAA).

the speed and direction of winds in cloud-free areas. This capability helps forecasters to predict the movement of weather systems developing over remote tropical seas, where other types of observations are sparse.

The National Oceanic and Atmospheric Administration (NOAA) routinely keeps a replacement GOES in space storage in case one of the operational geostationary satellites fails. Also in geostationary orbit are environmental (weather) satellites operated by Japan, China, South Korea, India, and a European group (see Figure 5.3), as well as hundreds of other satellites, most for communications purposes. In

fact, the "geostationary parking lot" is getting pretty crowded. Consider the time lapse image in Figure 5.5 which was created by a telescope atop Kitt Peak in Arizona on March 29, 2014. Each dot is a geostationary satellite—the array of about 40 dots represents only about 10% of the entire geostationary fleet.

How do we know that these dots are geostationary satellites? Well, when photographers take time-lapse images of the nighttime sky, the stars trace out trails as the earth rotates about its axis (revisit the discussion at the beginning of Chapter 3). This means that the bright, fixed dots in the midst of the belt of star trails

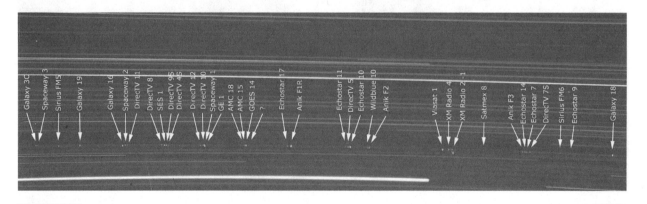

FIGURE 5.5 A time lapse image of a small portion of the geostationary orbit taken from atop Kitt Peak in Arizona on March 29, 2014. The long, slightly curved lines represent star trails, while dots (with corresponding labels) mark the positions of geostationary satellites (courtesy of Bill Livingston).

must be stationary with respect to the Earth (they obviously didn't move during the time-lapse photography). There's no way that the light reflected by these geostationary satellites would be sufficiently bright to see them clearly on just a single snapshot, but the long exposure of the time-lapse photograph allows them to leave a faint imprint.

Unlike equatorially-parked geostationary satellites, whose view of high latitudes is blocked or blurred by the highly curved Earth, **polar-orbiting satellites** "see all" as their orbits pass approximately over earth's poles (see Figure 5.3). Indeed, the entire surface of the Earth is observable by polar-orbiting satellites as they trace out paths that are nearly fixed in space while the Earth rotates beneath them.

But these satellites don't "see all" at once. On each pass (each polar-orbiting satellite makes 14 orbits a day), the satellite scans a swath of earth thousands of kilometers wide. The swaths are nearly adjacent at the equator but they progressively overlap with increasing latitude. Thus, regional polar-orbiting images are montages of images that were gathered at different times, a disadvantage compared to the full-disk views provided by geostationary satellites. However, polar orbiters, unlike geostationary satellites, can acquire undistorted images over high latitudes. Figure 5.6, which focuses on Alaska and surrounding regions, illustrates the telltale "swath" of data characteristic of polar-orbiting satellites.

Visible Cloud Imagery: Scanning the Globe

When television weathercasters say that they're looking at a "visible satellite picture" (such as the one in Figure 5.7a), some viewers may question their sanity. "Of course, it's visible, silly. We can see it, can't we?"

What weathercasters really mean to say is that the image of clouds, continents, ice caps, snow fields, and oceans pictured on viewer's screens has been created using visible radiation from the sun that back-scattered off these objects. Thus, much like photos you take with a camera or cellphone, these satellite images utilize back-scattered visible light.

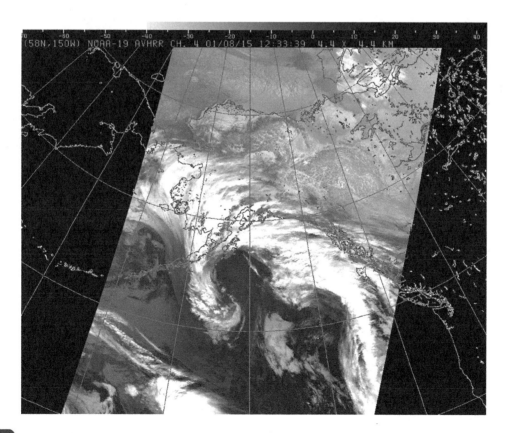

FIGURE 5.6   An infrared satellite image taken from NOAA-19, one of the polar-orbiting satellites in NOAA's fleet of Polar Operational Environmental Satellites, or POES. This image, from January 8, 2015, shows a typical swath of satellite data characteristic of polar orbiters (courtesy of NOAA).

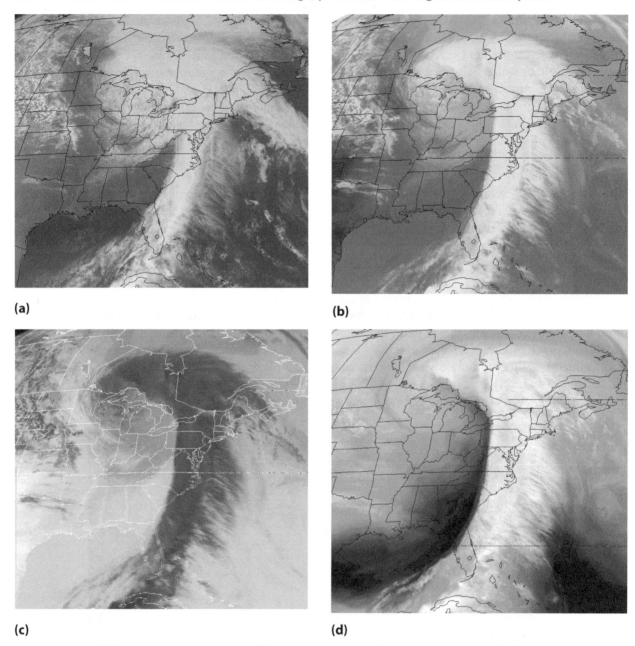

(a)

(b)

(c)

(d)

FIGURE 5.7 Different types of satellite images, taken at the same time; (a) Visible; (b) Infrared; (c) Infrared, but a negative of part (b), so the coldest areas are dark and warmest bright; (d) Water vapor (courtesy of NOAA).

Creating visible images from satellites requires that the visible light back-scattering off clouds and the Earth's surface is able to pass through the atmosphere without being extinguished. If the atmosphere totally or largely absorbed back-scattered light, visible satellite images would appear pitch black (at worst) or very faint (at best). In either case, such images would not shed much light on weather forecasts. Fortunately, back-scattered visible light travels through the atmosphere to space without much depletion (look back at Figure 2.19—and note that the atmosphere does not absorb visible light very well). Just think of the atmosphere as a soft, stained-glass window in a small church that allows sufficient sunlight to illuminate the chapel by day. In a similar way, sufficient visible light that's back-scattered by the earth and clouds passes through the atmosphere's glazed pane of gases and particles for the satellite to detect.

Onboard weather satellites, a sophisticated instrument called a **radiometer** intercepts visible radiation back-scattered by clouds and the Earth (satellite radiometers also detect other wavelengths of radiation—more on this later). More importantly, the sensitive radiometer distinguishes differences in the amount of back-scattered light reaching the satellite from various objects. In essence, the radiometer measures the albedo of clouds and the earth's surface by distinguishing between the varying intensities of sunlight back-scattered to space. For display purposes, contrasts in albedo are mathematically converted into appropriate shades, such as bright for clouds and dark for land and oceans (note the subtle darker shading of oceans compared to land in Figure 5.7a, indicating that the Earth's seas back-scatter less visible radiation than the continents).

There are also subtleties in the shades of brightness for clouds. As an example, note in Figure 5.7a the streamers of wispy clouds (likely **cirrus** or **cirrostratus clouds**) blowing off the tops of a line of tall thunderstorms moving off the Southeast and Mid-Atlantic Coasts. These high thin clouds appear a bit duller than the very bright white tops of **cumulonimbus clouds** because the albedo of wispy high clouds is typically less than 0.5, compared to an albedo of more than 0.7 for thick cumulonimbus clouds (thunderstorms). Thus, by using visible satellite imagery, meteorologists are able to distinguish between thick and thin clouds.

To understand why thin clouds have a lower albedo than thick clouds, recall the experiment we conducted for the Focus on Optics in Chapter 4 (The Color of Clouds). By way of review, we added increasing amounts of milk to a bottle of clear water. When we first mixed in a small amount, you could still see through the bottle, although not as clearly as you could when it contained only water. Thus, this thin mixture back-scattered only a modest amount of light from the window behind the bottle (the mixture transmitted enough light for us to still see the window blinds). In other words, the thin mixture had a relatively low albedo. And so it is with thin cirrus clouds: They back-scatter a modest amount of visible light from the sun. Thus, they appear white on visible satellite imagery, but not bright white like the tops of thunderstorms.

Again, we liken cumulonimbus clouds to the final mixture of milk and water in our Chapter 4 experiment. Recall that, in this case, we added a lot of milk and we couldn't see the window blinds behind the bottle. Indeed, the mixture back-scattered a lot of light from the window, and thus transmitted very little light to the camera. In other words, the thick

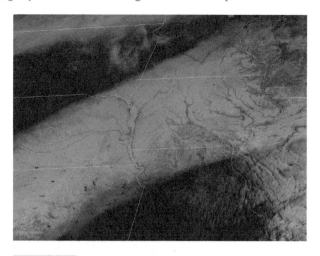

FIGURE 5.8 The relatively dark outlines of branching rivers stand out against the white backdrop of a stripe of snow across the Midwest (the rivers are ice-free or partially ice-free). Note that there are a few clouds over Illinois in the southeast corner of the image (courtesy of NOAA).

mixture had a high albedo. And so it is with thick cumulonimbus clouds. They back-scatter a lot of sunlight to space and their tops look bright white on visible satellite imagery.

Generally speaking, distinguishing between clouds and snow cover on a static visible satellite image is challenging. That's because the albedos of clouds and snow are in the same ballpark. When broken splotches of clouds move over snow-covered regions, meteorologists sometimes look at loops of visible images. Clouds will move, of course, but snow cover will not (unless it melts very rapidly, sometimes lending the false impression that moving or dissipating clouds are present). At other times, meteorologists will determine that there is snow cover over a region by spotting the unfrozen, dark outlines of rivers and their tributaries. Figure 5.8 is a striking example of ice-free (or partially ice-free) rivers of the Middle West standing out against the white backdrop of a stripe of snow cover left in the wake of a winter storm.

Like rivers against a backdrop of snow cover, deciduous and coniferous forests also appear darker on visible imagery. For example, check out Figure 5.9, a visible satellite image centered on New York State the day after an early season snowstorm. In this case, snow accumulated primarily in the higher elevations, including the Catskills and Adirondacks of New York and the Green Mountains of Vermont. The whiteness is certainly not uniform, in part because trees lose the snow that accumulates on their limbs fairly quickly

FIGURE 5.9   A visible satellite image of the northeastern United States in the aftermath of a snowstorm shows snow cover in the higher elevations of Pennsylvania, New York, Vermont, and several other states. The white is not uniform because of the dense forests that cover many of the higher elevations (courtesy of NOAA).

FIGURE 5.10   Back-scattered moonlight was used to capture a nighttime visible image of Hurricane Isaac at 1:57 A.M. CDT on August 29, 2012. Skies were mainly clear over much of the Florida peninsula at this time, so the lights of cities such as Tampa, Orlando, and Miami were detected as well (courtesy of NASA).

as winds increase in the wake of a storm. As a result, the dense trees that dominate these higher elevations mask the high albedo of the underlying snowpack, presenting a darker appearance to space than areas with fewer trees.

Because visible satellite imagery utilizes back-scattered sunlight, you might think that such imagery from weather satellites is only possible during the daytime hours. The launch of a new polar-orbiting satellite called Suomi (in honor of a pioneering satellite meteorologist) in 2011 dispels this notion—Figure 5.10 is a visible image taken in the wee hours of the morning of August 29, 2012 showing Hurricane Isaac in the northern Gulf of Mexico (Florida is seen on the right). Isaac's clouds were illuminated by just enough moonlight to be detected by the satellite's very sensitive low-light sensor. Also notice that where clouds didn't completely block the view, city lights were also detected by the satellite—in Florida, for example, the densely populated Tampa and Orlando areas and the stretch from West Palm Beach to Miami show up quite well from space at night.

Infrared Satellite Imagery: It Goes Out through the Atmospheric Window

Being a polar-orbiting satellite, Suomi can't provide the kind of continuous coverage of nighttime cloud cover that forecasters need. But as it turns out, meteorologists have had this capability for decades. As you learned in Chapter 2, Earth and its atmosphere (including clouds) continuously emit infrared radiation. Thus, infrared sensors aboard weather satellites give forecasters the ability to detect clouds, even at night, assuming that the infrared radiation can reach the satellite.

According to Figure 2.19, there are several ranges of infrared wavelengths in which the atmosphere has low absorptivity (which means radiation at these wavelengths can pass through the atmosphere to the satellite). Using our "window" metaphor (but in this case for infrared radiation), wavelengths near 4 microns and 10 microns, for example, correspond to infrared energy that preferentially passes through the window. In fact, the radiation band between 8 and 13 microns is sometimes called the **atmospheric window**, and infrared radiation in this band is commonly used to produce infrared satellite images. Most other infrared wavelengths get absorbed by gases (and clouds) in the atmosphere, and thus never reach the satellite.

Unlike visible satellite images, which allow meteorologists to distinguish between thick and thin clouds, infrared images give forecasters the ability to differentiate between high and low clouds. Consider Figure 5.7b, which is the infrared counterpart to Figure 5.7a. Notice the grayish clouds over Ohio, West Virginia, western Pennsylvania, and Michigan. In contrast, note the bright white clouds associated with the thunderstorms along the Eastern Seaboard and the cirrus clouds streaming east off these thunderheads. The gray color indicates that the Midwestern clouds are low (probably the ones we call stratus), with tops only a thousand meters or so above the ground. The bright white colors indicate that the tops of the thunderstorms and the blow-off cirrus are high.

How does infrared imagery differentiate between high and low clouds? Recall that temperature generally decreases with altitude. So, in Figure 5.7, the tops of the low-altitude Midwestern clouds are relatively warm compared to the tops of the Eastern Seaboard's high clouds. Where there are no clouds (over the northern Gulf of Mexico and much of Georgia, Alabama, and Mississippi, for example), the infrared radiation reaching the satellite comes from the Earth's surface, which is typically warmer than most low clouds. Now let's apply another principle—recall the Stefan-Boltzmann Law, which states that the amount of radiation emitted by a body (per unit area) is proportional to the fourth power of its temperature (expressed in Kelvins). Thus, more infrared radiation is emitted from the tops of warmer low clouds than

from the tops of colder high clouds. And the ground will emit more infrared radiation (typically) than the tops of low clouds. Computers translate these different intensities of emitted infrared radiation into shades of black, gray or white.

On infrared imagery taken by commercial photographers, cold objects are depicted in dark shades and relatively warm objects in white hues. For example, in Figure 5.11 (taken with a microbolometer camera tuned to wavelengths of 7 to 14 microns), a man opens a refrigerator door. Note that the cold interior of the refrigerator is dark in the infrared while the warm man appears white. Infrared satellite images are no different. High, relatively cold clouds should look dark in the infrared (like the inside of the refrigerator). But how weird would black clouds look to most people (see Figure 5.7c)? To prevent any confusion, meteorologists reverse the shading scheme, making high, cold clouds appear white (Figure 5.7b). Ah, that's better!

On infrared satellite imagery, the earth and its oceans often appear dark because they are relatively warm and emit relatively large amounts of infrared radiation. In spring and early summer, however, large bodies of water, especially those at higher latitudes, typically appear brighter than the land because water temperatures are slow to recover from their winter chill (at a time when the land more rapidly starts to heat up). For example, consider Figure 5.12, which is an infrared image centered on Pennsylvania on an early spring morning. The bright grayish appearance

FIGURE 5.11   On commercial infrared imagery, cold objects appear dark while warm objects appear white, illustrated by this image of a man opening a refrigerator door. The warm man looks white in the infrared, while the cold interior of the refrigerator appears dark (courtesy of FLIR Systems, Indigo Operations).

FIGURE 5.12 An infrared satellite image on an early spring morning in the Northeast when skies were mainly clear. The bright grayish appearance of water bodies indicates that surface waters were relatively cold. Darker splotches are the infrared footprints of relatively warm urban areas (courtesy of NOAA).

of Lake Erie and the Atlantic Ocean indicates that surface waters were cold, which is a hallmark of early spring in this region. As an aside, we call your attention to the darker "blemishes" in the image, especially noticeable in southeastern and southwestern Pennsylvania. These dark splotches are the infrared footprints of cities (urban "heat islands"), which were noticeably warmer than the surrounding rural areas.

Still focusing on Figure 5.12, we note that the fair-weather cumulus clouds (with relatively low tops) in northwestern Pennsylvania and southwestern New York are easy to distinguish from the warmer land. In contrast, detecting low clouds at night or just around sunrise on infrared satellite imagery is often challenging, particularly in winter when there's snow cover. That's because the temperature of a cold snowpack can be very similar to the temperature of the tops of low clouds. To meet these challenges, forecasters working the night shift must look closely at the sky conditions on the hourly surface observations, rather than solely trusting satellite imagery. In addition, there is also special satellite imagery, which relies on detecting radiation at wavelengths near 3.9 microns, that forecasters use for detecting fog and low clouds such as stratus.

Sometimes forecasters will color-enhance infrared images, typically to better reveal the highest, coldest cloud tops that might be associated with, for example, the tallest and most violent thunderstorms surrounding the eye of a hurricane. Figure 5.13a is a close-up visible satellite image of Hurricane Sandy off the East Coast of the United States at 18 UTC on October 28, 2012. Some of the thinner (less reflective) high clouds on the northwest side of Sandy had already moved over land, but clearly there were brighter (thicker) clouds offshore, particularly just off the coast of the Carolinas and near the center of the swirling storm. These thicker clouds are suggestive of tall cumulonimbus (thunderstorms). You can get a better idea of where the highest cloud tops are on the standard infrared image shown in Figure 5.13b—there, the brightest white, indicative of the coldest cloud tops, are right off the Carolinas. To seal the deal, Figure 5.13c is a color-enhanced infrared image, with shades of green corresponding to the lowest cloud-top temperatures, in the −60°C range.

Similar principles of color enhancement were used to create the infrared image of the space shuttle landing at Edwards Air Force Base in southern California (Figure 5.14). In this case, the warmest areas on the shuttle (which emitted the most intense infrared radiation) are shown in red and orange, and include the nose of the spacecraft and the tires, which were heated by friction as the shuttle touched down and decelerated on the desert runway. NASA discontinued the space shuttle program in 2011 after 135 missions.

By stringing a sequence of visible or infrared cloud images together (which is how satellite movies, or "loops," are produced), forecasters can even determine the direction the wind is blowing at various altitudes (high clouds can move one way while low clouds move in another direction). However, when there aren't any clouds, particularly over the oceans where there are precious few surface observations, what do forecasters do? They turn to a third type of satellite imagery at their disposal—water vapor imagery.

Water Vapor Imagery: Infrared Imagery, With A Twist

Water vapor imagery is actually a special type of infrared imagery that allows forecasters to qualitatively gauge the water vapor content in the middle and upper troposphere. Perhaps more importantly, water vapor imagery allows meteorologists to track atmospheric motions, even in regions where there aren't any clouds.

First, recall that the satellite imagery that we have considered so far (visible and standard infrared) rely on atmospheric windows—that is, the atmosphere is mostly transparent to the wavelengths of radiation

(a)

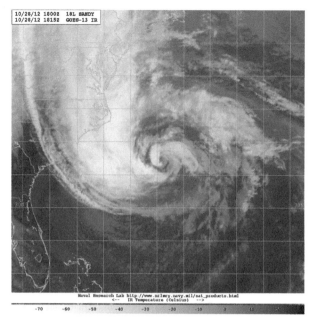

(b)

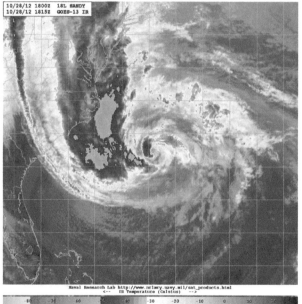

(c)

FIGURE 5.13 Geostationary satellite images of Hurricane Sandy at 18 UTC on October 28, 2012: (a) Visible image; (b) Infrared image (the temperature legend on the bottom gives estimated cloud-top temperatures); (c) Color-enhanced infrared image. Note the temperature legend at the bottom (courtesy of NOAA).

used in these types of imagery. These wavelengths are able to pass through the atmosphere and reach the satellite. In contrast, water vapor imagery—despite being a type of infrared imagery—utilizes wavelengths of radiation (near 6.7 microns, to be precise) which the atmosphere readily absorbs! You can even confirm that absorption by checking back to Figure 2.19. Why, you might ask, would meteorologists use wavelengths of radiation that have a difficult time getting out of the atmosphere? Well, it turns out that water (in all its phases) is the dominant

emitter of wavelengths near 6.7 microns. Thus, if a radiometer onboard a weather satellite is able to detect such wavelengths emanating from the atmosphere, there's a very good chance that the emitter was water in one of its forms.

But here's the hitch. Recall from Kirchoff's Law in Chapter 2 that efficient emitters of radiation at certain wavelengths also tend to be efficient absorbers of those wavelengths. In the context of water vapor imagery, this means that radiation of wavelengths near 6.7 microns emitted by water low in

FIGURE 5.14 A color-enhanced infrared image of the space shuttle landing at Edwards Air Force Base in southern California. Red represents areas of relatively intense infrared radiation emission, which mark the warmest parts of the spacecraft (courtesy of NASA).

the troposphere has very little chance of reaching the satellite because such radiation will likely be absorbed by water at higher altitudes. This means that radiation of wavelengths near 6.7 microns that is able to reach the satellite must have come from water at relatively high altitudes, where the chances of that radiation being absorbed on its way to the satellite is much less (there's not much water, water vapor, or ice at such lofty altitudes). Now honestly, in some extreme cases (for example, when the satellite scans the frigid, bone-day atmosphere over Siberia in winter), radiometers can indeed detect water at low altitudes. But this is truly the exception and not the rule. Generally speaking, a satellite radiometer attuned to radiation of wavelengths near 6.7 microns will only detect water residing at altitudes corresponding to the middle troposphere and above.

Recall that the temperature of an emitter can be deduced from the amount of radiation emitted (via the Stefan-Boltzmann Law—a higher temperature means that the object emits more intense radiation). When the upper reaches of the troposphere are dry, most of the 6.7-micron radiation reaching the sensor likely originates from water vapor residing at lower altitudes in the middle troposphere—source regions that are warmer than the upper troposphere (we assume here, reasonably, that the temperature of the air decreases with increasing altitude). The 6.7-micron radiation emitted from this "warmer water vapor" is then represented as darker blotches on a water vapor image (with a little help from a computer). These

darker blotches, though really a radiative snapshot of water vapor in the middle troposphere, represent a negative print of dryness in the upper troposphere.

Figure 5.15a is a water vapor image from GOES-East on March 8, 2015 (the green numbers are surface dew points—we'll come back to them later), while Figure 5.15b is the corresponding infrared image. Let's focus first on the island of Cuba, just to the south of Florida—from the infrared image, you can conclude that skies there are free of clouds. Note the dark appearance of the water vapor image

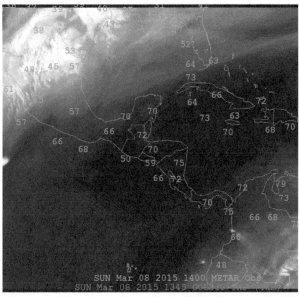

(a)

(b)

FIGURE 5.15 (a) A water vapor image, with surface dew points (in °F) superimposed; (b) The corresponding infrared image (courtesy of NOAA).

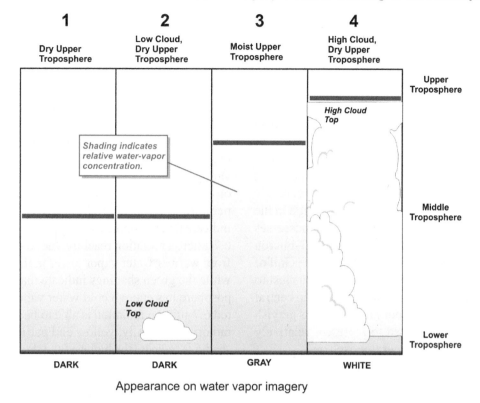

Appearance on water vapor imagery

━━━━━ The red bar represents the highest layer of water vapor (or high cloud top) whose emissions the satellite measures.

FIGURE 5.16 Sample columns of air (numbered 1–4 for ease of reference) with different moisture characteristics, and how each would register on a water vapor image. In this schematic, darker shading corresponds to higher concentrations of water vapor. The red bar represents the highest layer of water vapor detectable by the satellite. For Columns 1 and 2, the upper troposphere is dry. Thus, the satellite detects 6.7-micron radiation emitted by "warm water vapor" in the middle troposphere and the water vapor image appears dark over these regions. The low cloud in Column 2 does not register on the water vapor image because the 6.7-micron radiation it emits is absorbed by water vapor higher up. For Column 3, the upper troposphere is moist. There the satellite detects radiation emitted from "cold water vapor" and the water vapor image appears gray. In Column 4, the water vapor image appears white because only 6.7-micron radiation emitted by the very cold top of the cloud reaches the satellite.

over the island. If we examine a sample column of air in this region, it would resemble Column 1 in Figure 5.16. Essentially, the red bar marks the layer of air in the middle troposphere that has sufficient water vapor so that its emitted 6.7-micron radiation reaches the satellite and meaningfully registers on the radiometer (6.7-micron radiation emitted upward by water or water vapor molecules below the red bar is absorbed by water in the layer between those molecules and the red bar). Being in the middle troposphere, the water at the level of the red bar is relatively warm (compared to water at higher altitudes) and thus emits a relatively high amount of 6.7-micron radiation. In this case, the air column is represented as dark on the water vapor image, indicating that higher altitudes of the air column are dry.

While we're focusing on dark areas on the water vapor image, note that many surface dew points of 65°F or higher (values considered humid by human standards) were observed in these dark areas. Thus, we must conclude that water vapor imagery cannot be used to assess water vapor content near the Earth's surface. And scientifically this makes sense—6.7-micron radiation being emitted upward by this low-level water cannot reach the satellite because it's absorbed by water higher up. Nonetheless, trying to glean the moisture content of low-level air from water vapor imagery is probably the biggest mistake made by television weathercasters who use this specialized satellite imagery.

Now let's focus on another area—check out the infrared image (Figure 5.15b) from extreme southern Mexico southward through most of Central America. Here, the grayish tint that covers many areas of these countries suggests low clouds, yet this region on the water vapor imagery is just as dark as the region over Cuba. What's going on here? Consider Column 2 in Figure 5.16, which mimics the conditions in this region. The 6.7-micron radiation emitted upward by the low clouds is absorbed by water vapor higher up. In this way, the mid-tropospheric water vapor represented by the red bar in Column 2 prevents the low clouds from being detected by the satellite. Lesson learned: Low clouds typically do not show up on water vapor imagery.

Now let's find a region on Figure 5.15 that corresponds to Column 3 in Figure 5.16. Such a column is cloud-free, but the upper troposphere is relatively moist—thus, the red bar in Figure 5.16 is high in the column, marking a colder layer of air that possesses sufficient water vapor so that its emitted 6.7-micron radiation reaches the satellite. Portions of the Gulf of Mexico certainly fit that bill, as well as areas just to the east of southern Georgia and northern and central Florida. Note the water vapor image appears grayish in these areas, a shade used to represent relatively cold water vapor in the upper troposphere.

Finally, a complication arises in water vapor images when clouds are present at high altitudes—check out, for example, much of northern Mexico into Texas and western Louisiana in Figure 5.15b. Note that in these areas, the water vapor image mimics the infrared image pretty well (remember, a water vapor image is a special kind of infrared image; high, cold cloud tops match perfectly on both images). Here, the 6.7-micron radiation reaching the satellite originates from the tops of these cold clouds. So in these regions, the water vapor image is not really depicting

water vapor, but instead the water (mostly in the form of ice crystals) that are present in the high, cold tops of clouds such as cirrus or cumulonimbus. This situation is represented by Column 4 in Figure 5.16.

Water vapor images are sometimes color-enhanced, a technique which is particularly useful when clouds are present because the color enhancement better distinguishes variations in cloud-top temperature. Figure 5.17a is a color-enhanced water vapor image of Hurricane Katrina as it bore down on New Orleans on August 28, 2005. Note the temperature scale at the bottom—here, blue shadings indicate that the upper troposphere is dry (that is, the 6.7-micron radiation reaching the satellite originates from warmer water vapor lower in the troposphere), while the green shadings indicate that the upper troposphere has enough cold water vapor that its emitted 6.7-micron radiation is able to be detected by the radiometer. Finally, yellow and particularly maroon shadings indicate the highest (and thus coldest) cloud tops. Note the donut of dark maroon around the eye of Katrina, denoting very cold cloud tops associated with tall thunderstorms. This ring of strong thunderstorms is the fingerprint of a powerful hurricane.

Figure 5.17b is a similar color-enhanced water vapor image of Katrina, taken 12 hours later. On the western flank of the hurricane, the two narrow ribbons of blue and green shading indicate drier air circulating counterclockwise into the storm at higher altitudes. When you compare this image with Figure 5.17a, you can see that the western flank of the storm

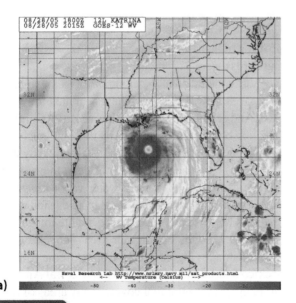 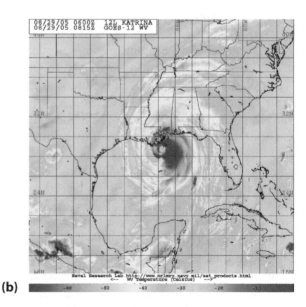

(a) (b)

FIGURE 5.17 (a) Color-enhanced water vapor image of Hurricane Katrina at 2015 UTC on August 28, 2005. Green shadings indicate that the upper troposphere is relatively moist, while blue shadings indicate that the upper troposphere is dry. Yellow and maroon shadings indicate high cloud tops; (b) Same image, 12 hours later (courtesy of NOAA).

TABLE 5.1 Guidelines for Interpreting Clouds on Satellite Imagery

Low Clouds (such as stratus)

- The base of a deck of low clouds typically looks rather dark, which means that a limited amount of visible light is getting through. Thus, a significant amount of sunlight must be reflected (back-scattered) back to space by low clouds. As a result, low clouds look bright on visible images.
- On standard infrared images, low clouds appear gray because their tops are relatively warm.
- On standard water vapor images, low clouds are typically not represented because they are too low in the atmosphere.

High Clouds (such as cirrus or cirrostratus)

- From the ground, a layer of thin high clouds usually looks white, meaning that sufficient sunlight is getting through to make the clouds appear rather bright to an observer on the ground. So, because ample sunlight is getting through, the amount of sunlight that is reflected (back-scattered) back to space by these thin clouds must be somewhat less than what is reflected back by more reflective low clouds. Thus, thin high clouds look just a bit off-white (compared to low clouds) on visible imagery.
- On standard infrared images, thin high clouds appear white because their tops are very cold (they are high in the atmosphere).
- On standard water vapor images, thin high clouds often appear white (sometimes a dull white or a very light gray) because their cold, high-altitude ice crystals emit just enough radiation at wavelengths near 6.7 microns to be detected by the satellite.

Clouds of Vertical Development (such as cumulonimbus)

- From the ground, the base of a cumulonimbus cloud usually appears very dark, which means that little visible light is getting through. Thus, such thick clouds must be reflecting (back-scattering) a large amount of visible light back to space, so tall cumulonimbus clouds appear bright white on visible images.
- The high tops of cumulonimbus clouds are cold and thus appear bright white on standard infrared imagery.
- The high tops of cumulonimbus clouds appear white on standard water vapor images because their cold, high-altitude ice crystals emit just enough radiation at wavelengths near 6.7 microns to be detected by the satellite.

"took a hit" as drier air eroded some of the thunderstorms in this sector of the hurricane. Forecasters at the National Hurricane Center in Miami, FL routinely look at enhanced water vapor imagery to better assess the current and future state of tropical systems. Indeed, these forecasters know that dry air in the upper half of the troposphere is poison to moisture-fueled hurricanes.

Water vapor imagery has another, arguably more crucial, use that derives from its ability to detect mid- and upper-level water vapor (which, remember, is invisible). By animating a sequence of water vapor images, meteorologists can get a sense of the direction and speed of mid- and upper-tropospheric winds, even in the absence of clouds. This capability is particularly important to tropical forecasters because these high-altitude winds can control the movement and strength of tropical storms and hurricanes. Often, during hurricane season, tropical systems are the only clouds over large stretches of tropical oceans, so there aren't many extensive cloud masses to give away the direction of upper-level winds. And there aren't many weather observers over the ocean to help—just a few ships, buoys, and airplanes. Without animations of water

vapor imagery, forecasters trying to assess the movement of tropical systems would have to rely solely on computer simulations.

Figure 5.18 is a water vapor image at 21 UTC on September 10, 2004, with wind barbs superimposed. These winds were derived by following features on sequences of water vapor imagery around that time to get a sense for the direction and speed of air motions at various levels (the different colors indicate different altitudes in the middle to upper troposphere). In this case, note the ghostly outline of Hurricane Ivan south of Cuba—can you see how nearly all of the upper-level winds appear to flow outward and away from the center? Such a divergent pattern of high-altitude winds is common in the vicinity of strong hurricanes because it helps to remove air from near the storm's center, keeping surface pressures low (hurricanes are low-pressure centers—more to come in Chapters 6 and 11).

Once moisture-laden tropical systems reach land, forecasters focus on where and how hard it's raining, using an active remote sensor called radar. Before we delve into the topic of radar, we direct your attention to Table 5.1, which is a summary of the guidelines you can use for interpreting satellite imagery.

FIGURE 5.18 Satellite-derived wind barbs superimposed on a water vapor image at 21 UTC on September 10, 2004. At the time, Hurricane Ivan (seen near the center of the image) was a Category-4 hurricane with 120-knot sustained winds (courtesy of the Cooperative Institute for Meteorological Satellite Studies at the University of Wisconsin).

RADAR: ECHOES FROM A STORM

Besides the vigilant satellite sentinels in space, meteorologists also rely on ground-based remote-sensing instruments that use invisible radiation to scan and probe the atmosphere. Of all these earthly electromagnetic endeavors, detecting precipitation is one of the most crucial to forecasters. Indeed, radar displays of precipitation such as the one shown in Figure 5.19 are not only staples of modern television weathercasts, but also routinely accessible to anyone with an internet connection or smartphone. This radar image, from 00 UTC on April 28, 2011, shows several broken lines of severe thunderstorms (in red) that produced killer tornadoes in the southeastern United States, particularly in Alabama, Mississippi, and Tennessee, during one of the largest and deadliest tornado outbreaks in recorded history. These thunderstorms also produced hail and heavy rainfall, while other colors on this radar image mark areas where rain of more modest intensity fell.

Given the easy accessibility of radar images these days and their colorful presentation, you might think that radar is a recently developed technology. As it turns out, radar has a long and illustrious history.

Reflectivity Mode: Boomerang Microwaves

During the 1920s and 1930s, ships navigating the Potomac River near Washington, DC, would inadvertently disrupt radio communication signals by passing between a transmitter on one side of the Potomac and a receiver on the other side. This radio interference eventually led to the realization that radio waves (a form of electromagnetic radiation) could be used to detect ships and airplanes, giving the military a distinct advantage during times of war. Early in World War II, the United States joined Great Britain in a cooperative effort to develop radar, which is an acronym for RAdio Detection And Ranging. Simply stated, radar detects objects and then determines how far the objects lie from the transmitter of the radio waves. Radar was used successfully in World War II, especially by the Allied Forces, though primarily for military, not weather, purposes. Nonetheless, when important weather was occurring, radar did get a meteorological workout (see Figure 5.20), though don't be fooled by the look of this historical radar image—it's not a satellite image—in the old days of radar, black and white ruled.

Given that radio waves have long wavelengths (typically 100 cm or longer), they required large transmitting antennae, which were relatively inaccurate

FIGURE 5.19 A mosaic of radar images at 00 UTC on April 28, 2011, during one of the largest and deadliest tornado outbreaks in recent history. Many of the areas shown in red were severe thunderstorms producing tornadoes and large hail from New York to Mississippi (courtesy of NOAA).

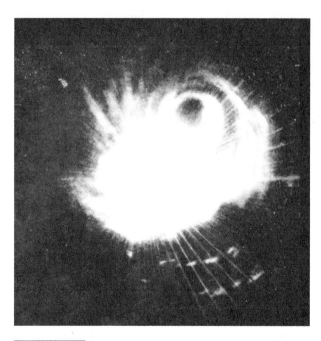

FIGURE 5.20 On December 18, 1944, a U.S. Navy ship captured this radar image of a typhoon (the western Pacific version of a hurricane) near the Philippine Islands. It was only the second tropical system ever observed on radar (courtesy of NOAA).

in fixing the positions of targets. However, rapid breakthroughs in technology allowed shorter-wavelength microwave radiation to replace the radio waves, paving the way for lightweight radars with much smaller antennae. Radars using shorter wavelengths and smaller antennae proved to be much more precise in pinpointing targets, providing clear advantage when the task at hand called for detecting enemy aircraft. After the war, surplus radar equipment was put up for sale to civilians. Some visionary weather enthusiasts bought military radars and aimed them at the sky, starting an era of radar research that would culminate in the sophisticated, high-tech, colorized displays available today. Like satellite imagery, radar is based on meteorologist's ability to "see" invisible radiation.

The first national network of weather radars became operational in the late 1950s. Compared to the nifty modern radar displays, early images from the WSR-57s (Weather Surveillance Radar, circa 1957) now seem like they came from the Dark Ages; they were black and white and static (no loops). Different intensities of rain were displayed in various shades of gray, and meteorologists outlined storm locations with a grease pencil on a plastic overlay. Using multiple overlays, forecasters could estimate the direction and speed of movement of the precipitation. Even though the coverage was rather limited in some areas of the country, and the radars had great difficulty detecting light precipitation (especially snow), these WSR-57 radars provided meteorologists with an unprecedented tool for short-term forecasting. The National Weather Service updated many of the radars in the 1970s, when the demand for animations and color radar images increased.

The current generation of government-run weather radar, and the software and hardware that interpret the radar data, are collectively known as NEXRAD, for NEXt Generation RADar. Weather forecasters often refer to one of these radars as a WSR-88D. Again, "WSR" is short for Weather Surveillance Radar, while the "88" refers to the year this type of radar became operational. The "D" stands for "Doppler," indicating the radar's added capability (over previous types of weather radar) to measure horizontal wind speed and direction relative to the radar. A discussion of this enhanced radar utility will wait until later in this section.

More than 160 NEXRAD systems are deployed throughout the United States and selected territories (including Puerto Rico and Guam). Each dot in Figure 5.21a marks the location of a WSR-88D, while Figure 5.21b shows a WSR-88D. There are also many privately operated Doppler radars, some owned by television stations. If forecasters require a wider view of precipitation (such as the entire Northeast or the entire country), radar data from multiple WSR-88Ds must be combined into one larger mosaic image. That's how the radar image in Figure 5.19 was created.

Weather radars are *pulsed,* which means that the radar transmits and receives short pulses of electromagnetic radiation. A pulse travels through the atmosphere and strikes airborne "targets" (such as raindrops, snowflakes, sleet, hail, birds, or insects). The targets then backscatter some of the transmitted energy to the radar's receiver (the targets scatter energy in all directions, but only the back-scattered component matters to radar imagery). The roundtrip time for the signal is typically very short, on the order of a thousandth of a second. Knowing this travel time and the velocity of the back-scattered energy (the speed of light!), meteorologists can determine a target's distance from the radar: travel time, multiplied by the speed of light, divided by two. The radar rotates through 360 degrees so it can detect all the precipitation falling around the radar site.

Initially, the radar beam's angle above the horizon, or **elevation angle**, is very small, approximately half a degree (just above the horizon). After the radar rotates through 360°, this elevation angle is adjusted upward, typically to 1.5°. Depending on the type of weather situation, a Doppler radar may use as many as 14 different elevation angles, the maximum being 19.5°. In this way, the radar captures the vertical structure of precipitation.

National Doppler Radar Sites

Alaska

Hawaii

(a) Guam

Puerto Rico

FIGURE 5.21 (a) Locations of the government-run NEXRAD Doppler radars in the United States and Puerto Rico; (b) A National Weather Service WSR-88D. The radar antenna is housed inside the large white dome (courtesy of the National Weather Service).

(b)

Now let's look more closely at the energy that is back-scattered to the radar. In a nutshell, the radar measures the rate of energy return, or power, which gets converted into a number called the radar reflectivity factor, commonly shortened to simply **reflectivity**. Computer software then creates images of reflectivity such as the one shown in Figure 5.22, which is centered on the WSR-88D at Mobile, AL, on a rainy March day in 2015. A single WSR-88D has a maximum range of about 240 km (about 150 mi), a distance indicated by the circle near the edge of the image. Figure 5.22a uses reflectivity values from only a single elevation angle (in this case, half a degree), a product that meteorologists call **base reflectivity**. In contrast, a **composite reflectivity** image shows the maximum reflectivity value measured from all elevation angles—Figure 5.22b is the corresponding composite reflectivity image to Figure 5.22a. Comparing the two images, note that reflectivity values higher than 35 dBZ, for example (the yellows, oranges, and reds), are more expansive on the composite reflectivity than on the base reflectivity. That's not surprising because the composite reflectivity is the maximum value chosen from a set of reflectivity values at different elevation angles.

So, what do all those colors in the legends of Figure 5.22 represent? It turns out that the power of back-scattered energy from precipitation has a very large range of values. For example, the power returned from softball-size hail is many orders of magnitude (multiples-of-ten) greater than the return from drizzle. To shrink the range of possibilities, meteorologists express reflectivity using a measure of this order of magnitude, a unit called dBZ (which stands for "decibels of Z," where "Z" stands for reflectivity and a "bel" is a unit of power ratio). This unit, dBZ, is used on radar reflectivity images, including those we've already shown in Figures 5.19 and 5.22. As a general guideline, the reflectivity of rain typically varies from 20 dBZ to 50 dBZ. And the larger the dBZ value, the harder the rain is falling and, thus, the greater the rainfall rate. Because it has so many practical applications (for example, forecasting flash floods), let's explore this last idea a bit more closely.

The power associated with the back-scattered energy that returns to the radar receiver depends on the number, size and composition of the targets. Thus, in the case of raindrops, the power of the returning signal provides meteorologists with an estimate of how hard it's raining (rain rates are commonly expressed in units such as in/hr or mm/hr). The working assumption here is that rainfall rates are high when large populations of relatively large raindrops (a lot of water, in other words) are detected by radar as these drops fall through a cloud. Thus, we make the assumption that the relatively high intensity of returning energy, or high reflectivity, corresponds to high rainfall rates. In Table 5.2, we show some dBZ markers to help you estimate rain rates from reflectivity images. We caution you that the relationship between dBZ and rainfall rate is quite variable; for example, in tropical systems, the rainfall rates

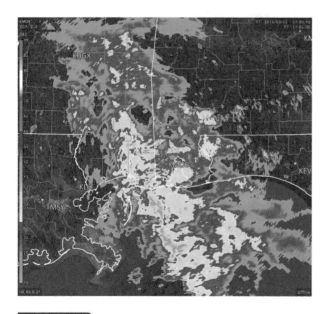
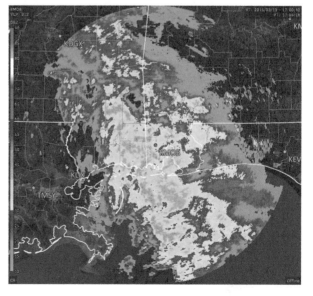

FIGURE 5.22 Radar reflectivity imagery from the WSR-88D at Mobile, AL, from around 1700 UTC on March 13, 2015: (a) Base reflectivity (at 0.5° elevation angle); (b) Composite reflectivity (courtesy of the National Weather Service).

TABLE 5.2	General correspondence between dBZ and rain rate.	
dBZ	Rain Rate (in/hr)	General Qualifier
65	16+	
60	8.00	
55	4.00	Extreme
52	2.50	
47	1.25	Heavy
41	0.50	
36	0.25	Moderate
30	0.10	
20	Trace	Light

"multi-sensor" image, meaning that data came from radar, precipitation gauges, and satellite estimates. Combined with other heavy rain that fell on adjacent days, this drenching from Tropical Storm Lee led to an historic crest of the Susquehanna River in Binghamton, NY (see Figure 5.24).

The presence of large hail in thunderstorms can complicate the process of inferring rainfall rates from radar reflectivity. Typically, radar reflectivity from a thunderstorm is greatest in the middle levels of the storm. That's because large hailstones start to melt as they fall earthward into air whose temperature is higher than

FIGURE 5.24 An aerial photograph of Binghamton, NY on September 7, 2011, shows a floodwall protecting a hospital from the swollen Susquehanna River during Tropical Storm Lee. The river reached a record crest of 25.73 ft the next day – flood stage there is 14 ft (courtesy of FEMA).

could be double what is given in Table 5.2, especially at the higher dBZ values. Further muddying the waters, values of 55 dBZ or higher may be associated with hail (details forthcoming).

Over the lifetime of a rainstorm, computers routinely convert radar reflectivity and the corresponding rainfall rates into an estimate of the storm's precipitation total. Figure 5.23 shows the estimated 24-hour rainfall ending at 12 UTC on September 8, 2011, over the Eastern Great Lakes and Mid-Atlantic, a period when Tropical Storm Lee soaked parts of Virginia, Maryland, Pennsylvania, and New York with more than six inches of rain. We note that Figure 5.23 is a

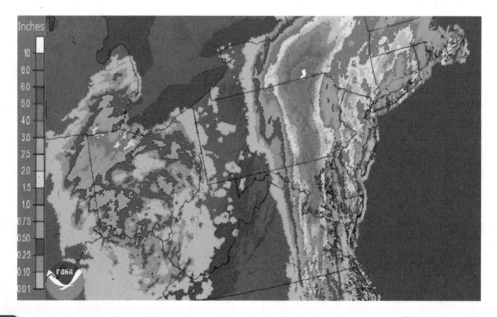

FIGURE 5.23 A multi-sensor estimate of 24-hour rainfall ending at 12 UTC on September 8, 2011, shows a swath of heavy rain associated with Tropical Storm Lee from Virginia to New York, with some areas receiving in excess of eight inches (courtesy of NOAA).

0°C (32°F), the melting point of ice. Covered with a film of meltwater, these large hailstones look like giant raindrops to the radar (more on "wet ice" and radar reflectivity in just a moment). So when large, "wet hailstones" are present in thunderstorms, rainfall rates inferred from the very high reflectivity of what the radar perceives as "giant raindrops" are unrealistic.

As an example, consider Figure 5.25 which shows reflectivity images of a powerful thunderstorm in central South Dakota on July 23, 2010, at four different elevation angles. This thunderstorm produced the largest (8 inches in diameter) and heaviest (1 pound, 15 ounces) hailstone ever measured in the United States. This giant was recovered at Vivian, SD about 50 km (30 mi) south of Pierre, the state capital. In Figure 5.25, values greater than approximately

60 dBZ, depicted here in purple, mark the position of large wet hailstones. This four-panel display is an excellent example of how multiple elevation angles can be used to infer the vertical structure of precipitation. Our main point here is that very high reflectivity does not guarantee that it's raining really hard. Rather, extremely large values of dBZ often signal the presence of wet hail.

During the cold season, the composition of the radar targets also plays a key role in reflectivity. For instance, the power of the return signal from raindrops is approximately five times greater than the return from snowflakes of comparable volume. In part, that's because there are usually tiny "holes" in snowflakes (such as the one shown in Figure 5.26). Forecasters worth their salt always keep the lower

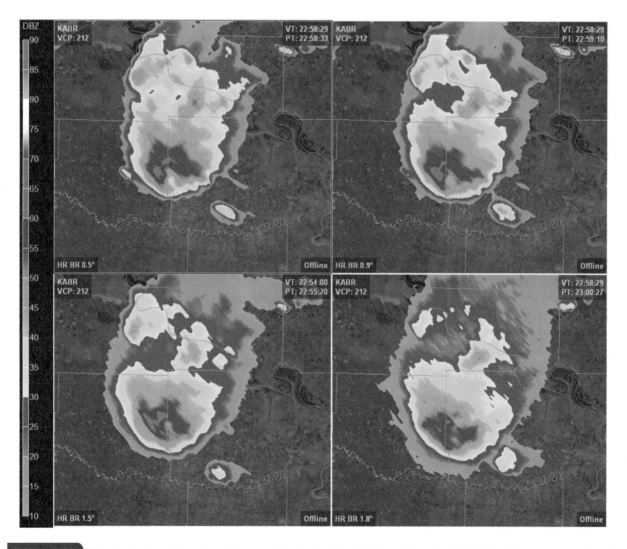

FIGURE 5.25 The thunderstorm that produced the record-breaking hailstone in Vivian, SD on July 23, 2010 is captured on radar reflectivity imagery at four different elevation angles (clockwise from upper left, 0.5°, 0.9°, 1.8°, and 1.5°). The dBZ values greater than approximately 60 dBZ, shown here in purple, mark the position of large wet hailstones (courtesy of National Weather Service, Aberdeen, SD).

FIGURE 5.26   A snow crystal (courtesy of Kenneth G. Libbrecht, California Institute of Technology).

return signal from snow in mind. Otherwise, it will be easier to underestimate the areal coverage and intensity of snowstorms.

There's another reason that the reflectivity of snow can be misleadingly low. Consider that layered, precipitation-bearing **nimbostratus clouds** are often not very tall, especially in winter ("nimbus" is a Latin word meaning rainstorm or rain cloud, so "nimbo" or "nimbus" in a cloud name indicates precipitation). If such layered clouds are relatively far from the radar, then the radar beam will often only intercept the drier upper portions of the clouds (or miss the clouds altogether), simply because the radar beam is slanted upward and thus its altitude increases with increasing

distance from the radar. Figure 5.27 shows the radar beam overshooting a shallow, snow-bearing cloud near the radar's maximum range. The bottom line here is that you need to be careful when interpreting radar reflectivity images during the cold season.

However, there is a type of radar image that makes the interpretation easier—an example is shown in Figure 5.28. Unlike the radar reflectivity images we've shown you so far, this "winter radar" image is primarily used to depict precipitation type. In this example, blue means snow, pink shades are used for mixed precipitation (which includes sleet and freezing rain), while the remaining colors (such as green, yellow, and orange) correspond to rain. For snow and mixed precipitation, the darker the corresponding color, the heavier the precipitation. Although surface observations of precipitation type are used to create colorful images such as Figure 5.28, gaps between observing sites mean that winter radar can be somewhat incomplete. To fill in the gaps, surface analyses of temperature and dew point are used in order to make an "educated guess" about the type of precipitation reaching the ground. We recommend that you use these "winter radar" images with a healthy dose of caution, especially near the boundaries between precipitation types, given the imperfect assumptions used to create these displays.

While detecting snow is high on the meteorologist's list of priorities in winter, detecting severe thunderstorms capable of producing tornadoes jumps to the front of the line during spring and summer. Like the "X" that marks the spot on a treasure map, some thunderstorms that spawn twisters leave a telltale reflectivity signature on radar that offers compelling evidence that there is rotation within the storm. And where there's rotation in thunderstorms, there can be tornadoes. Figure 5.29a is the reflectivity image of a severe thunderstorm in central Oklahoma

FIGURE 5.27   The radar beam can completely miss shallow, snow-bearing clouds that lie relatively far away from the radar.

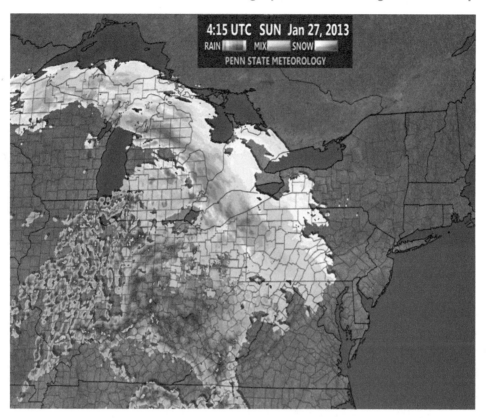

FIGURE 5.28   A "winter radar" image on January 27, 2013. Shades of blue depict snow, shades of pink depict mixed precipitation (including sleet and freezing rain), while other colors indicate rain of various intensities (Data source: WSI Corporation and Penn State University).

on May 20, 2013. About 20 minutes after this image was captured, a powerful tornado with winds near 200 mph slammed into the community of Moore, OK, causing widespread destruction and several dozen fatalities. The curlicue feature (just to the northwest of the dot labeled TOKC), which is on the southwestern flank of the severe thunderstorm, is formally called a **hook echo**. This feature is a clear sign that rotation was occurring within the thunderstorm. However, a hook echo does not necessarily seal the deal on the presence of a tornado. Indeed, weather forecasters require more solid evidence—those clues are found in Figure 5.29b and 5.29c. So let us introduce you to a few other products of a Doppler radar.

Velocity Mode: Building Additional Evidence for Tornado Detection

In the mid-nineteenth century, Johann Christian Doppler discovered an effect that today bears his name, an accomplishment that was acknowledged again in modern times by naming a revolutionary type of radar after him. Doppler's place in scientific history was secured when he noticed a shift in the **frequency** (or pitch) of sound as the source of the sound moved away or toward a stationary observer. Sound, like radiation, travels in waves—frequency is simply the number of waves passing a given point per unit time. The classic example of this **Doppler effect** involves a train speeding toward a crossing with its whistle blowing (see Figure 5.30). An observant bystander waiting at the crossing notes that the sound of the whistle is higher in pitch when the train approaches and lower in pitch when the train recedes (we make a distinction here that "pitch" is not the same as "loudness").

The same principle that applies to sound applies to electromagnetic radiation and, thus, to radar meteorology. The frequency of the radiation back-scattered by a moving target is not the same as it would be for a stationary target. Baseball scouts use this principle when they point radar guns at the fast balls of prospective pitchers. The frequency of the returning signal after it bounces off a baseball is quickly calibrated into a speed. Radar guns used by police to catch speeders operate in the same manner.

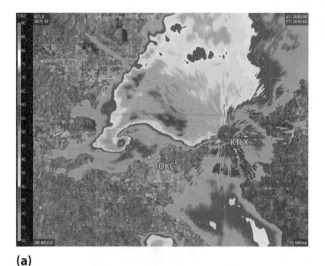

(a)

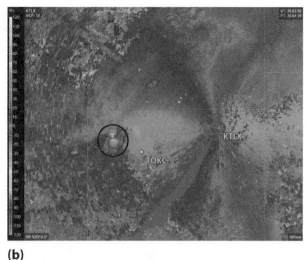

(b)

(c)

FIGURE 5.29   (a) Reflectivity mode of NEXRAD at Oklahoma City, OK, at 20 UTC on May 20, 2013, captures a hook echo associated with a tornadic thunderstorm (the radar location is labeled KTLX – TOKC is a nearby Doppler radar at the Oklahoma City airport). The hook just to the northwest of TOKC indicated rotation within the thunderstorm. A few minutes earlier, the thunderstorm's rotation provided the backdrop for a tornado to spin up. The tornado lasted almost 40 minutes and travel about 17 miles (27 km) on the ground through central Oklahoma; (b) Storm-relative velocity image from the NEXRAD at KTLX at the same time as the reflectivity image; (c) Correlation coefficient from the NEXRAD at KTLX (courtesy National Weather Service).

FIGURE 5.30   As a whistling train approaches a crossing, a stationary observer hears a higher frequency of sound than when the train is departing.

How does this principle apply to detecting tornadoes? First, let's revisit the reflectivity image of the severe thunderstorm in Figure 5.29a. Recall that this radar presentation of the storm showed a hook echo, which is a classic reflectivity feature that indicates

rotation within the thunderstorm. Why does this feature appear on images of radar reflectivity? In a nutshell, radar targets such as raindrops (and sometimes hail) move with the strongly rotating portion of the storm as they fall toward the ground. In turn, these targets act as tracers for the storm's circulating winds, allowing the radar to gather data for imaging the footprint of the storm's rotation.

Let's explore the process by which the velocity mode of Doppler radar measures these winds and displays them. Like a bystander at a train crossing who hears a change in frequency as a whistling train approaches and then passes, Doppler radar has the ability to sense the changes in frequencies between energy back-scattered from raindrops moving toward or away from the radar (see Figure 5.31).

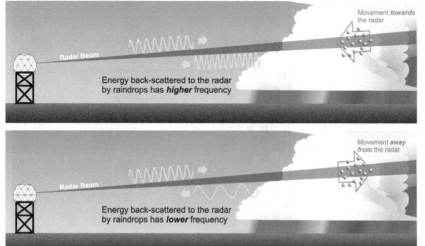

FIGURE 5.31 (Top) If microwaves transmitted toward a precipitating cloud by Doppler radar strike raindrops that are moving toward the radar, then the frequency of returning radiation is higher than the frequency of the initial pulse of energy; (bottom) If Doppler microwaves strike raindrops that are moving away from the radar, then the microwaves return at a lower frequency than the frequency at which they were initially transmitted.

Just like a ship sailing into the wind encounters and then reflects a greater frequency of incoming wind-driven waves, microwaves "reflecting off" raindrops moving toward the radar also return at a higher frequency. And the faster the speed of the ship or raindrop, the higher the frequency of the reflecting waves. Conversely, just like a ship sailing with the wind encounters and then reflects a lower frequency of wind-driven waves, microwaves "reflecting off" raindrops moving away from the radar also return at a lower frequency. Changes in frequencies are then translated, by computer, into wind directions (relative to the radar site) and also into wind speeds.

What do we mean by "relative to the radar site"? A single Doppler radar can only sense air motions directed toward or away from the radar. Such air motions are called **radial velocities** because lines drawn from the radar outward are called **radials**. For example, consider Figure 5.32, where the radar site is in the middle (red dot) and the wind at four points near the radar is represented by thick white arrows (notice that in this diagram, the wind is not blowing <u>directly</u> along a radial at any of the locations). The Doppler radar will only be able to detect the part (component) of the wind that blows along a radial. In this case, the radial velocities at each point are represented by yellow arrows. At points B and C, the radial velocity is directed toward the radar, while at point A the radial velocity is directed away from the radar. Notice that as the angle between the wind and the local radial gets smaller, the magnitude of the radial velocity gets larger. Finally, at point D, the wind is blowing perpendicular to the radial, so the radial velocity there would be zero.

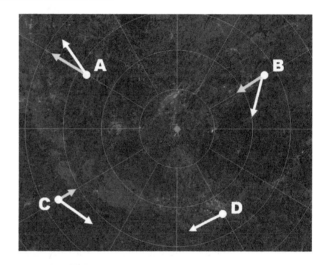

FIGURE 5.32 The white arrows at points A, B, C, and D represent sample winds at these points. The corresponding yellow arrows show the radial velocity at each point—essentially the component of the wind along the radial. The closer to parallel the wind is to the radial, the larger the magnitude of the radial velocity. Traditionally, a radial velocity away from the radar is positive, while a radial velocity toward the radar is negative.

So how are these radial velocities used to detect rotation within a thunderstorm? Figure 5.29b is a display of Doppler radial velocities, known as a **velocity image**, valid at the same time as the reflectivity image in Figure 5.29a (strictly speaking, this type of velocity image is a **storm-relative velocity image** because the motion of the thunderstorm itself has been subtracted out—essentially, all that's left is motion inside the thunderstorm that helps meteorologists better visualize the rotation). In this image, colors indicate motion toward (inbound) or away from (outbound) the radar at KTLX, measured in knots. Negative radial velocities, corresponding to motion toward the radar, are represented primarily by greens and blues, while positive radial velocities, corresponding to motion away from the radar, are represented primarily by red, pink, and orange colors. Note the juxtaposition of green right next to a few pixels of pink, just to the northwest of TOKC (inside the circle). This **velocity couplet** indicates a relatively small area in which fast radial velocities toward the radar (green) are adjacent to fast radial velocities away from the radar (pink). Note that the velocity couplet matches perfectly with the location of the hook echo on the reflectivity image—no coincidence there! In effect, this velocity couplet pinpointed the rotating part of the thunderstorm that spawned the tornado (essentially, meteorologists "fill in" the rest of the rotation to complete a circle—see Figure 5.33 for a schematic). Forecasters look for such velocity couplets during outbreaks of severe weather as telltale signs of rotation in thunderstorms, prompting the National Weather Service to issue tornado warnings. In fact, based on similar reflectivity and velocity images, a tornado warning was issued 20 minutes earlier at 2:40 PM CDT for several counties in central Oklahoma (see Table 5.3).

During the period 1986 to 2012 (which includes the time when the national Doppler radar network was deployed), the probability that meteorologists in the National Weather Service could detect a tornado before it formed increased from about 25 percent to 75 percent, while the average warning time for tornadoes increased from around 4 minutes to about 14 minutes. In the face of deadly tornadoes, every precious minute counts. Still, roughly three out of every four tornado

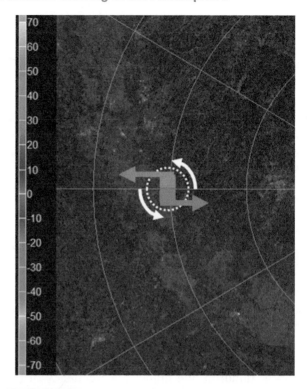

FIGURE 5.33 Schematic showing an idealized velocity couplet. Fast radial velocities toward the radar (represented by green) adjacent to fast radial velocities away from the radar (represented by red) are consistent with rotation. Meteorologists effectively "fill in" the remainder of the circulation (white arrows).

warnings issued by the National Weather Service are false alarms—despite rotation detected by Doppler radar, tornadoes do not form. So clearly, the combination of reflectivity and velocity imagery from Doppler radar is powerful, but it only goes so far.

TABLE 5.3 Tornado warning issued by the National Weather Service in Norman, OK, at 2:40 PM CDT on May 20, 2013.

National Weather Service Norman Ok

240 PM CDT MON MAY 20 2013

THE NATIONAL WEATHER SERVICE IN NORMAN HAS ISSUED A TORNADO WARNING FOR . . .

NORTHWESTERN MCCLAIN COUNTY IN CENTRAL OKLAHOMA . . .

SOUTHERN OKLAHOMA COUNTY IN CENTRAL OKLAHOMA . . .

NORTHEASTERN GRADY COUNTY IN CENTRAL OKLAHOMA . . .

NORTHERN CLEVELAND COUNTY IN CENTRAL OKLAHOMA . . .

• UNTIL 315 PM CDT

• AT 238 PM CDT . . . NATIONAL WEATHER SERVICE

METEOROLOGISTS DETECTED A SEVERE THUNDERSTORM CAPABLE OF PRODUCING A TORNADO. THIS DANGEROUS STORM WAS LOCATED NEAR NEWCASTLE...AND MOVING EAST AT 20 MPH. IN ADDITION TO A TORNADO . . . LARGE DAMAGING HAIL UP TO GOLF BALL SIZE IS EXPECTED WITH THIS STORM.

• LOCATIONS IMPACTED INCLUDE . . . NORMAN . . . MOORE . . . NEWCASTLE . . . BRIDGE CREEK AND VALLEY

Dual-Polarization Radar: Entering Another Dimension

In 2012, the National Weather Service began to upgrade its fleet of Doppler radars with a new capability called **dual polarization**. For starters, "dual pol" provides forecasters with the means to identify airborne debris from tornadoes. Used in conjunction with reflectivity and storm-relative velocities, dual polarization helps to reduce the false alarm rate (more on this in a moment).

Dual polarization also provides the means by which forecasters can distinguish between various types of precipitation particles (hailstones and raindrops, for example, or snowflakes and ice pellets). The bottom line here is that dual pol is an important tool that can be used to determine precipitation type during winter and to improve radar estimates of rainfall all year round.

So what is dual polarization, you ask? Remember that conventional Doppler radar sends out pulses of microwave radiation—you can think of these pulses as waves that propagate in the horizontal dimension (see Figure 5.34 and focus on the blue wave). With this perspective, conventional Doppler radar can only provide information about the size and shape of targets in one dimension (note the blue lines drawn across the precipitation particles in the schematic). In contrast, dual-pol radar transmits pulses of microwave energy in both horizontal and vertical dimensions (again, see Figure 5.34—the red wave represents this extra dimension). In this way,

dual-pol radar provides information about both the horizontal and vertical configuration of targets, giving meteorologists a better sense for their size and shape. That's a pretty valuable tool when forecasters are trying to distinguish between hailstones and large raindrops in strong thunderstorms, or between snowflakes and ice pellets in a winter storm, or when they're looking for additional, real-time corroboration that a tornado is causing damage.

How can the dual-pol capability provide evidence that a tornado has formed? The underlying premise is that twisters spew debris into the air—leaves, branches, trees, and pieces of buildings, for example. Such debris, which are much larger than most precipitation particles, vary in size and shape. Stated another way, there's not much "correlation" between the sizes and shapes of tornado debris. On the flip side, the sizes and shapes of raindrops in a thunderstorm are much more uniform and are more highly "correlated." When dual polarization scans a tornadic thunderstorm, it can essentially distinguish volumes of air with targets that are highly correlated in size and shape from those volumes of air containing targets that have a low correlation in size and shape.

Dual polarization radar quantifies this idea in a product called, appropriately, the **correlation coefficient** (CC), a number that varies from 0 to 100, with 100 representing a volume of air with perfectly uniform targets (sometimes a range of 0 to 1 is used).

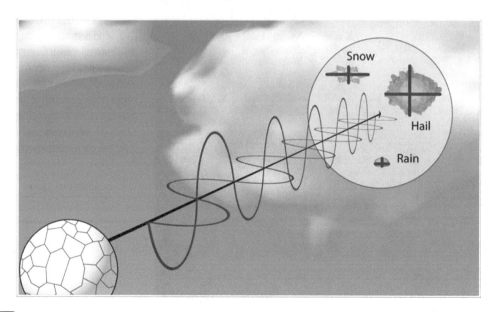

FIGURE 5.34 Conventional Doppler radar produces pulses, or waves, of radiation that allow only the horizontal dimension of precipitation particles to be sampled (see the blue wave and the corresponding blue lines across the targets). The upgrade to dual polarization (represented by the red wave and the red lines on the targets) allows the vertical dimension to be sampled as well, allowing meteorologists to better distinguish between different sizes and shapes of targets.

An example of this dual-pol product is shown in Figure 5.29c—looks like a mess, right? But let's identify some familiar markers. First, note the legend—values near 100 (corresponding to red and purple colors) indicate volumes of air filled with relatively uniform particles. Now compare the reflectivity image in Figure 5.29a with the correlation coefficient in Figure 5.29c—do you see a resemblance between where it's raining and the relatively large CC values? You should! This makes sense because those areas are filled with primarily raindrops. Now look at the very high reflectivity values in the hook echo on Figure 5.29a, and find the corresponding area on Figure 5.29c—the CC colors are now blue and gray, telling us that the targets there are not very uniform. What could possibly give both high reflectivity and a relatively low correlation coefficient, and be associated with a hook echo? It is airborne tornado debris, likely a jumble of tree and building pieces of widely differing shapes and sizes. The roughly circular area of relatively high reflectivity and relatively low CC associated with a tornado is sometimes referred to as a **debris ball**.

We should point out that most twisters do not have a debris ball—many tornadoes are so weak and short-lived, or form in open fields, that they loft little, if any debris. Also, most of the airborne debris from a tornado stays relatively close to the ground, so the radar beam may overshoot it. Still, the violent tornadoes that cause most of the loss of life and property typically pick up enough debris to register on a reflectivity image as a debris ball, especially in densely populated areas. In combination with a velocity couplet and a corresponding pocket of low correlation coefficient, the presence of a tornado is all but confirmed! With this in mind, we hope you can appreciate how using all three Doppler radar products shown in Figure 5.29 is the most effective plan for meteorologists tracking these destructive storms.

We're not done yet with Doppler radar. We have one more product we'd like to introduce.

Velocity Azimuth Display: Radar on the Spot

So far, the radar images we've shown are horizontal maps on which radar data are plotted over an area around the radar site at a given time. Now we would like to show radar data in a different format. It turns out that a Doppler radar can also produce vertical plots of weather data—specifically wind speed and direction—at various altitudes above the radar. By stringing together many such observations taken at different times, a diagram such as Figure 5.35, known as a VAD Wind Profile (VWP),

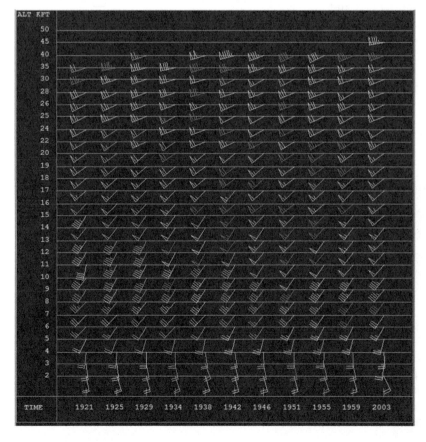

FIGURE 5.35 A VAD Wind Profile from Norman, OK shows wind barbs at 1000-ft intervals every 4 minutes from 1921 to 2003 UTC on May 20, 2013. Each wind barb actually characterizes a set of several measurements from which a representative observation was chosen, so each wind barb has some inherent error. The colors provide a sense of this error—red indicates a relatively high potential error, yellow a fairly reliable measurement, and green the most reliable measurements.

can be created. For the record, VAD stands for **velocity azimuth display**.

The "Wind Profile" in the name of this radar product is pretty obvious when you look at Figure 5.35—it shows wind barbs at various altitudes (vertical axis in thousands of feet) and times (horizontal axis in UTC) around and above the radar at Norman, OK, from 1921 to 2003 UTC on May 20, 2013, a period close to the time of the development of the Moore tornado. To get your bearings, the most recent time (2003 UTC) is traditionally found on the right side of VAD Wind Profiles. To interpret wind speeds and wind directions, use the conventions that you learned for the surface station models in Chapter 1.

The winds shown in Figure 5.35 are not instantaneous winds—instead, for each of the wind barbs, the Doppler radar took several measurements during a relatively short time period. After the wind data were collected, a representative wind direction and speed were then selected, so each wind barb has some inherent error—the colors provide a sense of this error. For example, notice that the wind barb at 18,000 feet at 1942 UTC in Figure 5.35 is red—this indicates that the wind changed dramatically during the short time period it was being measured, so the southwest wind at 60 kt has a relatively high potential error. A yellow wind barb, such as the one at 10,000 feet at 1951 UTC, indicates that the south-southwest wind

at 50 kt is a fairly reliable measurement, while green is used to indicate the most reliable readings. Lesson learned: some winds in a VAD Wind Profile should be used with caution because they can be unreliable.

Despite potential issues with reliability, the vertical profile of winds from the radar at Norman, OK, showed a significant change in wind direction and speed with altitude. For example, at 1946 UTC, winds at 2000 ft were southerly at 20 kt, and then transitioned to southwesterly at 50 kt at 11,000 ft. This vertical change in wind direction and speed is called vertical wind shear (more on this topic later). Significant vertical wind shear, coupled with a "streak" of fast winds in the middle troposphere (say, at 16000–18000 ft), can set the stage for the development of long-lived thunderstorms with rotating updrafts (called **supercells**) that have the potential to spawn tornadoes. And on this day, they did.

Before we address the connection between an upper-air "streak" of fast winds and severe thunderstorms, we have to lay the groundwork for interpreting weather maps and analyses at various altitudes above the ground. In Chapter 6, we introduce you to the patterns of pressure and winds on surface weather maps. Then, in Chapter 7, we will extend those concepts to the patterns of pressure and winds aloft.

Focus on Optics

Why Is Snow White?

Tiny ice crystals are the building blocks of snow-flakes (recall Figure 5.26). It should come as no surprise that a considerable amount of light is back-scattered when it strikes these mirror-like ice crystals—recall from Chapter 2 that snow has a relatively high albedo. So when a surface is covered by a freshly fallen layer of snow, there's a multitude of opportunities for visible light to be reflected. With all wavelengths of visible light back-scattered about equally, snow looks white.

But not always. Have you ever burrowed into a snow drift after a winter storm and noticed that the snow in these burrows takes on a bluish hue (see Figure 5.36)? It turns out that ice crystals absorb some visible light, particularly the longer wavelengths (such as those corresponding to red and orange). With that in mind, consider light that enters a hole in a snow bank. Surrounded on nearly all sides by snow, this light will undergo many more scatterings (each an opportunity for absorption) than what occurs at the surface of snow. With so many more opportunities for longer

wavelengths to be absorbed by snow on the sides of the hole, the light that eventually emerges from the hole has preferentially more of the shorter wavelengths in it. In short, it looks blue.

We hope you get our drift.

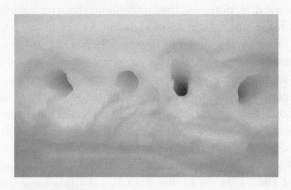

FIGURE 5.36 Holes burrowed into snow can look blue because the water in snow absorbs most of the longer wavelengths of visible radiation (courtesy of Kenneth G. Libbrecht, California Institute of Technology).

Surface Patterns of Pressure and Wind

LEARNING OBJECTIVES

After reading this chapter, students will:

- Understand that the total weight of an air column above a point is a proxy for the pressure at that point
- Understand the importance of using sea-level pressure instead of surface pressure to ascertain the high- and low-pressure systems that govern our weather
- Be able to identify troughs, ridges and closed high- and low-pressure systems on analyses of sea-level pressure

- Be able to identify regions with strong and weak pressure gradients and to qualitatively infer wind speeds
- Understand that the Coriolis force is an "apparent" force that arises from our earthly rotating frame of reference
- Be able to explain the counterclockwise (clockwise) circulation of air around low- (high-) pressure systems in the Northern Hemisphere
- Gain an appreciation for why the circulation around low- (high-) pressure systems in the Southern Hemisphere is opposite that of their counterparts in the Northern Hemisphere
- Understand why there must be divergence (convergence) aloft that offsets surface convergence (divergence) in order for a low- (high-) pressure system to develop or strengthen
- Understand why fronts are located in troughs and why they therefore mark elongated areas of surface convergence

On November 10, 1975, the Edmund Fitzgerald, a sturdy 222-m (729-ft) long and 23-m (75-ft) wide ship that was loaded to the hilt with iron ore, sank to the bottom of Lake Superior after surrendering to battering, angry waves stirred up by an intense storm (see Figure 6.1). The saga of the Edmund Fitzgerald became legendary in a popular song recorded by Gordon Lightfoot. His lyrics tell of the infamous "gales of November," a reference to the winds generated by intense low-pressure systems that often rage over the Great Lakes during the eleventh month.

Twenty-three years later to the day, another "gale of November," following a similar path to the storm that sunk the Edmund Fitzgerald, generated hurricane-force winds across parts of the Great Lakes (see Figure 6.2). At night, a buoy on Lake Michigan measured giant waves over 6 m (20 ft) high, while barometers in Minnesota nose-dived to set a new all-time state record for lowest air pressure.

If you have a barometer in your home, you probably are familiar with "inches of mercury" as a common unit of air pressure. If so, you may already know that decreasing air pressure often heralds stormy weather, while increasing air pressure is often the precursor of "fair weather" (although you might think that "fair" is a word that meteorologists use to hedge their predictions, "fair" generically describes a weather pattern characterized by a lack of precipitation, a sky whose coverage by low clouds is less than four-tenths, and the absence of any other extreme conditions of cloudiness, visibility, or wind).

FIGURE 6.1 The Edmund Fitzgerald in May 1975 (courtesy of Bob Campbell, Grand Ledge, MI).

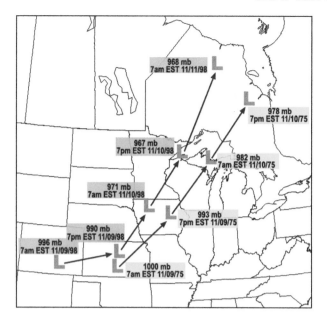

FIGURE 6.2 On November 10 in both 1975 and 1998, fierce gales lashed parts of the western Great Lakes as centers of low air pressure (marked by an "L") moved northeastward from the Great Plains. You will learn later in this chapter that the strongest winds associated with such low-pressure systems occur where the pressure gradient is largest (courtesy of Don Rolfson, National Weather Service, Marquette, Michigan).

With guidelines of "stormy" and "fair" on your home barometer in hand, you are well on your way to becoming an apprentice forecaster because no principle is more fundamental to weather prediction than the relationship between changes in air pressure over time and changes in the weather. In other words, it is crucial that forecasters know the rate at

which air pressure increases or decreases (sometimes expressed as "barometers rising or falling," with "rapidly" or "slowly" tacked on for good measure).

Relatively small variations of air pressure over horizontal distances are also crucial to weather forecasting, as we will soon learn. But first things first—we need a working model for pressure that will allow us to more easily understand the changes in air pressure, both in time and space, that play such a crucial role in weather.

PRESSURE AS A WEIGHT: HOW TO READ THE SCALES

We begin with a "working definition" of air pressure: The air pressure at any point can be closely approximated by the weight of the column of air that extends from the point to the "top" of the atmosphere (see Figure 6.3a). Note that we assume, for sake of simplicity, that the area of the base of the air column is one square inch.

Let's put this working definition to a test. Is it consistent with the standard definition of air pressure, which is the force (per unit area) exerted by air molecules? If we apply our working definition to an air column that extends upward from the earth's surface, it's pretty clear that the highest air pressure is always found at the ground where the column's weight is greatest. Now hold that thought for a moment. Let's next use the observation that air is compressible—in other words, it can be squeezed into a smaller volume. Just like a stack of facial tissues gets compressed as we place books on top of it, the air near the ground gets

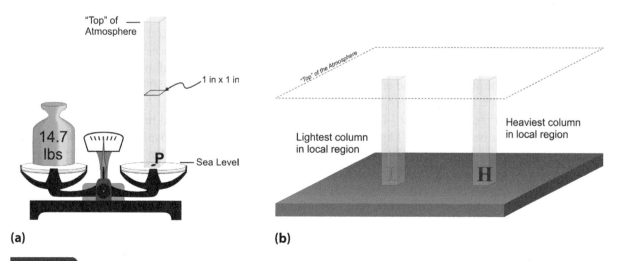

(a)   **(b)**

FIGURE 6.3 (a) One way to think about air pressure at a point P (in this case, P is at sea level) is to weigh a column of air with a cross-sectional area of one square inch. At sea level, it would weigh about 14.7 pounds; (b) The air columns associated with the centers of high and low pressure have the heaviest and lightest weights (respectively) compared to nearby air columns.

squeezed under the weight of all the air molecules piled on top of them. Indeed, the density of the air (mass of air molecules per unit volume) is always greatest near the ground (at sea level, it's about 1.2 kilograms per cubic meter). Talk about a squeeze play!

The bottom line here is that our working definition of air pressure is consistent with the standard definition of force per unit area. The weight of a column of air that extends vertically from the surface to the "top" of the atmosphere is greatest at the ground, where the number of air molecules per unit volume is also the greatest. And the greater the number of air molecules, the greater the force they exert per unit area.

Thus, our working definition of air pressure passes the test! Now applying it, we arrive at the inescapable conclusion that the column of air above a surface low-pressure system (represented by "L" on weather maps) is the lightest column in the surrounding region, while the column of air above a surface high-pressure system (represented by an "H") is the heaviest in the region (see Figure 6.3b).

Using weight as the basis for our working definition of air pressure might lead you to believe that there can be no air pressure on the International Space Station in the near weightlessness of space. Such a "far-out" notion ignores the universal definition of air pressure: force per unit area. There are indeed air molecules on board the Space Station (think of instances when astronauts appear on television without their space helmets—they are breathing air that contains oxygen). These air molecules have mass and speed and therefore exert a force per unit area. In other words, air molecules in the cabin of the Space Station exert air pressure, despite the near weightlessness of space.

Meanwhile, back on planet Earth, let's use our working definition to quantify sea-level pressure. On average, the weight of a column of air that extends from sea level to the top of the atmosphere and whose base has an area of one square inch is about 14.7 pounds. So average sea-level air pressure is about 14.7 pounds per square inch (psi), equivalent to about 1 kg/cm^2 (see Figure 6.3a). Such a pressure exerted on the exposed mercury in Figure 6.4 will cause the mercury to climb approximately 29.85 inches in the air-evacuated tube—thus, 14.7 psi ≈ 29.85 inches of mercury.

For ease of communication, meteorologists call this average sea-level pressure "one atmosphere." Now wait just a darn minute. There can't be two atmospheres of pressure, can there? Yes, under water! Keep in mind that one cubic foot of water weighs

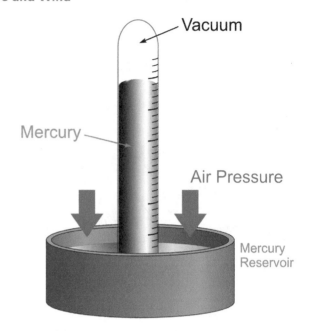

FIGURE 6.4 A mercury barometer. The weight of the mercury in the tube is balanced by the force of the air on the mercury in the dish.

about 62 pounds (1 cubic meter of water weighs 1000 kg). So the pressure exerted on a scuba diver is the weight of the column of water above him plus the weight of the column of air above his underwater position (the weight of the air pales in comparison to the weight of the water). For example, at the bottom of the Marianas Trench in the Pacific Ocean, where depths can reach almost 11 km (7 mi), the water pressure can reach 1100 atmospheres!

These crushing pressures at great depths make such waters off limits to scuba divers. But Alvin, the Deep Submergence Vehicle owned by the U.S. Navy and operated by the Woods Hole Oceanographic Institution (a research institution committed to the study of marine science), can dive to a maximum depth of 4500 m (nearly 15,000 ft). It takes about two hours for Alvin to reach this depth and another two hours to return to the surface.

So referring to "atmospheres of pressure" has its place under the sea. On the surface, "inches of mercury" is the standard choice on home barometers, most of which are aneroid (meaning "without liquid"). A typical aneroid barometer consists of a coil of hollow tubing from which air has been evacuated. The tubing is held apart by a spring. When air pressure increases, the tubing compresses, causing a needle to move. The movement is then calibrated to "inches of mercury."

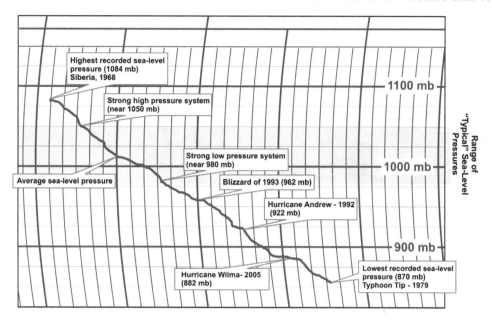

FIGURE 6.5 Typical and extreme values of sea-level pressure, in millibars.

Though "inches" is still a popular unit for air pressure, most meteorologists use a different unit, millibars (mb): For comparison, 1011 mb = 28.85 inches. Figure 6.5 shows average and extreme values of sea-level pressure on Earth, in millibars. On the millibar scale, nearly all sea-level pressures fall between 950 mb and 1050 mb, yielding a sea-level pressure variation across the globe of less than 10 percent. Mundane, day-to-day variations in sea-level pressure are even smaller. The lowest sea-level pressure ever recorded on Earth was measured inside the eye of Typhoon Tip in the western Pacific Ocean on October 14, 1979 (see Figure 6.6).

A careful eye might notice the reference to Siberia at the top of Figure 6.5 as the place where the highest sea-level pressure was measured. Now wait just a minute! Siberia isn't at sea level. What's going on here?

Correction to Sea Level: Getting Rid of Elevation

At locations above sea level, such as the Mile-High city of Denver, CO (see Figure 6.7), barometers measuring the local air pressure can only dream of someday registering pressures near 30 inches. That's because air pressure decreases with increasing altitude. In fact, the home barometer of a family

FIGURE 6.6 A visible satellite image of Typhoon Tip ("typhoon" is the name given to hurricanes that form over the northwest Pacific Ocean). On October 12, 1979, Tip's winds gusted as high as 190 mph as its pressure bottomed out at 870 mb (25.69 inches of mercury), setting the mark for the lowest sea-level pressure ever recorded. Winds circulating counterclockwise around Tip extended approximately 2200 km (1350 mi) in the east-west direction. Had Tip been placed over the continental United States, its wind circulation would have nearly spanned the western half of the country (courtesy of NOAA).

FIGURE 6.7 Denver, CO, lies at an altitude of 1615 m (5300 ft) at the base of the eastern foothills of the Rockies (courtesy of the Denver Metropolitan Convention and Visitors Bureau).

moving from Miami Beach to Denver might read 25.20 inches after unpacking! But what seems like a very low pressure relative to their old home in Miami Beach is no big deal in Denver—the scary drop in pressure is primarily a consequence of a natural loss in weight in the transition from weighing tall air columns over Miami Beach to shorter columns over mile-high Denver (see Figure 6.8).

And the rate of decrease in air pressure with altitude is fairly dramatic (see Figure 6.9). At an altitude of only about 5.5 km (3.4 mi), air pressure decreases to 500 mb, indicating that about half of the total weight of the atmosphere lies in the lowest 5500 m (18,000 ft) or so. If you look closely at Figure 6.9,

you can get a general idea of a typical pressure at Denver—about 850 mb. This is a far cry from average sea-level pressure and the difference is almost entirely due to elevation. To a meteorologist searching for pressure variations caused by the approach and retreat of low- and high-pressure systems, pressure differences due to elevation just muddle the issue.

To see what we mean, consider Figure 6.10a, which shows the pressure measured at the Earth's surface at 12 UTC on January 17, 2015. The large blue and purple blobs in the western United States and interior Mexico indicates pressures less than 840 mb, relatively low pressures that are consistent

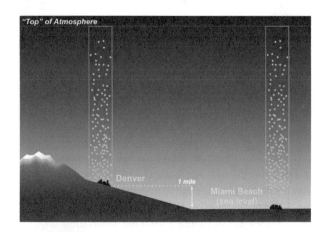

FIGURE 6.8 Comparing an air column above Miami Beach with one above Denver. The column above Denver will always be shorter and contain air of comparatively lower density, so the air pressure will always be lower there.

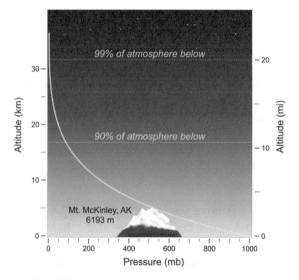

FIGURE 6.9 The decrease of air pressure with increasing altitude.

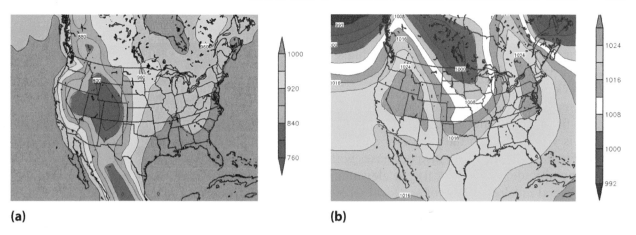

(a)
(b)

FIGURE 6.10 (a) Air pressure measured at the Earth's surface at 12 UTC on January 17, 2015. Lower surface pressures occur in the higher elevations (where air columns are shorter). In general, the highest surface pressure correlate with the lowest elevations (over the oceans, for example); (b) The corresponding chart of sea-level pressure. Given this pattern of sea-level pressure, an "L" (representing a center of low pressure) would be placed inside the 1000-mb isobar in central Canada, while an "H" (representing a center of high pressure) would be placed inside the 1024-mb isobars in the eastern and western United States (courtesy of NOAA).

with the high elevations in these areas. The highest pressures in Figure 6.10a are generally found at the lowest elevations—over the oceans. And unless there are dramatic changes to the laws of nature, maps of surface pressure will always exhibit this pattern—low pressure over the highest elevations and high pressure over the lowest elevations.

But don't the areas of high and low pressure you see on the evening weather map move, changing local weather conditions from day to day? Of course they do. Ultimately, what meteorologists need is a filter that removes the effect of elevation on pressure. Consider Figure 6.10b, which shows the surface pressures in Figure 6.10a after such a filter was applied—in essence, these are the pressures that would result if all points were at sea level. Figure 6.10b shows centers of high sea-level pressure over the Interior West and across the Northeast and Mid-Atlantic, and a center of low sea-level pressure over central Canada (and likely two others just off the northwestern and northeastern edges of the map). These high- and low-pressure systems would be "lost in the shuffle" if meteorologists used only maps of pressure at the Earth's surface. So how do meteorologists transform a map such as Figure 6.10a into one like Figure 6.10b?

In order to filter out the effect of elevation on pressure and thus allow direct comparison of pressure readings from different altitudes, air pressure readings are "corrected" to sea level. In other words, for any observing site located above sea level, the weight of the fictitious column of air that extends from the observing site to sea level is estimated using temperature observations at the station to assign an air density to the phantom column. The fictitious weight is then added to the weight of the real column of air above the observing site to arrive at "sea-level pressure" (an example is given in Figure 6.11). Making such "corrections to sea level"

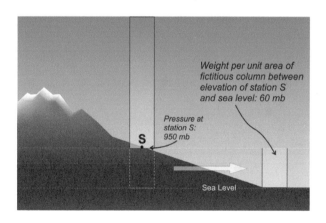

FIGURE 6.11 Computation of sea-level pressure. The observer at station S takes a pressure reading (assumed here to be 950 mb). By estimating the temperature (and thus the air density) of an imaginary air column that extends from sea level to the elevation of S, the observer can add a supplementary weight (air pressure) to the observed pressure reading at S. Note, for this example, that the pressure of the fictitious air column is 60 mb. In this way, the observer "corrects" the station pressure to sea level. Specifically, the station pressure corrected to sea level is 1010 mb.

is not perfect because errors often occur from making assumptions about density in the fictitious column. These errors are typically larger at higher elevations, particularly in mountainous regions such as the Rockies.

Speaking of assumptions made about air density, we first note that the vertical variation of air density is very similar to the profile of air pressure in Figure 6.9: Air density also decreases with increasing altitude. Most of us live at elevations within several hundred meters of sea level where air pressure is roughly 1000 mb and the molecules in a cubic meter of air collectively weigh about one kilogram (a representative value of air density at sea level is 1.2 kg/m³). In contrast, the air density at an altitude of 10 km (where air pressure is roughly 300 mb) is a little less than one third of its value at sea level. So, as pressure decreases to roughly one third of its sea-level value at 10 km, so does air density. The bottom line here is that the vertical profiles of pressure and air density are fairly similar.

Unlike pressure and temperature, however, we don't routinely measure air density with instruments. Rather, we calculate density from the **Ideal Gas Law**, an equation (derived from observations) that links the pressure, temperature, and density of an "ideal" gas. This fundamental relationship can be written as:

Pressure = Gas Constant × Density × Temperature

If temperature is expressed in Kelvins, density in kilograms per cubic meter (kg/m³), and the gas constant is taken as 287 (its units are m²/s²K), then pressure will be in units of Pascals (Pa), where 100 Pascals = 1 millibar. In an ideal gas, the atoms and molecules bounce perfectly off each other when they collide. This assumption does not necessarily hold in the real atmosphere, but for most meteorological purposes, the Ideal Gas Law works just fine.

In summary, the method of correcting to sea-level pressure allows meteorologists to compare oranges and oranges instead of apples and oranges. In other words, if pressures weren't corrected, comparing the pressure readings from different observing stations would mostly reflect differences in elevation. Once variations in pressure due to elevation are removed, forecasters still see small horizontal variations in pressure, and highs ("H") and lows ("L") of sea-level pressure can be located. Thus, the H's and L's that you see on the weather map are highs and lows of sea-level pressure.

Isoplething Sea-Level Pressure: Seeing the Forest from the Trees

Figure 6.12a shows abridged station models over the Upper Midwest during the evening of October 26, 2010, as a powerful storm moved through Minnesota and Wisconsin, setting records for lowest sea-level pressure in both states. The numbers to the upper left and lower left of the cloud-cover circle are already

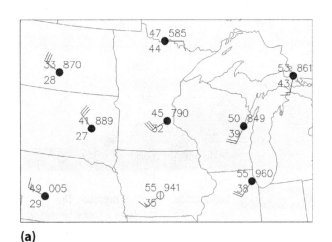

(a)

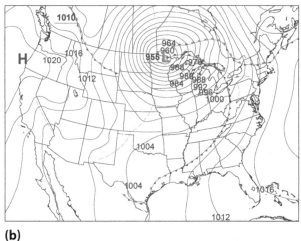

(b)

FIGURE 6.12 (a) Station models at 01 UTC on October 27, 2010, capture a record-setting storm passing through the Upper Midwest. The three-digit number to the upper right of the cloud circle is a coded sea-level pressure. To decode it, place either a "9" or a "10" in front of the three digits and place a decimal point in front of the last digit—the result will be the pressure in millibars. (b) An expanded view of the region showing isobars, or lines of equal sea-level pressure. At this time, the central pressure of the low over northern Minnesota was 955 mb, a record for the lowest sea-level pressure in that state (courtesy of NOAA).

familiar to you—temperature and dew point, respectively, in °F. The three numbers to the upper right of the circle represent a coded sea-level pressure. To decode this three-digit string, place either a "9" or "10" in front of these three numbers and then insert a decimal point before the last digit. Your decision between a "9" or a "10" should be based on the earlier assertion that nearly all sea-level pressures fall between 950 and 1050 mb.

For example, look at the station model for Minneapolis, in the southern part of Minnesota (coded pressure = 790). If you place a "10" before the three numbers, the sea-level pressure would be 1079.0 mb, which falls out of the typical range. Thus, the sea-level pressure must be 979.0 mb, which is fairly low but realistic. Next, consider Green Bay, WI, where the coded sea-level pressure is "849." Once again, if you place a "10" in front of these three numbers, the sea-level pressure would be unrealistically high, so the actual sea-level pressure is 984.9 mb. Finally, consider the station model at North Platte, NE (coded pressure = 005). Here, placing a "10" before the three numbers yields a realistic sea-level pressure of 1000.5 mb, whereas using "9" would have resulted in 900.5 mb. Although such a reading is possible in the eye of the most intense hurricanes, it has never been observed over land in the mid-latitudes.

Now we're ready to see the forest of sea-level pressure from all the station-model trees by schematically organizing pressure readings by drawing **isobars**, which are lines of constant pressure. Traditionally on a map of sea-level pressure, the 1000-mb isobar is drawn first, followed by isobars drawn at intervals of 4 mb (. . . 992, 996, 1000, 1004, 1008 . . .). When a more detailed analysis is needed (for example, during a winter storm or during summer when variations in sea-level pressures tend to be smaller), meteorologists often draw sea-level isobars at intervals of 2 mb.

We've emphasized "sea level" a great deal here, but we're sure you'll agree that it gets a bit cumbersome to always include this qualifier in any discussion about pressure patterns. From now on, any reference we make to "high- and low-pressure systems," "surface isobars," or even "surface pressure" for that matter, it's understood that we're talking about sea-level pressure.

Figure 6.12b, an expanded view of the storm in Figure 6.12a, shows the isobaric fruits of our labor. Isobars, in brown, appear at the traditional 4-mb interval, and they are labeled accordingly. With the pressure field now organized, the center of the strong

low-pressure system over northern Minnesota is marked by an "L," which, by convention, appears in red—its central pressure is 955 mb. Meanwhile, we label a center of high pressure near the West Coast with an "H," which is traditionally colored blue. In addition, a variety of fronts with their characteristic symbols appear on the map—locating fronts is more challenging and often requires a multi-prong approach, part of which will be revealed later in this chapter. In any case, by locating centers of high- and low sea-level pressure, we've now taken a big step toward making a weather forecast!

As one final example, consider Figure 6.13, which is an analysis of sea-level pressures over Asia on December 31, 1968, at 12 UTC. The 'H' over Siberia marks the core of the high-pressure system that set the world record for the highest sea-level pressure reading (recall Figure 6.5).

Notice in the sea-level pressure analyses in Figures 6.12b and 6.13 that, in most cases, isobars are not circular, but instead, they elongate in various directions from centers of high and low pressure. In other words, like spokes on a bicycle wheel, there are "spokes" of low and high pressure that extend outward from the hubs (centers) of lows and highs (see Figure 6.14a for this bicycle-spoke analogy). Some of these spokes are evident in the isobaric analysis shown in Figure 6.14b. The prominent bulges in the isobars extending to the southwest and northeast (in this example) of the center of the low are called **troughs**, which are elongated areas of low pressure.

The dashed lines running through the troughs mark the **trough lines**. Any two points, A and B, that lie on opposite sides of a trough line at some small distance measured perpendicular to the trough line, have pressures greater than the pressure of the corresponding point T that lies on the trough line (see Figure 6.14b). Thus, a trough line is indeed an area of elongated low pressure. And, like low-pressure systems with closed isobars around their centers, troughs can also produce clouds and precipitation. Completing the bicycle spoke analogy, we note that troughs can extend in any direction from the center of a low.

Similarly, the prominent bulges in the isobars in Figure 6.14b from the center of the high represent elongated areas of high pressure called **ridges**. The serrated lines running through the ridges mark the **ridge lines**. Any two points, C and D, that lie on either side of a ridge line at some small distance measured perpendicular to the ridge line, have pressures less than the corresponding point R on the

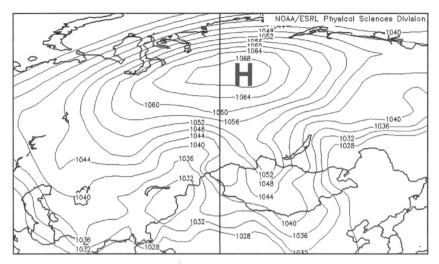

Sea-level Pressure (mb)   12Z December 31, 1968

FIGURE 6.13 The sea-level pressure over Asia at 12 UTC on December 31, 1968. Note the presence of the record-breaking high-pressure system centered over Siberia. Pressures greater than 1068 mb mark the core of the high (courtesy of NOAA).

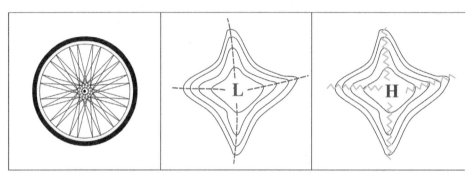

(a)

FIGURE 6.14 (a) Like spokes on a bicycle wheel, high- and low-pressure areas often elongate away from their centers; (b) An analysis of sea-level pressure, which shows troughs extending from low pressure in the East, and ridges extending from high pressure in the West. When crossing a trough perpendicularly, the lowest pressure will be found on the trough line. When crossing a ridge perpendicularly, the highest pressure will be found on the ridge line.

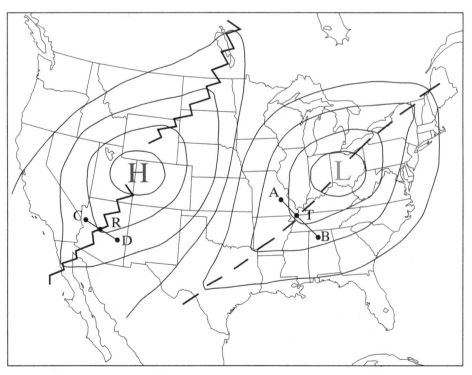

(b)

ridge line. Like high-pressure systems with closed isobars around their centers, ridges often bring fair weather. Ridges can extend in any direction from the center of a high.

Having completed our isobaric analysis, we are ready to determine the direction of the wind around areas of low and high pressure—another big step in making a weather forecast. With these wind patterns established, we will be able to see why lows and troughs are typically associated with clouds and precipitation, while highs and ridges favor dry, mainly sunny weather.

CIRCULATIONS OF AIR AROUND HIGHS AND LOWS: SOME ARE CLOCKWISE, OTHERS ARE NOT

In 1687, Sir Isaac Newton presented the three laws of motion in his book, *Philosophiae Naturalis Principia Mathematica*. These three laws mathematically relate the forces acting on an object to the motion of the object. The first law of motion states: A body at rest will remain at rest unless there is an unbalanced force acting on it. Moreover, an object in motion will continue to move at the same speed in the same direction unless there is an unbalanced force acting on it.

Let's put this law in a more meteorological context. Restated in our own words, there must be a net force (an unbalanced force) acting on an air parcel in order for the parcel to accelerate (change speed and/or direction with time). At the Earth's surface, there are essentially three forces that act on a parcel of air: the pressure gradient force, the Coriolis force, and the force of friction. The pressure gradient force, which derives from pressure differences over a horizontal or vertical distance, ultimately sets the air in motion. We'll examine it first.

Pressure Gradient Force: The Leader of the "Pack"

When you pump up a bicycle tire, the pump adds lots of air molecules that collectively increase the force per unit area inside the tire. In other words, an area of high pressure exists inside the tire (compared to the lower air pressure outside the tire). Moreover, there is a dramatic change in pressure in the short distance on either side of the nozzle of the tire, creating a substantial pressure gradient. If you depress the nozzle

of the tire, air flows rapidly from inside (higher pressure) to outside (lower pressure) the tire. The force that sets the air in motion is called the **pressure gradient force**, which acts from high to low pressure in a determined effort to erase the pressure gradient.

In effect, Mother Nature has a low tolerance for horizontal pressure gradients. Think about it. When a horizontal pressure gradient exists over the Northeast States, for example, there are air columns that weigh slightly more than others over the region. Trying to eliminate the imbalance, the atmosphere sends air from the heavier columns toward the lighter air columns. This transport of air is, of course, the wind.

To see Mother Nature in action, look at the four-part series of photographs of separated compartments of blue-colored water (Figure 6.15). In the first photograph, the compartment on the left has more water than the compartment on the right. Invoking the concept of pressure as a weight, it follows that the water pressure at the bottom of the chamber on the left is greater than the water pressure at the bottom of the compartment on the right. In other words, there is a horizontal pressure gradient. Mother Nature cannot even out this imbalance because a plastic divider separates the two compartments.

Now watch what happens as the divider between the two separated compartments is removed (photographs 2 and 3). Water moves from high pressure to low pressure. Eventually, Mother Nature achieves a balance (photograph 4) and all is well in this little world. Achieving a balancing act in the much larger theater of the atmosphere is not as easy. Let us be a little more definitive here. Try as it might, the atmosphere can never quite attain a perfect balance and erase all the horizontal pressure gradients. As a result, air flows relentlessly away from high-pressure systems toward low-pressure systems, creating persistent winds.

Figure 6.16 shows the average sea-level pressure across the globe. Note that, even when pressures are averaged over many years, there are favored regions for higher and lower pressure. Some of the largest horizontal pressure gradients at the surface lie in the two regions between the "Pacific High" and the "Aleutian Low" and between the "Bermuda-Azores High" and the "Icelandic Low." We will discuss these global semi-permanent pressure patterns later in the book in the context of the atmosphere's "general circulation." We will also show that the presence

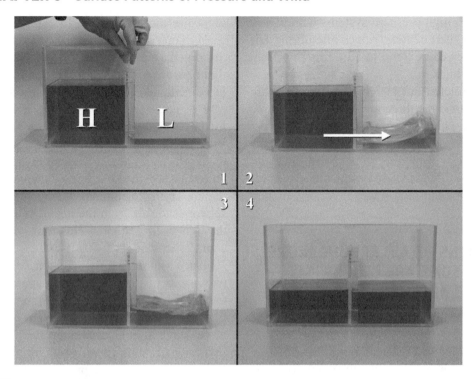

FIGURE 6.15 A series of photographs showing how two compartments of water of different heights, initially maintained by a divider, come to an equilibrium when the divider is removed. Initially, water flows from the higher-water compartment to the lower-water compartment.

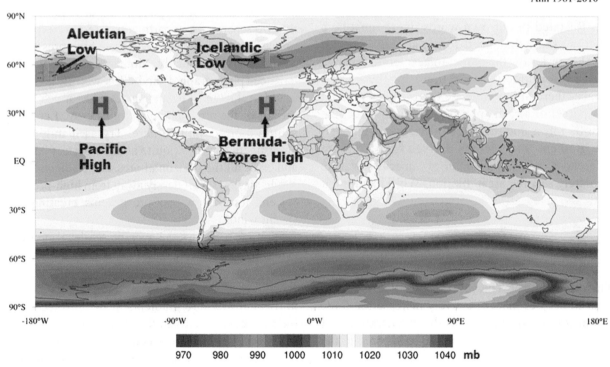

FIGURE 6.16 The long-term annual average of sea-level pressure around the globe. Note that, even after averaging over a long period of time, there are preferred areas of higher and lower pressure (courtesy of Climate Reanalyzer, Climate Change Institute, University of Maine, USA, https://ClimateReanalyzer.org).

of large surface pressure gradients over middle and high latitudes can be directly linked to the large north-south temperature gradients that characterize these regions.

For now, let's see if there is a relationship between the speed of the air and the magnitude of the pressure gradient force. To shed light on this matter, let's revisit the bicycle tire. Imagine inflating the tire with only two pumps of air (in effect, creating a small pressure gradient) and then depressing the nozzle to observe the resulting "wind." Now inflate the tire again with, say, eight pumps (creating a relatively large pressure gradient) and then depress the nozzle. Qualitatively compare the speed of the air in both cases—clearly, the larger the pressure gradient, the larger the pressure gradient force, and the greater the speed of the air.

On a typical weather map, sea-level pressure might change 20 mb over a horizontal distance of 1000 km (about 600 mi), yielding a pressure gradient of 0.02 mb/km (which doesn't seem all that large). Indeed, away from intense low-pressure systems, most sea-level pressure gradients are rather small. Yet even such seemingly paltry pressure differences can generate breezes that send flags flapping and street signs rattling.

Though we are presently concerned with horizontal movements of air (a.k.a., the wind), let's briefly consider the vertical direction: As a general rule, pressure typically falls from 1000 mb at sea level to nearly zero at 100 km (high enough to be, essentially, above the atmosphere). That's a vertical pressure gradient of 12 mb/km, 600 times as large as our example of a typical horizontal pressure gradient. Yet there is no mass exodus of air moving straight up from the ground into space. The key to explaining this apparent inconsistency is gravity, which offsets the vertical pressure gradient force (on average) over large spatial and long time scales (see Figure 6.17). It's where and when the two forces do not exactly balance that air rises (or sinks). And these vertical motions are crucial to forecasters because rising air leads to clouds and precipitation, while sinking air favors dry, clear weather (more on this later).

Okay, back to the horizontal. We can easily identify relatively large pressure gradients near the Earth's surface by inspecting isobaric analyses and noting where isobars are packed close together—for example, in the eastern Dakotas and most of Minnesota and Wisconsin in Figure 6.12b. Tight packing of

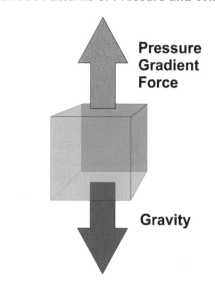

FIGURE 6.17 On average, the downward tug of gravity nearly balances the upward push from the pressure gradient force.

isobars indicates relatively large pressure gradients because pressure changes rapidly (relatively speaking) over a short distance (where distance is always measured at right angles to the isobars). Thus, like the rush of air from the eight-pump-inflated bicycle tire, we should expect winds to be relatively strong here—indeed, note in Figure 6.12a that winds in these areas were blowing at 20 to 30 knots. In regions where isobars are more loosely packed, such as over Florida or Texas in Figure 6.12b, we should expect lighter winds because pressure changes more slowly with horizontal distance (we would say the pressure gradient in these areas is weak).

Now let's consider wind direction. Here we have a problem. Observations show that air tends to flow away from higher pressure toward lower pressure, but not directly. In other words, air does not move from higher to lower pressure at right angles to local isobars. So, when it comes to determining the direction of the wind in the real atmosphere, the situation isn't quite as simple as the direct, outward rush of air from a rapidly deflating bicycle tire. We must withhold our final judgments on wind direction pending a discussion of other horizontal forces that are at work.

Friction: Whoa Nellie!

It may not be intuitive to you that air in motion near the earth's surface is slowed by contact with the ground. Perhaps this claim makes better sense if you accept the notion that even a hockey puck slows down as it skims across seemingly smooth ice. The ground is a lot rougher. Once the pressure gradient force puts air in motion, collisions between air molecules and the unyielding, rough ground cause air parcels to decelerate a bit, a slowing down of the wind that we attribute to a force called **friction**.

When a meteor or a spacecraft enters the earth's atmosphere at very high speeds, friction between the air and the object is great, causing intense heating (see Figure 6.18). Such dramatic examples of the effects of friction, though out of the realm of weather, allow us to rightfully conjecture that the magnitude of the force of friction increases with increasing speed—the faster surface winds blow, the greater the force of friction. It's sort of a check-and-balance system on the speed of the wind.

How does friction slow the wind? When the wind blows, the roughness of the surface causes mechanical eddies to form. Recall from Chapter 2 that mechanical eddies develop anytime the wind blows over the Earth's surface. These eddies mix relatively fast-moving air down toward the ground, where the air's momentum diminishes. Formally, **momentum** is the mass of an object multiplied by its speed. A 240-pound, brute fullback running the football at full speed has much more momentum than a baby turtle crawling on the bank of a pond.

Mechanical eddies also mix relatively slow-moving air next to the ground upward. As a result, wind speeds higher up in any well-mixed layer of air above the ground don't blow quite as hard. The bottom line here is that eddies serve as "middle men" to slow winds that blow near the surface. Note, in Figure 6.19, that wind speed (on average) increases with height above the ground. This classic vertical profile of wind speed illustrates that the greatest impact from friction (downward and upward mixing of momentum by mechanical eddies) occurs right next to the ground. One other general rule: The effects of friction over bodies of water, particularly when they're fairly smooth (no big waves), is generally less than over rougher land.

Let's take a moment and review. We have the pressure gradient force causing a parcel of air to accelerate and the force of friction slowing it down a bit. The only problem that remains is to explain why, in large-scale weather systems, air does not move directly from high to low pressure. This one is a bit more complicated because it involves a subtlety called an "apparent force."

Apparent Forces: Your Bottom Deceives You

Nature can sometimes lend false impressions. For example, the Earth revolves around the Sun, yet we have absolutely no sense that we're speeding through

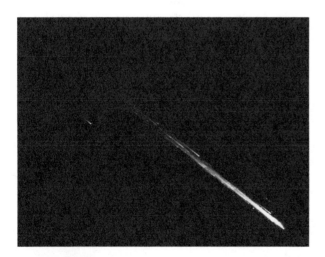

FIGURE 6.18 Friction with the Earth's increasingly dense atmosphere led to the fiery reentry and disintegration of the unmanned Cygnus resupply ship on August 17, 2014. This commercial cargo freighter had spent the last month delivering nearly two tons of supplies to the International Space Station (NASA/ESA/Alexander Gerst).

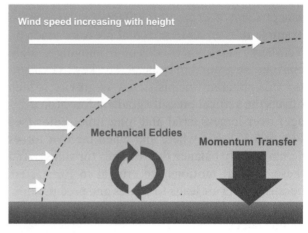

FIGURE 6.19 Mechanical eddies generated by friction transfer momentum from faster winds toward the ground where the momentum diminishes. As a result, wind speed generally increases with increasing height in a layer of air above the ground (here, the length of the wind vector indicates wind speed).

space at about 66,000 mph (29,700 m/s). We earthlings even perceive the Sun "moving" across the sky on a daily basis (a false perception, of course)—by merely standing on this planet and monitoring the Sun's position during the day, you might harbor the mistaken idea that the Sun is moving relative to the Earth.

It turns out that the notion of "false impressions" has implications to our perception about moving "objects" on Earth, including moving air (that is, the wind). To understand these implications, consider that wind velocity is a **vector**; it has a magnitude, or speed (25 mph is a breezy example) as well as a direction (from the northwest, for example). Similarly, the velocity of a car is also a vector—it has a magnitude (for example, 65 mph on the interstate), and it has direction (perhaps you're driving west).

Acceleration is a change in velocity over time. Thus, because velocity is a vector, acceleration must be a vector also. So there are two ways for an acceleration to occur: by changing speed (you hit the gas to pass a truck or you brake to avoid a squirrel) or by changing direction (you round a sharp curve in the road at constant speed, and your bottom slides outward as you negotiate the bend). In the latter example, the acceleration results from the continuous change in direction—the motion of your bottom is the evidence of a net force at work.

Let's think more about the acceleration that occurs when you round the bend. For you to successfully negotiate the curve and not continue in a straight line, your car must continuously accelerate toward the center of the curve. If there weren't some force directed inward, your car would surely run straight ahead off the road. It's that simple. You don't sense this inward acceleration, though. You only feel yourself sliding (accelerating) outward. So there seems to be a contradiction here.

The outward acceleration that you feel arises from your "frame of reference." By frame of reference, we mean the part of your immediate surroundings that you sense is unaccelerated (the interior of the car, in this case). Indeed, you perceive the car to be unaccelerated as it negotiates the curve at constant speed, even though it really accelerates toward the center of the curve. This perception leads you to falsely sense that some other force, which acts to accelerate you outward, is the "real" force at work. But it's not: This phantom force, called the **centrifugal force**, is only an "apparent" force (see Figure 6.20).

FIGURE 6.20 The centrifugal force pushes amusement riders into their seats when roller coasters negotiate inverted curves.

The key point we want to make here is this: Our sense of what is accelerated and what is not accelerated sometimes depends on our frame of reference. And this connection has an important consequence regarding our observations of the direction of the wind.

Coriolis Force: As the World Turns

To understand this consequence as it pertains to the direction of the wind, let's first start with the true (but not so obvious) premise that all locations at latitude 40°N are now racing east at about 794 mph (357 m/s), a consequence of the rotating Earth.

However, not all points on the planet move at the same speed. For points on the equator, the posted eastward speed limit is about 1036 mph (466 m/s). The difference in these rotational speeds is a result of the spherical shape of the Earth.

Let's explore this concept further. Consider two points at the same longitude, one at 40°N, the other at the equator (as in Figure 6.21). In a given time period, the point at the equator must move a larger distance (in other words, around a larger latitudinal circle) than the point at 40°N, whose latitudinal circle has a smaller circumference. The only way to accomplish this feat is for the point at the equator to move faster. If the equator is the "turbo-supercharged racer" of latitudes, then polar regions, where latitudinal circles become progressively smaller with increasing latitude, turn out to be golf carts. They just putter along.

As you might guess, points on different latitude circles moving east at different speeds can lead to some unexpected results. Imagine a projectile is launched from point Q (in Figure 6.21) northward toward point P. By way of preview, let's just say that the projectile never makes it to point P.

We can get a clue to what might happen to the projectile by heading off to the gridiron for a clinic on football (see Figure 6.22(a–c)). For those who don't follow the sport closely, the X's in Figure 6.22a are

offensive players and O's represent the defense. Let's say that the quarterback (the yellow "X") calls a pocket pass. In Figure 6.22b, he drops straight back (into the "pocket") and easily completes a pass to his receiver near the right sideline. The next play (Figure 6.22c), the quarterback calls a roll-out pass to the same receiver. This time, the quarterback runs fast toward the right sideline. While he's running, he fires the pass toward his receiver. But, because the quarterback (and the football) have momentum toward the right sideline, the football misses the receiver slightly to the right (relative to the quarterback rolling out to his right). Notice that the receiver has a few choice words (bleep, bleep) for his quarterback, who, while rolling out, failed to take into account his momentum toward the sideline.

Let's take the lesson learned by the quarterback and apply it to the projectile launched northward from point Q in Figure 6.21. At the instant before launch, the projectile was moving eastward at 1036 mph, courtesy of the rotating earth. Once airborne, the projectile retains its original endowment of eastward momentum. Think of the projectile's northward flight as the flight of the football after the quarterback released it during his rollout toward the right sideline. Until it hits the ground, the projectile will continue to move eastward with the approximate speed of the equator. But the projectile is also moving northward. With each passing moment, it moves over ground that has an eastward speed less than its own. In essence, the projectile surges into the lead in this eastward race with the slower-moving ground below (see Figure 6.23).

But there is no error in the projectile's flight telemetry. Nor has the projectile crossed over into the Twilight Zone. The apparent deflection is a natural consequence of living on a spherical, rotating planet. This deflection is the **Coriolis effect**, named after the French engineer and mathematician Gustav Coriolis. Though he is credited with discovering the Coriolis effect as it relates to the atmosphere, Coriolis actually studied rotating parts on machines and apparently never postulated what might happen to objects moving over a rotating planet. In all fairness, however, Coriolis should be recognized for laying the mathematical foundation for a force necessary to explain the direction of air motion in a rotating frame of reference—the **Coriolis force**.

Let's return to our projectile and launch it from another location in the Northern Hemisphere. If the projectile is launched southward from the North Pole,

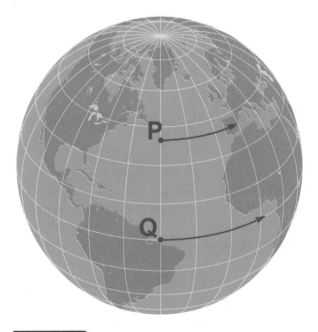

FIGURE 6.21 Point Q at the equator must travel faster around its larger latitudinal circle than point P at 40°N latitude.

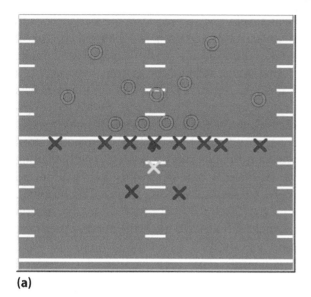

(a)

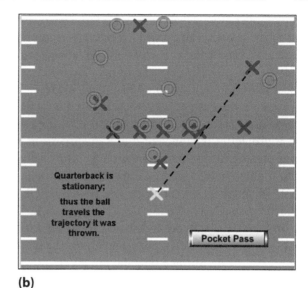

Quarterback is
stationary;

thus the ball
travels the
trajectory it was
thrown.

Pocket Pass

(b)

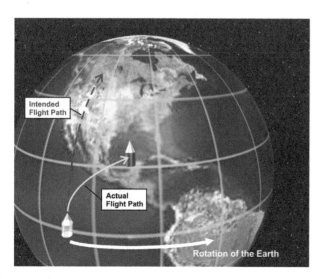

!$*?#

Quarterback has
momentum (red arrow);

thus the ball travels a
trajectory that is the sum
of the intended path
(yellow line) and his
momentum.

Roll-out Pass

(c)

FIGURE 6.22 Who would have thought that passing plays in football could help to explain what happens to a projectile launched northward from a point in the Northern Hemisphere? (a) Players line up; (b) The quarterback drops back into the "pocket," sets his feet firmly in the turf, and then throws a forward pass; (c) The quarterback rolls out and throws a pass on the run. To be a successful pass play, the running quarterback must take into account his motion toward the sideline. In this case, he didn't, and the pass was incomplete.

FIGURE 6.23 Like a football thrown by a quarterback running toward the sideline, a projectile fired northward from the equator retains its fast eastward speed. As it moves pole-ward, the projectile moves east ahead of the slower ground below it. In effect, the projectile curves to the right of its intended path.

it moves over ground that possesses a faster eastward speed than its own. Thus, it will lag behind the ground below and appear to veer westward. Relative to an observer at the North Pole, the projectile will again appear to deflect to the right of its intended path.

To tackle the case of a projectile moving in the east-west direction, check out Figure 6.24. Suppose that a projectile is launched due eastward from a tower on a platform floating over the Pacific Ocean (note the southernmost brown projectile in Figure 6.24). To understand what happens, consider that the projectile is actually moving faster to the east than the surface on the latitude circle below it (because the total eastward speed of the projectile equals the eastward speed of the surface plus the projectile's own firing speed). To appreciate the consequence of this difference in eastward speed, think about twirling a ball attached to an elastic cord: as you twirl faster, the ball goes outward (away from you). Similarly, the projectile gets thrown outward from the Earth's axis of rotation at a right angle—the solid black arrow in Figure 6.24 points the way to the projectile's new position at the dashed circle.

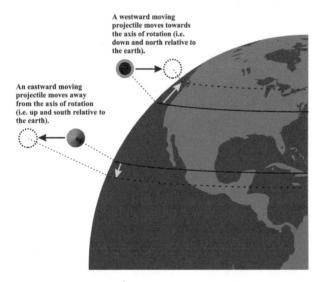

A westward moving projectile moves towards the axis of rotation (i.e. down and north relative to the earth).

An eastward moving projectile moves away from the axis of rotation (i.e. up and south relative to the earth).

FIGURE 6.24 A projectile launched east moves outward at a right angle from the earth's axis of rotation in response to a stronger centrifugal force. In so doing, it drifts upward and also deflects southward. Gravity offsets much of the upward movement, but the southward movement appears as a deflection to the right relative to an observer at the launch site. In response to a weaker centrifugal force, a projectile launched westward moves toward the earth's axis of rotation at a right angle. In so doing, it drifts downward and also deflects northward. Relative to an observer at the launching site, the projectile appears to deflect to the right.

Such a right-angled path from the Earth's axis of rotation causes the projectile to gain altitude, but it also takes the projectile south of its original latitude. Note that the dashed line drawn from the projectile's new position toward the center of the Earth intersects the surface at a latitude south of the projectile's original launch latitude (the yellow arrow indicates this change in latitude). Eventually, the projectile stops its outward fling as gravity reasserts itself, exerting a pull on the projectile in the direction back toward the center of the Earth (again, along the dashed line). But clearly, from the perspective of an observer on the launch platform, the projectile appears to deflect to the right.

Now consider a projectile launched due westward from a different ocean platform (in Figure 6.24, we're now talking about the northernmost projectile). Again, think about twirling a ball attached to an elastic cord, except now, twirl slower—the ball comes in toward you, analogous to the westward-launched projectile which is now moving opposite the direction of the Earth's surface on the latitude circle below. In this case, the projectile moves inward toward the Earth's axis of rotation at a right angle, which means it has to lose altitude and move north of its original latitude (see Figure 6.24). Again, from the perspective of an observer on the ocean platform, the projectile appears to deflect to the right.

So what do these arguments using projectiles have to do with the weather? Think of horizontally moving air parcels (that is, the wind) as "projectiles" and apply the same principles. In other words, air already in motion is subject to the Coriolis force. However, you may point out that our examples are special cases that consider only motion in the north-south and east-west directions. What about air moving northwestward, or east-southeastward? Recall from Chapter 5 that motion in any direction can be decomposed into components in the north-south and east-west directions, which then allows us to apply the arguments above to any wind direction. Thus, regardless of the direction of air motion, the deflection resulting from the Coriolis force in the Northern Hemisphere is to the right—formally, scientists say that the Coriolis force acts 90° to the right of the motion. And ocean currents feel its influence as well.

You may repeat all of the above arguments for the Southern Hemisphere by noting that if you are in space looking "up" at the South Pole, the sense of the earth's rotation is clockwise. Try this by holding a globe above your backward-tilted head and spinning the globe clockwise, which imitates the true sense of the Earth's rotation from this perspective. You should arrive at the result that deflections due to the Coriolis force are to the left of the observer in the Southern Hemisphere.

But the Coriolis effect does not dole out its deflecting influence evenly over the planet. Take a long, hard look at a globe. Concentrate on the latitude circles which girdle the globe near the equator. Though the largest of these circles lies right at the equator, nearby circles are almost as big. Hence, points rotating eastward on these circles move at comparable speeds. Thus, a projectile launched northward (or southward) from one low latitude to another low latitude will have almost the same eastward momentum as the ground over which it travels. In other words, the Coriolis effect is relatively small at low latitudes. Indeed, hurricanes (see Chapter 11), which need some contribution from the Coriolis effect in order to develop their characteristic spin, rarely form within 10° latitude of the equator because the Coriolis effect is simply too weak there. Figure 6.25 shows the origin points of tropical storms and hurricanes that formed in the period September 11–20 in the North Atlantic and eastern North Pacific Oceans, a period of time which is very close to the peak of hurricane season. Although waters are plenty warm

to support hurricane development in the tropics this time of year, note the relative lack of storms between the equator and 10°N latitude. Clearly, something puts the brakes on tropical development there—we will elaborate on this idea in Chapter 11.

Now shift your attention to the globe's high latitudes. Here, latitude circles are shrinking fast with decreasing distance from the pole. Points rotating eastward on these circles move at significantly slower speeds the closer you get to the pole. Thus, the eastward momentum of a projectile launched from one high latitude to another high latitude will be increasingly different than the eastward momentum of the ground below. In essence, the Coriolis effect becomes relatively large at high latitudes. Our conclusion: The Coriolis effect increases with increasing latitude.

The apparent deflection caused by the Coriolis effect also depends on the speed of the moving object. Suppose we launch two projectiles northward from a point at 40°N and impose the condition that they fly for five hours before they self-destruct. One projectile moves at an average speed of 200 mph. The other moves at 100 mph. In five hours, the projectiles have traveled 1000 mi and 500 mi, respectively. Because the first projectile travels farther north, it eventually flies over much more slowly

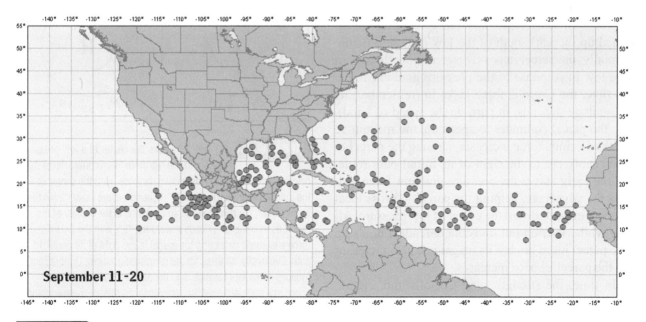

FIGURE 6.25 Red dots mark the points of genesis for tropical storms and hurricanes that formed during the period September 11–20 in the North Atlantic and eastern North Pacific Oceans (using data from 1851 to 2009). The dearth of development between the equator and 10°N latitude is related to the relatively weak Coriolis effect at low latitudes (courtesy of the National Hurricane Center).

moving ground than the second projectile. Thus, it will surge farther east than the second projectile. Its apparent deflection will be greater. Our conclusion: The faster a moving object, the greater the apparent deflection caused by the Coriolis effect.

Contrary to popular belief, the time and spatial scales of a flushing toilet are too small for the rotation of the earth to have any noticeable effect (water is the "projectile" in this case). Thus, claims that water spins counterclockwise down a toilet or the drain of a sink in the Northern Hemisphere and clockwise in the Southern Hemisphere because of the "rotation of the earth" are hogwash! The spin of the water has everything to do with how the toilet or sink was constructed, and nothing to do with the Coriolis force.

Armed with our understanding of the Coriolis force, we are now ready to put the finishing touches on the flow of air around surface high- and low-pressure systems.

Surface Air Flow: Diagrams of Football Plays in Meteorology

Consider a Northern Hemisphere weather map with a low-pressure system in the center, as shown in Figure 6.26a. The "L" marks the point with the lowest sea-level pressure in the local area. We depict this observation by writing "HIGHER PRESSURE" around a few isobars that surround the low. Like television sportscasters, let's isolate on a parcel of air lined up (at rest) south of the low in Figure 6.26b. Initially, the parcel runs a fly pattern directly toward the low in response to the pressure gradient force. In time, the Coriolis force kicks in (remember, its magnitude depends, in part, on the speed of wind). As the parcel continues to speed up, the magnitude of the Coriolis force increases, causing the parcel to noticeably cut to the right of its initial pass route. In the early stages, the pressure gradient force also has the advantage, helping the parcel to cross local isobars inward toward the center of low pressure. All the while, friction (an invisible defender) holds up the parcel just a bit, but its magnitude also increases as the parcel's speed increases.

In time, the strengthening Coriolis and friction forces will offset and eventually balance the pressure gradient force, allowing the parcel to proceed on a slant pattern without making any further cuts (changes in direction) or changes in speed (unless balance is upset by some external force or by a change in the pressure gradient force). So, the

intended pattern of this parcel is clear: straight off the line of scrimmage, cut to the right, and then go long.

Like the water-compartment experiment we conducted earlier in this chapter, the pressure gradient force continually adjusts as weights of air columns change in response to incoming and outgoing air parcels. Unlike the water-compartment experiment, a perfectly uniform air pressure is never reached in the atmosphere.

In Figure 6.26c, we completed the pass routes of several other air parcels around the low. Here's the bottom line: The circulation of air around a surface low-pressure system in the Northern Hemisphere is in the counterclockwise sense, with air moving slightly inward toward the low. More generally, the circulation around a low is **cyclonic** (lows are sometimes called **cyclones**).

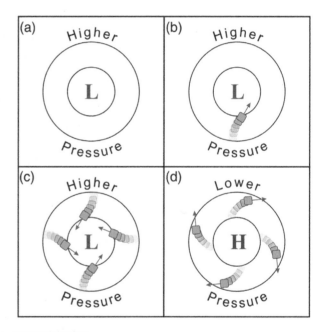

FIGURE 6.26 (a) The "L" marks the center of a low-pressure system in the Northern Hemisphere. Note that pressure increases with increasing distance from the center of the low. In other words, there's "higher pressure" away from the low's center; (b) If we release an air parcel in the vicinity of the low, it accelerates and heads toward the "L", with the Coriolis force eventually deflecting the parcel to the right of its original path. Eventually, all three forces (pressure gradient, Coriolis, and friction) come into balance, and the parcel moves at constant speed in a constant direction; (c) When we release three additional parcels in the vicinity of the low, their collective paths suggest a counterclockwise rotation around the low's center, with a tendency for air to converge toward the "L"; (d) The circulation around a high-pressure system in the Northern Hemisphere ("H" marks the center of highest pressure) is clockwise, with a tendency for air to diverge away from the high's center.

Weather Folklore and Commentary

Of Swallows and Bats

The link between cloudy, inclement weather and low-pressure systems can be observed in nature. The folklore "Swallows and bats fly close to the ground before rain" is generally true. Its veracity is based on the idea that swallows and bats have very sensitive ears, which are greatly affected by sudden changes in air pressure. When air pressure drops, as it would with the approach of a low-pressure system, bats and swallows will skim the surface, trying to get as close as they can in order to keep the pressure on their ears as high as possible (even though the air pressure decreases everywhere, the pressure will be highest next to the ground). After observing swallows or bats flying low to the ground, clouds and precipitation often arrive with the low, giving this folklore a ring of truth.

As a result, air converges in the vicinity of surface low pressure. A similar diagram for a high-pressure system in Figure 6.26d shows that the circulation of air around high pressure in the Northern Hemisphere is in the clockwise sense with air moving slightly outward from the high. More generally, the circulation around a high is **anticyclonic** (highs are sometimes called **anticyclones**). At any rate, air diverges in the vicinity of surface high pressure. Figure 6.27 is a new and improved Figure 6.14b because we added arrows to show how these rules are applied to more realistic highs and lows and their associated closed isobars. Note how the wind crosses the isobars—inward slightly toward the closed low (and toward the troughs) and a bit outward away from the closed high (and outward away from the ridges).

Let's take a closer look at the final heading that a parcel takes as it crosses the isobars toward lower pressure and away from higher pressure. Over land, the wind crosses the isobars at approximately 30 degrees or so, on average (see Figure 6.28). Over the ocean and other large bodies of water, the crossing angle is generally smaller, owing to less friction over the smoother water (compared to rough land).

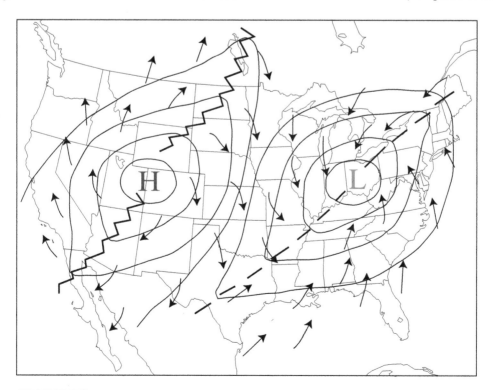

FIGURE 6.27 The analyzed map of Figure 6.14b, now with arrows showing the surface wind flow.

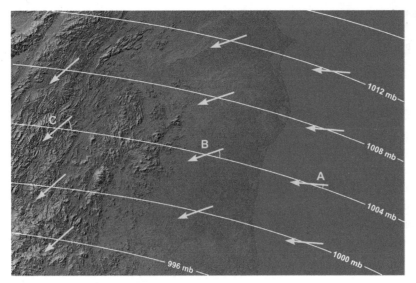

FIGURE 6.28 The angle at which winds cross local isobars depends, in large part, on the underlying surface. Smoother surfaces (such as water, on the right) provide less friction, allowing the wind to cross local isobars at smaller angles.

In mountainous areas, where the effects of friction are greater, the crossing angle is typically larger, sometimes 45 degrees or more.

It is the convergence in the vicinity of surface low pressure and troughs, and divergence in the vicinity of surface high pressure and ridges, that gives each feature its characteristic weather. The even-keeled atmosphere always tries to maintain fairly consistent weights for all of its air columns. It accomplishes this feat by avoiding major congestions and evacuations of air. That's why surface pressures at any location usually vary by less than 5 percent from day to day, week to week, and month to month.

To avoid big traffic jams, air converging toward the center of a low must rise to avoid a big congestion (see Figure 6.29). Rising air cools, and, if the air cools enough, clouds and precipitation can form. By the same token, air diverging away from the center of a surface high prompts air to gently sink over the high in order to avoid any large evacuation of surface air (see Figure 6.29). Sinking air warms and dries out, conditions unfavorable for clouds. So it's the vertical motions related to the low-level convergence and divergence that are ultimately responsible for the weather associated with highs and lows.

To support our claims about the link between vertical motions and high- and low-pressure systems, check out Figure 6.30 where isobars are superimposed on satellite and radar imagery on a day when a storm was affecting the eastern U.S. Notice the colorful radar reflectivity (where rain was falling) around the low in Indiana and in the vicinity of a trough extending south of the low. In contrast, notice the absence of clouds and precipitation associated with surface divergence and sinking air near high pressure ruling the Plains.

Using weight as a proxy for pressure, the column of air over the center of a surface low-pressure

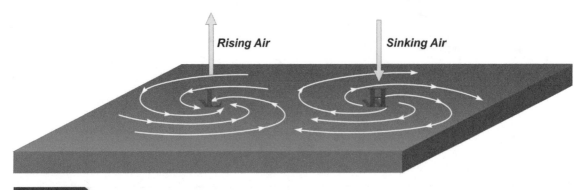

FIGURE 6.29 Convergence around surface lows sets the stage for rising air, while divergence in the vicinity of surface highs favors sinking air.

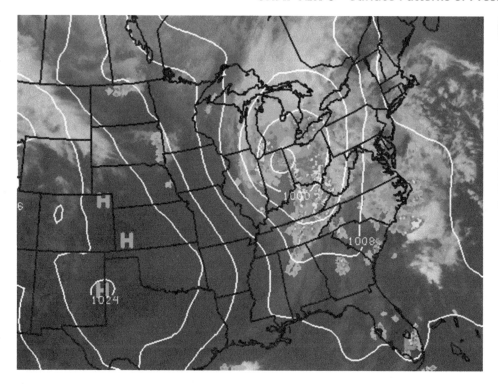

Isobars (in white) are superimposed on infrared satellite and radar imagery of the eastern two-thirds of the United States on an October afternoon. Note the close correlation between low pressure and clouds and precipitation, and between high pressure and dry weather (courtesy of NOAA).

system (high-pressure system) must be the lightest (heaviest) column in the general vicinity. But this observation creates a dilemma. If air converges toward the center of a low, doesn't this convergence of air constitute a weight gain? Shouldn't the pressure immediately start to increase at the center of a low? By the same token, if air diverges away from the center of a high, doesn't this divergence constitute a weight loss? Shouldn't the pressure immediately start to decrease at the center of a high? What's a developing low (high) to do? Go on a crash diet (weight-gain program)? Not quite. Let's investigate.

Controlling the Weights of Air Columns in the Middle Latitudes: A Well-Balanced Diet of Convergence and Divergence

As we just discovered, surface low-pressure systems are in a real bind. Low-level convergence adds weight, which is not good for weight-conscious lows. Fortunately, the atmosphere avoids any large congestion by forcing converging surface air to rise (recall Figure 6.29). But rising air still doesn't take the weight off. To understand what we mean, raise your arms above your head while weighing yourself on your bathroom scale. Did raising your arms cause your weight to decrease? Of course not! And so it is with air rising over the center of a low-pressure system: It still contributes to the overall weight of local

air columns. One answer to this pressure predicament is a pattern of divergence (a weight loss) at higher altitudes in local air columns at and near the center of low pressure. We'll return to this weight-loss plan a little later, but it suffices to say that if more air leaves the air column high over the center of low pressure than enters it via convergence near the surface, the low can maintain or lose weight and thrive.

Surface highs have a different problem. Diverging air away from the center of surface high-pressure systems constitutes a weight-loss plan (not good for weighty highs). Fortunately, air sinks in an attempt to avoid large evacuations of air away from the center of a high. Does sinking air raise the weights of local air columns around the center of a high-pressure system? To answer this question, weigh yourself again, this time starting with your arms over your head. Now lower your arms toward the floor while you're on the scale. Did your weight increase? Of course not! And so it is with air sinking over the center of a high-pressure system. One answer to this pressure predicament is a pattern of convergence (a weight gain) at higher levels in local air columns at and near the center of high pressure. We'll also return to this weight-gain plan a little later, but it suffices to say that if more air enters the air column over the center of high pressure than leaves it via divergence near the surface, the high can gain or maintain its weight and thrive.

You can infer the weight loss or gain of air columns by watching your barometer. As a low (high) pressure system approaches your town, local air columns start to lose (gain) weight and surface pressure starts to decrease (increase). In these situations, weather forecasters keep tabs on the **pressure tendency**, which is simply the change in pressure over a given time period. As a matter of practice, forecasters often choose three hours as the time period over which they measure pressure changes—that's generally long enough to capture the significant pressure rises and falls that accompany the movement of large-scale high- and low-pressure systems. A coded form of this three-hour pressure tendency is placed in the surface station model, just below the sea-level pressure. We note that the pressure tendency is coded in tenths of a millibar (decimal points are not included).

Some of the most dramatic pressure changes occur in the vicinity of hurricanes (which are strong low-pressure systems). Figure 6.31 shows a few station models over a portion of the northern Mid-Atlantic States at 15 UTC on October 29, 2012. At the time, Hurricane Sandy was approaching the New Jersey coast from the southeast. The "−75" at Atlantic City, NJ, indicates that sea-level pressure had fallen 7.5 mb in the previous three hours, a hefty pressure decrease. Pressures were falling dramatically elsewhere, even inland, a sure sign that the weather was heading downhill quickly. Philadelphia, where the "−69" indicates that pressure had fallen 6.9 mb in the previous three hours, went on to set its lowest sea-level pressure on record, 952.2 mb.

Figure 6.32 is a meteogram from Atlantic City, NJ, for the 24-hour period beginning at 12 UTC on October 29, 2012. Note that the pressure fell a whopping 40 mb in the first 12 hours of this period as Sandy approached, and then rose approximately the same amount in the next 12 hours as the storm moved away. Pressure tendencies of this magnitude only occur in the vicinity of strong tropical cyclones (and occasionally near rapidly developing winter storms). Also note that the sky was cloudy the entire period and precipitation was observed much of the time, two common characteristics of low-pressure systems. The meteogram also nicely illustrates that winds shift direction as a low-pressure system passes, in this case from northerly to southerly (although we caution you not to generalize this wind shift observation, given that hurricanes can move in many directions, while typical mid-latitude low-pressure systems generally move west to east). Nonetheless, the Sandy example should cement the idea that as a low-pressure system moves by, winds shift direction.

So far we have given all our attention to centers of low pressure, but highs deserve equal billing—large positive pressure tendencies are often the hallmark of approaching (or strengthening) high-pressure systems, especially during the winter. For example, on January 7, 2015, a formidable high-pressure system (with central pressure 1056 mb) associated with an Arctic air mass settled into the Central United States, setting records for sea-level pressure in Lincoln and Omaha, NE, Des Moines, IA, and Topeka, KS, to name a few places (see Figure 6.33a). South and east of this high, pressure was rising rapidly, as much as 5 mb or more in three hours over parts of the Plains, as seen in the station models in Figure 6.33b. In these areas of large positive pressure tendency, strong, low-level cold advection (which can be inferred from Figure 6.33b by noting the northerly winds and the large north-south temperature gradient) decreased

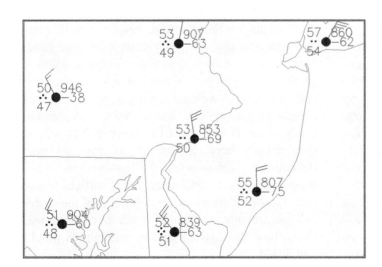

FIGURE 6.31 Station models at 15 UTC on October 29, 2012, showed dramatic pressure decreases as Hurricane Sandy moved toward the New Jersey coast. In the surface station model, the pressure tendency is placed below the sea-level pressure, and is in units of tenths of millibars per three hours (courtesy of NOAA).

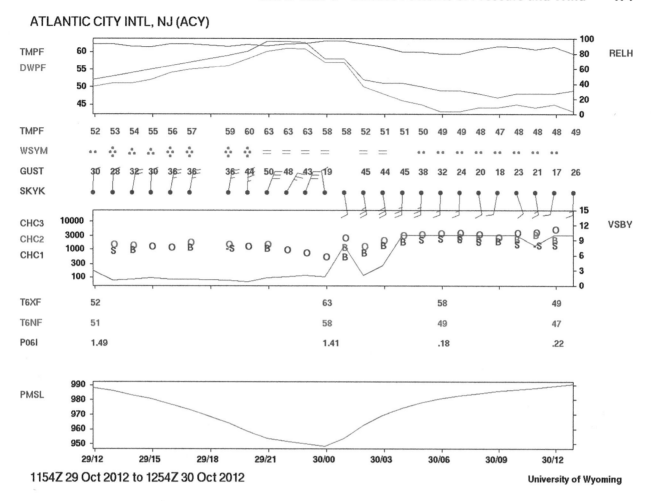

ATLANTIC CITY INTL, NJ (ACY)

1154Z 29 Oct 2012 to 1254Z 30 Oct 2012

University of Wyoming

FIGURE 6.32 A meteogram from Atlantic City, NJ for the 24-hour period beginning at 12 UTC on October 29, 2012 captures the landfall of Hurricane Sandy on the southern New Jersey shore. The bottoming out of pressure and the shift in wind direction around 00 UTC on October 30 coincide with the passage of the center of the storm (courtesy of the University of Wyoming).

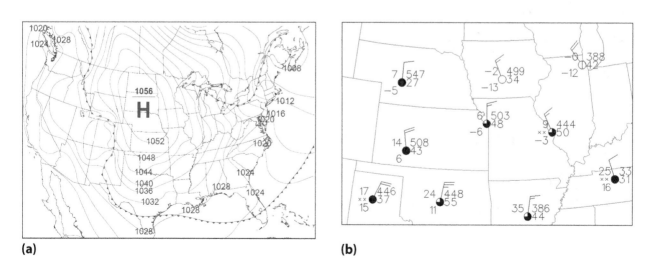

(a)

(b)

FIGURE 6.33 (a) The surface weather map at 15 UTC on January 7, 2015, showed a very strong high-pressure system (with a central pressure of 1056 mb) over the Plains, associated with a bitterly cold Arctic air mass; (b) With the high-pressure system moving to the southeast, surface station models at that time showed very large positive pressure tendencies, some in excess of 5 mb per three hours (courtesy of the National Weather Service).

the average temperature in local air columns. In turn, with molecules huddled closer together in the colder air, the average air density increased, so weights of local air columns and their corresponding surface air pressures also increased with time.

Our main point from these examples is this: pressure tendency is a key variable that forecasters routinely check, particularly in changeable weather situations. In fact, knowing the pressure tendency is arguably more useful from a forecasting perspective than knowing the value of the sea-level pressure because pressure tendency provides local clues when an important high- or low-pressure system is approaching. Given that surface highs and lows often bring with them characteristic weather, pressure tendency thus can have great predictive value.

To close this chapter, we will stay in the mid-latitudes and make the link between horizontal pressure patterns and fronts—a connection that helps forecasters to locate these important boundaries between warm and cold air masses.

FRONTS REVISITED: FITTING INTO THE PRESSURE PATTERN

With the circulations around surface pressure systems now established, we see for the first time how lows and highs circulate warm air poleward and cold air equatorward, thereby doing their fair share to help ease temperature contrasts between the poles and tropics. Of course, there are the inevitable showdowns as polar air masses knife equatorward and tropical air

masses stream poleward, both entering alien territory. A ruckus of stormy weather often results where battle lines are drawn between air masses.

You may be surprised to learn that the instigators of confrontations between air masses are areas of high pressure. To understand this claim, we take a step back and consider how huge masses of air become hot or cold, or dry or moist. An air mass acquires its temperature and moisture characteristics based on the latitude of the source region (polar or tropical) and on the underlying surface (land or water). What's also key, however, is that the air mass must linger long enough in the same general area to take on the characteristics of the underlying surface. In practical terms, this means that source regions for air masses are typically characterized by relatively weak pressure gradients and, thus, relatively light winds, which is a hallmark of high-pressure systems. It follows that the regions near the center of sprawling highs typically mark the cores of air masses.

To illustrate this relationship, consider Figure 6.34a, a map of isobars (brown lines) and station models at 03 UTC on January 3, 2015. At this time, a high-pressure system (central pressure 1048 mb) was centered over the Yukon Territories, to the east of Alaska, a common source region for continental polar and continental Arctic air masses. Note the weak pressure gradient and corresponding light winds in the vicinity of the high (in this case, within the confines of the first isobar or so). Here, weak winds, snow-covered ground, and limited daylight (owing to the high latitude) conspired to create a

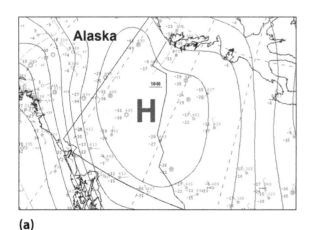

(a)

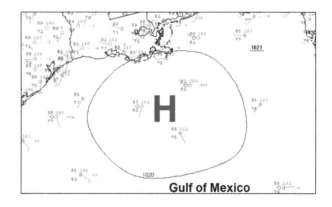

(b)

FIGURE 6.34 The centers of sprawling high-pressure systems, which feature weak pressure gradients and thus relatively light winds, typically mark the cores of air masses: (a) A bitterly cold continental Arctic air mass centered in the Yukon Territories on January 3, 2015; (b) A maritime tropical air mass over the northern Gulf of Mexico in mid-July of 2007 (courtesy of the National Weather Service).

(a)

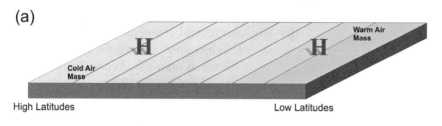

High Latitudes   Low Latitudes

(b)

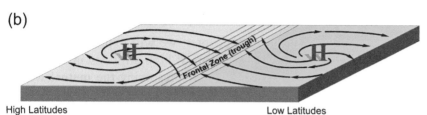

High Latitudes   Low Latitudes

FIGURE 6.35 (a) A hypothetical cross section from the North Pole to the equator, showing an area of cold high pressure to the north and an area of warm high pressure to the south. The initial north-south temperature gradient is uniform (solid lines are isotherms); (b) Diverging winds from the highs concentrate the temperature gradient in a frontal zone in the mid-latitudes, which corresponds to a trough of lower pressure.

favorable environment for the air mass to become bitterly cold and extremely dry. This Arctic air mass provided the seed for the blockbuster high-pressure system which, four days later (on January 7, 2015—recall Figure 6.33), set all-time pressure records in the central United States.

Similarly, Figure 6.34b shows a sprawling high-pressure system center over the northern Gulf of Mexico on a July day. Note the weak pressure gradient and corresponding prevalence of light winds around this high, allowing the air to take on the characteristics of the underlying warm water. Not surprisingly, the air mass associated with this high was warm and humid, or formally, maritime tropical.

With highs at the center of air masses and air diverging away from the center of surface highs, showdowns between contrasting air masses are inevitable. Check out Figure 6.35a which shows an idealized set-up of two stationary high-pressure systems: one over warm tropical latitudes and a second over cold polar latitudes. At some time later, after air spreads out from both high centers (arrows depict this divergence), a narrow zone of sharply contrasting temperature developed over the mid-latitudes (see Figure 6.35b). As mentioned in Chapter 3, the narrow zones where differing air masses meet and vie for control are called fronts.

Given that relatively tranquil high-pressure systems mark the cores of air masses, it makes sense that the outer edges of air masses must have relatively low pressure. Thus, any front that forms when two air masses meet must be located in a zone of relatively low pressure—a trough (see Figure 6.35b).

A trough line marks a distinct wind shift. For the example in Figure 6.36, winds blow from the north-northwest behind the trough axis, while winds blow

from the south-southwest ahead of the trough. These converging winds at the trough concentrate temperature (and moisture) gradients in the vicinity of the front. Converging air at the surface leads to rising air, often supporting clouds and precipitation in the vicinity of fronts. Sometimes, troughs associated with fronts are subtle, so convergence at the front is weak. In such cases, finding a front in the pressure, temperature, or moisture patterns is more difficult.

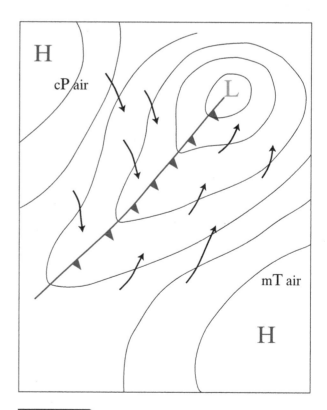

FIGURE 6.36 A cold front in a trough of low pressure between two high-pressure areas, with arrows showing the typical wind flow near the front.

A slight wind shift might be all that gives the position of the front away.

In summary, you can locate fronts on a weather map by looking for a trough, a relatively large temperature and dew point gradient, and/or shifts in the wind direction. A cold front that has all of these characteristics is documented in Figure 6.37. Note that the front (the thicker, black line) lies just at the leading edge of the large temperature gradient (in part (a)), just at the leading edge of the large dew-point gradient (in part (b)), in the pressure trough (in part (c)), and in the elongated zone where winds converge (in part (d)).

Because fronts lie in troughs and zones of typically large horizontal temperature contrasts, they make favorable breeding grounds for centers of low pressure to form and intensify. We will investigate this claim in detail in Chapter 12.

Reflecting Ahead and Looking Back: All About the Weights of Air Columns

In the next chapter, we will explore the pressure and wind patterns at high altitudes in an attempt to pave the way for resolving the mutual weight

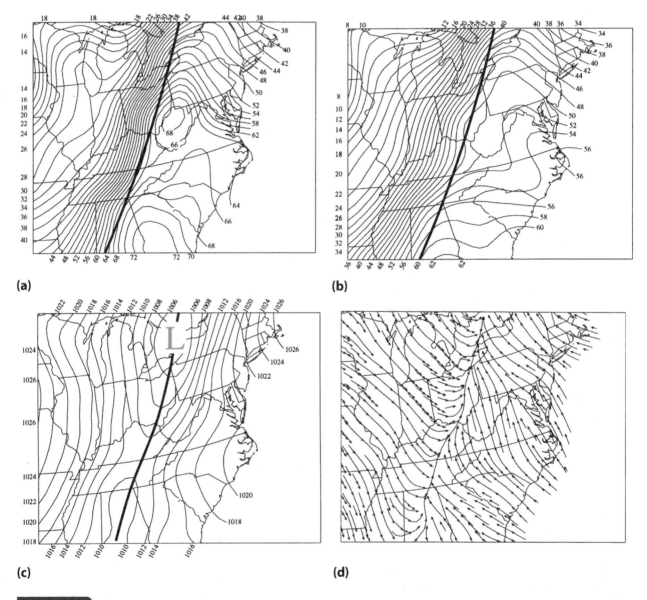

(a)

(b)

(c)

(d)

FIGURE 6.37 Weather data from a March day when a front stretched from a low near Lake Ontario to Alabama: (a) Isotherms (every 2°F); (b) Isodrosotherms (every 2°F); (c) Isobars (every 2 mb); (d) Streamlines, lines that are everywhere parallel to the wind. Note that the front (the dark bold line marks its position) lies on the warm side of the large temperature gradient, on the moist side of the large gradient in dew points, in a trough, and in a zone of converging winds.

issues of surface low- and high-pressure systems (remember that converging air adds weight to local air columns near the centers of weight-conscious lows, while diverging air takes weight off local air columns near the centers of weighty highs). Then,

in Chapter 12, we will link upper-air patterns to regions of convergence and divergence aloft, which, in turn, will have a bearing on the development of weight-conscious low- and high-pressure systems at the surface.

Focus on Optics

The Inferior Mirage: A Piece of the Sky

Visible light (and radiation of any wavelength) is always in a big hurry, zipping through thin air at about 300,000,000 m/s (186,000 mi/s). Light slows down a bit, however, when it is forced to travel through denser mediums such as water, ice, or glass. At only slightly reduced speeds, you might think that light would still be a contented traveler. But light is always looking at its watch, apparently preoccupied with travel time. The super-punctual nature of light was first recognized by the French mathematician Fermat, who arrived at the following conclusion.

Fermat's Principle. The path taken by light in traveling from one point to another is such that the time of travel is a minimum when compared with nearby paths.

While traveling through the reduced speed zones of glass, water, and ice, light gets in such a hurry that it will literally go out of its way to make its travel time as short as possible. Consider Figure 6.38, which shows possible paths for light traveling between two points, one in air (point A) and one in glass (point G). Point P_1 lies on a straight line from A to G, but impatient light will not take this path because the distance traveled in glass (a denser, slower medium for travel than air) is relatively great, so light will lose a lot of time while traveling at reduced speed.

So light looks for alternate routes. By taking a path slightly to the right of P_1, light, though it travels an overall greater distance, reduces its total travel time because it shortens its travel distance in glass (the slower medium). In other words, a path slightly to the right of the straight-line route takes less time because the time gained by traveling a shorter distance in glass more than compensates for the time lost while traveling a greater distance in air. Though the path requiring the least time is not apparent

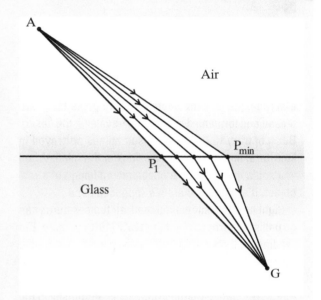

FIGURE 6.38   Light trying to get from point A in air to point G in glass will refract or bend, taking a course through point P_{min} that minimizes the total travel time.

from Figure 6.38, there is obviously some "trade-off" path that will yield minimal travel time. We will call the point on the glass surface corresponding to this optimal path P_{min}. This "bent path" through P_{min} will be the route that light from point A will take to reach point G.

Now let's generalize these ideas. When speed-conscious light travels from one medium to another that has a different density, it will bend or "refract." This refraction of light in the atmosphere is responsible for several magical optical phenomena.

For instance, assume you're driving down the road on a sunny, summer day. Up in the distance, the road shimmers and the pavement looks wet (see Figure 6.39). Yet, as you drive on, the "puddle" disappears, only to reappear farther in the distance. You are witnessing

(Continued)

(Continued)

FIGURE 6.39 The wet-puddle-on-dry-road mirage (courtesy of Alistair Fraser).

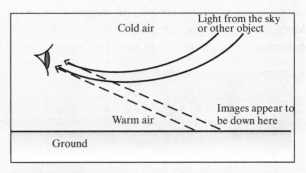

FIGURE 6.40 Over a sun-baked road, air temperatures decrease rapidly with height above the hot pavement. Rays of light from the sky or other lofty objects (such as clouds or trees) that pass through this layer bend so that the cooler, denser air lies on the inside of the curve. Curved light entering an observer's eyes is interpreted by the brain as having entered the eyes in a straight line. Thus, the brain "sees" the object on the road—below its actual position. For this reason, the displaced image of the object is called an inferior image.

a mirage, just like the fleeting watery pools that have teased and tormented many a lost traveler in the desert. But, contrary to how they're sometimes portrayed in literature, mirages are not illusions of the mind. When you see a mirage, you're observing an image of a real object. It's just not where it's supposed to be.

Right next to a sun-baked road, air temperatures can go ballistic, perhaps reaching 120°F (49°C) or more. But readings fall off rapidly with increasing height above the road. Because air density is a function of temperature (assuming pressure is approximately constant), there are relatively large contrasts in air density in the first foot or so above the road. As light from the sky (or a tree or anything higher up) enters this narrow zone of contrasting densities, it is bent (refracted), with the direction of bending such that the denser, cooler air higher above the road is on the inside of the bending ray of light (see Figure 6.40). This direction of bending, with the denser medium on the inside of the curving ray, is consistent with the earlier discussion we had with air and glass in Figure 6.38.

The degree of bending is determined by the temperature (density) gradient above the road: The greater the

rate of temperature (density) change with height above the pavement, the greater the bending. Thus, it should not be surprising that rays from the sky or trees can be bent strongly enough to reach our eyes, as Figure 6.40 suggests. Because our brains interpret incoming light rays as having reached our eyes in unbending, straight lines, our brain "sees" an image of the sky or tree on the road. Thus, the puddle on the road in Figure 6.39 is just an image of a piece of the sky.

It is a real image—not an illusion. The piece of the sky or tree we see on the road is simply not where it's supposed to be. This type of mirage is called an **inferior mirage** because we see the image below where it really is.

So the next time you're driving on a warm, dry day and a friendly passenger points out the wet road ahead, show sound meteorological judgment. Tell him it's just a piece of the sky.

Upper-Air Patterns of Pressure and Wind

LEARNING OBJECTIVES

After reading this chapter, students will:

- Gain an appreciation for the connection between topographic maps and maps of height contours on constant pressure surfaces
- Have a quantitative sense for the standard heights of mandatory pressure levels
- Be able to identify troughs and ridges from the patterns of height contours on constant pressure charts, as well as derive a qualitative sense for wind speeds
- Understand why pressure decreases relatively rapidly (slowly) in cold (warm) air

- Understand the close connection between high (low) heights on a constant pressure surface and high (low) pressure on a nearby constant height surface
- Understand why the wind tends to blow nearly parallel to height contours on any constant pressure surface that lies above the boundary layer
- Understand why winds increase markedly with increasing altitude in regions where there are relatively large horizontal temperature gradients, and then be able to identify the pressure altitude where the mid-latitude jet stream resides
- Understand the seasonal variations in the mean position and strength of the jet stream
- Be able to distinguish between high-amplitude and zonal flows and understand the general impacts to weather forecasting
- Be able to identify blocking patterns on analyses of 500-mb heights

We live at the bottom of an ocean of air. Like swells rolling along the sea surface, the waves of low and high pressure that cause the variety of weather we observe at the Earth's surface often ripple at much higher altitudes. These lofty weather systems are called upper-air troughs and ridges. We'll describe exactly what they are and how to identify them on weather maps later in the chapter. For now, it suffices to say that upper-air troughs produce divergence at high altitudes, leading to decreases in air pressure at the surface (see Figure 7.1). In response, there is convergence of air at the surface and upward motion, paving the way for clouds and precipitation.

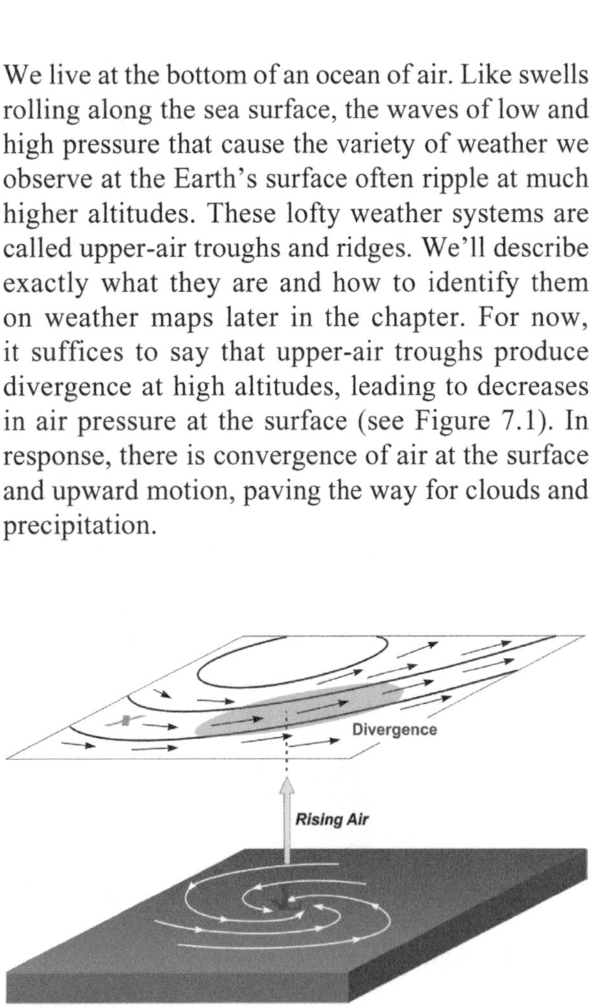

FIGURE 7.1 An upper-air trough ("X" marks the bottom, or "base," of the trough) produces an area of divergence at high altitudes, causing pressure to decrease at the surface. In response, low-level air converges and rises, setting the stage for clouds and precipitation. Note that the upper-air trough trails the surface low to the west (more on this later).

Detecting and charting upper-air troughs and ridges are crucial steps in weather forecasting. The process begins with observations gathered from the gaggle of radiosondes that are launched every day at 00 UTC and 12 UTC. These in-situ instrument packages probe the atmosphere up to heights of 15–30 km (9–18 mi), collecting measurements of temperature, pressure, humidity, and wind. Using these upper-air observations as well as others from aircraft and satellites, meteorologists plot maps of the weather above the surface. But they don't draw these upper-air weather maps at constant heights (for example, 2000 m, or 10,000 m). Instead, they plot weather data on **constant pressure surfaces**, which are surfaces on which the pressure (not the altitude) is constant.

Figure 7.2 shows two views of a hypothetical constant pressure surface. Note that it undulates in a series of rolling hills and valleys, indicating that a given pressure does not occur at the same height everywhere. As it turns out, charting these rolling hills and valleys is crucial for identifying upper-air troughs and ridges that help to produce "weather" observed at the ground.

That may sound like a pretty big challenge. Let us assure you that it's as easy as reading a topographic map. To see what we mean, check out Figure 7.3a, which is a visible satellite image of the Big Island of Hawaii. Skiing anyone? That's snow on the summits of Mauna Loa (center) and Mauna Kea (to its north). Yet, after skiing, you can also surf big waves on the island's northern shores. Clearly, there are large differences in elevation on the island. Figure 7.3b, a topographic map that shows contours of constant elevation (in feet) for the island, confirms our observation.

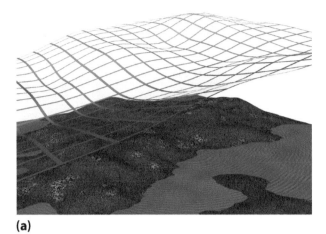

(a)

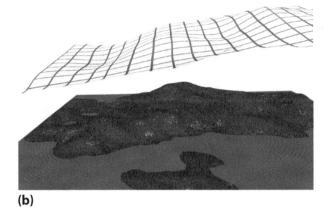

(b)

FIGURE 7.2 (a)–(b) Two schematic views of a constant pressure surface. Notice that such a surface is not flat, but rather has undulations, indicating that we can observe a specific pressure (500 mb, for example) at different altitudes.

For all practical purposes, constant pressure charts are very similar to a topographic map. In this chapter, you'll learn how to interpret upper-air weather charts and to identify upper-air troughs and ridges, giving you more skills in your apprenticeship as a weather forecaster.

CONSTANT PRESSURE SURFACES: LIKE CRUMPLED THROW RUGS

As we already mentioned, constant pressure surfaces are not flat. If we liken constant height surfaces to perfectly laid throw rugs, then constant pressure surfaces look a bit crumpled, as if someone wiped their feet and didn't bother to flatten the rugs out (see Figure 7.4). So think of constant pressure surfaces as nearly flat surfaces with some crinkles and crumples.

Why would meteorologists prefer crumpled constant pressure surfaces to nice and flat constant height

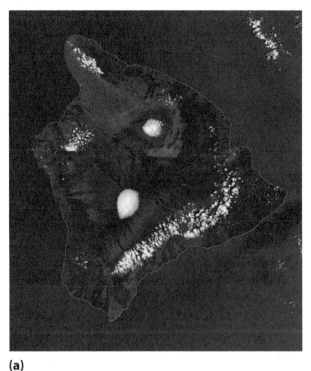

(a)

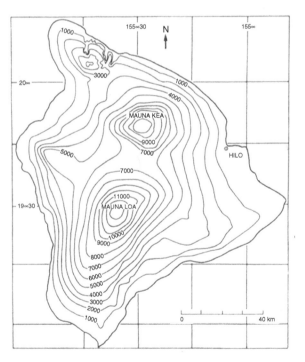

(b)

FIGURE 7.3 (a) A true-color visible satellite image of the Big Island of Hawaii. Note the snow-covered summits of Mauna Loa (center) and Mauna Kea (courtesy of NASA); (b) A topographic map of the Big Island shows contours of constant elevation (courtesy of the American Meteorological Society).

FIGURE 7.4 A constant pressure surface is like a crumpled throw rug. It has ridges and valleys as if someone wiped their feet and didn't bother to flatten it out.

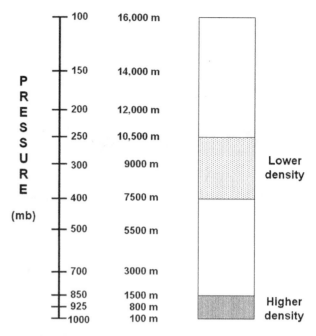

FIGURE 7.5 The mandatory pressure levels, with representative standard heights. At high altitudes, where air density is less, a thicker layer of air is needed to match the weight of a thinner layer of air at lower altitudes. Here, as an example, the vertical separation between 400 mb and 250 mb, representing a pressure difference of 150 mb, is much greater than between 1000 mb and 850 mb in the lower troposphere.

surfaces? That's a good question. As it turns out, the complicated mathematical equations that represent the behavior of the atmosphere are simpler when they're applied to constant pressure surfaces. So, if professional meteorologists use constant pressure surfaces to analyze and predict the weather, then we want you, as an apprentice weather forecaster, to use them too. Plus, nearly all of the upper-air maps you'll see on the Web are plotted on constant pressure surfaces.

The constant pressure surfaces that you'll deal with most frequently are called **mandatory pressure levels**. They are the ones at which radiosondes always take observations: 1000 mb, 925 mb, 850 mb, 700 mb, 500 mb, 400 mb, 300 mb, 250 mb, 200 mb, 150 mb, and 100 mb. We give the standard heights for each of these mandatory levels in Figure 7.5. The notion of a standard height is not in conflict with the idea that constant pressure surfaces have hills and valleys. Think of a standard height of a mandatory pressure level as a representative, or reasonable estimate, of the height at which that particular pressure would likely be found.

Figure 7.5 shows the vertical spacings between the mandatory pressure levels. Given that these pressure levels span from sea level to nearly 20 km, we had to mathematically scrunch the vertical axis in a way that accurately portrayed the relative spacing between successive pressure levels. We used the standard heights as guidelines to draw them. Notice that the vertical spacing between 1000 mb and 850 mb, for example, is noticeably less than, say, the spacing between 400 mb and 250 mb, though both intervals represent the same change in pressure. What's up with that?

Remember that air density decreases with increasing altitude. Also recall that the weight of an air column (per unit cross-sectional area) approximates air pressure. To match the weight corresponding to the 150 mb of pressure between 1000 mb and 850 mb (a layer in the relatively dense lower troposphere), a much thicker layer is required in the relatively low-density upper troposphere (see Figure 7.5). Thus, the vertical distance between the 400-mb and 250-mb levels is much larger than the vertical distance between 1000 mb and 850 mb.

Like the topographic map of Hawaii, constant pressure charts have contours of equal elevation. Figure 7.6 is a 500-mb map over the contiguous states on a winter day. The contours of constant elevation represent the height of the 500-mb pressure surface above sea level, in units of meters (m). Note that the 500-mb surface dipped lowest over central Canada (the purple region marks the lowest heights of the 500-mb surface above sea level at this time). Also note that the 500-mb surface reached its apex over the southeastern United States and tropical regions farther south (the red shading marks the highest heights of the 500-mb surface above sea level at this time).

Now that we've presented an example of data plotted on a constant pressure surface, we'll rewind and take you through the process of creating constant

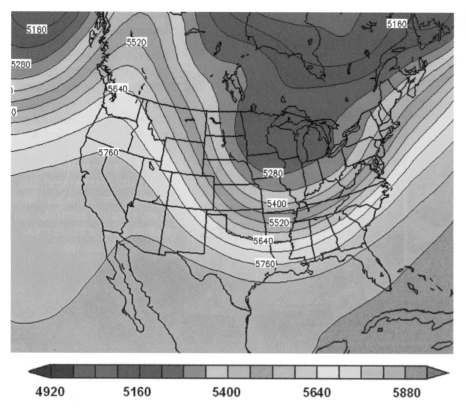

FIGURE 7.6 The 500-mb constant pressure map at 12 UTC on a January day. Isopleths of constant height above sea level are expressed in meters, with constant height lines drawn every 60 m (courtesy of NOAA).

4920   5160   5400   5640   5880

pressure maps. Then we'll teach you how to interpret them. If you have an idea of how topographic maps work, you're already ahead of the game.

Creating and Interpreting Maps on Constant Pressure Surfaces: Step by Step

Figure 7.7 shows three columns of air that extend from sea level (where we assume the pressure is 1000 mb) upward to 500 mb. The column on the left lies over the tropics. The center column resides in the middle latitudes, while the column on the right lies over polar regions. Although the pressures at the bottoms and tops of the three columns are all the same, the air columns vary in height.

The colorful shadings in the three air columns represent a graduated scale for temperature (dark red indicates the warmest air and dark blue corresponds to the coldest air). The variations of warm and cold air with latitude and altitude offer clues to how we can best explain the difference in the heights of the tops of the three air columns. Clearly, the tropics lay claim to the tallest air column, whose average temperature (from top to bottom) is the highest of the three columns. In contrast, the shortest air column over the polar regions has the lowest average temperature. We hope you agree that making a direct

connection between the average column temperature and the column's height seems promising.

Let's gather some empirical evidence. Consider Figure 7.8, which shows the results of an experiment conducted in one of the author's kitchen. He placed an empty soda bottle into a pan on the stove and

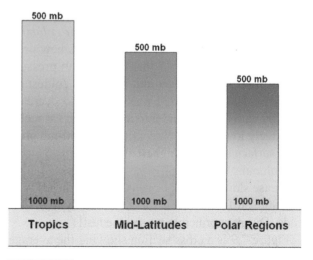

FIGURE 7.7 Columns of air over the tropics (left), mid-latitudes (center), and polar latitudes (right), stretching from sea level (where the pressure is assumed to be 1000 mb) to the 500-mb level. Colors correspond to different temperatures; warmest air is red, coldest is dark blue.

FIGURE 7.8 Like a cold column of air that deflates in the vertical, a balloon fastened over the neck of an empty soda bottle hangs limply after the bottle was immersed in ice for several minutes (left). Like a warm column of air that inflates in the vertical, the balloon expands upward as the bottle-balloon system heats up on a kitchen stove (right). This experiment should give you a sense of why a relatively warm column in the tropics is taller than its colder counterpart in polar regions.

covered the top of the bottle with a balloon (left). Then he turned on the burner and observed the balloon expanding upward (right), demonstrating that air increases its volume when heated (provided pressure is held constant). Applying the results of the kitchen experiment to Figure 7.7, it makes sense that warm tropical air columns are typically the tallest, while cold air columns that lie at high latitudes are the shortest.

We can now assert a guiding principle that serves as the first step in understanding maps on constant pressure surfaces. Looking at the three air columns in Figure 7.7, we deduce that *pressure decreases more rapidly with increasing altitude in cold air columns and more slowly with altitude in warm air columns.*

Let's stay with Figure 7.7 and test this new principle. Note in Figure 7.7 that the 500-mb pressure surface lies at the lowest altitude in the coldest air column. That means that pressure has to decrease rapidly with height in order to reach 500 mb at such a relatively low altitude. In contrast, the 500-mb pressure surface lies at the highest altitude in the warmest air column. What's the interpretation? Pressure has to decrease slowly with height in order for the 500-mb pressure surface to lie at such a relatively high altitude. Our new principle passed the test!

Figure 7.9 is a cross section showing the average height of the 500-mb surface from the equator to the North Pole. On this cross section, note that the 500-mb pressure surface slopes downward from the taller, warmer tropical air columns to the shorter, colder columns over polar regions, which is consistent

with the principle you just learned. On any given winter day (when temperature contrasts between the tropics and polar regions are large), the drop in altitude of the 500-mb surface from the equator to the North Pole can approach 1200 m or so. For reference, a representative 500-mb height during winter is 4700 m at the poles and 5900 m over the tropics.

Okay, let's take the next step in creating and interpreting weather maps drawn on constant pressure surfaces. Look again at the cross section of the average 500-mb heights in the Northern Hemisphere. Doesn't it remind you of hilly terrain that reaches from a valley (polar regions) to a plateau (tropics)? Just as we contoured elevations on the Big Island of Hawaii, we can make a topographic map of this "hilly terrain" that visually describes the cross section in Figure 7.9.

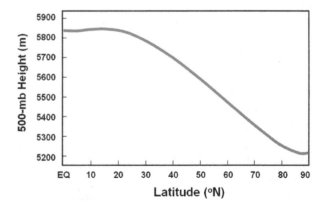

FIGURE 7.9 Long-term average height (in m) of the 500-mb surface from the equator to the North Pole.

By way of review, a topographic map is a flat, plan view (recall Chapter 1), even though it represents terrain that is not flat. The contours of constant elevation drawn on standard topographic maps allow us to locate terrain that is steep or relatively flat. What about a map of a constant pressure surface such as 500 mb? It is also a flat, plan view, even though the 500-mb constant pressure surface it represents is not flat. Once we draw contours of constant elevation (height above sea level) on a plan view of the 500-mb surface, we will be able to use this map to help us forecast the weather.

But we're getting a bit ahead of ourselves, so let's start from scratch. First, for sake of argument, let's suppose that the lines of constant elevation of the 500-mb pressure surface on a winter day line up perfectly with latitude (as in Figure 7.10a). Formally, we call these isopleths of equal elevation **height lines** or **height contours,** or more succinctly, just **heights** or **contours.** Again, the lowest heights on this 500-mb pressure surface lie to the north where air columns are coldest; the highest heights of the 500-mb surface lie to the south where air columns are warmest.

Of course, the tropics and poles are very territorial, with cold and warm air masses staking a claim to as much territory as possible. So cold air masses routinely forge southward and warm air masses advance northward (see Figure 7.10b). In other words, latitude markers have no control when these air masses wage war. To realistically reflect these territorial struggles, the idealistic height lines in Figure 7.10b must be distorted to look more like Figure 7.10c as pushy cold air dips southward and warm air bulges northward.

And indeed, some 500-mb patterns have pronounced dips and bulges. For example, the 500-mb map in Figure 7.11a shows a pronounced southward dip of heights over the Middle West, associated with cold air between the ground and 500 mb. There is also a prominent northward bulge in 500-mb heights over the western United States and western Canada, associated with warm air between the ground and 500 mb. Other 500-mb patterns (such as the one in Figure 7.11b) are not nearly as wavy—note the lack of pronounced northward bulges and southward dips of 500-mb heights. We'll talk more about these different types of upper-air patterns and their connections to weather a bit later in this chapter. For now, just note the variety of height patterns, and how this variety is linked to the average temperature of air columns.

Note the consistency of the labels on the heights for the maps in Figure 7.11. That's because there are

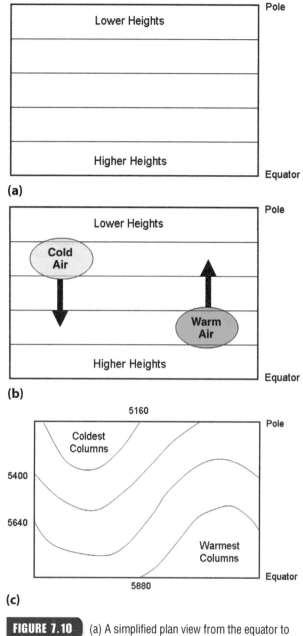

FIGURE 7.10 (a) A simplified plan view from the equator to the North Pole showing a few idealized 500-mb height lines; (b) Realistically, the atmosphere, which works to mitigate north-south temperature contrasts, routinely sends cold air equatorward and warm air poleward; (c) Thus, a more realistic pattern of 500-mb heights has dips and bends. Here, we have assigned realistic values (in m) to the height lines. In this case, the coldest columns of air (on average) between the surface and the 500-mb level are found north of the 5160-m height line, while the warmest columns of air (on average) are found south of the 5880-m height line.

conventions for drawing heights on 500-mb maps (and maps of other constant pressure surfaces). For example, the standard contour interval for 500-mb maps is 60 m. The conventional height contours on

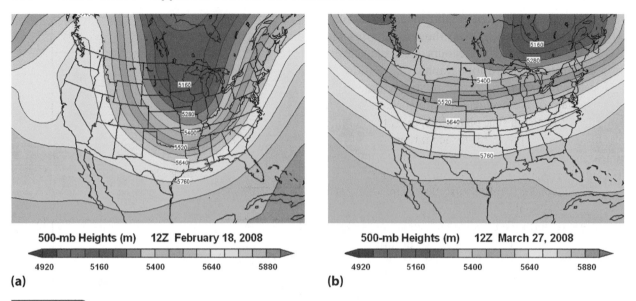

(a) 500-mb Heights (m)   12Z February 18, 2008

4920   5160   5400   5640   5880

(b) 500-mb Heights (m)   12Z March 27, 2008

4920   5160   5400   5640   5880

FIGURE 7.11 (a) The 500-mb map on a February day shows a pronounced northward bulge in heights over the western United States and western Canada, associated with warm air between the ground and 500 mb. Also note the pronounced southward dip in heights over the Middle West, associated with cold air between the ground and 500 mb; (b) The 500-mb map on a different day shows a lack of pronounced northward bulges and southward dips of 500-mb heights (courtesy of NOAA).

500-mb maps include . . . 4920 m, 4980 m, . . . , 5400 m, 5460 m, . . . , 5640 m, 5700 m, and so on. How high or low the contour labels go depends on season and geographic location. For example, the 5940-m height sometimes appears on 500-mb maps during the warm season over low latitudes, while the 4820-m height sometimes appears on 500-mb maps in winter over polar latitudes.

For this text, we'll talk primarily about three constant pressure maps, each representative of one level of the troposphere: 850 mb (lower troposphere), 500 mb (middle), and 300 mb (upper). The standard height and interval at 850 mb are, respectively, 1500 m and 30 m (see Figure 7.12a). At 300 mb, the standard height and interval are, respectively, 9000 m and 120 m (see Figure 7.12b).

Note that these 850-mb and 300-mb charts are from the same time as the 500-mb chart in Figure 7.11a, and the height patterns on the three charts share some similarities. This is not a coincidence. After all, height patterns on a constant pressure surface are a reflection of the average temperature of the air below that pressure surface, so temperature data that were used to create Figure 7.12a (the 850-mb chart) also are used to create the 500-mb chart, but they're combined with additional temperature data between 850 mb and 500 mb. Similarly, temperature data used to create the 500-mb chart are used to create the 300-mb chart, but they're combined with additional temperature data between 500 mb and 300 mb. Got the idea? Upper-air height charts are built on a foundation of temperature.

How are maps of heights on constant pressure surfaces (like the ones you've just seen) generated? The height data, as well as temperature and humidity observations, come from the network of radiosondes that are launched each day at 00 UTC and 12 UTC, from satellites, and from aircraft. In addition, a radiosonde can be tracked as it rises to provide information about the wind speed and direction aloft. The upper-air data observed at a particular pressure level are then organized into station models that resemble the surface station models we introduced in Chapter 1. Upper-air data can then be isoplethed with height lines, isotachs (lines of equal wind speed), and isotherms to identify important weather patterns at high altitudes. In fact, many upper-air weather charts used by operational meteorologists display multiple fields. For example, Figure 7.13 is a 300-mb map of the North Pacific Ocean and North America that displays heights (dark contours), wind barbs that indicate wind speed (in knots) and wind direction, and regions with the fastest 300-mb wind speeds (color-filled isotachs).

Though information about winds aloft can be estimated from radiosondes, weather forecasters can also infer the wind direction and speed from the

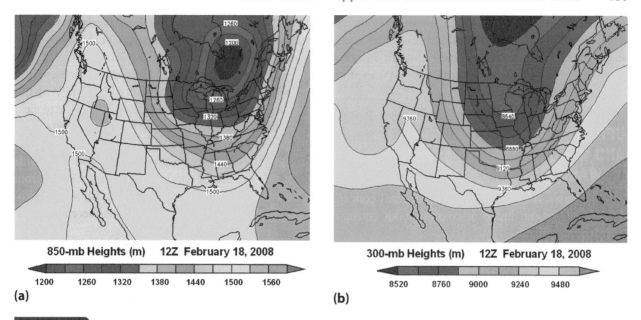

850-mb Heights (m)   12Z February 18, 2008

1200   1260   1320   1380   1440   1500   1560

(a)

300-mb Heights (m)   12Z February 18, 2008

8520   8760   9000   9240   9480

(b)

FIGURE 7.12 Upper-air maps on a February day: (a) Heights (in m) on the 850-mb pressure surface, with height lines drawn every 30 m (every other height line is labeled); (b) Heights (in m) on the 300-mb pressure surface, with height lines drawn every 120 m (every other height line is labeled).

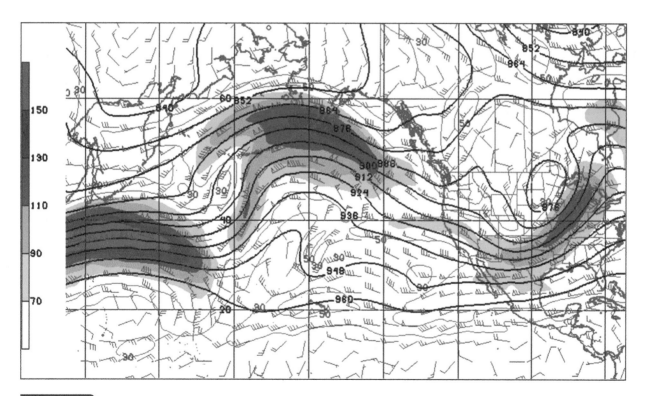

FIGURE 7.13 Maps at the mandatory pressure levels sometimes display other analyzed fields in addition to heights. For example, here is a 300-mb chart from mid-December 2015. The dark contours are 300-mb heights (labeled in tens of meters), while the wind barbs indicate wind speed (in knots) and wind direction at 300 mb. The color-filled isotachs indicate regions with the fastest wind speeds, using the color code on the left (courtesy of NOAA).

patterns of heights on constant pressure surfaces. This relationship is fundamental to our understanding of how atmospheric patterns aloft fit into the grand scheme of weather forecasting. Let's explore this topic further.

UPPER-AIR PATTERNS OF HEIGHT AND WIND: IT ALL BOILS DOWN TO WARM AND COLD

For emphasis, let's restate the fundamental concept you learned in our initial discussion about constant pressure surfaces. In a nutshell, relatively low heights typically lie at higher latitudes, where tropospheric air columns are relatively cold. And relatively high heights typically are found at lower latitudes, where tropospheric air columns are relatively warm. With this fundamental principle in hand, the next hurdle to clear is understanding the relationship between patterns of heights and upper-air wind directions and wind speeds.

Figure 7.14a shows the average 300-mb heights over an expanse of the North Atlantic Ocean from December 1, 2014 to February 28, 2015 (meteorological winter in the Northern Hemisphere). Note the

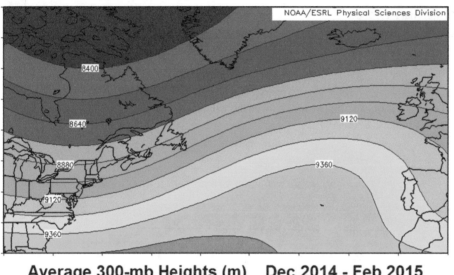

Average 300-mb Heights (m)   Dec 2014 - Feb 2015

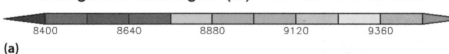

(a)

FIGURE 7.14 Average 300-mb heights (in m) over the North Atlantic Ocean from December 1, 2014 to February 28, 2015. The eastern United States is on the left and extreme western Europe is on the right; (b) The average 300-mb winds over the North Atlantic Ocean from December 1, 2014 to February 28, 2015. Arrows indicate wind direction, and wind speeds are color-coded in meters per second (m/s). To convert from m/s to mph, simply multiply by 2.2; thus, for example, 40 m/s is approximately 88 mph (courtesy of NOAA).

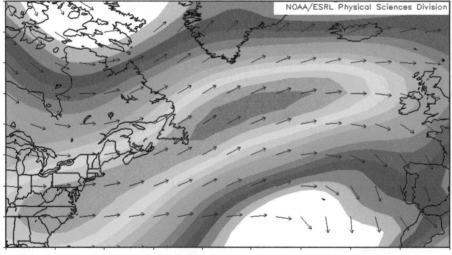

Average 300-mb Wind (m/s)   Dec 2014 - Feb 2015

(b)

predominantly west-southwest to east-northeast orientation of the 300-mb heights over the Atlantic. Also note the large gradient in 300-mb heights (where height lines are packed closely together) in a channel that extends from the East Coast of the United States to the British Isles.

Now compare these observations with Figure 7.14b, which shows the average 300-mb winds over the same region during the same time period. Note that, as a rule, 300-mb wind arrows are nearly parallel to the height lines, resulting in a general west-southwesterly average 300-mb wind direction over the North Atlantic region during these three winter months. Also note that the fastest winds, on average, are co-located with the largest height gradient, in a corridor that extends from the East Coast across the North Atlantic. This core of fastest 300-mb winds marks a portion of the jet stream, a narrow channel of relatively speedy upper tropospheric winds that encircles each hemisphere over the mid-latitudes. During the Northern Hemisphere's winter, the speedy jet stream typically flows across the North Atlantic Ocean, a pattern of upper-level winds that commercial airlines try to exploit on trans-Atlantic flights to Europe. We'll study the jet stream in greater detail a bit later in this chapter.

First, there are still two hurdles for you to clear on the fast track to understanding the usefulness of constant pressure surfaces. Why do winds tend to blow nearly parallel to height lines (this is particularly true on constant pressure surfaces that lie high above the earth's surface)? And why are the fastest wind speeds on constant pressure surfaces found in areas where height lines are packed closest together? Let's begin by clearing the first hurdle.

Interpreting Heights on Constant Pressure Surfaces: Clearing the First Hurdle

Now that we've established that heights on constant pressure surfaces resemble contours of constant elevation on standard topographic maps, the next logical question is this: How do we interpret, in the context of weather forecasting, the patterns of heights we observe? To answer this question, let's take a balloon ride (see Figure 7.15a). We promise that it will be "uplifting" in more ways than one. The first thing to remember is that pressure decreases more slowly with height in a warm air column than it does in a cold air column. Or, alternatively, pressure decreases faster with height in cold air than it does in warm air.

Okay, suppose you decide to take a balloon ride up through a cold air column. We'll take the balloon

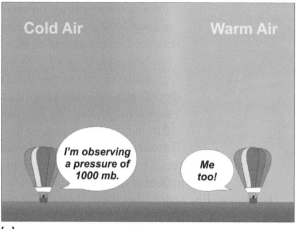

(a)

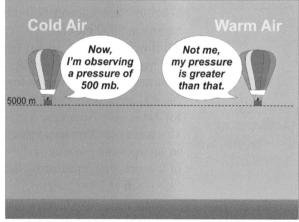

(b)

FIGURE 7.15 (a) Two hot-air balloons at the surface both register a pressure of 1000 mb; (b) The balloonist ascending in the cold air column decides to "hang out" at 500 mb, which we assume, for sake of argument, lies at an altitude of 5000 meters. The balloonist ascending in the warm air column reaches an altitude of 5000 meters, but the reading on his barometer is *greater* than 500 mb (remember that pressure decreases more slowly with altitude in warm air). Compared to the barometer mounted on the balloon in the cold air column (which reads 500 mb), the pressure at 5000 meters in the warm air column is *higher*. Thus, on the constant height surface at 5000 meters, there is relatively high pressure in the warm column and relatively low pressure in the cold column.

in the warm air column (thank you very much). Ready, on your mark, get set . . . . go! Rising from the ground, you get to 500 mb in the cold air column faster than we do in the warm air column. So, the 500-mb height in the cold air column is relatively low. Just for fun, let's say it's an even 5000 meters (see Figure 7.15b).

Now suppose you decide to just "hang out" at 5000 meters in the cold air column. Shivering, you look longingly southward toward the warm air column. Our balloon has also reached 5000 meters and we've decided to hang out there, too (it's not as cold). Unlike you, we haven't yet reached a pressure of 500 mb because, as you recall, pressure decreases more slowly in warm air. Indeed, the 500-mb level is still above us. Thus, as Figure 7.15b shows, the pressure we measure in the gondola attached to our balloon is greater than 500 mb. So, on the flat surface at 5000 meters, your gondola marks an area of low pressure (compared to the higher pressure measured in our gondola).

The bottom line here is that the 500-mb height in the cold air column is lower than the 500-mb height in the warm column. And the pressure you measured at 5000 meters is also lower than the one we measured. In summary, the cold column has a lower 500-mb height and a lower pressure at 5000 meters.

Thus, we have the following general result: On any constant pressure chart, a point that marks a region's lowest height corresponds to a center of low pressure on a flat surface at an altitude equal to (or approximately equal to) that height. Here's the working interpretation of this principle: You may treat a center of low height on any constant pressure surface as if it were a center of low pressure on a nearby flat (constant height) surface (such as 5000 meters).

We can mimic the argument for high heights and warm columns of air, yielding a second result: On any constant pressure chart, a point that marks a region's highest height corresponds to a center of high pressure on a flat surface at an altitude equal to (or approximately equal to) that height. In other words, you may treat a center of high height on any constant pressure surface as if it were a center of high pressure on a nearby flat (constant height) surface. From a practical standpoint, these are pretty powerful results because we know some things about circulations of air around high- and low-pressure systems.

We can go one step further. On constant pressure maps, heights packed tightly together represent a large height gradient. Given the close connection between patterns of height and patterns of pressure that we just developed, it follows that a large height gradient on a constant pressure surface can be interpreted the same way as a large pressure gradient on a nearby constant height surface; that is, a large height gradient implies relatively fast winds. Conversely, a small height gradient on a constant pressure surface can be interpreted the same way as a small pressure gradient on a nearby constant height surface. Looking back to Figure 7.14, you should now understand

why the fastest 300-mb winds are co-located with the largest height gradient, while winds tend to be weaker where the height lines are less tightly packed.

Hopefully, these results help you feel more comfortable working with upper-air maps. To close the deal, consider Figure 7.16a which shows a generic pattern of 500-mb heights over the United States. Note the equatorward dip in heights over the East, associated with southward-building, relatively cold columns, and the poleward bulge in heights over the West, associated with northward-building, relatively warm columns of air. Figure 7.16b shows what the corresponding pattern of isobars would look like at 5500 meters, a reasonable altitude for a pressure of 500 mb. Areas of high and low pressure at the 5500-m level are marked with H and L. We also indicate an elongated area of low pressure, or trough, with a dashed line, and an elongated area of high pressure, or ridge, with a serrated line (recall these conventions from Chapter 6). Also note that where the height gradient is relatively large in Figure 7.16a, the pressure gradient is relatively large in Figure 7.16b (for example, over the East), and vice versa.

Clearly, the similarity between the two patterns is striking. For practical purposes, these two charts are interchangeable—they tell the same general story. Lastly, in Figure 7.16c, we repeat Figure 7.16a, but now we indicate areas of relatively high and relatively low height at 500 mb with H and L, and the elongated zones of high and low heights with the same serrated and dashed lines that we used for ridges and troughs of pressure. Yes, these same terms apply to upper-air charts—these are ridges and troughs of height.

Having the means to identify centers of high and low pressure aloft, we're now ready to talk about upper-level winds.

Determining Wind Directions on Constant Pressure Surfaces: Buys-Ballot's Law

The next hurdle in learning how to use upper-air weather maps is to relate height lines to the direction of the wind. Of course, as you learned in Chapter 6, horizontal differences in pressure (that is, the pressure gradient force) drive the wind. However, on a constant pressure surface, there are no differences in pressure, and thus no isobars! That's where heights come in—they are the constant-pressure equivalent to isobars, and so on a constant pressure chart, the **height gradient force** drives the wind. Like the

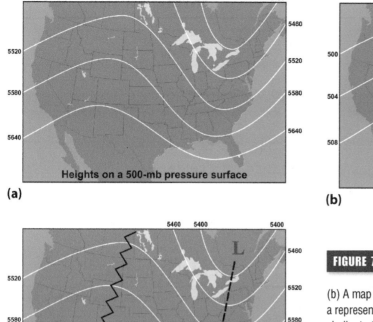

(a)

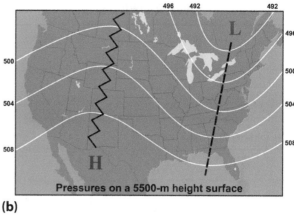

(b)

(c)

FIGURE 7.16 (a) A generic map of 500-mb heights with a trough in the East and a ridge in the West; (b) A map of pressure on a constant height surface at 5500 m, a representative altitude for a pressure of 500 mb. The map looks similar to the map of 500-mb heights in part (a), suggesting that high heights (low heights) on a constant pressure surface can be treated as if they were high pressures (low pressures) on a nearby constant height surface; (c) Highs and lows and ridges and troughs of height can thus be analyzed on a constant pressure surface just as highs and lows and ridges and troughs of pressure would be on a constant height surface.

pressure gradient force, the height gradient force acts from high to low heights in a determined effort to erase the height gradient.

With this in mind, turn your attention to the idealized map of 500-mb heights shown in Figure 7.17a (heights are the thin horizontal lines). Assume that we're in the Northern Hemisphere where an air parcel (the black circle) is initially restrained from moving toward lower heights in response to the height gradient force, which is represented by the red arrow.

Now, release the parcel. Immediately, it accelerates toward lower heights, but as soon as the parcel is in motion, the Coriolis force kicks in, acting 90° to the right of the motion. At some short time later (see Figure 7.17b), the parcel has made progress toward lower heights, but not directly so, having been deflected by the Coriolis force (represented by the green arrow). Note that the magnitude of the Coriolis force is also relatively small because it depends, in part, on the parcel's velocity. As the air parcel accelerates toward lower heights in response to the height gradient force, the magnitude of the Coriolis force steadily increases because the parcel's speed increases (Figure 7.17c). As a result, deflections to the right

become greater with time. The greater deflection always means a new course adjustment for the air parcel, and the Coriolis force shifts its direction accordingly so that it can continue to act 90° to the right of the new path of the air parcel. In a nutshell, as the parcel accelerates, the direction in which the Coriolis force acts steadily shifts like an hour hand on a clock, while the magnitude of the Coriolis force increases (see Figure 7.18).

Eventually, the height gradient force and the Coriolis force reach a balance, shown in Figure 7.17d—at this point, no net force acts on the air parcel. Based on the fundamental laws of motion (first proposed by seventeenth-century physicist and mathematician Isaac Newton), the parcel will stop accelerating and continue to move in a constant direction at a constant speed. In this case, the final direction of the air parcel is directly eastward (in other words, the wind blows from the west). Indeed, note that the velocity vector parallels the height lines. In fact, no matter how large or how small the initial height gradient is, the wind will always blow parallel to the height lines once the Coriolis and height gradient forces balance. The height gradient

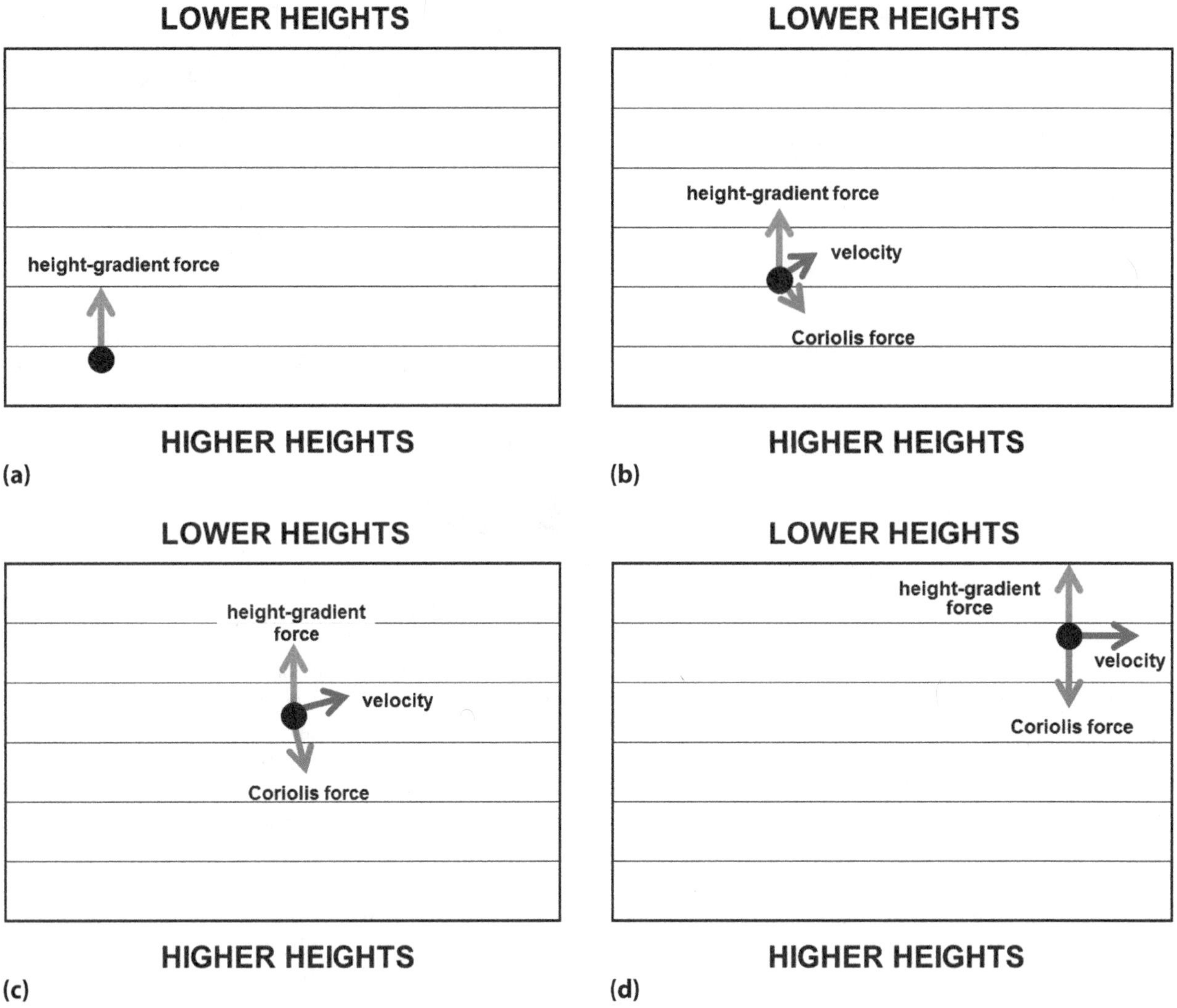

LOWER HEIGHTS

(a)

HIGHER HEIGHTS

LOWER HEIGHTS

(b)

HIGHER HEIGHTS

LOWER HEIGHTS

(c)

HIGHER HEIGHTS

LOWER HEIGHTS

(d)

HIGHER HEIGHTS

FIGURE 7.17 (a) An idealized 500-mb map showing an air parcel initially restrained from moving. The red arrow represents the height gradient (pressure gradient) force; (b) The parcel is released and accelerates, but the velocity of the parcel (blue arrow) is small at the start. Note that the Coriolis force (the green arrow) acts 90° to the right of the velocity. Also notice that the magnitude of the Coriolis force is small because it depends, in part, on velocity; (c) Some time later, the air parcel has continued to move toward lower heights and continues to speed up (because the forces are not in balance). At this point, the parcel's velocity has noticeably increased. Note that the magnitude of the Coriolis force has also increased in response to the parcel's increasing velocity; (d) Once the height gradient force and the Coriolis force balance each other, the parcel no longer accelerates, and it will now move parallel to the height lines at a constant velocity. The velocity arrow in this balanced state represents the geostrophic wind.

can only dictate the final speed of the parcel (more on this topic in just a bit).

The idea that the air parcel would end up moving parallel to the height lines rather than making a beeline directly toward the area of lowest heights may surprise you. Such a result shows the huge impact of the Coriolis force. Never underestimate the impact of the effects of the rotating earth.

In the interest of full disclosure, we should mention that, in reality, the height pattern in Figure 7.17 won't stay exactly the same as we go from part (a) to part (d);

instead, the heights will change slightly as air parcels move because their movement changes the weights of local air columns. However, this slight adjustment won't make any difference in the overall result.

The wind that results when the height gradient force (or pressure gradient force) and the Coriolis force come into balance is called the **geostrophic wind**. From the standpoint of word origins, *geo* means "earth" and *strophic* means "turning," a clear reference to the importance of the role of the Coriolis force in creating this idealized wind. Formally, the

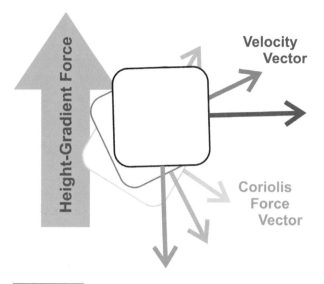

FIGURE 7.18 Green arrows represent the Coriolis force in parts (b)–(d) of Figure 7.17. Over time, the Coriolis force strengthens as the wind speed increases, eventually balancing the height gradient (pressure gradient) force.

state at which the height gradient force (pressure gradient force) and the Coriolis force are in perfect balance is called **geostrophy**.

Before we go any further, we want to emphasize that the geostrophic wind is an idealized wind. It results from a picture-perfect balance between the pressure gradient force and the Coriolis force. This balance is fine in theory, but often is, in practice, unrealistic, because air moving in the real world is subject to other forces such as friction. Yes, the real atmosphere can, at times, get close to geostrophy, particularly at high altitudes where friction with the ground becomes negligible. But realistically, you can't fly a kite in the geostrophic wind.

At the surface, winds are far from geostrophic. Recall from Chapter 6 that observations show that air moves across the isobars inward toward lower pressure at an average angle of about 30° over land—the crossing angle is smaller over water and often larger over rougher land. As an example, Figure 7.19a shows isobars of sea-level pressure and

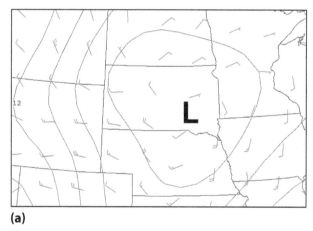

(a)

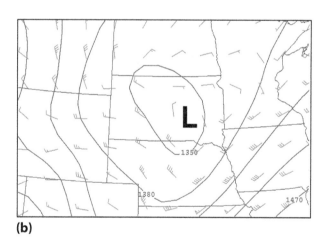

(b)

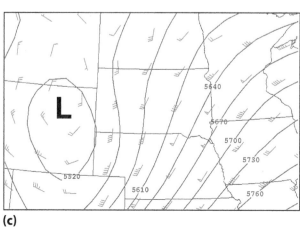

(c)

FIGURE 7.19 Weather data at 12 UTC on an October day. (a) Surface wind barbs (in knots) and isobars of sea-level pressure (in mb); (b) Corresponding 850-mb heights (in m) and wind barbs; (c) Corresponding 500-mb heights (in m) and wind barbs (courtesy of NOAA).

surface wind barbs associated with a low-pressure system over the Northern Plains. Notice that surface winds east of the low centered over South Dakota crossed isobars at relatively small angles. But on the western side of the low (over Wyoming, for example), crossing angles were noticeably greater, owing largely to the rough terrain of the Rockies. In Figure 7.19b, which shows the corresponding 850-mb heights and wind barbs, note that the winds east and south of the 850-mb low centered over central South Dakota were pretty much parallel to the height lines (the effects of friction were relatively limited there). But the crossing angles west of the low, where some of the terrain reaches up to at least 850 mb, were still rather large, indicating that friction was still hard at work. Here, winds are not geostrophic (not even close). Higher in the atmosphere, at the 500-mb level (see Figure 7.19c), winds everywhere more closely paralleled the height lines, indicating that friction played less of a role in the balance of forces at these higher altitudes. Lesson learned: The effect of friction with the earth's surface fades with increasing altitude. As a result, winds at higher altitudes tend to be closer to geostrophy than winds at lower altitudes.

With this lesson in mind, we can now simply look at the height lines on an upper-air map and immediately get a sense of the wind direction. How do we do this? First, always remember that winds in the middle and upper troposphere tend to be geostrophic and blow nearly parallel to the height lines.

This observation narrows down the possibilities for wind direction to essentially two choices. Now imagine standing on the map with the wind (which we assume is nearly geostrophic) at your back as in Figure 7.20. Then low heights (low pressure) should be on your left in the Northern Hemisphere,

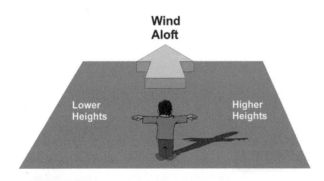

Wind Aloft

Lower Heights

Higher Heights

FIGURE 7.20 If you stand with the wind at your back on a constant pressure surface, lower heights (lower pressure) will be on your left in the Northern Hemisphere.

allowing you to make the correct choice for wind direction. This rule for determining wind direction on upper-air maps is called **Buys-Ballot's Law**, named for the Dutch meteorologist who published the first rigorous validation of the rule in 1857. Look back to Figures 7.14 and 7.19 to see Buys-Ballot's Law at work.

Although you already have an inkling of the connection between height gradients and wind speeds, we will now specifically address this concept.

CONTROLLERS OF WIND SPEEDS ALOFT: BECAUSE THEY ARE THERE

By the early 1920s, successful expeditions to the North and South Pole left only the so-called "Third Pole" unconquered. This last challenge was Mount Everest, whose icy summit protruded above 29,000 feet (see Figure 7.21). On June 6, 1924, two British climbers, George Mallory and Andrew "Sandy" Irvine, left their base camp high on the mountain for a final assault on the summit that Mallory believed would take three days. Mallory's motivation to climb the mountain has long been immortalized by his glib remark, "Because it is there."

John Odell, a mountaineering enthusiast who five years earlier had made the first public suggestion that Mount Everest should be climbed, acted as the expedition's cinematographer. On June 8, 1924, Odell, positioned farther down the mountain as a "support team," spotted Mallory and Irvine as two tiny black dots "nearing the base of the summit pyramid." Suddenly, clouds rushed in and a snow squall lashed the summit. Mallory and Irvine were never heard from again.

It should come as no surprise that the number-one concern of Everest climbers is always weather, with strong winds posing the greatest threat. For safety, wind speeds near the summit (where the pressure is about 300 mb) should not exceed 18 m/s (40 mph). As wind speeds increase beyond this threshold, climbers are exposed to dangerous wind chills (see Chapter 16) and exhaustion (after all, oxygen is scarce when the pressure is only 300 mb, so any steep climb in strong winds is even more exhausting). Heavy snow squalls, like the one that likely took the lives of Mallory and Irvine, rank a close second in the list of weather threats on Mount Everest.

The most common "window of opportunity" to climb Mount Everest is May, after the strong westerly winds of early spring weaken and retreat

FIGURE 7.21 Mount Everest as viewed from the south. The cloudless sky suggests perfect climbing weather, but the blowing snow at the summit indicates that winds might have been an issue. The summit of Everest towers to 8849 m (29,035 ft), a measurement determined by the National Geographic Society in 1999. However, the government of Nepal has not recognized this value, which is six feet higher than the elevation of 29,029 feet first determined by an Indian survey in 1955 (courtesy of Henry Firus, Flagstaffotos, http://www.flagstaffotos.com.au/gallery2/main.php).

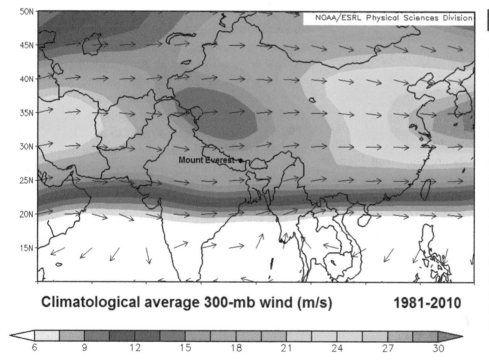

Climatological average 300-mb wind (m/s)   1981-2010

FIGURE 7.22 The climatological average wind speed and direction at the 300-mb level during May (in units of m/s) shows that the channel of fastest winds (the yellows and reds) has shifted north of Mount Everest, stretching from Iran and Afghanistan to eastern China and South Korea (this is true despite the relative minimum in southwestern China). Winds near the summit of Everest average about 18 m/s (40 mph) in May, right around the threshold for safety near the top of the world (courtesy of NOAA).

northward but before precipitation from Southeast Asia's summer monsoon begins (more on this feature in Chapter 10). Figure 7.22 shows the average wind speed and direction during May at 300 mb for a large swath of southern Asia centered on Mount Everest. Note that average wind speeds at this lofty height in May are about 18 m/s (40 mph) near Everest, the threshold for safety, so it's safe to assume that winds are below this value at least

part of the time in May. In fact, in recent years, about 95% of all summits in May have taken place between May 13 and May 26.

Why do winds tend to be so strong at high altitudes? What controls the speed of the wind on constant pressure surfaces such as 300 mb? Why do wind speeds increase from summer to winter, and what is the jet stream? Time to get to the bottom of these issues.

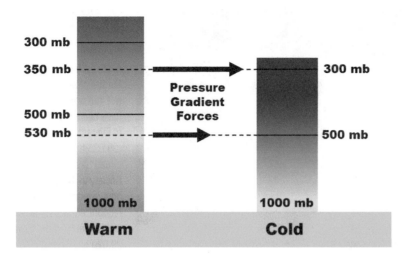

FIGURE 7.23 Horizontal temperature gradients create differences in heights between warm and cold air columns on constant pressure surfaces (here, the 500-mb and 300-mb levels are shown in both columns). As a result, there are horizontal pressure gradients, which, in turn, cause the wind to blow.

Leading to the Jet Stream: A Recurring Theme of Warm and Cold

The first question we'll tackle is why winds typically increase with altitude in the troposphere. Figure 7.23 shows two columns of air (one warm and one cold) extending from the surface to the tropopause. Note the disparities between the heights of the 500-mb and 300-mb surfaces in the two columns.

For sake of argument, suppose the 500-mb height and 300-mb height in the cold column are 5100 m and 8100 m, respectively. Imagine a horizontal line drawn from the cold column into the warm column at these altitudes (both lines are shown in Figure 7.23). Reasonable values for the pressures at these altitudes in the warm column might be 530 mb and 350 mb. Clearly, there are now horizontal pressure gradients at 5100 m and 8100 m (in other words, differences in pressure between the two columns over some horizontal distance). These horizontal pressure gradients are represented by the two black arrows— the length of each arrow qualitatively represents the magnitude of each pressure gradient force.

You've previously learned that we can treat gradients in heights on constant pressure surfaces as if they were horizontal pressure gradients. In light of Figure 7.23, this connection between height gradients and pressure gradients should ring true even more. Now we can go a step further. Note that the larger height gradient at 300 mb corresponds to a larger pressure gradient (compare the lengths of the two arrows). Thus, we would expect the wind at 300 mb to be faster than the wind at 500 mb.

To demonstrate that Figure 7.23 does indeed represent scientific truth, check out Figure 7.24a which shows 500-mb heights and wind barbs at 12 UTC on a September day in 2015. Where the heights are most tightly packed—for example, in western Canada—the wind barbs clearly depict the fastest winds, some in excess of 50 knots. In contrast, where height lines are loosely packed (for example, over the Pacific Northwest and Atlantic Ocean), wind speeds are much lower.

Now let's go higher in the atmosphere. Figure 7.24b shows the 250-mb heights and wind barbs over the same area at the same time. The values of the heights are greater, of course, but the pattern of 250-mb heights is fairly similar to the pattern at 500 mb. The major difference between the maps is that the winds are noticeably faster on the higher-altitude pressure surface (check out, for example, western Canada, where some wind speeds are now in excess of 100 knots). And, in general, winds increase with increasing altitude over the mid-latitudes in the troposphere, a notion that may not be unexpected, especially given our story about Mount Everest. That idea is nicely illustrated in Figure 7.25, which shows a typical wind profile over the mid-latitudes during the cold season, a time when north-south temperature contrasts are at their greatest and winds aloft are fastest. A close look at this vertical wind profile, however, shows that wind speeds reach a maximum in the upper troposphere (here, near 270 mb) but then actually <u>decrease</u> above that level, a result that may surprise you. Once again, the answer is all about temperature gradients.

The crux of the explanation begins with Figure 7.26a, which represents a typical cross section from north (cold air in blue) to south (warm air in red). The temperature pattern in this cross section is consistent with the wind profile in Figure 7.25.

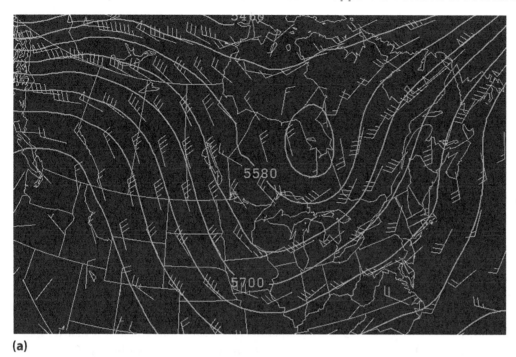

(a)

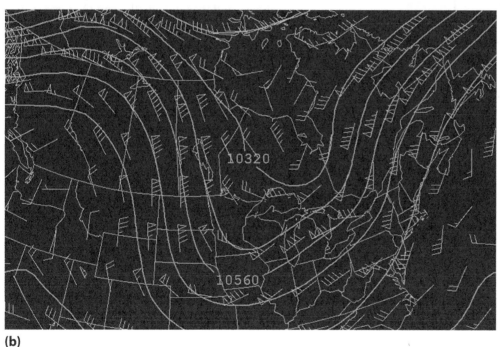

(b)

FIGURE 7.24 Upper-air weather charts from 12 UTC on September 11, 2015, showing heights and wind barbs: (a) The 500-mb map; (b) The 250-mb map.

Note that the vertical spacing between successive mandatory pressure levels is smaller in the colder air to the north than the warmer air in the south (for example, the yellow arrows highlight the disparity in the vertical spacing between 300 and 250 mb). This observation reflects the principle you learned earlier in the chapter: *Pressure decreases faster* *with increasing altitude in cold air than in warm air.*

Note how the constant pressure surfaces are relatively flat in both the cold air mass and the warm air mass, reflecting the horizontal uniformity of temperatures in each. But the transition zone between the warm and cold air masses, which lies over the

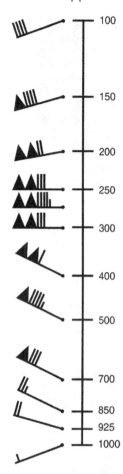

FIGURE 7.25 The vertical profile of winds measured by a radiosonde released at Caribou, ME, on a November day. Winds (both direction and speed) are shown. The station model convention for interpreting wind barbs applies here. Thus, for example, the wind at 500 mb is blowing from the west-northwest (about 290 degrees on the compass) at a speed of 95 knots.

mid-latitudes, is another story altogether. Given the differences in vertical spacing between successive pressure levels, each pressure surface must slant upward from the cold air to the warm air. Because the differences in vertical spacing between successive pressure levels in the cold and warm air masses magnify with decreasing pressure (increasing altitude), the slanted portions of the constant pressure surfaces that bridge the transition zone between the two air masses must get steeper with decreasing pressure (increasing altitude). This means that, in the zone where there is a transition from low heights in the cold air to high heights in the warm air, the height gradient on constant pressure surfaces increases with decreasing pressure (increasing altitude). And this means that wind speeds must increase with increasing altitude. Whew! Behold Figure 7.26b, where we have added the wind barbs

from Figure 7.25! With low heights to the north (in the cold air) and high heights to the south (in the warm air), winds blew from a general westerly direction—in fact, high-altitude winds over the mid-latitudes blow so persistently from a westerly direction that these winds are often simply referred to as the **westerlies**.

As it turns out, the westerlies reach their maximum speeds just below the tropopause. In the case shown by Figure 7.25, wind speeds reached a maximum (135 kt) at 270 mb (between 250 mb and 300 mb). Thus, it's reasonable to assume that the tropopause lay very close to 250 mb—for sake of argument, let's say that's the case. So the natural question to ask is this: why did wind speeds decrease at 250 mb and higher up?

Once again, the answer lies with temperature contrasts and the idea from Chapter 3 that the height of the tropopause varies with latitude (and therefore, temperature). Recall from Chapter 3 that temperatures remain constant with altitude just above the tropopause and then eventually start to increase with altitude (see Figure 7.27). With the decrease in temperature arrested in the cold air mass above about 250 mb, the vertical spacing between the 250-mb level and the 200-mb level is larger than it would have been had temperatures continued to decrease with altitude.

This is not the case in the warm air mass at these pressure levels because the tropopause is higher in the warmer air (maybe at the 150-mb level or so) and temperatures there continue to decrease with increasing altitude (see Figure 7.26c, where now we show higher-altitude pressure surfaces). As a result, in the zone where the two contrasting air masses meet, the slanted bridge on the 200-mb pressure surface is less steep than the bridge on the 250-mb surface. And the bridge on the 150-mb surface is less steep than the bridge on the 200-mb surface. Translation: At these lofty altitudes, the height gradient in the transition zone between the two contrasting air masses is weakening with increasing altitude. This reduced height gradient corresponds to a decrease in wind speed with increasing altitude, as demonstrated by the wind barbs in Figure 7.26c.

The bottom line here is this: when considering the vertical profile of winds in the transition zone between clashing cold and warm air masses, the fastest winds occur just below the tropopause, and tend to be westerly. Moreover, at this level of fastest wind, you will typically find a narrow channel where the height gradient is maximized and thus winds are fastest. This current of fast-moving air

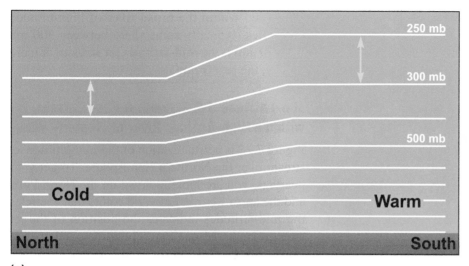

(a)

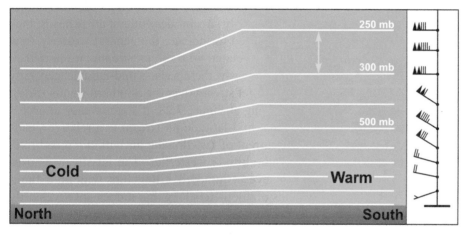

(b)

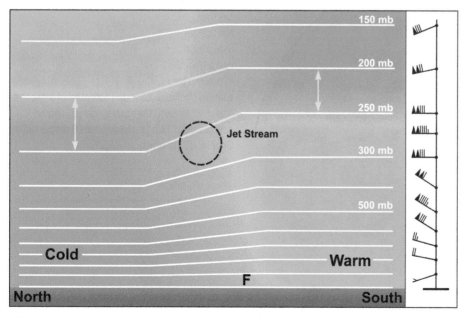

(c)

FIGURE 7.26   (a) The spacing between any two successive mandatory pressure levels is smaller in cold air (in blue) than in warm air (in red). This causes the slanted portions of the constant pressure surfaces that bridge the transition zone between the cold and warm air masses to get steeper with decreasing pressure. In other words, the height gradient on constant pressure surfaces in the zone where the air masses meet increases with decreasing pressure (increasing altitude); (b) As a result, winds blowing generally from the west increase in speed with increasing altitude in the troposphere; (c) The general decrease in temperature with increasing altitude in the troposphere slows at the tropopause (close to 250 mb) and then reverses in the stratosphere. As a result, the north-south height gradient on a constant pressure surface weakens (above about 250 mb, in this example). Schematically, the "bridge" connecting equal pressures in the warm and cold air masses becomes less steep above the tropopause, so wind speeds begin to decrease with altitude. Thus, a maximum in wind speed—the mid-latitude jet stream—will be found in the vicinity of 250 to 300 mb.

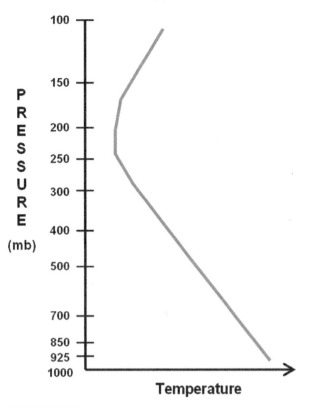

FIGURE 7.27 On average, temperatures in the troposphere decrease with increasing altitude. Above the tropopause, temperatures remain fairly constant with altitude. Temperatures generally start to increase with increasing altitude in the stratosphere.

embedded within the broad river of high-altitude westerlies (typically somewhere between 300 mb and 250 mb) is the **mid-latitude jet stream**, or more simply, the **jet stream**.

Mid-Latitude Jet Stream: A Fast Current within a High-Altitude River of Westerly Winds

Figure 7.28 is a visualization of the mid-latitude jet stream over the Northern Hemisphere. To get your bearings, you are looking down on the North Pole, and the narrow current of vivid colors that encircles the mid-latitudes indicates the position of the jet stream. Dark red qualitatively corresponds to the fastest winds while orange and bright yellow represent lower (but still relatively speedy) winds. The slowest winds are in blue. The bunched, narrow lines are essentially streamlines without arrowheads; they are intended to indicate the general westerly flow of air associated with the jet stream (we note that, simultaneously, there's also a mid-latitude, generally westerly jet stream over the Southern Hemisphere).

Looking at Figure 7.28, it's pretty easy to accept that the mid-latitude jet stream is a relatively narrow current of fast-flowing air within a broad river of high-altitude westerlies. The jet stream is usually several hundred kilometers wide and thousands of kilometers long, but it does not necessarily

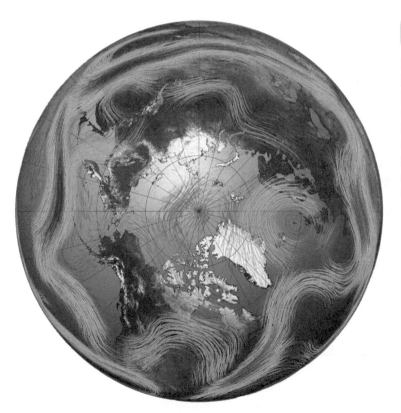

FIGURE 7.28 A visualization of the mid-latitude jet stream in the Northern Hemisphere as viewed from above the North Pole. The channel of reds and yellows indicates this current of fastest winds embedded within the mid-latitude westerlies. Note that, in places (for example, at low latitudes of the eastern Pacific Ocean), the footprint of a second major jet stream, the subtropical jet, can be seen as well. This jet stream will be discussed in Chapter 10 (courtesy of NASA).

wrap continuously around the globe. Note that, in Figure 7.28, the jet stream briefly loses its identity over the West Coast of the United States near the base of a wavy trough. So indeed, there are "breaks" in the jet stream, particularly during the warm season when jet-stream winds are much slower than they are in winter. To understand why, let's explore the seasonality of the jet stream.

Like a migrating bird, the jet stream undergoes seasonal shifts in position and speed. In the cold season, it migrates equatorward as cold air from high latitudes forges farther south into the mid-latitudes. Furthermore, because north-south temperature contrasts are largest in winter, pressure contrasts aloft can be very large as well, boosting jet stream speeds in the cold season. By way of example, Figure 7.29a shows a late winter day with a very wavy pattern of troughs and ridges in the jet stream—one trough had plunged into the Southeast states and another was just off the U.S. West Coast, while a ridge bulged into central Canada. For the record, wind speeds in the jet stream in winter and spring can exceed 150 knots (about 170 mph).

In contrast, cool air retreats northward during summer in the Northern Hemisphere, and the jet stream follows this poleward retreat. Given the greater number of daylight hours during summer, you should not be surprised that north-south temperature gradients decrease across the mid-latitudes during this time of year. As a result, jet-stream wind speeds dramatically slow in the warm season from their hectic winter pace. For example, check out Figure 7.29b which shows a modest current of jet stream winds flowing over the northern tier states and southern Canada on a late summer day. Notice the relative lack of reds on this image compared to Figure 7.29a, indicative of slower jet-stream winds during the warm season. Also, note that the jet stream is less wavy in Figure 7.29b, another characteristic difference between the two seasons.

We close this section with another reference to temperature. Many texts refer to the mid-latitude jet stream as the "polar jet stream" or "polar jet"—we avoid this usage because we worry that "polar" could give the mistaken impression that this jet stream resides exclusively near the poles, or at high latitudes. The term "polar jet" originated in the early 1900s when a group of Norwegian meteorologists first described the narrow battle zones that separated polar and tropical air masses. Using military terms, they dubbed this battle zone the **polar front**. Thus, "polar jet" was the logical choice to describe the ribbon of fast, high-altitude winds that form in concert with the polar front.

On a daily basis, the polar front manifests itself on mid-latitude weather maps as the interconnected family of fronts that often stretches from one low-pressure system to another (see Figure 7.30). When television weathercasters show the jet stream, it's often positioned directly above these fronts. However, that placement isn't quite right. Recall that surface fronts lie at the leading edge of the colder (or drier) air mass. For a perspective on this, take a look back at the cross section in Figure 7.26c—the leading edge of the cold air mass would be approximately at the point labeled F, so that's where any surface front would be. Notice that the jet stream lies on the cold side of this location. And that is true in general, as shown schematically in Figure 7.30.

(a)

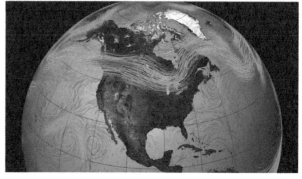

(b)

FIGURE 7.29 (a) The mid-latitude jet stream as it might appear in winter or spring, with a wavy pattern of troughs and ridges and relatively high wind speeds (represented by the reds); (b) A typical late summer mid-latitude jet stream—farther poleward, less wavy, and with lower wind speeds than its winter counterpart (courtesy of NASA).

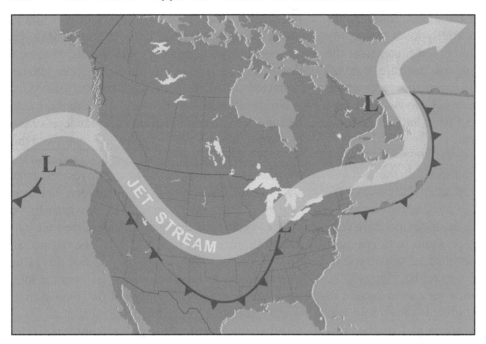

FIGURE 7.30 In general, the mid-latitude jet stream lies on the poleward, or cold, side of the polar front.

In this chapter, we've presented plenty of upper-air maps, some with heights, some with streamlines, and others with the wind barb portion of station models. These highly visual techniques make it easier for meteorologists to recognize different atmospheric patterns. And in fact, pattern recognition (especially on upper-air charts) is an important part of a forecaster's toolbox.

PATTERN RECOGNITION: PART OF THE ART OF FORECASTING

The world around us is full of patterns that we constantly mine for information. From offensive football coaches sitting in lofty stadium booths looking for patterns in the alignments of opposing defenses, to the media conducting national polls that gauge voting patterns in presidential elections, we constantly search for cues that help us make informed decisions and predictions. In a similar fashion, meteorologists use **pattern recognition** when they look for cues from patterns on constant pressure charts. Here we'll explore some of the more common patterns that forecasters routinely recognize in the height and wind fields on upper-air charts.

Jet Streaks: The White-Water Rapids of the Jet Stream

A fast-moving river sometimes has small sections where turbulent water flows very fast (for example, white-water rapids). And so it is with the mid-latitude

jet stream. It's easy to spot several pockets of relatively speedy winds, called **jet streaks**, on the upper-air wind charts we've shown on the last few pages. For example, in Figure 7.29a, one jet streak stretches from central Canada into the upper Midwest while another is impinging on the Pacific Northwest. As a general rule, jet streaks occur in concert with dramatically large height gradients at jet-stream level. This idea is nicely illustrated in Figure 7.31 which shows 250-mb heights (solid lines) and wind speeds (legend in color) on a late February day. At this time, a powerful jet streak was centered near the Southeast coast, with maximum wind speeds exceeding 150 knots (about 170 mph). Note how the jet streak is co-located with the zone of largest height gradient. Without a doubt, the most powerful jet streaks occur in winter, given the large horizontal temperature gradients (and thus large height gradients) that are a hallmark of the season. Such jet streaks are commonly the instigators of significant low-pressure systems (often along the East Coast), a topic that will be explored in more detail in Chapter 16.

Favored places for jet streaks to form are confluence zones, which are regions where two currents of fast-moving, high-altitude air merge together. To get a conceptual sense of what we're talking about here, consider Figure 7.32a, which shows the confluence of the Monongahela and Allegheny Rivers at Pittsburgh, PA, to form the Ohio River. Now consider Figure 7.32b, which shows one branch of the

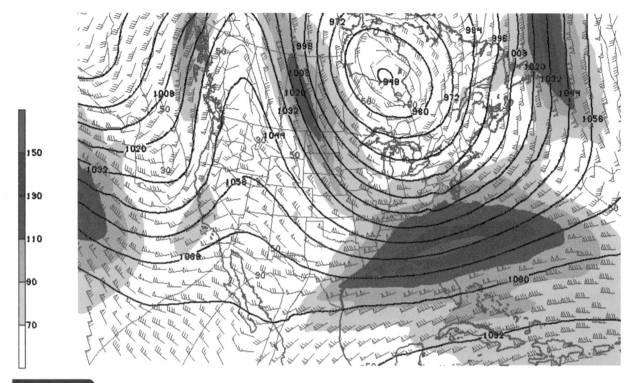

FIGURE 7.31 A map of 250-mb heights (solid lines, labeled in decameters), wind speeds (in color, in knots) and wind barbs on a late February day. Only wind speeds above 70 knots (about 80 mph) are colorized. A jet streak was positioned over the Southeastern United States at this time, co-located with the zone of largest height gradient (courtesy of NOAA).

(a)

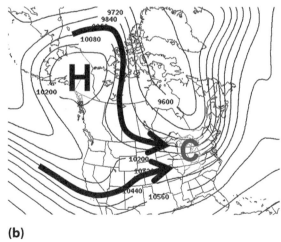

(b)

FIGURE 7.32 (a) The confluence of the Monongahela River (flowing in from the upper right) and the Allegheny River (flowing in from the upper left) at Pittsburgh, PA, marks the headwaters of the Ohio River (courtesy of the Army Corps of Engineers); (b) A 250-mb height map illustrating a confluence zone, marked with a "C."

jet stream directed from central Canada into the northern United States merging with another branch of the jet stream flowing from the Pacific Ocean. The area where they meet is a confluence zone, marked here with a "C." The large height gradient that marks this confluence zone is a result of the merging of the height lines associated with the two jet streams.

We'll delve into the role of jet streaks in the life cycle of mid-latitude lows later in the text. In the remainder of this chapter, we'll explore other classic upper-air features and patterns and show you how forecasters use pattern recognition to help them more accurately predict the weather.

Upper-Air Pattern Recognition: Wavy or Not?

The westerlies (and the jet stream embedded within them) occasionally flow almost directly from west to east, a pattern that meteorologists describe as **zonal**. In zonal flows, height lines tend to parallel latitude lines (see Figure 7.33). Meteorologists describe a zonal flow as a **low-amplitude pattern** because any upper-level troughs and ridges are subtle and not very wavy. In such a generally straight west-to-east flow, upper-air troughs and ridges tend to move fast, particularly during winter. These speedy and ill-defined upper-air features are the basis for the popular forecasting mantra "Fast flow, weatherman's woe," which is a reference to the difficulty in timing the arrival of clouds and precipitation (and the improving weather that follows) in zonal flows.

In contrast, the upper-level westerlies sometimes follow a much more serpentine pattern, carving out a wavy series of troughs and ridges. Meteorologists describe such patterns as

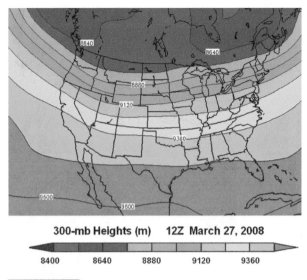

300-mb Heights (m)   12Z March 27, 2008

8400   8640   8880   9120   9360

FIGURE 7.33   Heights at the 300-mb level on a March day over the U.S. provide an example of a zonal flow (courtesy of NOAA).

meridional (see Figure 7.34a), because upper-level winds blow nearly parallel to meridians (longitude lines). Forecasters recognize these **high-amplitude patterns** as cues to the possible onset of extreme weather—big snowstorms and record chill in winter, outbreaks of tornadoes in spring and record heat

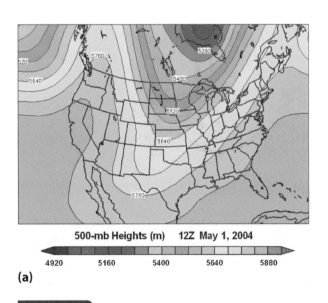

500-mb Heights (m)   12Z May 1, 2004

4920   5160   5400   5640   5880

(a)

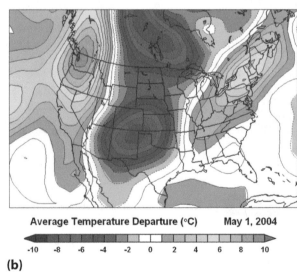

Average Temperature Departure (°C)   May 1, 2004

-10   -8   -6   -4   -2   0   2   4   6   8   10

(b)

FIGURE 7.34   (a) A high-amplitude 500-mb pattern on a May day. Note the sharp trough in the center of the country bookended by ridges near each coast; (b) Departures from the daily average temperatures (in °C) on this day. Note that these departures mirror the high-amplitude height patterns fairly closely, with higher-than-average temperatures in the eastern and western states (where there were highly wavy 500-mb ridges) and below-average temperatures over the middle of the country that coincided with a high-amplitude 500-mb trough (courtesy of NOAA).

in summer. You'll learn more about the connection between high-amplitude patterns and strong surface pressure systems in Chapter 12. For now, we'll illustrate how forecasters use this information about the amplitude of the upper-air pattern with a simple lesson in temperature forecasting.

When you observe any upper-air trough on a constant pressure surface, you know that local air columns are relatively cold from the ground to that pressure level. By way of review, pressure decreases relatively fast with increasing altitude in cold air columns, which translates to low heights (characteristic of a trough). We made a similar statement about ridges on constant pressure surfaces. For upper-air ridges, local air columns are relatively warm from the ground to that pressure level—pressure decreases more slowly with altitude in warm air columns, which translates to high heights (characteristic of a ridge).

Forecasters take these observations to the bank by making predictions about patterns of surface temperatures in high-amplitude 500-mb flows. Under a 500-mb trough in a high-amplitude pattern, surface temperatures tend to be lower than average, particularly during the daytime when clouds tend to fill the trough (we'll explain why in the next chapter). On the flip side, temperatures tend to be above average beneath a high-amplitude upper-air ridge.

Figure 7.34b shows the departures from average temperatures on the day corresponding to the 500-mb heights shown in Figure 7.34a. Note that the departures from the date's average temperatures closely mirror the high-amplitude 500-mb pattern of heights at this time, with temperatures below average in the trough in the middle of the country and above average in the ridges near the East and West Coasts. Though clouds can complicate the situation at night (can you remember how?), the temperature signal sent by a high-amplitude pattern is pretty clear. By simply recognizing that a high-amplitude 500-mb pattern will prevail, forecasters can make general predictions about future temperature trends.

High-amplitude patterns that stagnate can cause protracted sunny, dry weather in one region and rainy, cooler weather in neighboring regions. Such persistent, high-amplitude patterns are often associated with what meteorologists call "blocking highs."

Blocking Highs: Not an Offensive Lineman's Euphoria

When weather forecasters talk about blocking highs, avid football fans might conjure up images of the gridiron satisfaction experienced by an offensive lineman flattening a would-be tackler. But in the world of weather, a **blocking high** is a high-pressure system in the middle to upper troposphere that acts as an obstacle to the upper-level westerlies, forcing the swiftest current to split into two discrete branches that flow around the high. Figure 7.35 shows a

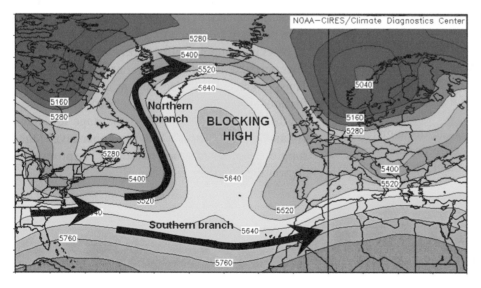

FIGURE 7.35 The 500-mb analysis at 00 UTC on February 25, 2004, showed a blocking high south of Greenland and Iceland. The blocking high split the relatively fast flow of air at 500 mb into northern and southern branches (remember that 500-mb winds blow nearly parallel to the height lines) (courtesy of NOAA).

FIGURE 7.36 A rock in a stream acts as a block to the water, forcing the flow to split into two distinct branches.

blocking pattern over the North Atlantic Ocean. You can think of a blocking high ("block," for short) as a large rock in a shallow fishing stream that causes flowing water to split and flow around it on both sides (see Figure 7.36).

In 1950, a meteorologist named D. F. Rex made the first climatological study of blocking highs, using upper-air data and surface weather maps. Rex (sometimes called the "father of blocking highs") required that the following criteria be met in order for a 500-mb high to qualify as a "block":

1. The fast core of 500-mb winds must split into two discrete branches.
2. Each branch must transport an appreciable mass of air.
3. A sharp transition from zonal flow to meridional flow must be observed at the split in the upper-air currents.
4. The pattern must persist for at least 10 days.

Although criteria for blocking vary from meteorologist to meteorologist, Rex's standards have served as a model for many studies on blocking highs.

Blocking high-pressure systems, like all upper-level highs and ridges, tend to suppress clouds, showers, and thunderstorms (details to follow in Chapter 8). As a result, abundant sunshine and the lack of rain within the domain of a long-lived blocking high (some have lasted more than a month) can lead to tinder-dry conditions, wildfires, and crop damage during the warm season.

Not all the headlines about blocking highs deal with dryness, however, given that a typical blocking pattern also includes stagnant, closed 500-mb lows (revisit Figure 7.35 and look to the west and northeast of the blocking high). Detached from the relatively strong steering winds associated with the northern and southern branches of the jet stream, these **closed lows** have nowhere to go and thus contribute to the overall blocking pattern. Unlike upper-level highs, closed lows promote clouds, showers, and thunderstorms. Thus, protracted blocking patterns can pave the way for flooding in the regions where closed lows develop.

Why do upper-level lows set the stage for showers and thunderstorms? Why do upper-level highs suppress clouds? All good questions. We promise to answer them in the next chapter.

Weather Folklore and Commentary

Aviation Risks

In December 1997, a United Airlines jet, while flying high over the Pacific Ocean from Japan to Hawaii, violently plunged 30 meters (about 100 ft) without warning, killing one passenger and injuring more than 100 others. The cause of the mishap was clear-air turbulence ("CAT", for short)—an aviation hazard that can occur in the vicinity of the jet stream. Clear-air turbulence can also occur over tall mountains such as the Rockies, where turbulent waves of air develop in response to winds flowing over lofty peaks.

When CAT is severe, flight attendants and passengers not buckled into their seats can be slammed against storage compartments and the cabin ceiling. More commonly, clear-air turbulence jostles food carts and whitens the knuckles of passengers.

What exactly is clear-air turbulence and why does it often form in the vicinity of the jet stream? Let us start by telling you what CAT is not. Sometimes you'll hear folks refer to "air pockets" with regard to in-flight turbulence, apparently suggesting that there are floating pockets devoid of air that presumably can cause the aircraft to suddenly lose lift and altitude. There are no such things as "air pockets," however. In other words, there are no vacuums in the atmosphere.

Here's the real scoop. As you learned in this chapter, winds increase in speed with increasing altitude below the jet stream. This change in wind speed with height, known as **vertical wind shear**, leads to a "rolling over" of the air. To understand what we mean by "rolling over," imagine a pencil resting on your left hand. This hand represents the slower wind below the jet stream. Rub your other hand forward on top of this pencil (this hand represents the faster jet stream). Note that the pencil rolls. Similarly, vertical wind shear in the vicinity of the jet stream can cause air to "roll over," creating a series of intense turbulent eddies like the ones depicted in Figure 7.37a. These turbulent eddies, which can be 20 to 200 m (65 to 650 ft) in width, create a series of updrafts and downdrafts that pilots and passengers experience as clear-air turbulence. Like the jet stream itself, clear-air turbulence is not typically continuous, but rather occurs in patches that are 100 km (60 mi) long or less.

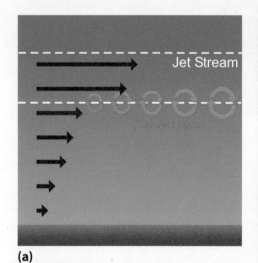

(a)

(b)

FIGURE 7.37 (a) Vertical wind shear can be strong near the jet stream, especially during winter when upper-level winds are routinely fast. In turn, strong wind shear can generate turbulent eddies, heightening the risk of severe clear-air turbulence; (b) Breaking waves of clouds called Kelvin-Helmholtz billows can give away the presence of turbulent eddies associated with strong vertical wind shear at lower altitudes (courtesy of Benjamin Foster, University Corporation for Atmospheric Research).

(Continued)

(Continued)

Clear-air turbulence can occur at lower altitudes as long as sufficiently strong wind shear is present in a layer of air (the stability of the layer also plays a role, and we will discuss this topic in Chapter 8). At altitudes of a few kilometers (where dew points are typically higher than at jet-stream level), clouds can sometimes form, allowing you to see the turbulent eddies. These "breaking waves" of clouds, which sometimes form to the lee of mountains or large hills, are formally called Kelvin-Helmholtz billows (see Figure 7.37b); they mark a layer of turbulence at an altitude of a few kilometers. Although it probably would be bumpy flying into these billows, this example does not qualify as "clear-air" turbulence simply because clouds are present. For turbulence to formally qualify as CAT, the air, of course, must be clear.

The Role of Stability in Weather

8

LEARNING OBJECTIVES

After reading this chapter, students will:

- Understand that an air parcel's buoyancy depends on the difference in temperature (density) between the parcel and its environment
- Be able to apply a stability test to an air parcel at rest, using the environmental lapse rate and the initial temperature of the parcel
- Understand that the dry and moist adiabatic lapse rates apply to the rates of change of temperature inside rising or sinking air parcels
- Have a quantitative sense for the amount of energy consumed or released as water changes from one phase to another

207

- Understand the impact of conditional instability on the buoyancy of an air parcel lifted from the earth's surface
- Understand the stability of temperature inversions and the instability of superadiabatic layers
- Understand the impact on stability from cooling or warming aloft
- Be able to describe how and where subsidence inversions typically form, and to convey their effects on weather and air quality
- Be able to distinguish between stratiform clouds (precipitation) and convective clouds (precipitation) and the environments in which they form
- Be able to describe the Bergeron-Findeisen process and then defend the assertion that a large portion of the precipitation that falls over the middle latitudes begins as snow at high altitudes
- Be able to take visual cues from smoke plumes and distinctive cloud patterns to qualitatively assess the stability of the atmosphere
- Be able to explain the sometimes breezy afternoon that follows a calm morning in a weather pattern characterized by a relatively weak surface pressure gradient

- anvil
- cirrocumulus cloud
- altocumulus cloud
- stratocumulus cloud
- fumigation
- lenticular cloud
- superior mirage

Larry Walters of Los Angeles always wanted to fly. After graduating from high school, he enlisted in the Air Force, but less-than-perfect eyesight prevented him from flying the friendly skies. Even after his discharge, Larry's dream never faded. One day in 1982, his long frustration finally prodded him to take matters into his own hands. With a plan reminiscent of the schemes of Wyle E. Coyote from the cartoon show *The Roadrunner*, Larry walked into the local Army-Navy store and bought several tanks of helium and 45 weather balloons. Each balloon, when fully inflated with the very light gas (only hydrogen is lighter), would measure about 1.2 m (4 ft) in diameter (see Figure 8.1).

Larry went home and attached this do-it-yourself flying kit to a lawn chair. Before inflating the balloons, he secured the lawn chair with an anchoring line (the "launching pad" was the roof of his girlfriend's house). Larry knew full well that the balloons, once inflated with featherweight helium, would have great buoyancy and rise up and away if precautions weren't taken (to understand buoyancy, think of an inflated rubber duck playfully submerged in a bathtub of water and then released – it buoyantly bobs quickly to the surface). Once all the balloons were inflated, Larry donned a parachute and climbed into the chair to get a feel for his trusty but eclectic craft, which he had dubbed "Inspiration 1."

Knowing that there wouldn't be any flight attendants or landing protocol, Larry had packed a large bottle of soda, containers filled with water for ballast,

a CB radio, an altimeter, a camera, and a pellet gun that he planned to use to pop some balloons for a controlled final descent. Without any countdown, Inspiration 1 jumped the gun and lifted off when sharp

FIGURE 8.1 A weather balloon (courtesy of the National Weather Service).

edges on the roof of his girlfriend's house severed the mooring line.

Instead of drifting lazily skyward, the helium-ballooned craft rocketed toward the heavens, quickly reaching a chilly cruising altitude of about 4875 m (16,000 ft). Later, Larry would say that the view was so amazing that he forgot to take any pictures. Floating in the relatively rarefied air three miles above the Los Angeles area, Larry began to feel cold and dizzy. As if his situation couldn't get any worse, Inspiration 1 then drifted into the primary approach pattern of the Long Beach Municipal Airport. Two pilots on commercial jets spotted him (now that must have been quite a sight!) and immediately radioed the control tower that they had just observed a guy on a lawn chair with a gun.

Pulling out his pellet gun, Larry shot several of the balloons, and Inspiration 1 started its long descent toward the ground. He tried to land on a section of a country club near Long Beach, but his craft got entangled in some high-voltage power lines a bit short of the "runway." Luckily, the plastic tethers connecting the helium balloons to the lawn chair prevented Larry's electrocution. To make a long story short, Larry's entire flight covered only ten miles, but he created quite a ruckus. The Federal Aviation Administration (FAA) later charged and fined him for "not establishing and maintaining two-way communications with the control tower."

This story is not an urban legend—it is true. If you harbor any doubt, consider that NASA uses helium-filled balloons to carry scientific instruments weighing the equivalent of three compact cars to altitudes of 40 km (25 mi)—near the top of the stratosphere (see Figure 8.2)! Flights by some of these Ultra-Long Duration Balloons have lasted as long as two weeks, giving scientists invaluable data to conduct their research.

The up-and-down movements of volumes of air (called **vertical motions**) are of keen interest to meteorologists simply because rising air cools, which can lead to net condensation (clouds) and precipitation. As a means of assessing vertical motions, meteorologists take a simple approach: They isolate a large volume of air and assume that it behaves, essentially, as a unit separate from the air around it. Such a blob of air is referred to as a **parcel**. This approach to understanding vertical motions requires us to consider the forces that act on air in the vertical direction. For example, given that air pressure decreases with altitude, there's always a large vertical pressure gradient force in the atmosphere, directed upward—thus, we

FIGURE 8.2 NASA's Ultra-Long Duration Balloons, which are made of polyethylene film that is about as thick as sandwich wrap, carry heavy scientific payloads to great altitudes, gathering invaluable data for research (courtesy of NASA).

might expect that air parcels would rise as dramatically as helium-filled balloons! If you buy into this idea, you are underestimating the "gravity" of the situation. Let's investigate.

ASSESSING THE POTENTIAL FOR VERTICAL ASCENT: THE SKY'S THE LIMIT

To get a sense of the magnitude of the vertical pressure gradient, consider that air pressure decreases from approximately 1000 mb at sea level to 250 mb at an altitude of about 10 km (6 mi). So that's a vertical pressure gradient of 750 mb in 10 km, or 75 mb per km.

How does this vertical pressure gradient compare to the magnitude of horizontal pressure gradients observed in the atmosphere? We'll consider the large pressure gradients associated with the fierce winds around the center of intense hurricanes. There, the horizontal pressure gradient can reach 1 millibar per kilometer just outside the eye, nearly an order of magnitude <u>smaller</u> than the atmosphere's vertical pressure gradient! But air clearly doesn't rush upwards at speeds faster than the strongest winds of the most powerful hurricanes. That's because gravity balances the vertical pressure gradient force (or nearly so), generally limiting the magnitude of vertical motions (see Figure 8.3). When these two forces balance, the atmosphere is said to be in **hydrostatic equilibrium**.

On large spatial and time scales, the atmosphere stays pretty close to hydrostatic equilibrium, so we should expect typical vertical velocities to be fairly small. Let's see just how small.

Consider Figure 8.4a which shows the vertical velocity at the 700-mb level (roughly 3000 m) at 00 UTC on October 27, 2010, a day when a powerful low-pressure system moved through the upper Midwest (we looked at this storm in Chapter 6). Note that the units for vertical velocity in this case are millibars per second (mb/s)—here, vertical velocity represents a "change in pressure per unit time" and not the more typical "change in distance per unit time" usually associated with a velocity. However, don't let this slightly different approach throw you for a loop. When an air parcel moves upward, its pressure decreases. Thus, a rising parcel has a negative vertical velocity when the units are mb/s. In contrast, if an air parcel subsides, its pressure increases with time, so sinking parcels have a positive vertical velocity when the units are mb/s.

To gauge the magnitude of the vertical motions in Figure 8.4a, focus your attention on Minnesota and North Dakota, where the dark purple indicates relatively large negative vertical velocities (and thus upward motion). The powerful storm (see Figure 8.4b for a satellite image of its large cyclonic circulation) set records for low-pressure readings in this part of the country—in fact, the pressure fell to 955.2 mb at Bigfork, MN, the second lowest barometric reading

on record for any non-tropical low-pressure system over the contiguous United States.

Given such a powerful storm, you might expect air to be rising rather quickly—we can assess this based on Figure 8.4a. To compute how long it would take for a parcel of air rising at a vertical velocity of −0.004 mb/s (the dark purple) to ascend from 850 mb to 700 mb (a vertical distance of about 1500 m), we would divide 150 mb by 0.004 mb/s. The answer is 37,500 seconds, or about 10.5 hours—that's quite a long time to move just 1500 m and certainly does not sound "fast" (it's equivalent

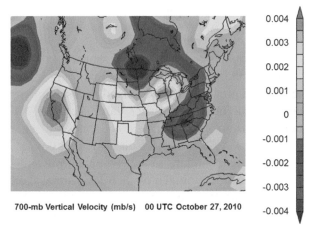

700-mb Vertical Velocity (mb/s)   00 UTC October 27, 2010

(a)

(b)

FIGURE 8.4 (a) The vertical velocity at the 700-mb level at 00 UTC on October 27, 2010 in units of millibars per second (mb/s), with negative values indicating upward motion and positive values indicating downward motion. Meteorologists commonly evaluate vertical motion at the 700-mb level because it lies approximately in the middle of the lower half of the troposphere, where most of the "weather action" occurs; (b) The relatively large upward vertical motions in the Upper Midwest were associated with a deep low-pressure system, its large counterclockwise circulation easy to see in this satellite image (courtesy of NOAA).

Pressure Gradient Force

Gravity

FIGURE 8.3 On average, the upward pull of the pressure gradient force is approximately balanced by the downward tug of gravity.

to about 470 feet per hour)! Similarly, the largest (in magnitude) sinking motions in the wake of this powerful low (in the vicinity of southern Lake Michigan) were slightly slower (the orange corresponds to +0.003 mb/s). These slow upward and downward motions indicate just how close to hydrostatic equilibrium the atmosphere typically operates on relatively large spatial and time scales.

Now compare that example to what goes on within the core of a severe thunderstorm, where air parcels sometimes can rise from near the ground (where the pressure is around 1000 mb) to the tropopause (about 200 mb) in perhaps 30 minutes or less. Now that's fast! Such speedy currents of rising air might occur within a powerful thunderstorm such as the one shown in Figure 8.5.

This example demonstrates that large departures from hydrostatic equilibrium can occur on small spatial and time scales such as those typical of a thunderstorm. To aid in predicting when thunderstorms might become severe (formally, producing large hail, damaging winds, or a tornado), forecasters must assess the stability of the atmosphere by determining how buoyant the air can potentially become. Let's explore this idea.

Positively Buoyant Air Parcels: In High Spirits

Helium-filled balloons are high fliers because the density of helium is considerably less than the density of

A severe thunderstorm on April 28, 2002, photographed from the window of a commercial jet. This severe thunderstorm spawned a killer tornado in LaPlata, MD (courtesy of Steven Maciejewski).

air (remember, density is mass divided by volume). For comparison sake, a representative density for helium is 0.17585 kg/m^3, while a typical density for air is about 1 kg/m^3 at sea level. Think of Larry's balloons before "launch" as squeaky rubber ducks submerged in a bathtub of water. Such a metaphor is appropriate here because the density of a typical rubber duck is much less than the density of water, which is about 1000 kg/m^3.

With the rubber-duck metaphor in mind, consider Figure 8.6. Here, one of the authors immersed a rubber duck in a container of water (left image). When he released it, the rubber duck immediately bobbed to the surface (image on the right). That's because there was a net upward force called the **buoyancy force** acting on the submerged duck. The idea that water exerts an upward force on the submerged rubber duck comes from **Archimedes' Principle**. Archimedes was an ancient Greek mathematician who left his scientific mark by shouting "Eureka" when he realized that his body displaced an equal volume of water in his bathtub. Formally, Archimedes' Principle states: *The total upward force on a body submerged or partially submerged in a fluid is equal to the weight of the fluid displaced by the body.*

In the case of the submerged bathtub toy, the buoyancy force pushed the rubber duck upward with a magnitude equal to the weight of water displaced by the duck (the author had to exert a downward force to keep it submerged, but once he released his grip, the buoyancy force did its thing). Obviously, the weight of the water displaced by the duck was much greater than the weight of the duck. In this situation, meteorologists would say that the duck was **positively buoyant**.

Larry's helium-filled weather balloons were overwhelmingly positively buoyant. As the balloons ascended into air of lower pressure, they expanded, but the decreasing density (same mass but a larger volume) of helium inside the balloon continued to be less than the decreasing density of the outside air. The balloons ascended until there was a balance between the buoyancy force and the combined weight of Larry, his lawn chair, and his provisions.

Can objects be **negatively buoyant**? Sure. A rock immersed in water sinks to the bottom once released. That's because the density of the rock is greater than the density of water. Thus, the weight of the water displaced by the rock is less than the weight of the rock.

Now let's get out of the water and get some air! In terms of rising parcels of just plain old air, the

FIGURE 8.6 A rubber duck is submerged in a tub of water (with blue dye added for effect). As soon as the duck was released, it rapidly bobbed to the surface in response to the buoyancy force, represented by the arrow in the image on the right (Photograph ©Ann H. Taylor. Duck courtesy of Joshua Bram and Cuyler Luck).

buoyancy force is related to the density difference between the air inside the parcel and the air surrounding the parcel (for simplicity, the air surrounding the parcel will be called the "ambient air" or, more simply, "the environment"). As long as the parcel is less dense than its environment, the parcel will remain positively buoyant and the "sky's the limit"!

Air density is difficult to measure, but fortunately, air density is related to air temperature. We have a working principle on which to fall back: Warm air is less dense than cold air, under the constraint that the pressure of a parcel moving vertically adjusts to the pressure of the surrounding air (a reasonable assumption given that a rising or sinking parcel expands or compresses as the ambient pressure changes). With this in mind, the critical test for buoyancy comes down to temperature differences between the air parcel and the ambient air.

Suppose a parcel of air initially in equilibrium with its surroundings at the ground is given a nudge upward. If the parcel finds itself warmer (and thus less dense) than its new surroundings, it will be positively buoyant (see Figure 8.7). The positively buoyant parcel will continue to rise until the temperature (density) of the parcel equals that of its surroundings at some higher altitude (more on this eventual equalization in just a bit). Conversely, if a parcel of air initially at rest is given a nudge upward and finds itself colder (and thus, more dense) than the ambient air, the parcel will be negatively buoyant and will sink back toward its original level. A

third possibility, called neutral buoyancy, is that the parcel, once nudged, is equal in temperature to its surroundings—in this case, the parcel would stay put.

The temperature difference between rising, positively buoyant parcels and the ambient air is typically slight, with most differences not exceeding about 1°C (about 2°F). However, in tall cumulonimbus clouds (like the one in Figure 8.5), which have vigorous currents of rising air near their centers, temperature differences between ascending parcels and the ambient air may approach or even

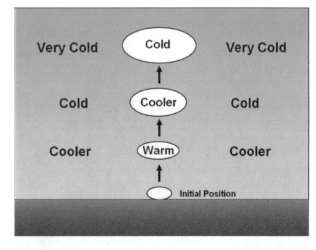

FIGURE 8.7 An ascending parcel of air will continue to rise as long as it remains positively buoyant (warmer than its environment).

exceed 5°C (9°F). And the larger the temperature difference, the greater the upward acceleration (and thus the faster the upward speed).

With our buoyancy test in mind, we are now ready to assess the stability of the atmosphere.

STABILITY: NUDGE AND JUDGE

To make our point about stability, let's try a simple experiment. Stand with both feet firmly planted on the floor and spread slightly beyond your shoulders, thus creating a broad base. Ask a friend to give you a firm but friendly nudge on the shoulder. Although your upper body will initially move, your feet will stay planted, and you can easily return to your original position. You might say that, before you were given the nudge, you were in a "stable" equilibrium. Now repeat the experiment, but start by standing on just one leg. Although somewhat awkward, you are still, nonetheless, in equilibrium. Again, ask your friend to give you a firm but friendly nudge. The result will be different here—you will likely hop away from your initial position—so you might say that you were in an "unstable" equilibrium.

Now we apply the same approach to test whether air parcels initially at rest are in "stable" or "unstable" equilibrium. Suppose a parcel of air in hydrostatic equilibrium is given a nudge upward. If the parcel is positively buoyant and keeps moving away from its initial resting position, then the parcel is said to be in **unstable equilibrium** (see Figure 8.8). In this case, the layer of the atmosphere in which the test took place is said to be unstable. If, however, the parcel is negatively buoyant and comes back toward its original position after the nudge, the parcel is said to be in **stable equilibrium** (see Figure 8.8). In other words, the layer of the atmosphere in which the test took place is stable. Finally, a parcel in hydrostatic equilibrium that is given a push upward and comes to a stop after a short rise is said to be in **neutral equilibrium**. In all these examples, we emphasize the word *equilibrium*—the test is no good if the parcel is already in motion. Thus, you would not apply this test to an air parcel already rising at the center of a powerful thunderstorm—it should have been applied before the thunderstorm developed.

With this stability test in hand, we are on the brink of making the jump from a general discussion to

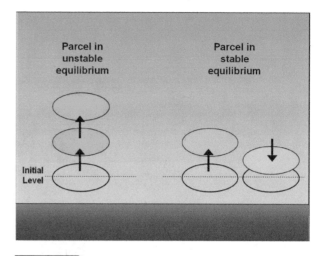

FIGURE 8.8 When an air parcel in unstable equilibrium is nudged upward, it will continue to rise because it is warmer than its environment. In other words, the parcel is now positively buoyant. When an air parcel in stable equilibrium is nudged upward, it will return toward its original position because it becomes colder than its environment. In other words, the parcel becomes negatively buoyant.

practical applications of stability and instability to weather forecasting.

Lapse Rates: Keeping Track of Temperatures Inside and Outside a Parcel

The key to determining whether an air parcel is positively, negatively, or neutrally buoyant at a particular altitude boils down to a simple comparison between the temperature of the parcel and the temperature of its ambient environment. While the temperature of the environment at any altitude depends on the prevailing weather pattern, the temperature changes that occur inside a rising or sinking air parcel are governed by a strict set of rules.

For the time being, let's assume that net condensation is not occurring in a rising air parcel, so the parcel is unsaturated. This assumption is reasonable (particularly in the lower troposphere) because most clouds form as the result of rising air, and the bases of most clouds are often a kilometer or more above the surface (see Figure 8.9). Such a rising parcel of unsaturated air cools at a rate of 10°C per kilometer of ascent (roughly equivalent to 5.5°F per 1000 feet). Although it may seem that this number "came out of the blue," recall that a parcel of air expands as it rises. To accomplish this expansion, the molecules inside the rising parcel must work to push out its "walls." The energy required to perform this molecular feat translates into an average cooling rate of about 10°C/km

FIGURE 8.9 A rare photograph of Andean condors a few kilometers above Peru's Colca Canyon provides evidence of invisible currents of rising air, an example of ascent without net condensation. Andean condors have a wing span of 3–4 meters (10–14 ft), enabling them to soar and glide over the canyon on these invisible thermals (courtesy of Dianne Lefty).

we're referring specifically to the rate of decrease in an unsaturated ("dry") parcel of air.

In the case of downward motion, a sinking parcel of unsaturated air warms at a rate of 10°C per kilometer of descent. As a parcel sinks, it moves into an environment of increasingly higher pressure. In the process, molecules in the higher-pressure ambient air compress the sinking parcel. The energy these ambient air molecules expend is not lost, however. Instead, the air molecules inside the parcel reap the rewards by gaining an equivalent amount of kinetic energy. This gain causes the temperature of the unsaturated parcel to increase to the tune of 10°C per kilometer of descent.

So why is this specific lapse rate called "dry adiabatic"? The implication of "dry" is easy—it simply means that the parcel is not saturated. How about the "adiabatic" part? In an **adiabatic process**, there is no exchange of heat energy between a parcel and its surroundings. Applying this restriction to an air parcel means that the parcel's temperature can only change if it expands or compresses. In this context, all other processes are not adiabatic. For example, any warming from the absorption of solar radiation disqualifies the process from being adiabatic because heat energy is added from outside the parcel. Similarly, the cooling that occurs when raindrops fall into a parcel and evaporate does not qualify as an adiabatic process because the parcel's decrease in temperature is the direct result of an exchange of mass (the raindrops) with the surroundings. When

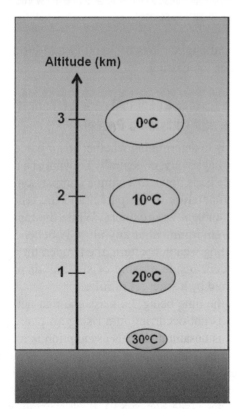

FIGURE 8.10 A rising, unsaturated air parcel cools at the rate of 10°C per kilometer (about 5.5°F per 1000 feet).

of ascent (see Figure 8.10), a value known as the **dry adiabatic lapse rate**. Recall from Chapter 3 that the general term *lapse rate* refers to the rate of decrease of temperature with increasing altitude—in this case, an unsaturated parcel rises and cools, the decrease in temperature inside the parcel results largely from a dry adiabatic process.

We should pause here, take a breath, and be careful not to mix apples and oranges. As we just argued, rising parcels of unsaturated air have their very own lapse rate, and it's always 10°C/km. On the other hand, the vertical temperature profile of the parcel's environment is a completely different matter. It is determined by a number of factors, including height above the heat-dispensing ground and temperature advections at different levels. On average, the temperature of the environment decreases at a rate of about 6.5°C per kilometer (6.5°C/km, approximately 3.6°F/1000 ft). Recall from Chapter 3 that this rate of temperature decrease with altitude is the average environmental lapse rate. It is important to remember that the value of 6.5°C/km is an average—the local environmental lapse rate varies from time to time, place to place, and atmospheric layer to atmospheric layer. That's why weather balloons have a job.

Meteorologists visualize the vertical variations in the temperature of air parcels and their environment on diagrams such as the one in Figure 8.11. The solid red line represents a sample temperature profile for the atmosphere over a given location (this data, for example, could have been collected from a weather balloon). Such a profile is called a **sounding**. The dashed line is a template path of a parcel that is

rising and cooling at the dry adiabatic lapse rate—the slope of the line is intended to represent a temperature decrease with altitude of 10°C/km. Such a line is called a **dry adiabat**.

At this point, perhaps you're wondering why both the dry adiabatic lapse rate and the average environmental lapse rate are positive numbers, even though they represent a rate of *decrease* of temperature with altitude. In a way, this convention is a transparent attempt to avoid working with negative numbers. In this context, it makes sense, because most of the time, temperature decreases with increasing altitude. So in order to make life easier, lapse rates are always taken to be positive. Admittedly, this convention may be in conflict with your intuition to use a negative sign when a variable decreases, but dropping the negative sign is standard practice in meteorology when referring to the rates of decrease in temperature with increasing altitude.

There are, however, times when the temperature in certain layers of the atmosphere remains constant with height or even increases with increasing height. For example, on clear nights with nearly calm winds and low dew points, the lower atmosphere tends to be coldest right near the surface, with air temperature increasing with height in a layer perhaps a few hundred meters thick. A layer of the atmosphere in which temperature increases with height is called an **inversion**. As you will soon learn, inversions also occur at higher altitudes. Indeed, there are inversions between 2 and 3 km in Figure 8.11 and also above 9 km. At other times, a layer of air can be **isothermal**, meaning that temperature stays constant with increasing altitude. Two isothermal layers are also shown in Figure 8.11.

Now back to the behavior of rising and sinking parcels of air. Imagine an unsaturated air parcel rising from the ground all the way to the tropopause and cooling at the dry adiabatic lapse rate of 10°C/km. Along the way, such a parcel would cool nearly 100°C (180°F) in polar regions and nearly 150°C (270°F) over tropical latitudes. In reality, a rising parcel that's cooling at this rate soon gets much colder and thus denser than its environment. That's because the temperature of the ambient air usually decreases at a rate less than 10°C/km (remember, the average environmental lapse rate is 6.5°C/km). Thus, rising parcels of unsaturated air would likely never make it to the tropopause. But we know that some parcels get there in strong thunderstorm updrafts (some penetrate into the lower stratosphere)! So there must be a mechanism that slows the rate of

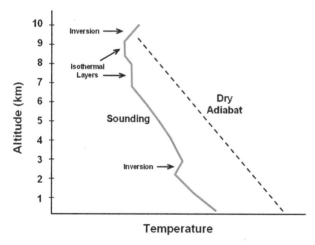

FIGURE 8.11 A graphical way to represent how the temperature of a rising, unsaturated air parcel and its environment vary with altitude. The solid line, called a sounding, represents the temperature of the environment (typically measured by a radiosonde). The dashed line is called a dry adiabat. It represents the rate of decrease (increase) in temperature inside a rising (sinking) parcel of unsaturated air.

cooling of rising parcels, increasing their chances of rising to great heights.

Moist Adiabatic Lapse Rate: Tapping Hidden Heat Energy

For the most part, the lower troposphere is sufficiently moist that air parcels near the earth's surface need only rise modest distances (sometimes less than one kilometer) before they become saturated. With just a tad more ascent and cooling, net condensation occurs and clouds form. Recall from Chapter 4 that the height at which this occurs is called the lifting condensation level, or LCL for short (see Figure 8.12). Above the LCL, rising parcels no longer cool at 10°C/km (the dry adiabatic lapse rate) but rather at a smaller rate of about 6°C/km. That's because net condensation releases latent heat (recall Chapter 4), slowing the rate of cooling from 10°C/km to about 6°C/km (3.3°F per 1000 feet). This altered rate of cooling for rising, saturated parcels is called the **moist adiabatic lapse rate**. You may wonder why we can still use the term *adiabatic* to describe this lapse rate, given that heat is added to the parcel during the process. But this heat energy does not come from the parcel's surroundings; rather, it was stored in the parcel's water vapor all along, and simply released during net condensation. Thus, we're safe calling the process adiabatic.

In fairness, we point out that the moist adiabatic lapse rate is not constant. It is typically less than 6°C/ km in tropical regions (as low as 4°C/km) where dew points in the lower troposphere are usually high—all that water vapor can translate into a significant release of latent heat when net condensation occurs, dramatically reducing the rate of cooling inside rising air parcels. By contrast, the moist adiabatic lapse rate associated with frigid, very dry Arctic air masses is close to the dry adiabatic lapse rate. That's because sparse supplies of water vapor translate to a relatively small release of latent heat during net condensation, only slightly reducing rates of cooling inside rising air parcels. Similarly, in the upper troposphere, the moist adiabatic lapse rate and the dry adiabatic lapse rate are nearly "one and the same" because almost all the water vapor has been depleted by the time parcels get to such high altitudes.

The bottom line here is that a rising parcel cools at the dry adiabatic lapse rate until net condensation begins at some altitude. Above this level, cooling proceeds at a reduced rate. Figure 8.13a is similar to Figure 8.11 except that we've added a dotted, curved line that represents the moist adiabatic lapse rate— such a line is called a **moist adiabat**. The changing slopes of moist adiabats are consistent with variable

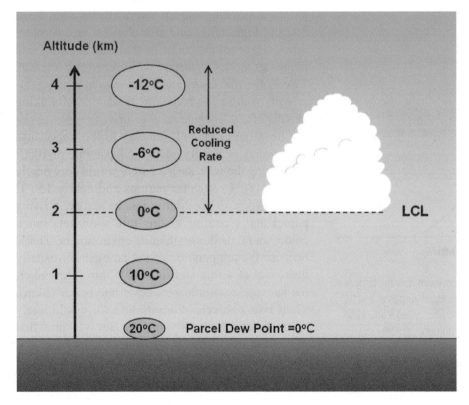

FIGURE 8.12 A parcel of air at the surface is forced to rise. Initially, we assume the parcel's temperature is 20°C and its dew point is 0°C. If we assume that the dew point stays nearly constant as the parcel ascends, the parcel will become saturated and net condensation will begin at an altitude of about 2 km (once the parcel has cooled 20°C). Above this lifting condensation level, the parcel cools at a reduced rate of approximately 6°C/km.

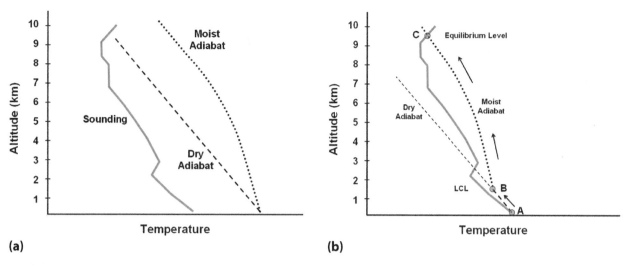

FIGURE 8.13 (a) Once net condensation begins, the decrease in temperature inside a rising parcel of air follows the local moist adiabat (dotted curve), sloped to represent a vertical cooling rate less than the dry adiabatic lapse rate; (b) Let's track the vertical changes in temperature inside the rising parcel of air. Initially, cooling occurs along a dry adiabat until the parcel reaches the lifting condensation level (here, assumed to be at the level of point B). As the parcel continues to rise, its rate of temperature decrease now follows the local moist adiabat. This reduced rate of cooling gives parcels a better chance of remaining warmer than the environment (temperature increases to the right in this diagram).

moist adiabatic lapse rates, which depend on the amount of water vapor available in rising air parcels. At low altitudes, where water vapor is more plentiful (and thus the potential release of latent heat is greater), the line is sloped so as to represent a decrease of about 6°C/km; at high altitudes, the slope is close to that of the dry adiabat.

So let's follow a parcel that is forced to rise from its original level (by a mountain, front, or low-pressure system, for example). Before ascent begins, we assume the parcel has the same temperature as its environment. So we'll start at point A on the sounding in Figure 8.13b. Initially, the cloudless parcel cools at the dry adiabatic lapse rate, and so its rate of cooling follows the local dry adiabat at point A (the dashed line). As a matter of practice, we can draw a local dry adiabat anywhere our test parcel begins its ascent (or descent). We follow this "path" upwards until net condensation occurs and a cloud forms. For sake of argument, let's agree that this happens at the level of point B. Now, because the rate of cooling inside the parcel will change from dry adiabatic to moist adiabatic, the parcel's rate of cooling will follow the local moist adiabat (dotted line) at point B. Note that we can draw a local moist adiabat anywhere our test parcel reaches the lifting condensation level. The parcel, now saturated, will follow this "path" until its temperature decreases to the environment's temperature. In this case, the "stop sign" is posted at the altitude of point C, the

equilibrium level (in reality, because the rising parcel has momentum, it will slightly overshoot this level).

Had the parcel remained on its initial dry adiabatic "path" and cooled at a faster rate, its equilibrium level would have been much lower. Lesson learned: With their cooling rate reduced, rising parcels of saturated air can remain positively buoyant longer. Keep in mind here that the vertical decrease in temperature of the ambient air is, on average, about 6.5°C/km, so rising parcels of moist air cool a bit more slowly than the average drop-off in temperature of the ambient air. In other words, saturated parcels have a better chance at remaining positively buoyant.

The operative word in the preceding sentence is "chance." Indeed, though the amount of moisture that parcels contain helps to determine how high they rise after they're given an upward nudge, the environmental lapse rate also comes into play. We can sum up this dual, somewhat fuzzy fate for rising parcels in one mouthful: **conditional instability**.

To assess whether a layer of the atmosphere is conditionally unstable, consider Figure 8.14 which shows a test parcel at a starting altitude of 2 km, initially in equilibrium with its environment (the atmospheric sounding is the red line). Notice that the environmental lapse rate in the layer just above the parcel's initial altitude lies between the local moist adiabat and the local dry adiabat. Now, give the parcel a nudge upward. If the parcel is saturated and

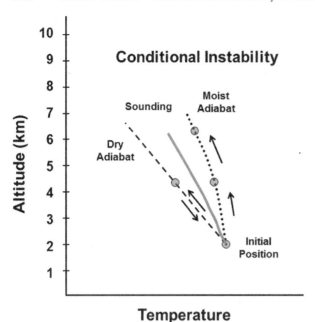

FIGURE 8.14 An example of a conditionally unstable layer. In this example, a saturated parcel (red circle) rising from an initial altitude near 2 km and cooling at the moist adiabatic lapse rate is positively buoyant because it is warmer than its environment. However, if net condensation were not occurring inside the parcel, its rate of cooling would equal the dry adiabatic lapse rate. In a hurry, the parcel would become cooler than its environment and thus negatively buoyant.

thus follows the moist adiabat, the parcel will remain warmer than the environment, positively buoyant, and keep rising. If, however, net condensation does not occur and the parcel follows the dry adiabat instead, the parcel will be colder than the environment, negatively buoyant, and sink back toward its initial

level. So, in this case, the stability of the layer depends on whether the rising parcel is saturated or unsaturated—hence the word "conditional."

In the end, when surface air is very warm and humid, the stage is set for prolonged ascent because parcels need to rise only a modest distance for net condensation to occur. Once that happens, parcels cool at the reduced moist adiabatic lapse rate which, in turn, increases the likelihood that they can remain positively buoyant to greater altitudes. That's what happened with the severe thunderstorm in Figure 8.5. As another example when conditional instability comes into play, keeping air parcels positively buoyant to great altitudes around the eye of a hurricane is pivotal for the storm to intensify (see Figure 8.15).

We'll have plenty more to say about tropical cyclones in Chapter 11. For now, let's explore the various ways to make a layer of the atmosphere more stable ("stabilize") or less stable ("destabilize").

WAYS TO CHANGE STABILITY: FROM THE GROUND FLOOR UP

While living in Atlanta over 30 years ago, one of the authors routinely rode his bicycle to graduate school, taking a route that included a stretch of dirt road. During summer dry spells, an early morning ride could get pretty gritty at times, with palls of choking dust hanging in the air whenever a car passed. But the afternoon ride home was noticeably less dusty. Although the occasional car whipped up dust, it quickly dispersed in the hot afternoon air. How could the personality of the environment change so dramatically in such a short time without any change

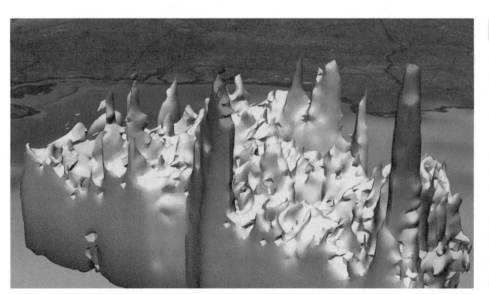

FIGURE 8.15 A radar cross section (acquired by satellite) through Hurricane Katrina in August 2005 shows two isolated hot towers (in red) indicative of positively buoyant parcels that have reached great altitudes. The tower near the center is about 16 km high—towers this tall near the eye of a hurricane are often an indication of intensification. In fact, Katrina became a Category 4 storm soon after this image was taken (courtesy of NASA).

in the prevailing weather pattern? Answer: It all boils down to stability.

Low-Level Stability and Instability: Grounded in Ground Temperatures

On a clear night with light winds and relatively low dew points, dramatic cooling of the earth's surface by a net loss of radiation sets the stage for a ground-based temperature inversion to form, appropriately called a **nocturnal inversion** (see Figure 8.16). When a parcel of air is nudged upward in such a layer, it always becomes negatively buoyant (colder than its environment) and sinks back to its original position. Lesson learned: Layers with a temperature inversion are highly stable (and, to a slightly lesser degree, so are isothermal layers). Indeed, such layers effectively snuff out any tendencies for the air to move up or down. In simple terms, stable layers of air at the earth's surface have relatively cold bottoms. That's why an early morning bicycle ride on a busy dirt road can get rather gritty as dust gets whipped into the air by a passing car and then just hangs there because the cool surface air is so stable.

That's also why a layer of ground fog usually appears to be stratified. In other words, the top of the layer is fairly flat, as you can see in Figure 8.17. For the record, **ground fog**, which is essentially just a cloud touching the ground, forms at night when the sky is clear, the surface air becomes saturated, and gentle eddies associated with light winds gradually help to thicken the nocturnal inversion. Fog can also form in a stable layer near the earth's surface when

FIGURE 8.17 A stratified morning ground fog (courtesy of Rebekah Labar).

warm, relatively moist air is advected over relatively cool waters (see Figure 8.18) or snow and ice covering the ground. These wind-aided fogs are called **advection fogs**. Both of these fogs tend to have relatively flat tops, indicative of the overall stable nature of the layer.

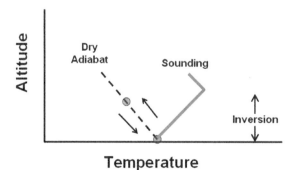

FIGURE 8.16 On a clear night with light winds, a nocturnal inversion forms near the ground. Even when a sample air parcel (red circle) is nudged upward, it is always colder than its environment and thus negatively buoyant, so it sinks back toward its original position. Lesson learned: Highly stable layers of the atmosphere snuff out any tendencies for air to move up or down.

FIGURE 8.18 Advection fog in San Francisco Bay during the early morning of September 26, 2009. The top of the Golden Gate Bridge is visible, with the San Francisco skyline in the background (courtesy of Mila Zinkova, Earth Science Picture of the Day).

Weather Folklore and Commentary

Making Dew

"When the dew is on the grass,
Rain will never come to pass.
When grass is dry at morning light,
Look for rain before the night."

Dew and ground fog are close cousins. Dew forms when water vapor in a thin layer of air next to the ground condenses into beads of water on grass as the temperature of the ground falls just below the dew point. Fog forms when water vapor in a thicker layer of air next to the ground condenses onto airborne condensation nuclei. Both require a clear, cool night to form, but slightly different wind conditions. Dew forms on a truly calm night when the greatest chill is confined to grasshopper level. For fog to occur, a very light wind is needed to spread the chill from the ground through a deeper layer of air.

The clear, cool, and tranquil conditions needed for dew (as well as ground fog) to form at night are typically fostered by an area of high pressure. Thus, the first part of this folklore has merit since an evening dew on the grass is consistent with the presence of a fair-weather high-pressure system. If the grass is dry at morning light, then the night was likely too windy, too cloudy, or just too warm. Since any of these conditions could occur ahead of an approaching low-pressure system that promises rain, the second part of the lore has some validity as well.

How do fogs that form in stable layers dissipate? To see one mechanism in action, consider Figure 8.19. One of the authors placed a Plexiglass cylinder into a bucket of ice (not shown) for 30 minutes to mimic a deep nocturnal chill on a clear night with light winds. A piece of hollow tubing was inserted into a small hole near the bottom of the cylinder. After removing the cylinder from the bucket of ice, the author lit a cigarette and gently blew smoke through the tubing into the cylinder. This gentle blowing simulated light winds that help to thicken the inversion (had he blown too hard, the smoke would have simply skyrocketed up through the inversion).

Next, the cylinder was placed on a hot plate. Convection began in quick fashion, indicating a rapid destabilization of the "atmosphere" inside the cylinder. In the picture on the right, note the hints of up-and-down motions associated with convective eddies. The downward circulation of the eddies mixed clear air toward the bottom of the cylinder, in effect, dispersing the smoke. The same process explains why the author's afternoon bicycle ride on the dirt road was not as dusty as the morning commute. Just like the eddies in the cylinder experiment, convective eddies in the warmer, less stable lower troposphere more quickly dispersed the afternoon dust kicked up by passing cars.

Ground fog dissipates in similar fashion after sunrise. Solar radiation penetrates the thinner, outer edges of a bank of fog and heats the ground. In turn, the heating produces convective eddies, which mix down drier air from above and foggy air upward, causing the outer edges of the fog to evaporate. Then the process works its way inward toward the thickest part of the fog. In fact, on animations of visible satellite imagery, meteorologists often observe a bank of fog dissipating from the outer edges inward.

This account of how ground fog dissipates also sheds light on how to destabilize the atmosphere near the surface: Heat the ground! The heated surface then warms the air directly above while imparting less energy to air higher up. Indeed, on sunny days, the environmental lapse rate near the earth's surface will tend toward the dry adiabatic lapse rate (10°C/km). Sometimes, a layer near the ground can even become **superadiabatic** (see Figure 8.20), with the temperature decreasing at a rate greater than 10°C/km. In such layers, convective eddies develop quickly and tend to mix cooler air toward the hot ground, helping to ease the extremely rapid temperature decrease with height. However, superadiabatic lapse rates can reform in response to subsequent heating by the sun. In essence, it becomes a battle between heating and mixing.

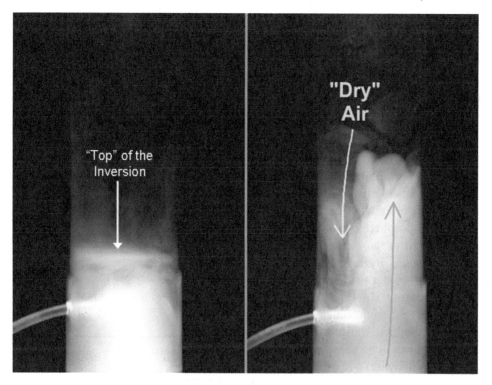

FIGURE 8.19 (Left) Cigarette smoke blown into a cylinder that had cooled in a bucket of ice for 30 minutes dramatically stratified, indicating the highly stable environment inside the cylinder. (Right) After placing the cylinder on a hot plate, convection quickly began, indicating a rapid destabilization of the "atmosphere" inside the cylinder. Note the up-and-down motions associated with convective eddies. The downward circulation of the eddies mixed smoke-free air toward the bottom of the cylinder, in effect, dispersing the smoke. Ground fog similarly dissipates after sunrise.

There are other ways to change the stability of the atmosphere near the surface. For example, on hot and humid days, afternoon thunderstorms sometimes erupt. As small raindrops evaporate, air temperatures decrease toward the dew point in response to evaporational cooling (see Figure 8.21). This cooling tends to stabilize the lower troposphere, in effect producing a "cool bottom." On the flip side, warm advection near the surface destabilizes the lower troposphere, creating a warm bottom layer and sometimes paving the way for convection.

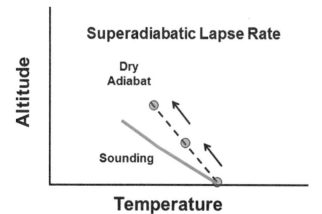

FIGURE 8.20 On sunny, warm days, the environmental lapse rate can become superadiabatic in a layer next to the ground. In such a case, an air parcel given a nudge upward is positively buoyant and thus continues to rise.

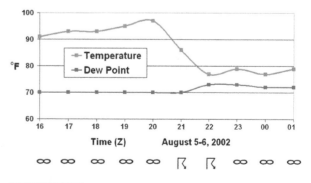

FIGURE 8.21 A portion of the meteogram from Washington DC's Dulles International Airport, beginning at 17 UTC on an August day. That afternoon, temperatures soared to 96°F (36°C) under hazy skies. Extreme heat and humidity paved the way for a thunderstorm, which began between 20 UTC and 21 UTC. In response, surface temperatures dramatically decreased via evaporational cooling, thereby stabilizing the lower troposphere.

Lest we leave you with the impression that the surface has full control over changes in stability, we will now look higher in the troposphere.

Stability Changes Aloft: The Opposite of What Occurs at the Ground

Although the chill that forms near the ground on a clear and relatively calm night stabilizes the lower troposphere, cooling at high altitudes tends to destabilize the troposphere, particularly during the daytime.

Figure 8.22 demonstrates what happens when a layer of the middle troposphere cools. Let's assume a saturated parcel of air at the bottom of this layer is given a nudge upward. Before any cooling occurred, the parcel was colder than its environment and thus negatively buoyant. However, after the mid-level cooling, note that the environmental lapse rate moved to the left of the local moist adiabat. In this case, the rising parcel would be warmer than its environment (positively buoyant) and thus continue to rise. So cooling aloft is a destabilizing process.

To see the destabilizing effects of mid-tropospheric cooling in action, consider Figure 8.23, an 18 UTC analysis of 500-mb heights and temperatures,

superimposed on radar imagery, on October 15, 2015. At the time, a closed 500-mb low and its associated pool of mid-level cold air was situated off the

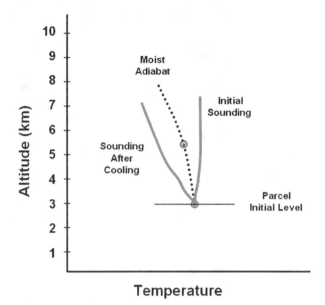

FIGURE 8.22 Cooling aloft destabilizes the troposphere with respect to moist ascent because a parcel rising into the cooled layer would become warmer than its new, colder environment.

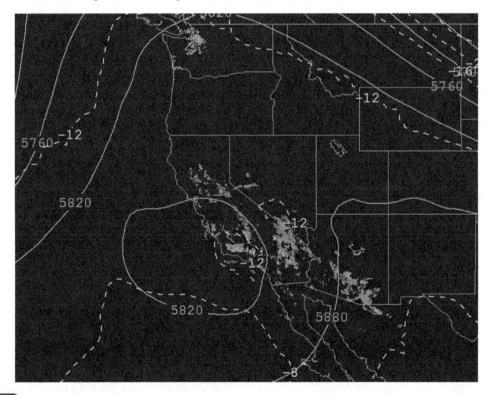

FIGURE 8.23 An analysis of 500-mb heights (green solid lines, in meters) and temperatures (yellow dashed lines, in °C) superimposed on a radar image at 18 UTC on October 15, 2015. A mid-level cold pool associated with a closed 500-mb low was situated near the southern California Coast, destabilizing the middle troposphere and paving the way for heavy showers and thunderstorms to develop.

coast of southern California (note the closed contour of −12°C air right near the coast). Meanwhile, heavy showers and thunderstorms had erupted in concert with this cold pool, causing flooding and mudslides serious enough to close parts of Interstate 5 near Los Angeles.

This example during October highlights a peculiarity of coastal California weather. Whereas most people associate thunderstorms with spring and summer, the cold season is actually the time of year when thunderstorms are most likely there (though they are still uncommon). That's because the period from roughly October to March is the time when invading pools of mid- to high-level cold air move over relatively warm Pacific waters. Granted, water temperatures off the coast are generally in the 5°C to 15°C range (41–59°F), but they are high compared to the frigid air that sometimes arrives at higher altitudes (in concert with 500-mb troughs). In such situations, there is a relatively warm bottom and a deep chill aloft, which together set the stage for relatively fast upward motions.

On the flip side, warming aloft tends to stabilize the middle troposphere (the opposite effect of warming at the ground). The simplest example is an upper-level ridge building over a region. As you recall from Chapter 7, an upper-level ridge marks the core of relatively warm air aloft.

Another example of the stabilizing effect of warming aloft occurs near and to the east of the center of surface high-pressure systems, where air tends to sink from high altitudes into the lower troposphere. We'll use Figure 8.24 to help to tell this story. During the day, this large-scale subsidence usually stops around the 850-mb level (air below that level often rises after being heated by the sun-warmed ground—see the left side of Figure 8.24). We've already established that sinking air warms by compression, but the warming of a column of subsiding air is not uniform. This uneven warming leads to the formation of an inversion that's typically found around the 850-mb level, appropriately called a **subsidence inversion**, given that it owes its existence to air that gently subsides.

It may not be immediately clear to you why an inversion would form just because a column of air subsides. To explain, we have to recall a fundamental result from Chapter 7: The spacing between successive pressure levels expands as altitude increases, particularly in the upper troposphere. With this idea in mind, you're ready to tackle subsidence inversions! We'll use Figure 8.25 as our guide. Let's suppose a

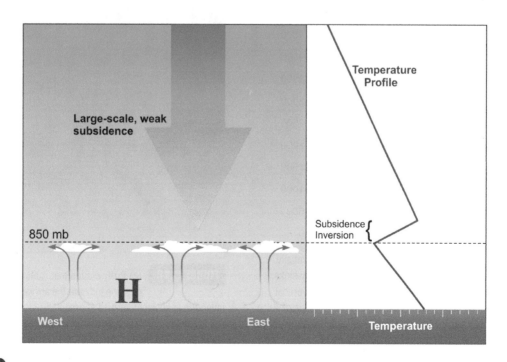

FIGURE 8.24 Over and east of the center of a surface high-pressure system, air sinks from great altitudes and warms. During the daytime, subsidence usually stops around 850 mb, and a subsidence inversion typically forms near this level. Below 850 mb, air often rises from the sun-warmed ground.

volume of air between 300 mb and 400 mb subsides so that it ends up between 700 mb and 800 mb (near the level where many subsidence inversions occur).

During descent, the rarefied volume of air is compressed, given that pressure increases with decreasing altitude. But the air is not compressed evenly. A parcel of air at the top of the volume at point T descends a greater distance than a parcel starting at point B on the bottom of the volume—note in Figure 8.25 that the vertical distance from T to T' is greater than the distance from B to B'. Keeping in mind that a sinking, unsaturated parcel of air warms at a rate of 10°C/km, it follows that the air arriving at T' will have warmed more than the air arriving at B'. As a result, the air at T' usually ends up warmer than the air at B'. This vertical profile of temperature corresponds to an inversion—more specifically, a subsidence inversion. These inversions routinely form near the centers and on the eastern flanks of most surface high-pressure systems.

There can be visual clues that a subsidence inversion is forming or is already present. During the daytime, puffy, fair-weather cumulus clouds flatten out as the center of a robust high-pressure system approaches. This flattening occurs because the tops of fair-weather cumulus steadily collapse in response to the strengthening subsidence inversion (see Figure 8.26a). Some meteorologists refer to these

(a)

(b)

FIGURE 8.25 In this illustration, a column of air initially between 300 mb and 400 mb gently sinks, settling between 700 mb and 800 mb. Given that the vertical distance between pressure surfaces is not uniform, air at the top of the original column subsides a greater distance than air at its bottom. Thus, the top of the column undergoes more compressional warming than the bottom, often leading to a temperature inversion between 700 mb and 800 mb. Inversions that form when air subsides (sinks) are appropriately called subsidence inversions.

FIGURE 8.26 (a) Puffy clouds that "flatten out" during the daytime indicate the arrival of a subsidence inversion associated with an approaching high-pressure system; (b) The flat top of this haze layer over Georgia formed near an altitude of 2000 meters (about 6500 ft), approximately corresponding to the bottom of a subsidence inversion. Some convective clouds in the distance have already broken through the inversion, spurred on by the strong Southern sun (courtesy of David Odum).

clouds as "pancake cumulus." As sunset approaches, these clouds typically crumble away, setting the stage for a clear, cool night. Another clue that a subsidence inversion is present is a sharp upper edge to a layer of haze. In Figure 8.26b, the top of the haze layer approximately marks the bottom of the subsidence inversion. Such images are more common in the eastern United States, where pollution from large industrial and urban areas provides ample particles for net condensation.

Although people often associate clouds and precipitation with instability, rain and snow routinely fall from clouds that form in a stable environment. What's up with that?

CATEGORIZING CLOUDS AND PRECIPITATION: STABLE VERSUS UNSTABLE

While the image of a towering thunderhead with heavy rain is consistent with the stormy implication of the word *unstable*, don't be fooled into thinking that precipitation can't fall in a stable atmosphere—it can, it does, and it can be heavy! In a typical thunderstorm, precipitation comes and goes in a short burst and affects only a small area (it can be raining on one side of town and not on the other). In contrast, clouds that form in a stable atmosphere have limited vertical extent and tend to cover a much wider area. And when they precipitate, the rain or snow tends to be steady, long-lived, and extensive in coverage. Such differences in cloud appearance and precipitation are profound, so it makes sense for us to categorize clouds and precipitation based on stability.

Clouds and Precipitation in an Unstable Atmosphere: Growing by Heaps and Bounds

Derived from Latin, the word *cumulus* means "heap," an appropriate description, especially when these clouds develop into towering **cumulonimbus clouds** (Figure 8.27a) and tall **cumulus congestus clouds** (see Figure 8.27b)—both resemble heaps of cotton. More diminutive fair-weather cumulus (see Figure 8.27c) are rather symmetrical—typical dimensions might be several hundred meters tall as well as wide. Even the vertical and horizontal scales of cumulonimbus and cumulus congestus are comparable: For example, a cumulonimbus cloud can reach to the tropopause (a vertical extent of perhaps 10 km or so) while its horizontal span is also on the order of kilometers.

Cumulus, cumulus congestus, and cumulonimbus clouds form by way of convection, a process which (in this context) refers to the vertical transport of heat and

(a)

(b)

(c)

FIGURE 8.27   Cumulus represents the family of clouds that resemble hills or mountains of clouds. For this reason, they are sometimes referred to as "clouds with vertical development." The relatively strong updrafts of air that promote cumulus clouds help to give them their distinguishing cauliflower-like appearance. Cumulus clouds range in size from (largest to smallest): (a) cumulonimbus; (b) cumulus congestus (courtesy of Tim Martin); and (c) fair-weather cumulus.

moisture driven primarily by positive buoyancy. These clouds are examples of **cumuliform clouds** (sometimes called "convective clouds") and may last only for minutes, quickly evaporating when surrounding dry air mixes into them via a process called **entrainment**. On the other hand, given an adequate supply of moist air, towering cumulus clouds that develop into long-lived severe thunderstorms can persist for hours. Updrafts of moist air that sustain cumuliform clouds vary in speed from tens of centimeters per second (in fair-weather cumulus) to tens of meters per second (in severe thunderstorms). The common summertime thunderstorms that develop in the heat and humidity of the afternoon are naturally convective, with rains that typically build to a downpour and then taper off in a matter of tens of minutes while affecting an area on the order of several kilometers in diameter.

The transition from a plain old cumulus to a precipitation-producing cloud is not a simple matter of cloud drops growing by net condensation until they're large enough to fall. A typical cloud drop is a million times or more smaller (in volume) than a typical raindrop (check back to Figure 4.13). As these tiny cloud drops grow, water vapor dwindles in surrounding cloudy air, limiting the further growth of cloud drops. In addition, the drag exerted by the air on tiny, lightweight cloud drops causes them to fall very slowly, while updrafts further slow their descent or even make them rise. Finally, more often than not, descending cloud drops easily evaporate in the air below cloud base. The conclusion from all this: Raindrops aren't just a bunch of oversized, condensation-formed cloud drops—other mechanisms are at play to make raindrops.

To provide some insight into one of the most important mechanisms, consider that the rain that falls from a summertime thunderstorm over the middle and high latitudes almost always begins as snow in the cold upper troposphere. In fact, almost all precipitation that develops poleward of the tropics starts as snow. Intrigued?

The Bergeron-Findeisen Process: There's Snow in Summer Thunderstorms

To understand why almost all precipitation that falls over the middle latitudes begins as snow (even in summer), you first need to know that some tiny water droplets can resist freezing down to temperatures as low as –40°C (–40°F). If that shocks you, consider the results of an experiment shown in Figure 8.28. One of the authors took an empty can, greased the bottom with a little oil, and placed nine drops of water on it using an eyedropper (left photograph). Then he placed it into the freezer for ten minutes with the temperature approximately –10°C (14°F). Lo and behold, four of the drops did not freeze (right photograph), showing that relatively small drops of water can indeed exist at temperatures well below 0°C. Drops of water that stubbornly resist freezing at such low temperatures are often described as **supercooled drops**.

Why the resistance to freezing? At very low temperatures in the middle to upper troposphere, supercooled droplets typically lack ice nuclei, tiny particles that jump-start the formation of cloud drops. In the case of the experiment conducted in Figure 8.28, the author boiled the water before placing the drops on the can. In this way, ice nuclei were likely removed from the water (although nuclei could have been reintroduced in the water drops by contact with the air or contact with nuclei on the bottom of the can).

At –40°C, however, the presence of ice nuclei becomes a moot point because all water drops freeze spontaneously at this very low temperature. Still, this leaves plenty of temperature leeway for ice, water,

FIGURE 8.28 (left) Drops of water on the bottom of an empty tin can; (right) After ten minutes inside a freezer, four drops stubbornly refused to freeze.

and water vapor to coexist high in the troposphere. That's part one of the solution to the puzzle of why most of the precipitation that falls over the middle latitudes begins as snow.

Second, you need to know that it's much more difficult for molecules rigidly bonded in the lattice of ice (see Figure 8.29) to squirm free and sublime to vapor than it is for more loosely bonded water molecules at the same temperature to evaporate. As a result, the equilibrium vapor pressure over ice at a temperature below 0°C is less than the equilibrium vapor pressure over water at the same temperature. We demonstrate this principle schematically in Figure 8.30a with a partitioned box containing water and ice at a temperature just below 0°C.

If we remove the barrier (see Figure 8.30b), water vapor molecules spread more uniformly throughout

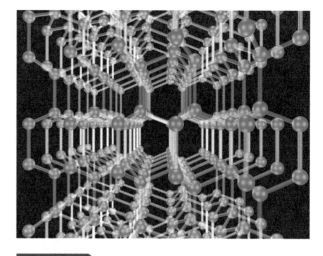

FIGURE 8.29 A computer simulation of the lattice structure of ice. The blue spheres represent water molecules.

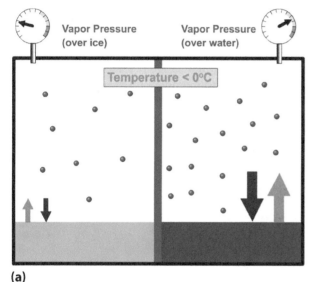

(a)

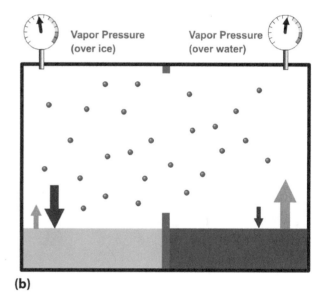

(b)

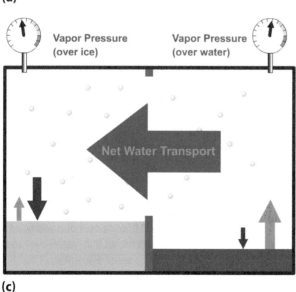

(c)

FIGURE 8.30 (a) Initially, more water vapor molecules reside in the air above the water because water molecules can more easily break their bonds with their neighbors than ice molecules can. In the water compartment (on the right), the rate of evaporation (red arrow) equals the rate of condensation (blue arrow). In the ice compartment (on the left), the rate of sublimation (red arrow) equals the rate of deposition (blue arrow); (b) After the partition between the two compartments of air is removed, the horizontal gradient in vapor pressure between the two compartments decreases as water vapor molecules spread more evenly throughout the container. The vapor pressure above the water is now less than its equilibrium vapor pressure, so the rate of evaporation exceeds the rate of condensation (the red arrow is larger than the blue arrow), and the water starts to evaporate. Meanwhile, the vapor pressure over ice is now greater than its equilibrium vapor pressure, so the rate of deposition exceeds the rate of sublimation (the blue arrow is larger than the red arrow), and the ice starts to grow; (c) In the end, ice grows at the expense of the water. In other words, there was a net transport of mass from the water to the ice.

the entire box. Over the water, there are now fewer water vapor molecules than needed for saturation, so net evaporation proceeds. Over the ice, there are now more water vapor molecules than before, collectively exceeding the equilibrium vapor pressure of ice. So net deposition occurs and the ice grows (Figure 8.30c). In other words, ice grows at the expense of water. Formally, this process is called the **Bergeron-Findeisen process** or the **ice-crystal process**.

The same rob-Peter-to-pay-Paul process occurs in cold clouds where all three phases of water coexist. We illustrate the process in Figure 8.31, where ice crystals in the shape of hexagonal plates and columns grow at the expense of surrounding water drops.

Figure 8.32 provides two examples of this process in action. Figure 8.32a is a close-up photograph of a portion of a car windshield on a cold, late autumn morning. Note the relatively dark, water-free areas surrounding the ice crystals. Essentially, the ice crystals on this otherwise wet windshield have grown at the expense of the surrounding water. Figure 8.32b shows a hole in a deck of clouds composed of supercooled water droplets. The Bergeron-Findeisen

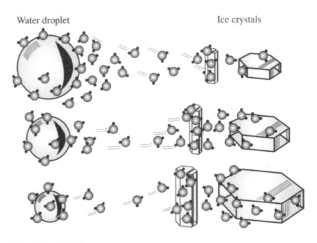

FIGURE 8.31 The Robin Hood code of conduct during the Bergeron-Findeisen process: Water vapor is robbed from vapor-rich cloud droplets in order to pay vapor-poor ice crystals. In this way, ice crystals grow at the expense of surrounding cloud droplets.

process was initiated in a portion of this cloud deck, and the hole formed as ice crystals grew at the expense of the surrounding cloud drops. Note the dark

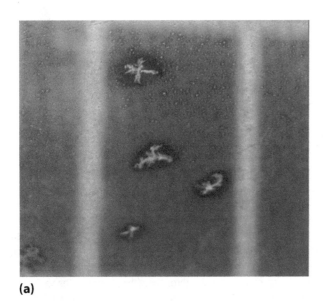

(a)

(b)

FIGURE 8.32 (a) Several ice crystals on an otherwise wet car windshield have grown at the expense of surrounding water; (b) The ice crystal process was initiated in a portion of this deck of clouds near 3000 m (10,000 ft) composed of supercooled water droplets. In all likelihood, ice crystals fell into the cloud from above, perhaps from the contrail of a jet aircraft that passed overhead. There's also the possibility that particles fell from the exhaust of an airplane and "seeded" the cloud with ice nuclei. Regardless of the source, the Bergeron-Findeisen process began and ice crystals grew at the expense of surrounding cloud drops, causing the hole to form (courtesy of Jeff Higgins).

trail of ice crystals falling below the hole. Assuming that the ice crystals sublimated or evaporated before reaching the ground, this dark trail is an example of **virga**, precipitation that evaporates or sublimates before reaching the ground (see Figure 8.33).

We will refer to a single crystal of ice as a **snow crystal** because it is the most basic form of a snowflake. As the weights of snow crystals suspended high in a cold cloud increase, they begin to fall. On descent, some snow crystals can splinter while others can bump into and stick to other snow crystals. At other times, snow crystals can fall through a cloudy layer filled with supercooled droplets. The droplets can rapidly freeze onto falling snow crystals, creating a **rimed** snow crystal.

A **snowflake** can be an individual snow crystal or a few snow crystals that are stuck together, sometimes called an **aggregate**. The photograph in Figure 8.34 documents these various kinds of snow crystals (more on wintry precipitation in Chapter 16). The helter-skelter splintering, bumping and sticking that snow crystals undergo within cold clouds helps to explain why "no two snowflakes are exactly alike."

Snowflakes that form high in summertime cumulonimbus clouds all share exactly the same fate. Unless they settle on the cold tops of tall mountains, they melt into raindrops before they reach the ground. In fact, in any season, much of the precipitation that falls over the middle latitudes begins as snow within cold clouds high above the ground. The type of precipitation that reaches the ground depends on the temperature structure in the lower troposphere, a few

FIGURE 8.34 Three different kinds of snow crystals.

kilometers below the altitude where the Bergeron-Findeisen process gave birth to snowflakes. We'll explore two other types of precipitation, sleet and freezing rain, in Chapter 16.

Though the Bergeron-Findeisen process is, undeniably, a dominant force for making precipitation over the mid-latitudes, it is not alone. Sometimes, raindrops can be so large that they immediately break up into much smaller droplets, creating a new generation of raindrop embryos that can set off what cloud physicists call a "warm-rain chain reaction."

A simple process is at the core of this chain reaction—collision and coalescence. One description of this so-called "warm-rain process" reads like a script for a classic boxing film: "The bigger they are, the harder they fall." In other words, bigger raindrops fall faster than smaller raindrops, thus overtaking and colliding with them and consequently growing as all or part of the smaller drops' water sticks or "coalesces" to the larger, speedier drops. In this way, the warm-rain process allows thunderstorms to rapidly grow raindrops by collision and coalescence, heightening the chances of a burst of heavy rain.

In summary, clouds that develop in a conditionally unstable (or unstable) environment are predominantly cumuliform. When these convective clouds produce precipitation, rain or snow tends to be spotty and short-lived. Even in summer, precipitating cumuliform clouds have a connection to winter. Indeed, raindrops falling from towering cumulonimbi typically began as snow crystals high above the melting level.

FIGURE 8.33 Streaks of virga beneath a cumulus congestus cloud over central Pennsylvania in May 2013 (courtesy of Marisa Ferger).

How about clouds forming in a stable environment? Read on.

Clouds and Precipitation in a Stable Atmosphere: Layered and Stratiform

Just because the troposphere is stable doesn't mean that precipitating clouds can't rain on your parade. But to overcome the resistance to upward motion that exists in a stable environment, air parcels require more than just a gentle nudge. Take, for example, the **cap cloud** shown in Figure 8.35 observed over a mountain near Breil-sur-Roya in southeastern France (for reference, the upslope flow of air in the photograph is from left to right). Clearly, the orographic lifting was strong enough to force air parcels to rise sufficiently for the cap cloud to form, despite the obvious stability of the environment. You can tell that the environment is stable because the cap cloud bends downward after reaching the summit, a sign that cloudy air parcels were negatively buoyant.

In light of the appearance of the "nuage chapeau" in Figure 8.35, clouds that form in stable environments bear little resemblance to their taller cumuliform counterparts. But cap clouds aren't your daily, run-of-the-mill type of "stable" cloud. The more common types form in layers, spreading out horizontally and covering large areas such as the leaden **stratus clouds** seen in Figure 8.36. Unlike their cumuliform counterparts, these **stratiform**

FIGURE 8.36 Stratus clouds (courtesy of Alistair Fraser).

clouds have limited vertical development. In fact, *stratiform* is derived from the Latin word *sternere* which means "to stretch." Fog is also a stratiform cloud, even though it usually forms in the absence of a lifting mechanism.

When it comes to mid-latitude weather, stratiform clouds routinely form on the cold side of the surface warm front associated with a low-pressure system (Figure 8.37a). On the chilly side of a warm front, surging warm air (often an mT air mass) glides up and over relatively dense cold air (often a cP air mass). In essence, like the air forced to flow up our mountain in France, relatively warm air is forced to climb up the "slope" of the retreating cold air mass. Warm air must make this climb because it simply cannot dislodge the denser, colder air near the surface. Appropriately, this lifting mechanism is called **overrunning** because warm air glides up and over cold air (see Figure 8.37b).

In typical overrunning situations (especially from autumn into early spring), our test for stability would show that air parcels ascending the wedge of cold air quickly become negatively buoyant. In other words, the lower atmosphere is stable here. So why do parcels continue to rise? Basically, because they have no choice! The lifting mechanism of overrunning, although not usually as strong as orographic lift, is indeed compelling. In fact, forecasters always look for situations when overrunning might produce stratiform precipitation whenever they prepare their forecasts. A classic example of overrunning is shown in Figure 8.37c. High pressure centered north of New England provided the chill, and a low-pressure system in the

FIGURE 8.35 A cap cloud forms as moist air is forced to rise over a mountain summit near Breil-sur-Roya in southeastern France. In French, the name of this cloud is "nuage chapeau," which translates to "hat cloud" (courtesy of Vincent Jacques).

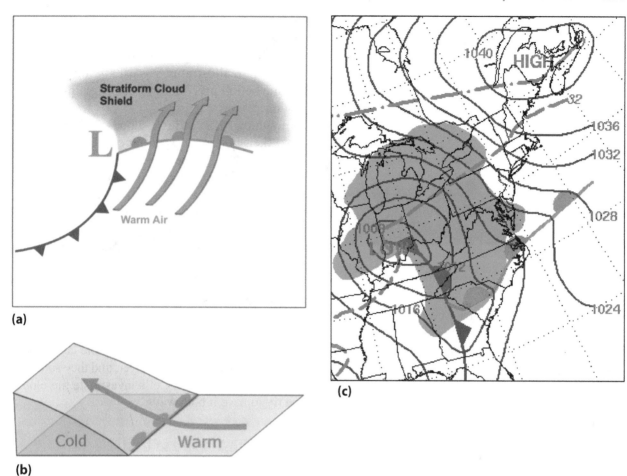

(a)

(b)

(c)

FIGURE 8.37 (a) Stratiform clouds commonly form on the cold side (typically north) of a surface warm front as relatively warm air surges northward ahead of a mid-latitude low-pressure system; (b) Like air forced to rise over a mountain, relatively warm air is forced to glide up the slope formed by the wedge of cold air. This process is called overrunning; (c) Surface analysis at 12 UTC on February 1, 2008, with green marking areas where precipitation was falling at the time. High pressure centered near Maine provided the low-level chill for widespread overrunning precipitation to form north of a warm front, which stretched from the Southeast into the western Atlantic Ocean (courtesy of NOAA).

Midwest drew warmer air northward over the cold wedge. An expansive shield of precipitation, much of it frozen, fell north of the warm front, where relatively warm parcels overran low-level cold air.

But these air parcels can only be forced to rise so far. That's because they become increasingly negatively buoyant as they steadily cool on ascent. Once rising air parcels reach an altitude where their negative buoyancy is simply too great, they essentially stop rising and spread out laterally. As a result, clouds that form on the cold side of a surface warm front tend to be rather shallow and layered. Only when the overrunning warm air is exceptionally warm and moist can cumuliform clouds (such as thunderstorms) form on the cold side of a surface warm front.

In addition to stratus, there are other members of the stratiform clan. We'll introduce them from lowest to highest in the troposphere, essentially in the order they would typically form north of a warm front. Stratiform clouds that produce steady rain or snow and cover a wide area are known as **nimbostratus clouds** (*nimbus* means "rain"). Nimbostratus are low, dark, and claustrophobic clouds, with bases typically several hundred meters off the ground (see Figure 8.38). The Bergeron-Findeisen process is often at work in nimbostratus clouds.

Continuing vertically up the wedge of cold dense air, we would next encounter **altostratus clouds** (see Figure 8.39), which form at altitudes between about 2500 and 6000 m (about 8000 and

FIGURE 8.38   Nimbostratus clouds (courtesy of Tim Martin).

20,000 ft). Though primarily composed of water droplets, they may contain some ice crystals, especially in winter at higher altitudes. Finally, **cirrostratus clouds** form at altitudes above about 6000 m (about 20,000 ft) and are primarily composed of ice crystals (see Figure 8.40). At times,

sunlight refracted by these ice crystals forms a colorful ring around the sun (or moon) known as a **halo**. If these high-altitude clouds lower and thicken into altostratus, it's usually only a matter of time before precipitation-bearing nimbostratus clouds arrive.

In summary, stratiform clouds have a layered look; their horizontal span far exceeds their vertical reach. And unlike their cumuliform brethren, stratiform clouds have both smooth bottoms and smooth tops. Slow and steady are the operative words for vertical motions in stratiform clouds; air parcels typically ascend at only a few centimeters per second or less. But when they produce precipitation (almost exclusively by the Bergeron-Findeisen process), it is usually steady and covers a wide area.

Sometimes, clouds that form in stable layers can show visual signs of some instability. On the flip side, clouds that develop in an unstable layer sometimes can encounter a stable layer, and they suddenly take on a layered look. Let's investigate the clouds that try to have it both ways.

Clouds with Stable and Unstable Components: Having It Both Ways

When powerful thunderstorms erupt, they often boil up to the tropopause where the temperature decrease with altitude slows to a trickle. Above the tropopause, the lapse rate becomes nearly isothermal, and the atmosphere is very stable. When the tops

FIGURE 8.40   A layer of cirrostratus clouds provides a background for a contrail left by a high-flying jet over Nevada in June 2014 (courtesy of John Kupersmith, www.jkup.net).

FIGURE 8.39   Altostratus clouds (courtesy of Hampton Shirer).

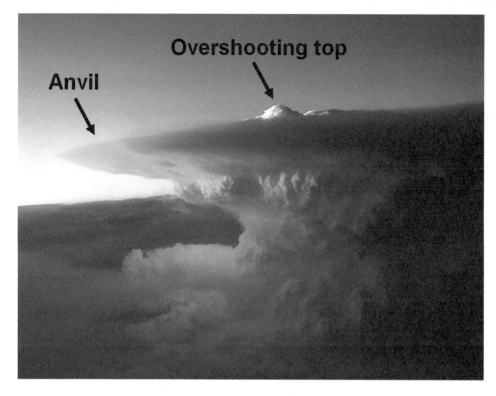

FIGURE 8.41 The over-shooting top of this severe thunderstorm indicates that rapidly rising air parcels in the speedy updraft in the core of the thunderstorm have overshot the stable tropopause—they had great momentum and couldn't stop on a dime. The anvil indicates that air parcels with less momentum hit the tropopause and were forced to move laterally (courtesy of Steven Maciejewski).

of thunderstorms bump up against the tropopause, there's a confrontation of sorts—instability versus stability.

Despite the stable layer in their path, rapidly rising parcels in the speedy updraft of a potent thunderstorm still have plenty of momentum. As a result, they often continue their ascent a small distance into the stable layer above, forming **overshooting tops** (see Figure 8.41). These overachieving parcels cannot, however, break through the inversion in the lower stratosphere, so even they eventually run out of gas.

Figure 8.41 also shows another interesting feature occasionally seen atop a powerful thunderstorm. Away from the thunderstorm's speedy core updraft, rising air parcels have less momentum. Upon reaching the tropopause, they quickly become colder than their surroundings and their ascent stops. They spread out horizontally and form an icy layered cloud called an **anvil**, so named because of its resemblance to the tool used by blacksmiths (see Figure 8.42).

Anvils can also form with convection during winter. Figure 8.43 shows a wintertime cumulonimbus cloud that produced a snow squall with thunder over central Pennsylvania in response to very cold air arriving at the 500-mb level. Note that the convective cloud has an anvil, which, in this case, is at a relatively low altitude (compared to the warm season).

So a cloud that forms in an unstable environment can develop a layered look after encountering a stable layer. Does it also work the other way? In other words, can a layered cloud that forms in a stable environment develop more of a convective look? The answer is again yes.

Throughout this discussion of stability, the key to any cloud makeover is the temperature structure of the layer in which the cloud develops. For example, consider what happens when the top of a stratiform cloud cools, perhaps by undergoing a net loss of radiation. This cooling has a destabilizing effect on the layer in which the cloud resides, given that

FIGURE 8.42 A blacksmith's anvil (courtesy of Smithsonian Institution).

FIGURE 8.43 A wintertime cumulonimbus cloud that produced a snow squall with thunder over central Pennsylvania developed an anvil.

FIGURE 8.45 Altocumulus clouds.

the lapse rate of the layer increases (the top cools). In this case, a layer of cirrostratus, altostratus, or stratus clouds can develop cumulus-like cells, particularly if the bottom of the layer warms a little (from radiation emitted by the ground below, for example). When convection develops within initially stable layers, **cirrocumulus, altocumulus,** or **stratocumulus clouds** can result (see Figures 8.44, 8.45, and 8.46).

Radiative cooling at the tops of clouds is the same type of cooling that can make a hot-air balloon lose buoyancy at night. Indeed, measures must be taken during transoceanic balloon crossings in order to avoid having the balloon sink after the sun goes down. Had his helium-ballooned lawn chair ventured over the Pacific at night, the same sinking fate might have awaited Larry Walters, a

FIGURE 8.46 Stratocumulus clouds.

man who seriously underestimated the importance of buoyancy.

We end this chapter with a discussion that focuses on a few ways that you can sense the stability or instability of the atmosphere.

OTHER SIGNS OF STABILITY AND INSTABILITY: FROM SMOKE TO SUMMER BREEZES

Earlier in this chapter, we reminisced about one of the authors assessing the stability of the lower atmosphere by observing the behavior of dust kicked up by cars on a dirt road. Besides the layered appearance of stratiform clouds and the heap-like structure of convective clouds, there are other visual signs that offer clues about the stability of the atmosphere.

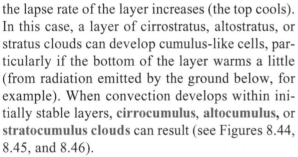

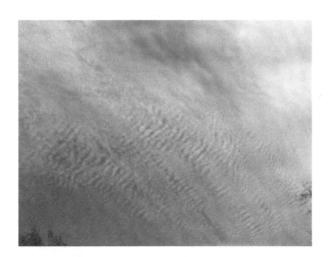

FIGURE 8.44 Cirrocumulus clouds (courtesy of Alistair Fraser).

Smoke: Fanning and Looping

When Mount Etna erupted on the Mediterranean island of Sicily in late October 2002, astronauts aboard the International Space Station captured the striking photograph in Figure 8.47. Note that the erupting plume spewed upward at first (it was quite hot and positively buoyant), but then it hit a stable layer and started to fan out horizontally (the process is actually called **fanning**), as winds carried the plume away from the volcano.

The plume of smoke from a chimney or smoke-stack can also send clues about stability. On a clear, calm night, rising smoke from a factory stack typically fans out horizontally in a low-level nocturnal inversion, much like the fanning of the plume of volcanic ash from Mount Etna. If the nocturnal inversion is fairly strong, pollutants get trapped in a layer of air next to the ground. Figure 8.48 is an early-morning photograph that visually makes our point (similarly, the effluent from vehicles, chimneys, and the like, gets trapped). Sometimes, pollutants spewing from a relatively high smoke-stack or cooling tower can spread out horizontally in a thin layer above the ground (see Figure 8.49a). After the nocturnal inversion breaks and the lower troposphere destabilizes, convective eddies may

dramatically distort the horizontal plume of smoke, causing it to loop up and down as it gets caught in alternating currents of rising and sinking air (Figure 8.49b). The downward arcs of the looping smoke plume can bring relatively high concentrations of pollution to the surface, a process called **fumigation**.

Cloud Clues: UFOs and Castles

A series of alternating up-and-down air motions can also indicate stability. When a westerly wind blows over a roughly north-to-south oriented mountain range, a series of waves will sometimes set up in the lee of the mountains (see Figure 8.50a). The theory of these lee waves is rather complicated, but it suffices to say that they typically have wavelengths of several kilometers and can extend downwind of the mountain barrier for tens of kilometers or more. When lee waves form in a layer of air that's fairly close to saturation, clouds can develop on the upswing and evaporate on the downswing, creating a "wave train" of clouds and, on occasion, turbulence that impacts aircraft flying near and through these mountain wave patterns. Figure 8.50b is a visible satellite image of wave clouds that formed downwind

FIGURE 8.47 The eruption of Mount Etna on the island of Sicily in October 2002, as seen from the International Space Station. Note how the hot, positively buoyant plume first rises, then fans out laterally as it encounters a stable layer (courtesy of NASA).

FIGURE 8.48 An early-morning photograph that shows how a nocturnal inversion can trap smoke near the ground (courtesy of Chuck Forsberg).

of three of Alaska's Aleutian Islands (at the time, winds were northwesterly). The long chain of small, volcanic islands have rocky coasts that rise abruptly to steep, forbidding mountains.

In some cases, mountain wave clouds can take on eerie elliptical shapes, sometimes resembling stacks of pancakes. In the past, these **lenticular clouds** (see Figure 8.50c) were sometimes reported as unidentified

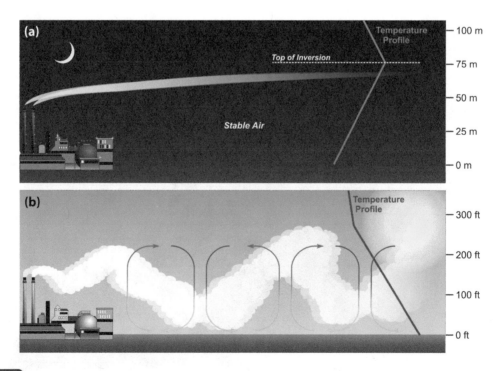

FIGURE 8.49 (a) On a clear, calm night after a fairly strong nocturnal inversion forms, pollutants from a relatively high smokestack may accumulate in a thin layer above the ground; (b) After the inversion breaks, convective eddies may cause the plume to loop up and down, transporting pollution to the ground in a process known as fumigation.

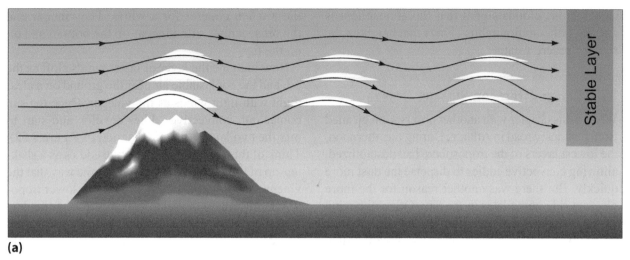

(a)

(b)

(c)

FIGURE 8.50 (a) When mountains force the prevailing flow of air to upslope, waves can form in the lee (downwind) of the mountains when the lower layers of the atmosphere are stable. If the air is close to saturation, a series of mountain wave clouds can form; (b) A visible satellite image on June 18, 2006 shows wave clouds forming downwind of three of the larger mountainous islands that comprise Alaska's Aleutian Islands. At the time, winds were northwesterly (courtesy of NASA); (c) Lenticular clouds over South Georgia Island, located to the east of the southern tip of South America in the South Atlantic Ocean (courtesy Lieutenant Elizabeth Crapo, NOAA).

flying objects. But they are far from alien: The bottom line here is that mountain wave clouds are a visual clue that the atmosphere is stable.

Other than the typical cumuliform clouds, altocumulus castellanous clouds (see Figure 8.51) indicate instability in middle levels of the troposphere, often at altitudes near 3000 m (about 10,000 ft). As their name suggests, these mid-level clouds resemble castles. Their appearance on a spring or early summer morning often portends the arrival of severe thunderstorms in the afternoon or evening. Indeed, if warm, moist surface air is lifted later in the day, perhaps by an approaching low-pressure system or front, the rising parcels will have no problem accelerating through the middle troposphere on their way to the tropopause. In this way, altocumulus

FIGURE 8.51 Altocumulus castellanous clouds.

castellanous clouds signal that the atmosphere is ripe for tall cumulonimbus clouds that heighten the risk for severe weather.

Wind Clues: Afternoon Summer Breeze

We end this chapter with another observation gleaned from that dusty road in Atlanta. During the afternoon, the lowest layers of the troposphere had destabilized, allowing convective eddies to disperse the dust more quickly. But there was another reason for the more pleasant ride home later in the day. Often, the wind was calm in the early morning, but, by afternoon, a breeze had kicked in, helping to disperse the dust kicked up by passing cars.

Where did these afternoon breezes come from? Consider the diagram on the left in Figure 8.52, which represents the vertical profile of wind speeds in the lower troposphere before dawn on a typical summer day. Not surprisingly, the winds near the ground are relatively light, but stronger winds lurk about a kilometer or so above the ground (note the prominent bulge to the right in this pre-dawn profile of wind speeds). These faster winds are often de-coupled from the light winds in the layer of surface air. To understand what we mean by "decoupled," think of oil and vinegar salad dressing sitting on the kitchen counter for a while. The vinegar and the oil separate, with vinegar on the bottom and oil on the top.

Now imagine the layer of faster winds aloft as the oil, and the cool, stable air near the ground on a clear night with light winds as the vinegar. Once the sun comes out, convective eddies develop and start to mix the two layers (middle diagram of Figure 8.52). Think of this mixing by convective eddies as a shaking-up of the salad dressing; in the same way that the vinegar and oil mix, wind speeds in the lower troposphere become more homogeneous during the day. With faster momentum air mixing downward, wind speeds near the ground increase and a noticeably gusty afternoon breeze develops (right diagram of Figure 8.52). In turn, this breeze helped to disperse dust on the dirt road in Atlanta.

Of course, other factors such as the approach of a low-pressure system can increase the pressure gradient and cause the wind to increase from morning to afternoon. But, in the absence of such a large-scale mechanism, an afternoon breeze following a calm morning is usually a sign that the lower troposphere destabilized during the day.

In the next chapter, we'll look more closely at the ultimate example of atmospheric instability—thunderstorms.

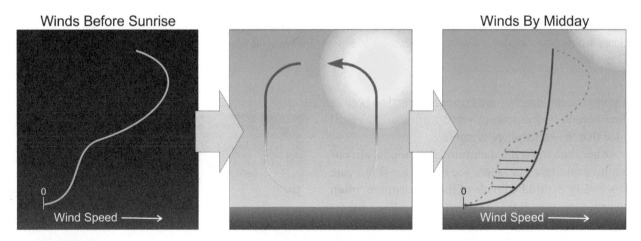

FIGURE 8.52   (Left) A typical profile of wind speeds in the lower troposphere before dawn during the warm season; (Middle) Sunshine acts like a giant spoon, mixing the lower troposphere by causing currents of rising and sinking air. As a result, momentum from faster winds aloft is mixed downward, energizing winds near the ground by the afternoon (and weakening initially faster winds higher up); (Right) The resulting vertical profile of wind speeds in the afternoon shows that winds near the ground have increased in speed, while wind speeds higher up have decreased. After mixing, the vertical profile of wind speeds during the afternoon is more homogeneous.

Focus on Optics

Superior Mirage: The Phantom of the Arctic

Suppose, on an adventurous excursion to the Arctic, you came upon an eerie mountain of ice like the one shown in Figure 8.53. What would you do? Perhaps you'd choose to turn around and go home. That's exactly what John Ross did in 1818, when, as a captain in the British Navy, he led a sailing expedition to Canada's Lancaster Sound in search of the "Northwest Passage" to the Pacific Ocean. After balking at such an Arctic spectacle, Ross returned to Britain, only to have his story and reputation ridiculed when his second-in-command, a fellow by the name of Perry, sailed through the Sound without a hitch a year later.

What Ross observed was not a daunting mountain of ice—it was likely a mirage made fearsome by upwardly magnified images of tiny pieces of snow or ice resting on the Arctic tundra. Unlike the inferior mirage discussed in Chapter 6 (an example of which is the "water on the road"), this type of mirage is called a **superior mirage** because images of an object appear above its actual location. A superior mirage forms when cold, dense air at the ground is overlain by warmer, less dense air (see Figure 8.54a), a common temperature-inversion structure over high latitudes in winter. More specifically, temperature first must increase slowly with height above the frigid tundra. Above this layer of slow temperature increase, temperatures increase more rapidly with height. Completing the temperature-inversion sandwich, the temperature increase returns to a slower rate higher above the ground.

As light from a point on the snowpack travels through the temperature-inversion sandwich, it can be refracted either strongly (while passing through regions

FIGURE 8.53 A superior mirage. Multiple images of points on the snow-covered ground appear above the snowpack, creating a mirage that looks like a mountain or wall of snow to the observer (courtesy of Robert Greenler).

(Continued)

(Continued)

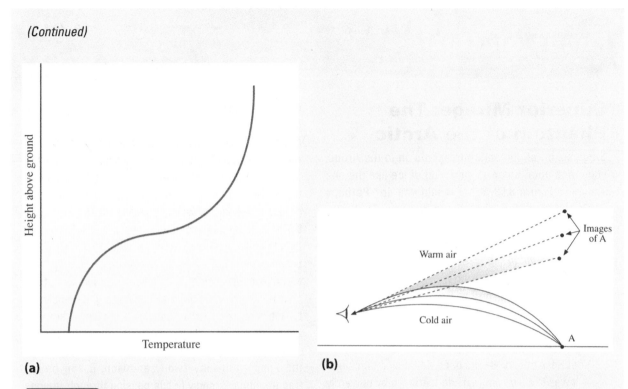

(a)

(b)

FIGURE 8.54 (a) The temperature-inversion sandwich that often leads to a superior mirage. Continuing with the metaphor of a sandwich, the two slices of bread are represented by the lowermost and uppermost inversion, which are both doughy and weak (temperature doesn't change very rapidly with height). The meat of the sandwich is the stronger inversion in the middle; (b) As light from a clump of snow on the ground at point A passes through the temperature-inversion sandwich, it can be bent in a variety of ways, allowing light to enter an observer's eyes at various angles. Thus, the brain "sees" several images of point A above point A. Each image is a superior mirage, and the collective of these images can take on the appearance of a mountain of snow as shown in Figure 8.53.

of strong vertical temperature gradients) or not so strongly (while passing through regions where vertical temperature gradients are weaker). Given that light can bend in a variety of ways while traveling through these fluctuating gradients, several distinct rays can reach an observer's eyes from a single point on the snowpack (see Figure 8.54b).

The rays shown in Figure 8.54b are concave down because light always bends so that denser, colder air lies on the inside of the curve. But now recall that our brains are programmed to interpret all the light waves that reach our eyes as having traveled in straight lines. Thus, using Figure 8.54b as a guide, the observer can see three distinct images of the snow at point A above the snowpack—and probably many more that aren't even shown. Each of these images of point A is a superior mirage because it appears above point A.

When superior mirages stack up on top of one another like this, a strange, vertical magnification results. Moreover, when there are small variations in the temperature-inversion sandwich, one point might be greatly magnified while its next-door neighbor might appear almost normal. The end result would be an image like the one in Figure 8.53—like the one that John Ross probably saw.

By the way, a similar spectacle can sometimes be seen over cool, mid-latitude lakes during summer (because the same type of temperature-inversion sandwich can occur there). So, if you're boating, be on the lookout for superior mirages suspended over the lake—they would likely appear as a series of towers or a wall with gaps.

If you see one, there's no need to do a John Ross impression and go home.

Thunderstorms

LEARNING OBJECTIVES

After reading this chapter, students will:

- Be able to explain the theory that describes how lightning forms in cumulonimbus clouds
- Be able to explain the underpinning science of thunder and to estimate the distance that lightning strikes from an observer
- Have a working knowledge of lightning safety
- Understand the concept of the level of free convection and its relevance to the development of thunderstorms

- Be able to articulate the differences in the development of elevated convection versus thunderstorms initiated by high-level heat sources
- Understand how differential heating between land and water can generate sea and land breezes (and associated convection along their leading edges)
- Understand the role that vertical wind shear plays in determining the type of convection in the context of single-cell, multicell and supercell thunderstorms
- Be able to explain the life-cycle of single-cell thunderstorms and how these storms can sometimes produce severe weather
- Be able to recognize weather patterns favorable for flash flooding
- Understand how hail forms in thunderstorms
- Be able to describe the physical differences between microbursts and macro-bursts and understand the processes that favor their development

- hail
- graupel
- wet-bulb zero
- precipitation loading
- downburst
- microbursts
- macrobursts
- rainbow

It's a situation we've all probably experienced at some point, typically in summer: your evening picnic or softball game is washed out by a drenching downpour, while just a short drive away, nary a drop falls. Such is the character of convection, especially in summer when thunderstorms typically have a spatial scale of 10 km (about 6 mi) and last just a few tens of minutes, on average. Those who have high standards for meteorologists should take these observations to heart. No wonder that expressions such as "scattered" or "isolated" or "hit-or-miss" and qualifiers such as "probability of precipitation" (POP) really get a workout in summer.

Formally, POP represents the probability that measurable liquid precipitation (that is, at least 0.01 inches of rain or melted snow) will occur at any specific location in the forecast area during the forecast period. So a 50% chance of rain tonight means there's a five in ten chance that any point in the region covered by the forecast (such as your front yard) will receive measurable precipitation by morning. Or, considering ten forecasts with a similar weather pattern, it should rain at your house five of the ten times. Essentially, the use of POPs signals that forecasters are uncertain that measurable precipitation will occur.

When forecasters expect precipitation to only affect part of the forecast area, they can use areal qualifiers such as those given in Table 9.1. For example, forecasters will predict "numerous" thunderstorms for a specific region when they expect thunderstorms to develop and affect 60% to 70% of the forecast area. Figure 9.1 is a satellite image on a day when such a forecast would have been appropriate for much of Indiana and Kentucky.

TABLE 9.1 Terminology used when forecasters expect that thunderstorms will develop but only affect a fraction of the forecast area.

Coverage	Term
10–20%	Isolated
30–50%	Scattered
60–70%	Numerous
80–100%	(no term used)

The thunderstorms seen in Figure 9.1 likely produced thousands of bolts of lightning. In fact, each year thunderstorms produce about 25 million cloud-to-ground lightning bolts in the United States. This estimate comes from the National Lightning Detection

FIGURE 9.1 A forecast of "numerous" thunderstorms would have been appropriate in Indiana and Kentucky on this August afternoon (courtesy of NOAA).

Network (NLDN), which consists of over 100 ground-based sensing stations that pinpoint the location of a cloud-to-ground lightning bolt within seconds of a strike. Figure 9.2a shows the cloud-to-ground lightning strikes during just one hour of a July afternoon over the southeastern United States. Figure 9.2b is the corresponding radar reflectivity image at the end of the hour. Note the excellent correlation between the locations of lightning strikes and the highest reflectivity values (particularly the orange and red).

The newest generation of GOES has the capability to detect lightning, the first geostationary weather satellite with this feature. The "Geostationary Lightning Mapper" can detect lightning flashes over

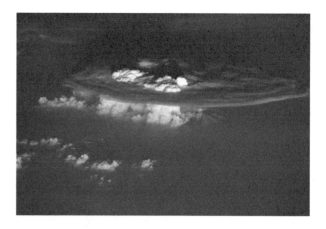

FIGURE 9.3 A thunderstorm viewed from the International Space Station in March 2016 (courtesy of NASA).

much of the Western Hemisphere, giving forecasters a heads-up when thunderstorms are forming or intensifying.

Lightning is but one of a thunderstorm's dangers, which also include hail, flooding rains, damaging straight-line winds, and tornadoes. But thunderstorms are a mixed blessing from Mother Nature: On the plus side, they provide a large amount of the precipitation needed for agricultural production and water resources over many parts of the world. In fact, in the Central Plains, the agricultural heartland of the United States, organized clusters of thunderstorms are the primary source of precipitation during the summer. Each thunderstorm is a boiling consequence of instability, a cumulonimbus cloud that can form, grow, mature, and die within an hour (see Figure 9.3).

In this chapter, we explore the variety, structure, life cycle, and human impacts of thunderstorms. We begin this exposé by explaining the traits that give them their name—lightning and thunder.

LIGHTNING AND THUNDER: BIRTHRIGHT OF THE THUNDERSTORM

Lightning has long captured the imagination of humankind. Ancient societies attributed bolts from the sky to angry or vengeful gods casting their wrath upon mortals below. In 1752, Benjamin Franklin offered a more scientific explanation based on the results of his famous kite experiment (see Figure 9.4). In essence, Franklin found that lightning was simply a large, visible electrical discharge—in other words, a spark in the air.

The prerequisite for lightning is an imbalance of electrical charge in the atmosphere, often referred to as **charge separation**. To understand charge

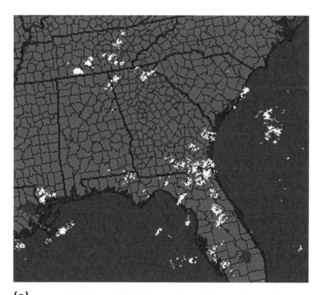

(a)

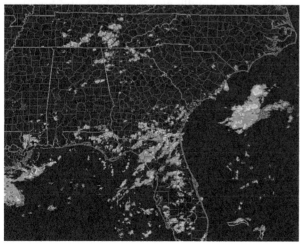

(b)

FIGURE 9.2 (a) A display of the cloud-to-ground lightning strikes during the hour ending at 18 UTC on July 29, 2008 (courtesy of WSI); (b) The corresponding 18 UTC composite of radar reflectivity (courtesy of NOAA).

separation, recall that all matter is composed of atoms. Atoms, in turn, are made of negatively charged electrons orbiting a positively charged nucleus. Charge separation occurs when there is a transfer of electrons between objects (in the case of lightning, between colliding precipitation particles). Atoms with excess electrons carry a negative charge; atoms short on electrons carry a positive charge. A common example of charge separation occurs in your clothes dryer. As clothes tumble and rub against each other, extra electrons can build up on some pieces of clothing. When shirts, pants, and socks are pulled from the dryer, **static electricity** crackles, signaling the dissipation of that charge imbalance.

When Benjamin Franklin's kite passed near a thunderstorm cloud, negative charges leaked onto the kite and raced down the conducting string and key, shocking his knuckles (see Figure 9.4). At the time, Franklin did not fully realize how dangerous his experiment was—a more powerful electrical discharge could have killed him. But his experiment demonstrated that electrified clouds, and therefore lightning, are akin to static electricity. Anyone who has strutted

across a carpet on a cold, dry winter day and built up excess negative charge through friction, only to release it on a doorknob by inducing a shocking spark, has "first-hand" evidence of how lightning works.

Lightning Formation: The Shocking Truth

Shuffling your feet across a carpet will cause negative charge to build up on your body, but a spark will not leap from your hand to a doorknob if the charge separation dissipates. Consider that you usually don't get shocked on humid days—that's because water freely conducts electricity, reducing charge separation and thus the likelihood of a spark.

Exactly how does charge separation occur in clouds? Of the "current" explanations for cloud electrification, the **precipitation theory** typically gets the biggest "plug" from meteorologists. This theory maintains that electrification is a contact sport between different **hydrometeors** (a general term for precipitation particles). For example, ice crystals, which are lightweights, often go one-on-one with **graupel**, which are rimed snowflakes. Though uncharged before colliding, graupel and ice crystals recoil from their in-cloud collisions with each having acquired a charge. Heavier graupel acquires a negative charge, while featherweight ice crystals are charged positively. Separation of these oppositely charged particles then proceeds as lightweights are whisked skyward by rising currents of air, or **updrafts**, within the thunderstorm. Meanwhile, gravity does its job on graupel, building up negative charge in the middle and low levels of a cloud. A

FIGURE 9.5 Charge separation in a cumulonimbus cloud sets the stage for lightning formation.

schematic of this two-layer model of charge separation is shown in Figure 9.5. Because like charges repel, the buildup of negative charge low in the cloud induces a positive charge on the surface beneath, particularly on tall objects such as buildings and trees. Having achieved charge separation, electrical sparks are ready to fly!

We should note that other research on thunderstorms suggests that this two-layer model of cloud electrification (in other words, positive on top, negative below, as shown in Figure 9.5) may be a bit too simple. By

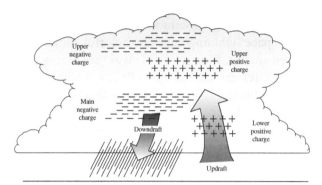

FIGURE 9.6 The charge distribution inside a thunderstorm, based on research conducted by University of Mississippi scientists. The electrification structure is more complicated than the traditional two-layer model of charge, with several pockets of negative charge and several pockets of positive charge (From "Electrical structure in thunderstorm convective regions: 2. Isolated storms" by Maribeth Stolzenburg, David Rust, and Thomas C. Marshall, Journal of Geophysical Research: Atmospheres, Volume 103, 1998, pp. 14097–14108. Copyright © 1998 by John Wiley and Sons. Permission conveyed through Copyright Clearance Center).

analyzing dozens of soundings of the charge distribution within thunderstorm clouds, researchers at the University of Mississippi found that a four-layer charge distribution similar to the one shown in Figure 9.6 is probably closer to reality. Regardless of the exact charge structure within the cloud, separation of charge is the essential prerequisite for lightning.

Lightning is possible between any two oppositely charged regions in and around a thunderstorm, provided the difference in charge is large enough to overcome air's natural resistance to the flow of electricity. Generally speaking, there are two principal types of lightning: strokes within a single cumulonimbus cloud (intracloud lightning) which account for about 80% of all lightning, and strokes from cloud-to-ground, which constitute most of the other 20% (see Figure 9.7). Lightning strokes between clouds or between a cloud and the surrounding air are infrequent, because the difference in charge between two clouds or between a cloud and the surrounding air is not nearly as great as the differences found inside and beneath a cumulonimbus cloud.

A typical cloud-to-ground lightning stroke begins with a discharge of electrons from the cloud base toward the ground. The torrent of negative charge proceeds in a series of segments called a **stepped leader**. Each segment is about 50 to 100 meters long and a few centimeters wide, and usually invisible to the human eye. As the stepped leader approaches the ground, it attracts positive charges from objects at the surface, particularly tall ones such as antennas and trees. Eventually, a traveling spark moves

FIGURE 9.7 A 15-minute composite of cloud-to-ground lightning near Forsyth, Missouri, on August 6, 2011 (courtesy of Allan Burch, Earth Science Picture of the Day).

upward from one of these ground-based points to connect the stepped leader to the surface. In the first instant after the spark meets the stepped leader, electrons in the lowest part of the channel surge into the ground. Subsequently, a brilliant **return stroke** propagates upward from the ground to connect with the downward-moving stepped leader. In effect, the return stroke completes a pathway so charge can be lowered to the ground.

Once the difference in charge between the cloud and the ground dissipates, the channel through which the lightning bolt traveled closes quickly and the stroke is complete. However, this dissipation in charge often requires successive transfers of electrons to the ground, each followed by a successive return stroke that is less brilliant than its predecessor. These subsequent transfers of electrons, called **dart leaders**, are separated by only fractions of a second. Dart leaders give some lightning strokes a flickering appearance.

A small percentage of cloud-to-ground lightning strokes are outlaws, originating not in the mid or low levels of a thunderstorm cloud but rather near the top, an area that typically carries a large positive charge. These **positive flashes**, which constitute less than 10% of all cloud-to-ground lightning, are particularly dangerous for several reasons. They frequently strike on the periphery of the thunderstorm, away from the rain core (see Figure 9.8). In this way, they can reach considerable distances (perhaps 20 km or more) to areas that most people do not consider a lightning risk. Also, the peak current of their return strokes is often much larger than the peak current of negative return strokes, so they are more lethal and can cause greater damage. It's believed that positive flashes cause a disproportionate percentage of forest fires and damage to power lines.

Recent research into lightning has revealed that some thunderstorms also produce other electrical phenomena that occur high in the atmosphere. Collectively called **transient luminous events** (TLEs), this "space lightning" goes by colorful names such as **red sprites, blue jets,** and **elves** (see Figure 9.9). Red sprites can extend up to 95 km (about 60 mi) above a thunderstorm and, because they are not very bright, can be seen only at night and usually only with highly sensitive cameras (see Figure 9.10). Blue jets, which have been seen by pilots, extend upward from a thunderstorm top to heights of 40–55 km (about 25–35 mi) and last only a fraction of a second. A low-light camera on the Space Shuttle first imaged elves, which are rapidly expanding, glowing discs of light that occur above areas of significant cloud-to-ground lightning. Elves last less than a thousandth of a second.

FIGURE 9.8 Positive flashes originate in the upper reaches of a thunderstorm and often strike relatively far from the main core of the thunderstorm (courtesy of NOAA).

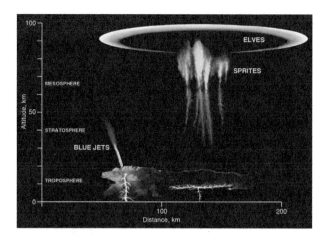

FIGURE 9.9 The collection of transient luminous events (courtesy of NSSL, adapted from Carlos Miralles [AeroVironment] and Tom Nelson [FMA]).

FIGURE 9.10 A red sprite, photographed from the International Space Station on August 10, 2015 (courtesy of NASA/JSC).

Thunder: Atmospheric Rumbles

Lightning heats the surrounding air instantaneously to extreme temperatures, likely near 30,000°C (about 50,000°F). In response to this rapid heating, the air immediately surrounding the bolt expands explosively, creating a sound wave that propagates away from the lightning in all directions at a speed of approximately 340 m/s (755 mph)—the speed of sound. On the other hand, the light from the bolt travels at—what else—the speed of light (which is about a million times greater than the speed of sound). As a result, you see lightning the instant it flashes, but unless the bolt strikes very close to you, there will be a slight delay until you hear the thunder.

Because of the huge difference between the speed of sound and the speed of light, you can use the popular "flash-to-bang" rule to estimate your distance from a lightning strike. Simply count the number of seconds between seeing the flash and hearing the thunder. At 750 mph, sound travels roughly one mile in five seconds. Thus, a five second delay means the lightning was about one mile away.

As a general rule, if the delay is 30 seconds or less, the lightning is close enough to be a hazard. Of course, if multiple lightning strokes are occurring at about the same time, it may be difficult to link peals of thunder to individual lightning flashes.

The delay between seeing lightning and hearing thunder also helps to explain why thunder sometimes lasts a few seconds ("rumbles" of thunder). Check out Figure 9.11, which shows a hypothetical vertical lightning bolt between Point 2 and Point 1. Assume the bolt is 2000 meters long and strikes the earth 2000 meters away from you (you are standing at Point O). Thunder originating from the nearest segment of the stroke (at Point 1) will arrive at your ears first, in

$$\frac{2000 \text{ meters}}{340 \text{ meters/second}} = 5.9 \text{ seconds}$$

But thunder emanating from the most distant segment of the stroke (at Point 2) will be heard in

$$\frac{2828 \text{ meters}}{340 \text{ meters/second}} = 8.3 \text{ seconds}$$

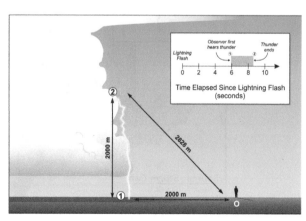

FIGURE 9.11 The distance between an observer (at Point O) and a lightning flash can be estimated by determining the length of time between seeing the flash and hearing the thunder (the length of the hypotenuse of the right triangle is computed using the Pythagorean theorem—if the legs of a right triangle have lengths a and b, and the hypotenuse has length h, then $a^2 + b^2 = h^2$). In this example, the hypotenuse (the line from the observer to Point 2) is 2828 m long.

The end result is that as sound produced by different parts of the bolt reaches your ears, you will hear a continuous rumble of thunder that begins about 6 seconds after you see the lightning and lasts for 2.4 seconds.

All lightning produces thunder, but the sound usually cannot be heard beyond 20 km (12 mi) or so, limited by the dampening effect that the atmosphere has on sound waves (thunder is the atmosphere's way of shouting, but even its stentorian voice can't be heard over great distances). Often, people call distant lightning that seems to have no thunder "heat lightning," suggesting that the lightning somehow owes its existence to daytime heating and not to a thunderstorm. That's nonsense. "Heat lightning" is simply lightning from a thunderstorm that is too far away for the thunder to be heard.

Lightning Safety: Avoiding Hair-Raising Situations

Even in these days of improved forecasting, increased technology, and heightened public awareness of weather hazards, lightning remains a killer. For the ten-year period 2007–2016, the average was 31 lightning fatalities per year in the United States,

according to the National Weather Service (see Figure 9.12). Many more are struck each year and survive. In fact, it's estimated that 90% of those struck by lightning live to tell about it. But many lightning survivors suffer a permanent affliction or disability, often from an injury to the nervous system. Moreover, most authorities agree that the actual number of injuries and fatalities is probably higher because some incidents, particularly those that occur in small, rural communities, do not get incorporated into national statistics.

The majority of lightning injuries and deaths occur during the months of June, July and August, between the hours of 2 p.m. and 6 p.m. Most people struck are outdoors in open fields and recreation areas, under trees, or in and near bodies of water. Though no place is absolutely safe from lightning, some are safer than others. The best shelter is a sturdy enclosed building, especially if a lightning rod is mounted on the roof. The rod is usually connected to an insulated conducting wire that channels the electrical discharge of a lightning strike to a long metal rod that is driven far into the ground. The lightning rod's elevated position increases the likelihood that it will be struck, sparing other parts of the structure or other nearby objects.

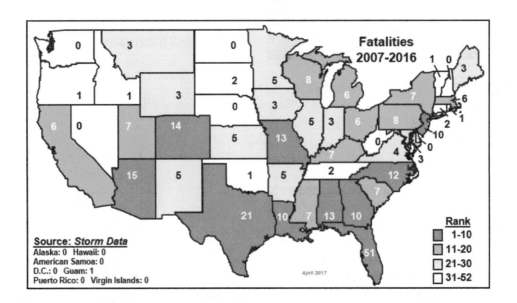

FIGURE 9.12 Lightning fatalities by state, from 2007 to 2016, compiled from data reports issued by the National Weather Service. Over the last few decades, trends in lightning casualties reflect a shift away from agricultural incidents to those involving recreation and urban activities in the more populous states of the West and South. When these statistics are normalized by population, the states with the most lightning fatalities per capita are Wyoming, Colorado, and Montana (courtesy Ronald L. Holle, Holle Meteorology & Photography).

When lightning strikes a building, the electricity typically seeks the path of least resistance to the ground. Thus, pipes and wires inside a home, which are efficient conductors, can pose a danger by serving as inviting conduits for electricity to follow. As a result, you should refrain from taking a shower, cooking, or using electrical appliances during a thunderstorm.

If there are no easily accessible buildings in which to seek shelter, your next best choice is a fully enclosed metal vehicle such as a car (no convertibles) or bus with the windows rolled up. Contrary to popular belief, the relative safety of vehicles is not related to the rubber tires. Rather, if lightning strikes a car, the metal "skin" acts to conduct electricity to the ground. As long as you are not in contact with metal inside the car (and there's not much inside modern cars), you are relatively safe.

If you are caught outdoors during a thunderstorm, stay clear of tall objects and elevated locations, in addition to materials that conduct electricity well. Avoid rooftops, hilltops, individual trees, telephone and electrical poles, and flagpoles because the chance of becoming a lightning target increases with height. Also avoid metal objects such as fences, bicycles, mowers, and golf clubs, which all conduct electricity. The same is true of bodies of water, so boaters and swimmers should carefully monitor threatening weather. Table 9.2 summarizes the most important guidelines for minimizing your risk from lightning.

If you are about to be struck by lightning, you may experience a tingling sensation on your skin or feel your hair stand on end (see Figure 9.13). Like clothes from a hot dryer that are held together by

FIGURE 9.13 This hair-raising photograph was taken at Moro Rock in California's Sequoia National Park. This person evacuated the top of the mountain soon after this picture was taken. Moments later another person walked up to that same location and was struck and killed by lightning (courtesy of NOAA/NWS).

static cling, positively charged hair can be attracted by the negative charges in the descending stepped leader. In such a situation, you should immediately squat low to the ground on the balls of your feet and tuck your head—essentially, assume an upright fetal position. This position minimizes height (and thus possible selection as the return stroke site) as well as contact with the ground. The latter is important because electricity from a nearby lightning strike can be conducted through the earth—so the more of your body touching the ground, the greater your risk.

Lightning strike victims do not retain an electrical charge and are safe to touch immediately after being struck. Typically a lightning strike will cause cardiac or respiratory arrest; many victims can be revived by immediate resuscitation and, if the victim has no pulse, CPR.

ORIGINS OF THUNDERSTORMS: TO BE, OR NOT TO BE

With an average diameter of perhaps 10 km (6 mi), a typical thunderstorm is a mesoscale system. Its

---

TABLE 9.2 Some Lightning Safety Tips

DO
- Seek shelter immediately in a building or car if you hear thunder.
- Get out of boats and away from water.
- Allow at least 30 minutes after thunder is last heard to resume normal activities.

DON'T
- Take shelter in small sheds, under isolated trees, or near fences or metal structures.
- Remain in the open outdoors.
- Use electrical appliances or take a bath or shower.

Weather Folklore and Commentary

Lightning Never Strikes the Same Place Twice

A lightning strike essentially wipes out the local imbalance of electrical charge between the ground and the parent cumulonimbus cloud along the path of least resistance. Thus, locations that are relatively high in altitude in comparison to their surroundings (and thus closer to the clouds) are more vulnerable to lightning strikes. For example, the Empire State Building was designed to act as a lightning rod for the surrounding area: The top of the building is struck, on average, about 100 times per year (see Figure 9.14). On one occasion, the building was struck 15 times in 15 minutes!

Indeed, all else being equal, the risk of being struck by lightning increases with height. For an antenna rising above flat surroundings, the average number of strikes per year as a function of the height of the antenna is given in Table 9.3. Clearly, the oft-repeated statement that "lightning never strikes the same place twice" is just a myth.

FIGURE 9.14 Lightning strikes the Empire State Building (courtesy of National Weather Service, Upton, NY).

Table 9.3 Lightning Strikes as a Function of Height

Average number of lightning strikes per year on an antenna above flat ground, as a function of the height of the antenna. As a basis for comparison, it is assumed that an antenna of height 91 m (300 ft) is struck by lightning once per year.

Antenna Height	Average Annual Strikes
183 m (600 ft)	3
244 m (800 ft)	5
305 m (1000 ft)	10
366 m (1200 ft)	20

tall turrets of bubbling convection are the visible manifestations of positive buoyancy. Compared to the near vertical calm in the surrounding hydrostatic atmosphere (where the vertical pressure gradient force and gravity nearly balance), a thunderstorm is a working system of vigorous updrafts and downdrafts that fly in the face of hydrostatic equilibrium. In a way, thunderstorms are atmospheric agents that help to mitigate small-scale temperature contrasts, both in the vertical and horizontal. For this meteorological powerhouse to develop, both instability and a lifting mechanism are required.

Instability and Lifting: Thunderstorm Preludes

Remember the test for instability in Chapter 8? If a parcel is nudged upward from its equilibrium state, it will continue to rise if it is warmer (and, thus, less dense) than its surroundings. Therefore, if a positively buoyant parcel of air rises from the lower troposphere into a deep chill above, it will likely continue to rise. If the parcel pushes above its lifting condensation level, the latent heat released will put the brakes on the parcel's rapid rate of cooling (from dry to moist adiabatic), improving the parcel's chances for continued ascent. Thus, warm and moist conditions in the lower troposphere, together with sufficiently cold air aloft, up the ante for thunderstorms to develop.

When meteorologists assess the potential for thunderstorms, they routinely check to see whether lifted parcels of air will be able to reach their **level of free convection**. By definition, the level of free convection (LFC, for short) is the altitude at which a parcel of air, after lifted dry adiabatically until saturation and then moist adiabatically thereafter, becomes warmer than its environment, setting the stage for an updraft to form and paving the way for deep convection (see Figure 9.15). Of course, this definition only makes sense if the atmosphere is conditionally unstable. Otherwise, the moist adiabat that a rising parcel follows after it reaches the lifted condensation level will always be to the left of the environmental sounding—in other words, the parcel will always be colder than its environment (revisit Figure 8.15 if you need a refresher on conditional instability). If a parcel of air can reach its LFC, the parcel can then accelerate into the upper troposphere, helping to generate a strong updraft that produces a thunderstorm.

Uneven solar heating alone, by warming the lower troposphere, is sometimes sufficient to get a parcel to its LFC, as Figure 9.16 illustrates. Typically, solar heating works in concert with surface boundaries (such as a front) which promote low-level convergence and additional lift. Even with a front, thunderstorms may not develop along the length of the boundary. In these situations, convergence may simply be too weak to initiate thunderstorms everywhere. Instead, there may be only a few select places where low-level convergence is strong enough to get air parcels to their LFC. In any case, thunderstorms whose updrafts originate near the ground are often described as **surface-based convection**.

Such a definition suggests that some thunderstorm updrafts have their origins higher in the troposphere. More specifically, a strongly stable layer near the surface inhibits surface-based convection, but instability may lurk above the stable layer. If air parcels can tap this instability, they may still become positively buoyant, initiating updrafts above the stable layer in what is termed **elevated convection**. A particularly striking example of elevated convection sometimes occurs in winter when surface air isn't very warm or very moist. In essence, think of it as "instability with a jacket on."

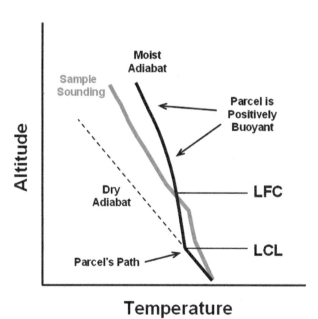

FIGURE 9.15 Determining the level of free convection, or LFC. A sample temperature sounding is shown in red; the parcel's path is the thick dark line. A lifted parcel cools dry adiabatically until it reaches saturation at its lifting condensation level (LCL). It cools moist adiabatically thereafter. The altitude above the LCL at which the parcel becomes warmer than its environment, and thus positively buoyant, is the Level of Free Convection (LFC).

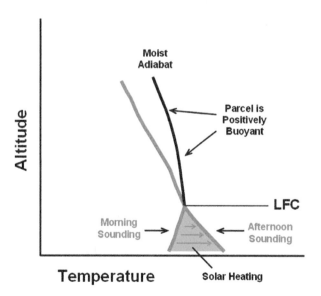

FIGURE 9.16 Solar heating warms the surface and shifts the temperature sounding in the lower troposphere to the right toward higher temperatures, with more warming occurring closer to the surface. If the sounding becomes dry adiabatic in the lower troposphere, parcels can sometimes reach their LFC, raising the odds for thunderstorms to develop.

For example, a crippling ice storm accompanied by thunder struck parts of the Southern Plains in early December 2007. Figure 9.17a shows an icy mold taken off the top of a municipal fire hydrant in Norman, OK. The ice was three-quarters to one inch thick (2–2.5 cm)! The photographer surgically separated this "ice sculpture" around the edge of the top of the hydrant, and then placed it on a white beverage cooler in order to better illustrate the impressive veneer of ice.

What meteorological setup led to this historic buildup of ice? The regional surface analysis at 00 UTC on December 10, 2007 (see Figure 9.17b) showed an area of high pressure building into the Southern Plains behind a cold front that had swept into Texas. The meteogram at Tulsa, OK, from 03 to 18 UTC on this day (see Figure 9.17c) indicated northerly winds had ushered in below-freezing air at the surface. Interspersed among the many symbols of freezing rain, note the thunderstorm icons at 08 UTC and 17 UTC, when surface temperatures were below 32°F. Given this low-level chill, no doubt these thunderstorms had their roots at higher altitudes. Indeed, warmer winds blowing from the southwest

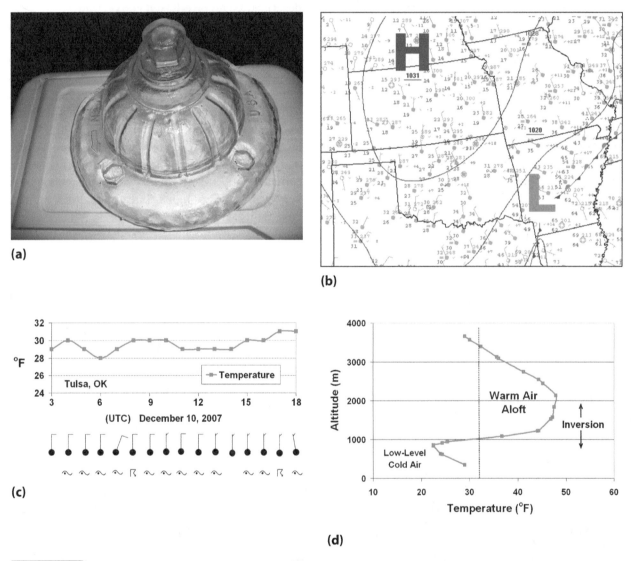

(a)

(b)

(c)

(d)

FIGURE 9.17 (a) An "ice sculpture" removed from the top of a fire hydrant in Norman, OK, after a crippling ice storm on December 10–11, 2007 (courtesy of Greg Carbin); (b) Surface analysis at 00 UTC on December 10, 2007, shows northerly winds and subfreezing temperatures across Oklahoma in the wake of a cold front that had swept south of the state (courtesy of NOAA); (c) Meteogram at Tulsa, Oklahoma, from 03 UTC to 18 UTC on December 10, 2007, shows persistent freezing rain with occasional thunderstorms; (d) The vertical profile of temperature over Norman, OK at 00 UTC on December 10. Thunderstorms at Tulsa on this day originated in the relatively warm layer of air just above 2000 m.

a few thousand meters above the surface overran the wedge of low-level cold air. The vertical profile of temperature over Norman, OK at 00 UTC on this day, given in Figure 9.17d, shows the two important pieces of the recipe for elevated convection. Note the thin wedge of sub-freezing air near the ground and, above that, a prominent temperature inversion that extends to about 2000 meters, the footprint of the warmer southwesterly winds aloft. Updrafts for the elevated convection likely originated just above this temperature inversion where some relatively warm parcels were able to reach their LFC, become positively buoyant, and accelerate upward to form thunderstorms.

No matter the season, the recipe for thunderstorms is always the same: get air parcels to their level of free convection and have them remain positively buoyant through a deep layer. The subtle message in this recipe is that thunderstorms never develop randomly, despite the use of common qualifiers such as isolated, scattered and numerous. Indeed, thunderstorms erupt at specific places and specific times for very sound meteorological reasons.

So why do meteorologists use such qualifiers? The simple explanation is that we just don't have the capability to accurately discern, in a timely fashion, the small-scale environments in which thunderstorms develop. Our observational network is just too coarse (in space and time). In other words, the small-scale environments that favor thunderstorms developing at a specific time often "fall through the cracks" of our observational network. And that's a good reason why there's always uncertainty associated with forecasting thunderstorms.

So much for the bad news. The good news is that, by looking at weather data over a long period of time, we can identify regions where thunderstorms are more likely to occur. Moreover, within these more active regions, we can identify and better understand the mechanisms that initiate thunderstorms. Forecasters also rely on sophisticated computer simulations that help them to better predict thunderstorms. In these ways, meteorologists reduce some of the uncertainty "that goes along with the territory" of forecasting thunderstorms. Let's explore these ideas further.

Thunderstorm Climatology: Blooming Where You're Planted

Figure 9.18 shows the frequency and distribution of lightning strikes (a proxy for average annual thunderstorm activity) within about 40° latitude of the equator. Thunderstorm frequency around the globe closely correlates with the low-level instability produced by strong solar heating of the continents, particularly over the tropics and subtropics. The highest annual frequency of lightning occurs in equatorial Africa, where tropical disturbances (clusters of thunderstorms) boost the tally. In concert with Southeast Asia's monsoon (which we'll explore in detail in Chapter 10), Indonesia also has a relatively high frequency of lightning. In some equatorial locations, thunderstorms occur, on average, on nearly half the days each year.

The average annual number of thunderstorm days for the United States is shown in Figure 9.19. Thunderstorm frequency is a maximum in the Deep South, particularly from coastal Louisiana, Mississippi, and Alabama into Florida. Here, the nearly inexhaustible

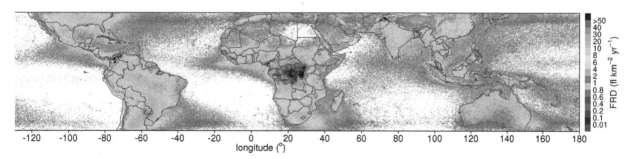

FIGURE 9.18 Average annual lightning flashes per square kilometer for the period 1998–2013. Areas with the fewest number per year are white and gray, while areas with the largest number of lightning flashes, as many as 50 per year per square kilometer, are dark. The Congo Basin in Africa has long been known as a global lightning hotspot, but recent research suggests that the Lake Maracaibo region in Venezuela may be the lightning capital of the world. The map is based on data collected by the Lightning Imaging Sensor aboard NASA's Tropical Rainfall Measurement Mission Satellite (courtesy of NASA).

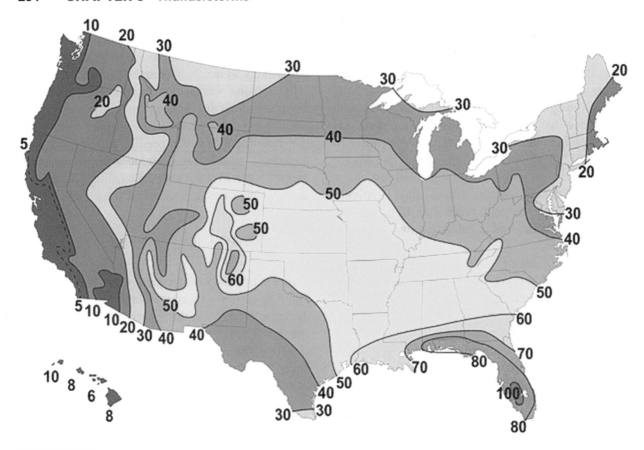

FIGURE 9.19 The average number of days per year with a thunderstorm (courtesy of the National Weather Service).

summer supply of water vapor made available by evaporation from the warm waters of the Gulf of Mexico and Atlantic Ocean serves as fuel for recurrent thunderstorms. The primary catalyst for many of the spring and summer thunderstorms that erupt along the Gulf Coast and over Florida is the mesoscale wind system called the **sea breeze**. At the most basic level, this local wind system is driven by the uneven heating of land and water. With ample morning sunshine, heights of constant pressure surfaces over land in the lowest few thousand feet begin to increase by late morning in response to greater solar heating, creating an area of slightly higher heights aloft (Figure 9.20). Air aloft over land flows from the weak high toward the water, causing surface pressure to decrease slightly over land (via divergence aloft) and increase over water (via convergence aloft). The bottom line is that a closed mesoscale circulation develops by early afternoon. The corresponding low-level flow of cooler maritime air toward land is the faithful sea breeze. We note that the sea breeze forms when the background large-scale pressure pattern is weak. Otherwise, a relatively strong large-scale wind pattern would overwhelm the sea breeze.

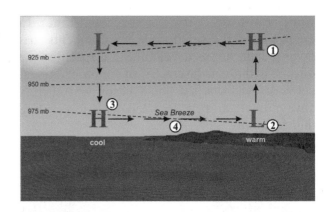

FIGURE 9.20 The meteorology of a sea breeze starts with the uneven heating of land and sea.
In response, the heights of constant pressure surfaces near 900 mb increase over land, promoting high pressure aloft (Step 1). As air aloft diverges from this relatively weak high toward the sea, surface pressure decreases over land (Step 2 – note the pressure surface dipping down toward the ground). Meanwhile, over water, air converging aloft causes higher surface pressure (Step 3). The flow from higher surface pressure toward lower surface pressure over land is the sea breeze (Step 4). For all practical purposes, these "steps" really occur at the same time, but we presented the concept as a sequence for ease of explanation.

FIGURE 9.21 Low-level convergence along the sea-breeze front (the leading edge of the sea breeze) helps to lift relatively hot, humid air over land, providing an organizing mechanism for showers and thunderstorms.

For areas near the water, the sea breeze makes hot, humid weather less stifling because it imports cooler air from the sea. But in the process, the leading edge of the sea breeze behaves like a "mini" cold front (called, appropriately, the **sea-breeze front**). For the record, a sea-breeze front is a mesoscale boundary that separates relatively cool maritime air from warmer continental air. Low-level convergence along the sea-breeze front helps to lift relatively hot, humid air over land, setting the stage for showers and thunderstorms to develop (see Figure 9.21). The trusty sea breeze is a major reason why, on a statewide basis, Florida, Louisiana, Mississippi, and Alabama battle with Hawaii for the title of wettest state. In particular, Florida's peninsular shape and low latitude combine to make the state one of the prime locations in the world for sea-breeze thunderstorms. Figure 9.22 is a close-up of sea-breeze thunderstorms along the Florida southeast coast. Note the lack of cloudiness over the relatively cool water bodies, especially Lake Okeechobee. In much the same way the sea breeze forms, a lake breeze often develops and flows outward from the shores of Lake Okeechobee. Not surprisingly, cumulonimbus clouds often develop along the corresponding lake-breeze front by the afternoon. And if the lake-breeze front from Lake Okeechobee meets the sea-breeze front from the Atlantic Ocean, look out, even more potent thunderstorms can erupt!

That's not the only collision of mesoscale boundaries that occurs in Florida during summer. At times, the sea-breeze front from the Atlantic meets the sea-breeze front from the Gulf of Mexico over the middle of the Florida peninsula (see Figure 9.23). This

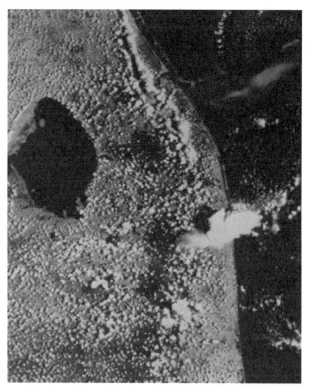

FIGURE 9.22 Thunderstorms developing along the sea-breeze front near the east coast of south Florida (courtesy of NASA).

collision of small-scale boundaries allows thunderstorms to merge and strengthen, bolstering Florida's claim to thunderstorm capital of the United States.

At night, the situation reverses: the land cools off faster than coastal waters, causing a **land breeze** to blow out to sea. In turn, showers and thundershowers sometimes form over offshore waters along the leading edge of the relatively weak land breeze.

Around the Great Lakes, the same process of uneven heating plays out in spring and summer, generating lake breezes and lake-breeze fronts. Though typically not as potent as their Florida cousins, these mesoscale boundaries can generate clouds and sometimes even lines of showers and thunderstorms.

In general, the number of thunderstorms decreases with increasing distance from the Gulf (check back to Figure 9.19). A secondary maximum of thunderstorm frequency is found in the vicinity of the Rocky Mountains, particularly near the Colorado-New Mexico border. Here, the mechanism for thunderstorm production, especially during summer, is a high-level heat source—the mountains.

How can mountains initiate thunderstorm development? Consider the situation shown in Figure 9.24a. In the morning, with the Sun rising in the eastern sky,

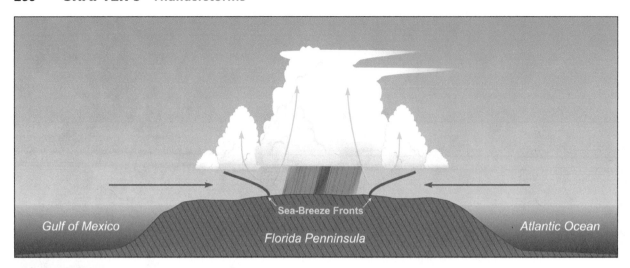

FIGURE 9.23 Sometimes, when the sea breezes from the Gulf of Mexico and the Atlantic Ocean converge over central Florida, updrafts associated with merging cumulus clouds strengthen, setting the stage for powerful thunderstorms.

solar energy strikes east-facing slopes more directly than the adjacent valley. In response, surface air in contact with these slopes warms, causing the local air density and, therefore, local air pressure, to decrease (relative to air at the same altitude that's not directly in contact with the ground). Note the impact in Figure 9.24a—constant pressure surfaces slope down as they approach the mountain slope, indicating a slightly lower pressure there. The resulting horizontal pressure gradient at any given altitude causes air to move toward the mountain where the terrain then forces it to rise. These horizontal and vertical motions are components of a mesoscale wind system known as a **mountain-valley circulation**.

To fully appreciate this mesoscale circulation, consider Figure 9.24b and focus on the two columns of air shown over the mountain slope. Because these columns got a head start on heating, by midday they are warmer than similar air columns over the nearby valley. As a result, pressure decreases more slowly in air columns over the mountain slopes. This means that pressure surfaces near mountaintop level bulge upward slightly, indicative of higher pressure aloft (the bulge is exaggerated in Figure 9.24b). In turn, the horizontal pressure gradient causes some air to flow away from the mountain at these higher elevations, eventually sinking over the adjacent valley. Taken altogether, these horizontal and vertical

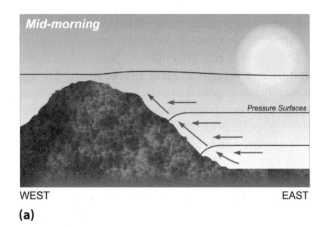

(a)

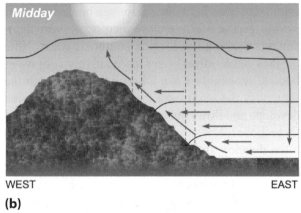

(b)

FIGURE 9.24 (a) Early in the morning, solar radiation more directly strikes east-facing slopes of mountains. The resulting horizontal pressure gradients cause air to move toward and up mountain slopes; (b) Because pressure decreases more slowly in warmer air columns over the mountain slopes, local pressure surfaces near the mountaintop level must bulge upward slightly, indicating higher pressure aloft. In turn, some air flows away from the mountain at these high elevations and eventually sinks over the valley, completing the mountain-valley circulation.

motions form the mountain-valley circulation. Like sea-breeze and lake-breeze circulations, a mountain-valley circulation develops when the large-scale pressure pattern is relatively weak; otherwise, stronger, large-scale winds would probably overwhelm its formation.

The bottom line here is that during the warm season (especially in the late morning and afternoon), some parcels of air will begin to rise high over the mountaintops, particularly if the upslope flow is warm and moist (air from the Gulf of Mexico, for example). In such a case, the stage is set for cumulonimbus clouds to develop. After they form, these thunderstorms often drift eastward with the prevailing westerly flow aloft. Figure 9.25 is a sequence of visible satellite images that shows the development of thunderstorms

(a)

(b)

(c)

FIGURE 9.25 A sequence of visible satellite images showing thunderstorms developing over the central Rockies during the afternoon of July 22, 2008: (a) 1603 UTC; (b) 1745 UTC; (c) 2115 UTC. Focus on the dark features in Colorado on the first image (the mountains) and then watch cumulonimbus clouds develop in the next two images (courtesy of NOAA).

generated by high-level heat sources (the peaks of the Rockies) from the late morning to the late afternoon on a typical July day.

While the United States east of the Rockies has access to maritime tropical air from the Gulf of Mexico, the western United States lacks a consistent source of warm, moist air needed for frequent thunderstorms. In fact, along the West Coast, the onshore flow of air chilled by the California current enhances low-level stability, squelching thunderstorm development. San Francisco International Airport averages only one to two thunderstorms per year (and they usually occur during winter). In fact, if there is such a thing as a thunderstorm season for coastal cities such as Los Angeles, San Diego, and Seattle, it's late autumn, winter, and early spring when cold upper-level troughs, in tandem with relatively warm Pacific air near the surface, destabilize the troposphere.

Although thunderstorm frequency is relatively low west of the Rockies, the lightning from the thunderstorms that occasionally form poses a disproportionately high risk of igniting forest fires, particularly in late spring and early summer. Before tropical moisture starts to invade the West in July, thunderstorms that develop in May and June tend to have high bases because dew points in the lower troposphere are still typically low, so air parcels rise higher before net condensation begins (see Figure 9.26). Raindrops falling from high-based thunderstorms have plenty of time to evaporate in their relatively long descent through dry air, thus accounting for "dry" thunderstorms in the western United States.

Lightning from thunderstorms often ignites wildfires, especially during the warm season and particularly in years when long-term drought and large rainfall deficits cause brush and forests to be tinder dry. For the period 2001–2015, lightning ignited, on average, 10,600 wildfires per year, burning, on average, 3.8 million acres annually (an area roughly the size of the state of Connecticut). Figure 9.27 shows the breakdown of lightning-induced fires by the 11 geographic areas that collectively cover the United States. On average, the Southwest geographic area reports the most wildfires caused by lightning (1790, on average, per year), while Alaska is the geographic area with the smallest average annual number (161). However, these lightning-induced Alaskan wildfires burn, on average, the largest area (1.4 million acres, on average).

But back to thunderstorm basics. To this point, we have discussed the recipe for thunderstorms from the

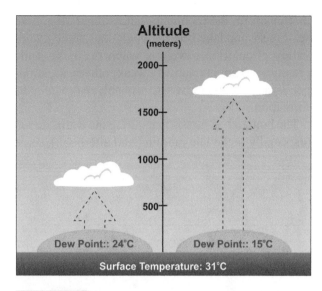

FIGURE 9.26 When low-level air is humid, thunderstorms have relatively low bases. A typical situation from Gulf Coast States in the summer (left) might feature surface air with a temperature of 31°C (88°F) and a dew point of 24°C (75°F). With parcels cooling at the dry adiabatic lapse rate, the lifting condensation level would lie at an altitude of about 700 meters. In contrast, western thunderstorms tend to have higher bases because low-level air is not as humid (right). For example, with a surface air temperature of 31°C (88°F) and a dew point of 15°C (59°F), the lifting condensation level would lie at an altitude of about 1600 meters.

standpoint of instability—essentially, get parcels to their LFC and have them remain positively buoyant as they rise through a deep vertical layer. However, observations clearly show that not all thunderstorms are created alike. Some form discretely, essentially separate from other storms, while at other times, thunderstorms develop into lines or clusters.

Regardless of variety or mode, the deciding factor in whether thunderstorms become severe—that is, produce large hail, damaging wind gusts, or tornadoes—is, to a large extent, the vertical variation of wind speed and direction in the environment in which the thunderstorms develop. The headline on this side of the thunderstorm recipe reads: vertical wind shear.

TYPES OF DISCRETE CONVECTION: WHEN THUNDERSTORMS GO IT ALONE

When thunderstorms are discrete, convection typically takes the form of a group of mostly detached storms. For an example of what we mean by "discrete," note the speckled appearance of the mostly

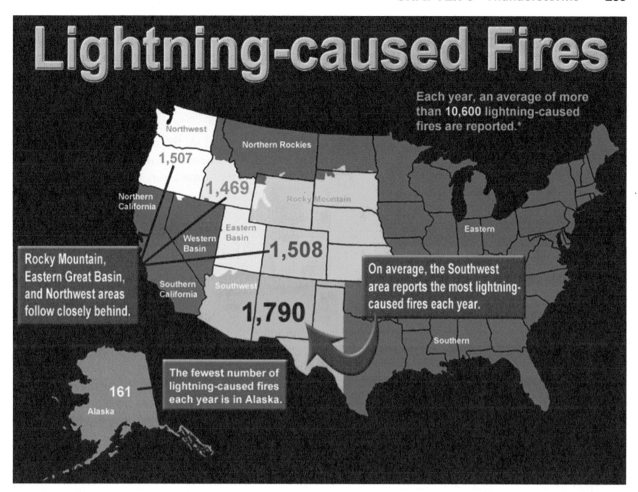

FIGURE 9.27 The number of lightning-caused wildfires in the United States by geographic area. Each geographic area has a coordination center that helps local agencies when wildfires get out of control. When the geographic area exhausts its resources, the coordination center then asks for help from the National Interagency Fire Center (courtesy of National Interagency Fire Center).

separate thunderstorms in Figure 9.28, which is a mosaic of composite reflectivity over part of the Southeast United States on a July day.

In contrast, consider Figure 9.29a, a composite reflectivity image from the early morning of November 17, 2015. Note the cohesive line of heavy thunderstorms stretching from Kansas to Texas. This line of storms formed in concert with a cold front along which a long line of relatively strong surface convergence helped to get air parcels to their level of free convection. You can see this long line of surface convergence in Figure 9.29b by noting the converging surface streamlines that pretty much match the location of the line of thunderstorms shown on radar.

There's a lesson to be learned here: When thunderstorms are discrete, there typically aren't any long lines or other expansive areas of strong surface convergence. Rather, surface convergence tends to be localized and isolated, and if this convergence helps to get air parcels to their LFC, the thunderstorms that develop are usually discrete. To be fair, we note that, at times, discrete thunderstorms can later "merge" into a larger, cohesive system of thunderstorms (a topic for Chapter 14). For now, however, we will only discuss thunderstorms that remain pretty much discrete throughout their lifetimes.

Generally speaking, there are three types of discrete thunderstorms: single-cell, multicell, and

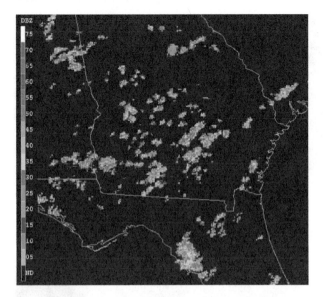

FIGURE 9.28 A composite of radar reflectivity at 1745 UTC on July 22, 2006, shows numerous single-cell thunderstorms in Georgia. Although their speckled appearance on radar might lend the impression of randomness, there were real meteorological reasons for when and where they formed. Some of the single-cell storms produced brief periods of severe weather in the form of hail and isolated strong wind gusts (courtesy of NOAA).

supercell storms. In this context, you can think of the suffix "cell" as referring to a towering cumulus cloud. Before we can discuss the distinguishing characteristics of these three types, we must first elaborate on the atmospheric factor that largely governs which type of discrete thunderstorm is favored in an environment primed for deep convection.

Vertical Wind Shear: Typecasting Thunderstorms

Without reservation, the magnitude of the vertical wind shear is one of the most reliable indicators used by weather forecasters to predict whether discrete thunderstorms can produce severe weather. For the record, **vertical wind shear** refers to a change in wind speed and/or wind direction with altitude. By convention, wind speeds are measured in knots, so forecasters routinely express vertical wind shear in knots. The vertical wind shear between the surface and a height of 6 km (around the 500-mb level) is a particularly robust predictor.

Figure 9.30a shows the vertical profile of winds in the lowest six kilometers at Nashville, TN, at 00 UTC on February 6, 2008 (the evening of February 5, when conditions were favorable for severe thunderstorms in Tennessee). If you do the math, the vertical wind shear between the surface and six kilometers was about 55 knots at this time. Such *strong* vertical wind shear on this day confirmed that ongoing and developing thunderstorms would be severe, whether discrete or organized into lines

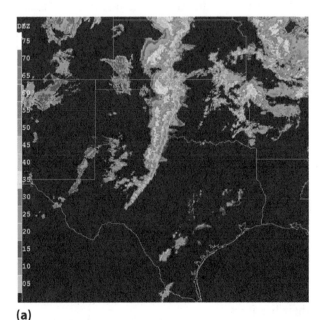

(a)

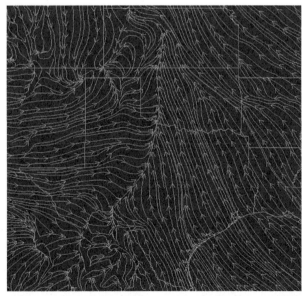

(b)

FIGURE 9.29 (a) Radar reflectivity at 06 UTC on November 17, 2015, shows a solid line of thunderstorms stretching from Kansas into Texas; (b) Surface streamlines at the time show a band of converging winds that correlates well with the location of the line of thunderstorms (courtesy of National Weather Service).

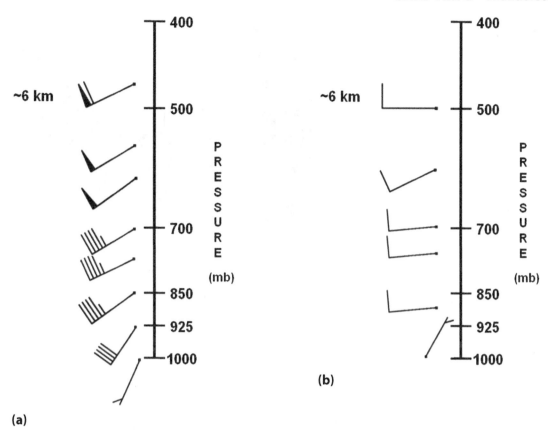

(a)

(b)

FIGURE 9.30 (a) The vertical wind profile over Nashville, TN, at 00 UTC on February 6, 2008 (the evening of February 5). At this time, the vertical wind shear between the ground and six kilometers was about 55 knots; (b) The vertical wind profile over Atlanta, GA, at 12 UTC on July 22, 2006. At the time, the vertical wind shear between the ground and six kilometers was weak, less than 10 knots.

or clusters. Indeed, this day marked the most prolific February tornado outbreak on record, with about 80 confirmed twisters and approximately 400 other reports of large hail and damaging winds (see Figure 9.31a). Several destructive tornadoes ripped through urban areas in Tennessee, including Memphis, Jackson, and the northeastern end of Nashville. At least 57 people were killed across four states, with hundreds of others injured. Figure 9.31b is a radar reflectivity image at 0330 UTC on February 6 during the peak of the tornado outbreak. Note the large thunderstorms (supercells) in north-central Tennessee and the two distinct lines of thunderstorms farther west. In this case, both the discrete thunderstorms and the lines of thunderstorms produced severe weather, aided and abetted by strong vertical wind shear.

Now compare the winds over Nashville in Figure 9.30a with the vertical wind profile over Atlanta, GA at 12 UTC on July 22, 2006, in Figure 9.30b.

At the time, the vertical wind shear between the surface and six kilometers over Atlanta was less than 10 knots (considered *weak* vertical wind shear by forecasting standards). Now look back to the radar reflectivity at 1745 UTC on July 22, 2006, in Figure 9.28. Note the speckles of radar returns in Georgia that represented relatively disorganized single-cell thunderstorms (we assume here that the vertical wind shear over Atlanta was representative of the environment where these thunderstorms developed).

Comparing these two events supports our claim that vertical wind shear is an essential tool that forecasters use to predict the type of discrete thunderstorms. In these examples, weak vertical wind shear was associated with single-cell thunderstorms (which, as you are about to learn, display relatively little organization). Generally speaking, the greater the vertical wind shear, the longer thunderstorms can last and the greater the chance

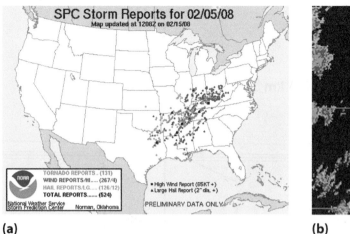

(a)

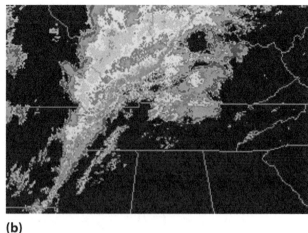

(b)

FIGURE 9.31 (a) Preliminary severe weather reports for the 24-hour period ending at 12 UTC on February 6, 2008. Red dots indicate tornado reports, blue dots represent damaging wind reports, and green dots indicate large hail reports. The actual number of tornadoes was ultimately less than the preliminary number because there were multiple reports of some tornadoes (courtesy of Storm Prediction Center); (b) Composite radar reflectivity image at 0330 UTC on February 6, 2008. The large thunderstorms (supercells) in north-central Tennessee and the two distinct squall lines of thunderstorms farther west are examples of the organized convection that can develop in the presence of large vertical wind shear (courtesy of NOAA).

that they will produce severe weather. As you will learn in this chapter, relatively strong vertical wind shear helps to keep a thunderstorm's updraft and downdraft working together to promote storm longevity (a **downdraft** is a current of descending air, usually accompanied by precipitation, associated with a shower or thunderstorm). As it turns out, the updraft and downdraft of a single-cell thunderstorm interfere with one another—thus, the longevity of single-cell thunderstorms is limited, and they pose a dramatically lower risk of producing severe weather. To understand the role that weak vertical wind shear plays in the relatively short lives of single-cell thunderstorms, we must first examine the three stages of this type of discrete convection.

Structure of Single-Cell Thunderstorms: A Life in Three Acts

Many commercial and military aircraft crashes during World War II resulted from encounters with thunderstorms. The need to better understand this nemesis to aviation led to the *Thunderstorm Project*, one of the first major meteorological field studies to probe the inner workings of thunderstorms. Using airplanes, radiosondes, radar, and an extensive network of observing sites over Florida and Ohio,

meteorologists collected and analyzed a mountain of weather data. The results of the project, published in 1949, produced a conceptual model of the structure and life cycle of single-cell thunderstorms that has changed little since.

Before we delve into this conceptual model for single-cell thunderstorms, we want to mention that many texts (including previous editions of this one) use the term "air mass" thunderstorm to refer to this ordinary type of thunderstorm. You may also hear these called "pulse" thunderstorms or "pop-up" thunderstorms. Generally speaking, these are simply less technical terms for single-cell thunderstorms, so the model we present applies to them as well.

For starters, single-cell thunderstorms have one updraft and one downdraft and tend to go through three distinct stages of evolution. In the initial, or **cumulus stage** of a single-cell thunderstorm, air parcels in updrafts reach their lifting condensation level, setting the stage for a young cumulus to develop in as little as 15 minutes. During this period, a single updraft emerges from trailblazing thermals.

Unlike idealized parcels that rise without interacting with neighboring parcels, buoyant thermals that become saturated and form cumulus clouds mix with their cooler, drier surroundings, a process called **entrainment**. As a cumulus cloud builds skyward above the lifting condensation level, this mixing of cooler, drier air into the sides of the developing cloud causes some cloud droplets to evaporate (see Figure 9.32a). This evaporation contributes to an overall cooling that steadily reduces the buoyancy of the thermals. In fact, by the time the initial cumulus cloud builds to a depth of about one and a half times its diameter at the LCL, it loses its buoyancy completely and stops developing vertically (Figure 9.32b).

But all is not lost for the budding thunderstorm. Sinking air at the edges of this "sacrificial" cumulus cloud encourages future clouds to develop in the same vertical column as the original. By following in the footsteps of earlier clouds, later cumulus feed off the water vapor that remains from the previous evaporation of cloud droplets. Entrainment still occurs, but the air mixing into the developing cloud is now more moist. Given an adequate supply of water vapor and enough differential heating, a parade of cumuliform clouds can grow in succession in a given vertical column, from the fair-weather cumulus to the towering cumulus congestus. In this initial stage

of a single-cell thunderstorm, only an updraft is present (Figure 9.33a).

The vigor of the updraft rapidly carries some water droplets above the 0°C (32°F) level of the cloud, which typically lies near an altitude of 4–5 km (2.5–3 mi) in mid-latitudes during the summer. As droplets cool below 0°C and become supercooled liquid water, the stage is set for the rapid growth of ice crystals by the Bergeron-Findeisen process. Once this process begins, the rate of production of precipitation increases rapidly.

The **mature stage** in the life-cycle of a single-cell thunderstorm begins when precipitation particles become sufficiently large to fall through the updraft, paving the way for rain to reach the surface. However, as precipitation falls, it reduces the positive buoyancy of the updraft (this **water loading** increases the density of ascending air parcels). And that's not the end of the bad news for the updraft. Falling precipitation drags some of the surrounding air *down* with it, forming a downdraft. As the mixture of precipitation and air descends, evaporation of water droplets and melting of ice crystals cool the air immediately surrounding the precipitation,

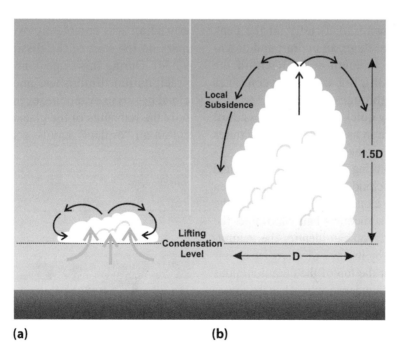

(a)   **(b)**

FIGURE 9.32 Entrainment plays an important role in the growth of cumulus clouds. (a) Drier air surrounding a developing cumulus cloud mixes into the cloud, increasing evaporation and reducing buoyancy; (b) Eventually, mixing between the cloud and the environment deprives parcels inside the cumulus cloud of their positive buoyancy, and the cumulus stops developing vertically. Subsidence immediately surrounding the cloud encourages further cloud development to occur in the same vertical column as the original cloud.

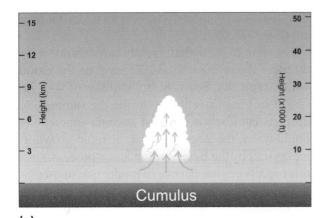

(a)

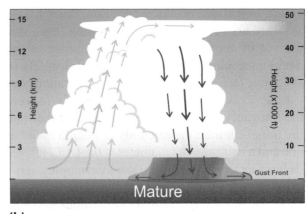

(b)

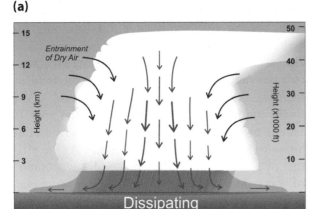

(c)

FIGURE 9.33 Single-cell thunderstorms have three distinct stages in their life cycle: (a) In the cumulus stage, buoyant parcels contribute to an updraft that successively promotes the vertical growth of a fledgling thunderstorm; (b) In the mature stage, both an updraft and downdraft coexist as precipitation falls to the surface at its strongest intensity; (c) A thunderstorm enters its dissipating stage as entrainment of drier air and drag from falling precipitation enhance a downdraft, leading to the demise of the updraft.

strengthening the downdraft. Briefly, at the peak intensity of the mature stage, an updraft and downdraft of comparable strength coexist within the cloud (see Figure 9.33b). But as drier air continues to mix into the cloud through entrainment, the pace of evaporative cooling quickens, and the downdraft can begin to dominate in as little as 15-30 minutes after precipitation first reaches the surface.

In addition to transforming the precipitation potential of the cumulonimbus cloud, the Bergeron-Findeisen process transforms the cloud's appearance. Just as the first wisps of gray in someone's hair can reflect the aging process at work, you can identify the maturation of a thunderstorm by its anvil of wispy ice clouds that begin to appear at the top of the cumulonimbus (see Figure 9.34). If a thunderstorm reaches the tropopause, these ice-crystal clouds spread out horizontally as updrafts hit the "lid" created by this stable layer. In mid-latitudes, anvils may be spread out by upper-level winds, rolling out a **glaciated** carpet of clouds downwind of the thunderstorm (in this context, glaciated means "made of ice crystals").

Once rain begins to fall, it is the beginning of the end of the single-cell thunderstorm. Eventually, the downdraft will completely overwhelm the updraft, marking the start of the **dissipating stage** (Figure 9.33c). During this stage, convection collapses, the precipitation diminishes, and the cumulonimbus cloud begins to evaporate, eventually leaving behind only the remnants of the glaciated anvil (sometimes called an "orphan" anvil—see Figure 9.35 for an

FIGURE 9.34 Astronauts aboard the International Space Station photographed this large glaciated anvil associated with a towering thunderstorm over Africa (courtesy of NASA).

FIGURE 9.35 An orphan anvil (courtesy of Harald Edens).

example). This final stage of a single-cell thunderstorm typically lasts 15-30 minutes, with the entire life cycle complete in the span of an hour or so.

Single-cell thunderstorms typically do not produce severe weather (large hail, strong straight-line winds, or a tornado). Occasionally, however, the updraft can abruptly strengthen, or "pulse," during the mature stage. Such pulse storms, like the ones that affected parts of Georgia during the afternoon of July 22, 2006 (revisit Figure 9.28), sometimes produce short-lived, localized severe weather in the form of penny-sized hail or wind gusts greater than 50 knots (58 mph). Pulse storms, as a general rule, do not produce tornadoes. Nonetheless, weather forecasters at the National Weather Service have their hands full trying to issue timely severe thunderstorm warnings for fleeting pulse storms.

How does vertical wind shear play a pivotal role in the life cycle of a single-cell thunderstorm? The short answer is that weak shear allows precipitation to fall through the updraft. Here's the whole story. Suppose, for sake of argument, that vertical wind shear is relatively strong. For all practical purposes, this assumption means that upper-level winds are strong. In such a case, precipitation particles would be carried away from the rather upright updraft, limiting any adverse impact on its positive buoyancy. In contrast, weak vertical wind shear translates into relatively weak upper-level winds. Thus, precipitation tends to fall through the updraft, leading to the eventual demise of the thunderstorm. The bottom line here is that weak vertical wind shear limits the life cycle of a single-cell thunderstorm.

There's more to the story about the connection between weak vertical wind shear and the relatively short life cycle of single-cell thunderstorms. Look back to Figure 9.33b, which depicts the mature stage, and focus on the feature labeled "gust front." For the record, a **gust front** is the leading edge of rain-cooled air that spreads away from a thunderstorm after the downdraft "splashes" down on the surface (like water from a kitchen faucet splashing down on the sink below). In essence, the leading edge of outflowing rain-cooled air is a mesoscale boundary that has a structure similar to that of a cold front. For that reason, a gust front is also called an **outflow boundary**.

Typically, a gust front (as its name implies) is accompanied by strong gusty winds. After the gust front passes, pressure typically increases briefly owing, in part, to the relatively high density of rain-cooled air.

When a thunderstorm forms in an environment characterized by weak vertical wind shear, there's little resistance to denser, rain-cooled air spreading outward from the storm, so the gust front moves rather quickly away from the thunderstorm (see Figure 9.36a). Given that a gust front acts sort of like a mini-cold front with its associated low-level convergence, warm and moist air from the surrounding environment typically rises over the gust front, often promoting cumulus clouds that may develop into new thunderstorms (see Figure 9.36b). However, with the warm, moist air rising to sustain this new convection, the fate of the original thunderstorm is sealed because its moisture supply is now cut off. Essentially, the same process that spawns new convection leads to the demise of the original single-cell thunderstorm. Downdrafts are the link between generations.

The character of thunderstorms changes, however, as the amount of vertical wind shear increases (as indicated earlier, meteorologists categorize vertical wind shear as *weak*, *moderate* and *strong*). In environments characterized by moderate to strong vertical wind shear, the downdrafts of thunderstorms don't fall directly through the updraft like they did with single-cell thunderstorms. That's because stronger upper-level winds carry precipitation particles farther downstream where they eventually fall and create downdrafts. In other words, as the amount of vertical wind shear increases, thunderstorm downdrafts don't interfere as much with updrafts, paving the way for storms to be more organized, more intense, and relatively long-lived. Time to introduce multicell and supercell thunderstorms.

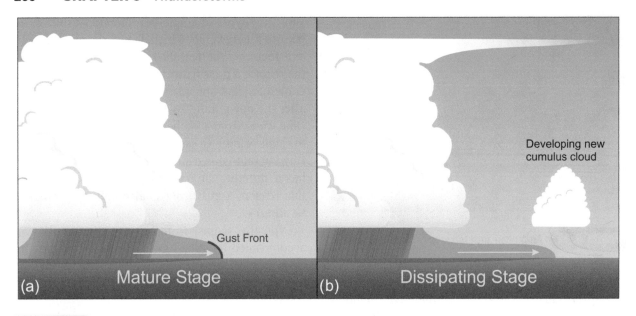

Developing new cumulus cloud

Gust Front

(a) Mature Stage

(b) Dissipating Stage

FIGURE 9.36 (a) In an environment with weak vertical wind shear, the gust front associated with a single-cell thunderstorm moves quickly outward away from the storm; (b) Low-level convergence along the gust front causes warm, moist air to rise, paving the way for new cumulus clouds that may develop into new thunderstorms.

Multicell Thunderstorms: All for One and One for All

You just learned that, in an environment characterized by weak vertical wind shear, a thunderstorm's updraft eventually loses access to its supply of warm, moist air. Thus, maintaining the updraft became physically impossible, sealing the fate of the short-lived single-cell thunderstorm.

It turns out that as vertical wind shear increases, gust fronts do not propagate away from the parent thunderstorm as quickly. As a result, the thunderstorm does not lose access to warm, moist, positively buoyant air right away. Meanwhile, in this environment of somewhat larger ("moderate") wind shear, stronger winds aloft carry precipitation particles farther away from the thunderstorm's updraft, insuring that there's less

interference between the downdraft and the updraft. The bottom line here is that as vertical wind shear increases, the stage is set for longer-lived, more organized thunderstorms. Most weather forecasters agree that a vertical wind shear of 20 to 35 knots between the surface and an altitude of six kilometers qualifies as moderate shear. In such environments, multicell thunderstorms can develop (assuming of courses that there's adequate instability and lift).

For the record, a classic **multicell thunderstorm** is a family (group) of single-cell thunderstorms in various stages of development (see Figure 9.37). Multicell thunderstorms, probably the most common type of deep, moist convection that erupts over the mid-latitudes, take shape as new thunderstorms develop along gust fronts. Figure 9.38 is a cross section showing the evolution of an idealized multicell

FIGURE 9.37 A multicell thunderstorm (courtesy of Henry Firus, Flagstaffotos, http://www.flagstaffotos.com.au/gallery2/main.php).

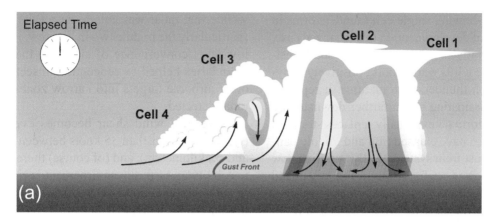

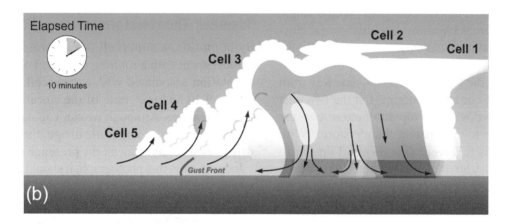

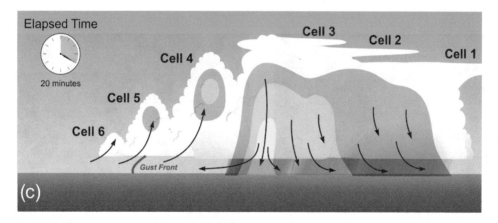

FIGURE 9.38 The evolution of a multicell thunderstorm during a time frame of 20 minutes demonstrates that a multicell thunderstorm is composed of single-cell thunderstorms at various stages of development. Colors indicate levels of radar reflectivity: (a) In the top panel, Cell #1 is dissipating while Cell #2 is mature (precipitation and downdraft have reached the ground). Cell #3 is nearing maturity and Cell #4 is in the cumulus stage; (b) In the middle panel (10 minutes later), the gust front has initiated a new cell (#5). Meanwhile, Cell #1 has almost completely dissipated. Cell #2 has entered the dissipating stage, and Cell #3 has matured. In effect, Cell #3 has "replaced" Cell #2 (and Cell #2 replaced Cell #1); (c) The third panel (10 minutes later) shows the gust front initiating a new cell (#6) and the continuing cycle of replacement.

thunderstorm, showing single-cell thunderstorms in various stages of development. In this idealized case, the gust front initiates new "cells" on its southwestern flank, where it lifts warm, moist, positively buoyant air. As each thunderstorm matures, it replaces the adjacent dissipating storm farther downstream. For example, storm #3 eventually replaces storm #2 after ten minutes have elapsed (top and middle panels). With the gust front systematically initiating new storms, the stage is set for a long-lived, organized multicell thunderstorm.

Sometimes rain-cooled air from clusters of thunderstorms (essentially multicellular thunderstorms) consolidates into a larger "cold pool" with a single gust front. Figure 9.39a shows essentially four clusters of thunderstorms and their "consolidated" gust fronts near North Platte, NE, at 01 UTC on June 11, 2007. Note that the gust fronts had moved away from the two middle clusters of storms, but the two gust fronts associated with the top and bottom clusters were still close to their parent thunderstorm. With their gust fronts relatively far away and unable to lift warm, moist air into the existing storms, the two middle clusters of storms weakened within 45 minutes (see Figure 9.39b). Meanwhile, the top and bottom clusters continued to thrive. In case you're

wondering, radar was able to detect the gust fronts (particularly the middle two in Figure 9.39a) because low-level convergence of air along these outflow boundaries helped to concentrate insects, dirt, and other airborne targets into narrow zones which the radar detected.

As vertical wind shear becomes even stronger (typically greater than 35 knots between the surface and six kilometers), and (of course) there's adequate instability and lift, the environment is ripe for what is, arguably, the granddaddy of all thunderstorms – the supercell – to erupt.

Supercell Thunderstorms: Heavy Hitters

By definition, a **supercell** is a discrete (separate) thunderstorm with a *rotating* updraft. To get a sense of rotation associated with a supercell, check out Figure 9.40 and take note of the storm's rounded, striated features. Although we can't actually observe rotating features on any static image, these rounded striations strongly suggest the presence of rotation.

Judging from this photograph, it should come as no surprise that supercells spark frequent lightning, with rates often exceeding 200 flashes per minute, some of the highest rates observed. And the impressive

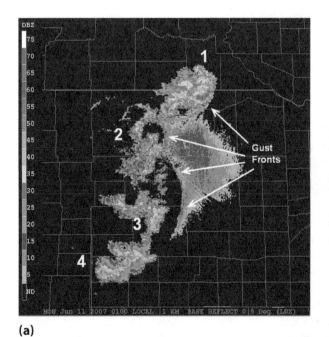

(a)

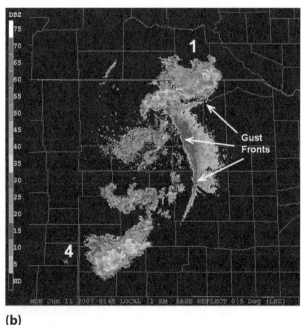

(b)

FIGURE 9.39 (a) Radar detected four clusters of thunderstorms and their associated gust fronts near North Platte, Nebraska, at 01 UTC on June 11, 2007; (b) Forty-five minutes later, as the two middle gust fronts consolidated and continued to move away from the two middle clusters of thunderstorms, these clusters weakened. Meanwhile, the top and bottom clusters continued to thrive because their gust fronts were still restrained (courtesy of NOAA).

FIGURE 9.40 Rounded striations give an overall sense of the rotation associated with this supercell thunderstorm near Grand Island, NE. Rotating updrafts are a hallmark of supercells (courtesy of Mike Hollingshead).

arsenal of supercells doesn't stop there. Indeed, these powerful thunderstorms produce nearly all of the strongest tornadoes (with maximum winds of 110 mph or higher) and the largest hail (diameters greater than two inches).

To get a sense of the highly organized nature of supercell thunderstorms, take a look at Figure 9.41, which is an idealized plan view of a classic supercell (the colors indicate varying degrees of radar reflectivity). On the southwestern flank of the thunderstorm, radar echoes appear to wrap around the **mesocyclone**. By definition, a mesocyclone is an area of low pressure that coincides with the supercell's rotating updraft; the presence of low pressure automatically suggests organization. Mesocyclones are on the order of a few to perhaps ten kilometers wide and at least half as tall as the depth of the cumulonimbus cloud. This means that mesocyclones can be seven to eight kilometers tall, assuming that supercells extend to the tropopause.

The interplay between the mesocyclone and strong vertical wind shear is really the key to the supercell's organization and longevity. The rotating updraft efficiently carries hydrometeors to high altitudes, where strong upper-level winds whisk them far downstream (ahead of the mesocyclone) where they eventually fall to the ground, forming a downdraft called the **forward-flank downdraft**. In a supercell thunderstorm, the strong vertical wind shear ensures that this downdraft does not interfere with the updraft (see Figure 9.42)—this separation between updraft and downdraft is key to the storm's

longevity and ability to produce severe weather. In addition, some of the precipitation gets caught in the circulation of the mesocyclone itself and eventually falls on the back side of the mesocyclone, forming the **rear-flank downdraft**. As you will learn in Chapter 15, if a supercell thunderstorm spawns a tornado, the rear-flank downdraft can play a pivotal role its formation. In such a case, the tornado would likely form near the point labeled "T" in Figure 9.41.

Note in Figure 9.41 that we have also drawn gust fronts (using the cold front symbol) representing the leading edge of the rain-cooled air from the two downdrafts, but we haven't drawn much in the way of convective clouds associated with the outflow boundaries. That's because upward motions in the belly of a supercell are usually so strong that compensating downward motion (subsidence) on the periphery of the supercell tends to suppress vertical development there (in part, this explains why supercells tend to be discrete). Thus, we arrive at a significant difference between multicell and supercell thunderstorms—the gust fronts do not play a major role in a supercell's organization.

Of the three types of discrete convection, supercells pose the greatest threat to produce severe weather. The rotating updraft in a supercell can suspend fledgling hailstones for long periods, allowing large hail to form. In fact, supercells produce a disproportionate percentage of hail greater than two inches in diameter. Perhaps more obvious, rotation in a supercell makes it easy to visualize how they might

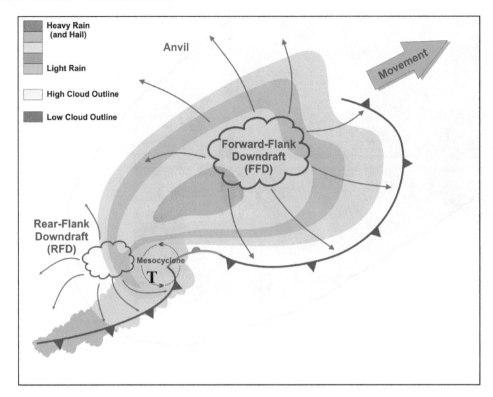

FIGURE 9.41 A plan view of a classic supercell. The colors represent various levels of radar reflectivity, while the cold front symbol is used here to denote gust fronts. Although cumulus clouds can form along the supercell's southwestern flank, the gust front there does not play a major role in the supercell's organization (note that we have used the cold front symbol here to represent the gust fronts).

FIGURE 9.42 In a supercell thunderstorm, fast winds aloft associated with strong vertical wind shear carry hydrometeors downstream, ensuring that downdrafts of supercells don't interfere with updrafts.

produce tornadoes—and in fact, supercells produce a disproportionate percentage of violent tornadoes (a topic we'll cover in a later chapter). For now, we take stock of the variety of other weather dangers posed by thunderstorms, in general: flash flooding, hail, and damaging straight-line (non-tornadic) winds.

TAKING THUNDERSTORM INVENTORY: A FULL STOCK OF WEATHER DANGERS

Thunderstorms are sometimes called "factories of weather" because they house virtually all types of meteorological phenomena, from snowflakes to lightning to hail. The inventory of thunderstorms includes some dangerous and damaging stock: We have already discussed lightning, while tornadoes, the most violent of all severe weather, will be covered in Chapters 14 and 15. We close this chapter by turning our attention to flash flooding, hail, and damaging straight-line winds.

Flash Flooding: The Number-One Thunderstorm Killer

The precipitation produced by thunderstorms plays a significant role in the hydrologic cycle of our planet. But heavy thunderstorms can also produce **flash floods**, which are rapidly rising surges of water that occur with little advance warning over a local area. Such concentrated downpours can overwhelm drainage systems, streams, creeks, and small tributaries of major rivers. According to the National

Weather Service, flash floods are the number-one storm-related killer in the United States over the long term (see Figure 9.43). In the 30-year period from 1987 to 2016, the national average of fatalities as a result of flooding was 84, exceeding that of lightning (47), tornadoes (70), and hurricanes (46). You'll note that the 10-year average tells a slightly different story, in part because the shorter averaging time allows individual extreme events to inflate the average, but flooding is still a formidable hazard to human health.

People are frequently caught off guard during flash floods because they often underestimate the power of moving water. Relatively shallow water that barely covers the hubcaps of a car can still carry a person away, while two feet of moving water can easily sweep cars downstream. Never drive into water of unknown depth—you may not get out! The National Weather Service reports that nearly half the deaths from flooding are "vehicle-related"—many occur in cars that are carried away by torrents of flood water. Follow the National Weather Service's simple advice about what to do when encountering flooding: "Turn around, don't drown."

Runoff from heavy, nearly stationary thunderstorms can be particularly dangerous in mountainous regions, where the sloping terrain rapidly funnels rainwater into usually sleepy mountain creeks and streams. Flash flooding is also especially dangerous in the deserts of the southwestern United States, where the lack of vegetation promotes rapid runoff that can, with little warning, fill typically dry

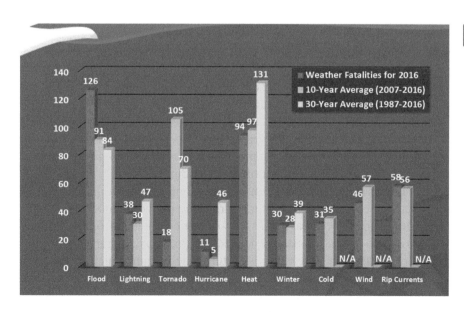

FIGURE 9.43 National Weather Service statistics for weather-related fatalities. The yellow bars indicate 30-year averages, the blue bars 10-year averages, while the red bars indicate the numbers for the year 2016 (courtesy of the National Weather Service).

(a)

(b)

FIGURE 9.44   (a) This flash flood in Maricopa County, AZ on October 13, 2003, was at its peak at 10:30 a.m. local time; (b) Only two and a half hours later, the offending wash had nearly emptied (courtesy of National Weather Service).

washes with fast, powerful currents of water (see Figure 9.44). Above all, the best defense against becoming a victim of flash floods is maintaining a healthy respect for the power of moving water.

For thunderstorms to produce flash flooding, typically either the upper-level wind flow must be weak or the upper-level winds must blow parallel to a line of thunderstorms. With weak upper-level **steering winds**, thunderstorms can stall or move very slowly, focusing heavy precipitation over a single area for an extended period of time. For the record, forecasters often estimate steering winds for single-cell thunderstorms by determining the average wind in the layer occupied by the cumulonimbus cloud (which, for all practical purposes, is the layer between 850 mb and 300 mb). On the other hand, when thunderstorms erupt over the same area and then move along a fixed path by steering winds that persistently blow from one direction, a series of individual thunderstorms can pass over the same location like railroad cars on a train track. This phenomenon, appropriately called **training**, can also produce flash flooding.

Individual, slow-moving thunderstorms that produce localized heavy rain and then dissipate are like lone thieves who rob banks one at a time. Losses from flash flooding can be great in these instances, but the scope of their crime is usually limited. Occasionally, there can be organized crime in the summertime (and other seasons as well) as the atmosphere conspires to widen the scope of flash flooding. Regardless, the telltale signs on the forecasting checklist for potential widespread flash flooding are:

- Weak vertical wind shear between 850 mb and 300 mb, with overall low wind speeds.
- High dew points near the surface and high moisture levels through a deep layer of the troposphere.
- A stationary or nearly stationary mechanism to promote rising air.

Obviously, weak winds aloft guarantee that thunderstorms will move slowly. With increasing wind shear (faster winds aloft), thunderstorms move more quickly. In addition, more and more dry air tends to get entrained into thunderstorms, thereby reducing the rain-producing efficiency of the storms.

To gauge the moisture content of the troposphere, meteorologists use a quantity called **precipitable water** (PWAT, for short), which is the amount of liquid that would accumulate if all the water vapor in an air column were forced to condense and fall as rain. Forecasters routinely use precipitable water to help them predict heavy rainfall rates and thus assess the threat of flash flooding.

Let's consider an example of the utility of precipitable water. During the period October 2–6, 2015, heavy rains led to historic flooding in South Carolina, in part associated with tropical moisture drawn from the vicinity of Hurricane Joaquin which was passing well offshore. Roughly a third of the state received

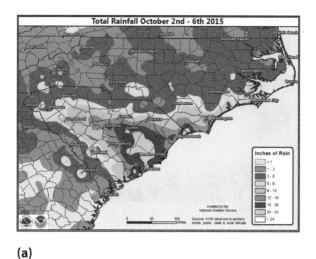

(a)

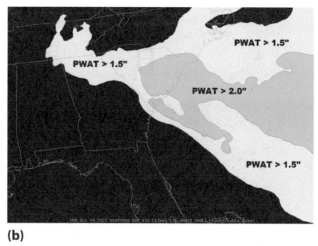

(b)

FIGURE 9.45 (a) Total rainfall for the period October 2–6, 2015, in South Carolina and surrounding states (courtesy of the National Weather Service); (b) An analysis of precipitable water at 06 UTC on October 4, 2015, showed a corridor of high PWAT values extending from the tropical Atlantic Ocean toward South Carolina.

more than a foot of rain, as shown in Figure 9.45a, as recurrent heavy showers and thunderstorms trained over the same areas. To solidify the tropical connection, consider Figure 9.45b which shows the precipitable water amounts at 06 UTC on October 4, 2015, when heavy rain was falling over much of South Carolina. A ribbon of relatively large PWAT (2.0 inches or higher) was pointing directly at the state, alerting forecasters that heavy rain would likely continue. Note that these PWAT values are nowhere near as large as the total amount of rain that fell—that's because PWAT only gives a snapshot of how much water vapor is in the air at a given time. In actual rainstorms, moisture is often continuously replenished from surrounding areas, allowing heavy rain to persist for long periods. Nevertheless, as a general rule, there is typically a good correlation between rainfall amounts from a given thunderstorm and precipitable water.

Of all the weather patterns associated with flash flooding, probably none are more dangerous than those that promote stationary thunderstorms over mountains. The background weather pattern almost always includes a persistent, upslope flow of moist air that gets air parcels to their level of free convection. Once thunderstorms erupt, the presence of weak steering winds encourages them to stall. The steep terrain then conspires to channel run-off from heavy rain rapidly into the valleys below. It is a recipe for disaster.

One of the most disastrous flash flooding events in U.S. history, which followed this recipe to a tee,

occurred on July 31, 1976, about 50 miles northwest of Denver in the Big Thompson Canyon, a popular camping destination. The tail end of a cold front had stalled over the Rockies, and moist low-level easterly winds in its wake were lifted up the east-facing mountain slopes (see Figure 9.46a). A weak trough at 500 mb approaching from the southwest provided additional lift and enhanced instability, setting the stage for impressive thunderstorms to develop (see Figure 9.46b). Steering winds in the mid-troposphere (around 500 mb) were very weak, generally less than 10 knots, allowing thunderstorms to remain virtually stationary for several hours. Over a foot of rain fell in the canyon, turning the normally placid Big Thompson River into a raging wall of water 19 feet high in some places. Boulders, vehicles and buildings were swept away. In two hours, the flash flooding killed 145 people, destroyed or damaged more than 700 houses and businesses, and caused more than $40 million in damages.

On average, flash flooding causes more than $2 billion in damage in the United States each year. Nationally, the annual cost of damage from hail approaches $1 billion, but, believe us, it's no bargain. Let's explore the topic of hail.

Hail: The White Plague

Onion-like stones of ice, called **hail**, can occur with single cell (pulse), multicell, and supercell thunderstorms. Their impact on agriculture can be significant—hail is often described as the "white plague" of crops.

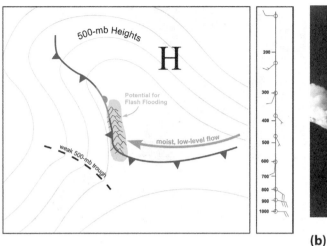

(a)

(b)

FIGURE 9.46 (a) The classic set-up for orographically induced flash floods, including the 1976 Big Thompson flood in Colorado. In the wake of a cold front, strong, moist low-level easterly winds (see vertical profile of winds on the right) rose abruptly up east-facing mountain slopes, allowing air parcels to reach their LFC and setting the stage for thunderstorms. A weak trough at 500 mb approaching from the southwest provided additional lift and cooling aloft, enhancing instability; (b) One of the only known photographs of the thunderstorms responsible for the Big Thompson flood (courtesy of John Asztalos).

Hailstones begin as meager frozen raindrops or small, soft snow-like particles called graupel. When strong updrafts suspend these particles above the 0°C (32°F) level of the thunderstorm, where the air is relatively rich in supercooled water, then new layers of ice will form on these fledgling hailstones. Research and observations have shown that, assuming a layer is conditionally unstable, the most favorable temperature range for hail growth is −30°C to −10°C (−22°F to −4°F). As a hailstone grows, it develops an onion-like pattern of ice that alternates between clear and opaque states. This variation is related to the non-uniform water content of the cloud as the hailstone gets tossed about within a turbulent cumulonimbus. In regions of the cloud of relatively low water content, air gets trapped as water freezes onto the hailstone, giving the layer a "milky" appearance. A high water content minimizes trapped air, which makes stones look clearer.

To assess the likelihood of hail, forecasters commonly compute the level of the **wet-bulb zero**. Essentially, this is an adjustment to the altitude at which the melting point of ice (0°C) is observed by taking into account the cooling effect that the evaporation of precipitation has on the air. When rain (or the meltwater on snowflakes or hail) evaporates, energy is drawn from the air, lowering surrounding air temperatures. Thus, when precipitation falls into relatively dry air, evaporational cooling will tend to lower the altitude of the melting level (in other words, the air gets colder at lower altitudes). This new altitude for the melting level, adjusted for the effects of evaporational cooling, is the level of the wet-bulb zero.

Observations suggest a preferred range of altitudes for the wet-bulb zero that makes hail more likely. If potent thunderstorms are expected to erupt, and the wet-bulb zero lies between 2200 and 2800 meters (about 7000 to 9000 ft), forecasters often predict hail. The chances of hailstones reaching the ground grow increasingly slim when the wet-bulb zero lies outside this thin envelope of altitudes. For instance, if the wet-bulb zero resides at an altitude higher than 2800 meters, then hailstones have a long, long way to fall through air at temperatures above 0°C, so they have little chance of surviving to the ground without melting. Such is the case in Florida in summer—the wet-bulb zero is often very high and, even though there are plenty of thunderstorms, very few produce hail that reaches the ground. On the other hand, if the wet-bulb zero is below 2200 meters, then forecasters know that relatively cold air is present in the lower troposphere. This low-altitude chill tends to stabilize the atmosphere, thereby suppressing the vigorous updrafts of air needed to support the growth of hailstones.

Average Number of 1" Hail Days per Year
1990-2015

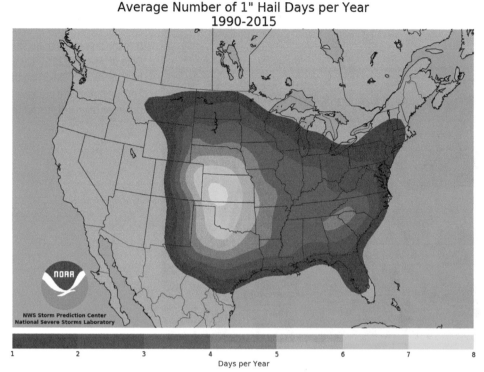

Days per Year

FIGURE 9.47 The average number of days per year when hail at least one inch in diameter occurs within 25 miles of a point (courtesy of Patrick Marsh, National Severe Storms Laboratory).

Of course, a seasoned forecaster always knows climatology. Figure 9.47 shows the annual average number of days during the period 1990–2015 with hail at least 1 inch in diameter (the size necessary for a thunderstorm to be considered severe). Clearly, a forecaster in Kansas and Oklahoma would be much more concerned about large hail than, say, a forecaster in Florida. Figure 9.47 suggests an elongated maximum in big hail stretching from northern Texas through the central Plains. This area of highest frequency corresponds to a region where updrafts in thunderstorms tend to be strong and the frequency of supercells tends to be high (more on this in Chapter 14). Figure 9.47 also suggests that large hail even occurs westward with a higher frequency along the Front Range of the Rockies, due in part to the relatively thin layer of ground-warmed air that typically overlies these higher elevations. Fast-falling hailstones defiantly zip through this thin warm layer without substantially melting. After heavy summer hailstorms along the Front Range, snow plows are sometimes called out to remove deep accumulations of ice off roads. In fact, Cheyenne, WY (elevation 1878 m) is one of the nation's most hail-prone cities.

For similar reasons, the area around Kericho, Kenya (elevation 2200 m) is one of the most-hail prone in the world, averaging about 50 days per year with hail.

The size of a hailstone depends, in large part, on the strength of thunderstorm updrafts—the stronger these upward motions, the longer that growing stones can wallow aloft in a bath of supercooled water. Generally, an updraft speed of 10–15 m/s (22–34 mph) can support the development of hailstones. Strong updrafts can suspend great volumes of hail and raindrops high in a thunderstorm, a process called **precipitation loading**. In such cases, vertical cross sections scanned by radar often reveal very high reflectivities that indicate the presence of large hail (see Figure 9.48).

Eventually, hailstones will fall toward the earth for one of two reasons. Hail can become too massive to be supported by the updraft, or the hailstone can find its way into either the downdraft region of the thunderstorm or be tossed into the clear air ahead of the thunderstorm, where it can then plunge to the surface. Hail produced by potent thunderstorms tends to fall in narrow streaks, often just a few kilometers wide and 10 or more kilometers long, that

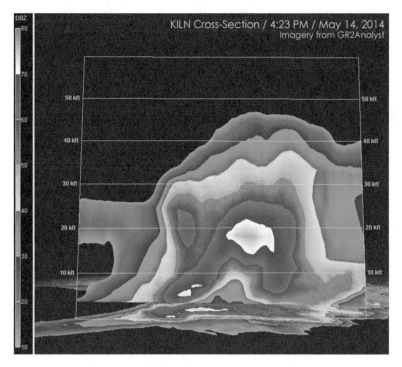

FIGURE 9.48 Cross section (vertical slice) of radar reflectivity through a severe thunderstorm in southern Ohio on May 14, 2014, shows reflectivity as high as 70 dBZ, indicating the presence of hail. This thunderstorm produced hail up to 2.5" in diameter (tennis-ball size) (courtesy of NOAA).

parallel the path of the storm (see Figure 9.49 for an example of such a "hailstreak"). When millions of hailstones simultaneously break free of the updraft, considerable damage can result. Hail can decimate farm fields, producing crop damage rivaling that of a tornado. In the High Plains east of the Rockies, hail typically damages five to six percent of the crops produced.

FIGURE 9.49 A hail streak near Airdrie, Alberta, Canada, on July 6, 2013. In just 15 minutes, many spots within the streak accumulated at least 3 inches of hail (courtesy of National Weather Service, Gaylord, MI).

Property losses from hail are also substantial. In just the Denver metropolitan area alone, seven hailstorms in the period from 1990 to 2004 each produced at least $100 million in insured damages. The Denver hailstorm on July 11, 1990, ranks as the second most destructive on record in North America, as softball-sized hail caused $625 million in property damage, mostly to cars and roofs. On April 29, 2012, severe thunderstorms pummeled parts of west Texas near Lubbock with baseball-sized hail, damaging hundreds of mobile homes and cars (see Figure 9.50). Hailstorms can also injure and sometimes kill. A hailstorm on May 5, 1995 near Dallas, TX, injured several hundred people, while the most deadly hailstorm on record killed 246 people in India on April 30, 1888.

In terms of weight and diameter, the largest hailstone ever documented in the United States was recovered in Vivian, SD on July 23, 2010 (see Figure 9.51). It weighed 0.88 kg (1 lb 15 oz) and was 8 inches (about 20 cm) in diameter. The record-setting hailstone fell as part of a barrage of large stones, many of them 6 inches (15 cm) or more in diameter, which left huge divots in the ground and damaged all of the town's 55 homes. Hailstones punched through many roofs and ended up in attics and even bedrooms. Notice the rather odd shape of the Vivian hailstone. While most smaller hailstones tend to be rounded, large hailstones sometimes freeze together into irregular lobed shapes.

FIGURE 9.50 Baseball-sized hail in west Texas in April 2012 heavily damaged many homes and automobiles (courtesy of National Weather Service).

As a heavy bundle of precipitation abruptly falls to earth, a strong downdraft often results. The stage is set for another type of severe weather: damaging straight-line winds from downbursts.

Microbursts: Dangerous Breaths from the Skies

On August 2, 1985, a passenger aircraft taking off from the Dallas-Fort Worth airport first encountered a strong headwind, then a sudden tailwind. This flight through changing wind directions (essentially, horizontal wind shear) as the plane climbed above the runway caused its airspeed to plunge from 207 to 138 mph (180 to 120 kts), leading to a loss of lift and a deadly crash. An investigation of this accident by severe storms expert Dr. Theodore Fujita (who also pioneered the scale on which tornado intensity is classified) led to startling revelations about another dangerous product of thunderstorms, the downburst.

For the record, a **downburst** (shown schematically in Figure 9.52) is the generic term for a strong downdraft of air which, upon reaching the earth's surface, causes an outrush of damaging, straight-line winds in the lowest one or two kilometers of the troposphere (at or near the ground). Upon striking the surface, the rain-cooled air "splashes" down like a strong stream of water from a kitchen faucet spreading out in all directions after hitting the sink below. Wind speeds associated with downbursts sometimes exceed 100 mph (87 kts). Downbursts often occur in concert

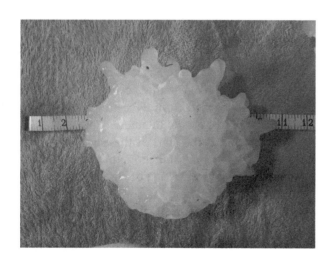

FIGURE 9.51 The largest hailstone on record in the U.S. (in terms of weight and diameter), documented by the National Climate Extremes Committee of NOAA, fell on Vivian, SD on July 23, 2010 (courtesy of National Weather Service, Aberdeen, SD).

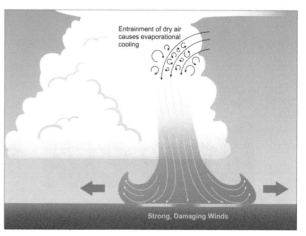

FIGURE 9.52 A cross section showing a downburst. When a powerful downdraft of air reaches the earth's surface, it spreads out in all directions, produces damaging, straight-line winds.

with strong downdrafts associated with heavy precipitation in thunderstorms. Downdrafts can further accelerate when they entrain dry air, which increases evaporational cooling and thus enhances the negative buoyancy and downward acceleration of air parcels plummeting to earth.

Downbursts come in two basic sizes. **Microbursts** are downbursts that typically last several minutes and whose outrush of winds extends no more than four kilometers in any horizontal direction. But don't judge a microburst by its somewhat limited duration and coverage. A powerful microburst on May 26, 2001, slammed the fiberglass skin of the WSR-88D radar shown in Figure 9.53. Winds estimated at 92–98 mph (80–85 kts) punched the dome inward and disabled the radar, the first time one of these radars had been so badly damaged. This photograph really puts the "micro" in microburst, wouldn't you agree?

FIGURE 9.53 This WSR-88D radar at Laughlin Air Force Base near Del Rio, TX, was badly damaged by the straight-line winds of a downburst shortly after 1 A.M. local time on May 26, 2001. Replacement radar domes are not easy to find, so the dome took weeks to replace (courtesy of National Weather Service).

Sometimes, the damage associated with microbursts is mistaken for that of a tornado: However, wind damage from microbursts occurs in straight lines, whereas a pattern of circulation or rotation is evident in the damage along the path of a tornado. For example, trees fall like dominoes from the powerful straight-line winds of a microburst, but form a twisted pattern (when viewed at a distance or from an airplane) after being felled by a tornado.

Macrobursts are downbursts whose outrush of winds exceeds four kilometers in horizontal scale; the upper limit is about ten kilometers. Macrobursts typically last several minutes to half an hour, and tend to occur when momentum from strong winds in the middle troposphere (or higher up) is transferred *downward* to the earth's surface (which, at face value, suggests a horizontal scale larger than that of a microburst).

On the evening of August 28, 2008, severe thunderstorms rolled across the metropolitan region of Phoenix, AZ, causing extensive straight-line wind damage. Figure 9.54a is a color-enhanced infrared satellite image over central Arizona at 9:30 p.m. MST. The bright red indicates cloud-top temperatures below –100°F (–73°C). Such extremely low temperatures translate to very tall thunderstorms that correlate closely to a heightened risk of severe weather. Figure 9.54b is a close-up display of Doppler velocities from the radar near Phoenix at 9:37 p.m MST, shortly after the satellite image in Figure 9.54a (the location of the radar is marked KIWA). Here, the most significant footprint of the macroburst is the pink area which indicates winds moving rapidly away from the radar at 60–70 kts (69–81 mph). The macroburst knocked down thousands of trees in the Phoenix metropolitan area (see Figure 9.54c); at the airport, winds gusted to 65 kts (75 mph) and temperatures dropped from 84°F to a daily record low of 64°F in just eleven minutes.

As mentioned at the start of this section, downbursts (especially microbursts) can be particularly dangerous to airplanes taking off or landing. Sometimes the threat to aviation is described as a "fist of wind" that drives an aircraft into the ground, but this is an oversimplification. The real threat from downbursts is the rapid change from a headwind to a tailwind. Consider, for example, Figure 9.55, which shows an aircraft encountering a microburst during landing. At Point 1, the plane feels a headwind and has a tendency to gain lift (because there is more air moving past the wings). In response, the pilot points

(a)

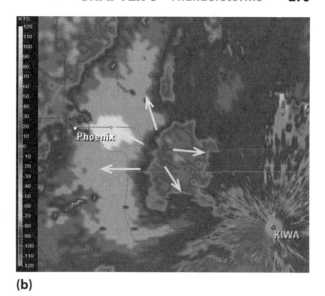

(b)

(c)

FIGURE 9.54 An enhanced infrared satellite image over central Arizona at 9:30 P.M. (MST) on August 28, 2008, shows severe thunderstorms with cloud-top temperatures as low as −100°F; (b) A display of Doppler velocities from the radar at Phoenix (radar located at the bottom right) captures the footprint of a macroburst. Pink indicates winds rushing away from the radar at 60–70 kts (69–81 mph), while green indicates winds rushing toward the radar at 20–30 kts (23–35 mph); (c) The macroburst downed trees in the Phoenix area (courtesy of NOAA).

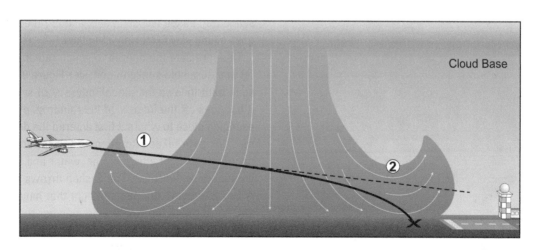

FIGURE 9.55 The microburst's threat to aviation. At Point 1, the aircraft encounters a strong headwind. Soon afterward, at Point 2, the aircraft encounters a strong tailwind. The relative motion of the tailwind to the aircraft can lead to loss of lift (plane not to scale).

the nose of the aircraft down. This makes the aircraft especially vulnerable when it suddenly encounters a tailwind on the opposite side of the microburst (at Point 2). Because the wind is moving with the aircraft in this area, the amount of air flowing over the wings and providing lift is severely reduced, causing the aircraft to rapidly lose altitude.

The threat of straight-line winds to aviation has led to the installation (at most major airports) of systems to detect the wind shifts associated with downbursts. Doppler radar is an integral part of these systems, given that its velocity component can detect the horizontal wind shear associated with downbursts.

POSTSCRIPT: LOOK SKYWARD

Thunderstorms come in many forms and impact society and nature in many ways. On a global scale, thunderstorms mediate the exchange of water, electrical charge, and energy between the atmosphere and the surface. Yet, the sight of a towering cumulonimbus cloud, the sound of thunder, and the cooling breezes of a gust front all bear witness to local effects as these meteorological marvels work to alleviate temperature contrasts on small spatial scales. Thunderstorms offer visible evidence of the inner workings of the atmosphere—they await your observation, with the appropriate respect for their power, of course.

Focus on Optics

Rainbows: Branding the Sky

Like ranchers in the old "Wild, Wild West" who branded their cattle with searing irons to protect their herds from rustlers, the sun leaves distinctive brands on the heavens, reminding us that it owns and operates the largest ranch on the big-sky prairie of atmospheric optics. Figure 9.56 shows the sun's most recognized optical brand, the **rainbow**, a colorful insignia that, like many of the arcs, rings, and spots that mark the sky, is a modified image of the sun itself. That's right, cowboys and cowgirls, the rainbow is simply reconstituted sunlight.

Raindrops are the agents that scatter incoming sunlight into an arcing deck of primary colors. To accomplish this optical feat, raindrops act like prisms (see Figure 9.57) and one-way mirrors, refracting and reflecting incoming sunlight. However, as mirrors, raindrops are flawed because most sunlight passes right through them—only a little is reflected off their back sides. But, as we will soon see, a little reflected light goes a long way in making a rainbow.

To appreciate the degree to which raindrops scatter and scramble sunlight, consider Figure 9.58a. The solid path follows the pinball odyssey of sunlight as it bounces off the interior of the raindrop, eventually escaping close to where it first entered the drop. Note that this path takes into account the refraction that occurs when light emerges from water and bends as it enters less dense air. The dashed arrows at Points A and B indicate the paths of light that have passed straight through the raindrop (remember the flawed, one-way mirror analogy). Note how these rays also bend as they move from water to air. Based on Figure 9.58a, the observation that raindrops scatter sunlight

FIGURE 9.56 A primary rainbow photographed in central Pennsylvania.

(Continued)

(Continued)

FIGURE 9.57 Visible light, upon passing through a prism, is separated into a rainbow of colors (courtesy of David Parker/Science Photo Library).

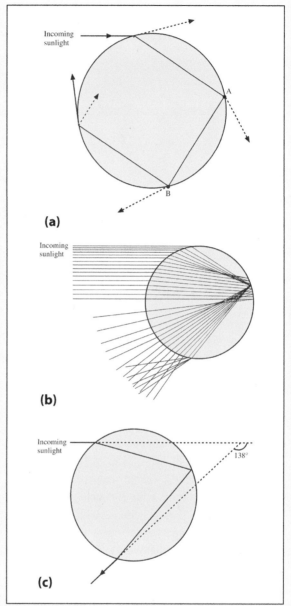

in many directions is irrefutable. Next, if we invoke the earlier assertion that raindrops also act like prisms to separate "white" sunlight into the primary colors, we then conclude that raindrops scatter rays of sunlight in jumbled fashion, thereby filling a large portion of the sky opposite the sun with many different colors.

At first glance, such a hodge-podge of colors scattered willy-nilly across the sky does not bode well for explaining a rainbow. But a closer examination of the spread pattern of sunlight around a single raindrop reveals more promising results (consider Figure 9.58b). First, however, realize that at each point of internal reflection, light on the inside of the raindrop grows weaker and weaker as more and more light "leaks" outside to the air. Thus, the best chance to see light bright enough to make a brilliant rainbow occurs when the number of internal reflections is limited to one. Oh, yes, there are rays that reflect more than once off interior walls, but the loss in the intensity of light at each reflection makes them less likely to make a bow. Two reflections can indeed fashion a "secondary rainbow" (see Figure 9.59). Note that it is much fainter than the

FIGURE 9.58 (a) Much of the sunlight that enters a "leaky" raindrop eventually passes through the drop and is scattered in many directions across the sky (dashed lines show this scattering). Only a small portion of sunlight takes the refractive, reflective tour on the inside of the drop before refracting again as it emerges back into the air (solid path); (b) Rays of sunlight that enter a raindrop and are reflected once off the drop's back side exit in many directions, but most rays tend to cluster around the "rainbow angle"—138°; (c) In other words, a typical incoming ray of sunlight that is internally reflected once off the back of a raindrop is deviated from its original path by an angle very close to 138°.

(Continued)

(Continued)

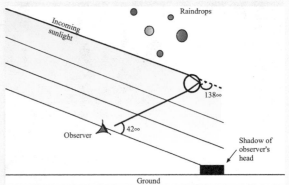

FIGURE 9.59 A primary rainbow with a fainter secondary rainbow.

FIGURE 9.60 Assuming that the shadow of the head of the observer lies 180° from the sun, then the light creating the rainbow must originate from raindrops that lie on a 42° circle centered around the shadow of the observer's head. Only these raindrops can send concentrated rainbow light to the observer's eyes. A person standing next to the first observer will see a different rainbow, since the raindrops on the 42° circle centered about the shadow of his head are, at any instant, different from the ones creating the rainbow for the first observer.

"primary rainbow" (also, the arrangement of its colors is opposite the primary rainbow).

Okay, hopefully we've convinced you why we're "primarily" interested in rays of light that have undergone just one internal reflection in Figure 9.58b. Now we need to resolve how the concentrated pattern of colors in the primary rainbow can emerge from the helter-skelter scattering of sunlight by raindrops. Though the incident rays at the top of the drop in Figure 9.58b are evenly spaced, those that emerge are not. Indeed, there is an angle of "minimum deviation" from sunlight's original path around which exiting rays congregate. The measure of this special angle is 138° (see Figure 9.58c). In other words, sunlight that enters a raindrop is deviated by at least 138° after it refracted, reflected, and refracted once again.

It is the clustering of light rays on departure from raindrops at approximately 138° from their original paths that provides the needed intensity and focus of light to make a rainbow. There is also an upper limit to the deviation of sunlight—it can be no more than 180°, which corresponds to a ray of sunlight striking the middle of the drop and doing an about-face. Thus, depending on the angle that incident light makes with a raindrop, the triple optical play of refraction, reflection, and refraction will throw incoming sunlight off course by an angle between 138° and 180°. But, as far as the primary rainbow is concerned, light rays deviated off course at angles close to 138° are the crucial ones.

Up until now, we've ignored you, the observer (please don't take this personally). Recall that raindrops are scattering and filling the sky with sunlight at angular distances between 138° and 180° from the sun. What does all this mean to the observer? The point that lies 180° from the sun is the shadow of your head (the sun is behind you in Figure 9.60) and the point corresponding to 138° from the sun lies 42° from the head of your shadow (you'll need to note that 180° − 138° = 42° while looking at Figure 9.60). Thus, at any instant, only those raindrops in front of you, which lie on a 42° circle centered around the shadow of your head, can send concentrated rainbow light to your eyes. The drops responsible for the rainbow can be at any linear distance from you as long as they lie on a 42° circle (actually, unless you are on a mountain or in an airplane peering down on a shower of rain, the greatest part of the rainbow circle that any observer can see is the half-circle above the horizon—the other half of

(Continued)

(Continued)

the rainbow circle theoretically lies below the horizon). And because raindrops are falling, there must be an uninterrupted supply of them in order for the rainbow to last more that a fleeting instant. Moreover, if any part of the 42° circle lacks raindrops or sufficient sunlight, then part of the bow will not be visible, thereby accounting for the partial bows that we sometimes see.

Finally, the order of colors in a primary rainbow is based on the observation that the longest wavelengths of visible light impinging and emerging from raindrops on the 42° circle are deviated through a slightly smaller angle than the shortest wavelengths of visible light (remember, raindrops act like prisms, and prisms refract red light the least). So the longest wavelengths of visible light that are destined for rainbow status must lie closest to the sun. Thus, red will be on the outside of the bow (which is closest to the sun) while violet will lie farthest inside the bow.

You are the center of "your" rainbow circle. No one else sees exactly the same rainbow that you see. Even if a person is standing next to you, there are different raindrops on a different 42° circle that are fashioning a different rainbow for him or her. Lest you get caught up in the issue of ownership of your rainbow, remember that "your" rainbow wears the distinctive brand of the sun—the biggest ranchero in the sky.

Tropical Weather, Part I: Patterns of Wind, Water, and Weather

10

LEARNING OBJECTIVES

After reading this chapter, students will:

- Gain an appreciation for the global energy budget and the role that the tropics play in helping to maintain the budget.
- Gain an appreciation for the history of the development of the model for the general circulation over the tropics and subtropics.
- Be able to identify the primary components of the general circulation over the tropics and subtropics, and to understand the underpinning meteorology associated with each feature.

- Gain an appreciation for the seasonal variations of the equatorial trough, subtropical high-pressure systems and the subtropical jet stream.
- Be able to identify the various impacts that the general circulation has on weather and climate in the tropics and subtropics.
- Be able to identify the primary features of Southeast Asia's monsoon and to understand the meteorology associated with each component.
- Be able to argue why the "summer monsoon" in the Southwest United States is not a classic monsoon.
- Understand the meteorology and oceanography associated with El Niño.
- Understand the role that the Southern Oscillation and the shifting Walker Circulation play in regional weather patterns during El Niño.
- Assess and understand the seasonal impacts of El Niño and La Niña on regional and global weather patterns.

The online Merriam-Webster Dictionary defines *idiosyncrasy* as "an individualizing characteristic or quality." Although the typical context of this word refers to peculiar traits in people, we feel compelled to use it to describe the overall personality of the tropics. In this introduction, we hope to pique your interest by calling attention to some of the rather quirky meteorological characteristics of the tropics.

First, think about what percentage of the planet's surface lies at low latitudes—for example, within 30° of the equator. Take a guess. Given that latitude values range from 90°N to 90°S, you might be tempted to speculate that the band from 30°N to 30°S doesn't occupy all that much of the total surface area of the globe. However, in reality, because the planet has such a wide girth at low latitudes, exactly half of the surface lies within 30° of the equator. If you reference Figure 10.1, where we have drawn the two latitude lines in question, this assertion may be easier to digest.

The tropics make up the lion's share of this low-latitude region. We will focus our attention on the area between the Tropic of Cancer (23.5°N) and the Tropic of Capricorn (23.5°S), although we acknowledge that these two major latitudes are not written in stone as absolute boundaries of tropical weather and climate.

Given that the tropics occupy such a large portion of the earth's surface, it should come as no surprise that they can have profound impacts on weather patterns in the middle and high latitudes. Figure 10.2, which shows the annual incoming solar and outgoing infrared energy as a function of latitude, captures another idiosyncrasy of the tropics. Without reservation, the

tropics have a radiation surplus while the high latitudes run a deficit. To keep this radiation imbalance from getting out of hand, heat energy and moisture are transported out of the tropics toward both poles. It's this poleward transport that extends the tropics' reach to higher latitudes.

The energy surplus in the tropics translates to relatively high annual average temperatures there. But, as it turns out, there are places in the subtropics

FIGURE 10.1 On a full-disk geostationary satellite image, it's easy to see Earth's wide girth in equatorial regions. Indeed, the region between latitudes 30°N and 30°S (which are shown) constitutes half of the surface area of the planet (courtesy of NOAA).

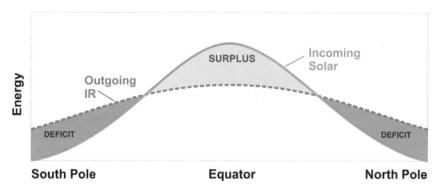

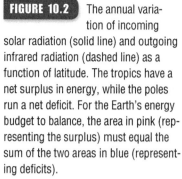

FIGURE 10.2 The annual variation of incoming solar radiation (solid line) and outgoing infrared radiation (dashed line) as a function of latitude. The tropics have a net surplus in energy, while the poles run a net deficit. For the Earth's energy budget to balance, the area in pink (representing the surplus) must equal the sum of the two areas in blue (representing deficits).

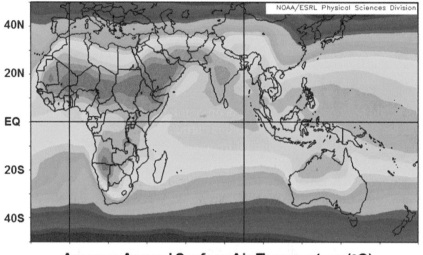

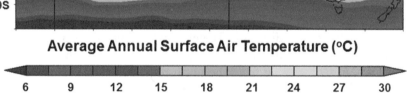

FIGURE 10.3 Annual average surface air temperatures, in °C, based on the period 1981–2010 (courtesy of NOAA).

that rival parts of the tropics for warmth (recall from Chapter 1 that the subtropics extend to approximately 30–35° latitude in each hemisphere). That's easy to see if you glance at Figure 10.3, which shows average annual temperatures for much of the Eastern Hemisphere. For completeness, we'll also include the subtropics and their role in global patterns of weather and climate in this discussion.

The laundry list of tropical idiosyncracies is rather long. Check out Figure 10.4, which compares average monthly high temperatures over the course of the year at Mumbai (formerly Bombay), India (latitude 19°N), with St. Louis, MO (a mid-latitude city). Note that St. Louis has a typical summer peak in mid to late July. But temperatures in Mumbai, like many cities in India, have a primary peak in May and then a secondary peak later in the year. This seemingly peculiar temperature trace arises because of the summer monsoon in Southeast Asia—a period from late May to October when recurrent rains inundate the subcontinent. There's also a winter monsoon in this region of

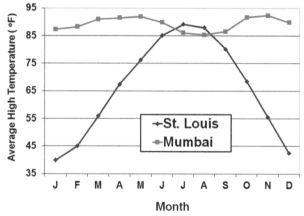

FIGURE 10.4 The annual trace of average monthly high temperatures at St. Louis, MO, and Mumbai, India.

the world, but hardly a drop of rain falls. Right away, you should get the impression that the word "monsoon" is not synonymous with "rain." We'll have plenty to say about monsoons later in the chapter.

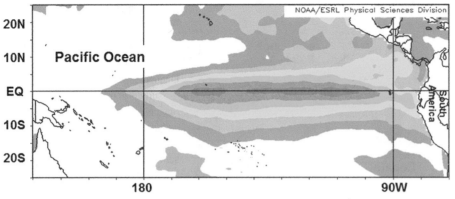

FIGURE 10.5 Departures from average sea-surface temperatures (in °C) for the period October 2015 to March 2016. Water temperatures during this period, which featured a strong El Niño, averaged more than 2°C (3.6°F) above average along the equator from the coast of South America to near the International Date Line (courtesy of NOAA).

Finally, no discussion of the tropics would be complete without addressing El Niño. For the record, **El Niño** is an anomalous warming of the waters of the eastern and central tropical Pacific Ocean, centered on the equator. The footprint of the strong El Niño of 2015–2016 is easy to see in Figure 10.5, which shows the departure from average of ocean water temperature from October 2015 to March 2016. El Niño, along with its flip side La Niña, typically has impacts on weather patterns far outside the tropics. We will dive deeper into El Niño and La Niña later in this chapter.

We haven't forgotten about hurricanes and tropical storms. In fact, we've reserved all of Chapter 11 to cover these compelling topics.

So you see, the tropics are full of hot topics!

WEATHER AND CLIMATE AT LOW LATITUDES: PERSISTENT PATTERNS

Over the long haul, weather observations across the middle latitudes indicate that surface winds tend to have a preferred direction at each town or city during a given month or season (or, for that matter, the entire year). These preferred directions depend on the specific location and occur despite the parade of transient high- and low-pressure systems that march from west to east across the middle latitudes.

To help visualize tendencies in wind speed and direction at a given location, meteorologists often draw a **wind rose**, a circular diagram with "petals" that summarizes wind speed and wind direction data over a given period of time (typically a month, season or year). Figure 10.6 is a wind rose for Rapid City, SD, incorporating many years of data from all months, so this particular wind rose represents average annual frequencies.

There are 16 compass directions on this wind rose: N, NNE, NE, ENE, E, ESE, SE, SSE, S, SSW, SW, WSW, W, WNW, NW, and NNW, and thus 16 petals. Each successive ring represents an increment of 4% in the frequency of a specific wind direction. To get your bearings, focus your attention on the longest petal (associated with winds blowing from the north-northwest). Its length from the center of the diagram indicates that such a wind direction occurs at Rapid City about 17% of the time. The lengths of each individual color in this particular petal indicate how that 17% is divided among various wind speed ranges (which are in m/s here—see the legend). For this example, NNW winds greater than 11 m/s (24 mph, the aqua color) have a frequency of about 4%, NNW wind speeds in the 8.5–11 m/s range (19–24 mph, the green) occur about 3% of the time, while NNW winds of 5.4–8.5 m/s (12–19 mph, the blue) were observed about 5% of the time.

Although every wind direction was observed at Rapid City (that is, there's a petal in all 16 directions), the relatively short length of the petals between north and east and between south and west indicates that winds from these directions occur with the least frequency. As we have noted, NNW is the preferred, or most common, wind direction, also sometimes referred to as the **prevailing wind**. To a large extent, the prevailing wind in mid-latitude locations has a westerly component. However, the situation is different in the tropics.

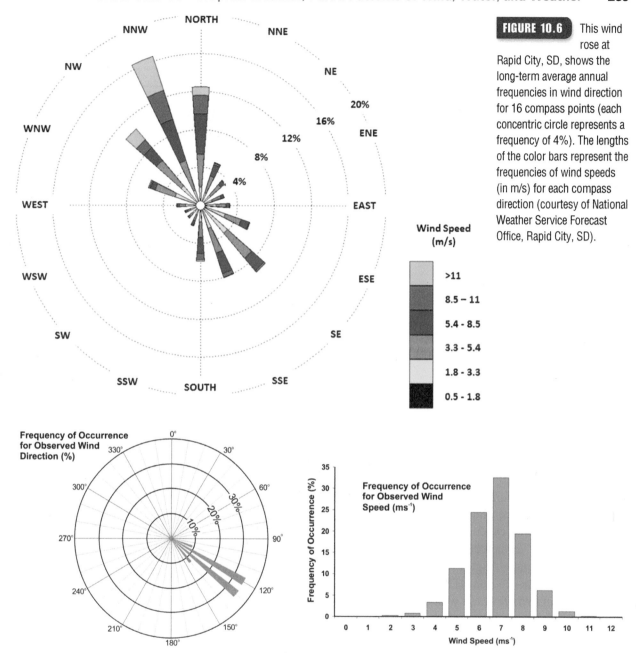

FIGURE 10.6 This wind rose at Rapid City, SD, shows the long-term average annual frequencies in wind direction for 16 compass points (each concentric circle represents a frequency of 4%). The lengths of the color bars represent the frequencies of wind speeds (in m/s) for each compass direction (courtesy of National Weather Service Forecast Office, Rapid City, SD).

Wind Speed (m/s)

	>11
	8.5 – 11
	5.4 - 8.5
	3.3 - 5.4
	1.8 - 3.3
	0.5 - 1.8

FIGURE 10.7 (Left) A wind rose for an ocean buoy moored at latitude 8°S, longitude 95°W, for the period 2001–2010. Note that winds blew almost exclusively from the southeast. (Right) A frequency diagram of the average daily wind speeds (in m/s) measured at the buoy.

Compare the Rapid City wind rose to Figure 10.7, a wind rose representing the 10-year period 2001–2010 for a buoy moored at 8°S, 95°W, in the tropical Pacific Ocean off the west coast of South America. The wind direction is overwhelmingly southeast, offering a clue that surface winds in the tropics are a lot more persistent than surface winds in the mid-latitudes. Rather than color-coding wind speeds directly on the wind rose, we choose to represent them in a frequency diagram (to the right of the wind rose). Clearly, these tropical southeast winds regularly blew with gusto—notice that the most common average daily wind speeds were 6–8 m/s (corresponding to about 13–18 mph).

This buoy was moored within the jurisdiction of the most consistent surface wind system on Earth, the reliable **trade winds**. In the Southern Hemisphere, the trades persistently blow from the

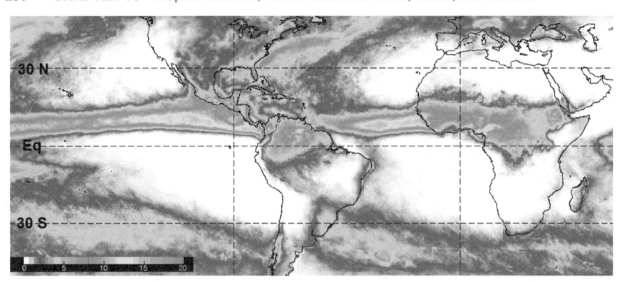

FIGURE 10.8 The average rainfall rates in August (in millimeters per day) based on satellite measurements from 1998 to 2010. The prominent ribbons of relatively high rainfall rates over the tropical Atlantic and the tropical eastern and central Pacific Oceans mark the mean position of the Intertropical Convergence Zone. Here, the northeast and southeast trade winds converge, promoting upward motion and recurrent showers and thunderstorms (courtesy of NASA).

southeast over tropical and subtropical oceans; in the Northern Hemisphere, they blow persistently from the northeast over low-latitude oceans. Figure 10.8 shows average rainfall rates in August during the period 1998–2010, based on satellite measurements. Note the prominent ribbons of higher average rainfall rates that stand out over the tropical Atlantic and the tropical central and eastern Pacific. These "strips" of enhanced rainfall rates correlate well with the average August position of the **Intertropical Convergence Zone** (ITCZ), a corridor over tropical oceans where trade winds converge, promoting upward motion and recurrent showers and thunderstorms (tropical disturbances such as tropical storms and hurricanes also contribute to these enhanced totals).

Both the trade winds and the ITCZ are part of the persistent circulation of air that tends to dominate the tropical oceans. Let's investigate.

The Hadley Circulation: In Control of the Tropics and Subtropics

In 1705, English astronomer Edmond Halley calculated that the bright comet he had observed in the heavens more than twenty years earlier was periodic and that it would return in 1758. His computations were right on target, although Halley didn't live long enough to see his prediction come true. In

FIGURE 10.9 Halley's Comet, on March 8, 1986, viewed from Easter Island (courtesy of NASA, photograph by W. Liller).

tribute to his great achievement, the comet, shown in Figure 10.9, now bears his name: Halley's Comet.

Halley was a scientific dilettante of sorts. In 1716, he proposed designs for diving bells that could be used for deep-ocean exploration. Halley also dabbled in cartography, naval navigation, and tropical meteorology. The trade winds caught his scientific eye. Used by Christopher Columbus on his voyage to the New World, these reliable winds were crucial to commercial sailing ships (hence the name "trades"). Ever inquisitive, Halley formulated a scientific theory that explained the trades.

Halley postulated that intense solar heating over equatorial regions caused rising currents of air that eventually hit the tropopause, spread poleward, and then subsided over latitudes noticeably farther from the equator. In 1735, English scientist George Hadley supported Halley's theory by envisioning a closed circulation of air with the trade winds as the return flow of surface air back toward the equator. Hadley saw his closed circuit of air at low latitudes as an integral part of the Earth's **general circulation**, which is the global system of average, large-scale horizontal and vertical air motions in the troposphere (see Figure 10.10).

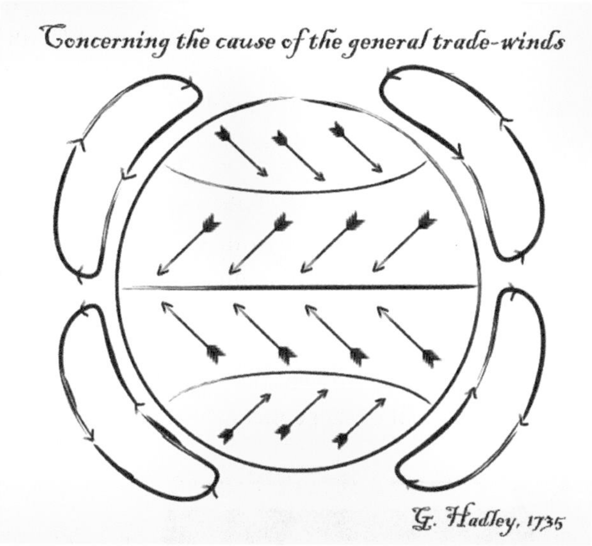

The strong signal of the general circulation in the tropics and subtropics comes from two relatively persistent convective circulations, essentially one in each hemisphere. These circulations are called **Hadley Cells**, named, of course, in honor of George Hadley. An idealized cross section of the two Hadley cells is shown in Figure 10.11. Note the deep convection at the ITCZ in the form of tall cumulonimbus clouds called **hot towers**. Within these, rising warm air hits the tropical tropopause and spreads poleward, eventually sinking near latitudes 30°N and 30°S. Some of the air then moves equatorward at the surface, creating the trade winds and completing a closed circuit.

Before we pursue a close-up look at each component of the Hadley Cells, it's worth addressing whether similar closed cells arise over the middle and high latitudes. Outside of the tropics and subtropics, transient surface high- and low-pressure systems typically bring an assortment of winds and up-and-down motions at any given location (this is especially true over mid-latitudes). As a result, the "average" horizontal and vertical motions outside of low latitudes are really just tendencies that only emerge over relatively long averaging periods—and these tendencies are relatively weak compared to the much stronger and persistent signal in the tropics and subtropics. As such, we don't believe these motions qualify as reliable closed circuits (cells) that can be used to explain day-to-day weather and its variability, certainly not to the extent of the Hadley Cells.

It's now time to dissect the Hadley circulations. Indeed, we will focus on each feature you see in Figure 10.11—in truth, there's plenty more than meets the eye.

FIGURE 10.10 In 1735, George Hadley proposed a model for the general circulation, which is the system of average horizontal and vertical motions of air in the troposphere. Hadley's description of the general circulation over low latitudes was pretty much right on target.

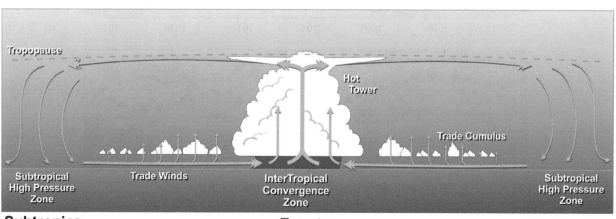

FIGURE 10.11 An idealized cross section of the two Hadley Cells which largely control weather patterns in the tropics and subtropics.

The Intertropical Convergence Zone: A Creature of the Sea

To piece together the puzzle we call the Hadley Cells, we start with the Intertropical Convergence Zone (ITCZ, for short). As its name suggests, convergence of ocean air is the mainstay of this surface component of the Hadley Cells. From what you've learned earlier, troughs of low pressure are the most common source for low-level convergence. For the record, the ITCZ is found in a trough—namely, the **equatorial trough**, formally an elongated belt of low pressure that, for all practical purposes, spans the entire globe at low latitudes.

Before you get the impression that the equatorial trough lies strictly along the equator (the name sort of suggests that, wouldn't you say?), consider Figure 10.12 which shows the long-term mean monthly sea-level pressures for January (top) and July (bottom) within 30° latitude of the equator. The dashed line indicates the average position of the equatorial trough. Note that, in July, the equatorial trough lies almost completely in the Northern Hemisphere, while during January, it's almost entirely in the Southern Hemisphere. In this way, the equatorial trough "follows the Sun," meaning that it lies mostly in the hemisphere where solar heating is greatest. This is especially true over large land masses such as Southeast Asia during the warm season, when surface air temperatures are very high. The close connection between the highest surface air temperatures and the position of the equatorial trough should make some sense—high air temperatures help to reduce air density and, therefore, reduce the weights of local air columns, which, in turn, help to maintain relatively low pressure at the surface.

The ITCZ lies within the equatorial trough over the tropical oceans where the northeasterly trade winds in the Northern Hemisphere converge with the southeasterly trade winds of the Southern Hemisphere. This low-level convergence at the ITCZ tends to add weight to local air columns, but the convergence is offset by upper-level divergence (revisit Figure 10.11 and focus on the diverging upper branches of the Hadley Cells). This high-altitude divergence removes mass from local air columns, limiting their weights and helping to maintain relatively weak low pressure at the surface. Thus, surface winds in the vicinity of the ITCZ are also generally weak. In fact, over tropical oceans where the ITCZ and the equatorial trough coincide, the corresponding

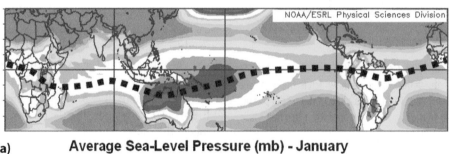

(a)   Average Sea-Level Pressure (mb) - January

| 1004 | 1006 | 1008 | 1010 | 1012 | 1014 | 1016 |

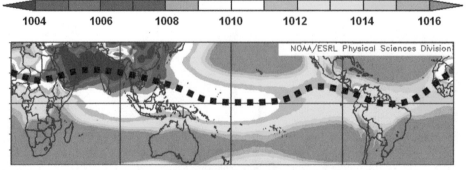

(b)   Average Sea-Level Pressure (mb) - July

FIGURE 10.12   A dashed line represents the average position of the equatorial trough in January (top) and July (bottom). The equatorial trough emerges from the polts of average sea-level pressure for these two months. Note how the equatorial trough tends to "follow the sun" into the summer hemisphere.

zones of weak surface winds is sometimes called the **doldrums**, a nautical term that refers to the light and variable nature of surface winds there.

Where there's low-level convergence and upper-level divergence, there's also upward motion (again, revisit Figure 10.11). So it should come as no surprise that the ITCZ corresponds to the ascending branch of the Hadley Cells. This upward motion commonly leads to the formation of a ragged belt of tall convective clouds marking the ITCZ (for an example, see Figure 10.13, a color-enhanced infrared satellite image from August 2015). As a result, recurrent showers and thunderstorms typically characterize the ITCZ. So in regions where the position of the equatorial trough shifts dramatically during the year, clear-cut wet and dry seasons emerge.

For example, the city of Fortaleza, located on the northeast coast of Brazil, has a wet season during the Southern Hemisphere's summer and autumn because the equatorial trough is nearby (see Figure 10.14). During the Southern Hemisphere's winter and spring, however, the equatorial trough shifts far north, setting the stage for a much drier period of months. Also, note the northward surge of the equatorial trough in July over India and Southeast Asia in Figure 10.12b. From May to October, the equatorial trough in this region of the world is called the **monsoon trough** (more on this breeding ground for prodigious rains later in the chapter).

Having dissected the ITCZ and the ascending branch of the Hadley Cells, let's shift our attention poleward to the subtropics.

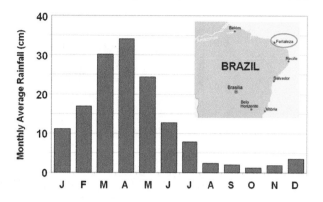

FIGURE 10.14 Average monthly rainfall (in cm) at Fortaleza, Brazil, shows a wet warm season and a dry cool season. Fortaleza is located on the northeast coast of Brazil (map courtesy of the Central Intelligence Agency).

Subtropical High-Pressure Systems: Merging Traffic from the Tropics

Figure 10.15 shows the distribution of average annual sea-level pressure. Note the corridor of relatively low pressure at low latitudes that marks the domain of the equatorial trough. The other features that might catch your eye are the belts of high-pressure systems that line up in both hemispheres near 30° latitude (note the belts of red that occupy much of the subtropical Atlantic, Pacific, and Southern Indian Oceans near this latitude). These time-averaged, semi-permanent features are called **subtropical highs**. The adjective "semi-permanent" implies that they tend to appear routinely on daily weather maps, particularly in the summer hemisphere. The long-term average position of these anticyclones lies near latitude 30° in both hemispheres, but subtropical highs tend to shift position with the seasons between approximately 20° and 40° latitude.

Now that we've established the existence of these semi-permanent subtropical highs, we can offer another explanation for the presence of the equatorial trough. For the sake of argument, focus your attention on the two belts of subtropical high pressure over the eastern Pacific Ocean—one in the Northern Hemisphere, one in the Southern Hemisphere. Logically, a zone of relatively low pressure—in other words, a trough (the equatorial trough)—must lie between these subtropical highs.

To understand why these highs populate the subtropical oceans, glance back to the cross section of the Hadley Cells in Figure 10.11. As lofty parcels of tropical air spread poleward just under the high tropical tropopause, they encounter a high-altitude traffic jam around 30° latitude. You can clearly see

FIGURE 10.13 A color-enhanced infrared satellite image at 15 UTC on August 11, 2015. The ITCZ is the broken band of convective clouds just north of the equator (courtesy of NOAA).

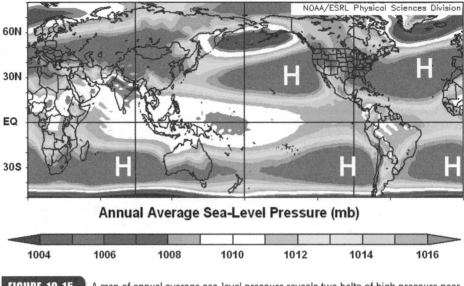

Annual Average Sea-Level Pressure (mb)

| 1004 | 1006 | 1008 | 1010 | 1012 | 1014 | 1016 |

FIGURE 10.15   A map of annual average sea-level pressure reveals two belts of high pressure near latitudes 30°N and 30°S (courtesy of NOAA).

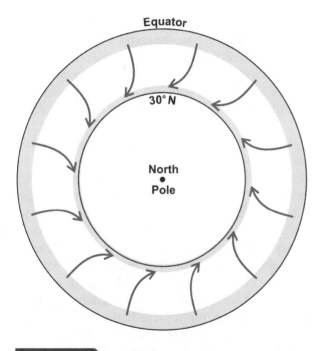

FIGURE 10.16   An idealized portrayal of the upper branch of the Hadley circulation. Air flowing poleward at high altitudes from the broad circle representing the equator is forced to "squeeze" into the smaller circle representing 30° latitude. With the Coriolis force deflecting air eastward, high-level air converges and starts to pile up.

inevitable traffic problems in Figure 10.16, which is a plan view (looking down on the North Pole) that shows the idealized upper branch of the Northern Hemisphere's Hadley circulation. Think of the outer ring as a latitude circle near the equator, while the inner ring represents latitude 30°N. Like a four-lane

highway narrowing to two lanes, parcels of air begin to pile up as they squeeze northward into increasingly smaller latitude circles. Note also how the Coriolis force starts to deflect parcels to the right, helping to halt their northward advance and thereby determining where the traffic jam gets bumper to bumper.

Indeed, the latitude at which the bottleneck develops is determined by the rotation rate of the earth (because the magnitude of the Coriolis force depends, in part, on the planet's rotation rate). If the rotation rate were significantly slower, the traffic jam would occur farther poleward. On the other hand, if the earth's rotation rate were much faster, the high-altitude traffic jam would take place closer to the equator. As it is, parcels piling up at high altitudes near latitude 30°N and 30°S creates convergence aloft, adding weight to local air columns and establishing the two belts of subtropical high pressure.

Like the nomadic nature of the ITCZ, the positions of the subtropical highs shift with the seasons. For example, the North Atlantic's subtropical high assumes an average position near the Azores Islands off the northern African Coast during winter (see the shadowy "H" in Figure 10.17). At this time of year, this anticyclone is commonly referred to as the **Azores high**. During the warm season, the high's average position shifts westward toward the middle of the Atlantic (see the representative analysis of sea-surface pressure in Figure 10.17). Sometimes, a "sub-center" of high pressure associated with the sprawling Atlantic subtropical high will take up residence near the island of Bermuda. As a result, this

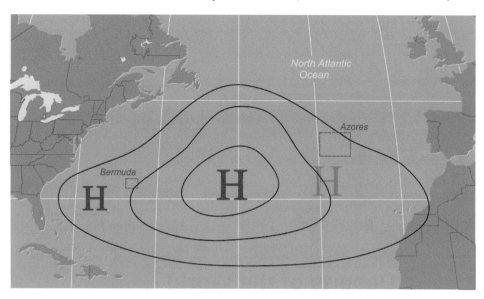

FIGURE 10.17 A schematic of average sea-level pressure showing typical summer positions of the Atlantic subtropical high-pressure system and a sub-center near the island of Bermuda (the Bermuda high). The shadowy "H" just to the south of the Azores marks the average winter position of the Atlantic subtropical high.

sub-center of the Atlantic subtropical high-pressure system commonly goes by the name **Bermuda high**. Those who live in the eastern half of the United States have probably heard weather forecasters invoke the Bermuda high while explaining very warm and humid conditions during summer. When its center lies off the East Coast of the United States, the Bermuda high's broad, clockwise wind circulation pumps maritime tropical air northward from the Gulf of Mexico.

How are the bottlenecks at 30° latitude resolved? After all, if parcels caught in the increasingly weighty traffic jams couldn't exit air columns near latitude 30°, surface pressures would increase without limit. The even-keeled atmosphere just doesn't allow such extremes. So, with their poleward paths halted and the barrier tropopause just above, parcels take the only off-ramp available—they sink in the centers and on the eastern flanks of the subtropical highs.

Recall that sinking parcels compress and warm as they descend toward the lower troposphere. High-altitude air parcels are typically unsaturated, given that any cirrus clouds streaming from the tops of ITCZ-generated thunderstorms have usually dissipated long before the upper branches of the Hadley Cells ever reach the subtropics. As a result, the warming rate of subsiding parcels is nearly dry adiabatic, and the air reaching 850 mb or so is very warm and very dry. The stage is now set for intense solar heating across subtropical latitudes and very high surface temperatures during the daytime.

Given the large-scale subsidence associated with subtropical highs, it should come as no surprise that precipitation in their vicinity tends to be sparse, as indicated by the minuscule average annual precipitation rates shown in Figure 10.18. On these satellite-derived measurements, the areas in white indicate relatively low precipitation rates (generally less than 0.5 mm/day,

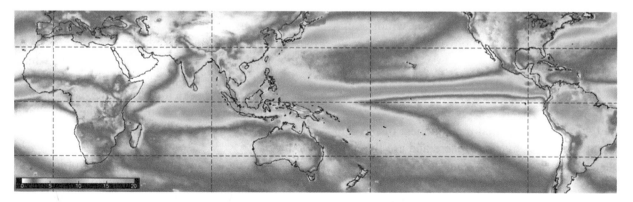

FIGURE 10.18 Average annual precipitation rates, expressed in millimeters per day, computed from the period 1998 to 2011 by the TRMM satellite. Areas of white, which are common over the subtropics in both hemispheres, indicate relatively low precipitation rates (courtesy of NASA).

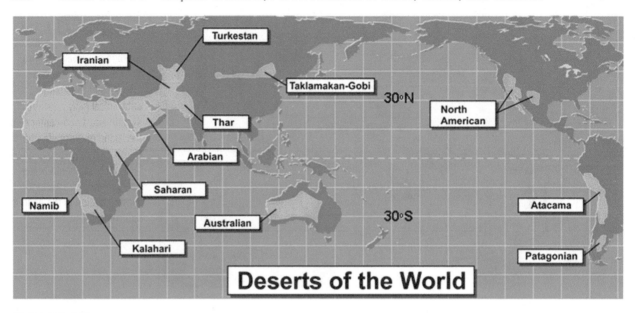

FIGURE 10.19   Many of the major deserts of the world lie near latitudes 30°N and 30°S.

which amounts to less than 8 inches/year), and tend to correlate well with the location of subtropical highs. Indeed, many of the major deserts of the world lie fairly close to latitudes 30°N and 30°S, as shown in Figure 10.19. The Sahara Desert in north-central Africa and the Great Australian Desert are prime examples.

On reaching the lower troposphere, some of the air subsiding in the vicinity of the subtropical highs then branches poleward and some branches equatorward (recall Figure 10.11). This equatorward flow completes the tropical travel tour of recycling air parcels in the Hadley Cells, while other parcels take a more extended tour toward higher latitudes. We'll discuss the impacts on mid-latitude weather patterns from air flowing poleward from the subtropical highs in Chapters 13 and 14.

The equatorward flow of air in the two Hadley Cells constitutes the system of trade winds. Let's explore the trades in a bit more detail.

Completing the Hadley Circuits: Jack-of-All-Trades

The trade winds are a major component of the general circulation of the atmosphere and are the most consistent surface wind system on the planet (particularly as it relates to wind direction)—no wonder that early explorers sought out the trades to navigate their sailing vessels.

The trades occupy most of the tropical oceans, blowing from the subtropical highs toward the ITCZ.

As with all winds, the direction of the trades is modified by the Coriolis force. Air moving equatorward from the subtropical oceans in the Northern Hemisphere is deflected to the right (which, in this case, is westward). Thus, the trade winds in the Northern Hemisphere are generally northeasterly. In the Southern Hemisphere, air moving equatorward from the subtropics is deflected to the left (also westward), resulting in generally southeasterly trade winds in that hemisphere.

Figure 10.20 shows annual average wind vectors (and wind speeds, color-coded in m/s) over the Atlantic and much of the Pacific Ocean within 40° of the equator, thus encompassing the tropics and subtropics. Even averaged over several decades, the footprint of the northeast trades in the Northern Hemisphere and southeast trades in the Southern Hemisphere is very evident, particularly between roughly 5° and 25° latitude in each hemisphere. Average speeds of 7 m/s (the yellow, about 16 mph) or higher, which most folks would describe as a decent breeze, cover large expanses of ocean in that latitude band, indicative of the persistence and strength of the trades (and the relatively weak friction) over these vast water bodies.

Initially, air flowing away from a subtropical high has low relative humidity, a consequence of warming during the prolonged descent associated with the anticyclone. Thus, as this air heads equatorward, there is a large vapor-pressure gradient between the warm sea surface and the initially "dry" trade winds.

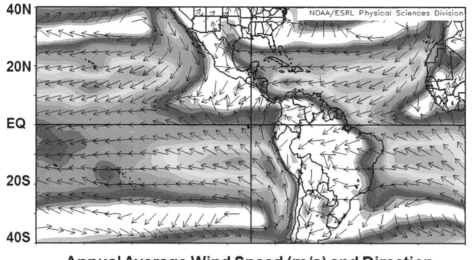

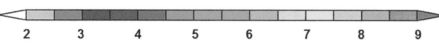

Annual Average Wind Speed (m/s) and Direction

FIGURE 10.20 Average wind direction and speed in the tropics and subtropics, derived from data observed during the period 1981–2010. Wind speeds are color-coded, in m/s. The persistent trade winds are easy to see over the oceans in both the Northern and Southern Hemispheres, blowing with greatest speeds between approximately 5° and 25° latitude in both hemispheres.

FIGURE 10.21 Hurricane Hunter aircraft, Miss Piggy, flies over trade-wind cumulus on the way to a developing storm (courtesy of NOAA's Hurricane Research Division).

In response, this air moistens via evaporation, and by the time parcels traveling in the trades reach the ITCZ, they are sufficiently moistened to fuel showers and thunderstorms along the ascending branch of the Hadley Cells.

Along the way, fair-weather cumulus clouds are routinely observed in the belt of trade winds, a visual clue of this moistening process (see Figure 10.21), and these moistened trades can also produce precipitation before they reach the ITCZ. The most substantial rains fall when the trades are forced to rise by topography. For example, consider the pattern of average annual rainfall on the Big Island of Hawaii, shown in Figure 10.22a. Note that the heaviest rain, in excess of 100 inches, falls on the northeast side

of the island where the sloping terrain (see a topographic map of the island in Figure 10.22b) ensures that the moist trades are abruptly lifted. Interestingly, the peaks of Mauna Loa and Mauna Kea are not the wettest areas—instead, the heaviest rain falls on the windward slopes of these mountains. By the time the winds reach the summits and areas to their leeward side, most of the moisture has been rained out.

At the start of this chapter, we mentioned that the tropics export heat energy to the mid-latitudes in an attempt to smooth out large temperature contrasts between the polar regions and low latitudes. The role that the trades play in this transport may be a bit surprising to you. That's because the trades actually transfer heat energy *toward the equator*.

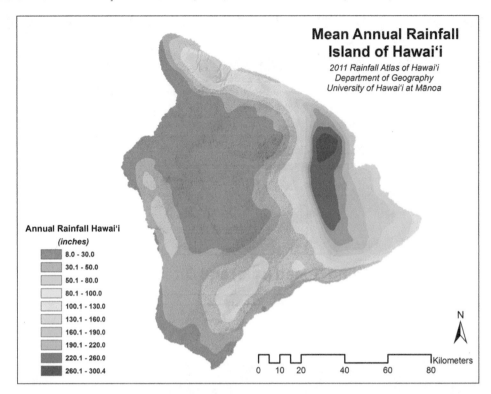

(a)

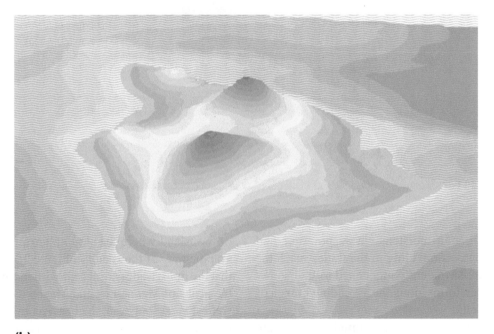

(b)

FIGURE 10.22 (a) Mean annual rainfall on the island of Hawaii (courtesy of Giambelluca, T.W., Q. Chen, A.G. Frazier, J.P. Price, Y.-L. Chen, P.-S. Chu, J.K. Eischeid, and D.M. Delparte, 2013: Online Rainfall Atlas of Hawai'i. Bull. Amer. Meteor. Soc. 94, 313–316, doi: 10.1175/BAMS-D-11-00228.1); (b) A three-dimensional topographic map of the island of Hawaii. Greens are the lowest elevations, transitioning to tan and brown as elevations increase. The peaks of Mauna Loa and Mauna Kea (which is the peak farther north) reach above 13,500 ft.

This transport is largely in the form of latent heat associated with the evaporation over the great expanse of oceans covered by the trades.

Certainly, at first glance, the assertion of "an equatorward transport of heat energy" by the trades seems to be completely at odds with an export of heat energy from the tropics. How could both be true at the same time? We'll begin our answer with the observation that the moisture picked up by the trades serves as fuel for the tall cumulonimbus clouds that populate the ITCZ. In turn, the large release of latent heat inside rising air parcels keeps them positively buoyant to great altitudes.

As high-flying air parcels start their poleward trek toward the subtropics, they are practically devoid of water vapor, having lost most of their moisture to net condensation during their long ascent. Thus, when they reach the subtropics and descend, they warm nearly dry adiabatically. In the end, they heat up so much that their temperatures become higher than they were at the same altitude during their ascent over the ITCZ. That's the power of latent heat! In turn, air circulating poleward on the western flanks of the subtropical highs transports this heat energy to the middle latitudes (where mid-latitude weather systems pick up the baton and continue the process of circulating warm air poleward—more details in Chapter 13).

So the most intriguing aspect of the trade winds' role in the global energy budget is that a large portion of the heat energy exported from the tropics can be traced to evaporation of warm ocean water into air parcels moving with the trades. Of course, ocean currents also transport heat energy out of the tropics. So do relatively compact hurricanes and tropical storms, although their contribution pales in comparison to other means of transporting heat energy.

In the final analysis, the reliable trade winds are truly a "jack-of-all-trades." Having completed the entire circuit of the Hadley Cells, there's still one other persistent feature worthy of our attention—the subtropical jet stream.

The Subtropical Jet Stream: An Ice-Skating Application

While flying westward toward Japan and other Pacific islands during World War II, American pilots reported ground speeds dramatically lower than the aircraft's indicated air speed. Flying at such reduced speeds relative to the ground could have meant only one thing—one lollapalooza of a headwind! Although World War II pilots made very little westward progress on some of their bombing missions, they had made a serendipitous discovery—the **subtropical jet stream** ("STJ", for short). As it turned out, the subtropical jet stream was the last major tropospheric feature of the general circulation to be discovered by direct human observation.

Figure 10.23 shows the average wind speed and direction at 200 mb (about 12,000 m) over Asia and the Pacific Ocean during Northern Hemisphere meteorological winter (December to February). You guessed it—the narrow ribbon of fast 200-mb winds near latitude 30°N marks the average winter position of the STJ. As Figure 10.23 indicates, the subtropical jet stream is very strong over the western Pacific region, where it sometimes achieves speeds over 90 m/s (200 mph) during winter.

To explain the existence of the speedy subtropical jet stream requires a principle known as the conservation of angular momentum. It's time for a brief sidebar.

You've probably used the term momentum before, but perhaps not in its formal scientific context. When

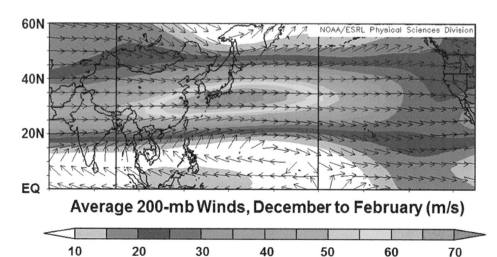

Average 200-mb Winds, December to February (m/s)

FIGURE 10.23 The average wind speed (color-coded in m/s) and wind direction (arrows) at the 200-mb level over Asia and the western Pacific Ocean during meteorological winter (December through February). Note the strong signal from fast winds near 30°N, marking the average position of the subtropical jet stream (courtesy of NOAA).

considering an object in motion, **momentum** is the product of its mass and its speed. With this in mind, let's head to the ice rink. A skater moving in a straight line has **linear momentum**. To perform a whirling spin on the ice, she transitions from this straight-line motion into a spiraling trajectory, converting linear momentum into **angular momentum**.

Like linear momentum, angular momentum depends on the object's mass and its speed, which in this case would be called angular velocity (the rate of spin). However, angular momentum also depends on the distance of the object from the axis about which it's rotating. In fact,

Angular momentum = constant × mass
× rate of spin × distance from axis of rotation

An ice skater starts the conversion from linear momentum to angular momentum with her arms outstretched. In quick fashion, she draws her arms inward close to her body, thereby increasing her rate of spin as her center of mass gets closer to the axis of rotation. The result is a crowd-pleasing, blurring spin.

Formally, angular momentum is conserved (that is, does not change) as long as there is no external force that exerts a net twist or "torque" on an object. Under that assumption, and applying the mathematical relationship above, we conclude that an ice skater's spin must quicken as the distance between her center of mass and her axis of rotation decreases.

So how does this apply to the subtropical jet stream? Consider parcels of air moving poleward in the high-altitude branches of the Hadley Cells. Figure 10.24, a view of the Northern Hemisphere looking down on the North Pole, shows the hypothetical trajectory of such an air parcel. Note that the trek toward the subtropics is not a direct flight. Instead, as the parcel moves away from equatorial regions, the Coriolis force kicks in and deflects the parcel to the right (eastward here). As a result, an air parcel may circle the globe at least once between its equatorial starting point and subtropical finish.

Okay, let's get back to the notion that air parcels tend to conserve angular momentum. This requires that an air parcel's angular velocity must increase as its distance from the Earth's axis of rotation decreases. On Figure 10.24, the Earth's axis of rotation passes straight down through the North Pole. Thus, as a high-altitude air parcel spirals away from the tropics toward the subtropics, its speed must increase. In fact, if we do the math, we find that air starting from rest (relative to the Earth's surface)

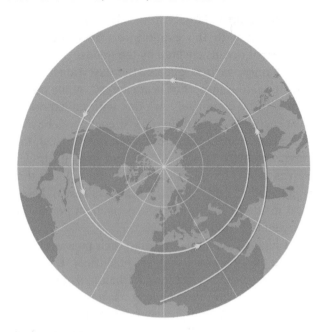

FIGURE 10.24 A typical trajectory of a parcel moving poleward in the upper branch of the Hadley Cell during Northern Hemisphere winter (December to February).

high over the equator will reach latitude 30°N with an eastward speed of about 134 m/s (300 mph)!

But the subtropical jet stream never attains such breakneck speeds. That's because parcels do not completely conserve their angular momentum. Tall mountains and towering cumulonimbus, for example, exert some drag on air parcels moving poleward in the upper branches of the Hadley Cells, slowing them down. Regardless, it is fair to say that air parcels *tend* to conserve angular momentum as they spiral inward toward the earth's axis of rotation.

What's so special about latitude 30°? As it turns out, the rate of rotation of the earth, manifested through the Coriolis force, largely determines the poleward extent of the STJ, so its *average position* is relatively fixed. If the earth's rotation rate were to increase, the subtropical jet would set up at a lower latitude (and vice versa).

Now let's address seasonal variations in the subtropical jet stream. Observations show that the STJ is much stronger during winter than summer. Figure 10.25 compares average 200-mb winds during Northern Hemisphere winter and summer in the latitude band between the equator and 50°N. The STJ is marked with a solid black arrow. As Figure 10.25b shows, for all practical purposes, the summer STJ is barely detectable, except for a modest streak of relatively weak 200-mb winds west of Africa.

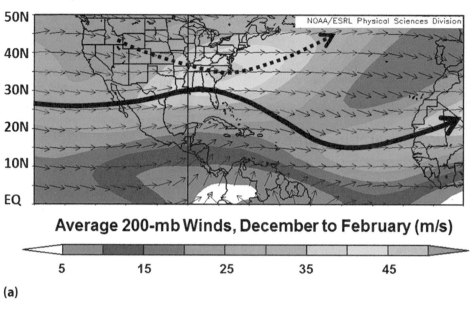

Average 200-mb Winds, December to February (m/s)

5   15   25   35   45

(a)

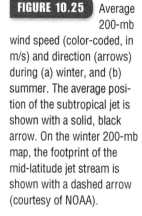

FIGURE 10.25 Average 200-mb wind speed (color-coded, in m/s) and direction (arrows) during (a) winter, and (b) summer. The average position of the subtropical jet is shown with a solid, black arrow. On the winter 200-mb map, the footprint of the mid-latitude jet stream is shown with a dashed arrow (courtesy of NOAA).

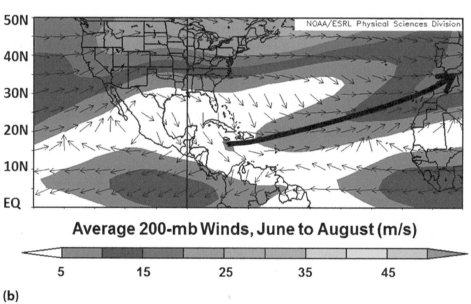

Average 200-mb Winds, June to August (m/s)

5   15   25   35   45

(b)

At this point, if you think back to the discussion about the subtropical jet and conservation of angular momentum, you might be saying "Whoa there, Nellie! Don't we have a contradiction on our hands?" Doesn't the upper branch of the Hadley Cell reach farther poleward in summer (compared to winter)? And wouldn't that mean the STJ would be faster in summer because parcels of air have spiraled closer to the Earth's axis of rotation?

We turn to temperature patterns to explain this apparent inconsistency (after all, fundamentally, temperature gradients produce pressure gradients and thus wind). During summer, intense solar heating over the land masses in the Northern Hemisphere's subtropical region means that these latitudes get much hotter, on average, than equatorial regions (see Figure 10.26). Thus, during the warm season in this latitude band, the typical north-south temperature gradient is reversed—that is, it's actually warmer to the north! And that's the key, because such a reversal in the temperature gradient would favor an easterly wind. In reality, what happens is simply that the westerly STJ is dramatically weakened during summer, so the STJ does not play as important a role in the overall weather pattern. During winter, however, the subtropical jet can push northward and play a role in powerful winter storms that affect the East Coast of the United States (more details in Chapter 16).

We have now seen many ways that uneven heating disrupts the symmetry of the earth's general

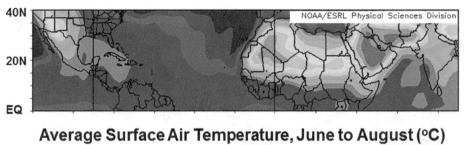

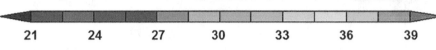

FIGURE 10.26 Average surface air temperatures during June, July, and August show that some of the hottest areas lie over land near latitude 30°N. Intense daytime heating so far from the equator disrupts the idealized Hadley circulation and weakens the subtropical jet during the summer (courtesy of NOAA).

circulation. Though rooted near the geographical equator, the equatorial trough and the Hadley Cells move north and south like migratory birds in pursuit of the summer hemisphere. Nowhere on earth does this migration lead to more dramatic consequences than the Indian subcontinent.

The Asian Summer Monsoon: A Gigantic, Long-Lived Sea Breeze

A quick glance at the Indian subcontinent shows that much of it lies in the latitude band between 10° and 25°N. Based on Hadley's original model (see Figure 10.10), we might expect this latitude band to be dominated by northeast winds. As it turns out, almost the exact opposite is true for much of the warm season.

Although India is hot during the summer, daytime temperatures at many locations across the country actually peak in May. Figure 10.27 shows average monthly maximum temperatures at Mumbai (western India,

latitude 19°N), Bhopal (central India, 23°N), and Kolkata (eastern India, 22°N). Note that, in general, average high temperatures decrease from May to August. This drop-off is particularly noticeable in Bhopal, where the average high temperature peaks at 102°F (39°C) in May and falls to 82°F (28°C) in August!

In response to this extreme heating across India (and much of the Asian subcontinent) during May, a gigantic sea breeze develops (see Figure 10.28). It is nothing like the typical sea breeze that forms at the shore during the day and then disappears at night. This sea breeze has a continental spatial scale and lasts for months, typically from late May to October. The Tibetan Plateau helps to enhance this continental-scale sea breeze because ocean air streams inland in order to replace the rising air currents over this expansive, high-level heat source [the plateau's average elevation is about 5000 meters (16,400 ft)].

Moist air streaming inland from the Indian Ocean, the Bay of Bengal, and the Arabian Sea

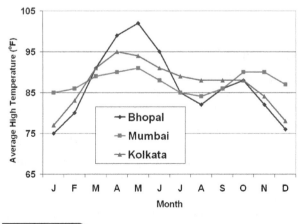

FIGURE 10.27 Average monthly maximum temperatures (in °F) at three Indian cities: Bhopal, Mumbai, and Kolkata. Note that average high temperatures tend to peak in May.

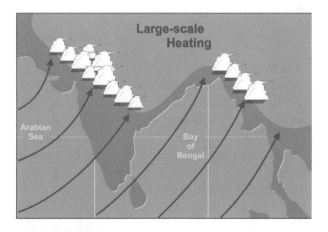

FIGURE 10.28 A gigantic sea breeze develops over India in late May or early June, paving the way for recurrent rains that are a hallmark of the summer monsoon.

fuels recurrent rains that are the hallmark of the **summer monsoon** of Southeast Asia. Figure 10.29 shows the long-term cumulative average rainfall at Mumbai, Bhopal and Kolkata. Note that, on average, the summer monsoon really gets going in June, and heavy rainfall really adds up in July and August. Mumbai averages more than 20 inches (51 cm) in June and July, and nearly that much in August! This wet period corresponds to the time when winds persistently blow inland from the warm surrounding sea. The average start date of the summer monsoon in India is shown on the left side of Figure 10.30. Spates of rain can come in heavy bursts, but it does not rain all the time. Indeed, there can be dramatic

spells of dry weather that last on the order of a week or more. Still, the dominant characteristic of the summer monsoon is recurrent rains, enough that Mumbai's average of 85 inches of rain per year is more than any city in the lower 48 U.S. states.

The heaviest rains associated with the summer monsoon typically occur in mountainous regions, where moisture-laden air is abruptly forced to ascend the sloping terrain. Rainfall can grow to biblical proportions in the mountains of northern India. Cherrapunji, located in the Khasi Hills of northeast India, owns the world record for most rainfall in one calendar year, 2299 cm (905.1 in), and the record for most rainfall in one month, 930 cm (366.1 in). Most of Cherrapunji's annual rainfall comes in a matter of a few months when the summer monsoon is at its peak.

For the record, the word *monsoon* does not mean "rain," but instead can be traced to an Arabic root that means "season." Thus, a pronounced seasonal shift in prevailing winds constitutes a monsoon. Judging from the running rainfall totals in Figure 10.29, the prevailing winds shift in October, heralding the start of the **winter monsoon**—a period when northerly (offshore) winds carry dry continental air across the Indian subcontinent and little rain falls. The average end date for the summer monsoon is shown on the right side of Figure 10.30.

Notice in Figure 10.28 that the gigantic sea breeze that provides the moisture for the summer monsoon across India blows generally from the southwest over the northern Indian Ocean. This prevailing

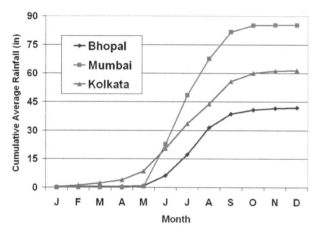

FIGURE 10.29 The running tally of average rainfall (in inches) at three Indian cities: Bhopal, Mumbai, and Kolkata. Note the steep climb from May to September, the signature of the summer monsoon.

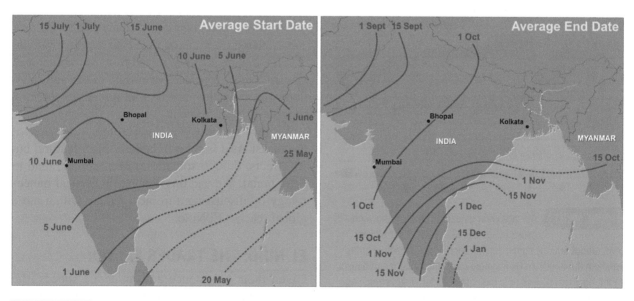

FIGURE 10.30 The average start (left) and end (right) dates for the summer monsoon.

wind has an interesting origin. The heating of the Indian subcontinent that sets the summer monsoon into motion is so intense that the equatorial trough (regionally, it's called the **monsoon trough**) shifts relatively far north to India (revisit Figure 10.12b). In the process, moist air associated with the south-easterly trades in the Southern Hemisphere gets drawn across the geographical equator (now that's a *gigantic* sea breeze). Once in the Northern Hemisphere, these moist southeast winds are deflected to the right (eastward) by the Coriolis force, creating a channel of relatively strong southwesterly breezes that supplies moisture for Southeast Asia's summer monsoon. Figure 10.31 shows the average wind direction and speed at the 925-mb level (about 800 meters) during June. The core of the fastest winds at 925 mb marks the **Low-Level Somali Jet**, a channel of low-altitude, speedy winds that helps to rapidly transport moist air toward India. This chart confirms that the southeasterly trades in the Southern Hemisphere are connected to the Asian summer monsoon. By the way, notice that the core of this low-level jet stream passes over its namesake, Somalia.

Sometimes, low-pressure systems form over the Bay of Bengal and move northwestward parallel to (and just south of) the monsoon trough. These **monsoon depressions** provide a lion's share of the rain in northeastern India, but also lead to some of deadliest flooding there. Monsoon depressions are typically a few hundred kilometers in diameter and, in addition to generating very heavy rains, feature maximum sustained winds of 30–35 mph. In a typical summer, six to seven monsoon depressions form over the Bay of Bengal and track westward across India. They rarely develop into the equivalent of a hurricane or tropical storm because the northern part of the Bay of Bengal is so narrow that most monsoon depressions spend relatively little time over warm water.

Compared to the big-time summer monsoon in India, the summer monsoon that occurs in the southwestern United States in July and August (more formally, the **North American Monsoon**) is strictly minor league. That's because there is no dramatic seasonal shift in wind. Granted, the southwestern United States heats up big-time during summer [for example, the average high in Phoenix in July is around 40°C (105°F)]. But the transition to southerly winds from the Gulf of California and the eastern tropical Pacific Ocean is rather subtle. In fact, meteorologists in Phoenix used to announce the onset of the North American Monsoon after the average daily dew point reached 13°C (55°F) or higher on three consecutive days. This somewhat "inside-baseball" criterion confused the general public, especially the people who mistakenly assumed that rain must fall in order for there to be a monsoon. Today, meteorologists in the Southwest United States simply define the *monsoon season* as June 15 to September 30.

With regard to the issue of rain, humid air invading the Southwest fuels showers and thunderstorms, particularly over high elevations (see Figure 10.32) where the uneven heating near mountaintops promotes rising currents of moist air and heavy rain. When moisture is drawn inland from the remnants of eastern Pacific tropical systems, showers and thunderstorms typically become widespread, heightening the risk of flash flooding.

Without reservation, the summer monsoon of Southeast Asia is a formidable tropical feature, but its effects, though sometimes catastrophic, tend to be regional. We end this chapter on tropical meteorology by discussing a tropical feature with a more global reach—El Niño.

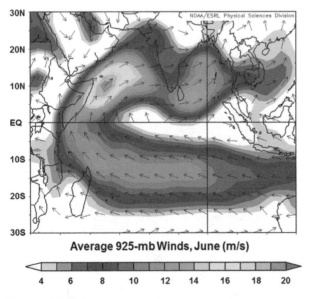

Average 925-mb Winds, June (m/s)

4   6   8   10   12   14   16   18   20

FIGURE 10.31 Average wind direction (arrows) and wind speed (color-coded in m/s) at the 925-mb level (about 800 m) during June. Note how southeasterly trade winds in the Southern Hemisphere cross the equator and deflect to the right (east). Also note the core of faster winds stretching across the Arabian Sea, a feature called the Low-Level Somali Jet (courtesy of NOAA).

EL NIÑO: THE TRADES FALTER

Given their persistence and moderate speeds, the trades move ocean water around at will. Over the vast tropical Pacific, these hard-working northeasterly and

FIGURE 10.32 Scattered thunderstorms erupt on August 14, 2014, over Arizona in response to a monsoonal flow of moist air from the Gulf of California and the eastern tropical Pacific Ocean. Meanwhile, easterly winds at higher altitudes can also transport moisture from the Gulf of Mexico over the Southwest States (courtesy of NOAA).

southeasterly winds pile up warm water over the western part of the basin near Southeast Asia, where sea levels are, on average, 20–30 cm (8–12 inches) higher than the levels in the eastern tropical Pacific. Figure 10.33 is a cross-section that is representative of the average state of the tropical Pacific Ocean. This average state is characterized by a mound of relatively deep and relatively warm water in the western Pacific.

At irregular intervals every few years, however, the trade winds slacken and this modest mound of relatively warm water sloshes back eastward, setting the stage for an important oceanic warming over the central and eastern tropical Pacific called El Niño.

The Equatorial Eastern and Central Pacific: Not as Warm as You Might Think

Because El Niño (and its counterpart La Niña) is defined in terms of anomalies of ocean temperature and surface winds, it's important to first discuss the average state of the tropical Pacific Ocean and overlying atmosphere.

The persistently westward-moving trades are in the water-moving business. The drag exerted by the trades on the ocean sets the topmost layer of water into motion, typically at a conservative pace of a few kilometers per day. As with any object moving over great distances for long periods of time, the Coriolis force gets into the act, deflecting this topmost layer of moving water to the right in the Northern Hemisphere (to the left in the Southern). Figure 10.34 shows the ensuing chain of events for the Northern Hemisphere.

In turn, water below the topmost layer is set in motion, but at a slower pace as a result of friction. Nonetheless, the Coriolis force deflects this slower-moving water to the right of the direction of the topmost layer. This domino effect continues to spread deeper and deeper. As each successive layer is set into motion, it moves to the right of the direction

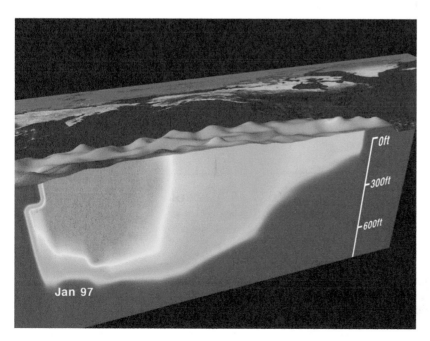

FIGURE 10.33 Average sea-surface heights and sea-surface temperatures with depth across the tropical Pacific Ocean during January 1997—South America is on the right, the far western Pacific on the left. Conditions that month were representative of average conditions in the tropical Pacific. Red indicates relatively warm water, while blue represents relatively cold water. Contrast the subtle mound of relatively high sea-surface heights and the deep layer of relatively warm water in the western Pacific Ocean with the relatively low sea-surface heights and relatively cool water in the eastern Pacific (courtesy of NASA).

Jan 97

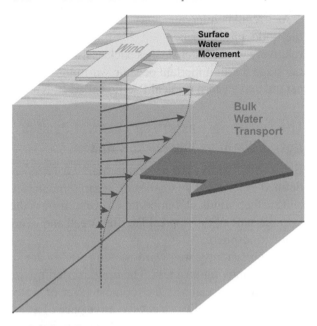

FIGURE 10.34 Surface water set into motion by the wind is deflected to the right of the direction of the wind by the Coriolis force (in the Northern Hemisphere). In turn, surface water sets a layer of water beneath it into motion. The Coriolis force deflects this slower-moving water to the right of the direction of the surface water. This domino effect continues to spread deeper until the initial energy imparted to the water by the wind is dampened out by friction at a depth of about 100 m. The resulting spiral of water with depth is called the Ekman spiral. The bulk of the water set into motion by the wind moves 90° to the right of the wind (in the Northern Hemisphere).

of the layer above it, creating a spiraling current of slow-moving water that weakens with increasing depth to about 100 m (about 330 ft), where the effects of the wind finally vanish.

This curling pattern of water motion with depth is called the **Ekman spiral** in honor of the Scandinavian physicist V. Walfrid Ekman, who published a scientific paper on this phenomenon in 1902. Ekman made calculations of the volume of the flow of water with depth and found that the bulk of water set into motion by the wind moves at an angle of 90 degrees to the right of the wind in the Northern Hemisphere (see Figure 10.34). This bulk-water transport is called the **Ekman transport**. In the Southern Hemisphere, the Ekman transport is 90 degrees to the left of the wind.

Figure 10.35a is a schematic of the Ekman transport (blue arrows) generated by the trade winds (tan arrows) over the equatorial Pacific Ocean. Note how the Ekman transport creates a pattern of divergence in the upper layers of the ocean near the ITCZ. The ocean, like the atmosphere, "abhors" unevenness and reacts by bringing colder, nutrient-rich waters up from the depths in order to replenish diverging surface waters (Figure 10.35b). This process is called **upwelling**.

The basic state of the tropical Pacific Ocean is often described by singling out the upwelling that

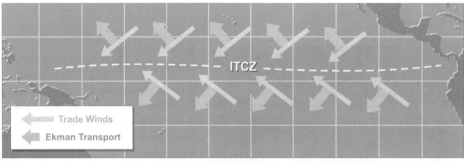

(a)

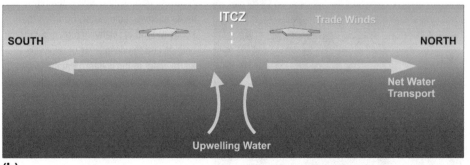

(b)

FIGURE 10.35 (a) The bulk-water transport (blue arrows) produced by the trade winds (tan arrows) acts 90° to the right of the trades in the Northern Hemisphere and 90° to the left of the trades in the Southern Hemisphere. (b) Water flowing away from the ITCZ via Ekman transport creates water divergence in the equatorial Pacific. In order to replace diverging surface waters, cold water upwells from beneath the ocean surface.

occurs in the far eastern part of the ocean, off the coast of western South America (specifically, off the coast of Peru). However, the trade winds also blow with gusto in the central tropical Pacific as you can see by referring back to Figure 10.20. Thus, in reality, there is a broad region of upwelling in the tropical Pacific, as suggested by Figure 10.35.

Still, the water that upwells off the coast of South America is typically cooler than the upwelling waters in the central Pacific. Revisit Figure 10.33 and note the temperature variations with depth in the eastern and central Pacific. Without reservation, the depth of the warm ocean layer is shallower in the eastern equatorial Pacific. Thus, the water that upwells to the surface from depths averaging 100 to 200 meters is coldest just off the coast of South America, as Figure 10.36 indicates, with sea-surface temperatures there only in the 21–27°C range (70–80°F), on average (a tad cool by tropical standards). However, this nutrient-rich chilly water fertilizes tiny ocean plants called phytoplankton, which serve as the base of an abundant and diverse community of marine life (and mammals that feed on marine life) in this region.

But these thriving communities teeming with life can change dramatically in a matter of months.

El Niño: Changing of the Guard

Although research aimed at identifying the precise cause of El Niño continues, its symptoms are well documented: The trade winds slacken in concert with warm water sloshing eastward from the western Pacific. Within months, the usually chilly, nutrient-rich surface waters off South America are replaced by warmer, nutrient-poor waters. In turn, plant, marine, and animal life starts to suffer. This anomalous warming of equatorial waters in the central and eastern Pacific is called **El Niño**, which means "The Child" in Spanish—a local name given to the temporary, warm reversal in the current of cold water off the Peru Coast that typically arrives shortly after Christmas each year. Local fishermen were keenly aware that this reversal occasionally developed into a robust, long-lasting warming.

Today, meteorologists typically use the term "El Niño" only to describe a major, protracted warming of the waters in the central and eastern equatorial Pacific Ocean. Figure 10.37 shows various regions in this part of the Pacific where sea-surface temperatures

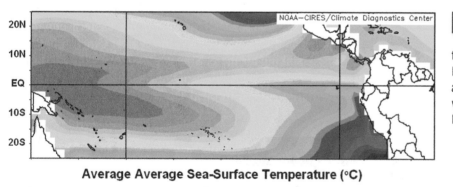

Average Average Sea-Surface Temperature (°C)

FIGURE 10.36 The annual average sea-surface temperatures (in °C) in the tropical Pacific. The coldest waters generally lie in the eastern Pacific and the warmest waters gather in the western Pacific (courtesy of NOAA).

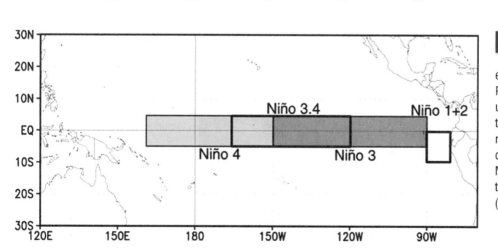

FIGURE 10.37 Regions of the equatorial central and eastern Pacific Ocean traditionally used to measure sea-surface temperature anomalies related to El Niño. An array of buoys covers these areas. Niño 3.4 is commonly used to define El Niño events (courtesy of NOAA).

are traditionally monitored. Niño 1+2, for example, is the area off the coast of South America in which local populations first recognized El Niño, while averages of measurements taken in Niño 3.4 are most commonly used to define El Niño events. During a strong El Niño, sea-surface temperatures can average 2.0–3.5°C (4–6°F) above average in Niño 3.4. To see an example, look back to the strong El Niño of 2015–16 documented in Figure 10.5.

Now you wouldn't want to "surf" the eastward slosh of warm water associated with the onset of El Niño. Indeed, this warm water undulates slowly eastward via an internal (submerged), slow-moving wave called an equatorially trapped Kelvin wave. Though the physics of this undulation are beyond the scope of this book, the end result is clear: In time, a deep layer of anomalously warm water spreads into the eastern equatorial Pacific Ocean (see Figure 10.38) Now, instead of the normal upwelling of cold, nutrient-rich water, there is weak upwelling of warm, nutrient-poor water. The detrimental impact on local ocean life can be staggering. Deprived of nutrients, phytoplankton soon diminish, and higher-order marine and mammal life leave the region in search of food.

El Niño's impact on weather needs some elaboration.

Weather Impacts from El Niño: Near and Far

In non-El Niño years, the mound of warm water in the western tropical Pacific keeps the overlying air warm, moist and buoyant, setting the stage for persistent currents of rising air and recurrent showers and thunderstorms. In contrast, upwelling relatively cool water in the eastern equatorial Pacific renders the overlying air cooler and more stable, resulting in a tendency for rainfall to be suppressed there. In addition, relatively dense air overlying cooler waters in the eastern tropical Pacific tends to maintain slightly higher surface pressures off the northwestern coast of South America (higher air density in the lower troposphere adds a little weight to local air columns). In contrast, low-level warm air overlying the warm water piling up in the western tropical Pacific helps to maintain slightly lower pressure there (see Figure 10.39a). In order to avoid any extreme increase (decrease) in sea-level pressure from unchecked surface convergence (divergence) of air around the low (high) shown in Figure 10.39a, the well-organized atmosphere adds upper and lower branches of horizontally flowing air, completing a cycle called the **Walker Circulation**.

We note that the Walker Circulation is an east-west circulation that is superimposed on the north-south circulation of the Hadley Cell. Given that the Hadley Cell favors rising air in equatorial regions (in the vicinity of the ITCZ), it's reasonable to look closer at what happens in the eastern equatorial Pacific simply because the Walker Circulation favors sinking air there. Sometime, the Walker Circulation just restrains the rising branch of the Hadley Cell in the eastern tropical Pacific rather than forcing the

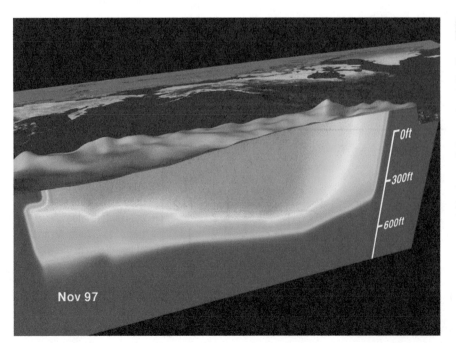

Nov 97

FIGURE 10.38 Sea-surface heights and sea-surface temperatures with depth across the tropical Pacific Ocean during November 1997, representative of conditions during a strong El Niño. Red indicates relatively warm water and blue represents relatively cold water. Note the higher sea-surface heights and the deep warm water in the eastern Pacific compared to the average state shown in Figure 10.33 (courtesy of NASA).

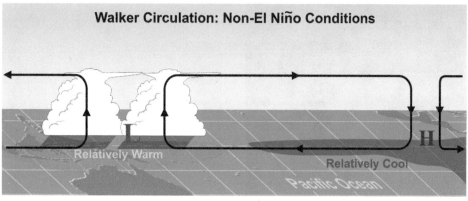

Walker Circulation: Non-El Niño Conditions

(a)

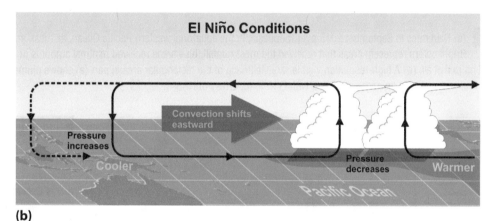

El Niño Conditions

(b)

FIGURE 10.39 (a) In non-El Niño years, relatively warm water in the equatorial western Pacific promotes lower surface pressure there, surface convergence, and rising air. Meanwhile, in the eastern equatorial Pacific, upwelling cold water fosters relatively high pressure, surface divergence, and a tendency for sinking air. Upper-level convergence over the eastern tropical Pacific and upper-level divergence over the western tropical Pacific complete the east-west Walker Circulation; (b) During an El Niño, surface pressures increase in the western tropical Pacific and decrease in the eastern Pacific. As a result, convection tends to shift eastward, and the Walker Circulation essentially reverses, aiding in the relaxation of the trade winds.

rising branch to reverse direction. However, in reality, the overall tendency during non-El Niño years is for rather gentle sinking motions to prevail over the eastern equatorial Pacific.

During an El Niño, however, the Walker Circulation dramatically reverses (see Figure 10.39b). This shift, called the **Southern Oscillation** (SO for short), leads to a conspicuous rising branch of air over the eastern tropical Pacific Ocean. As surface waters warm there, this rising branch is sustained by increasing low-level instability. Because the Southern Oscillation typically accompanies El Niño, the entire event is often called "ENSO" (short for El Niño Southern Oscillation). Regardless of name, ENSO is marked by a relaxation of the trade winds and an increase in showers and thunderstorms in the eastern tropical Pacific, where surface pressures tend

to decrease and evaporation rates and the exchange of heat between the sea and air increase. Upwelling continues off the coast of South America during an El Niño, but because of the greater depth of relatively warm water (revisit Figure 10.38), warm and nutrient-poor water upwells instead of colder, nutrient-rich water.

Meanwhile, over the western tropical Pacific where sinking currents of air dominate during an El Niño, rainfall dramatically decreases, heightening the risk of wildfires. During the autumn of 2015, with an El Niño in progress, anomalous dryness took hold in the western tropical Pacific. Figure 10.40a shows rainfall in September 2015 across Indonesia—bright colors represent areas that received the most rain while gray areas observed no rain at all that month. Figure 10.40b is

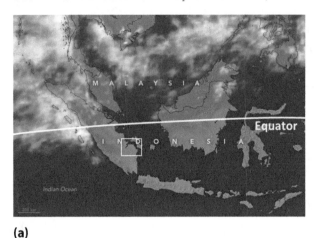

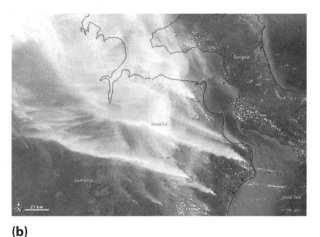

(a) **(b)**

FIGURE 10.40 (a) Rain rates in September 2015 over Indonesia, in the equatorial western Pacific Ocean, as measured by satellite. Bright colors represent areas that received the most rainfall, blue areas received minimal amounts of rain, while areas in gray observed no rain at all; (b) A high-resolution visible satellite image of the rectangular area in part (a) shows plumes of smokes from man-made fires worsened by dryness associated with El Niño (courtesy of NASA).

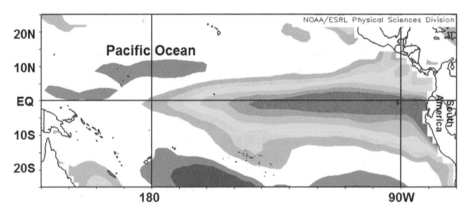

Sea-Surface Temperature Anomaly, Dec 1997 to Mar 1998 (°C)

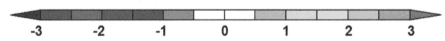

FIGURE 10.41 Departures from average sea-surface temperatures (in °C) for the period December 1997 to March 1998. Water temperatures averaged at least 3°C (5°F) above average in much of the central and equatorial Pacific Ocean, a clear signal of a major El Niño in progress (courtesy of NOAA).

a high-resolution visible satellite image that zooms into the rectangular area marked on Figure 10.40a. It shows plumes of smoke from wildfires over southern Sumatra (fire locations marked in red) in late September. Although some fires are part of typical land-clearing practices in this area, the dryness related to El Niño primed the region for an extreme fire season. In fact, by late October, six Indonesian provinces had declared a state of emergency due to the thick smoke and haze.

Because Indonesia is in the tropical Pacific, it's not surprising that changes in the Walker Circulation have a direct impact there. However, El Niño also has more far-reaching effects outside of the tropics that take the form of temperature and precipitation anomalies on seasonal time scales. One way that El

Niño's influence indirectly spreads to higher latitudes is via changes in the subtropical jet stream.

How does this happen? Given the anomalous warming of ocean waters in the central and eastern tropical Pacific Ocean during El Niño (especially a strong one), it should come as no surprise that the atmosphere over those waters also warms through a deep layer. This warming helps to enhance the north-south temperature gradient between the tropics and the mid-latitudes, and this helps to boost the strength of the subtropical jet stream. The strong El Niño of 1997–1998 provides a telling example of this ocean-atmosphere connection (Figure 10.41 shows sea-surface temperature anomalies during this El Niño). Figure 10.42a shows the long-term average (climatology) of 200-mb winds over the central and

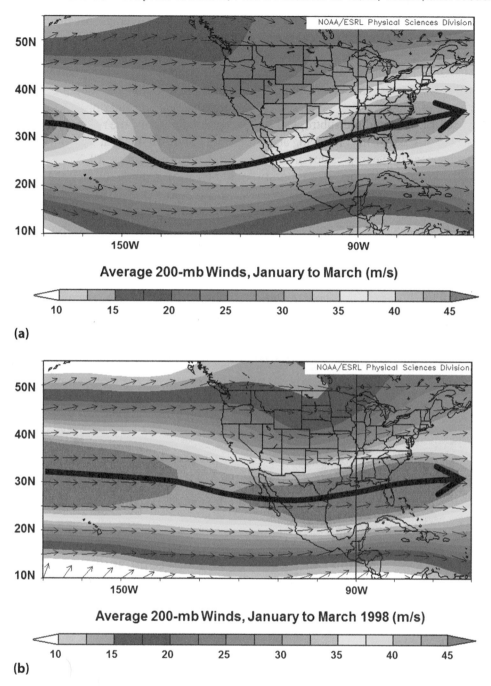

Average 200-mb Winds, January to March (m/s)

10   15   20   25   30   35   40   45

(a)

Average 200-mb Winds, January to March 1998 (m/s)

10   15   20   25   30   35   40   45

(b)

FIGURE 10.42 (a) The long-term average (climatology) of 200-mb wind direction and wind speed (color-coded in m/s) over the eastern North Pacific Ocean from January to March. The arrow shows the average position of the subtropical jet stream; (b) The average wind direction and wind speed at 200 mb from January to March 1998 when a strong El Niño was in progress. Again, the arrow marks the position of the core of the STJ. Note that wind speeds within the subtropical jet stream are stronger than the long-term average shown in part (a), and that the strongest winds of the STJ extend farther east toward the West Coast of North America (courtesy of NOAA).

eastern Pacific Ocean from January to March. The corridor of fast 200-mb winds indicated by the arrow represents the average position of the subtropical jet stream during these months. Compare this climatology to Figure 10.42b, which shows the average

200-mb winds from January to March 1998, during the strong El Niño. First, note that the wind speeds associated with this subtropical jet were generally higher than climatology. Second, notice how the core of the jet stream's fastest wind speeds reached farther eastward toward North America. This more active portion of the subtropical jet stream over the eastern Pacific helped to energize low-pressure systems off the California Coast, setting the stage for recurrent heavy rains (and heavy snow in the mountains) during that winter.

The link between El Niño and the subtropical jet stream demonstrates that, unlike the old saying about Las Vegas, what happens in the tropics doesn't necessarily stay in the tropics. Links between monthly and seasonal weather conditions in distant parts of the globe are called **teleconnections**, and some of the most newsworthy teleconnections start with El Niño. Consider Figure 10.43, which shows the typical temperature and precipitation anomalies that occur in Northern Hemisphere winter and summer when an El Niño is in progress. In North America, note the tendency for winters in central and western Canada to be warmer than average. Meanwhile, in the southern United States, the probability of cooler-than-average and wetter-than-average conditions increases during an El Niño winter—that's because the more active subtropical jet helps to energize low-pressure systems that develop and ripple eastward there, enhancing cloudiness and the risk of precipitation. Note also from Figure 10.43a that wetness is favored in southern California during El Niño winters—however, it's far from a given. In fact, the period from October 2015 to March 2016, during a strong El Niño, produced less than 60% of average precipitation in southern California. Clearly, from the teleconnection perspective, even all strong El Niños are not created equal (check out the commentary at the end of the chapter).

Figure 10.43b shows teleconnections with El Niño during the period June to August (meteorological

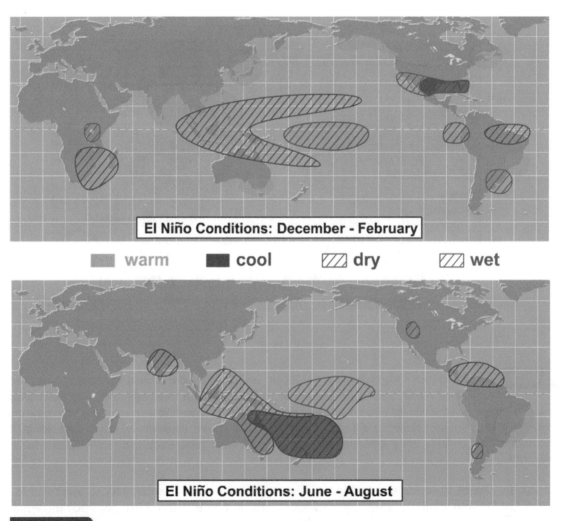

FIGURE 10.43 (Top) Typical large-scale temperature and precipitation anomalies that accompany an El Niño during Northern Hemisphere winter; (Bottom) Typical large-scale temperature and precipitation anomalies that accompany an El Niño during Northern Hemisphere summer (courtesy of NOAA).

summer in the Northern Hemisphere). Notice that nearly all of the hatched and colorized areas are in the Southern Hemisphere, where it's winter. In general, the impacts from El Niño in the summer hemisphere are not as pronounced as in the winter hemisphere, due in part to weaker temperature gradients in the summer hemisphere which reduces the overall response of the atmosphere to El Niño (for example, the subtropical jet stream is weaker in summer).

El Niños typically last from six months to a year. Then, for reasons still not completely understood, the waters of the central and eastern Pacific cool. The ocean may return to near its average state, or water temperatures there may, like a giant seesaw, swing the other way to cooler-than-average. The same huge patch of ocean that, months earlier, was several degrees warmer than average, may now be a few degrees below average. This flip side of El Niño is called **La Niña**.

Like El Niño, La Niña's cooler-than-average waters in the central and eastern tropical Pacific Ocean set in motion a rippling effect that can cause atmospheric temperature and precipitation anomalies in other parts of the world. Figure 10.44

shows La Niña teleconnections specifically for the United States during winter, for both temperature (Figure 10.44a) and precipitation (Figure 10.44b). The charts show the percentage increase in the risk of warm and cold extremes and wet and dry extremes. As an example of how to read these charts, note that in Texas, the risk of an unusually dry winter almost doubles when a La Niña is in progress, while in most of the Lone Star State, so does the risk of an unusually warm winter. Meanwhile, in northern North Dakota, La Niña brings an enhanced risk of an unusually cold and wet winter.

If you visit and explore the web site of NOAA's Climate Prediction Center, you'll discover that forecasters tend to lean heavily on established teleconnections from El Niño and La Niña when they craft their seasonal forecasts. When there is no signal from an El Niño or La Niña over the eastern tropical Pacific, the degree of difficulty associated with seasonal forecasting greatly increases.

Missing from this chapter on the tropics is an extensive discussion of hurricanes and tropical storms. As you will soon learn, they're in a class all by themselves.

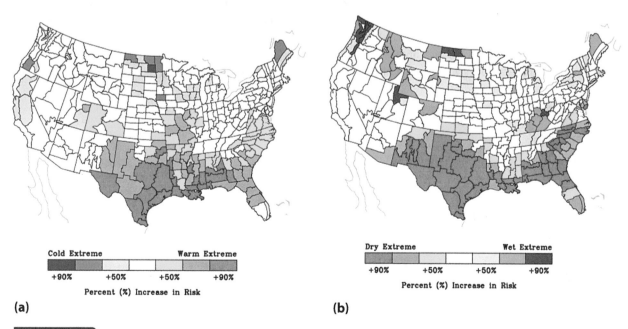

(a)   (b)

FIGURE 10.44   (a) Teleconnections in temperature in the United States during a La Niña winter. Below-average temperatures are favored over much of the upper Midwest and Northern Plains and parts of the West and northern New England, while chances increase for a warmer-than-average winter across the South; (b) Winter precipitation teleconnections in the United States during a La Niña winter. Dryness is favored in Texas and the Southeast, while there is an enhanced chance of above-average precipitation across several regions of the northern U.S. (courtesy of NOAA).

Weather Folklore and Commentary

What's in a Monster: The "Godzilla" El Niño:

In August of 2015, an oceanographer at NASA's Jet Propulsion Laboratory made news by calling the ongoing and potentially record-setting El Niño, "Godzilla." This tabloid headline was covered by the Los Angeles Times and instantly achieved a level of credibility. The media ran with it.

Such a sensationalistic description (NBC Nightly News crooned that "A monster El Niño is moving in our direction") fueled the public perception that El Niño was a "storm" that would send volley after volley of heavy rains into California, causing flooding, mudslides and other hardships similar to what happened during previous strong El Niños in the winter of 1982–1983 and 1997–1998. On the positive side, there was growing optimism that "Godzilla" would dramatically reduce the long-term drought conditions, particularly in southern California.

El Niños are not linear systems, however, meaning that the weather impacts from the El Niño of 2015–2016 wouldn't necessarily mirror the very stormy conditions in 1982–1983 and 1997–1998. Moreover, at the start of the rainy season for California (October–March), the state of sea-surface temperatures in the North Pacific Ocean was different than at the time of the two previous strong El Niños. Indeed, in mid-October of 2015, two additional "blobs" of relatively warm water—one to the south of Alaska's Aleutian Islands and the other to the west of Baja California—were present in the Pacific Ocean (see Figure 10.45). Such "competition" for control of the general weather pattern over the Pacific Ocean should have cast uncertainty on the seasonal forecast during the reign of the purported "Godzilla" El Niño.

However, it didn't. Not surprisingly, the early nicknaming of the 2015–2016 El Niño as "Godzilla" super-charged expectations, with the media disproportionately focusing on the rainy scenarios from past strong El Niños. However, a repeat was not in store. Measurements from 20 locations in southern California showed that less than 60% of the average precipitation fell from October 2015 to March 2016. In downtown Los Angeles, only 0.79 inches of rain were measured in February 2016, far below the monthly average of 3.20 inches and nowhere near the record 13.68 inches in February 1998.

Of course, hindsight is always 20-20. But here's some food for thought. Although teleconnections to El Niño and La Niña have proven useful for seasonal forecasting, ENSO is not the only game in town. Only a handful of strong El Niños have been observed and documented with modern-day tools, simply not enough information to provide a sound statistical basis for confidently making seasonal predictions, particularly of precipitation (which is notoriously much more difficult to predict than temperature).

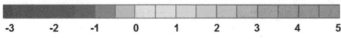

Sea-Surface Temperature Anomaly (°C), October 15, 2015

FIGURE 10.45 Sea-surface temperature anomalies (in °C) in mid-October of 2015 showed not only a strong El Niño (the relatively warm water in the central and eastern tropical Pacific) but several other areas of anomalously warm water in the Pacific that likely played a role in controlling weather patterns in the winter of 2015–2016 (courtesy of NOAA).

Tropical Weather, Part II: Hurricanes

LEARNING OBJECTIVES

After reading this chapter, students will:

- Be able to identify the ocean basins where tropical cyclogenesis occurs.
- Be able to identify the generic naming conventions used to describe the hierarchy of tropical cyclones in each of the basins (tropical depression, tropical storm, hurricane, etc.).
- Gain an appreciation for the history and protocol for generating the lists of specific names used in each of the ocean basins.
- Be able to assess the criteria for tropical cyclogenesis in the setting of real weather data.

- Understand the genesis, movement and weather associated with easterly waves.

- Gain an appreciation for why tropical cyclogenesis is rare in the South Atlantic Ocean.

- Understand how a positive feedback cycle plays a role in the development, maintenance and intensification of tropical cyclones.

- Understand the underpinning science of storm surge and its relationship to landfalling tropical cyclones.

- Understand the physics that explains the existence of a hurricane's eye.

- Understand the role of subtropical high-pressure systems in providing steering currents for tropical cyclones.

- Gain an appreciation for the role that the National Hurricane Center plays in warning the public, and be able to interpret some of the advisories and products issued by NHC.

At 06 UTC on October 23, 2015, a reconnaissance aircraft flew into Hurricane Patricia off the west coast of Mexico. Measurements indicated that Patricia's central pressure and maximum sustained winds were 879 mb and 180 kt (207 mph), respectively. Six hours later, satellite analysis suggested that maximum sustained winds had increased to 185 kt (213 mph), making Patricia the strongest hurricane on record in the eastern North Pacific Ocean (a visible satellite image of Patricia a few hours later is shown in Figure 11.1). This sustained wind speed ranked the hurricane as a Category 5 on the Saffir-Simpson Hurricane Wind Scale (see Table 11.1) which is

routinely used to rate the strength of hurricanes in the eastern Pacific and Atlantic Oceans (Patricia's 185-kt wind speed also exceeded the maximum sustained wind speed of any hurricane on record in the Atlantic Ocean). To be fair, regular monitoring by aircraft and the routine analysis of wind speeds using satellite imagery didn't begin until the early 1970s in the Atlantic Ocean and 1988 in the eastern Pacific, so there must be an asterisk by Patricia's record sustained wind of 185 kt.

Nonetheless, Hurricane Patricia was one for the history books. Around the time that the storm's sustained wind speed reached 185 kt, its central pressure was estimated to be 872 mb, the lowest sea-level pressure on record in the Western Hemisphere and the second lowest globally (the global record of 870 mb was set in 1979 in the western North Pacific Ocean by a tropical cyclone named Tip). For the record, **tropical cyclone** is the generic, universal

FIGURE 11.1 A visible satellite image of Hurricane Patricia at 1445 UTC on October 23, 2015, a few hours after it attained peak intensity (courtesy of NASA).

TABLE 11.1 Saffir-Simpson Hurricane Wind Scale		
Category/Description	Sustained Wind (knots; mph)	
1: Minimal	64–82	74–95
2: Moderate	83–95	96–110
3: Extensive	96–112	111–129
4: Extreme	113–136	130–156
5: Catastrophic	> 136	> 156

name given to low-pressure systems that form over warm tropical or subtropical seas—a **hurricane** is simply a tropical cyclone in the Atlantic or eastern Pacific Ocean with sustained winds of at least 65 kt (74 mph).

Tropical cyclones are a lot different than mid-latitude low-pressure systems (which, for contrast, are sometimes called **extratropical cyclones**). For starters, a tropical cyclone is a warm-core system, meaning that temperatures in the troposphere are higher at its center than on its periphery (in contrast, extratropical cyclones are cold-core). Indeed, aircraft reconnaissance flying into Patricia at the 700-mb level (a standard flight level for such aircraft) several hours after it reached peak intensity measured a temperature of 32.2°C (90°F), the highest 700-mb temperature ever recorded in a tropical cyclone.

In addition, thunderstorms tend to organize around the center of a tropical cyclone, a characteristic that's not typical for mid-latitude lows. Figure 11.2 shows the radar reflectivity of Hurricane Patricia at 1732 UTC on October 23, retrieved from a radar mounted on the reconnaissance aircraft. Note the ring of yellow and green (representing relatively high reflectivity) around Patricia's **eye**, which is the roughly circular precipitation-free island of generally light winds in the center of a hurricane. The ring of very tall thunderstorms around the eye is known as the **eye wall**.

Patricia's impressive legacy also included its rate of intensification. Although pressure and wind speed are commonly used to characterize tropical cyclones, pressure tendencies are often invoked in explosive storms such as Patricia. Over the 24-hour period ending at 06 UTC on October 23, Patricia's central pressure fell roughly 95 mb, and maximum sustained winds increased from 75 kt (86 mph) to 180 kt (207 mph). This 105-kt (121-mph) increase broke the all-time 24-hour intensification record of 95 kt (109 mph), set by Hurricane Wilma in the Gulf of Mexico in 2005. Over the 48-hour period also ending at 06 UTC on October 23, Patricia's maximum sustained wind speeds increased by nearly 150 kt (173 mph)! Indeed, Patricia is an extreme example of **rapid intensification**, a term that meteorologists sometimes use to describe explosively developing tropical cyclones. No wonder that at 06 UTC on October 23, aircraft data indicated that horizontal pressure differences near the center of Patricia were as large as 27 mb per mile, thought to be one of the largest horizontal pressure gradients ever recorded in a tropical cyclone.

By the time Hurricane Patricia made **landfall** (meaning that the storm's center crossed over land) later on October 23 along the coast of the Mexican state of Jalisco, near Playa Cuixmala, it had dramatically weakened. Still, in the historical database, the barometric reading of 932 mb is the lowest central pressure for a hurricane making landfall in Mexico. A graph of pressure with time on October 23 at Playa Cuixmala and other nearby weather stations is shown in Figure 11.3, vividly demonstrating the rapid fall and then rise of pressure as one of the

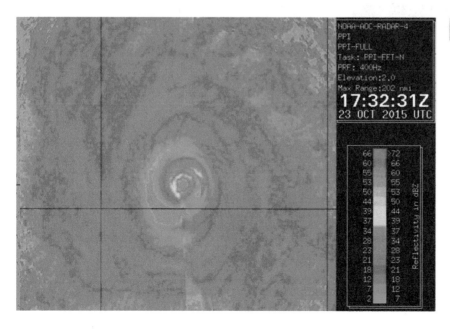

FIGURE 11.2 A radar reflectivity image of Hurricane Patricia near peak intensity at 1732 UTC on October 23, 2015, taken by a radar onboard a Hurricane Hunter aircraft (courtesy of NOAA).

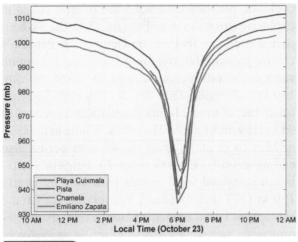

FIGURE 11.3 Graphs of sea-level pressure at four locations along or near the west coast of Mexico in the hours surrounding the landfall of Hurricane Patricia on October 23, 2015 (courtesy of NOAA).

planet's most powerful storms approached and then moved away.

Yes, Patricia was a remarkable storm from the standpoints of pressure, wind, pressure tendency, and other meteorological parameters. However, there is much more to learn about hurricanes. From space, the swirling, galactic-arm appearance on satellite images, together with an eye staring menacingly into space, absolutely invites investigation by the scientifically curious (see Figure 11.4). We will start with the locations on Earth where the greatest storms on Earth form.

TROPICAL CYCLONES: A GLOBAL PERSPECTIVE

There are seven ocean basins in which tropical cyclones routinely form: We recommend that you glance at Figure 11.5 while you read through the following list. To get your bearings, we point out that the shaded areas in Figure 11.5 mark the common breeding grounds for tropical cyclones. The red arrows indicate the paths that tropical cyclones typically follow within and beyond these breeding grounds.

1. Atlantic Basin (the North Atlantic Ocean, the Gulf of Mexico, and the Caribbean Sea)
2. Northeast Pacific Basin (from Mexico to the International Date Line)
3. Northwest Pacific Basin (from the International Date Line to Asia, plus the South China Sea)
4. North Indian Basin (includes the Bay of Bengal and the Arabian Sea)
5. Southwest Indian Basin (from Africa to about 100°E longitude)
6. Southeast Indian/Australian Basin (100°E longitude to 142°E longitude)
7. Australian/Southwest Pacific Basin (142°E longitude to about 120°W longitude)

Figure 11.5 contains a lot of information that we will explore in this chapter. One observation that we'd like to make right off the bat is that tropical cyclones do not form on (or very close to) the

(a)

(b)

FIGURE 11.4 (a) Some tropical cyclones bear a striking resemblance to spiral galaxies. This image of spiral galaxy Messier 101 is a composite of about 50 individual exposures from the Hubble Space Telescope (courtesy of NASA and ESA); (b) Hurricane Igor in the North Atlantic Ocean on September 13, 2010 (courtesy of NASA).

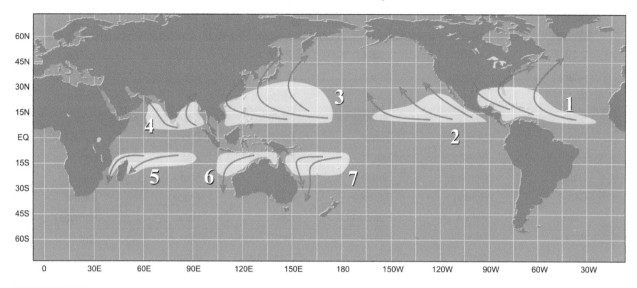

FIGURE 11.5 The primary global breeding grounds for tropical cyclones (shaded areas) and the typical paths they take (red arrows).

equator. Also note the absence of tropical cyclones in the South Atlantic Ocean.

We have already mentioned that the strongest tropical cyclones in the Atlantic and Northeast Pacific Basins are called hurricanes, but they go by different names in the other five basins. In the Northwest Pacific Basin, forecasters use **typhoon** to label a tropical cyclone whose maximum sustained winds have reached the hurricane threshold of 65 kt (74 mph), and **supertyphoon** if wind speeds reach 130 kt (150 mph). In the Southwest Pacific Ocean west of 160°E longitude and also in the Southeast Indian Ocean east of 90° longitude (gets a little complicated, eh?), strong tropical cyclones with hurricane-force winds are known simply as **severe tropical cyclones** (Australian forecasters refer to "weaker" hurricane-strength storms simply as cyclones). Continuing on our world tour, we note that forecasters call a strong tropical cyclone in the North Indian Ocean a **severe cyclonic storm**. Finally, in the Southwest Indian Ocean, the generic tropical cyclone is the chosen designator.

From "hurricane" to "typhoon" to "severe cyclonic storm," the labels that meteorologists attach to strong tropical cyclones vary widely across the globe. As it turns out, the lists of names that forecasters use to distinguish storms that form in a particular basin also vary across the globe. Before we delve into these naming conventions, we first need to trace the life cycle of a typical tropical cyclone because, in some ocean basins, assigning a name depends on the cyclone's stage of development.

Naming Tropical Cyclones: Ham with Chicken Livers and Mushrooms

On the way to becoming a hurricane, an intensifying tropical low-pressure system in the Atlantic and Northeast Pacific Basins evolves from a tropical depression to a tropical storm. A **tropical depression** is a tropical cyclone that has an observable cyclonic circulation on satellite imagery and whose maximum sustained winds are 33 kt (38 mph) or less. A **tropical storm** is a tropical cyclone with maximum sustained winds of 34–63 kt (39–73 mph). Of these different stages of development, hurricanes are distinguished by the characteristic eye in the center of the circulation (on rare occasion, a strong tropical storm can have an eye structure as well).

Globally each year, approximately 80 to 90 tropical cyclones reach tropical-storm intensity, with about sixty percent of these attaining the threshold of 64 kt (74 mph) to qualify as a hurricane (or typhoon, etc.). In order to efficiently communicate information about these systems and avoid confusion when more than one is active at the same time, meteorologists give them a number-letter tag when they become a tropical cyclone. For example, Tropical Depression 2-E would be the second tropical cyclone of the season in the Northeast Pacific Basin. This generic naming convention never changes.

Most tropical cyclones receive a name once they reach tropical storm intensity. Table 11.2 shows the six alphabetized lists of names currently used by the National Hurricane Center for the North Atlantic Basin. The lists of names are recycled.

TABLE 11.2 Atlantic Basin Tropical Storm and Hurricane Names					
2017	**2018**	**2019**	**2020**	**2021**	**2022**
Arlene	Alberto	Andrea	Arthur	Ana	Alex
Bret	Beryl	Barry	Bertha	Bill	Bonnie
Cindy	Chris	Chantal	Cristobal	Claudette	Colin
Don	Debby	Dorian	Dolly	Danny	Danielle
Emily	Ernesto	Erin	Edouard	Elsa	Earl
Franklin	Florence	Fernand	Fay	Fred	Fiona
Gert	Gordon	Gabrielle	Gustav	Grace	Gaston
Harvey	Helene	Humberto	Hanna	Henri	Hermine
Irma	Isaac	Imelda	Ike	Ida	Igor
Jose	Joyce	Jerry	Josephine	Julian	Julia
Katia	Kirk	Karen	Kyle	Kate	Karl
Lee	Leslie	Lorenzo	Laura	Larry	Lisa
Maria	Michael	Melissa	Marco	Mindy	Matthew
Nate	Nadine	Nestor	Nana	Nicholas	Nicole
Ophelia	Oscar	Olga	Omar	Odette	Otto
Phillippe	Patty	Pablo	Paloma	Peter	Paula
Rina	Rafael	Rebekah	Rene	Rose	Richard
Sean	Sara	Sebastien	Sally	Sam	Shary
Tammy	Tony	Tanya	Teddy	Teresa	Tomas
Vince	Valerie	Van	Vicky	Victor	Virginie
Whitney	William	Wendy	Wilfred	Victor	Virginie

So, for example, the 2017 list will be used again in 2023. The only exception occurs when hurricanes or tropical storms are extremely destructive or deadly. In these cases, their names are "retired" from the list because using the name again for a future storm would be inappropriate and/or insensitive. The World Meteorological Organization, a specialized agency of the United Nations headquartered in Geneva, Switzerland, then chooses a new name as a replacement. A recent example of a retired name is Sandy in 2012, which was replaced by Sara in 2018. Prior to the 2016 hurricane season, approximately 80 names had been retired from the North Atlantic list of names, including Andrew (1992), Allison (2001), Charley and Ivan (2004), Katrina and Wilma (2005), Igor (2010), and Joaquin (2015). At the time of this writing, Harvey, Irma, and Maria (2017) will likely be retired, given the extreme destruction these hurricanes inflicted. If that happens, it would mark the first time since 2005 (Rita and Stan) that back-to-back hurricane names were retired.

The custom of naming tropical storms and hurricanes has an interesting history. The practice dates back a few hundred years to the West Indies (the three main groups of islands that comprise the West Indies are the Bahamas, the Greater Antilles, and the Lesser Antilles). In the nineteenth century, islanders began to name hurricanes after saints. Indeed, when hurricanes arrived on a day commemorating a saint, locals christened the storm with the saint's name. For example, fierce Hurricane Santa Ana struck Puerto Rico on July 26, 1825, while Hurricane San Felipe (the first) and Hurricane San Felipe (the second) hit Puerto Rico on September 13, 1876 and September 13, 1928, respectively.

During World War II, Navy and Army Corp forecasters informally named Pacific storms after their girlfriends or wives. The spirit of this informal naming convention apparently started a trend in the United States. From 1950 to 1952, meteorologists named tropical cyclones in the Atlantic Basin according to the phonetic alphabet (Able, Baker, Charlie, etc.). Then, in 1953, the U.S. Weather Bureau switched to an alphabetized list of female names. In 1979, the National Weather Service amended their lists to also include male names.

FIGURE 11.6 A composite of visible images from the Japanese Himawari geostationary satellite shows Typhoons Chan-hom (left) and Nangka (right) on July 9, 2015 in the Northwest Pacific Basin (courtesy of CIMSS/University of Wisconsin).

FIGURE 11.7 A visible satellite image of Super Cyclonic Storm Gonu at 09 UTC on June 4, 2007. On this date, Gonu's maximum sustained winds reached 140 kt (161 mph), making it the strongest tropical cyclone ever to develop in the Arabian Sea (courtesy of MODIS Rapid Response Project at NASA/GSFC).

In 1959, forecasters monitoring the central Pacific Ocean started to use female names to designate tropical storms and hurricanes that formed near Hawaii. The rest of the Northeast Pacific Basin followed suit in 1960. In 1978, the lists of names for tropical storms and hurricanes in these basins were amended to include male names as well.

In the Northwest Pacific Ocean, forecasters began using female names for tropical cyclones in 1945. In tandem with the 1979 change in the United States, they also revised their lists to include male names. On January 1, 2000, however, in a dramatic departure from tradition, forecasters from the nations and territories in the Northwest Pacific Basin agreed to use a markedly different naming convention. From that day forward, the new lists did not (for the most part) contain male or female names. Instead, they included Asian words that referred to flowers, animals, birds, trees, foods, etc., while other names are simply descriptive adjectives.

For example, consider Typhoons Chan-hom and Nangka, which patrolled the Northwest Pacific Basin in July 2015 (see Figure 11.6—Chan-hom is on the left). Chan-hom (contributed by Lao PDR) is a type of tree while Nangka (contributed by Malaysia) means "jackfruit," which is a popular fruit tree in that country. Hungry for a tropical cyclone, anyone? Table 11.3 shows the names for this basin—you'll note that they do not appear in alphabetical order. Rather, the contributing nations are listed in alphabetical order. This list of countries determines the order in which names are assigned.

Before October 2004, tropical cyclones that formed in the North Indian Ocean were not named from a traditional list. For this basin, forecasters simply used a label consisting of a two-digit number and letter that the cyclone received once it attained tropical-depression status. For example, "Tropical Cyclone 02**A**" would be the second tropical cyclone of the season to form over the **A**rabian Sea. "Tropical Cyclone 01**B**" would be the first tropical cyclone of the season to form over the **B**ay of Bengal.

One of the most infamous *named* storms in the North Indian Ocean was Severe Cyclonic Storm Gonu, which developed over the Arabian Sea in early June 2007 (see Figure 11.7). On June 4, Gonu's maximum one-minute average sustained winds reached approximately 140 kt (161 mph), compelling forecasters to upgrade Gonu to a *super* cyclonic storm. As of this writing, Gonu still stands as the most powerful tropical cyclone ever to develop in the Arabian Sea. As Gonu moved toward the Arabian Peninsula, the storm weakened considerably as it drew dry air over the desert into its circulation.

The notion that dry air circulating into Gonu would cause the storm to weaken is not surprising, and is part of a much broader discussion about the conditions that are favorable (or unfavorable) for the genesis and development of tropical cyclones. Now that we've played the "name game," let's investigate these conditions.

TABLE 11.3 Northwest Pacific Tropical Storm and Typhoon Names

The names for tropical storms and typhoons that develop in the Northwest Pacific Basin are Asian words submitted by countries in the region. The names are compiled into "running lists": After all the names in one list are used, the name of the next storm that develops is simply the first word in the next list.

NORTHWEST PACIFIC BASIN

Contributor	I	II	III	IV	V
Cambodia	Damrey	Kong-rey	Nakri	Krovanh	Sarika
China	Haikui	Yutu	Fengshen	Dujuan	Haima
DPR Korea	Kirogi	Toraji	Kalmaegi	Mujigae	Meari
HK, China	Kai-Tak	Man-yi	Fung-wong	Choi-wan	Ma-on
Japan	Tembin	Usagi	Kanmuri	Koppu	Tokage
Lao PDR	Bolaven	Pabuk	Phanfone	Champi	Nock-ten
Macau, China	Sanba	Wutip	Vongfong	In-fa	Muifa
Malaysia	Jelawat	Sepat	Nuri	Melor	Merbok
Micronesia	Ewiniar	Mun	Sinlaku	Nepartak	Nanmadol
Philippines	Malaksi	Danas	Hagupit	Lupit	Talas
RO Korea	Gaemi	Nari	Jangmi	Mirinae	Noru
Thailand	Prapiroon	Wipha	Mekkhala	Nida	Kulap
U.S.A.	Maria	Francisco	Higos	Omais	Roke
Vietnam	Son-Tinh	Lekima	Bavi	Conson	Sonca
Cambodia	Ampil	Krosa	Maysak	Chanthu	Nesat
China	Wukong	Bailu	Haishen	Dianmu	Haitang
DPR Korea	Jongdari	Podul	Noul	Mindulle	Nalgae
HK, China	Shanshan	Lingling	Dolphin	Lionrock	Banyan
Japan	Yagi	Kaziki	Kujira	Kompasu	Hato
Lao PDR	Leepi	Faxai	Chan-hom	Namtheun	Pakhar
Macau, China	Bebinca	Peipah	Linfa	Malou	Sanvu
Malaysia	Rumbia	Tapah	Nangka	Meranti	Mawar
Micronesia	Soulik	Mitag	Soudelor	Rai	Guchol
Philippines	Cimaron	Hagibis	Molave	Malakas	Talim
RO Korea	Jebi	Neoguri	Goni	Megi	Doksuri
Thailand	Mangkhut	Bualoi	Atsani	Chaba	Khanun
U.S.A.	Barijat	Matmo	Etau	Aere	Lan
Vietnam	Trami	Halong	Vamco	Songda	Saola

RECIPE FOR HURRICANES: SIX INGREDIENTS IN JUST THE RIGHT MEASURE

So far we've mentioned that tropical cyclones have an observable cyclonic circulation on loops of satellite imagery. In addition, there are also organized thunderstorms concentrated around or near their centers. We will elaborate on the role that organized convection plays in the intensification of tropical cyclones later in this chapter. First, however, we will discuss the six ingredients required for the genesis of tropical cyclones. If you're interested in a tropical cookbook, here's the recipe:

1. Sea-surface temperatures of at least 26.5°C (80°F) and a relatively deep layer of warm water beneath the ocean surface
2. An environment primed to become conditionally unstable through a deep layer of the troposphere
3. Moist air in the middle troposphere
4. Weak vertical wind shear above the newly forming tropical cyclone

5. A genesis location that typically lies at least 5° latitude away from the equator

6. A group of loosely organized showers and thunderstorms, known as a **tropical disturbance**.

Although these ingredients might seem daunting at first, they all relate to the two characteristics that all tropical cyclones display—organized convection around or near their centers and an observable cyclonic circulation on satellite imagery. Keep these two traits in mind as we explore each of the six ingredients.

High Sea-Surface Temperatures: The Foundation for Organized Thunderstorms

Figure 11.8 shows average sea-surface temperatures (SSTs) over the North Atlantic Ocean from June 1 to November 30, the period that corresponds to official hurricane season in this basin. Note the corridor of water temperatures of 26.5°C (80°F) or higher that extends from the west coast of Africa across the tropical Atlantic to the Caribbean Sea and the Gulf of Mexico (roughly, the yellow, orange, and red). This corridor marks "hurricane alley," where tropical cyclones routinely develop during hurricane season.

The heart of Atlantic hurricane season runs from mid-August to mid-October, when there's a marked upturn in the frequency of tropical cyclones. Figure 11.9 shows the daily frequency of Atlantic hurricanes and tropical storms—the peak is during the second week of September. Not coincidentally, sea-surface temperatures in the North Atlantic are also at their highest during this month (see Figure 11.10).

Worldwide, summer and autumn are the seasons for tropical cyclones, given the vital importance of warm tropical seas. As a general rule, tropical season runs from July to October in the Northern Hemisphere and

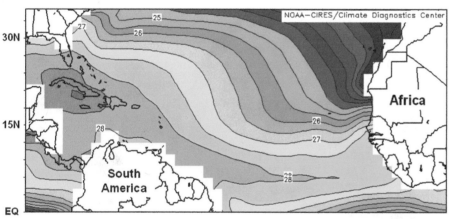

Average Sea-Surface Temperature (°C) – June to November

| 22 | 23 | 24 | 25 | 26 | 27 | 28 | 29 |

FIGURE 11.8 The long-term average of sea-surface temperatures (in °C) in the Atlantic Basin from June 1 to November 30 (courtesy of NOAA).

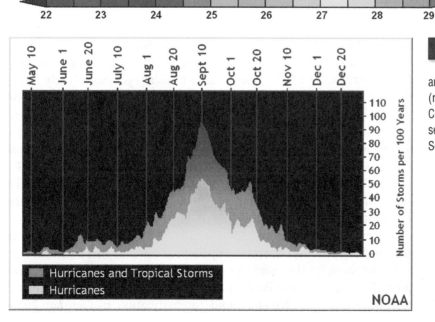

Hurricanes and Tropical Storms
Hurricanes

FIGURE 11.9 The daily frequency of hurricanes (yellow) and both hurricanes and tropical storms (red) in the Atlantic Basin, per 100 years. Climatologically, the peak of the tropical season in the Atlantic Basin is around September 10 (courtesy of NOAA).

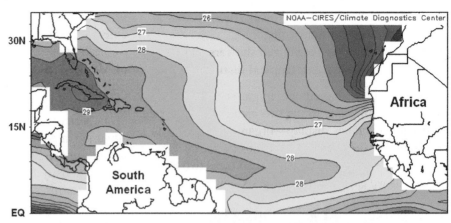

FIGURE 11.10 The long-term average of sea-surface temperatures (in °C) in the Atlantic Basin during September, the peak of Atlantic hurricane season (courtesy of NOAA).

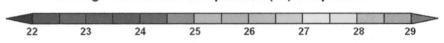

FIGURE 11.11 Sea-surface temperatures at 00 UTC on October 23, 2015, as Hurricane Patricia was nearing peak intensity. The hurricane's location at this time is marked with the black circle (courtesy of TropicalTidbits.com).

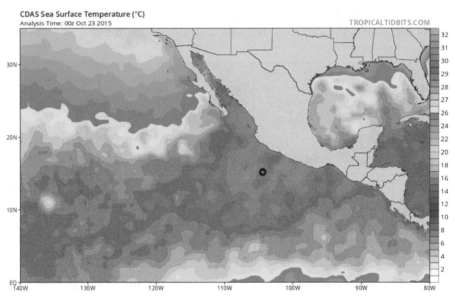

December through March in the Southern Hemisphere, though each ocean basin has its own idiosyncrasies. For example, typhoon season in the Northwest Pacific Ocean (Region "3" in Figure 11.5) essentially runs all year, the byproduct of persistently warm water. On February 23, 2019, the maximum sustained winds of Supertyphoon Wutip reached 160 mph (about 140 kt), marking the first Category-5 storm ever observed in the Northern Hemisphere during the month of February.

In early 2019, news of extreme tropical cyclones in the Southern Hemisphere also went viral. In March, long-lived Tropical Cyclone Idai in the Southwest Indian Ocean became one of the most catastrophic storms to impact Africa (specifically, Mozambique, Zimbabwe, and Malawi), ranking as the third deadliest storm ever recorded in the Southern Hemisphere. Then, in April, Tropical Cyclone Kenneth struck northern Mozambique with maximum sustained winds

over 140 mph (roughly 120 kt), the first time on record that two major tropical cyclones made landfall in Mozambique in the same season.

Without reservation, warm tropical seas were complicit in the extreme tropical cyclones of early 2019. High sea-surface temperatures ensure high evaporation rates and thus high dew points in the lower troposphere. High water temperatures also help to destabilize the lower troposphere by providing a "warm bottom." Any way you slice it, high SSTs pave the way for deep convection, which newly forming and intensifying tropical cyclones ultimately require to form around their centers.

As another example of the pivotal role that high sea-surface temperatures can play in the genesis and development of tropical cyclones, consider Hurricane Patricia (recall the beginning of the chapter). Patricia's rapid intensification in October 2015

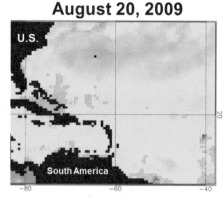

August 20, 2009

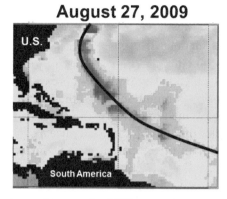

August 27, 2009

FIGURE 11.12 Sea-surface temperature anomalies (in °C) on August 20, 2009 (left) and August 27, 2009 (right—Bill's track superimposed) for a portion of the western North Atlantic Ocean traversed by Hurricane Bill. Sea-surface temperatures cooled by several degrees after Bill's powerful winds churned up ocean waters (courtesy of NOAA).

Sea-Surface Temperature Anomaly (°C)

occurred over an expansive region of anomalously warm ocean water, with temperatures approaching 31°C (see Figure 11.11), the highest sea-surface temperatures on record in mid-October in this part of the eastern Pacific south of Mexico.

Not only must sea-surface temperatures be high for tropical cyclones to form and develop, but a fairly deep layer of water below the surface must also be warm. Let's investigate.

Warm Water Below: When the Wind Stirs the Sea

A newly forming but slowly moving tropical cyclone can sputter and fade before it ever really develops. To see why, let's start on the high end of tropical cyclones by considering Hurricane Bill in August 2009, which stayed mainly over open Atlantic waters. For a time, Bill was a strong hurricane, even reaching Category-4 status on the Saffir-Simpson Scale for a day or so. Such powerful winds have an important impact on the water below—they agitate and mix the upper levels of the ocean, bringing cooler water to the surface. In response to this upwelling, sea-surface temperatures decrease. Figure 11.12, which shows sea-surface temperature (SST) anomalies (departures from average), illustrates upwelling caused by Bill's agitating winds. On the left are the SST anomalies on August 20, before Bill traversed this patch of ocean—nearly all of the water was warmer than average as indicated by the preponderance of yellow and orange. Compare this "before" view with the image on the right which shows the SST anomalies on August 27, after Bill moved through (Bill's track is superimposed in black). All along Bill's path, waters were colder than average (denoted by the blue), the result of cooler water upwelled from below the surface.

With this example in mind, you can imagine what might happen if a newly forming tropical cyclone encountered tropical seas where the layer of warm water beneath the surface was relatively shallow. Indeed, as winds began to strengthen, cooler water would quickly upwell to the sea surface, in effect limiting further development. So you can see why tropical meteorologists stipulate that high sea-surface temperatures must be accompanied by a relatively deep warm layer below the ocean surface in order to promote genesis of tropical cyclones.

Incidentally, hurricanes and tropical storms that pass over the cool wake of a previous storm can also lose strength. Again, the organized convection around the center of such a tropical cyclone would likely weaken, with the degree of weakening depending on how cool the water was and how long the storm stayed over the cool wake.

The Road to Conditional Instability: Paving the Way for Tall Thunderstorms

Truth be told, high sea-surface temperatures go hand-in-hand with a conditionally unstable troposphere. In other words, the two ingredients are not independent. Even if the tropical troposphere initially has neutral stability, high SSTs and increasing surface winds associated with a developing tropical cyclone promote increasingly higher evaporation rates, which, in turn, ensure that the air above the warm ocean is teeming with water vapor (the overlying surface air is also warm). In Chapter 8, you learned that warm, moist air parcels near the surface readily encourage net condensation after a relatively short ascent. Such rising parcels can remain positively buoyant to great altitudes because large releases of latent heat keep parcels warmer than their environment through a deep layer of the troposphere. Such ascent to great altitudes is pivotal for tall thunderstorms to form and organize around the center of a newly forming tropical cyclone.

Recall that the ring of very tall thunderstorms around the eye of a hurricane is called the eye wall. As ferocious as these thunderstorms might seem, the updrafts that sustain them are actually relatively tame (compared to, for example, the updrafts in a supercell thunderstorm). To explain this apparent paradox, recall from Chapter 9 that the strength of a thunderstorm's updraft depends largely on the degree of positive buoyancy—essentially, the difference between the temperature of a rising air parcel and the temperature of the environment. Over warm tropical seas, the environmental lapse rate in the middle troposphere is pretty close to the moist adiabatic rates that rising air parcels typically follow there. As a result, positive buoyancy is often surprisingly small inside eye-wall thunderstorms. Thus, the updrafts that sustain them tend to be rather weak.

Relatively weak updrafts in the eye wall help to explain the general lack of lightning around the eye of a hurricane. These weak updrafts are rather inefficient at separating electrical charges—that is, creating regions with large positive and large negative charges that are necessary to produce lightning. As an example, the eye wall of Hurricane Andrew, the last Category-5 hurricane to hit the United States (in South Florida), produced less than ten cloud-to-ground lightning strikes per hour between the time it was over the Bahamas and the time it made its second landfall in Louisiana.

But some eye walls are prolific lightning producers. Figure 11.13 is an infrared satellite image of Hurricane Irma as it approached Puerto Rico on September 6, 2017. At the time, Irma was a powerful Category-5

hurricane with maximum sustained winds of 160 kt (184 mph). Data from the Global Lightning Mapper onboard GOES-16 are superimposed on the infrared image and indicate substantial lightning around the core of Irma. Although these data were preliminary, the overall message is that eye-wall thunderstorms of very strong or rapidly intensifying hurricanes can produce a considerable amount of lightning.

So far, we have identified high sea-surface temperatures, a relatively deep layer of warm water below the surface, and eventual conditional instability through a deep layer of the troposphere, as primary ingredients for cooking up a tropical cyclone. There's one other thermodynamic (related to temperature and moisture) ingredient crucial to the development of a tropical cyclone.

Ingredient #3: A Moist Middle Troposphere

The color-enhanced water vapor image in Figure 11.14a shows Super Cyclonic Storm Gonu over the Arabian Sea at 18 UTC on June 4, 2007. At the time, Gonu's maximum sustained winds were 140 kt (161 mph), making it the strongest tropical cyclone ever to form over the Arabian Sea. In this image, the blue shading indicates dry air in the middle and upper troposphere. But as Gonu moved westward toward the Arabian Peninsula and started to draw dry, mid-tropospheric air into its circulation, the historic storm took a big hit. Just 18 hours later, Gonu's maximum winds had weakened to 105 kt (121 mph), and its presentation on water vapor imagery was much less formidable (see Figure 11.14b). The tropical cyclone continued to weaken as it ingested more and more of this low-humidity air from high over the deserts of the Middle East. The message here should be loud and clear—dry air in the middle troposphere is poisonous to hurricanes.

FIGURE 11.13 An infrared satellite image of Hurricane Irma at 0945 UTC on September 6, 2017, as the Category-5 hurricane bore down on the U.S. Virgin Islands and Puerto Rico. The red dots indicate lightning that occurred between 0942 and 0945 UTC. Yellow dots represent slightly older lightning flashes (courtesy of NOAA and Dave Santek, CIMSS/SSEC).

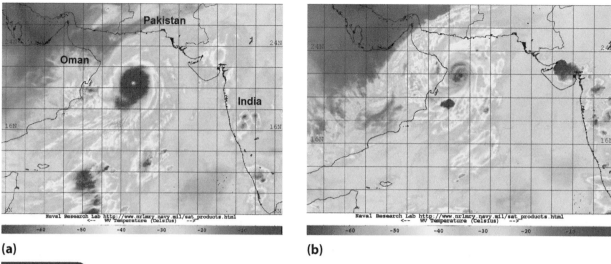

(a)   **(b)**

FIGURE 11.14 (a) A color-enhanced water vapor image of Super Cyclonic Storm Gonu over the Arabian Sea at 18 UTC on June 4, 2007 (from the Meteosat-7 geostationary satellite). Maximum sustained winds were 140 kt (161 mph) at the time. The blue indicates dry air in the middle and upper troposphere over the Arabian Peninsula and other parts of the Middle East; (b) A color-enhanced water vapor image of Gonu about 18 hours later. By this time, the storm had weakened after ingesting dry, mid-tropospheric air over the Middle East. Maximum sustained winds had dropped to 105 kt (121 mph) (courtesy of the Naval Research Laboratory).

Logically, if hurricanes take a hit when they ingest dry air in the middle troposphere, imagine the effect of dry, mid-level air on a newly forming and somewhat fragile tropical cyclone. It can't be good for development. To understand why, consider that mid-level dry air entrained into thunderstorms enhances evaporational cooling, so parcels become more negatively buoyant. This leads to stronger downdrafts, which are able to penetrate farther down into the lower troposphere, bringing drier air with them. This downward intrusion acts to snuff out new convection in the vicinity of the core of the disturbance. To the extent that new convection is pivotal for the development of the disturbance into a tropical cyclone, the die is cast and the disturbance fizzles (or at least doesn't develop any further).

Now that we've discussed all of the *thermodynamic* ingredients, let's move on to the *dynamic* ingredients (related to air motions) needed for the genesis and development of tropical cyclones.

Vertical Wind Shear: Exposing The Low-Level Circulation

In this chapter, you've learned that a tropical cyclone thrives when deep convection (tall thunderstorms) organizes around its low-level center of circulation. With this favorable organization in mind, consider Figure 11.15a, a color-enhanced infrared satellite image of Tropical Storm Julia early on September 15, 2016. The tropical storm's low-level circulation, indicated by the swirl of gray, relatively warm cloud tops

just off the South Carolina Coast, was clearly detached from its thunderstorms, which are marked by the cold, color-enhanced clouds to the east of the low-level circulation. In such situations, meteorologists sometimes say that "the low-level circulation is exposed."

Needless to say, Julia was not intensifying at this time. Indeed, the tropical storm was located in an environment of "high wind shear," meaning that vertical wind shear was strong enough to interfere with the organization of thunderstorms around Julia's low-level center. Let's investigate the impacts of vertical wind shear on tropical cyclones by addressing this question: How did Julia's low-level circulation become separated from its deep convection in this highly sheared environment?

We'll start with a couple "questionable" explanations sometimes offered by TV meteorologists—these include "strong upper-level winds rip the storm apart" and "strong upper-level winds push thunderstorms away from the center of the tropical cyclone." At first glance, these descriptions might seem at odds with our approach because they don't mention vertical wind shear at all.

Recall from Chapter 9 that vertical wind shear is the change in wind speed and/or wind direction with increasing altitude. Thus, computing vertical wind shear requires wind data from two levels—in the context of tropical cyclones, the 850 mb and 200 mb pressure levels are commonly used (the standard heights of these pressure levels are 1500 meters and 12,500

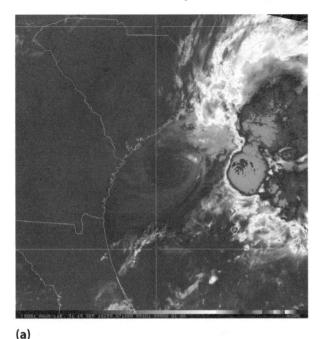

(a)

(b)

FIGURE 11.15   (a) A color-enhanced infrared image of Tropical Storm Julia at 10 UTC on September 15, 2016. Julia's low-level circulation, which is the swirl of clouds just off the South Carolina Coast, is displaced from the deep convection, marked by the colder cloud tops to the east (courtesy of NOAA); (b) Vertical wind shear between 850 mb and 200 mb at 06 UTC on September 15, 2016. White streamlines indicate the direction of the wind shear, while the magnitude is color-coded (in knots). The tropical storm symbol indicates the location of Julia at that time (courtesy of Cooperative Institute for Meteorological Satellite Studies, University of Wisconsin).

meters, respectively). To calculate the vertical wind shear, we simply subtract the wind vector at 850 mb from the wind vector at 200 mb. This isn't quite as easy as it sounds because it involves breaking down the winds at both levels into their east-west and north-south components (in other words, you need some trigonometry). Fortunately, in general, when 200-mb winds are relatively fast, the vertical wind shear between 850 mb and 200 mb is relatively large—essentially, fast upper-level winds are a proxy for significant vertical wind shear. So the two "questionable" explanations for Julia's satellite presentation (mentioned in the last paragraph) have at least some scientific merit.

The bottom line is that, in many cases, you only need to look at upper-level winds to get a rough, qualitative sense of the magnitude of vertical wind shear. If upper-level winds are relatively speedy, then vertical wind shear tends to be large, implying that the genesis and development of tropical cyclones will be disrupted. Having said this, it's prudent to have a quantitative rule to decide whether vertical wind shear is too large for tropical cyclones to thrive.

To this end, check out Figure 11.15b which shows the vertical wind shear between the upper and lower troposphere (for all practical purposes, between 850 mb and 200 mb) at 06 UTC on September 15, 2016.

To get your bearings, the white streamlines indicate the direction of the wind shear vector, while the magnitude of the wind shear is color-coded in knots. Why knots? Note that because wind shear is a difference between two wind vectors, it has the same units as wind speed – in this case, knots. At this time, Julia (its location is marked with the tropical storm symbol) was embedded in an environment with westerly vertical shear of magnitude approximately 30 kt.

As a general rule, forecasters have found that vertical wind shear less than 20 kt is favorable for tropical cyclones to develop or intensify. The wind shear that Julia experienced at this time exceeds this threshold, so conditions were unfavorable for intensification. But how did this vertical wind shear produce the separation between Julia's low-level circulation and its deep convection? Without getting too deep into the scientific explanation, strong vertical wind shear tends to displace the pocket of low-level convergence away from the tropical cyclone's central core. Because low-level convergence is the primary mechanism by which thunderstorms are initiated in tropical cyclones, deep convection then becomes separated from the center of the system's low-level circulation.

So, are the two "questionable" explanations we mentioned earlier scientifically sound? Not really. First,

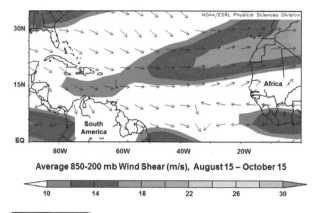

Average 850-200 mb Wind Shear (m/s), August 15 – October 15

| 10 | 14 | 18 | 22 | 26 | 30 |

FIGURE 11.16 The climatological average vertical wind shear between 850 mb and 200 mb during the period August 15 to October 15, the heart of hurricane season in the Atlantic Basin. Arrows represent the direction of the wind shear, while the magnitude is color-coded (in m/s). As a general rule, vertical wind shear of less than 10 m/s (which is about 20 kt, in white here) is sufficiently low to support tropical cyclone genesis and development (courtesy of NOAA).

strong upper-level winds really don't physically "rip" the tropical cyclone apart. The tropical cyclone still has organization—it's just not favorable for the system to thrive. And second, strong upper-level winds don't really "push thunderstorms away from the tropical cyclone's center"—in reality, the lifetime of tropical thunderstorms is too short for them to move all the way from the central core to a position far removed from the tropical cyclone's center of low-level circulation.

Figure 11.16 shows the long-term average 850–200 mb vertical wind shear over the Atlantic Basin during the heart of hurricane season, from August 15 to October 15. The areas in white mark the regions where vertical wind shear is less than 10 m/s, indicating that favorable wind-shear conditions prevail during this time (on average) over much of the basin. Given the region's favorable thermodynamic environment, tropical cyclones commonly form and intensify here during this time. Keep in mind that Figure 11.16 shows *average* conditions during a two-month period—wind shear over any specific region at any specific time can be more or less favorable for tropical cyclogenesis (tropical cyclone formation and/or intensification), depending on the prevailing weather pattern.

The bottom line here is that vertical wind shear over a tropical disturbance should be weak if the disturbance is ever going to develop into a tropical cyclone. This debate about vertical wind shear becomes moot, however, if the tropical disturbance is too close to the equator, which leads us to our next ingredient.

Tropical Cyclones and the Coriolis Force: Giving Storms a Little Latitude

Throughout the last few sections, we focused our attention on the two basic characteristics that all developing tropical cyclones display: an observable low-level circulation, and organized thunderstorms around the center of that circulation. We haven't yet given much ink to the cyclonic circulation itself, primarily because we defined a tropical cyclone using the words "low-pressure system," so the two already go hand-in-hand.

However, there's an underlying assumption being made here that's not obvious: A contribution from the Coriolis force is necessary for a circulation to form in the first place. Without a sufficiently strong Coriolis force, air would move almost directly toward the center of the low, rapidly adding mass to the air column and almost certainly causing air pressure to increase. Given that the magnitude of the Coriolis force goes to zero at the equator, a tropical disturbance too close to

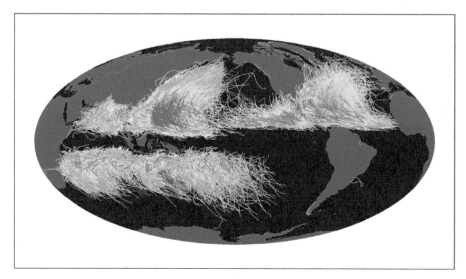

FIGURE 11.17 The tracks of all known tropical storms and hurricanes in the period 1851–2008. Light blue denotes tropical-depression status while green indicates tropical-storm intensity. Yellow through purple indicate hurricane intensities from Category 1 to Category 5 (courtesy of NOAA).

the equator would not have the opportunity to develop a low-level circulation and become a tropical cyclone.

Figure 11.17 dramatically illustrates our point. Here, the tracks of all recorded tropical storms and hurricanes in the period 1851 to 2008 are plotted—warmer colors indicate stronger winds. Note the lack of tropical cyclones on or near the equator. Indeed, as a general rule, tropical cyclones do not form within 5° latitude of the equator—the Coriolis force is simply too weak there—though there are exceptions to this rule. For example, here are the three tropical cyclones on record in the Northern Hemisphere that came closest to the equator while retaining hurricane strength:

- December 27, 2001: In the Northwest Pacific Ocean, Typhoon Vamei got within 1.5° of the equator.
- November 29, 2004: Cyclone Agni came within 2.7° of the equator in the North Indian Ocean.
- January 13, 2016: Hurricane Pali in the central Pacific Ocean reached 3.4°N before weakening into a tropical storm.

We will look more closely at Pali, which was the earliest (in the year) hurricane on record in the Central Pacific Ocean. Pali offers insights into the role that the Coriolis force plays in the fate of tropical

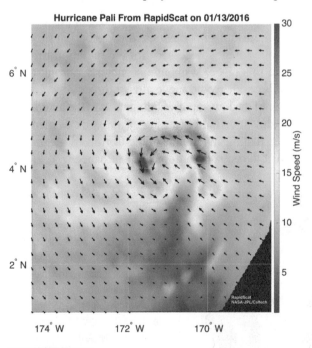

Hurricane Pali From RapidScat on 01/13/2016

FIGURE 11.18 Wind direction (arrows) and speed (color-coded, in m/s), obtained from the RapidScat instrument onboard the International Space Station, shows the surface circulation of Hurricane Pali as it neared its closest approach to the equator as a hurricane (courtesy of NASA JPL).

cyclones flirting with the equator. Figure 11.18 shows surface wind direction (arrows) and surface wind speeds (color-coded, in m/s) on January 13, 2016, the day Pali reached 3.4°N. By way of background, these wind measurements came from the RapidScat instrument, a special type of radar mounted on the International Space Station. RapidScat sends out pulses of microwave energy and then measures the return signals from the sea below. As it turns out, surface wind speeds and wind directions vary in a predictable way with the roughness of the sea surface. In this way, RapidScat captured the low-level cyclonic circulation associated with Pali, allowing forecasters to better gauge the strength of this tropical cyclone.

By that time, Pali had already been a named storm for six days—the storm was first upgraded to tropical cyclone status on January 7 when it was near 4°N latitude. In light of a weak Coriolis force so close to the equator, the obvious question is: How did Pali ever develop a low-level circulation in the first place?

For starters, there was an ongoing strong El Niño, so local water temperatures were as high as 29°C (85°F), plenty warm to support a tropical cyclone. However, that doesn't explain the circulation—for that, we turn to a phenomenon called the **Madden-Julian Oscillation** (MJO for short), a train of disturbances that moves slowly from west to east across the tropics, particularly the Indian and Pacific Oceans. First identified in the early 1970s, the MJO brings alternating periods of enhanced and suppressed convective rainfall, as well as accompanying shifts in surface wind direction. Typically, the prevailing easterly trade winds dominate ahead of an area of enhanced convection, while westerly winds take over once the disturbance passes (eventually, the wind direction returns to easterly, hence the word "Oscillation").

The westerlies that follow the passage of an area of enhanced MJO-driven convection can be quite strong, especially during an El Niño—so strong, in fact, that the abrupt transition to westerly winds is called a **westerly wind burst** (WWB). Such a WWB occurred during the period January 5–9, 2016, aiding in the development of Tropical Cyclone Pali. Figure 11.19 shows the anomalies (departures from the long-term average) of surface winds over the tropical Pacific Ocean during the period January 5–9, 2016. The area of dark purple marks the largest anomalies—the long arrows pointing in a general eastward direction inside this area are the footprint of the westerly wind burst. Just to the north of these anomalously strong westerly winds, note that some

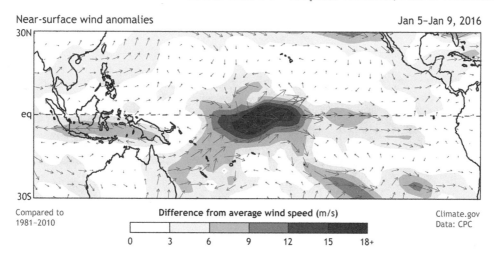

FIGURE 11.19 Anomalies in wind direction and wind speed for the period January 5-9, 2016, the period when Hurricane Pali formed and intensified. The area in purple in the central tropical Pacific Ocean with arrows pointing eastward represents a westerly wind burst associated with the MJO. Westward-pointing arrows just to the north of the purple area indicate strong-than-average easterly trade winds. The juxtaposition of these westerly and easterly winds provided a counterclockwise circulation that jump-started Pali (courtesy of NOAA).

arrows point westward—these represent stronger than average easterly trade winds. Together, the westerly wind burst associated with the MJO and the "boosted" easterly winds farther north created a counterclockwise rotation that served as the seedling circulation for Pali. In the final analysis, Hurricane Pali was a byproduct of the unusually strong El Niño of 2015–16 and a westerly wind burst associated with the Madden-Julian Oscillation.

With this insight into a rare tropical cyclone near the equator, you might wonder whether a tropical cyclone can ever cross the equator. To our knowledge, none has, but a few have come close, particularly in the North Indian Ocean. Is it possible? We would say "yes" because, at least initially, a tropical cyclone's large cyclonic circulation would not be affected by the weak change in the Coriolis force as the low crossed into the opposite hemisphere. However, there's another factor related to the variation of the Coriolis force with latitude that works against a crossover. Although this "other factor" (called the Beta effect) is beyond the scope of this textbook, we at least wanted to address the issue because we often get this question from inquisitive students.

With these extreme but interesting issues put to rest, let's talk generic. What is the primary source for the groups of loosely organized showers and thunderstorms that can move into thermodynamically favorable environments and increase the chance for tropical cyclones spinning up over the North Atlantic and Northeast Pacific Basins? Let's investigate.

Easterly Waves: Out of Africa

Recall that a group of loosely organized showers and thunderstorms over the tropics is formally called a tropical disturbance. Think of a tropical disturbance as a spark that can ignite a favorable thermodynamic environment and pave the way for a tropical cyclone to form.

Over the North Atlantic Ocean, approximately 60 percent of the tropical storms and "minor" hurricanes (Categories 1 and 2 on the Saffir-Simpson Scale) are initiated by tropical disturbances that move westward from Africa. These disturbances also initiate nearly 85 percent of all Atlantic hurricanes that reach Category-3 strength or higher, which the National Hurricane Center classifies as **major hurricanes**. If that's not enough to convince you that these tropical disturbances from Africa are worthy of our scientific scrutiny, consider that some meteorologists believe that nearly all the tropical cyclones that form in the Northeast Pacific Basin owe their existence to tropical disturbances from Africa.

Tropical disturbances that move westward from Africa are commonly called easterly waves. Formally, an **easterly wave** is a tropical disturbance that has a cyclonic circulation in the lower half of the troposphere, and that originates north of the equator over Africa and then moves westward across the tropical North Atlantic Ocean (and sometimes into the Northeast Pacific Ocean). Meteorologists sometimes call an easterly wave a **tropical wave**, though,

Aug–28–2008 21:01UTC

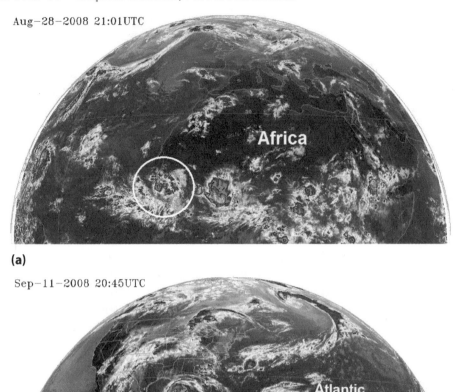

(a)

Sep–11–2008 20:45UTC

(b)

FIGURE 11.20 (a) An infrared satellite image at 21 UTC on August 28, 2008 that shows the eastern tropical Atlantic and Africa. An easterly wave just emerging into the Atlantic Ocean from Africa (circled in white) would provide the seedling low-level circulation for Tropical Storm Ike (copyright 2008 EUMETSAT); (b) Two weeks later, at 21 UTC on September 11, Hurricane Ike was churning westward in the central Gulf of Mexico (courtesy of NOAA).

more formally, it is an **African easterly wave.** The bottom line here is that easterly waves often serve as the catalyst for the seedling low-level cyclonic circulation that's needed for the genesis of tropical cyclones.

During the period from June to October, an easterly wave comes off the west coast of Africa into the eastern tropical Atlantic (north of the equator) every three or four days, on average. On August 28, 2008, a strong easterly wave emerged from Africa into the Atlantic, already generating plenty of high, cold cloud tops (see a colorized infrared image in Figure 11.20a). This tropical wave would eventually serve as the catalyst for the low-level circulation of Tropical Storm Ike, which formed on September 1. Two weeks after the image in Figure 11.20a,

Category-2 Hurricane Ike, packing winds of 85 kt (almost 100 mph), was moving westward through the Gulf of Mexico headed for an eventual landfall near Galveston, Texas (see Figure 11.20b).

On average, approximately 60 easterly waves emerge from Africa each year. Given that the long-term average annual number of named Atlantic storms is about ten (approximately six of which become hurricanes), it stands to reason that most easterly waves do not initiate a tropical cyclone. At the very least, some easterly waves produce clusters of thunderstorms that affect the Caribbean Islands or the Bahamas.

What causes easterly waves? This is a good question whose answer requires a deeper understanding of the dynamics of the atmosphere over northern

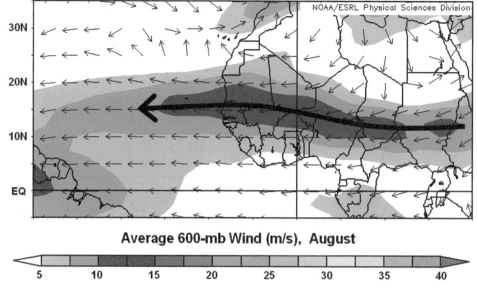

Average 600-mb Wind (m/s), August

FIGURE 11.21 Average wind direction (arrows) and speed (color-coded in m/s) at the 600-mb level (about 4000 m) during the month of August, showing the average position of the Middle Level African Easterly Jet as the black arrow (courtesy of NOAA).

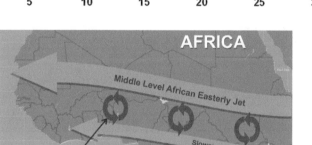

FIGURE 11.22 Easterly waves form on the cyclonic shear side (southern side) of the Middle Level African Easterly Jet.

Africa. Figure 11.21 shows the average wind direction and speed at the 600-mb level (an altitude of about 4000 meters) during the month of August. The narrow ribbon of fast 600-mb winds stretching across Africa and indicated by the black arrow is the **Middle Level African Easterly Jet** (MLAEJ for short); the 600-mb pressure level lies in the middle troposphere, hence the "Middle Level" in the name of this seasonal jet. During August, the core of the MLAEJ spans from roughly 30°E to 30°W longitude.

As it turns out, the MLAEJ is a major source for African easterly waves, particularly on the south side of the roughly east-west axis of the jet. That's where changes in the speed of the wind over a relatively short distance (measured north to south) impart a cyclonic circulation. To see how this **horizontal**

wind shear works, consider Figure 11.22 which shows the MLAEJ (the thick arrow) and also a current of slower easterly winds to the south of the MLAEJ. Imagine placing the center of a carnival pinwheel between these two streams of air. With faster winds on its northern side than on its southern side, the pinwheel will spin counterclockwise (that is, cyclonically in the Northern Hemisphere). Similarly, a newly minted easterly wave is a cyclonically circulating eddy that "spins up" from this shear. Given the pivotal role that easterly waves play in the genesis of tropical cyclones over the North Atlantic and Northeast Pacific Oceans, it behooves us to learn more about the MLAEJ.

During North Africa's hot summer, trade winds in the Southern Hemisphere cross the equator and penetrate to the southern edge of the Sahara Desert. As it turns out, this cross-equatorial flow is a component of Africa's summer monsoon (see Figure 11.23). At the southern edge of the Sahara, moist and relatively cool air associated with this monsoon meets hot and dry air flowing from the northeast over the Sahara Desert (these northeasterlies are the infamously scorching **harmattan winds**). Figure 11.24 shows average surface temperatures in August and a few arrows that represent these hot winds as well as a few arrows representing the cooler monsoonal flow. Note that average temperatures approach 40°C (104°F) over the western Sahara Desert. Also note the narrow ribbon of relatively large temperature gradient that stretches roughly east-west across Africa. Local African meteorologists call the boundary between these two contrasting air masses the **Intertropical Front**.

FIGURE 11.23 Dramatic seasonal shifts in wind direction occur in a zone from Africa to Southeast Asia and northern Australia, qualifying the highlighted area as the world's major monsoonal region. This region occupies a large portion of the tropical eastern hemisphere.

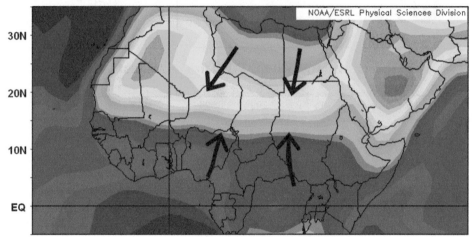

FIGURE 11.24 Average surface temperatures across Africa in August, in °C. Note the narrow ribbon of relatively large temperature gradient that marks the Intertropical Front. Arrows show how relatively cool, moist winds crossing the equator during Africa's summer monsoon meet hot and dry northeasterly winds from the Sahara Desert (courtesy of NOAA).

Average Surface Temperature (°C), August

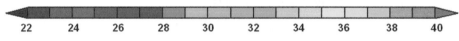

The bottom line here is that during summer, a front stretches east-west across Africa, with hotter air to the north and cooler air to the south—in essence, a reversal of the "normal" north-south temperature gradient. This front is relatively shallow, with north-south temperature gradients vanishing in the mid-troposphere. This lower tropospheric thermal gradient produces a height gradient that maximizes around 600 mb, with higher heights to the north and lower heights to the south at this pressure level (as shown in Figure 11.25). It follows that there are fairly strong winds that blow from the *east* just below the 600-mb level. These strong easterlies mark the Middle Level African Easterly Jet, a unique source for easterly waves.

Easterly waves don't become tropical cyclones overnight. It takes some time and distance from land for an easterly wave to develop into a hurricane. To our knowledge, only a handful of tropical cyclones have achieved hurricane status east of 25°W longitude in the deep tropics. That's about the longitude of Cabo Verde (formerly Cape Verde), an island archipelago located between about 15°N and

17°N latitude approximately 570 km (350 mi) off the coast of West Africa. Atlantic tropical storms and hurricanes that originate as easterly waves within 1000 km (600 mi) or so of Cabo Verde are called **Cabo Verde storms**. The season for Cabo Verde storms typically runs from August to early October, essentially the heart of hurricane season in the Atlantic.

With African easterly waves under our belt, we've taken a big step toward explaining a feature of Figure 11.17 that we ignored earlier—the almost complete lack of tropical cyclones in the South Atlantic Ocean. The Middle Level African Easterly Jet, which is the major source of tropical waves, is strictly a Northern Hemisphere phenomenon—it has no Southern Hemisphere counterpart. As a result, there aren't any easterly waves over the South Atlantic to provide the seedling low-level circulation to kick-start tropical cyclones.

Having said this, on rare occasion, tropical cyclones have formed in the South Atlantic, but they typically originate as mid-latitude lows (extratropical cyclones) that gradually acquire tropical characteristics.

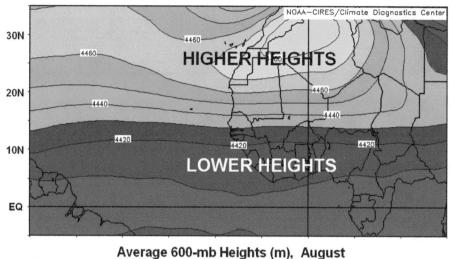

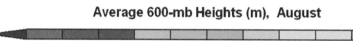

Average 600-mb Heights (m), August

4400   4420   4440   4460   4480   4500

FIGURE 11.25 Average 600-mb heights during August over Africa and the adjacent eastern Atlantic Ocean. With high heights to the north and low heights to the south (a reverse of the "normal" north-south gradient), a ribbon of strong easterly winds (the MLAEJ) sets up in the large gradient of 600-mb heights (courtesy of NOAA).

FIGURE 11.26 A visible satellite image of a rare hurricane off the east coast of Brazil on March 26, 2004. Dubbed "Catarina" by local meteorologists because it came ashore in the Santa Catarina province of Brazil, the storm destroyed 500 homes and damaged 20,000 others. The U.S. National Hurricane Center estimated its peak winds at 78 kt (90 mph), unofficially ranking "Catarina" as a Category 1 on the Saffir-Simpson Scale (courtesy of NASA).

One example of a rare South Atlantic hurricane that evolved this way was "Catarina" in late March 1994 (see Figure 11.26). Its track is the lone one you see in the South Atlantic in Figure 11.17.

With the six ingredients for the genesis and development of tropical cyclones in hand, it's now time to investigate and formalize the feedback processes by which tropical cyclones intensify.

STARTING AND REVVING THE HEAT ENGINE: RUNNING ON ALL CYLINDERS

Figure 11.27 is a visible satellite image showing Hurricane Kilo straddling the International Date Line in early September 2015. Because of this unique positioning, Kilo was in a kind of time warp, given that the local date for the eastern half of the storm was September 1 while the local date for the western half was already 24 hours ahead! For the record, a hurricane that moves west and crosses the International Date Line officially becomes a **typhoon** as the baton of jurisdiction passes from the Central Pacific Hurricane Center to the Joint Typhoon Warning Center on Guam.

This interesting twist of time aside, the legacy of Hurricane / Typhoon Kilo was that the storm retained tropical storm or hurricane strength for approximately three weeks, from August 20 to September 11 (see Figure 11.28), putting it in relatively exclusive company as one of the longest-lived tropical cyclones on record. During its long trek across the Pacific, Kilo spent about a day as a Category-4 storm—at its most powerful, it packed maximum sustained winds of 120 kt (138 mph) and registered a central pressure of 940 mb. Eventually, Kilo weakened before losing its tropical characteristics off the coast of Asia.

How do tropical cyclones like Kilo attain and maintain such low central air pressures, and what causes tropical cyclones to intensify (and then weaken)? Fundamentally, it's all about controlling the weight of air columns.

FIGURE 11.27   A visible satellite image of Hurricane Kilo at 1950 UTC on September 1, 2015. At this time, Kilo straddled the International Date Line (courtesy of JMA/NOAA/CIRA).

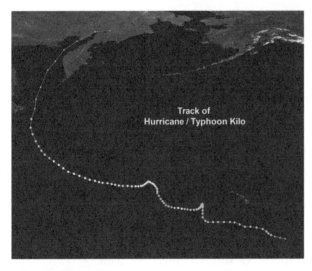

FIGURE 11.28   The track of Hurricane/Typhoon Kilo in the Pacific Ocean from approximately August 20 to September 12, 2015. Blue circles indicate tropical-storm strength, while tans, yellows, and oranges represent increasingly strong hurricane/typhoon strength (courtesy of Wikipedia).

Tropical Cyclone Health: Keeping the Weight Off

Once a tropical cyclone develops, its survival pretty much comes down to three characteristics of the local environments into which it moves:

- Is the underlying ocean water sufficiently warm?
- Is atmospheric wind shear relatively weak?
- Is the local atmosphere conditionally unstable?

Although all three factors are vital, we will begin by addressing the question of maintaining relatively low pressure in the context of conditional instability.

Earlier we pointed out that the development of a tropical cyclone hinged on thunderstorms organizing around the center of a tropical disturbance. What is critical about this particular pattern of organized convection?

For starters, high sea-surface temperatures translate to high evaporation rates, ensuring that the lower troposphere is teeming with water vapor. In turn, high humidity increases the likelihood that the local atmosphere is conditionally unstable. Recall from Chapter 8 that in conditionally unstable environments, the release of latent heat keeps rising saturated air parcels positively buoyant. As a result, thunderstorms clustering around the low-level center of a developing tropical cyclone typically reach very high altitudes. As they ascend, air parcels lose water vapor via net condensation, so by the time they reach such lofty heights, parcels are almost devoid of vapor.

Having reached the tops of thunderstorms, what becomes of these relatively dry air parcels? There are essentially two distinct routes: some diverge away from the center (more on this "upper-level divergence" in a moment), while others sink into the central core of the tropical cyclone (see Figure 11.29). Given that parcels are nearly devoid of water vapor, this subsiding air warms at a rate very close to dry adiabatic. As a result, the central air column in a developing tropical cyclone has the highest average temperature. For proof, check out Figure 11.30,

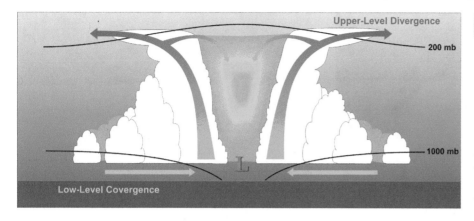

FIGURE 11.29 Here we track the routes of air parcels rising in thunderstorms around the center of a newly forming tropical cyclone. Some parcels reaching the top of thunderstorms sink into the central air column. Because they are practically devoid of water vapor, these sinking parcels dramatically warm at a rate close to dry adiabatic. Other air parcels diverge away from the center.

a cross section of temperature anomalies across Typhoon Kilo at 18 UTC on September 5, 2015—positive values correspond to air that is anomalously warm compared to the environment surrounding the storm. Kilo is easy to spot—the columns of green and especially yellow pinpoint its core. The message here is unmistakable: the average temperature of columns of air near the center of a hurricane is higher than in the surroundings. More generally (and in other words), tropical cyclones are "warm core."

Why is this temperature structure so important? Well, the warm anomaly at the core of a tropical cyclone reduces the average air density there—in turn, the weight of the central air column is the lowest in the environment of the tropical cyclone. This translates to a low air pressure at the center—and, as with all lows, that means winds converge at low levels.

But that can't be the end of the story. If air converging around the central core of a developing tropical cyclone was the only game in town, surface pressures would increase as converging air added weight to local columns. To offset this low-level weight gain, there must be divergence at high altitudes to evacuate air from atop the tropical cyclone's central core (revisit Figure 11.29).

How does this upper-level divergence occur? For the answer, we appeal to the notion that tropical cyclones are warm core. Recall that pressure decreases slower with height in warm air than in cold air. Thus, as the central core of the tropical cyclone warms, the rate of decrease in pressure with increasing altitude slows. As a result, constant pressure surfaces in the central core (which dramatically dip toward the surface in the lower troposphere) tend to flatten out with height and eventually bulge upward at high altitudes, the footprint of high pressure aloft (compare the 1000-mb and 200-mb pressure surfaces in Figure 11.29). This high pressure aloft leads to upper-level divergence

that helps to offset the weight gain from low-level converging winds. In fact, on animations of satellite images of hurricanes in the Northern Hemisphere, cirrus clouds (acting as tracers of the high-altitude wind) typically show a subtle pattern of divergence, with air circulating clockwise and outward away from the center. Sometimes even static satellite images, such as the one of Hurricane Isabel in Figure 11.31, suggest this upper-level divergence—in this case,

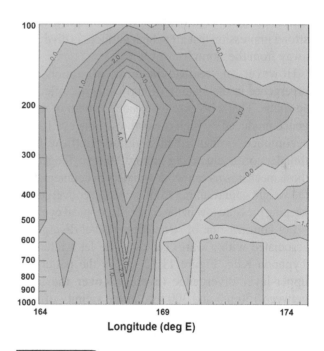

FIGURE 11.30 A cross section (from west to east) across Hurricane Kilo, showing the departure of temperature from the surrounding environment at 18 UTC on September 5, 2015. These measurements were taken by an instrument onboard a NASA satellite. The horizontal axis designates longitude (in degrees East), while the vertical axis denotes pressure (in mb). Positive values (greens and yellows) indicate warm anomalies and correspond to columns of air in and around Kilo's core (courtesy of NASA and CIMSS/University of Wisconsin).

FIGURE 11.31 A high-resolution visible satellite image of Category-4 Hurricane Isabel at 1640 UTC on September 10, 2003. At this time, Isabel was packing maximum sustained winds of 135 mph. The thin cirrus clouds on the hurricane's western and southern sides reveal a pattern of upper-level divergence (courtesy of MODIS Rapid Response Team, NASA/GSFC).

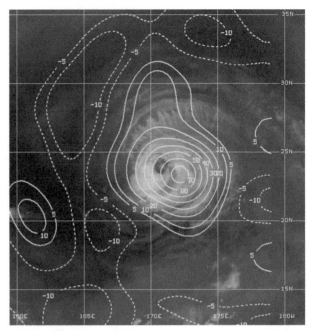

FIGURE 11.32 Isopleths of upper-level divergence superimposed on water vapor satellite imagery at 18 UTC on September 5, 2015, over the central Pacific Ocean in the vicinity of Typhoon Kilo. Positive values indicate divergence, while negative values represent convergence. Significant upper-level divergence was occurring over the storm, though the maximum divergence was not directly over the core (courtesy of University of Wisconsin/CIMSS).

the striated pattern of the thinner cirrus clouds on the western and southern flanks of the hurricane give the strong impression that air at high altitudes is diverging away from the storm.

However, in reality, patterns of upper-level divergence in the vicinity of hurricanes are often not so subtle. That's because the general pattern of divergence produced by high pressure over the eye has an accomplice—enhanced high-altitude outflow from thunderstorms clustering around the eye. Together, these mechanisms can work wonders in counteracting the weight gain resulting from low-level convergence, keeping pressure low in the center of a tropical cyclone.

For confirmation of strong upper-level divergence associated with a warm central core, let's return to Typhoon Kilo. Figure 11.32 shows the pattern of upper-level divergence (averaged over the layer between 150 mb and 300 mb) in the vicinity of Kilo at 18 UTC on September 5, 2015, superimposed on a water vapor image (this is about the same time as the cross section of temperature anomalies in Figure 11.30). Here, positive values indicate divergence, and although we won't delve into the details of the units, the values shown represent significant upper-level outflow. Notice that the maximum upper-level divergence in Figure 11.32 does not occur right over Kilo's center, a characteristic that's actually typical for tropical cyclones. Indeed, the pattern of upper-level divergence tends to be skewed toward the region with the strongest outflow from thunderstorms. Only rarely will you see a bullseye of upper-level divergence directly over the center of a tropical cyclone (though it can happen when the upper-level outflow is perfectly symmetric, with the emphasis on "perfectly").

This discussion should impress upon you the variety of interconnected processes occurring during the development phase of a tropical cyclone. We will now summarize the intensification of tropical cyclones in the context of a feedback loop so that you can get a better sense of how the heat engine of a tropical cyclone revs up. We purposely use the analogy of a heat engine here because, ultimately, the release of latent heat in thunderstorms clustering around a tropical cyclone's center drives the intensification process.

Amplifying the Storm: Positive Feedback

Let's review the amplification process of a strengthening tropical cyclone by starting at the beginning. As low pressure forms at the ocean surface, air converges toward the center. If the low develops far enough from the equator, the deflection by the

Coriolis force creates a cyclonic circulation. Meanwhile, as surface pressure decreases in response to compressional warming over the low's center, the horizontal pressure gradient strengthens. In response, wind speeds increase, especially lower-tropospheric winds around the core of the newly forming tropical cyclone where the pressure gradient tends to be largest. Stronger winds translate into increased low-level convergence and higher evaporation rates. As a result, the number and intensity of thunderstorms increase. In turn, high-altitude outflow strengthens, producing an increase in upper-level divergence and paving the way for the tropical cyclone to continue to deepen (as manifested by a lower surface pressure and faster low-level winds). The takeaway here is that upper-level divergence plays a pivotal role in the intensification process.

Now the stage is set for a positive (amplifying) feedback loop. Increased upper-level divergence associated with increased high-altitude outflow causes surface pressure to further decrease and low-level winds to increase. Increasing low-level winds further increase evaporation rates, the amount of water vapor and heat energy in the air, and low-level convergence. In turn, the number and intensity of thunderstorms organizing around the central core again increase, upper-level divergence increases, and the tropical cyclone finds itself caught in a positive feedback loop. This positive feedback loop involves elements of two theories used to explain the intensification of tropical cyclones: CISK, which stands for **Conditional Instability of the Second Kind**

(historically, an early theory to explain intensification), and a more contemporary theory known as WISHE, which stands for **Wind-Induced Surface Heat Exchange**. In this theory, the surface winds generated by tropical cyclones essentially drive the releases of heat energy and moisture from the ocean: the greater the wind speeds, the greater the fluxes of heat energy and moisture into the atmosphere.

Regardless of the theoretical details, the positive feedback and intensification cannot continue unchecked. Stronger vertical wind shear and mid-level dry air can disrupt the intensification process and cause tropical cyclones to weaken. And ultimately, even when the environmental conditions are ideal, sea-surface temperatures impose an upper limit on the potential intensity of hurricanes.

From a forecasting point of view, satellite imagery can provide clues that a tropical cyclone is intensifying. As you just learned, the intensification of a tropical cyclone goes hand-in-hand with an increase in the number and organization of thunderstorms around the central core. One sign of such an increase is the development of hot towers in the eye wall of a hurricane, towering cumulonimbus clouds that signal deep convection reaching the tropical tropopause. As an example, consider the hot towers that formed around the eye of Tropical Storm Wilma in October 2005, visualized in Figure 11.33 by a NASA satellite. These towering convective plumes were a clue to Wilma's upcoming intensification to a Category-5 hurricane as it approached Mexico's Yucatan peninsula.

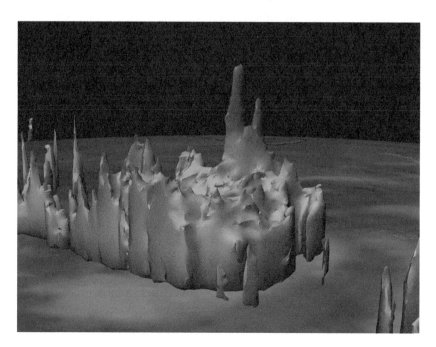

FIGURE 11.33 Radar onboard NASA's TRMM satellite (Tropical Rainfall Measuring Mission) measured the heights of precipitation columns around the center of Tropical Storm Wilma on October 17, 2005. Precipitation-sized hydrometeors were detecting by using a reflectivity threshold of 15 dBZ. The tallest precipitation columns (in deep red) were hot towers, which were the precursors to Wilma's rapid intensification (courtesy of NASA Goddard Space Flight Center's Scientific Visualization Studio).

FIGURE 11.34 A color-enhanced infrared satellite image of Super Cyclonic Storm Gonu at 18 UTC on June 4, 2007. The yellow shading, which represents cloud-top temperatures of –80°C (–112°F) or lower, marks the tops of tall thunderstorms in the eye wall of Gonu (courtesy of NOAA and the Naval Research Laboratory).

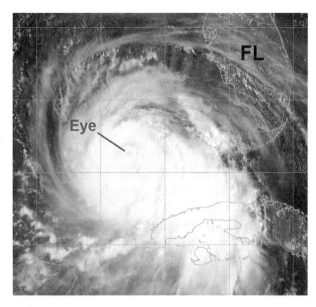

FIGURE 11.35 A visible satellite image of Hurricane Katrina over the Gulf of Mexico at 1515 UTC on August 27, 2005. Although you can see a faint outline of Katrina's center, thin cirrus clouds covering the eye prevent you from seeing low clouds in the eye or the underlying ocean surface (courtesy of NOAA and Naval Research Laboratory).

Color-enhanced infrared imagery, which visually conveys the temperatures of cloud tops, is also a tool used by tropical forecasters to gauge intensity. Figure 11.34 is a color-enhanced infrared satellite image of Super Cyclonic Storm Gonu at 18 UTC on June 4, 2007. The circular yellow shading around the eye of Gonu represents cloud-top temperatures of –80°C (–112°F) or lower. Such frigid cloud tops indicate very tall eye-wall thunderstorms, giving forecasters a clue about the formidable intensity of this tropical cyclone.

On the flip side, satellite imagery can also signal that a tropical cyclone may weaken. For example, the combing of cirrus clouds over a storm's eye on visible satellite imagery is a sign that the pattern of upper-level divergence is being disrupted, most commonly because the hurricane has moved into an environment of unfavorable vertical wind shear. Figure 11.35 is a visible satellite image of Hurricane Katrina on the morning of August 27, 2005. At this time, Katrina had re-intensified after crossing South Florida and was packing winds of 100 kt (115 mph) and registering a central pressure of 940 mb. However, despite warm Gulf waters, the intensification had been interrupted by strong upper-level winds (strong vertical wind shear) which had blown high-altitude cirrus clouds over the center (you can still

barely see the outline of the eye). In fact, Katrina's central pressure increased 9 mb in the three hours after this satellite image! Although the vertical wind shear then relaxed and Katrina began a rapid deepening, the message here is clear: If vertical wind shear is too strong (typically indicative of relatively speedy upper-air winds), surface pressures usually increase as the hurricane's sputtering exhaust system allows some air to pile up aloft—cirrus clouds blowing over the eye are a visible sign of this process.

EYE OF A HURRICANE: AT THE CENTER OF THE STORM

The eye of a hurricane is arguably one of the most captivating of natural phenomena. Figure 11.36 shows close-up visible and colorized infrared satellite images of the eye of Supertyphoon Vongfong on October 8, 2014, in the western Pacific Ocean. As stunning as the infrared image is, with the cold cloud tops of the eye wall (in red) ringing the relatively warm eye, the infrared image masks the complexity of the clouds in the eye. What, you say? Clouds in the eye of a hurricane? Indeed, despite what is commonly advertised, a hurricane's center is often not an oasis of clear skies. In fact, inside Vongfong's eye, you can see on the visible image

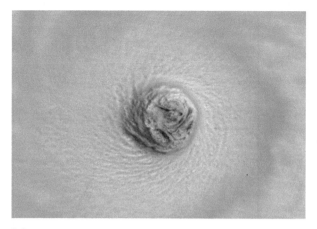

(a)

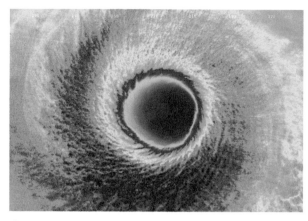

(b)

FIGURE 11.36 Visible (left) and colorized infrared (right) close-up satellite images of the eye of Supertyphoon Vongfong at 1640 UTC on October 8, 2014. The mesovortices seen on the visible image in the relatively warm eye are difficult to discern in the infrared image (courtesy of NOAA and the Cooperative Institute for Meteorological Satellite Studies, University of Wisconsin).

several **mesovortices**, which, in this case, are relatively small, cyclonic swirls of low clouds.

Nor is the eye necessarily calm. Momentum from fast-moving air in the eye wall can sometimes mix into the eye, creating swirls of turbulent clouds and gusty winds, especially on the periphery of the eye. Still, the central region of a hurricane's eye is, for the most part, an island of relatively light winds—truly a wonder of nature that, just 10 or so miles away in the eye wall, winds could be blowing over 130 kt (150 mph).

How can such a dramatic difference in wind speed occur over such a short distance? Why don't these strong winds penetrate all the way to the center? Or stated another way, why do hurricanes have an eye in the first place?

Features of the Eye: Tight Spirals and Football Stadiums

To help us answer these questions, we once again invoke the principle of conservation of angular momentum. In this case, it dictates that air parcels spiraling inward toward the center of a tropical cyclone must speed up. Figure 11.37 shows trajectories that air parcels take while spiraling toward the center of a hurricane.

Let's consider the problem in the context of the forces acting on these parcels near the center (see Figure 11.38). The inward-directed pressure gradient force is very large for sure—after all, this is a hurricane and the winds are howling! Yes, there's some Coriolis force, but on these small spatial scales, and compared to the pressure gradient force, Coriolis is small potatoes, so we'll disregard it. However, these air parcels take very curvy paths, so there's another force we can't

FIGURE 11.37 Trajectories of air parcels spiraling inward toward the center of a hurricane. Why do parcels stop short of the center? In other words, why is there an eye?

ignore—the outward-acting centrifugal force which gets larger as the speed of the inward-spiraling parcels increases. To understand why, think of your bottom sliding outward as your car goes around a curve— that's the centrifugal force at work, and the faster you negotiate the curve, the greater the outward force. As parcels spiral inward, they reach a point at which the increasing centrifugal force nearly offsets the hurricane's large, inward-directed pressure gradient force. As a result, air parcels stop crossing isobars before getting to the center, instead converging and rising in the stormy eye wall. Voila, an eye forms.

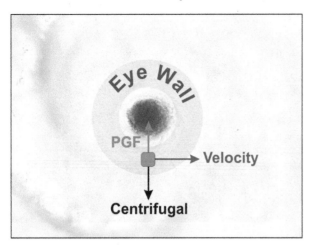

FIGURE 11.38 As an air parcel spirals inward toward the center of a hurricane, it reaches the eye wall where the outward-acting centrifugal force essentially balances the pressure gradient force. At this point, the parcel stops its inward spiral and rises in the eye wall (where many such air parcels converge). With parcels stopping their inward spiral in the eye wall, an island of relative calm—the eye—is left in the center.

In the interest of full disclosure, we should point out that near the ocean surface, friction between the wind and very turbulent seas substantially reduces the speed of inward-spiraling air parcels (essentially, near the surface, friction also plays a role in the balance of forces). If the truth be told, only air parcels above the boundary layer attain the near-balance of the centrifugal and pressure gradient forces – recall that the boundary layer is the layer in which friction with the surface is important, typically of depth 500-1000 m (about 1600-3200 ft). Figure 11.39 shows a typical vertical profile of wind speed in the eye wall of a hurricane. The data used to construct this figure were gathered via **dropsondes**, dispensable canisters of electronic weather instrumentation released from hurricane hunter aircraft (see Figure 11.40). Tethered to a parachute, a dropsonde measures temperature, pressure, and relative humidity as it descends toward the surface—wind speed and wind direction can also be retrieved by tracking the dropsonde using GPS. Note from Figure 11.39 that the "slowest" winds in the lower troposphere occur right at the ocean surface, where friction is greatest. Winds tend to increase with height as the effect of friction decreases, with maximum speeds typically near an altitude of 500 m (1640 ft).

Ordinarily, you would expect winds to keep increasing with height above the boundary layer (recall from Chapter 7 that this is particularly true in the mid-latitudes). But that's not the case here.

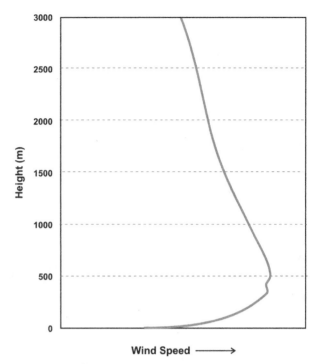

FIGURE 11.39 A plot of wind speed with altitude in the eye wall, using data gathered by dropsondes in more than 200 hurricanes, indicates that the maximum wind speeds in the eye wall occur about 500 meters (1640 ft) above the surface. Above the level of maximum wind speeds, winds in the eye wall gradually weaken with increasing altitude (adapted from Franklin, et. al., 2003, Volume 18, Weather and Forecasting).

Instead, downdrafts in eye-wall thunderstorms transport momentum downward into the boundary layer, putting the maximum wind speeds around 500 m (on average). This momentum doesn't really have an impact on winds closer to the surface, but does get transported down far enough to affect high-rise buildings. In fact, research suggests that during a hurricane, the upper floors of a high-rise building along the coast could experience winds a full Saffir-Simpson category stronger than on the ground floor.

The wind profile in Figure 11.39 also helps to explain an interesting visual feature of a hurricane's eye: its vertical structure is often reminiscent of a stadium, with eye-wall thunderstorms tilting outward with increasing altitude (refer back to Figure 11.29). To make the connection, we start with the fundamental link between the horizontal pressure gradient and wind speed—the decrease of wind speed in the eye wall above about 500 meters tells us that the magnitude of the horizontal pressure gradient across the eye wall must also decrease with increasing altitude. To see why this happens, consider Figure 11.41 which shows various constant pressure surfaces in

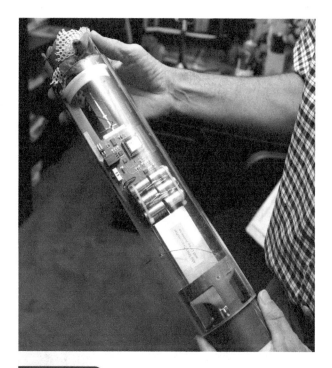

A dropsonde is a dispensable canister of electronic weather instruments released from Hurricane Hunter aircraft. Tethered to a parachute, a dropsonde measures temperature, pressure, and relative humidity as it descends toward the ocean surface. By tracking the motion of a dropsonde using the Global Positioning System, scientists can also retrieve wind speed and direction (courtesy of the U.S. Air Force).

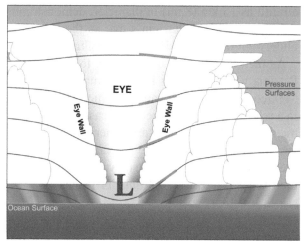

FIGURE 11.41 A schematic of various constant pressure surfaces in the eye and eye wall of a hurricane. The horizontal pressure gradient between the eye and the outer edge of the eye wall is greatest at the surface. Because pressure decreases relatively slowly in the eye's warm air column, the horizontal pressure gradient decreases with increasing altitude (indicated by the flattening of the constant pressure surfaces with height).

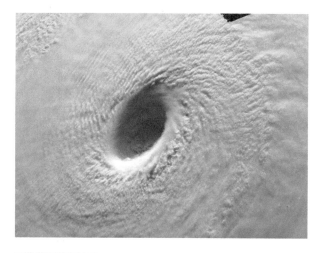

FIGURE 11.42 A close-up photograph of the eye of Typhoon Maysak on April 2, 2015, taken from the International Space Station, illustrating the stadium effect (courtesy of Astronaut Samantha Cristoforetti).

the eye and eye wall. Because pressure decreases more slowly with height in the eye's warm column, the constant pressure surfaces become "flatter" with increasing height—in other words, the magnitude of the horizontal pressure gradient decreases with increasing altitude. As a result, the inward spiral of air parcels relaxes with increasing height (and even reverses at very high altitudes where high pressure forms over the eye), and the vertical structure of the eye takes on the look of a stadium. Figure 11.42 is a close-up high-resolution visible image of Typhoon Maysak in the western Pacific Ocean on April 2, 2015, illustrating this so-called **stadium effect**. Staring at this photograph from this angle, you get the overwhelming sense that the eye is widest at the top and narrows with decreasing altitude, giving the impression that it is shaped like a stadium.

The size of a hurricane's eye can often provide vital clues to the current and future health of the storm. On average, the diameter of a hurricane's eye is 40 km (25 mi), but there is wide variation. For rapidly intensifying hurricanes, the eye wall usually contracts inward in response to the increasing pressure gradient, reducing the eye's diameter. Take Hurricane Patricia, for example, which rapidly intensified from a minimal hurricane early in the day on October 22, 2015, to a Category-5 with winds of approximately 180 kt (207 mph) in just 24 hours. Hurricane Hunter aircraft flying into the storm captured radar imagery on October 22 (Figure 11.43a) and reported an eye diameter of 22 km (about 14 mi). About 24 hours later, radar imagery showed a much smaller eye, reported by the Hurricane Hunter aircraft as 13 km (8 mi) in diameter, demonstrating the contraction of Patricia's eye as the hurricane underwent rapid intensification (see Figure 11.43b).

As mentioned in the introduction to this chapter, Patricia weakened a bit before it made landfall in Mexico, though it still caused tremendous damage in a relatively localized area. Forecasts of the movement of tropical cyclones are generally quite good, and that was the case with Patricia. As a result, the preparation for Patricia came down to understanding the risks posed by a landfalling hurricane.

HURRICANE ALERT: ASSESSING THE RISK

Figure 11.44 shows the tracks of all known hurricanes in the North Atlantic Ocean (1851–2013) and Eastern North Pacific Ocean (1949–2013). Although many Atlantic hurricanes clearly have come ashore

on the Gulf Coast or Atlantic Coast of the United States, most of the tracks completely miss land. For example, during the period 1950–2015, only about 27% of all hurricanes that formed in the Atlantic Basin made landfall in the United States.

How do forecasters know where a hurricane is headed? And what are the primary risks for life and property in the path of the storm? We'll tackle the steering question first, and then deal with the impacts of landfalling tropical cyclones.

Hurricane Movement: Subtropical Highs at the Steering Wheel

Figure 11.45 shows the tracks of all hurricanes and tropical storms that formed in Atlantic Basin in 2014. Except for two weak tropical storms in the Caribbean Sea and western Gulf of Mexico, the tracks of the

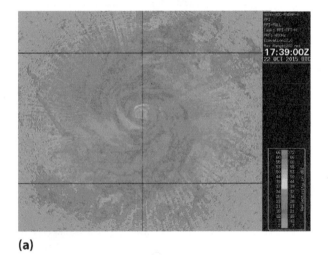
(a)

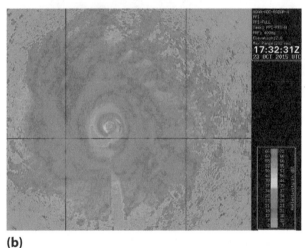
(b)

FIGURE 11.43 (a) Radar reflectivity from the NOAA Hurricane Hunters of the core of Hurricane Patricia at 1739 UTC on October 22, 2015 when the storm's maximum sustained winds were near 115 kt (132 mph). At this time, the diameter of the eye was approximately 22 km (14 mi); (b) About 24 hours later, at 1732 UTC on October 23, 2015, the eye had shrunk to about 13 km (8 mi) in diameter and maximum sustained winds approached 180 kt (207 mph) (courtesy of NOAA).

FIGURE 11.44 Tracks of all known hurricanes in the North Atlantic Ocean for the period 1851–2013, and in the Eastern North Pacific Ocean for the period 1949–2013 (courtesy of NOAA).

other named storms all exhibited a gradual clockwise arc over the North Atlantic Ocean. Such clockwise motion was orchestrated by the large-scale, anticyclonic circulation of air around the sprawling North Atlantic subtropical high—recall from Chapter 10 that this feature of the general circulation is a semi-permanent characteristic of the weather pattern over the North Atlantic Ocean, especially during the warm season when tropical cyclones thrive.

The bottom line here is that the expansive Atlantic subtropical high steered most of the 2014 named storms. More generally, the resident subtropical high in each tropical basin plays a major role in steering tropical cyclones. Other weather systems (such as upper-level troughs) can influence the paths that tropical systems take, but for the most part, subtropical highs rank at the top of the list for helping to steer many tropical cyclones.

How do forecasters determine the steering winds for tropical cyclones? For starters, it's not as simple as using the winds at a single altitude to move storms along. That's because tropical cyclones have depth—thus, it's more accurate to think about a steering *layer* of air instead of a single steering altitude. Indeed, the mean wind speed and the mean wind direction in such a steering layer provide forecasters with a reasonable estimate for storm movement.

How are the top and bottom of this steering layer determined? Research has shown that the depth of the steering layer increases as tropical cyclone intensity increases. That's because a more intense tropical cyclone will have a deeper, more well-developed vortex which, in turn, will be guided by a deeper steering flow. Figure 11.46 shows the various depths of the steering layer as a function of a tropical cyclone's minimum pressure (which is correlated with intensity). Let's try an example to see how this chart works.

On August 29, 2016, Category-3 Hurricane Lester, with minimum central pressure of 957 mb, was swirling over the eastern Pacific Ocean (see Figure 11.47a). In its 12 UTC forecast discussion, the National Hurricane Center noted: "The initial motion estimate remains 270/13 (moving westward at 13 kt). A strong, deep-layer ridge to the north of Lester should steer the cyclone on a continued westward course during the next three days."

So, is the National Hurricane Center's estimate and analysis consistent with the chart? Based on Figure 11.46, the steering layer for Hurricane Lester extended from 850 mb to 300 mb. In light of this steering layer, Figure 11.47b shows the mean wind direction (depicted by white streamlines) and the mean wind speed (color-coded in knots) for this layer. As the National Hurricane Center noted, a deep-layer ridge of high pressure north of Lester (essentially, a subtropical high) was steering the hurricane at this time. Indeed, note that the streamlines depicting the mean wind direction were pointing westward and that the mean wind speed in the steering layer was in the

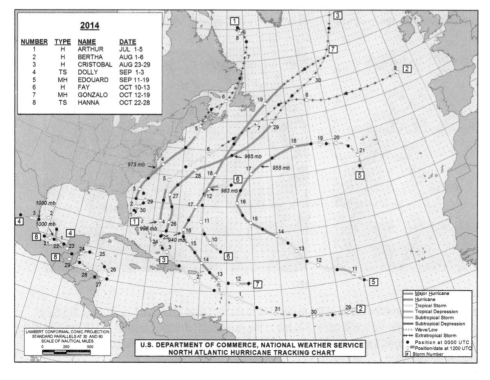

FIGURE 11.45 Tracks and intensities of all North Atlantic tropical cyclones in 2014 (courtesy of the National Hurricane Center).

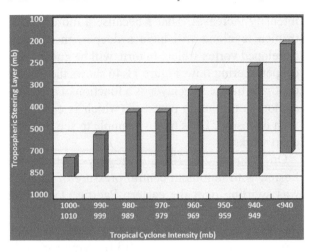

FIGURE 11.46 Typical steering layers for tropical cyclones as a function of minimum central pressure (adapted from CIMSS/University of Wisconsin).

neighborhood of 10 kt, both of which are consistent with the National Hurricane Center's analysis.

When the mean wind in the steering layer is weak, tropical cyclones tend to move slowly and sometimes erratically. Or they may simply stall. Either way, weak steering currents set the stage for lethargic tropical cyclones to produce protracted heavy rain and flooding.

The poster storm for slow-moving, prodigious rain-makers is Hurricane Harvey, which made landfall along the Texas Coast near Corpus Christi on August 25, 2017. Harvey was a Category 4 at landfall, so powerful winds and storm surge were part of its arsenal. However, Harvey will forever be known for its cata-

strophic rainfall and flooding. Southeast Texas bore the brunt of the rain, with some areas receiving over 40 inches (102 cm) in just 48 hours! Near Nederland, TX, about 80 mi (129 km) east of Houston, 60.58 inches (154 cm) of rain fell from the storm, setting a new single-storm record for North America. Figure 11.48 shows Harvey's extraordinary rainfall over Texas and Louisiana for the period August 24-31, 2017.

To produce such extreme rainfall, a tropical cyclone must stall and linger over the same region for days, which is exactly what Harvey did. Though the storm approached the Texas Coast at speeds of 5-9 kt (6-10 mph), the steering environment all but went calm after the storm moved inland. Figure 11.49 shows the mean steering currents at 15 UTC on August 26. At this time, Harvey's central pressure was 984 mb with maximum sustained winds of 65 kt (74 mph), so the steering layer (according to Figure 11.46) was 850-400 mb. The white streamlines indicate the direction of the mean steering wind, while the speed is color-coded (in kt). Note that the steering winds were very weak at this time – the National Hurricane Center estimated that Harvey was moving to the north-northwest at 2 kt (2.3 mph). The steering currents remained very weak for several days, which was a recipe for historic flooding.

On September 1, 2019, Category-5 Hurricane Dorian, packing sustained winds of 160 kt (184 mph), made landfall on the island of Great Abaco in the Bahamas. To make matters much worse, the mean wind in Dorian's steering layer essentially collapsed, so the storm moved only about

(a)

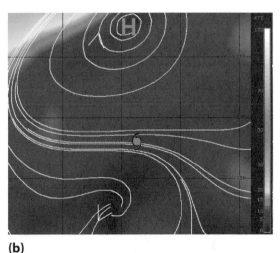

(b)

FIGURE 11.47 (a) Visible satellite image of Hurricane Lester in the eastern North Pacific Ocean on August 29, 2016 (courtesy of NOAA); (b) Mean flow in the layer from 850 mb to 300 mb in the vicinity of Hurricane Lester at 12 UTC on August 29, 2016. With an easterly mean flow in this steering layer, the National Hurricane Center predicted that Lester would continue heading westward (courtesy of CIMSS/University of Wisconsin).

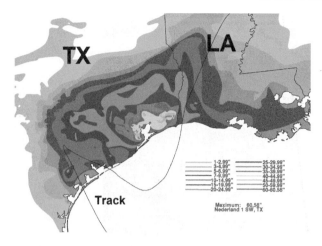

FIGURE 11.48 Rainfall (in inches) produced by Hurricane Harvey for the period August 24–31, 2017. Areas in yellow received more than 40 inches (102 cm) of rain. The thin black line shows Harvey's track (courtesy of the National Weather Service, Lake Charles, LA).

FIGURE 11.49 The steering layer for Hurricane Harvey at 15 UTC on August 26, 2017, was 850 mb to 400 mb. The mean wind direction in this layer is indicated by white streamlines, while the speed of the mean wind is color-coded in knots. With such weak steering currents for Harvey at this time, the National Hurricane Center estimated its forward speed at 2 kt (2.3 mph) (courtesy of CIMSS).

30 miles (48 km) over the course of a day. As a result, the eyewall of the hurricane slammed parts of the island (and Grand Bahama Island to its west) with Category-5 winds and catastrophic storm surge for 22 hours. Winds of major hurricane strength persisted on Grand Bahama for an incredible 36 hours. In the aftermath, portions of the Bahamas looked like a war zone. Since records began over 100 years ago, no other Atlantic hurricane has battered a populated area with such extreme conditions for such an extensively long time.

In the next section, we'll delve deeper into the topic of storm surge. We'll also discuss how some landfalling hurricanes can produce tornadoes. Read on.

Hour of Danger: Landfall and Storm Surge

Much like Paul Revere warned colonists that "The British are coming," gusty showers and thunderstorms organized in bands on the fringes of a hurricane often herald its approach. Revisit Figure 11.4 and recall that some tropical cyclones bear a striking resemblance to spiral galaxies. Taking this similarity one step further, we point out that the outskirts of a hurricane are dominated by its aptly named **spiral bands**. These tentacles of thunderstorms that pinwheel cyclonically around the center of a tropical cyclone correspond to areas of enhanced surface convergence and strong updrafts. Figure 11.50a shows that several spiral bands of Tropical Storm (previously Hurricane) Hermine on September 2, 2016

were prominent southeast of the center of the storm. "Squally" best describes the weather associated with spiral bands, with fitful but usually temporary rains, very gusty winds, and sometimes tornadoes.

Indeed, when hurricanes make landfall along the Southeast Coast or states that border the Gulf of Mexico, the Storm Prediction Center (SPC) in Norman, OK routinely issues tornado watches that frequently include parts of the spiral bands. In fact, SPC issued four tornado watches while Hermine was over land, including the one shown in Figure 11.50b (which was issued shortly after the radar image in Figure 11.50a). In this area ahead of the storm, easterly surface winds coming off the ocean were slowed by friction as they moved over land, while speedy winds aloft were, for the most part, unencumbered. In effect, friction created a layer in the lower troposphere where wind speeds increased rapidly with height. In turn, the potential existed for "horizontal rolls" to form, much like a pencil held in one hand (representing slower winds) rolls when your other hand rubs over it (representing faster winds aloft). In turn, thunderstorm updrafts can tilt the roll into an upright position, generating a tornado. Hurricane Ivan in September 2004 holds the record for most tornadoes spawned by a tropical cyclone—123 as it moved from Florida to Pennsylvania.

If you are in the direct line of fire of a hurricane, bigger problems than gusty bands of showers and thunderstorms are on the way. Rain typically becomes heavier and steadier as the core of a hurricane

(a)

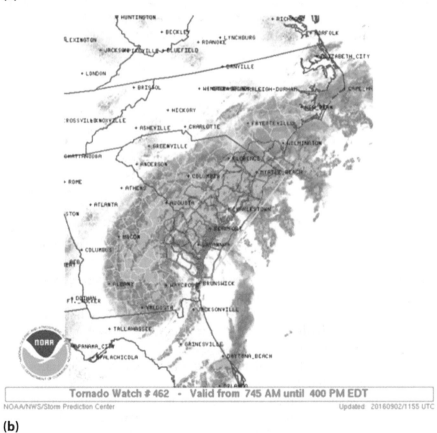

Tornado Watch # 462 - Valid from 745 AM until 400 PM EDT

NOAA/NWS/Storm Prediction Center   Updated 20160902/1155 UTC

(b)

FIGURE 11.50 (a) Composite radar image of Tropical Storm Hermine at 0854 UTC on September 2, 2016. Several spiral bands are prominent, particularly southeast of the center of the storm; (b) Areas outlined in red were part of a tornado watch issued by the Storm Prediction Center the morning of September 2, 2016, a few hours after the radar image in part (a) (courtesy of National Weather Service).

(a)

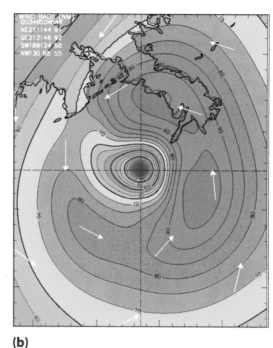

(b)

FIGURE 11.51 (a) A satellite image from the afternoon of August 28, 2005, just after Hurricane Katrina had intensified into a Category-5 storm (courtesy of MODIS Rapid Response Project at NASA/GSFC); (b) Observed surface winds around Hurricane Katrina at 09 UTC on August 29, 2005. Wind speeds are color-coded in knots, and arrows indicate wind directions. Measurements came from a variety of sources including aircraft reconnaissance, buoys, and Doppler radar. The hurricane was moving north at the time (courtesy of NOAA Hurricane Research Division).

approaches, and the winds dramatically increase as the eye nears. Typically, the strongest winds associated with a hurricane are not evenly distributed around the eye—Hurricane Katrina, from August 2005, provides a good example. On August 28, Katrina had intensified into a Category-5 hurricane with maximum sustained winds reaching 150 kt (173 mph)—the visible satellite image in Figure 11.51a shows Katrina in the northern Gulf of Mexico at this time. Fortunately for Gulf Coast residents, Katrina weakened into a Category-3 before making landfall the next day. Figure 11.51b shows surface wind speed and direction around the eye at 09 UTC on August 29, just two hours before Katrina's first landfall in Louisiana. The hurricane was moving almost due north at this time. Note that the maximum wind speeds—above 95 kt (109 mph) in this case—are to the north and east of the eye. If you take into account the direction of movement of the storm, this area represents the quadrant of the hurricane that is in "front" and to the right of the path, commonly known as the **right-front quadrant** (see Figure 11.52). This is typically where the strongest winds are found in a Northern Hemisphere tropical cyclone.

What's unique about this quadrant? Given that winds around a Northern Hemisphere tropical

cyclone circulate counterclockwise, the storm's forward motion is in the same direction as the winds in this quadrant of the hurricane—essentially the storm motion enhances the wind speed. To better understand this boost in observed wind speed, consider a robber in the Old Wild West running toward the engine on top of a moving train—the robber's total speed equals his running speed plus the train's forward speed. In light of this analogy, we infer that the strongest sustained winds of a hurricane typically occur in the right-front quadrant. The left-front quadrant is typically weaker because, in this quadrant, the hurricane's forward speed subtracts from the local wind speed. All bets are off, however, if the hurricane is stationary. In this case, winds in the left-front quadrant can be just as strong as in the right-front quadrant. In other words, a stationary hurricane's most destructive winds can extend over a much larger area. In September 2019, Hurricane Dorian essentially stalled over a portion of the Bahamas, inflicting widespread catastrophic destruction.

The strong, onshore winds in the right-front quadrant of a hurricane are primarily responsible for the **storm surge**, the rise in ocean level that occurs as a tropical cyclone makes landfall. The storm surge is

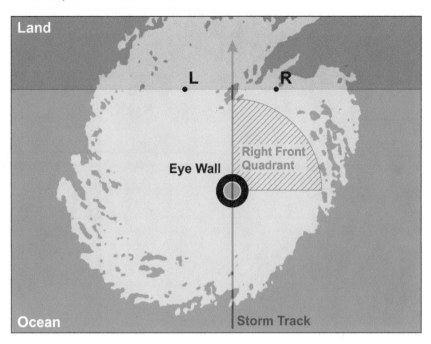

FIGURE 11.52 The fastest winds of a hurricane and the highest storm surge lie in the right-front quadrant. In this region, winds push water toward shore (point R is in this quadrant). At point L, offshore winds can actually lower water levels at the hurricane approaches.

magnified around the time of high astronomical tides and instances when sea water is forced to squeeze into narrower bays and rivers near the coast.

Despite Katrina's weakening just before landfall, the radius of hurricane-force winds on its east side was expansive, extending outward at least 130 km (80 mi) from its center (check back to Figure 11.51b). The large swath of strong onshore winds pushed the sea toward land, causing water to pile up in coastal shallows and leading to a devastating inland surge of water. Observations of high-water marks indicated that the storm surge was 24–28 ft (7.3–8.5 m) in a 32-km (20 mi) swath along the Mississippi Coast, 10–15 ft (3.0–4.6 m) along the coast of western Alabama, and 6 ft (1.8 m) along the western panhandle of Florida.

Large waves contributed to these high-water marks. Within the 24 hours prior to landfall, Katrina generated large ocean swells while it was a Category-5 hurricane. According to the National Hurricane Center, a buoy located about 110 km (70 mi) south of Dauphin Island, AL, reported a significant wave height of 30 ft (9.1 m) at 00 UTC on August 29 (the **significant wave height** is the average height of the highest one-third of the waves). Eleven hours later, the same buoy measured a significant wave height of 55 ft (16.8 m), the largest ever measured by a buoy operated by the National Data Buoy Center. Figure 11.53 shows the impact of the storm surge in south Plaquemines Parish near Katrina's first landfall.

Another significant storm surge occurred to the west of the track of Katrina (see Figure 11.54). There,

along the southern shores of Lake Pontchartrain, strong northerly winds produce a surge of 10–14 ft (3.0–4.3 m), causing some of the levees protecting New Orleans to fail and flooding many parts of the city situated below sea level.

Although wind often gets most of the attention in hurricanes (arguably, in large part because the Saffir-Simpson Scale is based solely on wind), storm surge historically causes most hurricane-related fatalities and damage. Worldwide, the low lands of Bangladesh, located along the northern shores of the Bay of Bengal, are probably the most vulnerable of all—in 1991, a 20-ft (6.1 m) storm surge produced by a strong tropical cyclone over-whelmed southeastern Bangladesh, killing 138,000 people and leaving an estimated 10 million people homeless. In the United States, New Orleans is arguably the most vulnerable city to storm surge, with Tampa and Miami at the top of that list as well. Sandy, in late October 2012, was not officially a hurricane at landfall, but nonetheless drove a cata-strophic storm surge into the New Jersey and New York coastlines causing an estimated $71 billion in damage, a large portion due to storm surge in those two states (see Figure 11.55).

The impacts from a tropical cyclone's winds don't just come from storm surge—hurricane-force winds can also inflict plenty of damage directly. Figure 11.56 shows some of the destruction in South Florida from the winds of Hurricane Andrew as it made landfall as a Category 5 in the wee hours of the

FIGURE 11.53 Storm surge pushed a large ship onshore in south Plaquemines Parish near the point where Hurricane Katrina first made landfall in Louisiana (courtesy of NOAA).

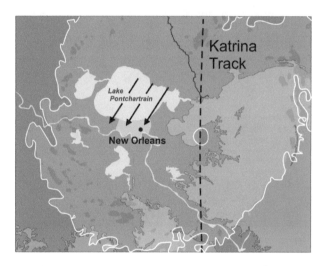

FIGURE 11.54 Northerly winds to the west of the track of Hurricane Katrina produced a large storm surge along the southern shores of Lake Pontchartrain, causing levees that protected New Orleans to fail.

morning on August 24, 1992. When most hurricanes come ashore, sustained winds usually decrease, but Andrew briefly intensified around landfall. The likely mechanism was an increase in low-level convergence in response to greater friction over land. That enhanced frictional convergence boosted updrafts in eye-wall thunderstorms, which in turn produced some 2000 streaks of damage attributed to this temporarily intensified convection.

Florida gets the lion's share of very strong hurricanes. On October 10, 2018, Hurricane Michael made landfall along the Florida Panhandle as a Category 5 storm, making it the strongest hurricane to strike the continental United States since Hurricane Andrew (see Figure 11.56b). Michael's minimum pressure at landfall was estimated to be 919 mb, which is the third lowest pressure ever recorded in the United States (only the Labor Day Hurricane in 1935 and Hurricane Camille in 1969 had lower pressures).

We close with some final thoughts about forecasting tropical cyclones.

THE ROLE OF THE NATIONAL HURRICANE CENTER (NHC): ITS MISSION AND VISION

NHC's Mission: *To save lives, mitigate property loss, and improve economic efficiency by issuing the best watches, warnings, forecasts and analyses of hazardous tropical weather, and by increasing understanding of these hazards.*

NHC's Vision: *To be America's calm, clear and trusted voice in the eye of the storm, and, with our partners, enable communities to be safe from tropical weather threats.*

In the last few decades of the 20th century, a majority of the fatalities caused by hurricanes and tropical storms were a consequence of inland flooding from heavy rains. But in 2005, Hurricane Katrina served as a grim reminder that storm surge still poses the greatest threat from landfalling tropical cyclones.

FIGURE 11.55   An aerial view of storm-surge damage to the Casino Pier in Seaside Heights, NJ, on October 30, 2012, in the aftermath of Hurricane Sandy (courtesy Master Sgt. Mark C. Olsen/U.S. Air Force/New Jersey National Guard).

(a)   **(b)**

FIGURE 11.56   (a) Hurricane Andrew (1992) inflicted serious wind damage in southern Florida (courtesy of NOAA). (b) An aerial view of Mexico Beach, FL, in the aftermath of Hurricane Michael's destructive storm surge and fierce winds. Michael made landfall near Mexico Beach at 1730 UTC (12:30 PM CDT) on October 10, 2018 (courtesy of NOAA).

Although the final death toll is not certain, approximately 1500 people lost their lives to *direct impacts* from Katrina (according to NHC). Without reservation, storm surge likely caused most of the estimated 1300 deaths in Louisiana and the 200 deaths in Mississippi (the fatalities in Mississippi occurred largely in three coastal communities).

The National Hurricane Center in Miami, FL, has worked closely with coastal communities to develop evacuation plans based on the predicted heights of storm surges. The height of the storm surge depends on the wind strength (category) of the hurricane and the location where the storm makes landfall. It also depends on the topography of the coastline and the adjacent ocean floor. Indeed, if the slope of the ocean floor adjacent to the coast is relatively gentle, water that's pushed toward land by hurricane-force winds piles up more readily, setting the stage for a higher storm surge compared to an ocean floor which has a steeper slope away from land.

To help communities formulate evacuation plans, the National Hurricane Center provides local emergency management officials with output from a computer model called SLOSH (which stands for **S**ea, **L**ake and **O**verland **S**urges from **H**urricanes). Figure 11.57 shows the SLOSH model's prediction

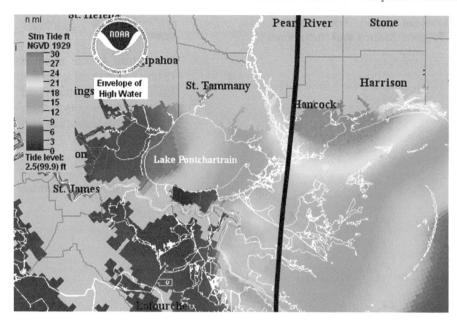

FIGURE 11.57 The SLOSH model's prediction for the storm surge of Hurricane Katrina (generated by the 03 UTC computer simulation on August 29, 2005). The height of the storm surge is color-coded in feet (red indicates 25 feet and higher). The black line indicates the National Hurricane Center's predicted path of the eye of the hurricane (courtesy of NOAA).

for the storm surge of Hurricane Katrina, generated by the 03 UTC computer simulation on August 29, 2005. The black line is NHC's predicted path of the eye. To get your bearings, the height of the storm surge is color-coded in feet. Note that the SLOSH model predicted a monstrous 25-foot surge into Hancock and Harrison counties in Mississppi, what turned out to be a very good forecast.

The National Hurricane Center alerts communities to the threat of hurricanes with two levels of advisories. NHC forecasters issue a hurricane watch for a coastal area when hurricane conditions are *possible* within the next 48 hours. The second, more urgent advisory, is a hurricane warning, which NHC forecasters issue when they *expect* hurricane conditions to arrive within the next 36 hours. When issuing these advisories, the National Hurricane Center must weigh conflicting societal impacts. NHC must allow enough lead time to evacuate densely populated areas, especially on barrier islands or other locations where evacuation routes can be cut off by rising water. On the flip side, evacuations are expensive. It costs time and money to board windows and sandbag homes and businesses in preparation for winds and waves. Businesses lose money when people evacuate shore areas. And it costs a lot of money just to facilitate the evacuation. These preparation costs increase each year as coastal populations and coastal property values increase.

Advances in technology, research, remote sensing, and forecasting techniques continue to improve the skill and accuracy of forecasts issued by the National Hurricane Center. Indeed, forecasts of the tracks of hurricanes and tropical storms are increasingly more accurate. Figure 11.58 shows the average annual error in the 24-, 36- and 48-hour forecasts for Atlantic hurricanes and tropical storms during the period 1970–2015 (it also includes the average annual error in the 96- and 120-hour forecasts from 2001 to 2015). For one, two, and three-day track forecasts, the improvement averages between one and two percent per year. Based on the trend lines in Figure 11.58, three-day track forecasts today are as good as one-day track forecasters were in the early 1990s, while five-day track forecasts are now as good as three-day forecasts in the early 1990s— so basically, two days of predictability have been gained in about the last 25 years. Still, the average track error at 72 hours is about 100 miles (160 km). Because of this uncertainty, NHC track forecasts always include a "cone of uncertainty" that widens with forecast time. Figure 11.59a shows NHC's track forecast of Katrina issued the evening of August 26, 2005—the "M" inside the black circles stands for "Major" hurricane. About 500 mi (800 km) of the Gulf Coast lay in the cone of uncertainty. To demonstrate how track forecasts have improved since then, Figure 11.59b shows the forecast that NHC would have likely issued had Katrina occurred in 2015. The cone at landfall narrows to about 300 mi (483 km) of coastline, a reduction of 40 percent in just a decade.

In contrast to *track* forecasts, NHC's forecasts of the *intensity* of hurricanes and tropical storms have improved only slightly since 1970. One of the primary reasons that intensity forecast improvements

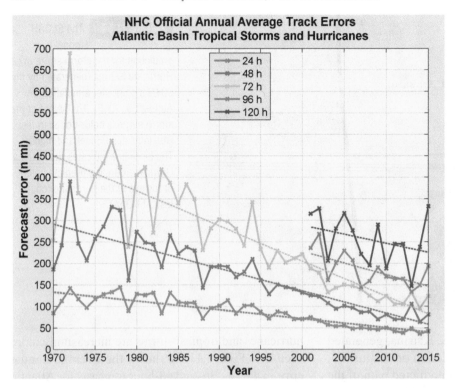

FIGURE 11.58 Annual average errors in 24-hour (red), 48-hour (green) and 72-hour (yellow) track forecasts for Atlantic Basin tropical storms and hurricanes, for the period 1970 to 2015, in nautical miles (1 nautical mile equals 1.15 miles). Errors in 96-hour and 120-hour forecasts are shown from 2001 to 2015 (courtesy of the National Hurricane Center).

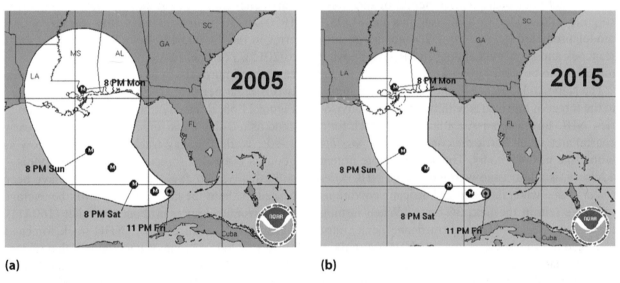

(a)   **(b)**

FIGURE 11.59 (a) Forecast of the track of Hurricane Katrina issued by the National Hurricane Center late in the day on Friday, August 26, 2005. The black dots indicate the official forecast of the center, while the white swath represents the uncertainty in the track forecast. Note that this "cone of uncertainty" widens with time, indicating the growing uncertainty in the track forecast with time. The "M" indicates that Katrina was forecast to be a "Major" hurricane; (b) Had Katrina occurred in 2015, the cone of uncertainty would have been significantly narrower, owing to improvements in track forecasts since the turn of the century (courtesy of NOAA).

have lagged those of track forecasts is that complex processes such as eye-wall replacement cycles are the primary controllers of intensity changes in major hurricanes, and these cycles are very difficult to accurately model. For the record, an **eye-wall replacement cycle** (ERC) occurs in major hurricanes when a spiral band completely encircles the existing eye wall. This spiral band robs the inner eye wall of moisture, so the inner eye wall collapses and the coiling spiral band forms a new outer eye wall. During the replacement cycle, the hurricane weakens, but, after the cycle is complete, the hurricane may become as strong (or stronger) as it was before the replacement cycle began.

(a)

(b)

FIGURE 11.60 (a) A satellite image showing the deep convection (in yellow and red) associated with Hurricane Wilma around 07 UTC on October 19, 2005. At the time, Wilma was a Category-5 hurricane, with maximum sustained winds of 150 kt (173 mph). Note the spiral band starting to wrap around the eye wall, signaling the start of an eye-wall replacement cycle. The eye of Wilma was only about 2.3 mi (3.7 km) in diameter, and the satellite could not resolve (detect) the "pinhole eye"; (b) A satellite image at 1845 UTC on October 20 shows the deep convection associated with a new (outer) eye wall. As the spiral band encircled the inner eye wall, it intercepted moisture destined for the inner eye wall. As a result, the inner eye wall collapsed and the now circular spiral band took over as the new (outer) eye wall. Wilma's maximum sustained speed at this time was 125 kt (144 mph) (courtesy of NASA and the Naval Research Laboratory).

Figure 11.60a is a satellite image that shows deep convection (yellow and red shades) associated with Hurricane Wilma at 07 UTC on October 19, 2005. For the record, the satellite sensor passively detected microwaves emitted by precipitation in the eye wall and spiral bands of Wilma. The eye was a "pinhole" at this time, consistent with Wilma's Category-5 status, with maximum winds of 150 kt (173 mph). Note the spiral band starting to wrap around the eye wall, signaling the beginning of an eye-wall replacement cycle. Figure 11.60b shows the end result of the eye-wall replacement cycle, with the spiral band having formed a new (outer) eye wall. At the time, the slightly weakened Wilma (after the ERC) was too close to land (the Yucatan Peninsula) to regain any strength—maximum sustained winds were 125 kt (144 mph).

So what's our point here? Given that such complex processes are difficult to accurately model, it's no wonder why NHC's advance in intensity forecasting has lagged behind improvements in track forecasting.

We end this section with some sobering thoughts. As of 2015, about 60 million people live in the 185 coastal counties abutting the Atlantic Ocean and Gulf of Mexico, an increase of about five million (nearly nine percent) over the previous decade. These numbers swell dramatically during peak holiday or vacations periods when millions more visit the nation's shores. Coastal populations continue to increase at a rate of four to five percent per year, on average. Indeed, because the lure of the ocean is so strong, people will always be attracted to live in harm's way. Education and preparation are the keys to preventing another disastrous loss of life from a future tropical cyclone.

Weather Folklore and Commentary

What a Difference a Day Makes

It has been said that the most beautiful day to be seen is the one that precedes a hurricane. This tidbit of tropical folklore is confirmed by satellite imagery of most hurricanes—just check a few of the images in this chapter.

The old adage "What goes up, must come down" helps to explain why skies surrounding a hurricane are relatively clear. Within a hurricane, rising currents of air sustain the thunderstorms that drive the storm's heat engine. Once updrafts reach the top of the storm, air spreads out and diverges, eventually sinking on the outer edges of the hurricane (of course, some air also sinks into the eye). This compensating subsidence often creates a rather sharp transition from cloudy to clear conditions, with cumulus cloud development suppressed hundreds of kilometers ahead of the storm. As careful observers dating back to Christopher Columbus have noted, weather conditions the day before a hurricane arrives are often beautiful. But there are typically subtle clues of the impending danger. Feathery cirrus clouds streaking the sky fan out ahead of the storm (from the tops of thunderstorms within the hurricane). The barometer starts to fall and the ocean gets rougher as swells roll out ahead of the ocean-agitated storm. All of these clues belie the blue skies above, foretelling the tempestuous approach of a hurricane.

Mid-Latitude I: Linking Surface and Upper-Air Patterns

12

LEARNING OBJECTIVES

After reading this chapter, students will:

- Gain an appreciation for the role that mid-latitude low-pressure systems have in mitigating horizontal temperature contrasts between the tropics and poles.
- Understand the role that upper-air divergence and upper-air convergence play in the life cycle of mid-latitude low- and high-pressure systems.
- Understand earth (planetary) vorticity and its variation with latitude.
- Understand the curvature and horizontal wind shear components of relative vorticity.

- Understand that the time rate of change in 500-mb absolute vorticity is related to horizontal mass divergence and convergence at this level.
- Learn to identify short-wave and long-wave troughs on analyses of 500-mb heights and absolute vorticity and to understand the role that short-wave troughs play in producing significant weather.
- Be able to identify jet streaks on analyses of upper tropospheric heights and isotachs.
- Understand the four-quadrant model for straight jet streaks in terms of vertical motion and upper-air divergence patterns, and why its direct application to real-time weather often fails.

The winters of the late 1970s were very cold over much of eastern North America. Not even Florida escaped the chill, as the winters of 1976–1977 and 1977–1978 tied for second coldest on record in the Sunshine State. On January 19–20, 1977, an unseasonably cold air mass brought record daily low temperatures as far south as Miami, while 0.2 inches of snow fell in Tampa on January 19 (see Figure 12.1), the only measurable snowfall there since the dawn of the twentieth century.

For such significant chill to penetrate deep into the Florida peninsula, the wind trajectory must keep southward-moving cold air mostly over land, bypassing the moderating effects of the Gulf of Mexico and the Atlantic Ocean. Figure 12.2 shows the 12 UTC surface analysis on January 19, 1977. The isobars suggest that the center of a potent low-pressure system was just off the eastern side of the map (note the tail end of a trailing cold front through central Cuba), while a high-pressure system was centered near the Texas/Louisiana border. This "couplet" of high and low pressure worked in tandem to drive unseasonably cold air southward over the Florida peninsula.

More generally, the circulation in the wake of a low-pressure system (and ahead of the trailing high-pressure system) transports cold air equatorward. Meanwhile, a low draws warm air poleward on its eastern flank, aided by the circulation of the downstream high. As you have learned, cold and warm air masses transported from polar and tropical latitudes (respectively) meet in zones called fronts that are marked by relatively large horizontal temperature gradients. These fronts are found in troughs and commonly attach to the centers of their associated low-pressure systems.

One of the basic tenets of meteorology is that mid-latitude low-pressure systems strengthen as they derive energy from these sharpening temperature contrasts. As more isobars encircle and expand

outward from the centers of deepening lows (see Figure 12.3), they extend their northward and southward reach, drawing together increasingly warm and cold air and thereby magnifying temperature contrasts at their associated fronts. In turn, this increased temperature gradient paves the way for further intensification—essentially, a positive feedback loop gets underway. We'll discuss this strengthening process in greater detail in Chapter 13. For now, just keep in mind that the more intense a low-pressure system is, the more effectively it transports warm air poleward and cold air equatorward.

Another way to think about mid-latitude lows is that they are atmospheric relay switches—they receive and process signals of warmth from the tropics and signals of chill from the poles. They then amplify these signals and transmit warm air farther toward polar regions and cold air farther toward

FIGURE 12.1   Residents of Tampa, FL, had to scrape snow off their car windshields on January 19, 1977 (courtesy of the Special Collections Department, University of South Florida).

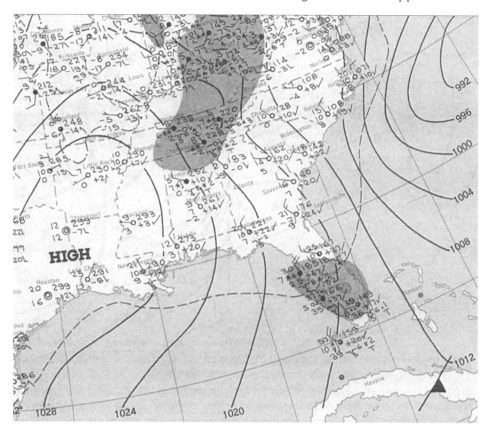

FIGURE 12.2   The surface analysis from the National Weather Service at 1200 UTC on January 19, 1977. Note that the north-northwesterly winds over Florida in the wake of the offshore low-pressure system and ahead of the trailing high over the Deep South essentially by-passed the moderating effects of the Gulf of Mexico and the Atlantic Ocean. Snow was falling in Tampa at this time, and temperatures near 20°F (–7°C) had penetrated all the way to the Gulf Coast (courtesy of NOAA).

tropical regions. In this way, lows act as important relay switches in the general circulation's intricate circuitry (including Hadley "Cells") that is designed to mitigate temperature contrasts between the poles and tropics.

In order for mid-latitude lows and highs to fulfill their commitment to the general circulation, they need a little help. The purpose of this chapter is to shed light on the role that upper-air patterns of wind and pressure play in aiding the intensification of surface lows and highs, allowing them to more effectively send cold air equatorward and warm air poleward.

LOW- AND HIGH-PRESSURE SYSTEMS: WEIGHT WATCHERS VERSUS BODY BUILDERS

Once a surface low-pressure system forms, air near the surface starts to converge around the center of the low, as shown by the cross section in Figure 12.4a.

Because the atmosphere works to minimize local congestions of air, the converging air must rise. If low-level convergence and upward motion were the only processes at work here, the air column over the center of the low would gain weight and the surface pressure would increase. That's not very promising for a newly formed low-pressure system. Indeed, by definition, the column of air at the center of a low-pressure system must always be the lightest column compared to other air columns in the general vicinity (remember that the air column extends from the ground to the top of the atmosphere). Obviously, there must be a "weight-reduction plan" in the upper half of the troposphere that allows the air column over the center of the low to maintain a relatively low weight. This weight-reduction plan is upper-level divergence.

By the same token, air near the ground diverges away from the center of a high-pressure system (see Figure 12.4b). In turn, air from above must sink toward the earth's surface to avoid a local depletion of air. Given

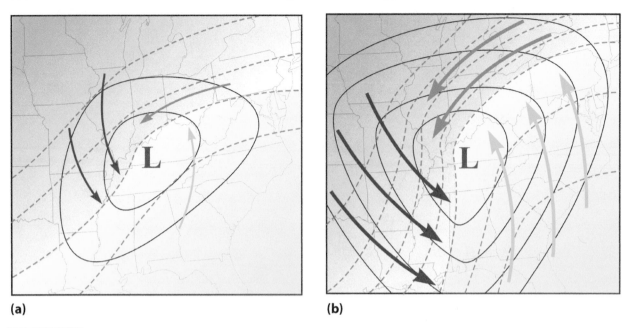

(a)   **(b)**

FIGURE 12.3 (a) The circulation around a low-pressure system in the Northern Hemisphere sends warmer air northward on the low's eastern flank and colder air southward on its western flank (solid contours are sea-level isobars and dashed lines are surface isotherms).Given that surface winds (arrows) cross isobars inward toward low pressure, there is convergence. As a result, temperature gradients increase in the vicinity of the low; (b) A low derives energy from these increasing temperature gradients and, thus, intensifies (its central pressure decreases). In response, more isobars close off around the deepening low, its counterclockwise circulation expands, winds strengthen, and the low more effectively sends warm air toward the pole and cold air toward the equator.

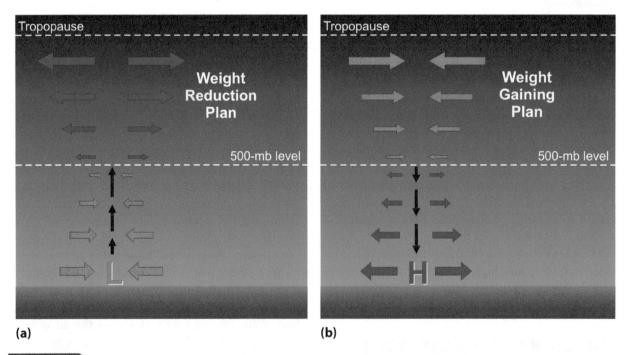

(a)   **(b)**

FIGURE 12.4 (a) The pattern of convergence in the lower half of the troposphere above the center of a developing low-pressure system. For a low to deepen (surface pressure decreases with time), there must be a weight reduction plan in the upper half of the troposphere to compensate for convergence in the lower troposphere; (b) The pattern of divergence in the lower half of the troposphere above the center of a developing high-pressure system. For a high to strengthen (surface pressure increases with time), there must be a weight-gaining plan in the upper half of the troposphere to compensate for divergence in the lower troposphere. In both diagrams, the thickness and length of the arrows qualitatively indicate the magnitudes of convergence, divergence, and vertical motions.

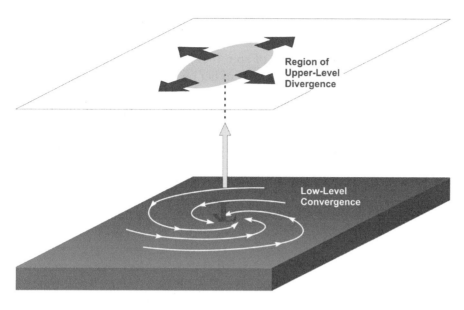

(a)

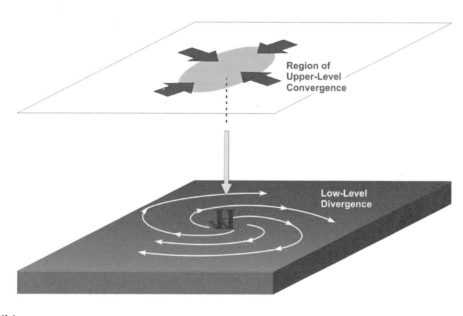

(b)

FIGURE 12.5   (a) For a surface low to deepen (central pressure decreases with time), upper-level divergence must more than offset the mass gained by low-level convergence; (b) For a surface high to intensify (central pressure increases with time), upper-level convergence must more than compensate for low-level divergence.

this loss of weight from the air column, there must be a weight-gaining process occurring in the upper half of the troposphere for the high-pressure system to maintain a relatively high weight. This weight-gaining plan is upper-level convergence.

Let's cut right to the chase. For a surface low to deepen (its central pressure decreases with time), divergence in the upper half of the troposphere (Figure 12.5a) must exceed low-level convergence. In other words, there must be net divergence above the center of a deepening low that allows the air column's weight and surface air pressure to decrease with

time. Similarly, over the center of a strengthening high-pressure system, there must be convergence in the upper half of the troposphere that exceeds low-level divergence (see Figure 12.5b). In other words, there must be net convergence above the center of a strengthening high that allows the air column's weight and surface pressure to increase with time.

The give-and-take between the upper and lower troposphere in controlling the weights of air columns is a close one. More succinctly, upper-air divergence typically exceeds low-level convergence by just a small amount in a developing low, while upper-air convergence barely exceeds low-level divergence in a strengthening high. The upshot of these checks and balances on weights of air columns is that the difference between sea-level pressures in lows and highs is not very great. Indeed, the change in barometric readings between a formidably stormy 980-mb low and a tranquil 1030-mb high is only about five percent. Yet, such a small change in barometric readings can translate into a drastic difference in weather. Amazing!

But how do meteorologists go about the task of assessing upper-air divergence and upper-air convergence? Read on.

Assessing Upper-Air Convergence and Divergence: Air Parcels Become Figure Skaters

To begin to answer the previous question, we first revisit the principle of the conservation of angular momentum. For starters, let's once again consider a figure skater executing a whirling spin. We take our discussion back to the ice because there's a clear relationship between the rate of spin and the position of the skater's arms: When a skater wants to

spin faster, she must draw her arms inward close to her body. In other words, there must be convergence of mass in order for there to be an increase in the rate of spin (her arms and hands converge toward the center of her body). When she wants to slow the rate of spin, her arms must extend outward. In other words, there must be divergence of mass in order for there to be a decrease in the rate of spin (her arms and hands horizontally spread away from the center of her body). So, in figure skating, there is clearly a connection between changes in a skater's rate of spin and horizontal divergence and convergence of mass.

Now, if the same kind of relationship between a change in rate of spin and horizontal divergence and convergence exists in the atmosphere, we're in business! And we're in luck—the ice-skater analogy does indeed have atmospheric relevance because most parcels of air have some spin around their local vertical axis. As Figure 12.6a shows, the **local vertical** at any point is approximately the line drawn from the center of the earth through that point. Operational meteorologists are interested in assessing changes in the rate of spin of air parcels about their local vertical because they reveal patterns of horizontal convergence and

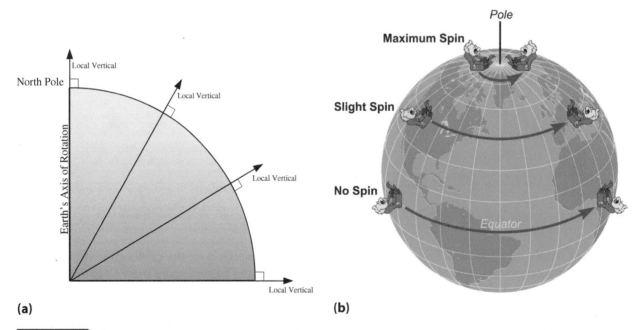

(a)   **(b)**

FIGURE 12.6   (a) The local vertical for a point is defined as the line drawn from the center of the Earth through the point; (b) The spin around the local vertical imparted to a parcel by the Earth's rotation is maximized at the North Pole, where the local vertical aligns with the Earth's axis of rotation. Quantitatively, the "planetary spin" imparted to a parcel above the North Pole equals one rotation in 24 hours. This planetary spin decreases to zero at the equator, where the local vertical becomes perpendicular to the Earth's axis of rotation. In the Southern Hemisphere, the spin imparted to a parcel above any given latitude is simply the negative of the value at the corresponding latitude in the Northern Hemisphere. In this context, "negative" implies a spin in the opposite (clockwise) direction.

divergence in much the same way that changes in the rate of spin of an ice skater about her local vertical axis are related to the horizontal position of her arms.

How can an air parcel acquire spin around its local vertical? One way is to be "onboard" a nearby object that is already spinning—in our case, the rotating Earth. To visualize how this can happen, let's go for a hike—a long hike north to the North Pole. Exhausted upon arrival, you sit on a lounge chair and quickly fall asleep (see Figure 12.6b). You're so tired that you sleep for 12 hours without changing the position of your head (now that's a sound sleep). When you awaken, your eyes are looking (relative to your local surroundings) in the same direction they were when you closed your eyes because both you and your surroundings rotated eastward with the Earth. However, from the perspective of an observer in space, you're actually facing 180° opposite the direction you were looking when you fell asleep (check Figure 12.6b). Of course, you don't sense this change in direction because both you and your local surroundings rotated with the Earth.

The bottom line here is that even though your position relative to your surroundings did not change, you rotated or "spun" 180° counterclockwise around the local vertical which, at the North Pole, essentially corresponds to the Earth's axis of rotation. If you had slept for a full 24 hours, you would have rotated a full 360°.

Like you and your lounge chair, the atmosphere rotates eastward with the Earth. So at the North Pole, an air parcel also gains counterclockwise "spin" as it rotates eastward with the Earth—one full rotation per day. At the South Pole, the spin gained by rotating around the local vertical would be in the clockwise sense.

What about at other latitudes? Is the spin gained by air parcels rotating with the Earth the same? To answer this question, imagine that you follow up your North Pole hike with a relaxing tropical cruise. Your ship drops anchor along the equator in the Pacific, and you again fall asleep for 12 hours—this time, on a lounge chair on the main deck (revisit Figure 12.6b). Again, it's a sound sleep, and you keep your head in exactly the same position as you snooze. When you awaken, your eyes are looking in exactly the same direction locally as they were when you fell asleep because—again—both you and your local surroundings rotated eastward with the Earth. This time, the same is true from the perspective of an observer in space—you're still looking in exactly the same direction.

The bottom line in this case is that, at the equator, you did not rotate at all around the local vertical. As a result, you gained zero spin while rotating eastward with the Earth. Even if you slept for 24 hours, the result would be the same. You won't gain any spin around the local vertical on the equator as you rotate eastward with the Earth.

So summarizing our results so far: At the North Pole, you rotate 360° around the local vertical in 24 hours, but at the equator, you don't rotate at all around the local vertical. What about intermediate latitudes? For sake of argument, let's choose 45°N.

As Figure 12.6b indicates, after sleeping soundly for 12 hours at 45°N latitude, you will wake up facing (as viewed by an observer in space) in a direction about 90° from the direction you were facing when you first went to sleep. Clearly, the counterclockwise spin that you gained here at 45°N is noticeably less that what you gained at the North Pole.

With these three results in mind, we arrive at this inescapable conclusion: The spin of an air parcel around the local vertical caused by the rotation of the Earth increases with increasing latitude, from zero at the equator to a maximum at the Poles.

The spin imparted to air parcels by the rotating earth is part of the parcel's **vorticity**, which is simply a measure of its rotation (here, we're only interested in rotation around the local vertical). Formally, the spin imparted to air parcels by the rotating earth is called **earth vorticity** (alternatively, planetary vorticity). By convention, we assume that earth vorticity in the Northern Hemisphere (counterclockwise spin) is positive.

While earth vorticity is spin imparted *by* the rotating earth, **relative vorticity** measures the spin of air parcels *relative to* the rotating earth. Such spin is routinely produced in the atmosphere by wind patterns associated with transient weather systems such as mid-latitude lows and highs. Figure 12.7 shows that in the Northern Hemisphere, their respective counterclockwise and clockwise circulations create positive relative vorticity (lows) and negative relative vorticity (highs).

Almost all parcels of air have both earth and relative vorticity. To picture how these two sources of spin can be present at the same time, consider Figure 12.8, which shows a tilt-a-whirl ride at an amusement park. When the ride is in motion, the platform on which the cars sit is rotating, akin to earth vorticity—all the cars spin in the direction the platform is spinning. However, each car also

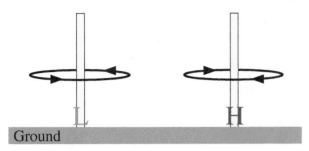

FIGURE 12.7 The circulation around low- and high-pressure systems (shown here for the Northern Hemisphere) also imparts spin around local verticals. The spin imparted to air parcels by a low is in the counterclockwise sense, or positive (the same sense as earth vorticity). The spin imparted to air parcels by a high is in the clockwise sense, or negative (opposite the sense of earth vorticity).

can rotate individually around its own axis, which is analogous to relative vorticity.

The sum of the two sources of spin about a vertical axis (in other words, the sum of relative and earth vorticity) is called **absolute vorticity**. Over mid-latitudes in the Northern Hemisphere, absolute vorticity is almost always positive because earth vorticity is the dominant source of parcel spin—so nearly all Northern Hemisphere parcels spin counterclockwise about their local verticals.

With the concept of absolute vorticity in hand, we can now reinforce the connection between spinning parcels and spinning ice skaters. If the absolute vorticity of a parcel in the Northern Hemisphere decreases with time, its counterclockwise spin slows and so it must undergo mass divergence. To understand the connection between an air parcel's slowing spin and mass divergence, think again of a

whirling ice skater (see the drawing on the left side of Figure 12.9). Now imagine that skater stretches her arms out horizontally, as shown in the drawing on the right-hand side of the figure. In response to the extending of the arms (akin to divergence), her rate of spin decreases. And so it is with an air parcel. Next, imagine the process in reverse. The skater begins with arms extended and then draws her arms inward. Her rate of spin then increases. And so it is with an air parcel.

We're almost home. All we have to do now is discuss how all of this fits together to make a pattern of upper-air divergence and convergence that sustains surface low- and high-pressure systems.

Upper-Air Vorticity Patterns: Putting a Different Spin on Ridges and Troughs

Okay, let's review where we've been and where we want to go. Once a low-pressure system forms, air near the ground converges toward the center of the low. In order to avoid local traffic jams (extensive piling up of air), the converging air rises. But rising air still contributes to the overall weights of air columns at and around the low's center (raise your arm above your head; while you're raising it, you still weigh the same). With air converging into air columns at and around the center of the low, divergence must occur in the upper half of the troposphere to offset the weight gains in the lower half of the troposphere. Of course, we can make a parallel summary for high-pressure systems, but with divergence near the surface and convergence in the upper half of the troposphere.

FIGURE 12.8 Each car on the Tilt-A-Whirl rotates (spins) with the platform, which is analogous to earth vorticity. Each car also spins around its own axis, which can be likened to relative vorticity (Darryl Brooks/Shutterstock.com).

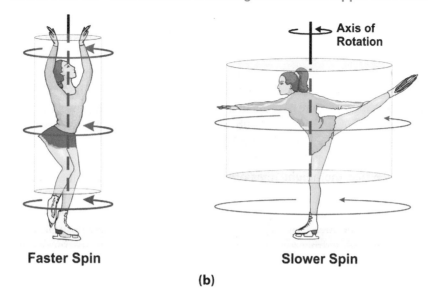

Faster Spin

(a)

Slower Spin

(b)

FIGURE 12.9 Suppose that we liken an air parcel shaped like a cylinder to a spinning ice skater. The dashed line is the axis of rotation (the local vertical). When the ice skater stretches out her arms horizontally, her rate of spin decreases in response to divergence (note the changing shape of the cylindrical parcel—it shrinks vertically and expands horizontally). If the process works in reverse, the air parcel undergoes convergence and the rate of spin increases.

That's where we've been. Where do we want to go? In a nutshell, the looming question now becomes: How do you determine where areas of divergence (or convergence) will occur on upper-air maps? As a matter of practice, weather forecasters routinely look at maps of height and absolute vorticity at the 500-mb level to identify the areas of divergence and convergence aloft. So let's start there.

Figure 12.10a shows height lines that outline a 500-mb ridge and trough (assume you're in the Northern Hemisphere). We assume here that air

parcels, by moving with the relatively fast 500-mb wind, travel right through the relatively slow-moving ridge-trough system (more on this later). As parcels of air (which we've shaped like medicine capsules for effect) negotiate the anticyclonic turn in the ridge, they acquire clockwise spin around their local vertical from the curved height lines. In other words, parcels attain negative relative vorticity as they round the anticyclonic curve associated with the ridge. Meanwhile, parcels of air negotiating the trough acquire counterclockwise spin around

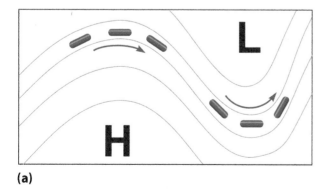

(a)

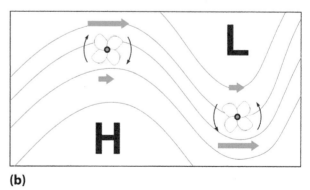

(b)

FIGURE 12.10 (a) Parcels of air traveling through upper-air troughs and ridges acquire spin (relative vorticity) by virtue of the curvature of the height lines (note that the parcel would also change shape as it moves through the trough/ridge system because its absolute vorticity changes—see Figure 12.9—but we do not show the convergence and divergence here because the focus here is the change in spin). Meteorologists refer to this type of relative vorticity as curvature vorticity; (b) Because winds at the base of a trough and the top of a ridge tend to be faster than winds near the center of the trough and ridge, air parcels, which we represent here by pinwheels, develop spin in response to this horizontal wind shear. Meteorologists refer to this component of relative vorticity as shear vorticity.

their local vertical from the curved height lines. In other words, parcels develop positive relative vorticity as they round the cyclonic curve associated with the trough. These contributions to relative vorticity are called **curvature vorticity** because they result from the cyclonic and anticyclonic curvature of the height lines.

Now shift your attention to Figure 12.10b where we show the same height pattern as in Figure 12.10a. But now we'll focus on the height gradient, starting in the ridge. Notice that the height lines are a bit more tightly packed at the top (northernmost part) of the ridge compared to farther south in the ridge. This is consistent with the idea that larger temperature contrasts exist where warmer air has penetrated farthest into territory usually occupied by cold air. As a result, the wind is usually faster near the top of a ridge than closer to the ridge's center. Now imagine placing a stationary pinwheel in the ridge, as shown in Figure 12.10b. Because winds near the top of the ridge are faster, the pinwheel will tend to spin in a clockwise direction, generating negative (anticyclonic) vorticity. Because this vorticity results from horizontal differences in wind speed (in other words, horizontal wind shear), this type of relative vorticity is called shear vorticity.

Similarly, height lines at the base (southernmost extent) of a trough, where cold air aloft meets warmer air equatorward of the trough, are usually a bit closer together compared to those farther poleward in the trough (see Figure 12.10b). As a result, the wind near the base of the trough is usually a bit faster, so a stationary pinwheel placed in the trough will tend to spin in a counterclockwise direction, generating positive (cyclonic) **shear vorticity**.

So, if we consider both vorticity generated by curvature and vorticity generated by shear, a 500-mb ridge is a source of negative relative vorticity and a 500-mb trough is a source of positive relative vorticity. Now we're ready to add earth vorticity which, you may recall, is positive (in the Northern Hemisphere).

For the sake of argument, it's reasonable to assume here that there is a single, representative value of earth vorticity in the latitudinal band that contains the ridge-trough system (there's not much variation in earth vorticity over the north-south span of the ridge-trough system). By adding this relatively large and positive value of earth vorticity to the negative value of relative vorticity in the ridge and to the positive value of relative vorticity in the trough, we find that absolute vorticity, in general, is positive in the

Northern Hemisphere. The center of the ridge marks a minimum in absolute vorticity (a "vort min") while the center of the trough marks a maximum in absolute vorticity (a "vort max"). The dashed lines in Figure 12.11a are **isovorts**, which are isopleths of vorticity.

Now let's follow a parcel of air as it moves through this ridge-trough system (see Figure 12.11b). As the parcel leaves the ridge, it moves toward a "straight-away," where negative relative vorticity due to both curvature and shear decreases. Thus, the parcel's relative vorticity increases in time. In fact, this increase in relative vorticity will continue until the parcel reaches the base of the trough. With the parcel's relative vorticity increasing as it moves from the ridge to the trough, its absolute vorticity increases as well (we're assuming here that earth vorticity is essentially constant over the north-south span of the ridge-trough system). In other words, the parcel's counterclockwise spin increases and, like an ice skater bringing her arms inward to increase her spin, the parcel undergoes convergence (note in Figure 12.11b that the parcel's horizontal area shrinks as it moves from the ridge to the trough).

The upshot here is that the region between an upstream ridge and a downstream trough favors upper-air convergence which helps to increase the weight of local air columns. This upper-level convergence paves the way for increasing surface pressure and often, as a result, developing or strengthening high pressure at the surface. Given that the axes of upper-level ridges and troughs typically run north-to-south (like those in Figure 12.11, and for that matter, Figure 12.10), a reasonable inference is that upper-level convergence occurs to the east of upper-level ridges.

A similar argument for a parcel leaving a 500-mb trough would lead to a decrease in absolute vorticity to the east of a trough. This decrease with time corresponds to a decreasing rate of spin, which can only be accomplished by divergence (like a slower-spinning ice skater who extends her arms away from her body, the air parcels east of the trough in Figure 12.11b expand outward). This upper-air divergence removes weight from local air columns and thus paves the way for decreasing pressure at the surface, and as a result, developing or strengthening surface low pressure. Thus, a reasonable deduction to make is that upper-level divergence occurs to the east of upper-level troughs.

Figure 12.12 summarizes these ideas. Note that a pocket of upper-level divergence, shaded in purple, lies between the upstream trough and downstream

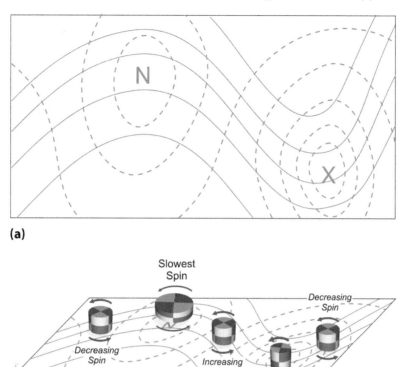

(a)

(b)

FIGURE 12.11 (a) A ridge marks a location of minimum absolute vorticity, which is denoted by the closed isovorts with the "N" in the center (short for vort miN). A trough marks a location of maximum absolute vorticity, which is denoted by the closed isovorts with the "X" in the center (short for vort maX); (b) As a parcel leaves a 500-mb ridge and heads toward the next trough, its absolute vorticity increases. The increase in vorticity with time means an increased rate of spin, which must be accompanied by convergence. Thus, the region to the east of a 500-mb ridge is favorable for surface high-pressure systems to develop. In contrast, as a parcel leaves a 500-mb trough and heads toward the next ridge, its absolute vorticity decreases with time. The decrease in vorticity is accompanied by a decreased rate of spin, and thus upper-level divergence. Thus, the region to the east of a 500-mb trough is favorable for surface low pressure to develop.

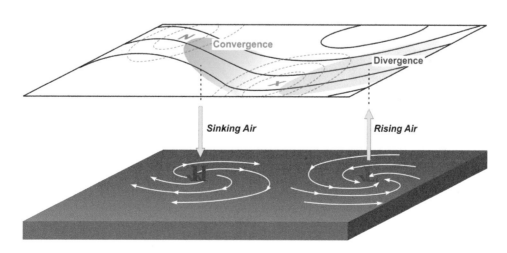

FIGURE 12.12 Upper-air convergence east of a 500-mb ridge promotes surface high pressure, while upper-air divergence east of a trough fosters surface low pressure.

ridge in a region where the spin of air parcels moving through the flow is decreasing. Beneath this upper-level divergence, low pressure tends to develop or strengthen. Meanwhile, a pocket of upper-level convergence, shaded in red, is favored east of the upper-level ridge. In this region, surface high pressure tends to develop or strengthen.

So there you go! That's the big-picture connection between vorticity patterns at 500-mb and surface pressure systems. But we can actually say a little more. As it turns out, the 500-mb troughs with the greatest potential to spawn powerful low-pressure systems tend to be relatively compact (we call them "short" waves). In this context, large, sprawling 500-mb troughs (the "long" waves) really can't compete with their more compact counterparts. Time to investigate the differences between "short-wave" and "long-wave" troughs.

ANTICIPATING BIG STORMS: FAVORABLE UPPER-LEVEL PATTERNS

When trying to predict significant mid-latitude storms, meteorologists typically first study the 500-mb pattern, looking specifically for the location of troughs that mean business. As it turns out, more compact troughs tend to produce the deepest low-pressure systems. But how do you measure "compact"?

Long and Short Waves: Don't Judge a Wave by its Size

Big waves always attract surfers looking to get big thrills (see Figure 12.13), but weather forecasters routinely look at relatively small 500-mb waves as potential catalysts for big (and powerful) low-pressure systems, especially during winter. Let us explain.

Meteorologists classify an atmospheric wave (a trough-ridge pair) according to its wavelength. Like the electromagnetic waves you studied in Chapter 2, the wavelength of an atmospheric wave is the distance corresponding to one complete cycle of a trough-ridge pair (see Figure 12.14). Waves with relatively short wavelengths (from several hundred to about one thousand kilometers) are called **short waves**. Weather forecasters use terms such as **short-wave trough** and **short-wave ridge** when they refer to individual components of a short wave. In contrast, long waves have wavelengths of several thousand kilometers.

Figure 12.15a shows a long wave at 500 mb that spans the entire nation (long-wave ridge in the West, long-wave trough in the East). The solid lines are 500-mb heights, and the dashed lines are isovorts. Given the gradual curvature and fairly loose packing of the height lines that form the long wave, the values of relative vorticity in the long-wave trough and long-wave ridge are comparatively small. In this case, the isovorts of absolute vorticity tend to follow the general pattern of the height lines; there aren't any closed isovorts associated with the long-wave trough and ridge.

In contrast, note the short wave over the Central States in the "straightaway" between the long-wave ridge and long-wave trough. Without reservation, the height lines associated with the short-wave trough

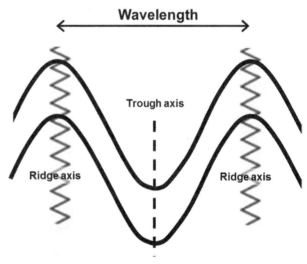

FIGURE 12.14 An atmospheric wave consists of a trough and a ridge, and its wavelength is the distance from one trough axis to the next trough axis (or one ridge axis to the next ridge axis). The official symbols for a trough axis (a dashed line) and ridge axis (a serrated line) are shown.

FIGURE 12.13 Surfer Alfredo Villas-Boas rides a big wave off the island of Maui in Hawaii (courtesy of Alfredo Villas-Boas and photographer Ricardo Cufi).

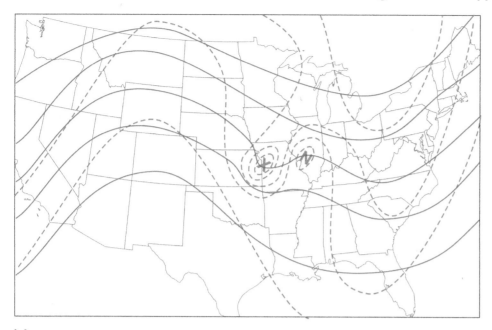

(a)

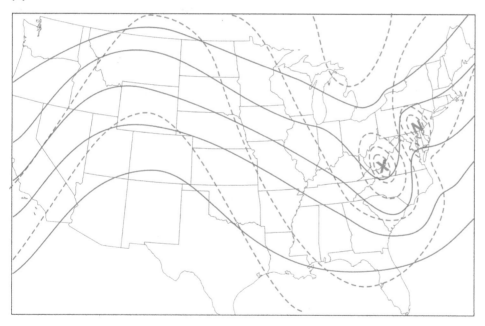

(b)

FIGURE 12.15   Two schematic 500-mb charts, with part (b) 24 hours after part (a). Solid contours represent 500-mb heights, while dashed contours are isovorts (isopleths of absolute vorticity). The heights indicate a long-wave ridge over the West and a long-wave trough in the East. To a large extent, the isovorts associated with the long wave follow the general pattern of the height lines. Notice there are no closed isovorts associated with the long wave trough and ridge. In contrast, the vort max associated with the more compact, curvier short wave in the Central United States is more conspicuous, indicated by several closed and tightly packed isovorts. Note that in part (b), the long-wave pattern didn't change much in shape or location, but the short-wave trough has noticeably progressed eastward, demonstrating that short waves move faster than long waves.

and short-wave ridge have noticeably more curvature, thereby enhancing local relative vorticity. Moreover, height lines are packed more tightly, particularly at the base of the short-wave trough and near the crest of the short-wave ridge. As a result, relative vorticity generated by horizontal wind shear is greater in the short wave. So, the bottom line to Figure 12.15 is this: The isovorts of the vort max and vort min that mark the short-wave trough and ridge are much more conspicuous than the isovorts associated with the long wave.

Figure 12.15b shows the 500-mb pattern 24 hours later. Note that the long wave has moved very little—it's still fair to say that there's a long-wave ridge over the West and a long-wave trough over the East. However, the short wave has progressed noticeably to the east, suggesting that atmospheric waves don't all move at the same speed. In fact, short waves almost always move faster than long waves. Indeed, a long wave makes relatively little eastward progress during the course of a week or so (some very long waves can actually drift westward), but a short wave can streak across the country in a couple of days during winter. This means that short waves can "travel through" long waves. In other words, short waves can round the top of a long-wave ridge, dive southeastward toward the long-wave trough, and then round the base of the long-wave trough and head northeastward toward the next downstream long-wave ridge.

We stated earlier that short-wave troughs, in general, are capable of spawning more formidable low-pressure systems than long-wave troughs, which often can barely muster a weak surface low. The 500-mb clue that helps explain this observation involves the difference between the pattern of isovorts associated with short-wave and long-wave troughs. Before we tackle this topic, let's review the link between the change in spin of an ice skater and patterns of divergence and convergence. For starters, remember the ice skater coming out of a fast spin. In order to markedly slow down her spin, a confident professional figure skater stretches out her arms in a clear act of divergence. But a shy, novice skater, who can only spin slowly, doesn't have to extend her arms very far to stop her spin. So, in the case of the whirling professional skater, large divergence corresponds to the most noticeable change in the skater's spin (from fast to slow in a short period of time). In the case of the novice, only small divergence occurs with a slight change in spin (from slow to slower).

If we jump from the ice-skating rink to spinning air parcels traveling through a short-wave trough at 500 mb, the same sort of conceptual approach still applies. In other words, large divergence occurs in concert with the most dramatic changes from fast to slow spin in a relatively short time. This happens downstream of a short-wave trough where isovorts are tightly packed (see Figure 12.15a) and, thus, parcels moving through the trough are slowing their spin rapidly with time. Like an ice skater slowing her spin

by outstretching her arms, air parcels undergo relatively large divergence, setting the stage for surface pressure to noticeably decrease. In contrast, in the case of a long-wave trough, the spin of an air parcel leaving the base of the trough does not change very much with time (the isovorts are loosely packed). Thus, upper-level divergence is rather weak downstream of a long-wave trough.

These same arguments can be applied to parcels of air moving through short-wave ridges and long-wave ridges. In both cases, parcels spin faster as they emerge from the ridge, but parcels leaving the short-wave ridge increase their spin faster with time than those leaving a long-wave ridge. Thus, upper-level convergence tends to be larger downstream of a short-wave ridge than downstream of a long-wave ridge.

So to this point, we have laid the groundwork for connecting patterns of divergence and convergence at the 500-mb level with surface pressure patterns. And, in fact, the 500-mb level has, over the years, become the go-to upper-air map for operational meteorologists. But as you learned in Chapter 7, the fastest winds occur at jet-stream level which, over the mid-latitudes, tends to lie around 300 mb. For meteorologists searching for the greatest upper-air divergence and convergence (which would support strong low- and high-pressure systems at the surface), it makes some sense to explore the level where winds blow the fastest. This jet-stream-level analysis is complementary to the 500-mb analysis—essentially, we are acknowledging that upper-level divergence and convergence typically does not occur just at one level, but instead in a layer of the upper troposphere (recall Figure 12.4). Nonetheless, a method for analyzing jet-stream winds has proven useful over the years by helping forecasters identify pockets of large upper-level divergence and convergence. Say hello (again) to jet streaks.

Jet Streaks: Big Storm Makers

In March 1993, one of the most powerful winter storms in modern history rapidly developed over the northern Gulf of Mexico and then roared up the Eastern Seaboard (see Figure 12.16), producing severe thunderstorms, a lethal storm surge, high winds, and, of course, prodigious snowfall (details to come in Chapter 16).

At the height of the storm, around 00 UTC on March 14, the central pressure bottomed out close to 960 mb (check out the map on the right in Figure 12.17). At

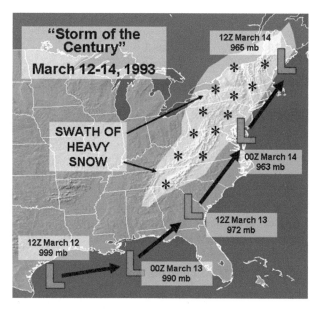

FIGURE 12.16 The track of the low-pressure system known as the "Storm of the Century" in March 1993.

the time, the core of a 300-mb jet streak, with maximum winds greater than 150 knots (170+ mph), was moving northward over Atlantic coastal waters just to the east of the path of the low (see the chart on the left in Figure 12.17). As it turned out, the intense surface low was under a region of the jet streak that favored strong divergence at 300 mb. Formally, this divergence-rich region is called the left-exit region of the jet streak (stay tuned for all the details). For now, it suffices to say that the storm's low barometric pressure was no accident. Let's investigate the relationship between deep low-pressure systems and jet streaks.

By way of review, **jet streaks** are pockets of relatively fast winds embedded within the mid-latitude or subtropical jet streams. These typically oval-shaped wind maxima are on the order of several hundred kilometers long and less than a few hundred kilometers wide. You can use the 300-mb chart in Figure 12.17 to get a sense of the spatial scale of a jet streak.

Jet streaks are somewhat of a paradox. In winter, they can house winds that blow as fast as 200 kt (230 mph). Yet they physically move slowly through the jet stream, typically advancing at a relatively slow-poke pace of 30–50 kt (35–55 mph)—typically much slower than the speed of the wind anywhere in the jet stream. Thus, parcels of air flowing rapidly along within the jet stream usually blow right through a jet streak. Parcels entering a jet streak accelerate as they pass isotachs of progressively higher values (see Figure 12.18). Air parcels continue to accelerate in this "entrance region" all the way to the core of the jet streak, where the fastest winds reside. In quick fashion, parcels then slow down as they move away

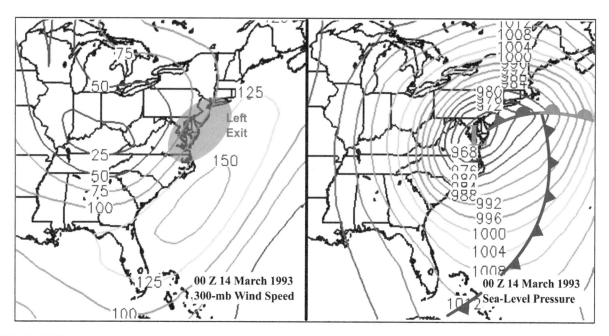

FIGURE 12.17 (Left) The 300-mb analysis of isotachs (in kt) at 00 UTC on March 14, 1993. The center of the 150-kt closed isotach over Atlantic coastal waters marks the core of a jet streak. The rose-colored oval marks an area of significant upper-level divergence associated with the left-exit region of the jet streak; (Right) The analysis of sea-level pressure at 00 UTC on March 14. At the time, the deep low lay under the divergence-rich left-exit region of the jet streak (courtesy of NOAA).

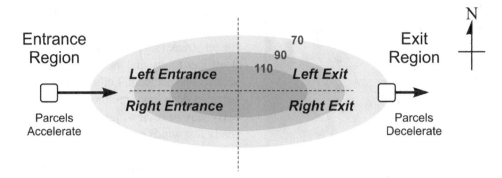

Isotachs representing a west-east oriented jet streak at the 300-mb level. Parcels of air accelerate when they enter a jet streak (the "entrance" region) and decelerate when they exit a jet streak (the "exit" region). A jet streak is also divided into "left" and "right" sides relative to an observer moving through the jet streak with an air parcel.

from the core of a jet streak in the "exit region," passing isotachs of progressively lower values. Meteorologists also identify regions of a jet streak as either "left" or "right," relative to an observer moving through the jet streak with an air parcel, yielding four quadrants: right-entrance, right-exit, left-entrance, and left-exit. The reason for this partitioning will soon become apparent.

The speed-ups and slow-downs of air parcels as they blow through a jet streak create patterns of divergence and convergence in the upper troposphere (such as the area of 300-mb divergence shaded on Figure 12.17). In tandem, surface low-pressure (and high-pressure) systems strengthen when they are properly positioned relative to jet streaks. Let's delve into the details.

In Figure 12.19a, the three arrows represent the direction and speed of 300-mb winds north of, south of, and in the core of a straight (west-to-east) jet streak (the longer the arrow, the faster the wind). This profile of wind around the core of a jet streak creates horizontal wind shear, which, in turn, generates cyclonic relative vorticity north of the core and anticyclonic relative vorticity south of the core (consistent with the sense of spin imparted to the two pinwheels in Figure 12.19a). Given that earth vorticity does not change very much over the limited north-south span of the jet streak, these areas of shear vorticity result in a vort max north of the core of a jet streak and a vort min south of the core of a jet streak (see Figure 12.19b).

Now imagine a parcel moving eastward away from the vort max. It loses absolute vorticity with time. In other words, it loses cyclonic spin with time. Like a spinning figure skater who slows her rotation by stretching out her arms, the air parcel undergoes divergence in the left-exit region of the jet streak.

Similarly, a parcel moving eastward away from the vort min gains absolute vorticity with time. In other words, the parcel's cyclonic spin increases with time. Like a spinning ice skater who quickens her spin by drawing her arms inward, the parcel undergoes convergence in the right-exit region of the jet streak. You can make similar arguments to show that there is convergence in the left-entrance region and divergence in the right-entrance region of a straight jet streak. Figure 12.19c summarizes this four-quadrant conceptual model of a straight jet streak.

Notice in Figure 12.19c that we have also included references to preferred vertical motions occurring in the various quadrants. For example, "upward motion" is indicated in the right-entrance and left-exit regions. The implication is that, in these two quadrants, upper-level divergence in local air columns is linked with low (or decreasing) surface pressure and, thus, convergence at low levels. Such a pattern of low-level convergence paired with upper-level divergence translates into upward motion throughout that air column (revisit Figure 12.4a). Similarly, in the left-entrance and right-exit regions, "downward motion" is consistent with low-level divergence and upper-level convergence in local air columns (see Figure 12.4b).

We referred to this four-quadrant model of a jet streak as "conceptual" for a reason. The model is a basic template for the pattern of upper-level divergence and convergence and vertical motions that occurs in the vicinity of a jet streak. In the real atmosphere, jet streaks are rarely this simple. The most important complicating factor is that most jet streaks are embedded in flows that are curvy—that is, within the realm of troughs or ridges. Figure 12.20, which shows the 300-mb pattern on an April day several years ago, provides an excellent example—note the

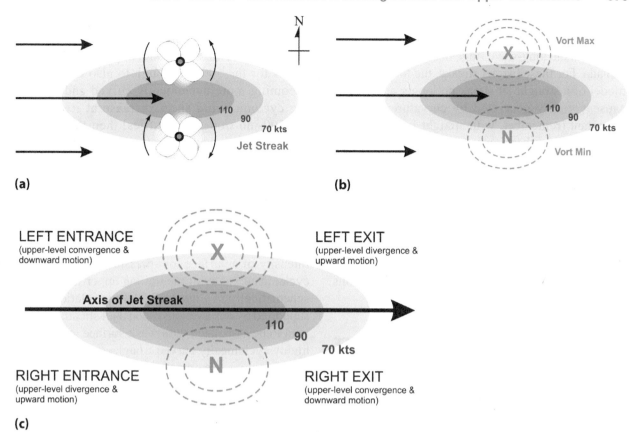

(a)

(b)

LEFT ENTRANCE
(upper-level convergence &
downward motion)

LEFT EXIT
(upper-level divergence &
upward motion)

Axis of Jet Streak

110
90
70 kts

RIGHT ENTRANCE
(upper-level divergence &
upward motion)

RIGHT EXIT
(upper-level convergence &
downward motion)

(c)

FIGURE 12.19 (a) The strongest winds associated with a jet streak lie at its core (represented by the longest arrow). Winds are not as strong away from the core (shorter arrows); (b) The horizontal wind shear associated with a straight jet streak creates cyclonic shear vorticity north of the core of the jet streak and anticyclonic shear vorticity south of the core of the jet streak. Thus, a vort max/vort min pair always accompany a straight jet streak; (c) A parcel moving eastward away from the vort max associated with a straight jet streak loses absolute vorticity with time, yielding divergence in the left-exit region. A parcel moving eastward away from the vort min gains absolute vorticity with time, yielding convergence in the right-exit region. Similar arguments yield convergence in the left-entrance region and divergence in the right-entrance region.

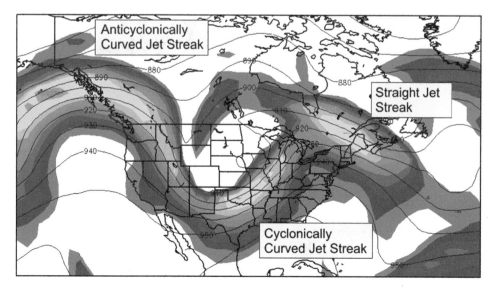

Anticyclonically Curved Jet Streak

Straight Jet Streak

Cyclonically Curved Jet Streak

FIGURE 12.20 A 300-mb pattern showing heights and isotachs. Three jet streaks are shown—one in a ridge (so it is anticyclonically curved), one in a trough (so it is cyclonically curved), and one in an area where heights are generally straight. The preferred areas of divergence and convergence associated with a jet streak depend on the curvature of the flow in which the jet streak is embedded.

jet streak embedded in the long-wave trough over the central United States and a second jet streak embedded in the long-wave ridge that arcs into western Canada. For these two jet streaks, the four-quadrant model of a straight jet streak no longer applies because vorticity generated by curvature becomes important (remember, with straight jet streaks, we only considered shear vorticity). In contrast, the four-quadrant model would work reasonably well for the relatively straight jet streak that stretches from Quebec into New England.

As it turns out, the left-exit and right-entrance regions of jet streaks often play a significant role in the life cycle of powerful mid-latitude low-pressure systems, a topic that we will address in the next chapter. For now, let's summarize the main points of this chapter.

STAYING LEAN: KEEPING THE WEIGHT OFF

In this chapter, we've established that a favorable area for surface pressure to decrease lies to the east of an upper-air trough, where upper-air divergence occurs. For a surface low to continue to deepen (intensify), this upper-air divergence must increase in time. One way for this to happen is for the 500-mb trough to also intensify by becoming more cyclonically curved and by having cyclonic wind shear increase at its base. If the 500-mb trough intensifies, then the trough's cyclonic relative vorticity (and hence its absolute vorticity) will continue to increase. Then, when a parcel moves through the stronger trough, divergence will be greater (the parcel will go from a fast spin in the trough to a markedly slower spin in the straightaway east of the trough).

We illustrate the relationship between the intensity of a surface low-pressure system and the cyclonic curvature of the trailing upper-level trough schematically in Figure 12.21 (this relationship is supported by observations). Zonal jet stream patterns tend to be associated with weak surface lows, while highly cyclonically curved troughs (in other words, meridional jet stream patterns) typically are associated with more intense surface low-pressure systems. Clearly, the cyclonically curved trough looks like the

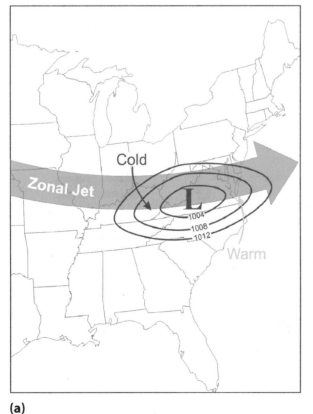

(a)

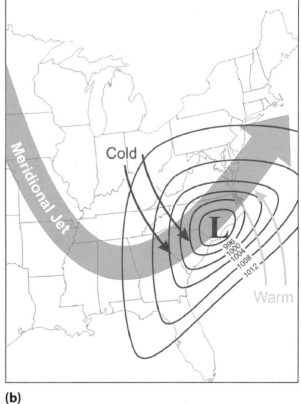

(b)

FIGURE 12.21 All other factors being equal, low-pressure systems that form in concert with highly curved jet streams, as in (b), typically have lower pressures than those that form in zonal flows, as in (a). In addition, the circulations around deep lows are broader and stronger and, thus, more efficient at mitigating north-south temperature contrasts.

bigger weather-maker. Moreover, the more expansive and stronger circulation of the deeper surface low, which sends warm air farther poleward and cold air farther equatorward, is much more efficient in helping the atmosphere achieve its goal of mitigating the large temperature contrasts between polar regions and the tropics.

Figure 12.21 can be interpreted another way: It shows two stages in the life cycle of a low-pressure system. As a low strengthens, it also alters the upper-level pattern, so the relationship is not just a one-way street. This synergy is but one of the interesting topics that awaits you in Chapter 13!

Mid-Latitude II: The Cyclone Model

13

LEARNING OBJECTIVES

After reading this chapter, students will:

- Understand the role of a 500-mb trough in initiating cyclogenesis along a stationary front.
- Gain an appreciation for the two-dimensional representation of frontal zones.
- Be able to identify the types of fronts on surface weather analyses and to gain an appreciation for the lifting mechanisms associated with each frontal zone.
- Be able to recogize the sequence of weather events typically associated with the approach and passage of cold and warm fronts.

- Understand the distinction between an anafront and a katafront.
- Be able to identify, on a surface analysis, a thermal ridge associated with a mid-latitude cyclone, and to discuss its role in initiating convection along and just ahead of the low's cold front.
- Understand the role of self-development in the life cycle of a mid-latitude low.
- Gain an appreciation for how a developing cyclone carves out a niche of negative pressure tendencies that allows the low to reconstitute itself downstream of its center.
- Understand the reasons why a surface low moves back into the cold air during the occlusion stage of its life cycle.
- Understand the role of the warm, cold, and dry conveyor belts in shaping the cloud and precipitation shields associated with a mature mid-latitude cyclone.

By way of review, we start with key observations from the last chapter. The sea-level pressure at the center of a powerful mid-latitude low-pressure system might read 980 mb, while a strong high-pressure area might barometrically top out at 1040 mb. That there's only about a five percent pressure difference between stormy lows and placid highs over the mid-latitudes illustrates the atmosphere's dedication to a system of checks and balances that keeps local air columns from getting too heavy or too light.

Controlling the weight of a column of air is a matter of (metaphorically speaking) exercise and diet. First, let's review the exercise program for a slimfast low. In the upper troposphere, aerobic jet stream winds perform calisthenics of divergence, a high-altitude outstretching that removes mass from the column of air above the low. In a spirit of check-and-balance, surface winds converge toward the center of the low, thereby tending to add weight to the air column in an attempt to offset the loss of mass aloft. But as long as upper-level divergence slightly exceeds low-level convergence, there is a net loss of mass in the air column and surface pressure should fall.

We say "should" because upper-level divergence must also compensate for another process that tends to add weight to air columns: the rising air that characterizes low-pressure systems. Recall that air cools during ascent, and this cooling tends to raise the average density of the air column which serves to slightly increase column weight. So, in the final analysis, upper-level divergence over a low must also offset this factor associated with cooling from ascent.

That's a tall order for upper-level divergence. Logically, as the low strengthens and convergence increases near the ground and cooling from ascent quickens, upper-level divergence has to increase in order to handle the increasing demand for maintaining a relatively light column weight.

How does upper-level divergence meet these demands? When does the demand become too great for upper-level divergence to keep pace? And what happens just ahead of the low's path? Put another way, does a low carve out a low-barometric-pressure niche so that it feels right at home once it arrives, or does it just move along like a helpless stick in a stream of steering winds? The answers to these questions are crucial to predicting the evolution and life cycle of mid-latitude low-pressure systems. As a forecasting tool, meteorologists have encapsulated the common characteristics and behaviors of mid-latitude lows into a template called the **cyclone model** (cyclone is a generic name sometimes given to low-pressure systems).

So attention all mid-latitude lows: This is your life!

BIRTH OF A LOW: WHICH COMES FIRST, THE CHICKEN OR THE EGG?

Recall from Chapter 6 that the cores of tropical and polar air masses are marked by high pressure. Thus, a front that marks the boundary between the outer edges of two contrasting air masses must lie in a zone of relatively low pressure (in other words, a trough). What better place for a mid-latitude cyclone to be hatched?

The First Domino Falls: Upper-Level Divergence

Commonly, stationary fronts embedded in a trough between two contrasting air masses serve as breeding grounds for mid-latitude cyclones (see Figure 13.1a). Such temperature contrasts can also exist naturally, driven by land-ocean boundaries, for example. In part, that's why cyclogenesis often occurs along or near the U.S. East Coast in winter—there, cold continental air rubs elbows with relatively warm air offshore. The bottom line here is that zones of relatively large temperature contrast are important in the life cycle of a mid-latitude low-pressure system.

Given a pre-existing zone of relatively large temperature gradient, what is the trigger for a center of mid-latitude low pressure? For all practical purposes, low-level convergence along a pre-existing stationary front cannot provide the spark for a low to form. Such convergence would cause an increase in the weight of local air columns, which is not exactly what an aspiring low needs. Instead, an upper-level short-wave trough must provide the necessary divergence aloft that leads to an initial loss of column weight and the formation of a fledgling center of low pressure (see Figure 13.1b). Once a counterclockwise circulation develops around this center, cold air begins to advance and retreat. Where cold air advances, a cold front is born; where cold air retreats, a warm front forms.

Think of the life of a low as a long, intricate series of dominoes, cleverly designed so that tumbling energy can set other contraptions into motion. In order to set off a chain reaction leading to the development of a full-fledged low-pressure system, upper-level divergence is required to tumble the first domino and thereby begin the process of **cyclogenesis**. The next domino to fall in the development of a low-pressure system requires low-level winds to advect (transport) warm and cold air northward and southward, respectively. Patterns of cold and warm air advection then tumble other atmospheric dominoes, ultimately leading to the intensification of the low.

The first step toward getting our dominoes all lined up is a closer look at warm and cold advection in the context of the frontal zones associated with mid-latitude low-pressure systems.

Cold and Warm Air Advection: A Follower and a Leader

Figure 13.2a shows a typical mid-latitude low-pressure system in the Northern Hemisphere in the middle stages of its life cycle, with its associated cold and warm fronts. Notice that here, unlike previous depictions, the fronts are shown as surfaces slanting upward from the ground.

The depiction of fronts as surfaces stems from the definition of air masses as volumes. The boundary between two volumes is not just a line on a horizontal map, but instead a surface. In the atmosphere, these

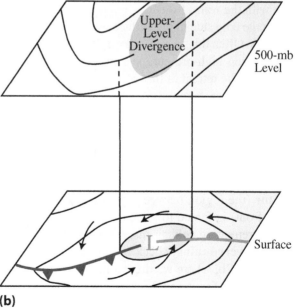

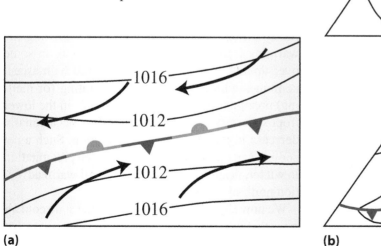

(a)

(b)

FIGURE 13.1 (a) A stationary front often serves as generic breeding grounds for a mid-latitude low-pressure system. Pressure is already relatively low there because the front lies in a trough between two highs; (b) An upper-level short-wave trough typically provides the upper-level divergence necessary for the initial lowering of surface pressure along the stationary front.

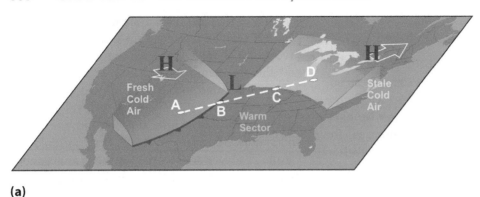

(a)

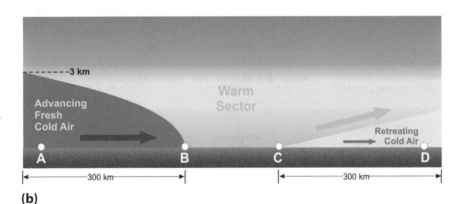

(b)

FIGURE 13.2 (a) For a typical mid-latitude low-pressure system, a warm front extends to the east of the low's center and a cold front trails to its southwest. Think of the fronts as surfaces that slant upward into the atmosphere. Relatively dense, cold air always lies below these frontal surfaces. (b) In a cross section from point A to point D, the cold air masses appear as wedges. At the warm front, the cold air retreats, leading to a gentler slope of the front. At the cold front, the cold air is advancing, so the slope of the front is steeper.

frontal surfaces slope upward over colder, denser air. Given that fronts lie in troughs, the associated low-level convergence implies upward motion along these boundaries, paving the way for clouds and precipitation. We will now investigate the details, starting with the warm front.

On the cold side of the surface warm front, cold air retreats. Often this cold air mass is "stale," given that a departing high-pressure system essentially marks its core. The retreat of this cold air mass allows less dense warm air to move poleward, so the warm front essentially advances by default. Typically, however, the cold air is slow to give ground, so warm fronts tend to move slower than cold fronts (where cold air is advancing).

As a result, the advancing warm air from the **warm sector** (the region between the warm front and the cold front) is forced to glide up the "ramp" formed by the retreating cold air mass. To get a better sense of this **overrunning**, check out the cross section in Figure 13.2b, where we drew a line between points A and D in Figure 13.2a and then took a vertical slice of the troposphere through this line. Now you can clearly see the overrunning at work, setting the stage for widespread stratiform clouds and precipitation to form on the cold side of the surface warm

front. While overrunning may be a visually descriptive term, it essentially is equivalent to warm air advection in this context because winds blow from a region of higher temperatures (the warm sector) toward a region of lower temperatures (the retreating cold air).

We note that the "wedge" formed by the retreating cold air mass has a very gradual slope, perhaps one kilometer in the vertical for every 300 kilometers in the horizontal. Although this slope is relatively small, rising warm air still cools, partially offsetting the warming influence from warm advection. In some overrunning events, cooling associated with ascent combines with cooling from evaporating (or melting) precipitation to cause temperatures in the lower troposphere (say around 850 mb) to occasionally decrease in the face of warm advection. Such a net cooling can tip the scales toward frozen precipitation in winter, despite the overall pattern of warm advection north of the warm front.

We now shift our discussion from the stale cold air mass that is retreating to the "fresh" cold air mass that is advancing behind the low's cold front. On average, this fresh cold air mass thickens by about one kilometer for every 100 kilometers of horizontal distance back into the cold air (look again at Figure 13.2b), so cold fronts tend to be steeper than warm fronts. With winds in the fresh cold air mass blowing in the general direction of the cold front, the entire span from the core of the cold air mass to the cold front is marked by cold advection. In other words, winds behind the cold front transport increasingly cold air into a given location once the cold front passes.

Armed with this background, we're now ready to see how the evolution of warm and cold air advection aids in the intensification of a low-pressure system.

SELF-DEVELOPMENT: DETERMINING THE DESTINY OF LOW PRESSURE

We have continually stressed the fundamental reason why low-pressure systems exist: They are agents employed by the general circulation of the atmosphere that help to lessen the temperature contrasts between polar regions and the tropics. But, in a way, they are actually "double agents" because, in an apparent act of double-cross, lows draw warm and cold air together over the mid-latitudes, steadily increasing temperature contrasts in narrow zones which we call fronts.

As we already hinted, however, lows derive energy from these contrasts, intensifying as they carry on their energy-feeding frenzy. In turn, more and more isobars surround the low's center, and its counter-clockwise circulation of air expands. In this way, expansive lows are able to propel warm air farther poleward and send cold air farther equatorward. In other words, they are able to produce warm and cold advections over an increasingly large area. In the final analysis, intense lows, by feeding off temperature gradients in frontal zones, effectively accomplish the mission of helping to even out larger-scale temperature contrasts.

In essence, low-pressure systems, by concentrating temperature contrasts in narrow zones along fronts and by promoting temperature advections, play a major role in their own development. In fact, this process is so common that meteorologists give it a name: **self-development**. To understand this positive feedback, we will discuss the details of the cyclone model, tracing the life of a low from birth to death.

Cold Advection: Dig That Trough!

We will now look more closely at the evolution of cold air advection behind the cold front to see how it affects the trailing 500-mb trough. If low-level cold air advection somehow sharpens the trough and

increases upper-level divergence over the low, the low will intensify. So that's exactly what we're after here—a connection between cold advection and the strengthening of the 500-mb trough.

Let's assume that a short-wave trough at 500 mb has already approached a stationary front (recall Figure 13.1b), helping to lower surface air pressure by removing mass from local air columns. In concert with the low-level, cyclonic circulation of air around the fledgling low (see Figure 13.3a), cold air starts to flow equatorward to the west of the low (cold advection) and warm air starts to flow poleward (warm advection) to its east. To make the link between cold air advection and the 500-mb trough, we will choose a point (labeled C in Figure 13.3) and then follow what happens to the height of the 500-mb surface at this point.

In time, the cold front moves eastward past point C, setting the stage for cold advection there (Figure 13.3b). Because pressure decreases more rapidly with altitude in cold air, the infusion of low-level cold air into the air column above point C causes the height of the local 500-mb pressure surface to lower. Figure 13.3b shows the surface low, cold front, and 500-mb height pattern at some time after cold air advection began at point C. Note that the 500-mb height at point C has decreased. Overall, the 500-mb trough has sharpened (become more cyclonically curved). An amplified, more curved trough has a higher vorticity maximum (in large part due to greater curvature), and thus produces more divergence aloft on its eastern flank. It follows that the central pressure of the low should decrease.

In response to decreasing surface pressure, more isobars encircle the deeper surface low, increasing pressure gradients (see Figure 13.3c). Thus, surface wind speeds also increase. Now a stronger wind blowing from the west-northwest behind the cold front quickens the transport of cold air, which increases the temperature gradient behind the cold front. Thus, cold air advection increases at point C. In turn, the 500-mb height further decreases there and the 500-mb trough continues to amplify. In concert with the trough's increasing cyclonic curvature, the vort max strengthens, which, in turn, augments the upper-level divergence over the surface low, causing the low to further deepen.

Hopefully, the appropriateness of the metaphor of dominoes to describe the development of a low-pressure system has become apparent. Initial upper-level divergence over a stationary front generates

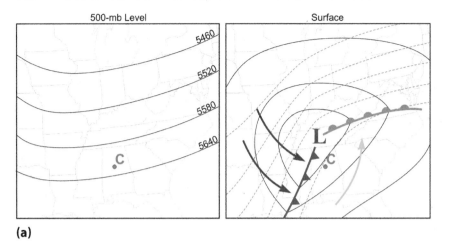

(a)

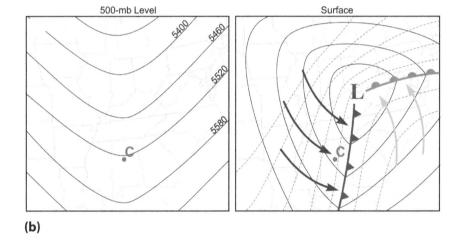

(b)

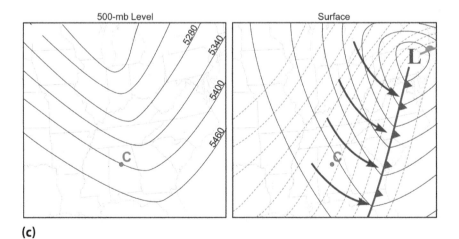

(c)

FIGURE 13.3 The self-development process of a low-pressure system. On the surface maps, isobars are solid lines while isotherms are dashed. (a) A developing low benefits from upper-level divergence associated with the 500-mb trough to its west; (b) The low strengthens, increasing cold advection to its west, further lowering 500-mb heights, which in turn sharpens the 500-mb trough; (c) A deeper trough means a more potent vorticity maximum (not shown) that translates to enhanced upper-level divergence to its east, which further deepens the surface low.

a fledgling low, which causes cold air advection behind a newly defined cold front. In turn, 500-mb heights fall, causing the 500-mb trough to sharpen and become more "curvy." In response, the vort max in the trough strengthens and upper-level divergence increases to its east. In turn, the surface low intensifies, pressure gradients increase, and surface winds speed up. With faster winds, cold advection increases, thereby setting the stage for a long chain of atmospheric dominoes to tumble and for the system to continue to strengthen via this self-development feedback loop.

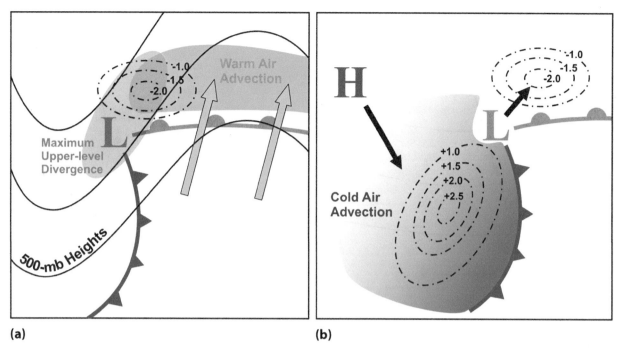

(a)   **(b)**

FIGURE 13.4 (a) The overlapping areas of warm advection and maximum upper-level divergence associated with a developing surface low-pressure system work in tandem to create a bull's-eye of negative pressure tendencies (units shown here are millibars per three hours); (b) By carving out a niche of maximum negative pressure tendencies, the low reconstitutes itself north and east of its center. Similarly, the trailing center of high pressure tends to move toward the bull's-eye of positive pressure tendencies that typically forms behind the cold front.

Ultimately, however, this self-development process can't continue unchecked. All low-pressure systems must eventually dissipate. But before the death knell sounds, we need to add one more piece of the puzzle in the life of a low.

Lows (and Highs) on the Move: Not Helpless Sticks in a Fishing Stream

In keeping with its industrious spirit of self-development, a low also has something to say about where and how it moves. Lows are not merely swept along in a current of upper-level steering winds like a stick in a fishing stream. Rather, a low continually reconstitutes itself, resourcefully expanding its portfolio of low pressure by wisely investing in warm air advection and upper-level divergence to the east of its center (and north of its warm front). Through these two processes, a low attempts to carve out an environment favorable for strengthening.

Figure 13.4a is a plan view that shows the areas of warm advection and maximum upper-level divergence associated with a developing low-pressure system. We already know that upper-level divergence

removes mass from local columns of air, thereby reducing column weights and surface pressures. Warm air advection, which is particularly hard at work on the cold side of the surface warm front, serves to lower the average density in local air columns, thereby reducing column weights and, in turn, reducing surface pressure.

The resulting falls in surface air pressure (meteorologists call them "negative pressure tendencies") from the combined effects of upper-level divergence and warm air advection ahead of a developing low-pressure system typically average a millibar or two every few hours (see Figure 13.4a). This bull's eye of maximum pressure falls is precisely the environment in which the low can reconstitute itself. So, like a self-made millionaire, the low moves in the direction of its self-made area of maximum pressure falls (note arrow in Figure 13.4b). Also note that the trailing center of high pressure tends to move toward the bull's eye of positive pressure tendencies that typically forms behind the cold front as cold advection helps to increase the average density of local air columns there.

By way of review: When a low-pressure system intensifies, pressure gradients and wind speeds

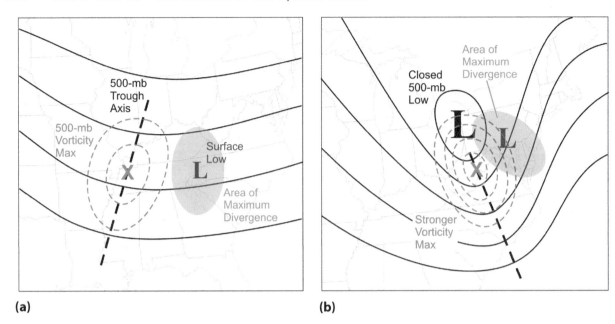

(a)

(b)

FIGURE 13.5 (a) A 500-mb trough and associated surface low-pressure system at the start of self-development; (b) A 500-mb trough and associated surface low near the end of self-development. The upper-level trough is now sharper, and a closed 500-mb low may have developed. The stronger vorticity maximum generates greater 500-mb divergence over and just west of the surface low.

increase in the lower troposphere. As a result, warm and cold advections increase. In turn, the upper-level trough sharpens, increasing upper-level divergence. Greater upper-level divergence and warm air advection promote greater falls in surface pressure ahead of the low, paving the way for more intensification as the low further reconstitutes itself east of its center.

But there's eventually a price to be paid for increasing divergence aloft and intensifying warm air advection: The resulting faster upward motion leads to greater cooling in and around the region of maximum pressure falls. Eventually, this cooling from ascent starts to arrest the tide of falling pressure east of a low's center, signaling the start of **occlusion**, the final stage in the life cycle of a low.

Occlusion: The End of Weight-Watching

Occlusion is a paradox because the low reaches its greatest intensity early in this stage, only to falter dramatically and dissipate later. To paraphrase a well-known quotation from Charles Dickens' *A Tale of Two Cities*, the occlusion stage is "the best of times" for a low, but it is also "the worst of times."

To begin to see why this applies, let's now compare the intensity and shape of a 500-mb trough at the start of self-development (Figure 13.5a) and near the end of self-development (Figure 13.5b). There are several key differences. Note how the upper-level trough has evolved—it has developed a closed height contour around a 500-mb low (meteorologists say that the 500-mb low has now "closed off"). Also, the tilt of the 500-mb trough has shifted from its original northeast-southwest orientation to a northwest-southeast tilt. Moreover, note how much more sharply curved the 500-mb trough has become during self-development, resulting in a stronger vort max (there are indeed more closed isovorts). Not only does the stronger vort max in Figure 13.5b generate greater divergence aloft, but the area of maximum upper-level divergence also shifts closer to the 500-mb low as the vort max makes a sharp left-hand turn around the more highly curved height lines.

Now let's turn our attention to changes in the vertical motion associated with the surface low during the evolution shown in Figure 13.5. As the surface low strengthens, so too does upward motion in the region of maximum pressure falls. However, stronger upward motion means increased cooling (remember, air cools when it rises), and with time, this cooling increasingly offsets the warming influence of warm advection. Essentially, as warm advection steadily loses its grip on helping to keep pressure tendencies negative during the process of occlusion, the

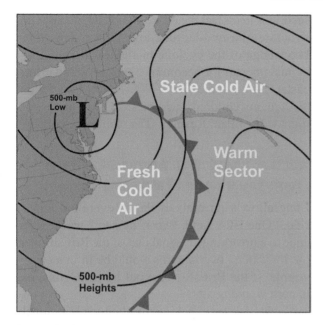

FIGURE 13.6 During the occlusion stage, the surface low moves closer and closer to a position directly underneath the 500-mb low. As this happens, the cold front appears to overtake a portion of the warm front, forming an occluded front. When this happens, the surface low loses contact with the warm sector, virtually ending the low's dependence on warm advection.

surface low increasingly follows the area of 500-mb divergence (which is shifting closer to the 500-mb closed low). In essence, the divergence-hungry surface low shifts back toward the cold air that lies on its western flank. As a result, the system gradually shifts to a "vertically stacked" alignment, with the surface low moving closer and closer to a position directly underneath the 500-mb low.

How does this realignment manifest itself in terms of the surface fronts? First, recall that advancing cold air typically moves faster than retreating cold air (another way of saying that cold fronts tend to move faster than warm fronts). As the surface low shifts back toward colder air, the cold front appears to overtake a portion of the warm front, forming an **occluded front**, shown in purple in Figure 13.6. Based on this generic frontal analysis, an occluded front separates the "fresh" cold air advancing behind the cold front from the "stale" cold air that is retreating north of the warm front. For all practical purposes, the surface low starts to lose direct contact with the warm sector, signaling the inevitable end of the low's reliance on warm advection. The point where the occluded front meets the low's warm front and cold front is called, appropriately enough, the

triple point. A new low-pressure center sometimes develops at the triple point, evidence that access to the warm sector is vital to a surface low's health.

While the cold front does indeed overtake the warm front during occlusion, this overtaking does not cause occlusion—rather, formation of the occluded front results from the reorientation of the divergence aloft and the response of the surface low moving back into the cold air to feed off this divergence (recall Figure 13.5).

As the low moves back into the cold air at the start of the occlusion stage, its fate is sealed, even though it is at or near full throttle. For once the system becomes perfectly vertically stacked (in other words, once the surface low lies directly beneath the 500-mb low), the surface low will lose its lion's share of divergence aloft. Strong low-level convergence and, to a lesser extent, cooling due to ascent, will be the primary mechanisms still operating in the column of air above the surface low. Convergence and cooling will add mass and increase the average density of the air column, leading to a weight gain and a rise in surface pressure. Now the low "fills" and its life cycle nears an end. But the low has served its purpose well, sending cold air equatorward and warm air poleward, thereby helping to ease temperature contrasts between the tropics and polar regions.

Now that you have an appreciation of the complicated and full life of a surface low-pressure system (the entire life cycle is also summarized in Figure 13.7), we will explore the myriad of weather that a low produces.

CYCLONE MODEL WEATHER: AS THE LOW GOES

Often, the first signs of an approaching warm front associated with a mid-latitude low-pressure system are patches of wispy cirrus clouds streaking across an otherwise deep blue sky (see Figure 13.8). Cirrus clouds (and their cousin cirrostratus) can be thought of as the "Paul Revere" of the atmosphere—they gallop ahead of the storm to herald the coming of thickening stratiform clouds and precipitation. Hooked-like cirrus are sometimes called "mare's tails" (after all, the midnight ride of Paul Revere couldn't have happened without his trusty horse).

Layered clouds and stratiform precipitation produced by overrunning are just one facet of the weather produced by a low-pressure system. The aim of this

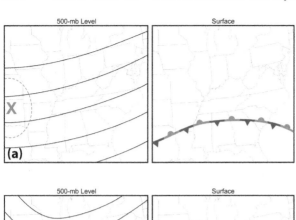

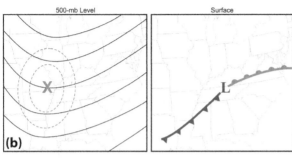

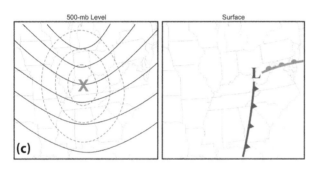

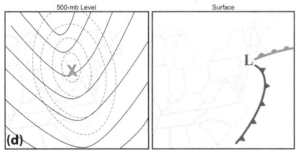

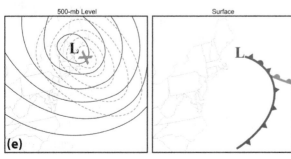

FIGURE 13.7 The life cycle of a mid-latitude low, at both 500 mb (left column) and the surface (right column).

section is to dissect the entire cloud and precipitation structure of a mid-latitude low, thus adding to our knowledge of the cyclone model. Such knowledge is fundamental for becoming a novice weather forecaster. So, to quote Henry Wadsworth Longfellow, "Listen, my children, and you shall hear, of the midnight ride of Paul Revere." The clouds are coming!

Weather Ahead of an Approaching Warm Front: Lowering and Thickening Clouds

Longfellow's lyric about Paul Revere's midnight ride, "One if by land, two if by sea," was a reference to a prearranged signal that, if the British "went out by water," two lanterns would be lit in a church steeple. If the British "went out by land," then one lantern would be lit.

Ahead of an approaching mid-latitude low, streaking cirrus clouds are a lantern of warning to weather

FIGURE 13.8 Two examples of wispy cirrus clouds. Made of ice crystals, these clouds are often the first signs of an approaching warm front.

forecasters who know, based on experience, exactly how the ensuing cloud invasion will evolve. First, cirrus will give way to a veil of cirrostratus that will cover the sky and dim the sun or moon. On occasion, ice crystals that compose cirrostratus, while acting like tiny prisms, bend incoming visible light from the sun or moon and separate the light into the primary colors. In such cases, a faint halo of light appears around the sun or moon, with red on the inside and blue or violet on the outside (see Figure 13.9). The appearance of such a halo through cirrostratus, coupled with decreasing pressure, is often followed by precipitation within a day, lending credence to the folklore "Ring around the moon or sun, rain before the day is done."

As the maturing low approaches, cirrostratus clouds thicken as their bottoms lower toward the middle troposphere. Eventually, a layered deck of altostratus arrives overhead, with the sun barely visible through the mid- and high-altitude overcast. In time, the sky darkens and precipitation starts to fall from low-hanging nimbostratus (Figures 8.38 and 8.39 show these stratiform clouds).

From space, the cloud shield associated with a mature mid-latitude cyclone and its attendant fronts typically has distinct patterns. One approach that has been successful in describing and explaining these patterns uses **conveyor belts**, streams of air of different temperature and moisture characteristics that help to sculpt the overall appearance of a low's cloud shield on satellite imagery. We now introduce the warm, cold, and dry conveyor belts.

Warm Conveyor Belt: The Subtropical Connection

As a mid-latitude low matures, its expanding circulation draws warm, moist air poleward from increasingly lower latitudes—in fact, some stronger mid-latitude lows can tap into air from as far away as the subtropics. This infusion of maritime tropical air into the broad cyclonic circulation of a deep mid-latitude low is actually rather orderly. Air originating in the lower troposphere over warm latitudes flows poleward in a well-developed channel called the **warm conveyor belt** (WCB). This stream of relatively warm and moist air gently rises as it flows poleward, eventually forming a layered shield of clouds mainly on the cold side of the surface warm front. In Figure 13.10a, an idealized WCB is represented by the rather broad, orange arrow originating ahead of the cold front. Notice how the warm conveyor belt gently rises as it flows over and north of the low's warm front.

To see a real warm conveyor belt in action, we will consider a mature mid-latitude low-pressure system in early winter a few years ago. On the surface analysis shown in Figure 13.10b, it's hard to miss the deep area of low pressure centered over western Illinois. There can be little doubt about the low's maturity, given that the low is occluded and its central pressure is 985 mb (for the record, this low produced blizzard conditions over the Upper Midwest). To get a sense for how the warm conveyor belt associated with this mature low appeared on satellite imagery, we will choose a water vapor image for reasons that will become apparent as our story unfolds.

Figure 13.10c is a color-enhanced water vapor image at the time of the surface analysis in Figure 13.10b. The location of the surface low in western Illinois is marked. Look to the east of the low and note the channel of light green that runs approximately south-to-north—based on the legend, this color represents temperatures of about −50°C. Recall from Chapter 5 that mid-level and especially high cloud tops typically show up on water vapor images. Thus, over the Ohio Valley and Great Lakes, this light-green color indicates cirrus and cirrostratus clouds that formed as the warm conveyor belt overran cold air to the north of the low's surface warm front—essentially, this corridor of clouds is the

FIGURE 13.9 A halo forms in a deck of cirrostratus clouds on August 21, 2016, silhouetted by Rainbow Bridge in southern Utah. The appearance of cirrostratus clouds and a halo around the Sun or Moon often indicates the approach of a warm front and the possible eventual arrival of precipitation (credit: Ray Boren and Earth Science Picture of the Day).

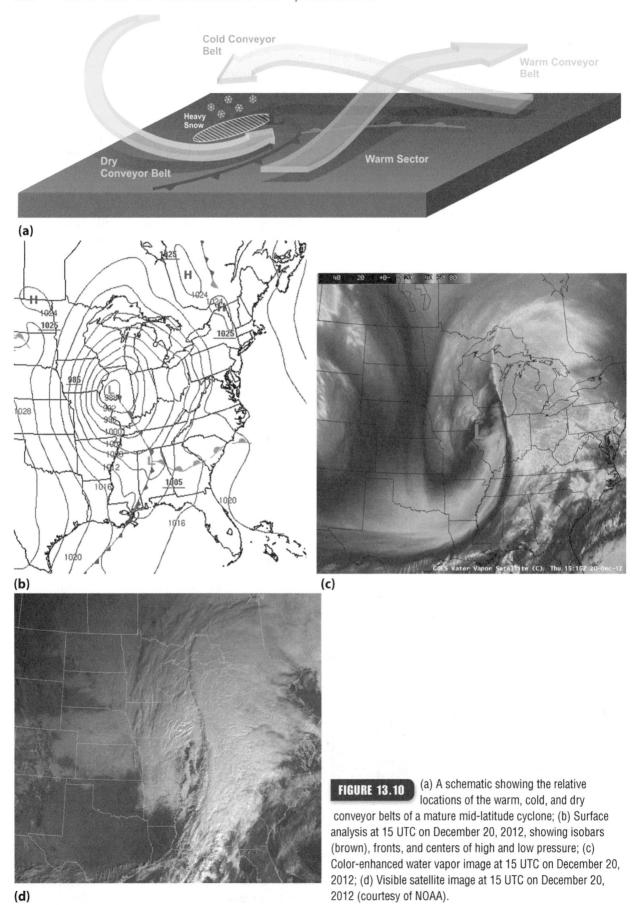

(a)

(b)

(c)

(d)

FIGURE 13.10 (a) A schematic showing the relative locations of the warm, cold, and dry conveyor belts of a mature mid-latitude cyclone; (b) Surface analysis at 15 UTC on December 20, 2012, showing isobars (brown), fronts, and centers of high and low pressure; (c) Color-enhanced water vapor image at 15 UTC on December 20, 2012; (d) Visible satellite image at 15 UTC on December 20, 2012 (courtesy of NOAA).

footprint of the warm conveyor belt. As an interesting aside, note the distinct western edge of the shield of light-green color—this sharp western edge of the WCB is called a **limiting streamline**. Air just to the right (east in this case) of this line typically possesses the highest moisture content in the warm conveyor.

In Figure 13.10c, we can only see the tops of the clouds associated with the WCB, but rest assured that thicker stratiform clouds were beneath this shield, including some nimbostratus (in fact, snow was falling over the central Great Lakes at this time). Typically, some of the precipitation falling from these nimbostratus clouds evaporates below cloud base, moistening lower-level air in the process. In response, the ceiling lowers, sometimes descending low enough to obscure mountain ridges and the tops of tall buildings.

This humidifying of the air below this part of the warm conveyor belt also plays a key role in shaping the moist personality of another of the cyclone's conveyor belts—the cold conveyor.

Cold Conveyer Belt: On the Up and Up

As a mid-latitude cyclone matures and evolves toward occlusion, isobars continue to encircle and close off around the intensifying center of low pressure. In response, a channel of cold, initially dry low-level air begins to circulate westward just to the north of the surface warm front. Revisit Figure 13.10a to see an idealized depiction of the **cold conveyor belt** (CCB) which is moistened by evaporation (and melting) of some stratiform precipitation that falls from the warm conveyor belt. Evaporation from the surface also contributes. Because the cold conveyor belt originates mostly in the lower troposphere, any low clouds associated with the CCB are not detectable by the water vapor image in Figure 13.10c. For our purposes, however, that's not a problem because we're more interested in what happens to the cold conveyor belt as it approaches the maturing low-pressure system.

At this point, you may have the impression that the cold conveyor belt takes the low road while the warm conveyor takes the high road. That's only part of the story. Indeed, after passing underneath the warm conveyor belt, the newly moistened cold conveyor approaches the surface low and begins to feel the effect of its large-scale rising motions (revisit Figure 13.10a). As the CCB ascends and emerges from beneath the warm conveyor belt, it often can be detected on water vapor imagery—in fact, Figure 13.10c shows the cold

conveyor's signature cyclonic swirl to the west of the surface low. In this enhanced water vapor image, the CCB shows up as bluish shades (water vapor) intermingled with white features (higher cloud tops) to the north and west of the low. From a weather forecasting point of view, the cyclonic swirl of the cold conveyor belt coincides with a "sweet spot" for precipitation. During winter, the heaviest snow typically falls to the north and west of the surface low where relatively cold, moistened air carried westward by the cold conveyor belt rises vigorously (see Figure 13.10a).

We have yet to address what is arguably the most noticeable feature on Figure 13.10c—the corridor of yellow and orange that swirls in a cyclonic fashion from the Southern Plains northward toward Lake Michigan. Although picturesque in a color-enhanced water vapor image, this conveyor belt actually signals the eventual demise of the cyclone.

Dry Conveyor Belt: Fools Rush in Where Angels Fear to Tread

Judging from the temperature scale in Figure 13.10c, we can deduce that the cyclonic swirl of yellow and orange on the water vapor image is relatively warm, a sign that the upper troposphere is relatively dry in that region (that is, devoid of clouds and much in the way of water vapor)—in other words, the radiometer is detecting vapor and clouds that are lower in the atmosphere, and thus warmer. For the record, a swirl of relatively warm and dry air that appears on water vapor imagery between the cold and warm conveyor belts is referred to as a **dry conveyor belt** (DCB). As the idealized model of conveyor belts in Figure 13.10a suggests, the DCB has its roots at very high altitudes—in fact, in the upper troposphere or lower stratosphere.

Given that sinking air warms by compression, this yellowish swirl is also a sign of downward motion. How can there be sinking air so close to the center of the surface low? It turns out that this sinking air is primarily in the middle and upper troposphere. Recall that an upper-level ridge typically follows on the heels of the upper-level trough that supports a surface low. The upper-level convergence east of the ridge promotes downward motion in the upper half of the troposphere. Sinking air warms, lowering the relatively humidity of the air. So the bright swirl on the enhanced water vapor image in Figure 13.10c also indicates a dry upper troposphere, which is consistent with what you learned in Chapter 5. As this air sinks it also moves southward (so the descent is essentially "slantwise"), and by the time the dry conveyor belt reaches the middle troposphere, some of its dry air is drawn into the cyclonic circulation of the 500-mb low (or trough) associated with the mid-latitude cyclone.

As the dry conveyor belt descends further (perhaps even reaching the 600-mb level or lower), it scours out clouds, creating a **dry slot**. Figure 13.10d is a visible satellite image at the same time as the enhanced water vapor image in Figure 13.10c. Note the partial clearing and thinning out of clouds from central Illinois southward—this dry slot corresponds closely to the same general area where we observed the dry conveyor belt on the enhanced water vapor image. The dry slot is sometimes referred to as a "fool's clearing," an old aviation term that refers to the clearing sky following the passage of a cold front. But as Figure 13.10d clearly shows, clouds associated with the trailing cold conveyor belt still lie to the west, so any clearing due to the dry slot could be short-lived.

Now step back and consider the visible satellite presentation of all the clouds associated with the low-pressure system (those from both the warm and cold conveyor belts). With a little imagination, we hope you see the resemblance to a comma (yes, the punctuation mark). Helping to give the cloud mass this likeness is the relatively cloud-free area (dry conveyor belt) between the rounded "comma head" (cold conveyor belt) and the "downward stroke" of the comma cloud (warm conveyor belt). This classic comma shape is an important visual cue to alert weather forecasters that the corresponding mid-latitude low-pressure system is in the mature stage of its life cycle.

Cold Front Weather: Curdling Clouds above a Wedge of Cheese

There's still one area of cloudiness that we haven't discussed yet—the clouds associated with a mid-latitude cyclone's cold front. As an example, see Figure 13.11a, which is a radar reflectivity image from a late spring morning. A line of showers and thunderstorms stretched from the upper Midwest to the Southern Plains at this time. This line of deep convection had recently produced tornadoes in Iowa, Missouri, Nebraska and Kansas. Figure 13.11b, the surface analysis from 12 UTC on this day, reveals that this convection formed along and ahead of a cold front. Later that day, additional severe thunderstorms developed from Missouri to Michigan as this cold front swept through, promoting vigorous currents of rising air.

Whereas the precipitation associated with a mature low's warm and cold conveyor belts is mostly stratiform—steady and covering a large area—precipitation associated with a cold front often is cumuliform, with convective showers and thunderstorms (or sometimes snow showers or squalls in winter) that tend to be hit-or-miss and short-lived. Recall that a cold front is located in a trough of low pressure, so as it approaches, local pressure decreases and there is enhanced low-level convergence, lifting increasingly warm and humid surface air and tipping the scales toward showers and thunderstorms (especially during the warm season).

The reason we refer to the air gathering ahead of a cold front as "increasingly warm and humid" is that surface winds at a given location in the warm sector tend to blow from a southerly direction once the front gets within several hundred kilometers (see Figure 13.12a). Thus, although it may surprise you, the warmest weather often occurs just before a cold front passes, where winds in the warm sector have blown from the south for the longest time. In turn, these persistent southerly winds cause isotherms to bulge northward, as shown in Figure 13.12b. This elongated area of higher temperatures that often develops just ahead of a cold front is called a **thermal ridge**. Typically, there's also a similar northward bulge in isodrosotherms ahead of the cold front. With relative maxima in both temperature and dew point, local air parcels are more likely to reach the level of free convection. Thus, it should come as no surprise that showers and thunderstorms often form along and just ahead of a cold front.

Behind the cold front, winds typically blow from the northwest, advecting drier and colder air southeastward. These winds, in tandem with developing currents of sinking air associated with high pressure building eastward in the wake of the low, help to evaporate clouds just behind the cold front. Thus, after a cold front passes, the sky often clears and air pressure rises. When the weather associated with a cold front behaves in this fashion, the cold front is called a **katafront**, which means that air sinks on the cold side of the surface front.

On occasion, however, a swath of stratiform clouds and precipitation will form in the wake of a cold front. In such a case, the cold front is termed an **anafront**, meaning that air rises on the cold side of the surface front. Figure 13.13 is a schematic that shows a typical scenario when a cold front acts as an anafront. Often, in such a case, the cold front is

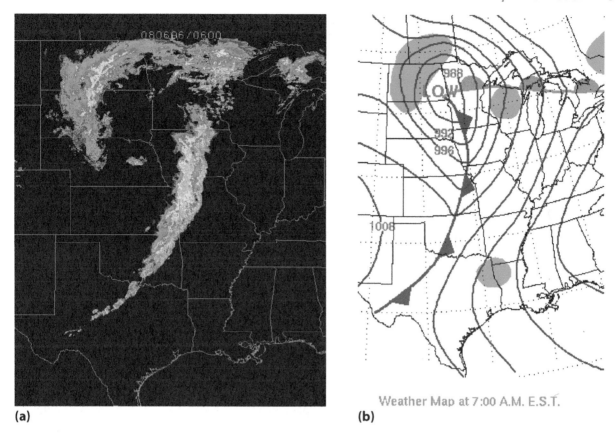

(a)   (b)

FIGURE 13.11   (a) Radar reflectivity image at 06 UTC on June 6, 2008, shows convection in the form of a line of showers and thunderstorms from Minnesota to Texas; (b) Surface analysis at 12 UTC on June 6, 2008, shows that this convection formed along and ahead of a cold front (courtesy of NOAA).

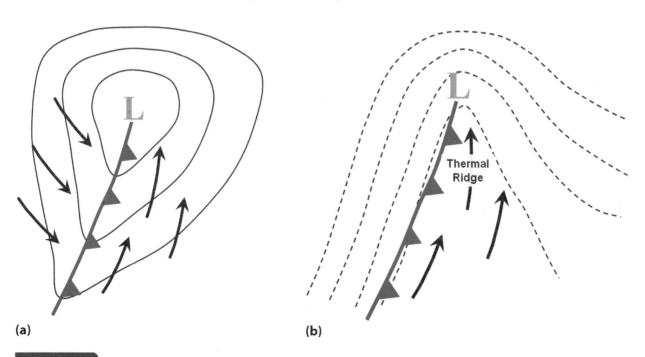

(a)   (b)

FIGURE 13.12   (a) A cold front is located in a trough of low pressure, so pressure decreases as the front approaches and increases after the front passes; (b) Just ahead of a cold front, southerly winds persistently draw warm air northward, producing a thermal ridge that appears as a poleward bulging of the isotherms (dashed lines).

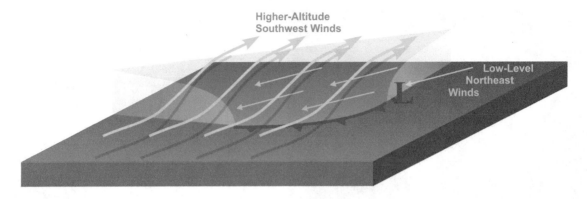

FIGURE 13.13 A cold front sometimes acts as an anafront. In a classic case, cold northeasterly winds behind the front "undercut" milder southwesterly winds higher up, forming a swath of stratiform clouds and precipitation behind the front.

sluggish, making very little progress to the south and east. Chilly surface winds behind the lethargic cold front typically blow from the northeast, while winds higher up blow from the southwest. In such a case, cold surface air wedging southwestward "undercuts" and lifts relatively warm, moist air arriving from the southwest higher above the ground. To visualize undercutting by cold air, think of sliding a snow shovel along a driveway covered with a newly fallen inch of snow. The shovel undercuts and lifts the snow, forcing it to rise slantwise onto the shovel.

Look again at the satellite image in Figure 13.10d. We've now accounted for the entire cloud shield associated with this mature cyclone by categorizing each area as stratiform or convective and by linking it to a lifting process and/or conveyor belt.

FINAL THOUGHTS

To close our discussion of the cyclone model, recall that a low-pressure system intensifies as it draws energy from horizontal temperature contrasts. Intensification takes place while cold and warm air, though close to each other, are separated by a sharpening front. In time, however, a low always succumbs as contrasting air masses encircle the cyclone and gradually mix during the latter stages of occlusion.

Figure 13.14 is an enhanced water vapor image of an occluded cyclone in the Gulf of Alaska in April 2012 that vividly illustrates this idea. Here, a spiraling ribbon of dry air in the middle and upper troposphere (the blue shades) has wrapped around and into the cyclone's center, giving it the look of a cinnamon bun. Like a boa constrictor squeezing the life out of its prey, these coils of different but gradually mixing air masses signal the eventual end of the low itself.

The Gulf of Alaska is a fairly common area where occluding lows meet their demise. In fact, climatologically speaking, there are favored areas in the North Pacific and North Atlantic Oceans where occluding lows go to die. To locate these favored regions, refer back to Figure 6.16, which shows the annual average of sea-level pressure over the globe. Note the "semipermanent" areas of relatively low pressure southwest of Alaska and south of Greenland—these are called, respectively, the **Aleutian Low** and the **Icelandic low**. They are, essentially, "graveyards of storms." Lows that decay here, at this high latitude, have certainly fulfilled their destiny—to act as a relay switch in the atmosphere's general circulation, transporting warmer air poleward and colder air equatorward and, thus, helping to mitigate north-south temperature contrasts.

FIGURE 13.14 Enhanced water vapor image of a highly occluded cyclone over the Gulf of Alaska. Here, blue indicates relatively dry air in the middle and upper troposphere (courtesy of CIMSS/NOAA).

Focus on Optics

Sundogs: Heeling at Their Master's Side

Dogs are people's best friends, faithfully standing at their masters' sides. The sun has optical "pets" too—bright and colorful spots that obediently heel on one or both of their master's sides (see Figure 13.15). These bright spots, popularly called **sundog**, form when certain ice crystals often found in cirrostratus clouds act like glass prisms to refract, or bend, sunlight. Like attentive puppies that have graduated from obedience school, sundogs (also called mock suns or parhelia, which means "with the sun") always appear at the same angle above the horizon as the sun.

We can further pinpoint the positions where sundogs sit beside their master. Remember the compass that helped you to draw circles in geometry class? Imagine a very large compass placed at the center of the sun. Sundogs, when they appear, will lie at an angular distance of about 22° on either side of the sun. To get a sense for this angular distance, stretch out your arm, open your hand, and cover the sun with your thumb—your little finger will fall at roughly an angular distance of 22°.

For students of geometry, the angle of 22° is not just pulled out of thin air. Hexagonal plates can act as 60° prisms (remember: 360° divided by 6 sides = 60° per side for hexagonal solids). For light passing through a 60° prism, the minimum deviation angle is 22° (see Figure 13.16). Thus, there is a clustering of bent light rays that emerge from 60° prisms at angles very close to 22°. When hexagonal columns (which can also act as 60° prisms) and plates are small and randomly oriented, any of their locations that lie 22° from the sun will be able to bend and direct sunlight to an observer's eye (see Figure 13.17). In this case, the observer will see a circle of light around the sun or moon, a ring called the 22° halo (look back to Figure 13.9).

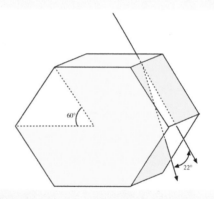

FIGURE 13.16 Like their hexagonal column cousins, ice crystals in the shape of hexagonal plates can act like 60° prisms. As a result, light shining through these prisms is deviated from its original direction (that is, refracted) by an angle of 22° or greater.

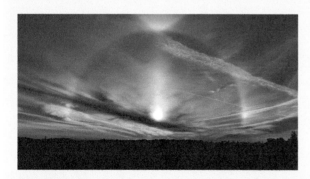

FIGURE 13.15 Sundogs appear at the same altitude to the left and right of the Sun in this photograph taken February 23, 2016, in California. You can also see a 22°-halo, a sun pillar (see Chapter 3), and an upper tangent arc (at the top) (courtesy of Mark Ritter and Earth Science Picture of the Day).

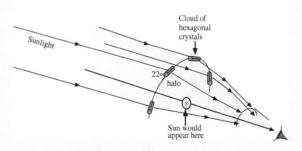

FIGURE 13.17 A 22° halo is a ring of light that surrounds the sun or moon at an angular distance of 22°. Such a halo can appear when the sky is filled with small, randomly oriented hexagonal ice crystals that act like 60° prisms to refract incoming sunlight or moonlight.

(Continued)

(Continued)

However, when ice crystals in a cold atmosphere are large hexagonal plates, a 22° halo will not be seen. Instead, only sundogs will be observed at an angular distance of approximately 22° on either side of the sun. But if both small, randomly oriented crystals and large, horizontally oriented plates exist in the atmosphere at the same time, we can see sundogs and the 22° halo together, as in Figure 13.15.

Often, sundogs will dazzle you with a vivid sequence of colors, with red on the inside (relative to the observer) and violet or blue on the outside. This color arrangement occurs because the longest wavelengths of visible light are refracted through a smaller angle than the short wavelengths of visible light. So red will lie closer to the sun (the source of the light).

Mid-Latitude III: Spawning Severe Weather

14

LEARNING OBJECTIVES

After reading this chapter, students will:

- Understand the three primary criteria for severe weather
- Be able to apply the checklist for predicting the development of supercell thunderstorms
- Recognize the conditions that are favorable for persistent supercells to develop along dry lines
- Understand that strong, widespread dynamic lift favors thunderstorms evolving into mesoscale convective systems

- Understand the role that the nocturnal low-level jet plays in the development of mesoscale convective systems

- Be able to distinguish a mesoscale convective complex from an "ordinary" mesoscale convective system

- Recognize the scenario for the development of squall lines

- Understand the life cycle of bow echoes and the role played by the rear-inflow jet

- Be able to apply the method of Larko's Triangle to determine the region at risk from severe thunderstorms associated with an approaching mid-latitude low-pressure system

- Understand the differences between tornado and severe thunderstorm watches and warnings, and the agencies responsible for issuing each product

On May 3, 1999, one of the most prolific tornado outbreaks in Oklahoma history left communities in the central part of the Sooner state looking like a war zone. Figure 14.1 shows an example of the destruction inflicted by one of the powerful twisters. For the first time in recorded history, a tornado ranked F5 on the Fujita damage assessment scale (F-Scale) struck within the city limits of Oklahoma City. At the time, an F5 rating corresponded to "incredible damage" associated with winds in the 261–318 mph range. As it turned out, the devastation caused by the Oklahoma City tornado resurrected the debate about

FIGURE 14.1 Destruction from a powerful tornado that struck central Oklahoma on May 3, 1999 (courtesy of National Weather Service, Norman, OK).

the wind estimates on the F-Scale being too high. In February 2007, the Enhanced Fujita Scale (EF-Scale) replaced the Fujita Scale (F-Scale) as the accepted standard for estimating a tornado's maximum wind gusts based on the degree of damage (more on this story in Chapter 15).

The tornado outbreak on May 3, 1999, totaled 74 twisters in Oklahoma and Kansas. The human toll was staggering: 46 dead and 800 people injured. The tornadoes damaged or destroyed more than 8000 homes, with the total loss in property approaching $1.5 billion.

During the outbreak, a few people seeking shelter beneath highway overpasses were killed when powerful tornadoes scored direct hits (see Figure 14.2), likely marking the first such documented cases of overpass fatalities and offering contrary evidence to the popular (and mistaken) notion that overpasses provide blanket safety against tornadoes. Such a notion probably arises from the false perception that a tornado is more like the hose attachment on an electric sweeper, whose nozzle merely vacuums objects up. Indeed, though the public typically thinks that the greatest threat from tornadoes involves being "sucked up" into them (that is, a vertical threat), the real danger comes from the violent winds and the flying debris (that is, a horizontal threat).

Formally, a **tornado** is a violently rotating column of air in contact with both the ground and the base of the parent cloud (see Figure 14.3). Tornadic winds can be very fierce, with a violent inflow of air that can hurl large and small objects at great speeds. Such a barrage of airborne missiles accounts for many injuries and deaths from tornadoes (broken glass propelled at tornadic speeds is lethal). Moreover, the

(a)

(b)

FIGURE 14.2 (a) An aerial view of the track of the powerful tornado that struck the suburbs of Oklahoma City, OK, on May 3, 1999. The brownish swath that begins on the bottom center of the photograph marks the path of the tornado. Interstate 44 runs from the lower right to the center of the photograph, where 16th Street (in the town of Newcastle, OK) passes over the interstate. A person seeking protection under this overpass was killed (courtesy National Weather Service); (b) The Newcastle overpass from ground level. Note the mud spattered against the steel girders and the concrete support, indicating that the tornado pummeled people under the overpass with debris slung through the air at lethal speeds (courtesy of Dan Miller, NOAA).

FIGURE 14.3 A tornado near the town of Stecker (southwest of Oklahoma City) on May 3, 1999. This twister was produced by the same parent thunderstorm that eventually spawned the F5 tornado that struck Oklahoma City. A single, long-lived severe thunderstorm can produce a family of tornadoes—one twister touches down and dissipates and then another develops and continues on a similar track (courtesy of the National Severe Storms Laboratory).

people seeking refuge on embankments under overpasses in the May 1999 tornado outbreak may have been subjected to even greater winds because they were elevated (Figure 4.2b) and exposed to the tornado's strongest winds (tornadic winds are strongest just above the surface).

Tornadoes are not the only weapon in the atmosphere's severe weather arsenal. Severe weather can also come in the form of large hail or damaging, straight-line winds associated with intense downdrafts in powerful thunderstorms. Our aim in this chapter is to show how mid-latitude weather patterns can create environments that pave the way for severe thunderstorms.

BREEDING GROUNDS FOR SUPERCELLS: A LOT OF SHEAR, A LITTLE LIFT

According to the National Weather Service, a thunderstorm officially becomes "severe" if it produces one or more of the following:

- A tornado
- Hail at least one inch (2.5 cm) in diameter, about the size of a quarter. Prior to January 2010, the threshold for severe hail was three-quarters of an inch in diameter, or about the size of a penny.
- Straight-line winds exceeding 50 knots (57.5 mph).

Of all the types of thunderstorms, supercells account for a disproportionate share of severe weather reports. That's really saying something because globally, supercells are probably the least common thunderstorm type. Part of the explanation for the prolific severe-weather producing capability of supercells is longevity. Indeed, supercells typically last one to four hours, ample opportunity to produce severe weather (some supercells have persisted for as long as eight hours). Such long durations help give credence to the "super" in "supercell."

However, longevity does not tell the whole story. Not only do supercells account for a disproportionate fraction of severe weather reports, they are also responsible for some of the most dangerous severe weather. Indeed, almost all strong and violent tornadoes, and hail two inches or more in diameter, are attributed to supercells. And although electrical activity is not a severe weather criterion, supercells are also prodigious producers of lightning, with flash rates as high as 200 flashes per minute!

That supercells would be the least common type of thunderstorm suggests that they require a special recipe with select atmospheric ingredients. So let's first look at the classic recipe for supercells, and then we'll examine each ingredient more closely.

Cooking up a Supercell: A Recipe with a Seemingly Secret Ingredient

In Chapter 9, we discussed a few of the atmospheric ingredients required to initiate supercells. For example, we mentioned that strong vertical wind shear helps to keep the rotating updrafts of supercells separated from thunderstorm downdrafts, which supports longevity. In this section, we will reveal the entire recipe first, including a seemingly "secret ingredient," and then examine each ingredient in more detail.

1. A thick layer of relatively warm, moist air at low levels, topped by a thin stable layer several thousand feet above the ground. The stable layer of air acts as a "lid" to initially inhibit or delay deep convection. Often, over the Great Plains, this stable layer is a temperature inversion related to the presence of relatively warm, dry air that "caps" the moist layer below.

2. Modest large-scale lifting (the secret ingredient). By "modest," we mean strong enough in spots to help overcome the stable layer's opposition to deep convection. It also helps some air parcels to reach the level of free convection (LFC) as the lid weakens. More importantly, modest lifting favors discrete convection.

3. A "steep" (large) lapse rate above the LFC. Such a lapse rate is typically between the moist and dry adiabatic lapse rates (and thus would qualify the layer as conditionally unstable). Such steep lapse rates above the LFC allow positively buoyant parcels to rapidly accelerate upward through a deep layer of the troposphere. In other words, there is large amount of "energy" available to buoyant air parcels above the LFC. Forecasters refer to this energy as Convective Available Potential Energy, or CAPE for short. Thus, Ingredient #3 boils down to large CAPE.

4. Strong vertical wind shear.

Ingredient #1: Getting All Your Ducks in a Vertical Row

To see how the ingredients in this recipe come together, we will consider the historic outbreak of tornadic supercells over portions of the Southeast United States during the late afternoon and evening of April 27, 2011. For the record, the National Weather Service confirmed 122 tornadoes over this region, with 62 twisters in Alabama alone. On this tragic date, 319 people lost their lives, marking the deadliest tornado day since March 18, 1925. The most destructive twister (rated an EF-4) moved across central Alabama, devastating parts of Tuscaloosa and Birmingham. This violent tornado packed winds of up to 190 mph (165 kt) and was 1.5 mi (2.4 km) wide at times. The tornado traveled approximately 80 mi (129 km), but the parent supercell traveled roughly 380 mi (612 km) from Mississippi to North Carolina (see Figure 14.4). As mentioned earlier, long-lived supercells such as this one have ample opportunity to produce severe weather. Indeed, this supercell spawned several additional tornadoes along its lengthy track.

At 18 UTC on April 27, 2011, the surface analysis showed a cold front approaching the area from the west (Figure 14.5a), with a dry line stretching southward ahead of the cold front. Surface temperatures in the warm sector to the east of the dry line climbed past 80°F (27°C) in spots. To add to the developing volatile mix, surface dew points rose above 70°F (21°C) across the region as southerly winds (indicated by the streamlines in Figure 14.5b) transported Gulf moisture inland.

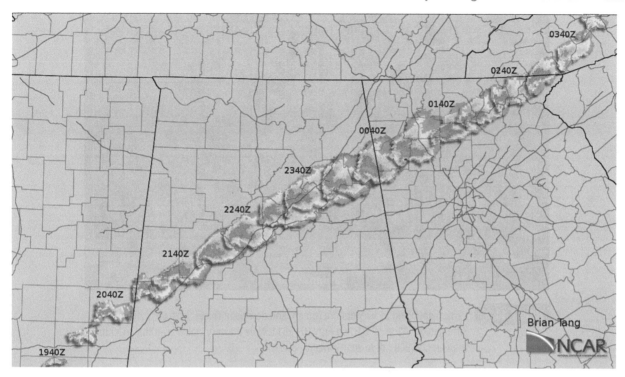

FIGURE 14.4 A montage of radar reflectivity images following a single supercell thunderstorm that traveled from Mississippi to North Carolina on April 27, 2011 (courtesy of Brian Tang, NCAR).

That takes care of surface conditions. How about the rest of the lower troposphere? Figure 14.6 shows soundings of temperature (in green) and dew point (in blue) at Birmingham at 12 UTC on April 27 (both are in °C). Dew points were 15°C (68°F) or higher from the surface to just below the 850-mb level, so moist air wasn't just confined near the surface. However, note the start of a temperature inversion just below 850 mb. Recall from Chapter 8 that an inversion is a very stable layer characterized by temperatures increasing with height. Furthermore, just above the temperature inversion was another very stable layer in which temperature remained essentially constant with height.

The bottom line here is that, despite the high low-level dew points, surface-based thunderstorms were improbable at this time, given the stable layer topping the relatively moist lower troposphere. In

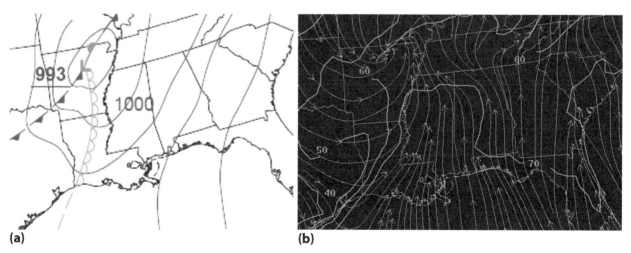

(a)   (b)

FIGURE 14.5 Surface data at 18 UTC on April 27, 2011; (a) The surface analysis. The boundary stretching southward from the low in Arkansas into Louisiana is a dry line; (b) Surface streamlines (blue) and surface dew points (green, in °F) show that dew points above 70°F (21°C) had overspread most of Mississippi and Alabama (courtesy of NOAA).

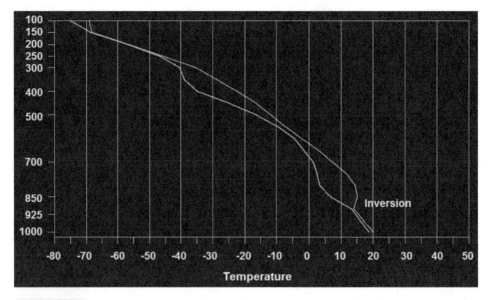

FIGURE 14.6 Vertical sounding of temperature (green) and dew point (blue), both in °C, from Birmingham, AL at 12 UTC on April 27, 2011. Note the very stable layer that extends from around 900 mb to approximately 800 mb.

these situations, the lower troposphere is said to be "capped," indicating that the stable layer aloft acts as a lid to inhibit deep convection.

Although it may seem counterintuitive, such a "lid" is pivotal to the development of severe thunderstorms later in the day. Think about it—if thunderstorms occur early in the day, the rain-cooled air they produce would tend to stabilize the lower atmosphere and possibly inhibit deep convection later on. However, with a cap in place, thunderstorm development is delayed, allowing a powder keg of heat and humidity to build in the lower troposphere (particularly on a day with ample sunshine and increasingly moist air—in this case, from the Gulf of Mexico). If the lid erodes later in the afternoon, the powder keg of heat and humidity can "explode" with very powerful thunderstorms. In essence, the lid is akin to a long fuse—once at its end, explosively deep convection can occur.

How does a lid form? To answer this question, revisit Figure 14.6. Note that, from the bottom of the temperature inversion to roughly 800 mb, dew points decreased dramatically. This signature of dry, warm air (essentially, a continental tropical air mass) gives us a clue to its origin: farther west over the high ground of west Texas and perhaps northern Mexico. The arrival of this cT air mass over a layer of moist air from the Gulf of Mexico is a classic set-up for a lid that inhibits deep convection.

With Ingredient #1 in the recipe for supercells covered, let's tackle Ingredient #2. In the process, we'll reveal how the lid erodes and allows the powder keg of heat and humidity to erupt into severe thunderstorms.

Ingredient #2: The Seemingly Secret Part of the Recipe

It's reasonable to think that in order to get a "big" thunderstorm (such as a discrete supercell), "heavy lifting" is required. In practice, things are not quite that simple. When lifting is sufficiently strong to initiate deep convection over a relatively large area, thunderstorms tend to merge into an organized group called a **mesoscale convective system** (MCS). Figure 14.7 is the radar reflectivity presentation of a typical mesoscale convective system. Before assigning the "MCS" designation to an organized group of thunderstorms, forecasters look for a solid area of precipitation with embedded thunderstorms that spans at least 100 km (62 mi) in at least one horizontal direction. The example in Figure 14.7 certainly fits this size criterion. Figure 14.8 provides another example that emphasizes a mechanism to organize the convection: here, strong low-level convergence along a cold front provided sufficiently strong lift that initiated a continuous line of thunderstorms.

In the spirit of true irony, the low-level lift required for discrete supercells is generally not as strong or

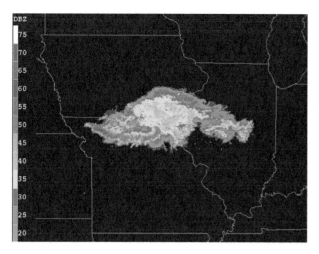

FIGURE 14.7 Radar reflectivity presentation of a mesoscale convective system (MCS) over the Midwest in June 2008 (courtesy of NOAA).

as widespread. In such situations, the lift can overcome the inhibition of the lid only in isolated spots where air parcels manage to get to their level of free convection.

To demonstrate our point, let's return to the historic outbreak of tornadic supercells during the afternoon and evening of April 27, 2011. For starters, focus your attention on Figure 14.9, the radar reflectivity at 2038 UTC from the Columbus Air Force Base in Mississippi (KGWX). You should recognize the classic hook-echo shape of the mostly discrete supercells that were in progress at this time over parts of Mississippi and Alabama.

To seal the deal, revisit Figure 14.5b, this time paying sole attention to the pattern of surface streamlines over Mississippi and Alabama. Notice the slight tendency

for these streamlines to crowd together over those states. This subtle confluence ("coming together") of surface streamlines is consistent with rather modest low-level convergence. Indeed, compare this pattern of convergence to the noticeably stronger convergence taking place in northern Arkansas, where a surface low was located at 18 UTC (recall Figure 14.5a).

Thus, modest and rather isolated lift is the second ingredient in generating discrete supercells. Because this may seem counterintuitive, we decided to refer to the second ingredient as "secret." Hopefully, it's not a secret to you any longer.

Now that we're talking about lift, we can finally explain how the lid erodes (in this case, on April 27, 2011). It's typically a gradual process—rising (and thus cooling) parcels of moist air penetrate the lid and begin to mix with the relatively warm, dry air. Some of the lift on April 27, 2011 came from solar heating at the ground, but the low-level convergence helped as well. By 18 UTC, the temperature inversion aloft and the nearly isothermal layer above it were gone (compare Figure 14.10 with Figure 14.6). Indeed, the lid had weakened considerably!

Also note in Figure 14.10 another consequence of sustained upward motion: the low-level moist layer was now much thicker. Indeed, look at the dew points, ranging from about 21°C (70°F) at the surface to slightly less than 10°C (50°F) near 700 mb. And surface temperatures were now above 26°C (80°F). So the low-level moist layer was now deeper, moister, and warmer. Talk about a powder keg of heat and humidity!

In the final analysis, the lower troposphere in northern Alabama was primed for discrete supercells

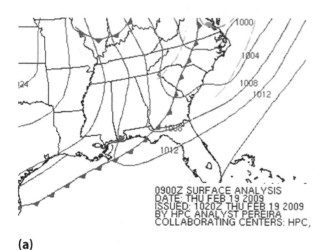

(a)

(b)

FIGURE 14.8 Weather data at 09 UTC on February 19, 2009: (a) Surface analysis shows a cold front located in a distinct trough over the Southeast; (b) Radar reflectivity shows an organized line of powerful thunderstorms (courtesy of NOAA).

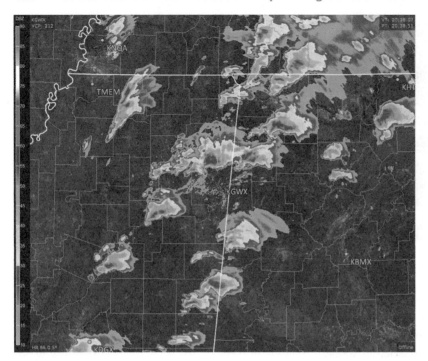

FIGURE 14.9 Radar reflectivity image from Columbus Air Force Base in Mississippi at 2038 UTC on April 27, 2011. Numerous discrete supercells are apparent (courtesy of NOAA).

in the afternoon. But what about conditions in the middle and upper troposphere? Time to consider Ingredient #3.

Ingredient #3: It All Comes Down to CAPE

Eventually, the lid can no longer suppress the pent-up powder keg of heat and humidity beneath it. At this point, solar heating and low-level convergence work to get parcels to their level of free convection, and the stage is set for explosive development of supercells.

Our use of the phrase "explosive development" refers to the idea that once air parcels reach the LFC, they accelerate upward like gangbusters because the parcels are positively buoyant. To see what we mean, check out the idealized sounding in Figure 14.11

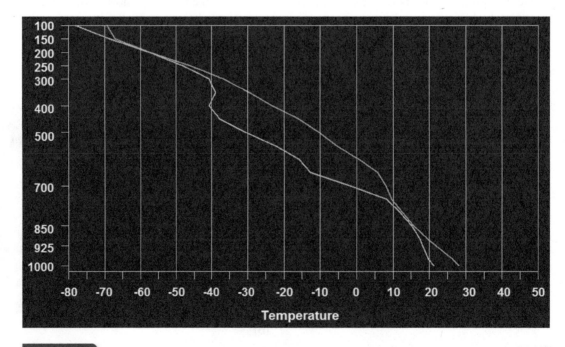

FIGURE 14.10 Vertical plot of temperature (green) and dew point (blue), both in °C, from Birmingham, AL, at 18 UTC on April 27, 2011. The inversion aloft (lid) that was present six hours earlier is now gone.

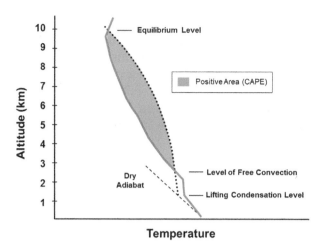

Temperature

FIGURE 14.11 A hypothetical temperature sounding is shown in red. If we assume that a parcel lifted to its LCL and then to its LFC follows a moist adiabat to the right (on the warm side) of the environmental temperature sounding, the area between the moist adiabat and the temperature sounding is called "positive area"—here shaded in green. Positive area is a graphical way to represent the total energy available to the positively buoyant parcel. Meteorologists refer to this energy as CAPE, for **C**onvective **A**vailable **P**otential **E**nergy—the greater the CAPE, the greater the upward acceleration of air parcels above the LFC.

which shows an air parcel dramatically warmer than its environment between the level of free convection and the equilibrium level (the level at which the rising parcel becomes colder than the environment). The region shaded in green between the parcel's moist adiabat and the environmental temperature sounding is known as "positive area" because the parcel is positively buoyant in this layer. Think of "positive area" as a measure of the total amount of energy available to the positively buoyant parcel. Specifically, this **Convective Available Potential Energy** (CAPE, for short) controls the magnitude of the upward acceleration of the parcel above the LFC. When CAPE is large, positively buoyant parcels above the LFC experience large accelerations—the kind associated with the explosive development of supercells after the lid erodes. Essentially, CAPE is a measure of the potential for strong updrafts, and more generally, it is an overall indication of instability in the troposphere. It's important to note, however, that thunderstorms won't necessarily develop just because CAPE is high—air parcels must first reach the LFC to tap into the CAPE.

CAPE is largest when the low levels of the troposphere are warm and moist and the troposphere above the LFC is unstable (typically, conditionally unstable). Moist air at low levels means that ascending parcels saturate at lower altitudes. Once saturated, they cool at

the reduced moist adiabatic lapse rate, thus increasing the likelihood that parcels will stay warmer than the environment through a deeper layer above the LFC. A relatively large environmental lapse rate above the LFC also increases the chances that rising parcels remain warmer than the environment (and thus positively buoyant) from the LFC to the tropopause.

So far, we've looked at the first three ingredients in the recipe for discrete supercells. Assuming that CAPE is relatively large and air parcels accelerate upward above the LFC, we have the makings of a thunderstorm updraft. All that has to happen at this point in order for a discrete supercell to develop is for the rotating updraft to stay separate from any thunderstorm downdraft—that leads us to the fourth ingredient, strong vertical wind shear.

Ingredient #4: Vertical Wind Shear

On April 27, 2011, vertical wind shear between the ground and an altitude of about 6 km was quite large over the Southeast States. Above Birmingham, AL, for example, low-level winds blew from the south, while wind direction gradually changed to southwesterly by an altitude of 6 km. This change in wind direction with height played a role in the large vertical wind shear from the surface to 6 km, but the change in wind speed was much more noteworthy. Figure 14.12 shows the 18 UTC plot of wind speed (in knots) as a function of pressure at Birmingham. Roughly speaking, wind speeds at the surface were

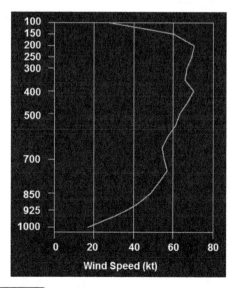

FIGURE 14.12 Vertical plot of wind speed (in kt) from Birmingham, AL at 18 UTC on April 27, 2011. The vertical axis is pressure (in mb). Vertical wind speed shear was substantial at this time.

less than 20 kt, while wind speeds at 6 km (near 500 mb) were about 70 kt. Such large vertical wind shear kept updrafts and downdrafts separated (see Figure 14.13), setting the stage for long-lived supercells and the historic outbreak of tornadoes on April 27, 2011.

Arguably, the most surprising aspect of this discussion about the recipe for supercells was the seemingly secret ingredient (Ingredient #2). As a general rule, modest low-level lift tends to go hand in hand with dry lines, boundaries between maritime tropical and continental tropical air masses. So let's next explore the dry line as an initiator of supercell thunderstorms.

Dry Lines: Initiating Discrete Supercells

As a general rule, dry lines form over the Southern Plains during spring and summer, when maritime tropical air (very moist) from the Gulf of Mexico spreads northward and northwestward toward the western Great Plains. Meanwhile, continental tropical air (very dry) flows eastward or northeastward from the deserts of western Texas, New Mexico, and northern Mexico. The proximity of these warm, dry areas to the warm waters of the Gulf sets the stage for an inevitable meeting of these two air masses. On such occasions, the dry line is the boundary that separates them.

Low-level convergence along the dry line is generally modest, with 500-mb short-wave troughs sometimes also adding upper-level divergence. In such situations, the dry line can initiate supercells that can persist as discrete storms. As an example, on April 29, 2009, the weather pattern was a classic set-up for the dry line to initiate discrete supercell thunderstorms in western Texas. Figure 14.14a shows the 21 UTC surface analysis of this area, and Figure 14.14b shows the 2130 UTC radar reflectivity, supporting our claim that dry lines often initiate discrete supercells.

Supercells tend to persist as discrete storms (and not merge or organize) when winds through a deep layer of the troposphere blow at right angles (or nearly so) to the dry line. In these situations, upper-level winds from the west (the dry line typically spans north to south) carry hydrometeors directly eastward. This means that rain from one supercell will likely stay separate from neighboring supercells, at least for awhile. In this configuration, there's little chance for the pools of rain-cooled air associated with neighboring supercells to merge and organize. Thus, supercells tend to remain discrete when upper-level winds blow at a right angle (or a fairly large angle) to the dry line.

Supercells that form in association with a dry line often remain distinct and separate, and rarely merge. However, in other situations, thunderstorms combine and organize, which brings us to the topic of mesoscale convective systems.

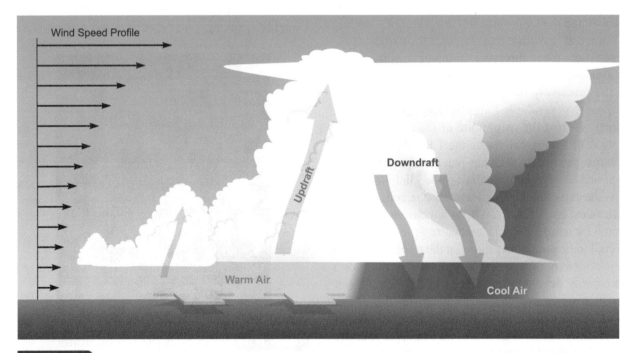

FIGURE 14.13 In an environment with large vertical wind shear, fast upper-level winds carry precipitation particles downstream away from the supercell's updraft, a separation that paves the way for longevity and an increased threat of severe weather.

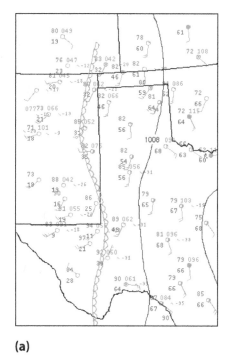

(a)

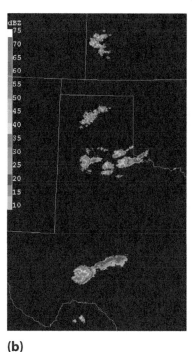

(b)

FIGURE 14.14 (a) The surface analysis at 21 UTC on April 29, 2009, indicates a dry line over eastern New Mexico and west Texas. A dry line appears on a surface analysis as a brown boundary with unfilled semicircles; (b) Radar reflectivity at 2130 UTC on April 29, 2009, shows that a few discrete supercells developed along and just ahead of the dry line (courtesy of NOAA).

MESOSCALE CONVECTIVE SYSTEMS: THUNDERSTORMS GET ORGANIZED

Forecasters at the Storm Prediction Center (SPC) in Norman, OK, routinely face challenges that the public might take for granted. First, forecasters must predict whether thunderstorms will develop at all. If the answer is "yes," they must predict when and where they will initiate, a task made more difficult given that local conditions often vary subtly across a region. Once thunderstorms develop, SPC forecasters must then predict the mode (type) of thunderstorms and the evolution over time. One reason that it's important to predict the correct mode is that tornadoes and large hail are much more likely when persistent, discrete supercells develop. On the other hand, squall lines are more likely to produce damaging wind gusts.

For the record, a **squall line** is a narrow, linear legion of thunderstorms that are often severe. The line of thunderstorms stretching from the eastern Carolinas into the northern Gulf of Mexico in Figure 14.8b qualifies as a squall line. As it turns out, a squall line is a type of mesoscale convective system. For your convenience, we'll repeat the definition of an MCS from earlier in this chapter: A mesoscale convective system is a relatively organized group of thunderstorms that produces a generally solid area of precipitation spanning at least 100 km

(62 mi) in one or more horizontal directions, so you can see how a squall line qualifies as an MCS.

An MCS can have humble beginnings, sometimes developing from relatively isolated thunderstorms that merge (see Figure 14.15). In such cases, pools of rain-cooled air associated with the gust fronts of individual storms eventually merge into a single, larger-scale "cold pool." This single pool of rain-cooled air can then initiate new thunderstorms that help to sustain the MCS or allow it to expand horizontally.

At other times, an MCS can develop almost immediately after thunderstorms erupt. Such rapid development often occurs in concert with strong overrunning associated with a nocturnal low-level jet (more coming soon about this feature). Figure 14.16 is a schematic weather map that depicts such a jet interacting with a synoptic-scale stationary front on an early summer night. Moisture rapidly transported north of the stationary front by this speedy nighttime wind maximum set the stage for an MCS to rapidly develop. The radar reflectivity associated with this MCS is superimposed to illustrate where the MCS developed in relation to the front.

We will now explore the nuts and bolts of nocturnal low-level jets so that you can get a better sense for the "dark side" (nocturnal nature) of many mesoscale convective systems.

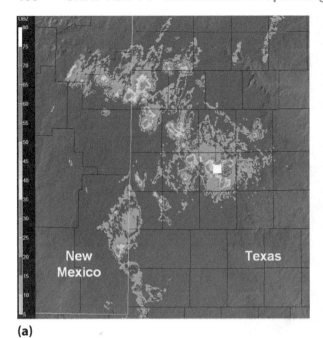

(a)

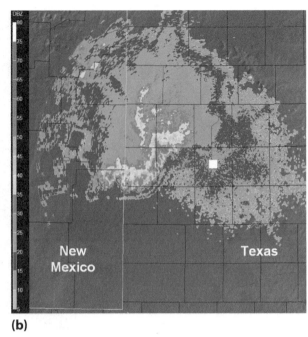

(b)

FIGURE 14.15   A sequence of radar reflectivity images from the Lubbock, TX radar site (marked with a white square), on May 27, 2007: (a) Relatively disorganized thunderstorms at 00 UTC; (b) By 0330 UTC, the isolated thunderstorms have merged to form an MCS (courtesy of NOAA).

Nocturnal Low-Level Jet: Acceleration After Dark

Figure 14.17a shows wind barbs retrieved from a wind profiler at Lathrop, MO (in the northwestern part of that state) from 00 UTC to 15 UTC on June 25, 2008. Time on this image advances from right to left, while the vertical axis is altitude, in meters. This period includes the nighttime hours when the MCS mentioned at the end of the last section developed (revisit Figure 14.16), so we should expect to see the footprint of the nocturnal low-level jet.

Indeed, around 06 UTC (1:00 AM CDT), southwesterly winds had increased to 50 kt (58 mph) at altitudes near 1000 m, marking the biggest footprint of the low-level jet over Missouri (see wind barbs inside the white oval in Figure 14.17a). For the record, the **nocturnal low-level jet** is a narrow ribbon of fast winds at altitudes that generally lie between 500 m and 1500 m (approximately 1600–4800 ft). Figure 14.17b shows winds on the 900-mb pressure surface (which is near an altitude of 1 km) at 06 UTC on June 25, 2008. Note the corridor of relatively fast winds stretching from southern Texas northward into Missouri. This corridor marks the core of the low-level jet, a speedy wind maximum which typically develops at night over the Central States during the warm season.

How does this nighttime low-level jet develop? For starters, recall what commonly occurs during the daytime when convective eddies stir the lower

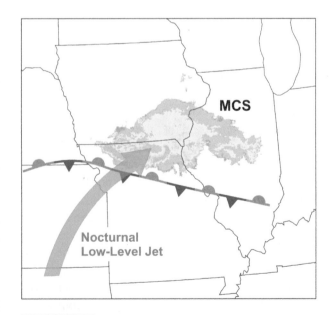

FIGURE 14.16   A nocturnal low-level jet rapidly transported abundant moisture north of a stationary front in the early morning hours of June 25, 2008, setting the stage for the rapid development of a mesoscale convective system. The radar reflectivity associated with this MCS is superimposed.

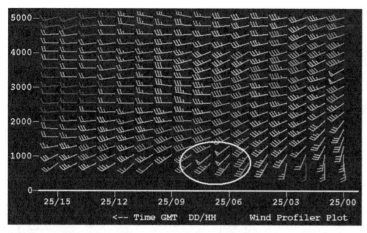

(a)

(b)

FIGURE 14.17 Wind-profiler data at Lathrop, MO, from 15 UTC on June 24, 2008 (right side of horizontal axis) to 15 UTC on June 25, 2008 (left side of horizontal axis). The vertical axis indicates height above the ground; it extends from the surface to 5000 meters. Around 06 UTC (1:00 a.m. CDT) on June 25, southwest winds accelerated to 50 knots at altitudes near 1000 meters (see wind barbs inside the white oval), marking the footprint of the nocturnal low-level jet over northwestern Missouri (courtesy of NOAA); (b) Streamlines and wind speeds (color-coded in kt) on the 900-mb surface at 06 UTC on June 25, 2008, right around the time the nocturnal low-level wind speed maximum occurred. Here, that corridor of speedy winds extends from the western Gulf of Mexico into Illinois.

layer above spans from 500 m to 1500 m in altitude. Because there are no longer any convective eddies to mix momentum from slower winds near the ground upward and faster winds aloft downward, the winds in the 500–1500 m layer accelerate (see leftmost image in Figure 8.52), setting the stage for a nocturnal low-level jet to form.

The sloping terrain of the Great Plains enhances the nocturnal low-level jet over this region. Consider that the land generally slopes upward from east to west in this part of the country (see Figure 14.18). Air at any given pressure altitude in the lower troposphere tends to cool more at night over the western High Plains because the air is closer to the surface, promoting the loss of energy by conduction to the ground. Thus, pressure surfaces tend to slope downward from east to west, creating a pressure gradient force directed to the west at low levels. The Coriolis force then turns the flow to the right, setting the stage for an enhanced southerly nocturnal low-level jet. Over this region, the jet can sometimes blow at speeds of 60 kt (nearly 70 mph) or higher.

On occasion, the nocturnal low-level jet helps to fuel the king of all mesoscale convective systems—the mesoscale convective complex. Read on.

The Mesoscale Convective Complex: King of Mesoscale Convective Systems

The advent of infrared satellite imagery allowed weather forecasters in the mid-1970s to assess the cloud-top temperatures of large nocturnal mesoscale systems over the Plains and Midwest now known as **mesoscale convective complexes**, or MCCs. Before we delve into the conceptual model of an MCC, we will first look at a specific event so that we can observe some of its basic characteristics.

Figure 14.19a is a color-enhanced infrared satellite image of a large MCS that formed over Texas in the early morning hours (around 09 UTC, or 4 A.M. local time) of June 12, 2016. Cloud-top temperatures as low as –83°C (–117°F) were observed on infrared satellite imagery. Such cold cloud tops are consistent with very tall thunderstorms that produced heavy rain, hail, and very strong wind gusts. The radar reflectivity of this MCS just 15 minutes later, shown in Figure 14.19b, certainly indicates, at the very least,

troposphere, creating a well-mixed boundary layer. If you look back to the rightmost image in Figure 8.52, you'll see that wind speeds in a well-mixed boundary layer are fairly uniform. Once the sun sets, however, cooling near the ground stabilizes the lowest few hundred meters as convective eddies dissipate and vertical mixing ceases. In effect, the boundary layer separates into a stable layer right next to the ground and another layer right above it (a process called "decoupling"). For the sake of argument, we'll assume the stable layer reaches to 500 m, while the

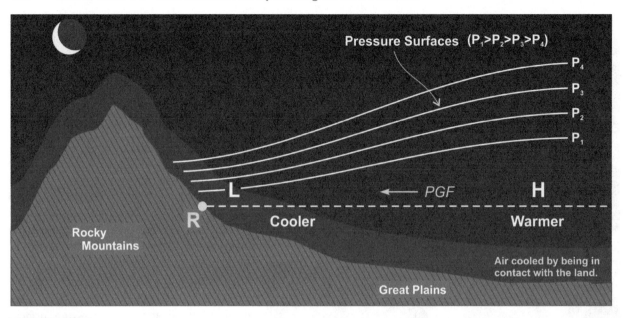

FIGURE 14.18 A cross-sectional view of the western High Plains that illustrates the contribution of sloping terrain to the enhancement of a nocturnal low-level jet. At night, air over the higher elevations cools faster than air at the same altitude farther east. As a result, pressure decreases faster with height over the western High Plains (note the tighter packing of pressure surfaces there compared to the looser packing to the east). The resulting slope of the pressure surfaces (exaggerated for effect) leads to a horizontal pressure gradient force directed from east to west, which (after accounting for the Coriolis force) enhances a low-level southerly jet.

heavy rain. In addition, over its lifetime, this large MCS produced over 5000 cloud-to-ground lightning strikes. In fact, the radar at Midland, TX was struck by lightning and temporarily knocked out of service. Meteorologists describe a large MCS such as this one as a mesoscale convective complex.

But size alone doesn't automatically mean that a mesoscale convective system qualifies as an MCC. In practice, a large MCS must satisfy several rigid criteria involving its cloud canopy in order to qualify as an MCC. Formally, a large MCS is designated an MCC if it satisfies all of the following:

- The spatial extent of the cloud shield with cloud-top temperatures of –32°C (–26°F) or lower must be at least 100,000 km², roughly two-thirds the size of the state of Iowa (such an area corresponds to a circle of diameter approximately 350 km)
- The spatial extent of the coldest cloud tops with temperatures of –52°C (–62°F) or lower must be at least 50,000 km²
- These size criteria must persist for at least six hours
- At the time of maximum spatial extent, the cloud shield must be roughly circular in shape. By "maximum extent," we mean that the shield of cold cloud tops (temperature of –32°C or lower) reaches its maximum size

To be perfectly frank, we're not big fans of using this very specific list of seemingly arbitrary criteria to distinguish between an MCC and a very large MCS. Heavy rain, flash floods, and severe weather have the same societal impact, regardless of what we call the offending mesoscale system. Still, such specificity emphasizes the unusual nature of MCCs, and their uniqueness doesn't end there—MCCs have much more of a nocturnal nature than the rest of the spectrum of mesoscale convective systems.

The radar presentation of an MCC typically differs markedly from its look on satellite imagery. Indeed, beneath an MCC's nearly circular cirrus canopy, thunderstorms (the convective region) can be linearly organized. You can certainly see that linear organization in the highest reflectivity values in Figure 14.19b. Also note the area of stratiform rain trailing the line of thunderstorms. This area of steady rain is also a standard feature of all MCCs (and some other mesoscale convective systems as well). Still, even considered in its entirety, the full precipitation shield of an MCC is not circular in shape like its corresponding satellite presentation. Such a structural contrast is yet another quirky characteristic. The bottom line here is that both the satellite presentation of an MCC and its radar presentation confirm that these large mesoscale convective systems have some underlying organization.

We mentioned earlier that another odd characteristic of MCCs is their overwhelmingly nocturnal nature, suggesting that the nocturnal low-level jet plays a pivotal role in their development and maintenance. Figure 14.19c is an analysis of 850-mb streamlines and isotachs (in knots) at 07 UTC on June 12, 2016. The footprint of the low-level jet is the relatively narrow corridor of faster 850-mb winds originating in the western Gulf of Mexico and moving northward through west Texas—in the red-shaded regions, wind speeds exceeded 30 kt (35 mph). The rapid transport of moisture by the LLJ was instrumental in initiating and maintaining the line of thunderstorms as the MCC moved slowly southeastward.

After sunrise, MCCs start to dissipate as the nocturnal low-level jet weakens. Logically, you might think that this signals that the MCC's story over. But not necessarily. While the MCC's convection was at full throttle during the nighttime, the large release of latent heat causes a drop in pressure in the middle troposphere. This relatively low pressure can, in turn, generate a cyclonic circulation of air aloft called a **mesoscale convective vortex** (MCV). The MCC that developed in west Texas on the night of June 12, 2016, left behind an MCV which you can readily see on Figure 14.19d, a visible satellite image later that afternoon. The presence of an MCV in the following daytime hours is further

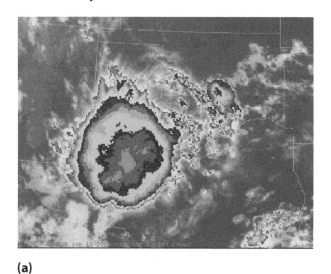

(a)

(b)

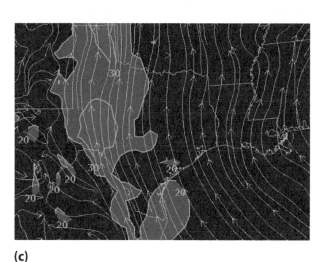

(c)

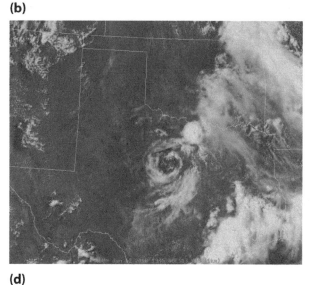

(d)

FIGURE 14.19 Various weather data from June 12, 2016, in Texas and nearby states: (a) A color-enhanced infrared satellite image of a large nighttime mesoscale convective system over west Texas at 0915 UTC; (b) Radar reflectivity of the MCS at 0930 UTC; (c) An analysis of 850-mb streamlines and isotachs (color-coded in kt) at 07 UTC shows a low-level jet aimed at west Texas. The core of the LLJ there is marked by the closed 30-kt isotach that's color-filled in red; (d) Visible satellite image at 1945 UTC shows a leftover mesoscale convective vortex (courtesy of NOAA).

evidence that an MCC is an organized mesoscale convective system. More importantly, a mesoscale convective vortex can help to initiate new afternoon thunderstorms, one of which you can observe just to the northeast of the MCV's center of circulation in Figure 14.19d. Finally, considering both Figure 14.19a and 14.19d, it's easy to understand why MCCs are sometimes called "land hurricanes."

Now that we've given you a specific example of an MCC and described its most prominent characteristics, we'll now generalize some concepts. For starters, we emphasize pattern recognition as playing a key role in understanding and forecasting MCCs. In other words, what kind of weather pattern favors their development?

To begin our closer look at MCCs, we note that they typically form in environments characterized by a ridge of high pressure in the middle to upper troposphere, a synoptic-scale surface front, and (no surprise here) the nocturnal low-level jet (see Figure 14.20). An upper-level ridge may seem counterintuitive, but MCCs tend to develop ahead of a 500-mb short-wave trough moving through the ridge. Meanwhile, the nocturnal low-level jet rapidly imports moist air—if the MCC is in the Central United States, the source of moisture is the Gulf of

Mexico. As the moist air moves north over a stationary front (or, more generally, a low-level relatively stable layer), the stage is set for an MCC (or a very large MCS) to develop. In addition, vertical wind shear must be modest for MCCs to form; their nearly symmetric cirrus canopies would not prevail in an environment of strong wind shear. Modest wind shear also implies relatively weak upper air winds, so MCCs tend to move slowly, increasing the risk for flash flooding.

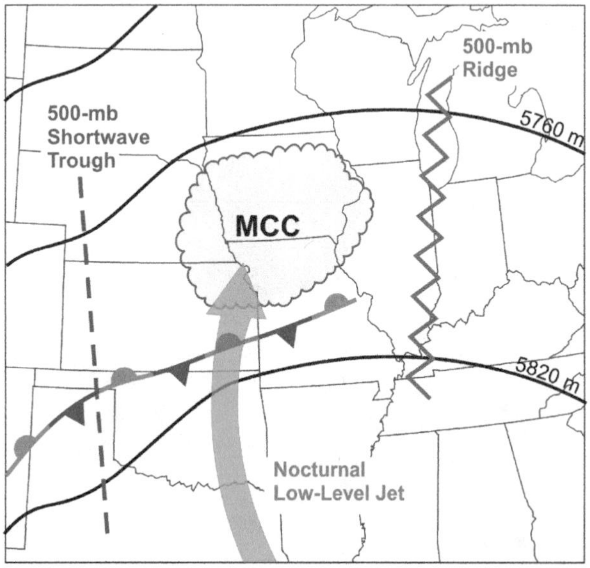

Now that we've described the typical weather pattern that favors the development of MCCs, let's look at a cross-sectional view that summarizes their structure.

Figure 14.21 is a cross section of a typical MCC. New thunderstorms form in the "convective region" near the edge of the complex facing the nocturnal low-level jet. In this region, gust fronts help to lift inflowing warm, moist air. However, observations show that the slope of a gust front associated with a newly formed MCC tends not to be very steep. That's because the gust front often moves slowly and its associated cold pool of rain-chilled air is relatively shallow. Acting alone, the shallow gust front would be too weak to lift air parcels to their LFCs. So air parcels require a little "boost." With MCCs mostly on the prowl at night, it should come as no surprise that this "boost" often comes from a nocturnal low-level jet that overruns a synoptic-scale front. You can think of the LLJ as a garden hose whose nozzle sprays a fast stream of warm, moist air that glides up and over cooler air near the ground, helping air parcels to reach their LFCs. In this way, the nocturnal low-level jet helps to initiate, feed, and maintain thunderstorms (at least for a while until the cold pool gets thicker).

In addition to the development of a nocturnal low-level jet, net radiative cooling near the tops of thunderstorms also helps MCCs to thrive at night. This cooling at high altitudes, coupled with warming in the middle troposphere associated with the release of latent heat in thunderstorms, further destabilizes the middle and upper troposphere. Indeed, MCCs (and MCSs, in general) tend to intensify in the wee hours of the morning. In fact, in a large region of the central United States stretching from Minnesota southward into northeastern Oklahoma, the average hour (local time) of thunderstorm initiation is between midnight and 3 A.M., owing to the frequency of nocturnal MCSs in this region.

FIGURE 14.20   Mesoscale convective complexes tend to form on the western flank of a mid-level ridge of high pressure as a 500-mb short-wave trough tries to move through the ridge. Meanwhile, a nocturnal low-level jet rapidly transports moisture over a stationary front, helping to initiate convection and completing the recipe for an MCC.

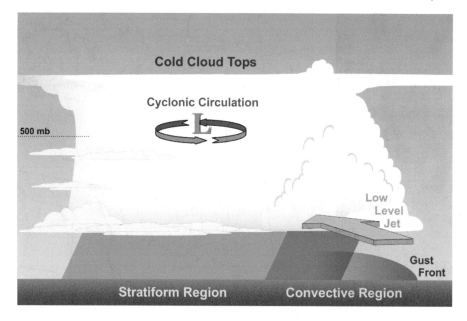

FIGURE 14.21 A schematic cross-sectional view of an MCC. Note that an MCC has a convective region that develops in response to the nocturnal low-level jet. There's also a stratiform region where steady, less intense rain typically falls. Given that MCCs are large and long-lived, a cyclonic circulation associated with lower pressure develops in the middle troposphere in response to the release of latent heat from convection. Here, an "L" marks the center of this mesoscale convective vortex which can sometimes be observed on visible satellite imagery after the MCC dissipates in the morning hours.

With the approach of dawn, both the nocturnal low-level jet and the MCC typically weaken. However, leftover outflow boundaries produced by MCC thunderstorms often serve as catalysts for additional thunderstorm development during the daytime hours. And with a weak cyclonic circulation left behind in the middle troposphere and soggy ground from the previous night's rains (water that can rapidly re-moisten the lower troposphere), the stage is set for possible re-ignition of an MCC the next night.

Although MCCs can spawn severe weather and heighten the risk of destructive flash flooding, their rains provide much of the water needed to grow the bounty of crops in America's "breadbasket." Indeed, these systems cross the corn and wheat belts in the Middle West during the growing season, when widespread rains offset evaporation from farm fields. Over the life cycle of one MCC (perhaps 12 to 16 hours), rainfall can exceed one trillion gallons—no wonder there's often flash flooding! Worldwide, MCCs also form in interior portions of Europe, Australia, China, Argentina, and Sahelian Africa, for example, where they provide the same vital link in the hydrologic cycle.

When we revisit the pattern of radar reflectivity in Figure 14.19b, we are again reminded that the nearly symmetrically round anvil observed on satellite imagery of this MCC completely disguised the linear nature of the convective region shown on radar reflectivity. We now move on to look more carefully at another type of mesoscale convective system that has linear characteristics: squall lines.

Squall Lines: Nature's Blitzkreig

To this point, we've described a squall line simply as a line of thunderstorms. Let's add some meat to the bones of this conceptual model. Figure 14.22a is a radar reflectivity image at 1150 UTC on May 2, 2008, showing a squall line that formed ahead of a cold front moving east toward the central Mississippi Valley (the corresponding surface analysis is shown in Figure 14.22b). Note the standard symbol that forecasters use to designate a squall line on the surface analysis. As it turns out, a squall line (labelled "SQLN") can be solid or there can be breaks in the line. By "breaks," we mean areas where the radar reflectivity associated with the squall line drops below roughly 50 dBZ. In this case, the squall line was solid from northeastern Missouri to west-central Arkansas, but there were breaks in the high reflectivity from west-central Arkansas across southeastern Oklahoma. But, as long as weaker reflectivity still connected the intense echoes (indicating that precipitation was falling between the heavy thunderstorms), then this portion with breaks was still an integral part of the squall line.

Note that there was an area of convective and stratiform rain behind the most intense storms, along the leading edge of the squall line. Such a structure indicates that squall lines qualify as a type of mesoscale convective system. Indeed, forecasters sometimes refer to a squall line as a quasi-linear convective system (QLCS, for short). Now that's a mouthful!

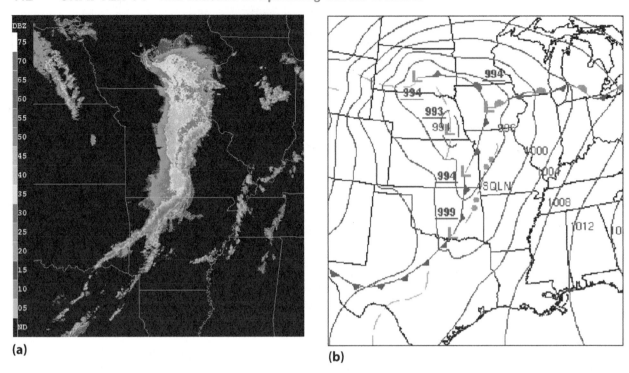

(a)

(b)

FIGURE 14.22 (a) Radar reflectivity at 1150 UTC on May 2, 2008, shows a squall line approaching the central Mississippi Valley. The squall line produced severe weather, primarily in the form of damaging straight-line winds; (b) The surface analysis at 09 UTC on May 2, 2008, indicates that the squall line had formed earlier ahead of a strong cold front. The symbol with dashes and two dots is the standard symbol for a squall line on a surface analysis (courtesy of NOAA).

Some squall lines develop bowing segments like the one over south-central Missouri in Figure 14.22a. We'll discuss these "bow echoes" in just a moment, but it suffices to say here that the greatest threat of severe weather posed by squall lines is damaging straight-line winds (especially when there are bowing segments).

How do squall lines form? Earlier in this chapter, you learned that discrete supercells tend to erupt in environments with large CAPE and modest large-scale lift in rather isolated spots. The most favorable environments for squall lines are those with large CAPE and *strong*, widespread lift, particularly when that lift is associated with low-level convergence along a sharp cold front (see Figure 14.23a) or along a surface trough ahead of a cold front, which meteorologists refer to as a **pre-frontal trough** (Figure 14.23b). Both surface patterns tend to favor squall lines, especially when strong upper-level divergence associated with a 500-mb short-wave trough works in tandem with the low-level convergence to create lift through a deep layer in the troposphere.

These two "best-case scenarios" for squall lines to form (a sharp cold front or a highly convergent prefrontal trough) occur more frequently from early

to mid-spring in the Southeast and Southern Plains and from mid-spring to early summer in the Central and Northern Plains, when southward-moving continental air masses are most likely to clash with maritime tropical air streaming north from the Gulf of Mexico. Quasi-linear convective systems can also evolve from thunderstorms initiated along outflow boundaries within the warm sector of a mid-latitude cyclone. QLCSs that affect the Great Plains sometimes develop from thunderstorms initiated by high-level heat sources (such as the Front Range of the Rockies). Generally speaking, quasi-linear convective systems can occur at almost any time during the warm season (primarily east of the Rockies), though winter is not out of bounds either.

Winds that blow from the southwest through a deep layer of the troposphere (nearly parallel to the initiating cold front or pre-frontal trough) contribute to the linear structure of developing squall lines. That's because these high-altitude winds carry hydrometeors produced by thunderstorms northeastward along the initiating boundary, where they eventually fall and help to consolidate the pool of rain-cooled air (like the portion of Figure 14.23a where winds aloft were predominantly southwesterly). Indeed, squall lines

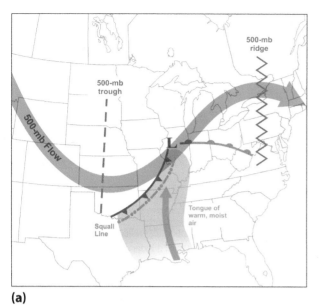

(a)

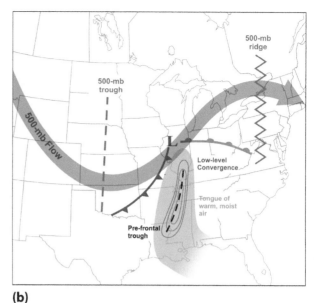

(b)

FIGURE 14.23 Squall lines tend to form in concert with strong low-level convergence associated with (a) cold fronts, or (b) pre-frontal troughs.

tend to have a single gust front and a relatively deep cold pool that helps to initiate and maintain intense storms along the leading edge (see Figure 14.24). Here, the gust front lifts warm and moist air parcels to their LFCs.

When squall lines approach, their position may be marked by a **shelf cloud**, a low, wedge-shaped cloud associated with a thunderstorm's gust front. Figure 14.25 is a striking photograph of a shelf cloud observed over eastern Wisconsin in July 2008. A shelf cloud is attached to the parent thunderstorm (or thunderstorms), and signals to the observer that the approaching gust front means business.

Some squall lines are more formidable than others. The type of squall line that produces a large-scale, damaging windstorm is called a derecho (pronounced "dey-RAY-cho").

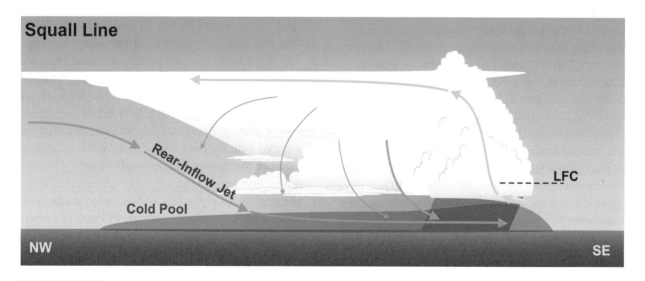

FIGURE 14.24 The pool of rain-cooled air associated with a squall line is usually sufficiently deep to lift air parcels to their LFC. As long as the gust front doesn't race out ahead of the squall line, the gust front will initiate new deep convection and thereby serve to maintain the squall line. The inflow of air in the mid-troposphere into the rear of the squall line can play a role in producing damaging winds at the surface.

FIGURE 14.25 A shelf cloud along a portion of a gust front associated with a squall line that produced severe weather in eastern Wisconsin on July 16, 2008 (courtesy of Doug Raflik).

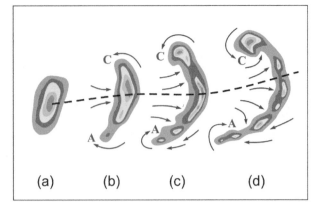

FIGURE 14.26 Schematic radar reflectivity showing the typical evolution of a radar echo associated with a line of thunderstorms into a bow echo. The dashed line marks the axis of greatest potential for downbursts. The letters "A" and "C" at the southern and northern ends of the bow echo represent areas of Anticyclonic and Cyclonic rotation.

Bow Echoes and Derechos: Radar Footprints of Widespread Windstorms

If you look back to Figure 14.24, you'll notice an inflow of mid-tropospheric air into the back of the squall line. This rear inflow develops in concert with an area of mesoscale low pressure (or "mesolow") that forms in the middle troposphere behind the leading edge of the squall line. This mesolow forms in response to the release of latent heat in the somewhat backward-tilted updrafts of the thunderstorms along the squall line's leading edge. As this rear inflow accelerates, a **rear-inflow jet** develops.

Some rear-inflow jets remain elevated, but some descend to the surface, setting the stage for a portion of the squall line to bulge forward. This forward bulge in the squall line appears on images of radar reflectivity as a bow echo (see Figure 14.26). By definition, a **bow echo** is a crescent-shaped radar echo often associated with a portion of a quasi-linear convective system (bow echoes get their name from their resemblance to an archery bow). Given that the rear-inflow jet descends to the ground whenever bow echoes appear on images of radar reflectivity, it should come as no surprise that bow echoes are a radar footprint of damaging straight-line winds. As a general rule, the portion of a QLCS associated with a bow echo spans roughly 40–120 km (25–75 mi) in length. The strongest winds are usually observed at the eastward-arching apex of the **surge region** of the bow echo (the dashed line in Figure 14.26 indicates the axis of the greatest potential for downbursts). Think of the surge region as the arrow on an archer's

bow. It is here that the rear-inflow jet, which has descended to the surface, causes the squall line to surge forward in tandem with a splash of damaging straight-line winds.

To add insult to injury, the now descended rear-inflow jet associated with a bow echo can also spawn tornadoes. In Figure 14.26(b-d), the "A" (**Anticyclonic**) and the "C" (**Cyclonic**) denote the sense of circulation of two **bookend vortices** that form on opposite ends of a bow echo. After several hours, which is a sufficient period of time for the Coriolis force to kick in, the cyclonic vortex becomes dominant, giving the northern end of the bow echo a comma-shaped appearance (Figure 14.26d). This cyclonic vortex can sometimes spawn a tornado, so recognizing this comma-shaped pattern on radar imagery is pivotal for forecasters at the National Weather Service to issue timely tornado warnings (more on watches and warnings later). In a bit of a positive feedback, the cyclonic and anticyclonic circulations associated with the bookend vortices work in tandem to enhance the descending rear-inflow jet.

Bow echoes tend to form in environments with large CAPE and relatively strong vertical wind shear, with 0–6 km shear sometimes reaching 45 knots (52 mph) or more. Strong winds aloft associated with relatively large vertical wind shear may help to intensify the rear-inflow jet.

In favorable environments, the northern end of a squall line forging out ahead of a strong low-pressure system can sometimes evolve into a series of long-lived, damaging downbursts. Such protracted windstorms are called **derechos**, a Spanish word, which, in this context, translates to "straight-ahead" or "direct." On radar, derechos have a telltale signature of one or more bow echoes, as shown in Figure 14.27.

A squall line that develops one or more long-lived bow echoes on radar can be upgraded to a derecho. There are several criteria for such a "convective windstorm" to qualify as a derecho. For our purposes, we will cite two of the most straightforward (no pun intended):

1. There must be numerous wind-damage reports and/ or reports of wind gusts greater than 50 kt (58 mph) over an area with a major axis (measured along the windstorm's forward path during its lifetime) whose length is at least 400 km (250 mi).
2. Some of the straight-line wind damage must be at least as severe as the damage inflicted by a weak tornado, or there must be several reports of winds in excess of 65 kt (75 mph).

Given such stringent criteria, it's probably no surprise that derechos are relatively rare. Figure 14.28 shows their frequency in the United States, one of the most derecho-prone countries in the world. On average, any given location east of the Rockies experiences one derecho every year or so, while significant

derechos (those that cause extreme damage and/or casualties) occur even less frequently.

For the record, roughly 75% of the derechos that impact the United States occur between April and August, when thunderstorms are most prevalent east of the Rockies. As it turns out, there are two principal types of derechos: serial and progressive. Figure 14.29 shows the idealized 500-mb pattern associated with derechos. Typically, a serial derecho evolves from a squall line that develops along or ahead of a cold front associated with a surface low-pressure system moving east from the Rockies—such a surface low is supported by a 500-mb shortwave trough, as shown in Figure 14.29. In contrast, progressive derechos tend to form along the northern fringe of a very warm 500-mb ridge of high pressure. In practice, the primary axis for progressive derechos in the United States stretches from the upper Mississippi Valley southeast toward the Ohio Valley, Great Lakes, and Mid-Atlantic States.

A schematic radar reflectivity signature of a serial derecho is shown in Figure 14.27. The radar presentation is typically a connected series of bow echoes that are part of a rather long squall line (usually several hundred miles long). The word "serial" is appropriate here because one of its definitions is "taking place in a series." In radar terminology, the series of bow

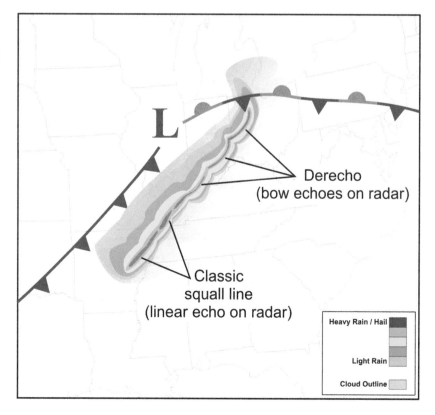

FIGURE 14.27 A schematic radar signature of a serial derecho associated with a squall line ahead of a strong mid-latitude low-pressure system. Note the bow echoes in the radar pattern east of the cold front. The red in the radar signature of the squall line and bow echoes marks the heaviest thunderstorms, while the gray indicates general cloud coverage.

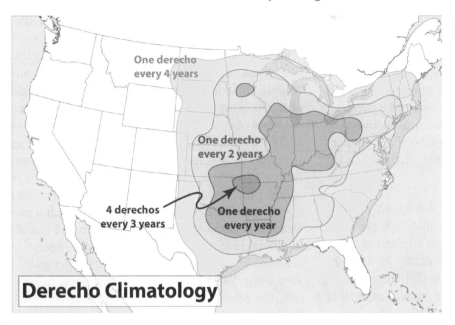

FIGURE 14.28 A climatology of U.S. derechos shows they occur, on average, once every year or few years east of the Rockies, with a relative maximum from northeastern Texas into the Ohio Valley (courtesy of Steve Corfidi and Dennis Cain, National Weather Service/Storm Prediction Center).

echoes associated with a serial derecho is sometimes called a **line-echo wave pattern** (LEWP, for short). For forecasting purposes, we'll stick with the term, serial derecho.

As Figure 14.27 indicates, a serial derecho occupies a portion of a long squall line associated with a mid-latitude cyclone and its attendant cold front. As you recall from Chapter 6, strong pressure gradients (at the surface and aloft) are the hallmark of intense, mid-latitude low-pressure systems, so it stands to reason that deep lows provide favorable environments for serial derechos because they tend to generate strong winds. Near the surface, strong

winds favor a line of strong low-level convergence along or ahead of the low's cold front. In addition, strong winds aloft are a hallmark of large vertical wind shear. As it turns out, large vertical wind shear helps to produce rear-inflow jets, which cause thunderstorms to bow forward in places.

Deep low-pressure systems and their strong cold fronts that initiate long, violent squall lines tend to occur in spring and fall, when north-south temperature contrasts become large. As a result, spring and autumn are the primary seasons for serial derechos.

A second kind of derecho, called a progressive derecho, typically forms over northern states east of

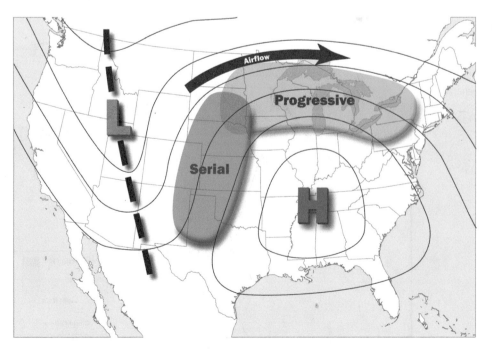

FIGURE 14.29 Synoptic upper-level setup for the two types of derechos, serial and progressive (courtesy of Steve Corfidi and Dennis Cain, National Weather Service/ Storm Prediction Center).

the Rockies during late spring and summer. Unlike their serial counterparts, progressive derechos tend to be a much shorter squall line (as short as perhaps 40 miles long). Moreover, progressive derechos can take the form of a single bow echo, as was the case for the historic derecho of June 29–30, 2012. Figure 14.30a is a radar montage showing the position of this progressive derecho at three-hour intervals starting at 15 UTC on June 29 and ending at 03 UTC on June 30. The corresponding storm reports are shown in Figure 14.30b.

Before we address the favorable environments for progressive derechos, we can't let our use of the word "historic" pass without comment. For starters, this very destructive windstorm was responsible for introducing the term "derecho" into the public's weather

vocabulary. That's because it impacted nearly every major metropolitan area from Indianapolis to Columbus to Baltimore and Washington, DC. This windstorm got widespread media attention, and everybody was talking about the "derecho." And for good reason. Very high wind gusts over 65 kt (75 mph) caused extensive damage and power outages. High winds blew roofs off homes, businesses, and schools, and overturned tractor-trailers. The derecho killed 22 people and seriously injured many more.

The progressive derecho of June 29–30, 2012, was not well predicted. In stark contrast, serial derechos have much greater predictability because deep low-pressure systems and strong cold fronts that generate violent squall lines are difficult to miss. In the case of the fast-moving progressive derecho of

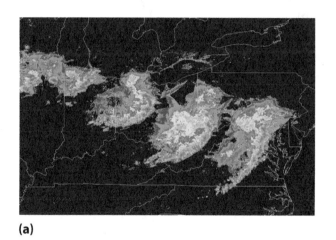

(a)

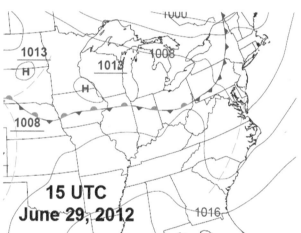

(c)

(b)

FIGURE 14.30 (a) A montage of radar reflectivity images every three hours from 15 UTC on June 29, 2012, to 03 UTC on June 30, 2012; (b) The area affected and storm reports associated with the June 29, 2012 derecho. Reports are for the 24-hour period beginning at 12 UTC on June 29. Some of the storm reports in Iowa and Illinois resulted from convection before the system took on derecho-like characteristics, or from a subsequent thunderstorm complex. Small blue squares indicate wind damage or wind gusts ≥ 50 kt (58 mph), large black squares with yellow centers indicate estimated or measured wind gusts ≥ 65 kt (75 mph), small green squares indicate hail ≥ 0.75 inches in diameter, large green triangles indicate hail ≥ 2.0 inches in diameter, and small red squares indicate tornadoes; (c) Although the surface analysis at 15 UTC on June 29, 2012, showed a front across the Ohio Valley, pressure gradients and pressure systems were weak (courtesy of NOAA).

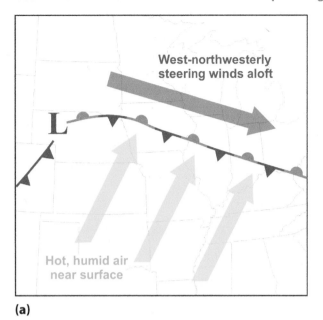

(a)

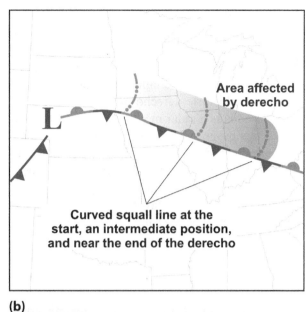

(b)

FIGURE 14.31 (a) The idealized synoptic setup conducive for warm-season derechos includes a stationary front that stretches generally from west to east and mid-level winds that blow approximately parallel to the front; (b) The idealized bow echo of a progressive derecho at several stages of its life, and the total area affected by the derecho over its lifetime (shaded).

June 29–30, 2012, however, there weren't any strong surface weather systems to grab the attention of weather forecasters (see Figure 14.30c). Such a relatively benign surface weather pattern is actually typical of progressive derechos. And when we say "fast-moving," we really mean it. All in all, the derecho covered approximately 700 miles in 12 hours, an average breakneck speed of 60 mph! Talk about "progressive!" Needless to say, forecasters were scrambling to stay on top of the streaking derecho.

Perhaps the key to understanding the environments favorable for progressive derechos is to note that, in the aftermath of the derecho of June 29–30, 2012, many thousands of people suffered without electricity for several days in a sweltering heat wave that lingered after the historic windstorm. Indeed, a hot air mass is a key ingredient in the recipe for progressive derechos. Keep this fact in the back of your mind as you read on.

The generic idealized setup for a progressive derecho is shown in Figure 14.31a. Virtually all progressive derechos form along a stationary front that stretches generally west to east across the northern states east of the Rockies. Such was the case with the progressive derecho of June 29–30, 2012—compare Figure 14.30c with the generic setup in Figure 14.31a. In addition, some of the most intense progressive derechos occur with a hot and humid air mass south of the front. On June 29, 2012, temperatures hit 100°F south of the front from Illinois and Missouri to Georgia and South Carolina during the afternoon, and dew points were 70°F or higher.

Let's return to another feature of the generic setup in Figure 14.31a. When a weak low-pressure system along the stationary front draws some of this hot, moist air northward over "cooler" air north of the front, warm advection sparks a system of thunderstorms on the "cool" side of the front. When progressive derechos develop in this kind of weather pattern, meteorologists have noted that high dew-point air also tends to "pool" along or near the portion of the stationary front where a progressive derecho spends its life.

Figure 14.31b is a schematic that shows the progressive derecho (here represented by a single bow echo) and the total area affected by the derecho over its lifetime. Note that progressive derechos tend to follow the stationary front. That's because the mean wind in the layer between the bottom and top of thunderstorms (the steering wind) tends to blow from the west-northwest, nearly parallel to the front.

Of all the characteristics of a progressive derecho, its forward speed is probably the most startling (as forecasters discovered on June 29–30, 2012). Indeed, the forward speed of a progressive derecho defies the rather lethargic summer weather pattern from which they sprout. On average, progressive derechos move at an average speed of 40–50 kt (46–58 mph), typically much faster than serial derechos.

From the original list of criteria for severe weather, only the topic of tornadoes remains for discussion. We devote all of Chapter 15 to explore this important subject. We close this chapter with a perspective on forecasting severe weather.

FORECASTING SEVERE WEATHER: LARKO'S TRIANGLE AND FINAL THOUGHTS

We hope that you've gained a deeper appreciation for the challenges that forecasters routinely face when the weather pattern favors an outbreak of severe weather. First, they must anticipate where thunderstorms will erupt. Then they must predict the mode, or type, of deep convection: For example, will thunderstorms develop as discrete supercells or in the form of a squall line? Then forecasters must predict how the storms will evolve with time. As we mentioned earlier, these forecasting issues are crucial because tornadoes and large hail are more likely when persistent, discrete supercells develop, whereas squall lines are more likely to produce damaging wind gusts.

In many cases, forecasts for severe thunderstorms can get pretty complicated. Take, for example, the outbreak of severe weather that occurred on April 10, 2009. Figure 14.32a is the 1630 UTC radar reflectivity over the lower Ohio and Tennessee Valleys on that day. At the time, a line of semi-discrete supercells had erupted from western Kentucky to northeastern Mississippi. We say "semi-discrete" here because we can distinguish individual supercells in the patterns of radar reflectivity, even though radar echoes connected them.

One hour later, the 1730 UTC radar reflectivity (Figure 14.32b) indicated that most of the semi-discrete supercells in western Tennessee had evolved into a more classic squall line, gradually shifting the primary threat posed by these storms from tornadoes to damaging straight-line winds. Despite the presence of the squall line, there were still discrete and semi-discrete supercells around, particularly the thunderstorm in southwestern Kentucky. Meteorologists describe such a mix of discrete, semi-discrete and organized thunderstorms (a squall line, in this case) as a *mixed mode*.

The idea here is to demonstrate that predicting the evolution of the mode of thunderstorms after they develop can be just as challenging as anticipating their initial mode. Both require forecasters to monitor CAPE, vertical wind shear, radar imagery, and a litany of other tools. For the outbreak that occurred on April 10, 2009, we suspect that predicting where the thunderstorms would develop was not quite as difficult as forecasting their evolving modes.

Figure 14.33 is the 18 UTC surface analysis on April 10, 2009. Notice that the thunderstorms formed mostly in an area bounded by a mid-latitude low-pressure system's cold and warm fronts, and the second closed isobar around the low's center. This

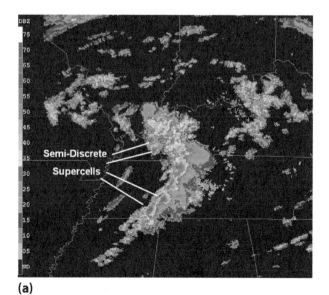

(a)

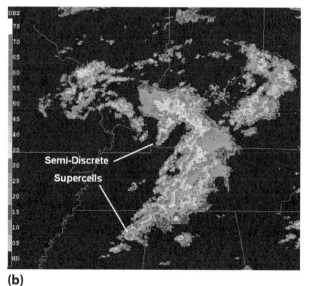

(b)

FIGURE 14.32 A pair of radar reflectivity images on April 10, 2009, shows how a series of semi-discrete supercells in western Tennessee evolved into a more classic squall line: (a) 1630 UTC; (b) 1730 UTC (courtesy of NOAA).

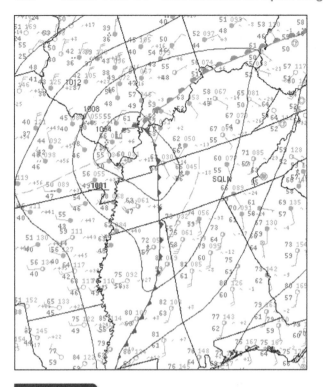

FIGURE 14.33 The 18 UTC surface analysis on April 10, 2009. A squall line formed earlier within the triangular region bounded by a mid-latitude low, its warm and cold fronts, and the second closed isobar around the low's center (courtesy of NOAA).

sector often marks the region where the synoptic-scale pattern can promote the initiation of severe thunderstorms. This triangular region is unofficially called **Larko's Triangle**, after the meteorologist who developed this forecasting tool (see Figure 14.34). This area constitutes the target where the atmosphere brings its biggest guns to bear. Within Larko's triangle, surface convergence around the low-pressure system and its cold front, coupled with upper-level divergence associated with the supporting 500-mb trough, sets the stage for severe thunderstorms. That's assuming, of course, that there is sufficient CAPE and a capping inversion that can eventually be breached.

The bottom line here is that Larko's Triangle is a tool that some forecasters use to pinpoint outbreaks of severe thunderstorms involving classic low-pressure systems. In other situations, forecasters look for mesoscale boundaries and complex terrain to initiate severe thunderstorms (examples include gust fronts, sea-breeze fronts, high-level heat sources, or dry lines).

The Storm Prediction Center (SPC) in Norman, OK, is charged with monitoring the country for conditions that promote severe thunderstorms. When conditions are favorable for an outbreak of organized severe weather, SPC will issue a **severe thunderstorm watch** that typically covers parts of several states. If conditions are especially favorable for severe thunderstorms to produce tornadoes, forecasters at SPC will instead issue a tornado watch. A typical severe thunderstorm or **tornado watch** is in effect for six to eight hours, but watches can be cancelled, replaced, or reissued as required. During a severe weather outbreak, it's not unusual for several watches to be in effect at the same time, especially during late afternoon and evening in the spring and early summer.

Although watch areas are often approximated by parallelograms, watches are actually issued by counties. As an example, Figure 14.35 shows three adjacent tornado watches issued by SPC during a mid-winter severe weather outbreak in the Deep South on January 21–22, 2017. Traditionally, tornado watches are shaded in red, but here the watch centered in Alabama is shown in purple, indicating a "Particularly Dangerous Situation" (PDS), a designation reserved for situations when SPC forecasters expect multiple strong or violent tornadoes. Note also that, superimposed on the watches, are indications of locations where severe weather was actually observed—nearly all of the green and black triangles and blue squares occurred within the watch areas.

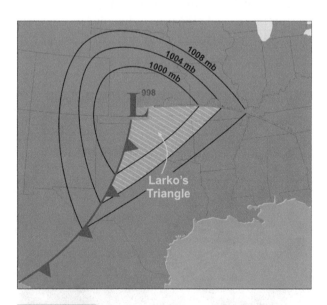

FIGURE 14.34 For synoptic patterns that favor outbreaks of severe thunderstorms, the forecasting tool called Larko's Triangle, defined as the triangular region bounded by the cold front, warm front, and first (or second) isobar around the low, can be used to identify the general region at risk of severe weather. In some cases, the area at risk expands to include the third closed isobar around the low.

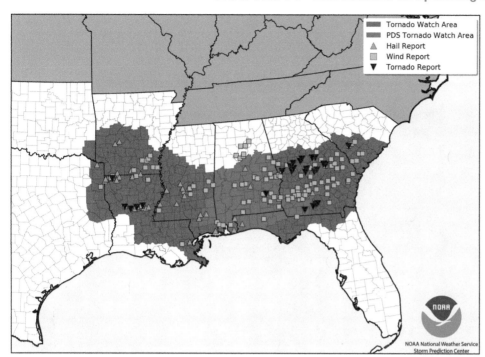

FIGURE 14.35 The three tornado watches issued by the Storm Prediction Center during the severe weather outbreak of January 21–22, 2017. The tornado watch that covered southern Alabama and parts of adjacent states was a "PDS watch," indicating a "Particularly Dangerous Situation." Colored triangles and squares indicate where severe weather (large hail, damaging straight-line winds, or a tornado) was reported (courtesy of Storm Prediction Center).

When severe thunderstorms are observed, or when radar reveals classic signs of severe weather, a local office of the National Weather Service will issue a **severe thunderstorm warning** or **tornado warning** for a specific county or portion of a county. Warnings are typically in effect for 30 to 60 minutes. In most cases, these warnings are issued far enough in advance to give people in areas potentially affected by severe weather at least several minutes to take appropriate safety measures.

Severe Thunderstorms: The Media and Chicken Little

When a severe thunderstorm or tornado watch has been issued for your area, there is no doubt that you should be prepared to take action and protect yourself. However, the truth is, in many severe weather outbreaks (not all, of course), only a relatively small fraction of a watch area will directly experience large hail, damaging winds, or a tornado. Thus, to some extent, intense media coverage of a potential severe weather outbreak is akin to a bit of "Chicken Little."

To understand the basis for this claim, suppose that two large tornadoes with relatively long damage paths, and several smaller twisters, develop within a given watch area. Further suppose that there are several dozen streaks of hail damage and wind damage associated with microbursts. Though this would be considered a very active day for severe weather, the area directly affected would typically amount to only a fraction of one percent of the original watch area. There are exceptions, of course, but the idea that severe weather only strikes a tiny fraction of a watch area is generally valid. Yes, you should be prepared to take quick action when severe thunderstorm and tornado warnings are issued. But during any given outbreak of severe weather, most people in a watch will not directly experience severe weather.

The one big exception is a derecho. When such a large windstorm develops, upwards of twenty percent of the original watch area can be affected by severe weather (mostly high winds). In this case, and cases such as the massive outbreak of tornadoes in central Oklahoma in May 1999, the media's intense coverage undoubtedly helps to save lives.

A Closer Look at Tornadoes

15

LEARNING OBJECTIVES

After reading this chapter, students will:

- Gain an appreciation for the climatology of tornadoes in the United States and other countries
- Distinguish between tornado vortex signatures and "ordinary" velocity couplets on images of Doppler velocities
- Understand the role of low-level wind shear in the development of the rotating updrafts in supercells

- Gain an appreciation for the Enhanced Fujita Scale to estimate forensically the wind strength of tornadoes
- Understand that only a relatively small portion of a tornado's damage path usually determines its assigned EF-rating.
- Gain an appreciation for tornadoes that develop from the ground up (landspouts)
- Understand the role that the rear-flank downdraft likely plays in tornadogenesis
- Gain a working knowledge of tornado safety
- Be able to dispel some common tornado myths
- Understand the science behind other whirlwinds that form in nature such as dust devils and fire funnels

Though most Americans will never be directly impacted by a tornado, these awesome whirlwinds are a source of fascination and awe for many people. Before 1950, in an era when relatively little was known about tornadoes compared to today, the use of the word "tornado" in official forecasts was discouraged and, for a time, even forbidden, for fear that it would cause public panic. Today, our once cloudy understanding of tornadoes has cleared somewhat, yet there are facets of tornadoes that meteorologists still don't understand, and misconceptions about them that still abound. In light of these considerations, we'll take a closer look at tornadoes in this chapter in order to make you better-informed weather consumers.

"Toto, I've a feeling we're not in Kansas anymore," marveled Dorothy after a fierce tornado swept her and her little dog to Munchkinland in *The Wizard of Oz*. Contrary to their Hollywood image (more recently in *Twister*, for example), most tornadoes are not giant monsters. The tornadoes that command attention on the evening news because they cut swaths of total destruction are the exception rather than the rule. Indeed, few "twisters" (as tornadoes are sometimes called) are powerful enough to completely destroy a house, or big or long-lived enough to cause a major disaster.

It was once thought that the whirling winds of a tornado blew at speeds comparable to the speed of sound (350 m/s, about 790 mph). But we now know that top wind speeds in even the most violent tornadoes seldom exceed 135 m/s (about 300 mph). Meteorologists chasing the violent twister that struck Oklahoma City on May 3, 1999 (see Figure 15.1a) used mobile Doppler radar to measure winds of 142 m/s (318 mph) just above the ground, the highest winds ever measured near Earth's surface (we point out, however, that this measurement indicated an instantaneous speed and arguably was not representative of the sustained power of the tornado—the twister's maximum sustained speed was probably lower). Even in this modern era that features armies of storm chasers with instruments mounted on vehicles, direct measurements of ground-level winds in a tornado are rare.

Tornadoes come in a variety of shapes and sizes (see Figure 15.1), but most share a common physical feature: a whirling, funnel-shaped cloud, appropriately called a **funnel cloud**, that lowers toward the ground from the base of a severe thunderstorm. As its name suggests, a funnel cloud rotates, but its circulation stays aloft (and thus does not cause any damage). If the circulation associated with this condensation cloud reaches the ground, then a tornado is born, though sometimes the funnel cloud itself can stay aloft (see Figure 15.2). In such cases, the tornado's circulation is often revealed by dust and dirt kicked up by the wind. In contrast, over grassy fields where there is little exposed dirt to be picked up, tornadoes sometimes become almost invisible, moving like whirling phantoms across the countryside. Caution: These tornadic ghosts are still very dangerous. Tornadoes can even occur without funnel clouds. This happens when weak ascent inside the twister can't sufficiently cool the air to produce net condensation, and/or when the air below the cloud base is too dry.

Tornadoes have an average diameter of about 100 m (330 ft), so there's no mistaking them for hurricanes, which possess diameters on the order of several hun-

(a)

(b)

(c)

FIGURE 15.1 A gallery of tornadoes: (a) May 3, 1999, near Oklahoma City, OK (courtesy of Jason Lynn and Steve Strum); (b) June 4, 2015, an anticyclonic tornado, near Sima, CO (courtesy of Brian Morganti); (c) A "Stovepipe" tornado near Matheson, Colorado, on June 4, 2015 (courtesy of Brian Morganti).

dred kilometers. The width of the damage path of some powerful tornadoes can sometimes exceed 2 km (about 1.3 mi), though the strongest wind speeds typically lie just inside the edge of the dust cloud that

FIGURE 15.2 A tornado near Sima, CO, on June 4, 2015. Note that the condensation funnel did not reach the ground, but the circulation associated with it did, picking up dust and dirt. This is the same tornado shown in Figure 15.1b, but before the condensation cloud reached the ground (courtesy of Brian Morganti).

often outlines the lower part of the twister. A tornado on May 31, 2013, reached an unprecedented 4.2 km (2.6 mi) in width as it passed through rural areas of central Oklahoma (see Figure 15.3). This large and violent twister, known as the El Reno tornado, killed four storm chasers, the first documented deaths since chasing tornadoes became popular.

Like the tornado in *The Wizard of Oz*, many twisters, when shown on video, appear as still life against the backdrop of a distant horizon, seemingly motionless as they corkscrew violent winds into the ground. In reality, tornadoes are often streakers, racing southwest to northeast (the typical direction of motion of the severe thunderstorms that spawn them) at forward speeds as high as 27 m/s (60 mph).

The traits of tornadoes mentioned briefly in this introduction are merely a sampling of the discussions to come. And tornadoes are not the only whirlwinds in nature. Let's now depart on a whirlwind tour to the Land of Whirlwinds. Like Dorothy and her companions Toto, the Tin Man, the Lion, and the Scarecrow, all we have to do is follow the yellow brick road.

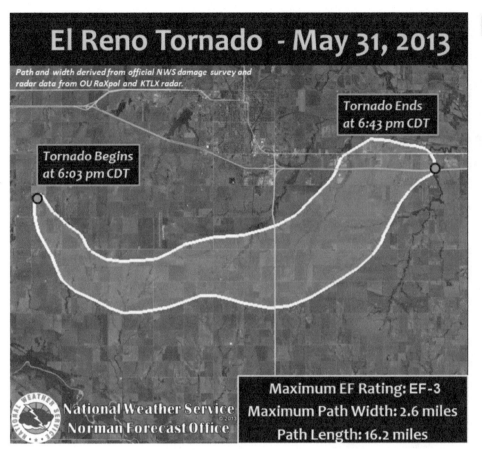

El Reno Tornado - May 31, 2013

Path and width derived from official NWS damage survey and radar data from OU RaXpol and KTLX radar.

Tornado Begins
at 6:03 pm CDT

Tornado Ends
at 6:43 pm CDT

National Weather Service
Norman Forecast Office

Maximum EF Rating: EF-3
Maximum Path Width: 2.6 miles
Path Length: 16.2 miles

FIGURE 15.3 The outline of the path of the "El Reno" tornado that ripped through central Oklahoma on May 31, 2013. The tornado was 4.2 km (2.6 mi) wide at one point, the greatest width of a tornado ever documented (courtesy of National Weather Service, Norman, OK).

LAND OF TORNADOES: THERE'S A TIME AND A PLACE

It was no accident that *The Wizard of Oz* was set in Kansas. The Jayhawk State lies in a corridor known informally as **Tornado Alley** (see Figure 15.4), a broad region of the central United States where tornado frequency is relatively high (especially for strong tornadoes). Figure 15.5 shows the average annual number of tornadoes observed in each state for the 30-year period 1985–2014. Overall, the United States averaged more than 1100 tornadoes per year during that time, with more than 10% of those twisters in Texas. Part of the reason Texas gets so many tornadoes is grounded in meteorology, but part is simply geography—Texas is a big state! In terms of numbers, Kansas is firmly in second place, so with regard to the story line in *The Wizard of Oz*, the state was cast very well.

The oranges and yellows in Figure 15.6 identify other regions of the world where conditions are often favorable for severe thunderstorms that can produce tornadoes. With a few exceptions (such as tornadoes spinning up when tropical storms or hurricanes come

ashore), the tornado-active regions of the world are positioned to frequently satisfy the four criteria for the development of tornadic supercells: high convective available potential energy (CAPE), strong vertical wind shear, modest synoptic-scale lift in rather isolated spots, and a breakable capping inversion.

Though the United States does not hold exclusive rights to tornadoes, there is no other region of the world where these ingredients come together more often than in Tornado Alley. Here, maritime tropical air from the Gulf of Mexico is readily accessible. So are elevated layers of warm, dry air from the Southwest and the plateaus of Mexico that form a capping inversion. Moreover, Tornado Alley lies along the beaten path of traveling low-pressure systems and their attendant fronts (and dry lines that form in concert with these weather patterns). Upper-air troughs add synoptic-scale lift that primes local environments for outbreaks of discrete supercells.

The primary season for tornadoes in the United States depends, for all practical purposes, on latitude. The ingredients for tornadic supercells typically start to come together from late January to early March over the Gulf States as burgeoning

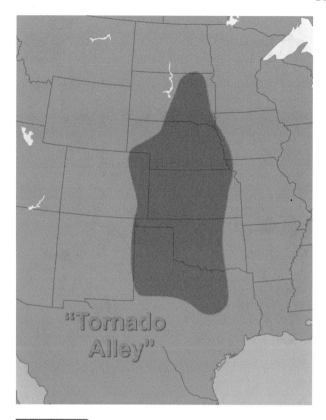

FIGURE 15.4 Tornado Alley generally stretches from Texas to South Dakota, but because tornado frequency can be measured in many ways, not all Tornado Alley maps will look alike. Here, the measure is based solely on the occurrence of "significant" tornadoes, defined as twisters having maximum winds of at least 51 m/s (113 mph) (adapted from *Climatological Risk of Strong and Violent Tornadoes in the United States*, by P. Concannon, H. Brooks, and C. Doswell, www.nssl.noaa.gov/users/brooks/public_html/concannon/).

warmth and moisture (mT air masses) from low latitudes interact with low-pressure systems in late winter and early spring. In fact, a second, unofficial tornado alley, often called "Dixie Alley," runs from Arkansas through Louisiana, Mississippi, Alabama, and Georgia (the numbers in Figure 15.5 certainly support this assertion). In this region, the period from January to May is prime season for strong tornadoes, owing, in large part, to the relatively easy access to warm, moist air and the prevalence of relatively strong wind shear.

The primary season for tornadoes then starts to spread northward during March, April and May as maritime tropical air continues to extend its reach. Peak time for tornadoes in the northeastern United States often waits until late May, June and July, when moisture from the Gulf of Mexico is more accessible. Though the jet stream generally weakens as it retreats toward Canada in summer, it can still dip southward over the Northeast States, allowing modest short-wave troughs to prime the troposphere for supercells by helping to erode capping inversions.

Figure 15.7 shows average monthly tornado frequency for the United States, based on data from 1991 to 2010. On average, May is the most active tornado month, though the record for most confirmed tornadoes in any month in the United States is held by April 2011, with 753 (May 2003 is second with 543).

On a daily basis, the most favorable time of the day for tornadoes to develop is the period from noon to sunset, peaking around 4 to 6 P.M. local time. Not coincidentally, this prime time for tornadoes corresponds to the most likely period of the

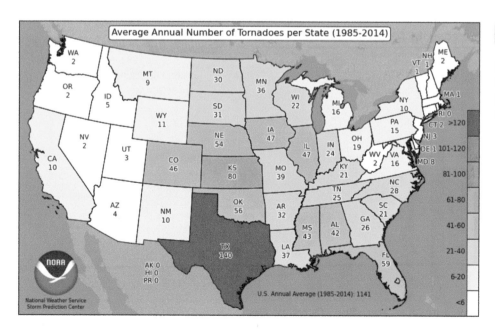

FIGURE 15.5 Average annual number of observed tornadoes for each state for the period 1985–2014 (courtesy of Storm Prediction Center).

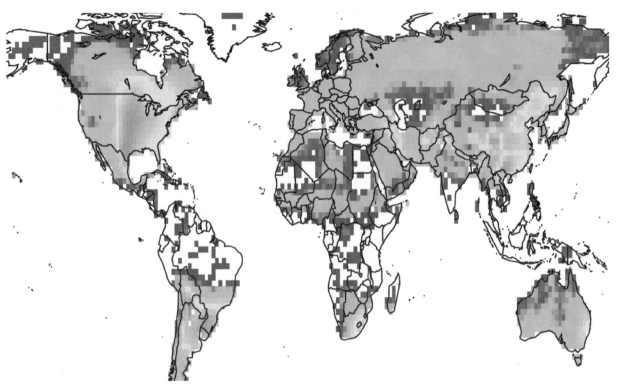

FIGURE 15.6 Regions of the world where conditions are, on average, most favorable for tornadic thunderstorms are shaded in orange and yellow. Besides the United States, other locations where tornadoes are relatively common include Canada, southern Brazil, Argentina, and eastern and southern China (courtesy of Harold Brooks, National Severe Storms Laboratory).

day for the lid to break after repeated assaults from vigorously rising currents of air set into motion by solar heating and synoptic-scale lift. Only about 20 percent of all twisters occur between midnight and noon.

If the road to the Emerald City in *The Wizard of Oz* is paved with yellow brick, then the road that leads to tornadic thunderstorms must be cobbled with wind shear. We will now take a second look at vertical wind shear and its role in tornadic supercells.

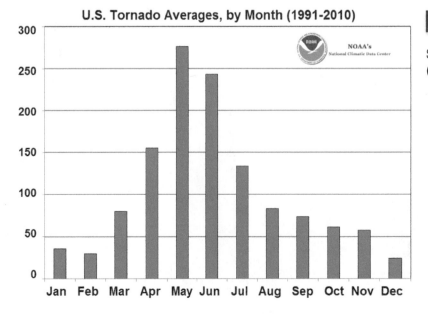

FIGURE 15.7 Average monthly tornado frequency for the United States during the period 1991–2010 (courtesy of NOAA).

PUTTING THE SPIN ON SUPERCELLS: WE'RE ON A ROLL

As we just reminded you, supercells require large vertical wind shear to keep updrafts separated from downdrafts. To fully understand what we're saying, check out Figure 15.8, which shows a unique view of the **storm-relative** inflow of low-level air and the storm-relative outflow of high-level air associated with a supercell. By "storm relative," we mean that we've subtracted out the thunderstorm's forward motion from the winds in and around the storm. For all practical purposes, we're treating the supercell as if it were stationary. In this example, low-level winds blow from the southeast relative to the supercell, while high-level relative winds blow from the southwest. For the record, storm-relative inflow can be quite strong. At a distance of 8-16 km (5-10 mi) from a tornadic supercell's updraft, storm-relative inflow can attain speeds of 65 kt (75 mph), while within a mile of a tornado, storm-relative inflow can be even higher! In this configuration (revisit Figure 15.8), high-altitude relative winds carry hydrometeors to the northeast, away from the updraft. Here, in this forward region of the supercell, the drag exerted on the air by falling raindrops produces the **forward-flank downdraft**. This mutually exclusive arrangement, which is a consequence of vertical wind shear, promotes longevity and increases the chances that the storm will produce severe weather. In a nutshell,

the stage is set for a mesocyclone (rotating updraft) to develop.

The origin of the mid-level rotation that characterizes a mesocyclone largely depends on the vertical change in the direction of the inflowing storm-relative winds that feed into the developing supercell's updraft. To visualize this, consider the vertical profile of storm-relative winds in the environment of an idealized supercell, shown in Figure 15.9. Notice that these storm-relative winds turn clockwise with height, known as **veering** to meteorologists (in this case, the wind direction changes from southerly to southwesterly to westerly with increasing height). In the lower troposphere, this veering with height generates some spin around a horizontal axis that's parallel to the stream of air flowing into the supercell (see Figure 15.10). Formally, meteorologists call this spin around a horizontal axis **streamwise vorticity**.

The generation of streamwise vorticity is difficult to envision, so consider this analogy: Think of streamwise vorticity as a tightly spiraling football. When a quarterback throws a tight spiraling pass, the fingertips impart spin to the football upon release. The football then spins around a horizontal axis that's aligned with the direction the football is thrown. In a way, storm-relative winds that veer with height in the lower troposphere are like a quarterback's fingertips: They impart spin around a horizontal axis.

Now we're ready to discuss the nuts and bolts of the origin of mid-level rotation in a developing

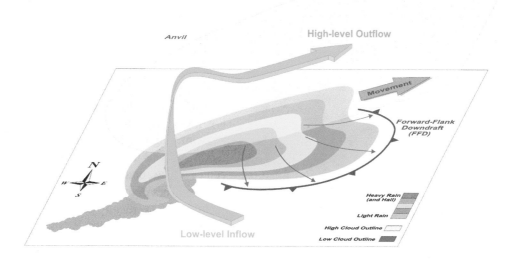

FIGURE 15.8 The storm-relative flow of air through a supercell. Here we subtracted out the storm's forward motion from the winds in and around the storm. For all practical purposes, we're treating the supercell as if it were stationary. In this case, low-level winds blow from the southeast *relative* to the supercell. Then they rise in the updraft. Eventually, high-level winds blowing from the southwest *relative* to the storm, carry hydrometeors to the northeast, where they fall and create the forward-flank downdraft. The colors represent the idealized radar reflectivity associated with the supercell.

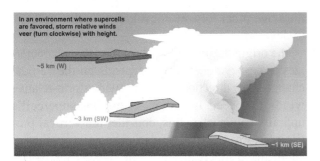

FIGURE 15.9 The storm-relative flow of air into a supercell at altitudes of 1 km, 3 km, and 5 km. Note that storm-relative winds veer with height.

mesocyclone. As a developing updraft ingests the now spinning inflow (again, storm-relative) of low-level warm, humid air, the spinning, storm-relative inflow (metaphorically, a spiraling football) is tilted upright (revisit Figure 15.10). This tilting converts low-level streamwise vorticity (which is typically on the southern flank of the developing supercell) into a mid-level, counterclockwise rotation about a vertical axis. A young mesocyclone has formed.

The rotating column of air (now a mesocyclone) is also associated with relatively low pressure (see Figure 15.11). To understand this, think of air in the spinning column as clothes in a washing machine during the spin cycle—just as clothes tend to pile up on the walls of the washer, air tends to migrate away from the center of the column, lowering the weight in the center and thus reducing the pressure.

FIGURE 15.10 When streamwise vorticity (created by the vertical change of low-level, storm-relative winds) gets tilted vertically by the updraft of a developing supercell, the updraft acquires cyclonic spin about a vertical axis. In effect, this spin about a vertical axis gives the mesocyclone its characteristic rotation (courtesy of Jessica Arnoldy).

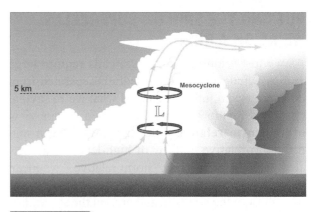

FIGURE 15.11 A supercell's rotating updraft, called a mesocyclone, is associated with low pressure. The width of a mesocyclone is on the order of a few to perhaps 10 kilometers (the parent cumulonimbus cloud has a diameter on the order of 30 km (19 mi) or so).

Now that we've established the rotation associated with the mesocyclone, we are in position to discuss how tornadoes form.

Stretching the Mesocyclone: Atmospheric Aerobics—"Twist, One, Two, Three"

Once the mesocyclone forms, a second process helps to intensify and transfer the spin to smaller scales. This process, called stretching, gives the rotating updraft a faster spin. To describe stretching, we once again invoke the tale of Olympic figure skaters. Some skaters finish their routines with a dizzying spin that begins as they draw their arms in toward their bodies, sometimes "stretching" their arms high above their heads to achieve maximum rotation. The quickening spin of skaters with upwardly stretched arms is a consequence of the conservation of angular momentum.

Remember that many severe thunderstorms form within the slow, cyclonic circulation of a synoptic-scale low-pressure system, so air feeding these storms already has some angular momentum. In the same way that an ice skater spins faster, low-level air flowing toward the center of the rotating mesocyclone will gain angular momentum. As the mesocyclone contracts inward, it also stretches vertically (see Figure 15.12). In essence, the contraction and stretching of the mesocyclone lead to an increase in its rate of spin. However, even at this stage of development, the rate of spin is still much slower and the spatial scale much larger than that of a tornado. Thus, there must be a way to further increase the rate of spin on a smaller spatial scale.

Truth be told, the last nuts and bolts needed to fully describe the process are still a bit loose; in

Slow Spin   Fast Spin

FIGURE 15.12 As low-level air converges toward the center of the mesocyclone, the column of air associated with the circulation of the mesocyclone stretches vertically. The column contracts and low-level air increases in speed as it moves closer to the center of rotation.

other words, the precise way that tornadoes are born is not completely understood. The most likely scenario is that the mesocyclone, which first develops at altitudes around 5–6 km (3–4 mi), begins to descend earthward and intensify, hitching a ride on a thunderstorm downdraft. If some of this downward-commuting air is then drawn back into the stream of humid air that is flowing toward the center of the mesocyclone, its angular momentum will further increase as the ring of inflowing air contracts. This creates faster cyclonic spin that can generate spin on a spatial scale worthy of a tornado.

There's a caveat to this entire discussion. Research suggests that some vortices destined to become tornadoes begin at the ground and work their way up rather than starting aloft and working their way down. Such a mechanism involves low-level convergence near the surface beneath a thunderstorm and the resulting vertical stretching and spin-up of the local air column. This mechanism likely produces **landspouts**, a term for relatively small and weak tornadoes that develop in concert with non-supercell severe thunderstorms (see Figure 15.13). Landspouts often form on the High Plains east of the Rockies, and are continental cousins to waterspouts (which will be discussed later in this chapter).

Twisters on Radar: Look for a Hook

Figure 15.14 shows the idealized radar reflectivity of a classic supercell. Note that we indicated the counterclockwise rotation associated with the mesocyclone on the southwestern flank of the supercell (for the record, mesocyclones can spin clockwise in the

FIGURE 15.13 A landspout formed a few miles west of American Falls, ID, during the afternoon of July 15, 2001. Landspouts are tornadoes that develop from the ground upwards. They are the continental cousins of waterspouts (courtesy of Norm and Linda Payne).

Northern Hemisphere, but they tend not to spawn anticyclonic tornadoes). Precipitation that is drawn into a corkscrew pattern around the periphery of the mesocyclone creates the classic hook echo feature on radar reflectivity. The motivation for the word "hook" is fairly obvious, given the curlicue shape of the precipitation wrapping around the mesocyclone. The word "echo" indicates that the precipitation pattern has been observed on radar.

When radar shows a hook echo and a tornado is present at the time, the twister usually spins up in the precipitation-free area on the southern side of the mesocyclone ("T" marks the spot in Figure 15.14). Figure 15.15a shows the radar reflectivity of a tornadic supercell that erupted near Oklahoma City, OK, on May 20, 2013 (the radar site is marked KTLX). This supercell spawned a violent tornado that struck Moore, OK with peak winds estimated above 200 mph, killing 24 people (this is the same Moore, OK that was struck on May 3, 1999).

The appearance of a hook echo on radar reflectivity does not guarantee the presence of a tornado or even that a tornado will develop. A relatively strong signal in the Doppler velocities near the hook echo confirms the presence of the corresponding mesocyclone. As you learned in Chapter 5, forecasters call this Doppler signature of a mesocyclone a velocity couplet, which is a juxtaposition of relatively fast inbound (toward the radar) and outbound (away from the radar) velocities. However, even a velocity

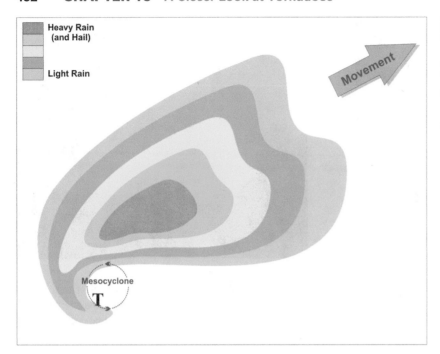

couplet does not necessarily indicate the presence of a tornado (or its imminent development).

However, if the difference between the maximum inbound and maximum outbound velocities—sometimes called the **gate-to-gate shear**—reaches a predetermined threshold, confidence runs much higher that a tornado has formed (or is about to form). For example, if the velocity couplet is within 55 km (34 mi) of the radar site, the National Weather Service uses 90 kt (104 mph) of horizontal shear as

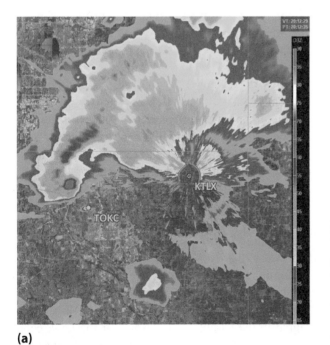

(a)

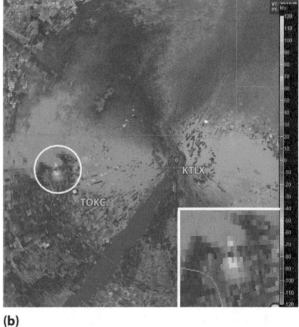

(b)

FIGURE 15.15 Radar data at 2012 UTC on May 20, 2013, showing a tornadic supercell that erupted near Oklahoma City, OK (the radar site is marked KTLX): (a) Radar reflectivity showed a hook echo on the southwest side of the supercell. The extremely high reflectivities in the hook echo indicate a debris ball (airborne debris); (b) Doppler-derived storm-relative velocities for the Moore, OK supercell. The inset is a close-up of the velocity couplet associated with the supercell's mesocyclone. In this case, the velocity couplet is a TVS (courtesy of NOAA).

a threshold. For a velocity couplet between 55 km (35 mi) and 100 km (62 mi) from the radar site, the threshold is 70 kt (80 mph). In these cases, the velocity couplet graduates to a **tornado vortex signature** (TVS for short). A TVS is circled on Figure 15.15b which shows the Doppler-derived storm-relative winds associated with the May 20, 2013 tornado near Oklahoma City (the inset shows a close-up of the TVS). Using the color-coded key on the right, we can estimate the horizontal wind shear across the velocity couplet. Light blue indicates inbound (negative) velocities of 60–70 kt (69–80 mph), while just to the north, bright pink corresponds to 60–70 kt (69–80 mph) of outbound (positive) velocities. Thus, the gate-to-gate shear is 120–140 kt (138–160 mph), easily large enough to satisfy the TVS criteria.

Recall from Chapter 5 that the radar beam widens as it gets farther from a radar site, and the returning signal is averaged over the width of the beam. As a result, when a tornado forms relatively far from a radar site, its rotating winds get averaged with the surrounding, larger-scale circulation of the mesocyclone. Thus, the Doppler-derived wind speeds associated with a mesocyclone's velocity couplet will not capture the true strength of the associated tornado (assuming one is present). This "watering down" of Doppler-derived wind speeds becomes more of an issue as distance from the radar site increases. In practice, the optimal range for detecting a TVS is about 60 km (36 mi) or so from the radar site.

So, if you *really* want to *accurately* sample the winds of a tornado with Doppler radar, you need to get "up close and personal." Figure 15.16a is a Doppler reflectivity image of the tornadic thunderstorms that exploded over Oklahoma on May 3, 1999. Note the supercell near the center of the image, particularly the curlicue at its southwestern tip (the white arrow points toward it). This hook echo indicates rotation within the parent supercell and, thus, the presence of a mesocyclone. Figure 15.16b is a Doppler velocity image focused on this supercell. The image came from a "Doppler on Wheels" (DOW), a special radar unit mounted on a truck that allows meteorologists to collect radar data from a tornado from a short distance away (see Figure 15.17). The TVS signature is circled in Figure 15.16b—here, the reddish pixels indicate fast

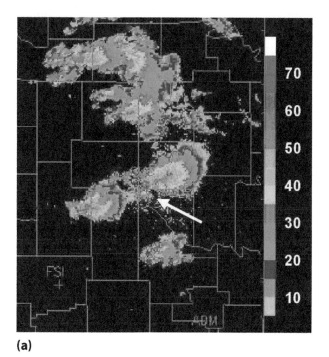

(a)

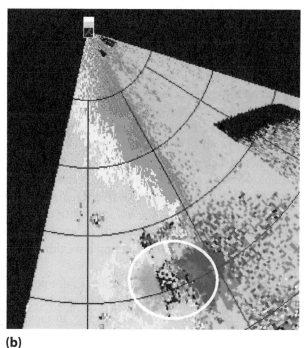

(b)

FIGURE 15.16   (a) The reflectivity radar display of supercells over Oklahoma on May 3, 1999. The hook echo at the southwestern tip of the supercell thunderstorm near the center of the image (see white arrow) marks the powerful twister that struck the town of Moore just south of Oklahoma City; (b) A Doppler radar velocity image of the tornado that struck Moore on May 3, 1999. The image came from a "Doppler on Wheels," a special portable radar unit. The TVS is circled. The center of the tornado is located between the reddish pixels, which indicate fast winds moving away from the radar (the DOW was stationed in the upper left), and the purplish pixels, which indicate fast winds moving toward the radar. Based on this image, tornadic winds were estimated at 142 m/s (318 mph) (courtesy of National Severe Storms Laboratory and the University of Oklahoma).

An example of a Doppler on Wheels deployed in the field (courtesy of Paul Markowski).

winds moving away from the radar, while purplish pixels indicate fast winds moving toward the radar (the DOW was stationed at the upper left of the image). The Doppler on Wheels measured a near-surface wind speed of 142 m/s (318 mph) in this tornado, the highest measured tornadic wind speed (that's why we decided to show these historic DOW radar products, even though DOW technology has advanced since then).

In the spring of 2009, Doppler on Wheels was one of a fleet of approximately 40 instrumented vehicles that gathered weather data for VORTEX2, the largest, most ambitious field experiment ever conducted on tornadoes. The **V**erification of the **O**rigins of **R**otation in **T**ornadoes **EX**periment **2**, which involved nearly 100 scientists and students, included ten mobile radars as part of its array of scientific instruments. The purpose of the fleet of vehicles was to deploy and create a relatively dense observational network around potentially tornadic thunderstorms. VORTEX2 documented the complete life cycle of a tornado spawned by a supercell in southeastern Wyoming on June 5, 2009, yielding a wealth of scientific data for future research.

Tornado detection by Doppler radar is not perfect. As we have already indicated, some velocity couplets come and go without a tornado forming. In fact, the false alarm rate on National Weather Service tornado warnings is currently about 75 percent. Studies have shown that, depending on the criteria used to define the presence of a mesocyclone, the fraction of supercells that spawn tornadoes can be as low as five percent or as high as thirty percent. On the flip side, tornadoes (especially weak ones) can form without registering a TVS—the frequency of tornado vortex signatures

is pretty low because they typically only appear on Doppler velocity images only when large tornadoes pass relatively close to the radar. Despite the high false alarm rate and the low frequency of tornado vortex signatures, Doppler radar has been an indispensable tool for detecting and predicting tornadoes.

As Figure 15.14 suggests, a tornado that forms in association with a classic supercell is often found in the precipitation-free portion of the supercell's southwestern flank. This position means that hail and rain often stop before a tornado arrives, lending a false impression that the storm is over. Then, without warning, the trailing tornado can barge in, unannounced by rain, hail, lightning, or thunder. The precipitation-free chamber of a tornadic thunderstorm is sometimes called the **vault**, the mesoscale strongbox of a supercell that often serves as a depository for tornadoes.

Though the vault is relatively "safe" from rain and hail, precipitation falls around the periphery of the mesocyclone. Raindrops and hailstones hold the combination to a second downdraft called the **rear-flank downdraft** that forms on the back side of the mesocyclone (see Figure 15.18a). This downdraft maintains a steady supply of evaporatively cooled air behind the advancing gust front extending south of the storm. Figure 15.18b shows a full-blown supercell thunderstorm, with the forward-flank downdraft (right) and the rear-flank downdraft (middle left) corresponding to the shafts of precipitation.

Recent research suggests that rear-flank downdrafts that are relatively warm near the surface may be more favorable for **tornadogenesis** (the production of tornadoes) than rear-flank downdrafts that are relatively cold. Why is the temperature of the rear-flank downdraft so important? For starters, previous studies indicate that some of the air drawn into the incipient tornado comes from the rear-flank downdraft (we hinted at this idea earlier). Though the jury is still out on the precise role that the temperature of this downdraft plays in tornadogenesis, if the air from the rear-flank downdraft is too cold, it would resist being lifted. As a result, vertical stretching would be limited in the column of air where tornadogenesis might occur, which in turn, would discourage the spin-up of a tornado.

How would forecasters get a handle on the temperature of the rear-flank downdraft? As it turns out, the temperature of this downdraft goes hand in hand with the altitude of the lifting condensation level (LCL). When the LCL is relatively low, the rear-flank downdraft tends to be relatively warm

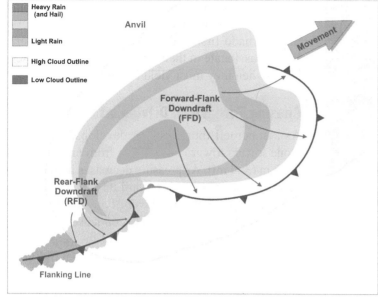

(a)

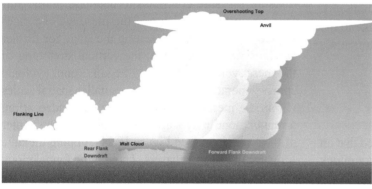

(b)

FIGURE 15.18 (a) A schematic of a classic supercell, showing the idealized pattern of radar reflectivity (color-coded) and the overall structure and position of the forward-flank and rear-flank downdrafts; (b) Looking southwest at a supercell thunderstorm, the position of the forward-flank downdraft corresponds to the shafts of rain beneath the thickest dark portion of the cumulonimbus. The vault is the brighter area to the left of the forward-flank downdraft, while the rear-flank downdraft lies just to the left. The apron of cloud hanging just below the cloud base between the downdrafts is called a wall cloud.

5–12°C (9–22°F). In this case, tornadoes became increasingly unlikely as LCL altitudes increased above 1200 m.

The bottom line here is that the altitude of the lifting condensation level is a fairly reliable predictor of whether or not severe thunderstorms will spawn tornadoes. In a nutshell, a relatively low LCL indicates an environment that's more favorable for tornadoes to form (assuming, of course, that conditions favor the development of supercell thunderstorms).

The gust fronts that mark the leading edge of a supercell's outflows of rain-cooled air sometimes spawn whirlwinds called **gustnadoes**. Here, the generating mechanism is strong horizontal wind shear—a change in wind direction and/or speed in the horizontal—though these vortices also sometimes form along the gust fronts of other types of thunderstorms. Note that we've resisted calling "gustnadoes" tornadoes. Indeed, gustnadoes seem to occur so frequently that they are *not* counted in any official tornado tally.

Gustnadoes are typically not strong enough to pick up and hurl debris through the air, which is why visual observations of these whirlwinds tend to be limited to dry places where swirls of dust near the ground typically outline their circulation. Figure 15.19 shows a gustnado associated with a thunderstorm over the Painted Desert of Arizona. The gustnado is the faint whirl just to the right of the center of the photograph. The dark shaft on the right, emblazoned by a flash of lightning, is a shaft of rain, while the pendulous cloud on the left side of the photograph marks the footprint of a microburst. Capturing both a gustnado and individual microburst in the same photograph is extremely rare.

when it reaches the ground. Indeed, a low LCL limits the production of cold air in the rear-flank downdraft because there is a limited vertical space in which falling raindrops can evaporate and cool the air. In contrast, when the LCL is relatively high, the rear-flank downdraft tends to be relatively cold because there's ample vertical space for cooling from evaporation of falling raindrops. The proof is in the pudding because temperature differences between the environment and the rear-flank downdraft of tornadic supercells are often less than 3°C (5°F). Such differences are consistent with relatively warm rear-flank downdrafts and limited evaporative cooling. Research has shown that tornadoes tend to form in environments with the LCL below 1200 m (about 4000 ft). The average height of the LCL for strong to violent tornadoes was about 700 m (2300 ft). When the LCL was higher than 1200 m, temperature differences between the environment and the rear-flank downdraft of a supercell ranged from

FIGURE 15.19 A striking photograph of a gustnado (the faint whirl to the left of the lightning) and a microburst (on the left side of the photograph) in the Painted Desert of Arizona (courtesy of Armand Scavo).

Of course, photographs of tornadoes are more commonplace, given the growing armies of both professional and amateur storm chasers who take to the highways each year to catch a glimpse of a tornado's fury. Let's go along for the ride.

THE CHASE: CLOSE ENCOUNTERS OF THE TORNADIC KIND

Circa 1970, a Notre Dame engineer suggested that an army tank be driven into a tornado in order to unlock some of the secrets of wind and pressure inside the twister's core (obviously, the engineer had the Fighting Irish spirit). Over the years, serious attempts at probing tornadoes have been made by launching instrumented rockets and remote-controlled airplanes at twisters. Since the late 1970s, researchers have built and attempted to deploy a variety of instrument packages in the paths of tornadoes to take meteorological measurements. Today, some storm chasers employ a heavily fortified "Tornado Intercept Vehicle" to take measurements and video near twisters. But tornadoes do not easily surrender their secrets.

Experienced tornado chasers use all the tools at their disposal, including Doppler radar, computer models, and surface and upper-air observations, to identity the areas where supercells and tornadogenisis are most likely to occur. When the rubber meets the road, however, chasers motoring around the countryside rely heavily on visual clues from cumulonimbus clouds to signal that a tornado is about to form. Think of these clouds as clues that help to guide storm chasers toward their goal of capturing a tornado on camera and video. A lot of chases are futile because storms that show signs of becoming tornadic eventually fizzle. Yet the promise of observing a tornado firsthand drives chasers to persevere. While passionate in their pursuit, all prudent chasers know when they must yield the right-of-way.

Chasing Tornadoes: Driving into a Wall

As a supercell thunderstorm develops, a mesocyclone lowers toward the ground from an altitude of 5–6 km (3–4 mi). As pressure decreases in the young mesocyclone, air drawn toward its center is cooled, causing water vapor in a layer of air directly below the base of the thunderstorm to condense. This cloud apron that hangs below the base of the cumulonimbus is called a **wall cloud**—one is shown in Figure 15.20. Because the wall cloud is part of the mesocyclone, it, too, shows rotation, a telltale sign to chasers that a tornado may soon form.

As air pressure continues to decrease, moist surface air on the periphery of the mesocyclone races inward, rises, and cools, setting the stage for further condensation. Meanwhile, the vertical stretching of the mesocyclone has already begun. Soon, cyclonic spin is generated on smaller and smaller spatial scales within the mesocyclone. With further condensation, a funnel cloud may start to descend from the center of the wall cloud (see Figure 15.21), upping the ante for gambling storm chasers. Though the funnel cloud might not reach the ground, the cyclonically spinning column in which it formed may have already touched down. Indeed, some weak tornadoes are characterized by a truncated funnel cloud. In these cases, the presence of the tornado is confirmed by the dirt and debris

FIGURE 15.20 A wall cloud descends below the base of a cumulonimbus cloud on May 24, 2004, near the town of Hebron, NE (courtesy of James Marquis).

FIGURE 15.21 From the wall cloud, a funnel cloud lowers toward the ground. Though the funnel cloud might not reach the ground, a tornado is born once the rotating column of air that houses the funnel cloud touches down (courtesy of Paul Markowski).

FIGURE 15.22 This tornado, which formed on May 22, 2008, in Kansas, stretched into a "rope," marking its final stages (courtesy of Doug Raflik).

swept into the air by strong, whirling winds. When powerful tornadoes form, pressure and temperature drops in the core of the mesocyclone are usually large enough to allow the funnel cloud to reach the ground.

As surface air is drawn toward a tornado, it cools by expansion as its pressure decreases, allowing moisture to condense into streamers of inward-moving clouds. These cloud streamers act like tracers, revealing the low-level convergence around a tornado. Though such streamers may be present, they are often hidden by flying dirt. However, streamers can sometimes be seen when tornadoes move over open grassy fields, where the amount of airborne dirt is limited.

Some tornadoes die very young, with a lifetime measured in seconds. A few long-winded twisters become senior citizens, living as long as several hours. Most tornadoes have a life span on the order of minutes, however, with durations averaging about 10 minutes. Occasionally, chasers will get a sign that the short life of a tornado is about to end: The tornado stretches into a "rope," a strung-out version of the twister (see Figure 15.22). As the parent thunderstorm pulls away from the tornado it spawned, the twister is literally stretched into oblivion.

Besides tornado chasers, trained tornado spotters also look for telltale signs of wall clouds and funnel clouds below the bases of supercells and other severe thunderstorms. When a spotter observes a tornado, the spotter's sighting becomes the basis for the local National Weather Service Forecast Office to issue a tornado warning. In practice, the National Weather Service aims to detect the rotation of a mesocyclone with Doppler radar *before* a tornado develops, providing valuable lead time for warning the public. In fact, the average warning time for tornadoes is now about 15 minutes.

Weather Instruments versus Tornadoes: Getting Blown Away

Over the past several decades, tornado researchers have attempted to place a variety of instrument packages in the path of a tornado to get measurements from within the twister's core, with varying degrees of success. Obtaining such in-situ observations is, at best, difficult, given safety considerations, the relatively narrow paths of most tornadoes, and the hostile environment such instruments would have to withstand.

The first in-situ surface observation system for tornadoes was developed in the early 1980s and christened TOTO, which stands for TOtable Tornado Observatory (see Figure 15.23). Between 1981 and 1985, researchers attempted to deploy this 400 kg (880 lb) industrial-strength instrument package directly in the path of tornadoes. Unlike Dorothy's Toto in *The Wizard of Oz* (who was actually carried up into a tornado), the instrumented TOTO was only brushed once (in April 1985). During that close encounter near Ardmore, TX, TOTO measured winds near 30 m/s (67 mph) and a small drop in air pressure (the instrument package named "Dorothy" in the 1996 movie *Twister* was inspired by TOTO).

FIGURE 15.23 During the early 1980s, researchers from the National Severe Storms Laboratory and the University of Oklahoma attempted to deploy TOTO in the path of oncoming tornadoes. TOTO recorded measurements of temperature, pressure, relative humidity, and wind on tape inside the 55-gallon drum. Only one tornado ever hit TOTO—a twister just brushed it in April 1985 (courtesy of NOAA Photo Library).

TOTO was retired in 1986, about the same time that a smaller portable instrument package called a "turtle" was developed at the National Severe Storms Laboratory. These squat, armored sensor packs, only about 35 cm (14 in) in diameter, measured temperature, pressure, and humidity, and were placed on the ground every few hundred meters in the path of a tornado. Another type of instrumented probe was successfully deployed in the path of a violent tornado that destroyed the town of Manchester, SD, on June 24, 2003 (see Figure 15.24). This HITPR device (Hardened In-situ Tornado Pressure Recorder) measured a pressure drop of 100 mb in a period of about 15 seconds as the tornado passed directly overhead. More recently, on April 21, 2007, a private storm-chasing vehicle equipped with weather instru-

FIGURE 15.24 The Manchester, SD, tornado of June 24, 2003. About seven minutes before this photograph was taken, an instrument package deployed in its path measured a pressure drop of 100 mb in a period of about 15 seconds (Photo © 2003 Jason Politte / http://www.onthefront.ws/)

ments measured a record pressure drop of 194 mb in a matter of seconds in a tornado near Tulia, TX.

Chasing tornadoes in order to deploy sturdy, in-situ instruments has scientific merit because traditional methods for measuring the vital signs of tornadoes have proven to be impractical. For starters, the probability of a twister striking any predetermined, fixed point is very, very low, even in Tornado Alley, unless one plans to wait a thousand years or more. Plus, most conventional barometers and anemometers would be ripped apart by a tornado's fierce winds. In addition, ordinary instruments are not designed to capture the rapid fluctuations in pressure and wind that likely take place inside a twister. Moreover, wind speeds are not necessarily uniform inside a tornado. Thus, any measurement at a fixed point may not be representative of a twister's maximum winds. One example of extreme winds within the powerful wind field of a tornado is the suction vortex.

Some of the most violent tornadoes contain smaller whirls that orbit around the center of the parent tornado, guided by the parent's counterclockwise spin. Such **multi-vortex tornadoes** have smaller, whirling offspring called **suction vortices**. Figure 15.25 shows the original conceptual model of a multi-vortex tornado created by Dr. Theodore Fujita in the early 1970s at the University of Chicago (Dr. Fujita, often called "Mr. Tornado" by his associates and the media, conducted ground-breaking research on a variety of severe weather phenomena). The path of a suction vortex is not circular, given that the parent tornado moves in a

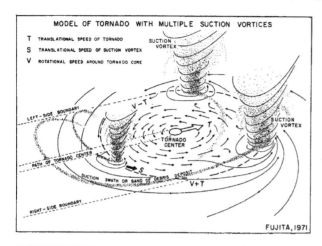

FIGURE 15.25 The original conceptual model of a multi-vortex tornado with multiple suction vortices orbiting the tornado center (courtesy of Dr. Theodore Fujita, 1971, Proposed Mechanism of Suction Spots Accompanied by Tornadoes. Preprints, *Seventh Conf. on Severe Local Storms,* Kansas City, MO, Amer. Meteor. Soc., 208–213).

relatively straight path across the landscape. The combination of circular and straight-line movements by a suction vortex creates a looping path for the whirling offspring. If you look closely at Figure 15.26, you'll see the **cycloid** pattern associated with the looping path of a suction vortex. Most of the field is out of the loop, but within the loop, damage would have been extreme; that's consistent with the notion that there was a narrow swath of extreme winds within the broader area of powerful winds associated with the tornado.

The reason that suction vortices produce extreme damage is that the whirling winds of the parent tornado and the winds of a suction vortex effectively

FIGURE 15.26 The combination of circular and straight-line movements by suction vortices creates a looping damage path called a cycloid. This cycloid pattern was created by an EF-4 tornado near Fairdale, IL (northwest of Chicago) on April 9, 2015 (courtesy of Jessyca Malina).

combine at some points within the cycloid path. The same principle applies to a train robber in the Old West running atop the moving cars toward the engine: The robber's total speed equals his running speed (relative to the train) plus the forward speed of the train.

In the same way, the twister's forward speed can also contribute to boosting its maximum winds. The forward speed adds to the speed of tornadic winds at places where they blow in the same direction as the twister's motion. In contrast, the forward speed of a tornado reduces the speed of tornadic winds at places where they blow in the opposite direction of the twister's movement. Our point here is simple: If an industrial-strength anemometer that was in place in the cornfield before the tornado arrived had not been hit by a suction vortex, the anemometer would have likely registered a wind speed much less than the tornado's maximum wind speed. Tornadoes guard their inner secrets fiercely.

Assessing Tornado Speeds: Giving Up the Chase

Over the years, movie photogrammetry, Doppler radar, and a forensic technique that utilizes tornado damage have provided indirect measurements of wind speeds in a tornado. In movie photogrammetry, for example, frame-by-frame analyses of flying debris in a tornado yield the distances that objects move over a specific period of time. Wind speeds can then be calculated from these distances and times.

Meteorologists can also estimate the wind speeds of a tornado by visually assessing the extent and degree of damage in the twister's aftermath. In 1971, Dr. T. Theodore Fujita of the University of Chicago developed a scale intended to relate the degree of tornado damage to the speed of tornadic winds. His **Fujita Scale** (F-Scale, for short), given in Table 15.1, was used for more than three decades to rate tornado intensity. In theory, the F-Scale enabled trained meteorologists to forensically estimate the wind speeds that caused the damage. The lowest ranking on the scale was F0, which applied to the weakest tornadoes whose maximum winds were less than hurricane strength. The F5 classification was reserved for only the strongest twisters.

In practice, estimating the winds of a tornado using the Fujita Scale was flawed. For example, the relationship between damage and wind speed had not been verified with any scientific calculations; to a large extent, Dr. Fujita relied on his intuition.

TABLE 15.1 The now-defunct Fujita Tornado Damage Scale (F-Scale)

	Wind Estimate (mph)	Typical Damage
F0	<73	**Light damage.** Some damage to chimneys and antennae; branches broken off trees; crops heavily damaged; shallow-rooted trees pushed over; sign boards damaged.
F1	73–112	**Moderate damage.** Mobile homes pushed off foundations or overturned; outbuildings blown down; shingles and parts of roofs peeled off; windows broken; some trees snapped.
F2	113–157	**Considerable damage.** Roofs torn off frame houses, leaving strong walls standing; weak structures and outbuildings destroyed; mobile homes demolished; large trees snapped or uprooted; cars blown off highways.
F3	158–206	**Severe damage.** Roofs and some walls torn off well-constructed houses; trains overturned; most trees in forest uprooted; heavy cars lifted off the ground and thrown.
F4	207–260	**Devastating damage.** Well-constructed houses leveled; structures with weak foundations blown away and disintegrated; cars thrown and large missiles generated; trees debarked by flying debris.
F5	261–318	**Incredible damage.** Strong frame houses lifted off foundations and disintegrated; steel-reinforced concrete structures badly damaged; automobile-sized missiles thrown 100 meters or more; trees completely debarked.

In addition, winds that cause comparable damage can vary dramatically in speed. That's because the extent of damage depends on factors such as the quality of construction and the direction and duration of the wind. Not accounting for these kinds of subtleties likely led meteorologists to incorrectly estimate the strength of some tornadoes in the past.

To address these shortcomings, a team of meteorologists and wind engineers developed an updated, or **Enhanced Fujita Scale** (EF-Scale, for short), that was implemented in the United States in February 2007. Like the original F-Scale, the updated EF-Scale ranges from EF0 to EF5; the higher the number, the greater the damage potential. What's new is that the assignment of an EF ranking begins with 28 *damage indicators* that represent different types of structures or objects that could be damaged by a tornado. These indicators include farm buildings, residential homes, shopping malls, transmission towers, flagpoles, and hardwood and softwood trees. For each of these damage indicators, specific guidelines link the degree of damage to an estimated range in wind speeds.

Table 15.2 compares the ranges in wind speeds for each category on the original F-Scale with the corresponding category on the updated EF-Scale. One notable difference between the two scales occurs in the higher categories, where wind speeds associated with EF3, EF4, and EF5 tornadoes are considerably lower than they were on the original F-Scale. These changes not only reflect a more reliable way to quantify the effects of wind on various structures, but they also indicate that winds slower than those estimated from the F-Scale can cause comparable degrees of

damage. For example, the EF-Scale classifies an EF5 as a tornado with winds at, or above, 200 mph, that can inflict damage previously ascribed to the lowest wind speed of the F5 range (261 mph).

The National Weather Service does not plan to re-evaluate past tornadoes using the EF-Scale. From the get-go, conformity between the two scales was paramount, and so a mathematical relationship was developed to correlate the ratings of the original F-Scale to those of the EF-Scale. This scheme allows a tornado to have the same rating on both scales, preserving consistency in the historical tornado record. So, for example, a tornado that was rated F4 using the original scale would be equivalent to an EF4 on the new scale, although the associated wind speeds would be different.

The first tornado to receive an EF5 rating moved through Greensburg, KS, during the evening of May 4,

TABLE 15.2 Comparison of Wind Speed Estimates from the F-Scale and the EF-Scale

F-Scale		EF-Scale	
F	Wind Speed (mph)	EF	Wind Speed (mph)
0	<73	0	65–85
1	73–112	1	86–110
2	113–157	2	111–135
3	158–206	3	136–165
4	207–260	4	166–200
5	261–318	5	>200

2007, damaging or destroying about 85 percent of the structures in the town. The costliest tornado on record (as of 2017—see Figure 15.27) moved through Joplin, MO on May 22, 2011, producing a swath of EF5 damage (see Figure 15.28). Figure 15.29 shows the National Weather Service's survey of the tornado damage path. Note the varying EF-Scale ratings along the tornado's track. The survey team rated this tornado an EF5 even though only a portion of the damage it produced was consistent with that of an EF5. By convention, the EF-Scale rating of a tornado (and, in the past, the F-Scale rating) is assigned according to the strongest winds (greatest damage) at any point along its path. This practice can give a false impression about the size of the area suffering damage consistent with the EF rating. As an example, studies of past tornadoes suggested that, in a typical F3 tornado, only about two percent of the area in the path of the twister actually experienced F3 damage. The rest of the damage was more characteristic of an F0, F1, or F2.

During the twenty-year period from 1995 to 2014, the United States averaged about 1240 tornadoes per year, but there was great year-to-year variability (look ahead to Figure 15.32). Fortunately, nearly 80 percent

FIGURE 15.28   EF5 damage in Joplin, MO after the tornado of May 22, 2011 (courtesy of National Weather Service, Springfield, MO).

of the twisters in the United States are relatively weak, earning a rating of (E)F0 or (E)F1, and about 95 percent of all U.S. tornadoes are below (E)F3 intensity. In contrast, less than one percent of all tornadoes receive an (E)F4 or an (E)F5 rating, with an annual average of just a few. Although rare, these monster twisters have been responsible for about two-thirds of all tornado-related fatalities since 1950. On March 3, 2019, an EF4 tornado ripped through eastern Alabama and western Georgia, killing 23 people and injuring at least 100. The death toll was more than twice the number of tornado fatalities in the entire United States for all of 2018. This killer twister was the first violent tornado (EF4 or EF5) in the nation in almost two years and the deadliest since the EF5 that devastated Moore, OK, on May 20, 2013.

Between 1950 and 2016, only 59 tornadoes have been rated (E)F5—the numbers in Figure 15.30 chart their locations in chronological order. Numbers 51 and 52 mark the Greensburg, KS, and Parkersburg, IA, EF5 tornadoes, while Number 59 is the May 20, 2013 Moore, OK tornado.

In terms of fatalities, the nation's worst tornado disaster occurred on March 18, 1925, when a powerful tornado produced a 353–km (219-mi) path of destruction from southeastern Missouri across southern Illinois into southwestern Indiana (see Figure 15.31). This "Tri-State" tornado (which may have been multiple twisters at some points along its path) was responsible for nearly 700 deaths and more than 2000 injuries during its more than three hours on the ground.

In the United States, the only structures that are built to withstand the powerful winds of the strongest

	DATE	LOCATION(S)	ACTUAL $	INFLATION ADJUSTED* $
1	22 May 2011	Joplin MO	2,800,000,000	2,921,780,000
2	27 April 2011	Tuscaloosa AL	2,450,000,000	2,556,550,000
3	20 May 2013	Moore OK	2,000,000,000	2,086,980,000
4	8 Jun 1966	Topeka KS	250,000,000	1,811,130,000
5	11 May 1970	Lubbock TX	250,000,000	1,512,380,000
6	3 May 1999	Moore/Oklahoma City OK	1,000,000,000	1,408,900,000
7	27 Apr 2011	Hackleburg AL	1,290,000,000	1,346,100,000
8	3 Apr 1974	Xenia OH	250,000,000	1,190,270,000
9	6 May 1975	Omaha NE	250,603,000	1,093,346,420
10	10 Apr 1979	Wichita Falls TX	277,841,000	898,283,680

FIGURE 15.27   The 10 costliest U.S. tornadoes since 1950 showing the actual dollar amount and an inflation-adjusted (to 2015 dollars) cost. The National Institute of Standards and Technology recommended changes in building codes nationwide based on their study of the Joplin tornado (courtesy of Roger Edwards/Storm Prediction Center).

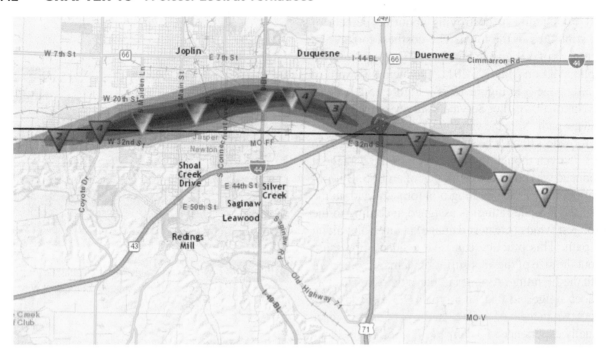

FIGURE 15.29 Survey of the track of the EF5 tornado that moved through Joplin, MO on May 22, 2011. The gray-shaded swath represents the extent of the tornado damage, almost one mile wide at its maximum. Only a small portion of the damage path was rated EF5 (courtesy of National Weather Service, Springfield, MO).

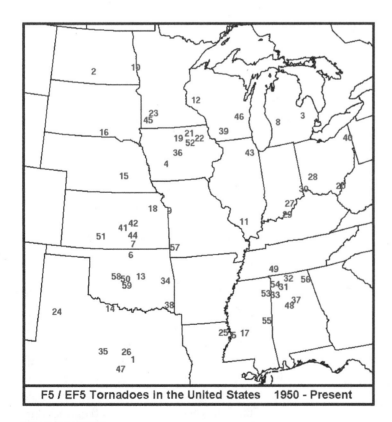

F5 / EF5 Tornadoes in the United States   1950 - Present

FIGURE 15.30 Locations of the F5 and EF5 tornadoes that have occurred in the United States from 1950 to 2016. The tornadoes are numbered chronologically (courtesy of the Storm Prediction Center).

tornadoes are the cores of nuclear reactors. Increasingly, many residents of tornado-prone areas are building "safe rooms," small, reinforced spaces in the interior of a home that are fortified by concrete and/or steel to offer extra protection. After the tornado outbreak on May 3, 1999, the state of Oklahoma launched an initiative to promote and support the construction of safe rooms through a rebate program, and more than 6000 shelters were built. When an F4 tornado hit Oklahoma City on May 9, 2003, many residents rode out the storm in these rooms, and no lives were lost. Though seeking shelter in a safe room may not be an option for you, there are other measures you can take that could save your life if a tornado approaches.

Tornado Safety: What to Do When a Twister Chases You

Let's start with what *not* to do. The worst course of action to take if a tornado approaches your house or building is to frantically run around opening all

FIGURE 15.31 Ruins of a school in Murphysboro, IL, where 17 children were killed on March 18, 1925, by the Tri-State tornado. This tornado holds several U.S. records, including most fatalities (695), longevity (3 hours, 3 minutes), and longest path length [353 km (219 mi)] (courtesy of NOAA Photo Library).

the windows. Many people still believe that a house will "explode" as pressure differences between the inside and the outside of a house become large as a tornado passes. This belief is scientifically unfounded (see the "myths" in the next section). Opening windows is useless and a waste of precious time. Most importantly, windows will likely be broken by flying debris, and glass propelled at high speeds is horribly lethal. Stay away from windows!

A basement or underground shelter is the safest place to take cover; get under a solid-oak desk, a heavy-duty work bench, a mattress, or something sturdy enough to protect you if the first floor caves in. If your dwelling has no basement, seek shelter on the lowest floor in an interior room (preferably one without windows!). An interior bathroom provides added safety, given that pipes and plumbing in the walls can give the room additional strength. If you're in a mobile home, get out and try to find a well-constructed building. If there are no buildings nearby, seek shelter in a ditch or low spot and protect your head as much as possible.

The same advice is appropriate if you're in a car and a tornado closely approaches. Vehicles are notorious death traps in tornadoes because they can be tossed and destroyed. In open country, if the tornado is far away and traffic allows, the best option is probably to drive out of its path. Table 15.3 summarizes some safety rules to follow if a tornado approaches.

Chasing Tornado Myths: Elusive Hearsay

Let's dispel the myth of the "exploding house" right off the bat. This misconception is based on the idea that air pressure lowers dramatically outside a house as a twister draws near, creating a pressure gradient large enough for the house to literally explode. In order to protect a house against tornado detona-

TABLE 15.3   Safety Rules When a Tornado Approaches

- In HIGH-RISE OFFICE or APARTMENT BUILDINGS: Go to the basement or a small interior room, preferably on a lower floor, and avoid windows. Protect yourself under a sturdy object.

- In BRICK or FRAME HOMES: Take shelter in the basement under sturdy items such as laundry tubs, work benches, pool tables, desks, staircases, or adjacent to furnaces or water heaters—these objects will break the fall of debris. If there is no basement, take cover in an interior bathroom or closet (do not lock the doors) or under heavy stuffed furniture in the center of the house. Keep away from windows and outside walls.

- In SCHOOLS and CHURCHES: Avoid rooms with large ceiling areas, such as auditoriums, gymnasiums, and churches. Go to an underground shelter, or a small interior room. Avoid windows and long hallways.

- In FACTORIES and SHOPPING CENTERS: Take shelter in a basement or a small room in the interior. Avoid windows and rooms or covered walkways with large roof areas. If there is only one large room, take cover under heavy objects away from walls.

- In MOBILE HOMES: Leave the mobile home and go to a shelter area or well-constructed building. If there are no buildings nearby, seek shelter in a ditch and protect your head as much as possible (though research suggests that taking shelter in a vehicle may be safer in some cases).

- In OPEN AREAS: Seek shelter in a nearby building or in a ditch. Protect your head as much as possible.

Weather Folklore and Commentary

Are There More Tornadoes Today?

Minutes after many weather disasters occur, television takes us to the scene. Such instant reporting often gives the impression that earth's weather has gone haywire. A cursory glance at yearly tornado statistics certainly supports this notion. As seen in Figure 15.32, the total annual number of confirmed tornadoes has climbed in recent decades—for example, 1103, 1098, and 1691 tornadoes were confirmed in 2006, 2007, and 2008, respectively, compared with the average of 820 twisters per year in the 1980s and about 400 per year in the 1950s. Are there really more tornadoes today? From relative tornado tallies, the answer is yes. In the absolute sense, however, probably not.

One thing is for certain: Compared to earlier times, today's annual number of twisters is inflated. There are now more people living and working in more areas, providing more opportunities to observe tornadoes. Moreover, many people own video equipment and most have cellphones, affording the citizenry a better chance to record tornadoes. Such evidence provides irrefutable confirmation. Perhaps even more importantly, the Doppler radar network that became operational in the early 1990s now allows meteorologists a better shot at identifying potentially tornadic thunderstorms and pinning down where to survey their damage.

Supporting the premise that improved observations are responsible for the higher tornado tallies in recent decades is that the number of tornadoes rated (E)F2 or higher (sometimes called *significant* tornadoes) shows no real trend. These stronger tornadoes, because of the destruction they often leave behind, have been well documented over the years (few have eluded detection). The lack of an upward trend in these stronger tornadoes reinforces the notion that the weaker (E)F0 and (E)F1 tornadoes that escaped detection in the past are now better observed.

Several decades ago, there were fewer people in fewer places with less technology. Is it any wonder, then, that modern-day tornado totals might be a bit inflated, given the no-frills way of counting tornadoes in bygone years?

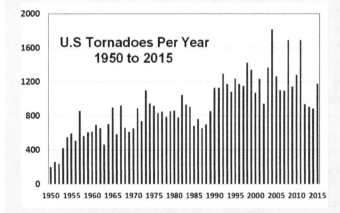

FIGURE 15.32 Total number of tornadoes observed in the United States each year from 1950 to 2015 (data courtesy of the Storm Prediction Center).

tion, it was once thought that windows should be opened so that air pressures could equalize rapidly. In reality, the last place you want to be as tornadic winds approach is near a pane of glass.

This myth at least gets partial credit, however. It is indeed true that air pressure decreases over the roof and outside walls of a house as speedy tornadic winds arrive—this is an application of Bernoulli's principle, named after the Swiss physicist who in 1738 postulated the relationship between increasing winds and lowering air pressure over a fixed object. Theoretically, there should be a large pressure gradient between the inside and outside of a house as a tornado passes. But houses are porous. They are not airtight. Thus, pressure differences between the inside and outside of a house quickly equalize as air rushes through nooks and crannies in the structure. In this way, the development of a large pressure gradient between indoors and outdoors is averted.

Roofs are like aircraft wings, especially if they have wide eaves. When subjected to increasing winds as a tornado approaches, aerodynamic roofs lift up a corner (or corners) of a house, weakening the structure and readying the house for severe damage once the tornado arrives.

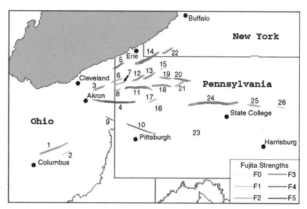

FIGURE 15.34 The tracks of 26 of the 41 tornadoes that struck Pennsylvania and surrounding areas on May 31, 1985. The tornado labeled "24" was an F4 that navigated the ridges and valleys of central Pennsylvania, downing an estimated 35,000 trees per minute. It was on the ground for approximately 100 km (60 mi) and was about 1000 meters (3300 ft) wide (courtesy of the Pennsylvania State Climatologist).

Before tossing Bernoulli's principle aside, you should know that it plays a role when tornadic winds flatten a house. As a tornado approaches and wind speeds increase over the nearest corner of a house (often the southwest corner because many tornadoes approach from this direction), air pressure decreases over that part of the roof, creating lift and causing that corner of the house to rise up slightly. The lift is augmented if the house has eaves (see Figure 15.33). With a corner of the house loosened from its moorings, the structure is in a weakened and vulnerable state. Then, when the mighty winds around the core of a strong tornado arrive, the house can be flattened in an instant.

A second tornado-related myth is the notion that tornadoes are somehow discriminating, able to skip over one house and completely destroy the next. Often, the explanation for this haphazard damage lies with suction vortices. Given the nature of their corkscrew paths, it's not hard to imagine a scenario in which a looping suction vortex misses one house but levels the next.

Finally, we dispel the myth that tornadoes cannot form in mountainous regions or don't hit large cities. Mountains and hills can indeed interrupt the flow of humid surface air into thunderstorms, thereby preventing the spin-up of tornadoes. However, once a tornado forms, it can go up and down mountains like an expert climber. During a historic tornado outbreak in Pennsylvania and surrounding states on May 31, 1985, some twisters literally went over hill and dale, climbing and descending mountains in Pennsylvania, all the while devastating everything in their paths (see Figure 15.34). Tornadoes can also occur in the Rocky Mountains, although they are infrequent there and typically weak.

Direct tornado hits on the downtowns of major cities are rare, but not for any meteorological reason. Rather, downtowns cover tiny land areas relative to the entire nation, and thus are small targets. Still, since 1995, tornadoes have hit some part of the central business district of Miami, Nashville, Little Rock, Salt Lake City (see Figure 15.35), Fort Worth, Atlanta, and Jacksonville, to name a few. Both Waco and Lubbock, TX have been hit by F5 tornadoes, in 1953 and 1970, respectively, while at least four tornadoes have struck St. Louis. And the notion that large buildings somehow can prevent or destroy a violent tornado is also a myth. The total circulation of a large tornado dwarfs even the largest skyscrapers in size and volume.

We end this chapter with a short discussion of other types of whirlwinds.

WATERSPOUTS, FIRE FUNNELS, AND DUST DEVILS: WHIRLWIND COUSINS

For an atmospheric whirlwind to be formally classified as a tornado, it must be in contact with both the ground and the base of a cumuliform cloud. Some of the other vortices in earth's atmosphere

FIGURE 15.35 An F2 tornado moved through the metropolitan Salt Lake City area on August 11, 1999. The tornado lasted 10 minutes and caused one fatality and more than 80 injuries. With damages in excess of $170 million, it was the most destructive tornado in Utah's history. The orange fireball is an electrical substation exploding (courtesy of ABC4 KTVX-TV, Salt Lake City, Utah).

FIGURE 15.36 Multiple dust devils in southern New Mexico on the evening of July 26, 2002 (courtesy of Randy and Kathleen Hensley).

satisfy this criteria, while others do not. We finish this chapter with a tour of a few of these whirlwind cousins. As you will see, they are not exclusive to our planet.

Dust Devils: Tasmanian Devil

It was Sunday, September 10, 1995. One of the authors took to the field for his weekly softball game on the outskirts of State College, PA. It was an autumnal day, with crystal-blue skies and temperatures near 21°C (70°F). A dry, cool breeze was blowing from home toward the outfield fence. It was a hitter's day.

In the sixth inning, a gust of wind suddenly blew up. The ground was very dry, and dirt and dust quickly filled the air between home and first base. Without much warning, a **dust devil**, a rotating column of air that forms near the ground and kicks up dust and dirt, spun up out of the disorganized wall of airborne dirt. With a diameter of 10–20 meters (33–36 ft) and a column of swirling dust extending more than 10 meters above the ground, the dust devil first moved across the author's bench, allowing him to estimate its wind speed to be at least 16 m/s (35 mph). The dust devil then moved toward the bleachers, where it lifted wooden seats off their runners. Unfortunately, the dust devil knocked over a little girl who fell through an opening in the bleachers. She was frightened and bruised but otherwise unharmed. The author had never experienced a dust devil like this in his life.

Dust devils are very much like the Tasmanian Devil, who would whirl and buzz across the countryside in hot pursuit of Bugs Bunny. Though on occasion he could rustle Bugs's ears, the Tasmanian Devil was, for the most part, harmless. Dust devils are typically small whirls that spin up on hot, sunny days over a dry landscape (see Figure 15.36). They too are short-lived and usually harmless, though occasionally they can cause minor damage. Wind speeds in a dust devil occasionally can top 20 m/s (45 mph).

Dust devils are not tornadoes in any formal sense because they do not connect to a cloud base. They form when strong heating by the sun promotes fast, rising currents of air near the ground. Just as the wind blowing past a corner of a building in autumn can cause a whirl of leaves to form, wind blowing past hills and mountains can impart a spin to currents of rising air, generating a dust devil. Even a semi-truck rolling along a highway can generate a dust devil. In fact, any horizontal wind shear can create a dust devil over dry ground on a sunny day. Dust devils aren't picky about what kind of horizontal wind shear: They can rotate clockwise or counterclockwise.

Apparently, dust devils aren't particular about which planet they form on either. Figure 15.37 captures a dust devil making its way across the Amazonis Planitia region of northern Mars in March 2012. Martian dust devils are actually fairly common and even made an appearance in the 2016 movie *The Martian*.

FIGURE 15.37 A dust devil on Mars, captured by a high-resolution camera onboard NASA's Mars Reconnaissance Orbiter on March 14, 2012. The orbiter is approximately 300 km (186 mi) above the surface, while the dust devil is about 70 m (225 ft) wide and approximately 20 km (12 mi) high (courtesy of NASA/JPL-Caltech/UA).

FIGURE 15.38 A waterspout off the Florida Keys, photographed from an aircraft (courtesy of NOAA Photo Library, Dr. Joseph Golden).

Waterspouts: Tornadoes of the Sea

Basically, a **waterspout** is a tornado over water (see Figure 15.38). There are two kinds of waterspouts. The first type forms in concert with a severe thunderstorm (a supercell, for example). They can begin as a tornado over land and then move offshore, or they can simply form over water (and sometimes move onshore). The second kind of waterspout is the non-supercellular variety, which is simply the maritime cousin of the landspout. These two varieties are more commonly called tornadic and fair-weather waterspouts (although we admit that most recreational boaters would hardly consider any waterspout an example of "fair weather").

Fair-weather waterspouts develop at the water surface and build skyward within the humid environment of the lowest few hundred meters of the troposphere. They are called "fair-weather" because many of them form underneath unassuming cumulus clouds. Fair-weather waterspouts typically require very warm water to initiate updrafts of air. Thus, shallow southern seas, such as those surrounding the Florida Keys, are fertile breeding grounds for fair-weather waterspouts. It's estimated that several hundred occur each year in the Keys, with August and September as the peak season.

Fair-weather waterspouts preferentially form along lines where surface air converges. A prime example is the leading edge of a land breeze off the Atlantic Coast of South Florida. Such surface convergence provides a favorable environment for rising, moist air to generate lines of cumulus and cumulus congestus clouds beneath which waterspouts can spin up. Wind speeds in fair-weather waterspouts rarely exceed 32 m/s (70 mph), so they're less dangerous than their tornadic counterparts. But if you're out on the water, avoid any confrontation: Waterspouts have been known to capsize boats. Fair-weather waterspouts are usually short-lived (with a life span on the order of a few tens of minutes), but they can come ashore and produce minor damage.

Funnels of Fire: Whirls of Hot Embers

When wildfires strike forests in the western United States during the summer, firefighters sometimes

FIGURE 15.39 A fire whirl. These funnels of fire are very dangerous and can spread fire beyond containment lines, starting new blazes (courtesy of United States Marine Corps, Camp Lejeune, NC).

encounter tornadoes of fire (see Figure 15.39). These fiery whirlwinds develop because the small-scale contrasts in air temperature between a fire's environment and the not-yet-torched surroundings can become very large. These large temperature gradients create, in turn, large pressure gradients, with several "mini" low-pressure areas developing over hot spots within the fire.

In response, a maelstrom of swirling winds starts to blow strongly around the wildfire, fanning flames, spreading fiery embers, and importing new supplies of oxygen (without replenishing the oxygen consumed by the fire, the fire would die out). These swirling winds rapidly converge around some hot spots, and as columns of air stretch vertically, tornadoes of fire can spin up – funnels of flames and hot embers that are very dangerous to firefighters.

POSTSCRIPT

Within the broad, slow circulation of a synoptic-scale low-pressure system, a rotating thunderstorm develops. Inside the rotating thunderstorm, a mesocyclone stretches vertically and quickens the tempo of spin. Like an aircraft's landing gear being lowered on approach to a runway, a rotating wall cloud descends below cloud base. From a wall cloud's undercarriage, a whirling circulation of air touches down on the earth below. Within the taxiing tornado, small suction vortices orbit around the center, producing looping swaths of extreme damage.

This cascade of whirling energy down to the scale of suction vortices is an example of the extreme measures that the atmosphere takes to mitigate large temperature contrasts. The idea that energy in the atmosphere cascades to smaller and smaller spatial scales, eventually even to the molecular (or viscous) level, was summarized by the physicist L. F. Richardson, a pioneer in numerical weather prediction, when he wrote:

> "Big whirls have little whirls that feed on their velocity.
> Little whirls have lesser whirls, and so on to viscosity."

The tornado is a quirky beast. It forms in a region of strong updrafts, yet its funnel cloud lowers toward the ground. It can flatten a house, yet leave the dwelling next door virtually unscathed. A twister can crumple a large, steel tank as if it were a plastic toy, yet it can pick up animals and deposit them unharmed far from their grazing pasture.

"Toto, I've a feeling we're not in Kansas anymore."

Mid-Latitude IV: Winter Weather

LEARNING OBJECTIVES

After reading this chapter, students will:

- Be able to distinguish the differences in the vertical profiles of temperature favorable for sleet, freezing rain, rain, and snow
- Be able to use the 0°C isotherm at 850 mb and the 5400-meter contour of 1000–500 mb thickness to estimate the position of the rain-snow line
- Understand how the jet stream's configuration determines the variety of winter storm tracks

- Gain an appreciation for the role that the natural temperature contrasts between land and ocean play in cyclogenesis along the East Coast of the United States
- Understand why intense storms along the U. S. East Coast qualify as nor'easters, and gain an appreciation for the role that high-pressure systems play in these storms
- Be able to identify patterns of cold-air damming on surface analyses, and to understand how such a pattern can increase the risk of frozen precipitation east of the Appalachians
- Understand the underpinning meteorology that leads to lake-effect snow
- Be able to distinguish weather patterns that promote multiple and single bands of lake-effect snow
- Understand the challenges of measuring the amount of snow
- Gain an appreciation for the shortcomings of wind-chill temperature in quantifying how a person's exposed skin "feels" in windy, cold weather

A snow-covered landscape is one of the most beautiful and tranquil scenes in nature, a winter wonderland of which greeting cards and childhood memories are made (see Figure 16.1). Perhaps winter's white coat makes such a lasting impression because it fundamentally changes the appearance of our surroundings. But winter weather can also be very disruptive, with brutal cold, howling winds, damaging ice, and blinding snow interrupting the rhythms of our daily lives (despite the gleeful mantra of avid skiers—"Let it snow, let it snow").

Although astronomical winter arrives in the Northern Hemisphere around December 21, meteorological winter is defined as the three coldest months of the year—December, January, and February. Of course, sometimes the atmosphere doesn't recognize the Gregorian calendar, given that incorrigible wintry weather often steps out of bounds and stakes a claim to late autumn and early spring. Some of the fiercest winter storms on record, such as the Blizzard of 1888 and the "Storm of the Century" in 1993, occurred in March, technically outside the domain of meteorological winter.

When describing the impacts of such extreme storms, forecasters sometimes refer to "white-outs," a term used to describe extremely low visibility caused by falling and blowing snow. When human behavior runs true to form, there often is a "whiteout" of sorts even before a snowstorm hits, with grocery stores selling out of the essential whites—milk, bread, and toilet paper. There's no doubt that when forecasters predict a big snowstorm, the anxious public braces for the worst. So issuing a forecast for a major winter storm is serious business.

In this chapter, we will explore the weather patterns and parameters that meteorologists analyze when forecasting winter storms. We'll examine the origins and impacts of winter weather, from the formation of fledgling snow crystals to the measures taken to combat aircraft icing. Along the way, we'll tackle ice storms, lake-effect snow, and atmospheric rivers, among many other topics.

WINTER PRECIPITATION: A PRODUCT OF THE ENVIRONMENT

Even in the summer, winter is close at hand. In the mid-latitudes, just a few kilometers above the surface, ice crystals can grow rapidly in an environment

FIGURE 16.1   A snow-covered landscape in eastern Pennsylvania in February 2017 (courtesy of Ben Reppert).

below 0°C (32°F). Through the Bergeron-Findeisen process (recall Chapter 8), ice crystals grow at the expense of neighboring supercooled water droplets, a give-and-take process that accounts for much of the precipitation that falls to earth in the mid-latitudes, regardless of the time of year. But what determines the type of precipitation that reaches the ground? To answer this question, we must understand the role that vertical profiles of water vapor and temperature play during the descent of water-gobbling ice crystals to the earth's surface.

The Snow Crystal: The Crown Jewel of Winter Precipitation

After a 25 cm (10 in) snowfall, a typical driveway might be covered with 1000 kg (more than a ton) of snow, enough weight of white to send even the most diehard shoveler to the neighbors to borrow the snowblower. Indeed, although winter can bring a potpourri of precipitation, snow is most readily linked with the season.

A **snow crystal** is a single crystal of white or translucent ice (see Figure 16.2). A **snowflake** is more encompassing. That's because a single ice crystal, several ice crystals stuck together, or giant puffs of snow crystals that sometimes fall in early spring all qualify as snowflakes.

The formation of individual snow crystals requires the presence of ice nuclei in a saturated

FIGURE 16.2 Microphotograph of a snow crystal, superimposed on a penny for a perspective of spatial scale. Typically, snow crystals range in size from about 1 to 5 mm (courtesy of Kenneth G. Libbrecht).

environment where temperatures are at or below 0°C. **Ice nuclei** are minuscule particles (naturally occurring mainly as tiny clay minerals) that have a microscopic crystal lattice structure resembling that of ice. This structure is inviting to sluggish water vapor molecules, which, upon deposition, begin the process of making snow crystals. But compared to the scores of cloud condensation nuclei (which serve as sites for the formation of cloud droplets), only a relative handful of ice nuclei are present in the atmosphere. Moreover, the ability of ice nuclei to attract water vapor molecules is temperature-dependent. For example, at a temperature of –20°C (–4°F), there are 1000 active ice nuclei per cubic meter of air (on average), but this number increases by a factor of ten with each subsequent 4°C (7°F) decrease in temperature. And, as temperature decreases, the increasing sluggishness of water vapor molecules makes them more likely to fuse with an ice nuclei and form a snow crystal. So, does this mean that "colder is better" for snow crystal formation? Well, not quite.

Snow crystals form (and grow) as water vapor molecules in the vicinity of supercooled water drops migrate toward, and deposit onto, ice nuclei (at temperatures below 0°C, of course). This "migration" is the result of a difference in equilibrium vapor pressure between liquid water and ice (revisit Figure 8.30). This difference is maximized at temperatures near –15°C (5°F), so the production and growth of snow crystals are highly favored near this temperature, despite the limited availability of ice nuclei.

As snow crystals descend toward the surface, there are several ways in which they can be modified and still qualify as a form of snow. First, snow crystals can collide and stick together, forming an "aggregate snowflake" (revisit Figure 8.34). Aggregate snowflakes are more likely to form at temperatures near 0°C when snow crystals are coated by a thin film of water, helping them to efficiently stick together. Alternatively, when snow crystals collide, they can break up into tiny ice crystal fragments, a process appropriately called **ice fragmentation**. These ice fragments can serve as additional sites for the growth of snow crystals. Finally, snow crystals can become coated with soft shells of ice during descent when they pass through an environment of tiny supercooled water droplets. During this process, called riming, a soft, opaque shell forms around the crystal. When snow crystals are completely coated with rime (similar to raisins that are coated with chocolate),

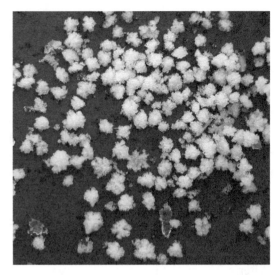

FIGURE 16.3 Two examples of graupel, or snow pellets. These white, opaque, nearly round particles are typically about 2–5 mm (.08–.20 in) in diameter. They form as a result of accretion of supercooled droplets on falling snow crystals.

they are called snow pellets or graupel (see Figure 16.3). Just as the raisin is hidden by the chocolate, the snow crystal is obscured by its shell of rime.

Given the many possible combinations of temperature and dew point at which a snow crystal may form, and the haphazard, helter-skelter nature of a crystal's journey toward the surface, the chances that two snowflakes will form in exactly the same way and, therefore, will be identical, are very small indeed. Although two snowflakes can "look" alike, the crystals will not be identical under a microscope.

Brigades of falling snowflakes sometimes charge earthward in short, heavy bursts, while, at other times, they settle toward the ground in a relatively steady descent. Not surprisingly, then, there are several terms used to describe the rate and duration of falling snow. A **snow squall** is a heavy, brief burst of snow, commonly occurring in concert with lake-effect snow or in convective bands within a broader shield of steady snow. A **snow shower** is a moderate but brief period of snow, which often results in spotty, light accumulations. On the light end, **snow flurries** refer to a very light and brief snow shower that produces only a trace of precipitation (no measurable accumulation). Note that these definitions share the word *brief:* They describe snowfall of short duration that is typically produced by convective, cumuliform clouds.

Longer-duration snows usually fall from stratiform clouds. For these steadier snows, meteorologists characterize intensity as light, moderate or heavy, depending on the horizontal visibility:

Horizontal Visibility	Snow Intensity
≥1 km (⅝ mi)	Light
½ km–1 km	Moderate
< ½ km (⁵⁄₁₆ mi)	Heavy

The media often invokes the term **blizzard** anytime there's a big snowstorm and the wind blows hard. In this context, the use of "blizzard" is colloquial. In a strict meteorological context, however, a blizzard must satisfy rather stringent criteria: at least three consecutive hours with sustained winds or frequent gusts of 30 knots (35 mph) or greater and falling and/or blowing snow that reduces visibility to less than 400 meters (0.25 mi). In reality, such conditions are very hard to come by. Note that falling snow is not a requirement for a blizzard: Snow already on the ground can be whipped up by strong winds, creating a **ground blizzard.**

Of course, a snowflake does not always end its skydive from clouds as snow. Let's investigate.

Freezing Rain and Ice Pellets: Children of Low-Level Chill

For a falling snowflake to reach the ground intact (that is, as a snowflake), the most favorable environment features temperatures below 0°C (32°F) throughout the troposphere. Even if such a "perfectly cold" troposphere is tainted by a thin layer of above-freezing air near the surface, snowflakes can still survive, especially if the snow falls rapidly and the flakes are large. Snow falling into a thin layer of relatively warm air next to the ground can occur in early spring as increasingly intense solar energy penetrates clouds and warms the surface (which, in turn, warms the overlying air). Snowflakes falling into this thin layer don't have much time to melt, allowing snow to reach the surface with air temperatures as high as 5°C (41°F) or even a bit higher. As a general rule, it takes a layer of above-freezing air at least 300 meters (about 1000 ft) thick to ensure that a parachuting snowflake ends its journey to the ground with the splash of a raindrop.

However, the atmosphere occasionally complicates matters by making a "warm-air sandwich." This "delicatessen special" features a layer of relatively mild

air above the ground (the meat in the middle) wedged between the deep chill at higher altitudes (top slice of bread) and a thin layer of subfreezing air near the ground (bottom slide of bread). This situation, shown in a cross section in Figure 16.4a, commonly occurs in winter on the cold side of a surface warm front as warm air aloft surges poleward ahead of an advancing storm. Cold, dense air near the ground is typically stubborn to retreat, so the arriving milder air runs up and over the low-level chill. The cold air is especially difficult to dislodge when it is trapped in deep valleys such as those in the Appalachians or the Columbia River Gorge in the Pacific Northwest. Regardless of geographic location, the greatest increase in temperature associated with warm advection often occurs initially in a layer centered near 850 mb, where temperatures can rise above 0°C. Snowflakes that melt as they fall into this relatively warm

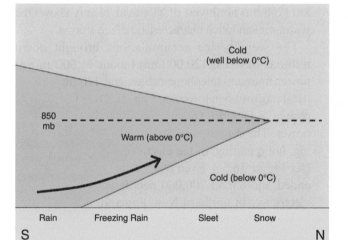

(a)

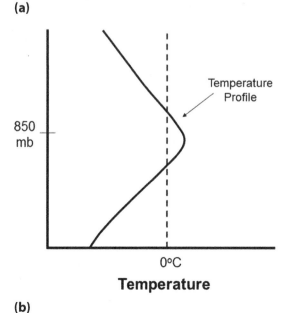

(b)

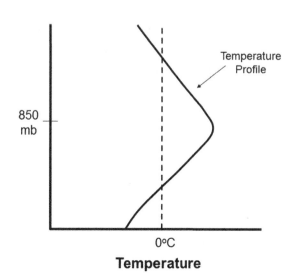

(c)

FIGURE 16.4 (a) An overrunning pattern marked by mild air from the south rising over a wedge of cold air near the ground, with the deeper chill of the mid and upper troposphere above the intruding warm layer; (b) A vertical temperature profile consistent with sleet. Snowflakes partially (or totally) melt as they fall through warm air aloft (≥0°C), and then refreeze into pellets of ice in the relatively thick layer of cold air (<0°C) next to the surface; (c) A vertical temperature profile consistent with freezing rain. Snowflakes completely melt as they fall through a thick layer of warm air aloft. The layer of low-level chill next to the surface is too thin for raindrops to freeze in the air, so they freeze after contacting cold, untreated surfaces such as trees, sidewalks and electrical lines.

layer are given a second opportunity to return to the flock of winter precipitation as either sleet or freezing rain when they fall through the subfreezing air below.

Let's assess their chances. Consider the temperature sounding in Figure 16.4b. As snowflakes originally in the thick crust of the top cold layer fall into the thin layer of above-freezing air centered around 850 mb, they either partially or totally melt. The process is not complete, however, because these raindrops (or partially-melted snowflakes) still must fall through the bottom crust of cold air next to the ground. If this cold layer is at least 300 meters thick, they freeze while airborne into **ice pellets** (also known as **sleet**). Ice pellets create a symphony of pinging and tapping on windowpanes, and can sting your face when driven by strong winds.

Note that sleet does not qualify as small hail because the two forms of icy precipitation form in completely different ways. Sleet belongs to the family of nimbostratus (stratiform clouds), while hail is a by-product of strong cumulonimbus (convective clouds). Calling sleet "hail" (or vice versa) is tantamount to comparing roast beef to Swiss cheese. Yes, both come from cows and both go on sandwiches. But that's about where the similarity ends. Hail and sleet are both ice and both fall from the sky. Hail's devotion belongs primarily to the warm season, while sleet's loyalty lies with winter.

Now back to the other possible outcome. If snowflakes completely melt as they fall through the above-freezing layer, and the bottom layer of cold air is thin, there's not enough time for the raindrops to refreeze in the air. Instead, supercooled drops freeze into a film of ice after impact with trees, electrical wires, and untreated surfaces whose temperatures are below 0°C. This infamous form of wintry precipitation is **freezing rain**, one of winter's most dangerous weapons. Figure 16.4c shows a classic temperature sounding that would prompt forecasters to predict freezing rain. In extreme cases, freezing rain can occur with near-surface air temperatures of −10°C (14°F) or lower! You might think that only snow could fall at such low temperatures, but the wedge of cold air next to the ground can be very shallow, with temperatures increasing rapidly with height above the thin slice of Arctic air. When forecasters talk about an "ice storm," their real concern is freezing rain, not sleet. Weighty accumulations of freezing rain can bring down wires and tree limbs, while numerous auto accidents make freezing rain a significant issue in the car insurance business.

One of the most severe, long-lasting ice storms in recent decades in North America occurred in January 1998 in northern New England and parts of Ontario, Quebec, and New Brunswick provinces in Canada. Over a five-day period, 7–11 cm (3–4 in) of ice accumulated in parts of southeastern Canada, affecting more people than any previous weather event in Canadian history.

Figure 16.5a shows the average pattern of sea-level pressure for the period January 6–9, 1998. An Arctic high-pressure system spanned west to east across Canada, providing a persistent northeast flow of cold air into New England. Meanwhile, the combination of relatively low pressure in the Ohio Valley and high pressure in the Atlantic sent warm, moist air northward to overrun the low-level chill. On average, a front set up across the Northeast States. North of the front, a prolonged icing event ensued. Figure 16.5b, a temperature sounding at 00 UTC on January 7 at Maniwaki, Quebec, about 240 km (150 mi) northwest of Montreal, clearly shows the warm-air sandwich characteristic of ice storms.

The weighty ice accumulations brought down millions of trees, 120,000 km (about 75,000 mi) of power lines and telephone cables, and 130 major electrical transmission towers in Canada. Over four million people in Ontario, Quebec, and New Brunswick lost power. The damage was so severe that major rebuilding, not repairing, of the electrical grid was necessary (see Figure 16.5c). Even two weeks after the ice storm ended, more than 700,000 people were still without electricity. In northern New England, half a million customers lost power, including 80 percent of Maine's population. Overall, more than 40 fatalities were blamed on the ice storm, and damages approached $5 billion in Canada and at least $1.4 billion in the United States.

Forecasting precipitation type during the cold season poses a major challenge to meteorologists, particularly when a low-pressure system moves up the East Coast of the United States. In this case, the transition zone between frozen and liquid precipitation often coincides with the corridor of large metropolitan areas from Washington, DC to Boston, so tens of millions of people are affected. To the east, warmer ocean air that circulates inland can tip the scales towards rain, while western portions of the region receive snow. A slight shift to the east or west in the track of the center of low pressure can mean the difference between all snow and a mix of freezing rain, sleet, and snow (or just plain rain).

Figure 16.6 shows the distribution of precipitation types associated with a classic winter storm. A shield

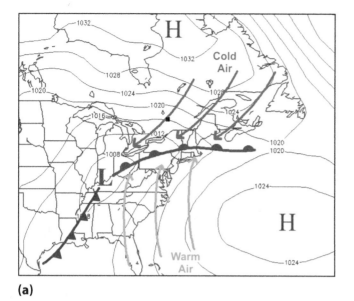

(a)

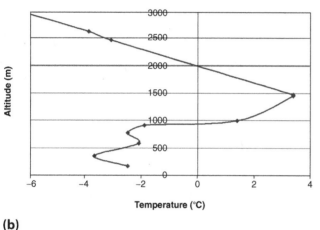

(b)

FIGURE 16.5 (a) Average sea-level pressure from January 6–9, 1998. Persistent high pressure over eastern Canada supplied cold, low-level air, while the combination of lower pressure over the Ohio Valley and high pressure offshore directed warm, moist air northward. The combination led to an ice storm of historic proportions in northern New England, eastern Ontario, and southern Quebec; (b) A temperature sounding at 00 UTC on January 7, 1998, at Maniwaki, Quebec, about 240 km (150 mi) northwest of Montreal, clearly shows the warm-air sandwich characteristic of ice storms. Maniwaki's location is marked with a black square in part (a); (c) A scene near Montreal on January 9, 1998. Ice accumulations brought down a total of 130 major electrical transmission towers in eastern Canada (courtesy Canadian Meteorological Centre).

(c)

of precipitation extends to the north of the warm front as moist, relatively mild air from the warm sector rides the warm conveyor belt and overruns a denser continental polar or maritime polar air mass. The heaviest snow typically falls to the northwest of a mature surface low, where its moist cold conveyor belt dramatically ascends. In this configuration, for a low moving to the northeast, the greatest snowfalls occur in a southwest-to-northeast oriented band.

Glance back to the temperature soundings for sleet and freezing rain in Figure 16.4(b–c). Notice that both show an 850-mb temperature above 0°C, indicative of the warmer air aloft. Forecasters often look to the temperature at the 850-mb level to pin down their forecast of precipitation type. By finding the 0°C isotherm on an 850-mb map, they separate locations more likely to see snow (850-mb temperature below 0°C) from those more likely to see sleet, freezing rain, or rain (850-mb temperature above 0°F).

In addition, forecasters often gain insight into the patterns of lower tropospheric temperature by analyzing the differences in heights between two mandatory pressure levels. Recall from Chapter 7 that pressure decreases faster in colder air than in warmer air. Thus, the vertical separation between two given pressure surfaces, formally known as the **thickness**, is related to the average temperature of the layer of air between them—the greater the thickness, the warmer (on average) the air between the two pressure surfaces.

Meteorologists commonly use the thickness between the 1000-mb and 500-mb pressure surfaces to assess precipitation type. When the **1000-500 mb thickness** is less than 5400 m, forecasters in the Eastern States lean toward snow reaching the ground (particularly at locations near sea level). In forecaster's jargon, 5400 m is the "critical thickness" for the 1000–500-mb layer. Statistically, this critical

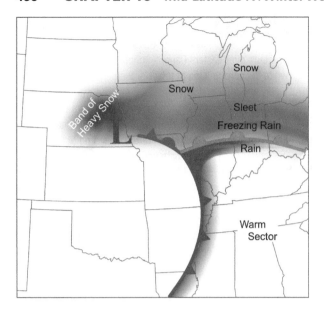

FIGURE 16.6 The idealized distribution of precipitation type associated with a mid-latitude cyclone during the cold season. On the cold side of the low's surface warm front, precipitation changes from freezing rain to sleet to snow with increasing distance from the front as low-level cold air thickens and the layer of invading warm air aloft becomes thinner.

thickness corresponds to a 50-50 chance of snow versus rain at sea level (in mountainous areas, the critical thickness is typically larger than 5400 m).

It's worth noting that a 1000–500-mb thickness of 5400 m corresponds to an average temperature of −7°C (19°F) between these two pressure levels. Given a typical temperature decrease with altitude in a column of air, such an average column temperature is low enough to allow the Bergeron-Findeisen process to be active in the upper part of the layer and low enough to give falling snowflakes a reasonable chance (about 50%) of reaching the ground.

Contours of 1000–500 mb thickness routinely appear on charts with sea-level isobars (see Figure 16.7). During the cold season, forecasters focus on computer predictions of the 5400-meter thickness contour to help them get a handle on the future position of the transition zone between solid and liquid precipitation. They also look carefully at computer predictions of the 0°C isotherm at the 850-mb level. Both of these contours typically shift northward with time in response to warm advection along and to the north of a low's warm front. It is here that

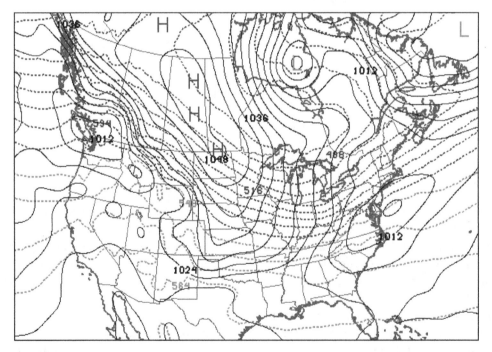

FIGURE 16.7 Forecasters routinely look at analyses of 1000–500 mb thickness such as this one from 12 UTC on March 10, 2017. On this particular forecast chart, isopleths of 1000–500 mb thickness are dashed and labeled in decameters, with contours greater than 540 decameters in red and contours less than or equal to 540 decameters in blue. The standard contour interval is 6 decameters (60 meters). Forecast maps of 1000–500 mb thickness typically also include sea-level pressure isobars (the thin, solid isopleths here).

snow can change to sleet or freezing rain as the air aloft warms. In a classic scenario, there's a narrow band of icy precipitation that roughly parallels the surface warm front on its cold side (see Figure 16.6). North of this icy zone, there is an expansive area of steady snow. Such a transition prompts meteorologists to issue forecasts such as "snow changing to sleet and then freezing rain" for towns slated to lie in the immediate path of the system's warm front.

FORECASTING WINTER STORMS: PATTERN RECOGNITION

During the cold season, weather forecasters rely on pattern recognition to anticipate the big guns of winter weather—biting chill, heavy snow, damaging ice, and stinging wind. Although every winter storm has unique characteristics, there are common surface and upper-air patterns that experienced forecasters recognize as telltale signs of developing winter storms. They then fashion their forecasts based on outcomes from similar patterns that occurred in the past and guidance based on computer model simulations. In this section, we will explore the techniques of pattern recognition in the context of winter storms.

Jet Stream Patterns: Laying the Storm Tracks

Perhaps the most recognizable jet stream configurations during winter are the high-amplitude patterns marked by a long-wave ridge and trough that span the contiguous states. In Figure 16.8a, the jet stream rides over the Rocky Mountains and plunges into the eastern United States. In such a highly wavy pattern, a surface low-pressure system that forms in concert with a short-wave trough moving through the long-wave trough would work in tandem with the trailing high-pressure system to transport cold air southward. This jet-stream configuration made frequent appearances during the winter of 2009–10, leading to a big chill and frequent storms in the eastern United States. All-time seasonal snowfall records were set that winter in many eastern cities, including Philadelphia (78.7"), Washington DC (56.1"), and Baltimore (77"). At the other extreme, the amplified jet-stream pattern shown in Figure 16.8b would favor cool, wet weather in the West and a milder spell in the East.

Troughs in the mid-latitude jet stream, coupled with horizontal temperature differences between the continents and oceans, produce a set of climatologically favored winter storms tracks over the United States, as

(a)

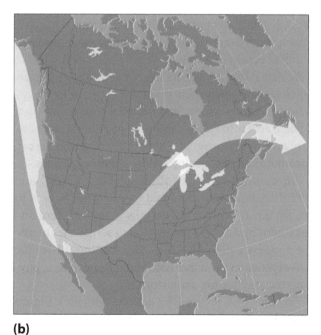

(b)

FIGURE 16.8 A highly meridional mid-latitude jet stream over the United States typically translates to cold waves or unusually warm winter weather in the East or the West, depending on the positions of the upper-level trough and ridge: (a) A winter jet-stream pattern that typically brings unseasonably low temperatures to the eastern United States and unseasonably high temperatures to the West. (b) In contrast, a jet-stream pattern that typically brings unseasonably mild weather to the East and unseasonably cold weather to the West.

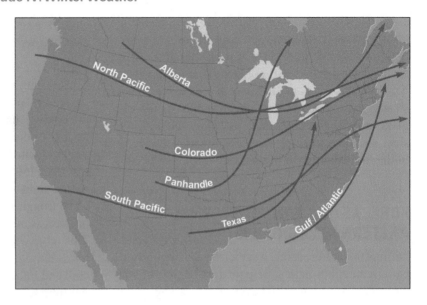

FIGURE 16.9 Climatological storm tracks over the continental United States. The most active tracks followed by cyclones in any particular winter are determined by the position of the mid-latitude and subtropical jet streams.

shown in Figure 16.9. In general, lows following more inland paths have limited access to moist air from the Gulf of Mexico and Atlantic Ocean, and, thus, usually produce modest amounts of snow. For example, low-pressure systems that take the "Alberta" track, appropriately called **Alberta Clippers**, typically yield a few to several fluffy inches. These storms, which originate in western Canada over or near the province of Alberta, are typically fast movers. Their name has a nautical origin. Indeed, clippers were 19th century sailing vessels that had tall masts and sharp lines; they were built for speed. Given their relatively quick movement and limited access to moisture, you'll understand why we say that Alberta Clippers are not typically big snow makers, although their potential to produce heavy snow increases as they approach the Northeast Coast, where access to Atlantic moisture and the natural horizontal temperature gradients between land and sea can promote rapid deepening of low-pressure systems.

In contrast, storms following the "Texas" and "Gulf/Atlantic" tracks have much easier access to moist air. In both cases, troughs in the southern and northern branches of the jet stream sometimes merge (a process called **phasing**) to create major-league divergence in the upper half of the troposphere, paving the way for big snowstorms.

To see this process in action, consider Figure 16.10a, an analysis of 500-mb heights (the solid green lines lines) and 500-mb absolute vorticity maxima (the thin dashed lines and color shading) at 18 UTC on February 7, 2010, during the record-setting winter of 2009–2010. Here, we will use the 500-mb level as a proxy for jet-stream level, using vorticity maxima to help discern important troughs. Note one

vort max over the Southwest United States in Figure 16.10a associated with a well-defined trough, and another vort max just north of the border between Canada and North Dakota affiliated with an even curvier flow. These vorticity maxima are embedded in separate branches of the jet stream—note how the height lines are relatively close together in the vicinity of each vort max (indicating relatively fast flow), but between the two vort maxes the height lines are relatively far apart. In this case, the mid-latitude jet stream had split over the eastern Pacific Ocean, with one branch heading north into Canada and the other branch taking a more southern route (sometimes the subtropical jet stream assumes the role of the southern branch).

Now consider the situation 72 hours later, at 18 UTC on February 10, 2010, shown in Figure 16.10b. The vorticity maxima (troughs) in the two distinct branches of the jet stream had merged, or phased, over West Virginia, forming a closed low with a very potent vort max. This upper-level disturbance energized an East Coast snowstorm that dumped almost 20 inches of snow on Baltimore and 16 inches on Philadelphia, two locations of many that set seasonal snowfall records that winter.

With this example of phasing under our belts, we will now take a deeper dive into the details of an East Coast winter storm.

Winter storms along the East Coast: Rapid Cyclogenesis and Nor'easters

Figure 16.11a is a water vapor image on January 23, 2016 at approximately 12 UTC. There's no denying

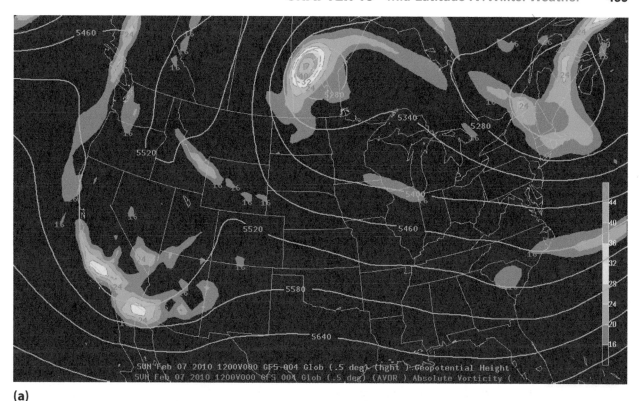

(a)

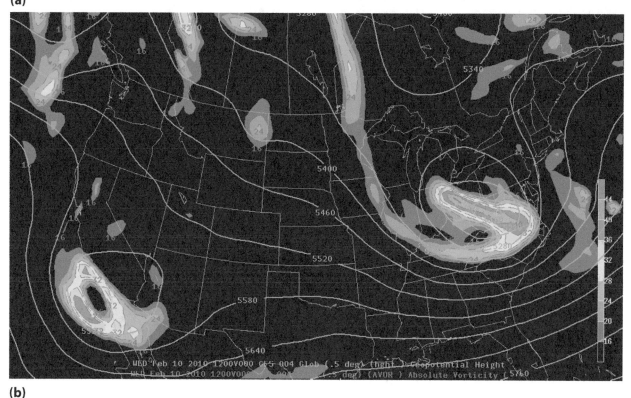

(b)

FIGURE 16.10 (a) A map of 500-mb heights (the darker solid lines, in meters) and 500-mb absolute vorticity (thinner dashed lines, in 10⁻⁵ s⁻¹, and color shading) at 18 UTC on February 7, 2010. The heights are drawn every 60 m and labeled in decameters, while the isovorts are drawn every 4×10^{-5} s⁻¹. Vorticity maxima are shaded in red and yellow. Note that the vorticity maxima (troughs) in the Southwest United States and in south-central Canada along the U.S. border are embedded in separate branches of the jet stream; (b) The same charts 48 hours later, at 18 UTC on February 9, 2010. The two troughs have merged, and the resulting upper-air disturbance would go on to energize a potent East Coast winter storm.

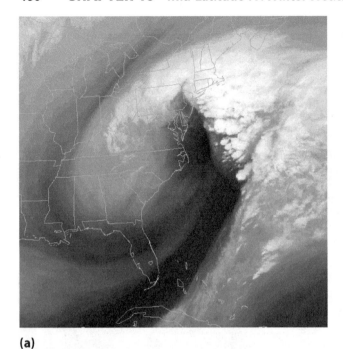

(a)

FIGURE 16.11 (a) Water vapor image at 12 UTC on January 23, 2016, shows the classic comma-shaped cloud that is characteristic of mature mid-latitude cyclones; (b) Surface analysis at 12 UTC on January 23, 2016 (courtesy of NOAA).

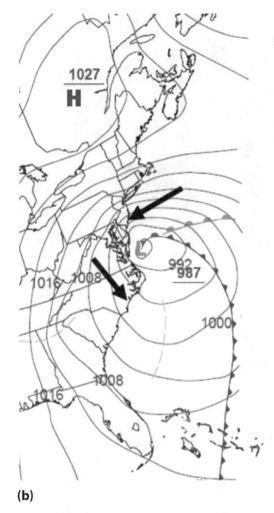

(b)

that the screaming message from this image is the classic comma-shaped cloud pattern associated with a mature mid-latitude low-pressure system, in this case situated along the Mid-Atlantic Seaboard. For confirmation, Figure 16.11b is the corresponding surface analysis with a few arrows added to emphasize the counterclockwise circulation of air around the storm.

This low-pressure system produced an historic snowstorm across highly populated parts of the Mid-Atlantic and Northeast States. Figure 16.12 shows the snowfall from this storm during the period January 22–24, 2016. Though snow fell from Arkansas to Massachusetts, the region marked "Northeast" took the brunt of the storm. There, over 20 inches of snow fell on several highly populated areas where more than 20 million people live. Meanwhile, in the "Southeast" region, the storm dumped over 10 inches of snow on approximately 5.6 million people.

The regional designations shown in Figure 16.12 are used in rating winter storms on a scale known as the **Regional Snowfall Index,** or RSI (see Table 16.1). In the Northeast, the winter storm of January 22–24, 2016 received a rating of 5, the highest on this scale. The Regional Snowfall Index is similar

to the Enhanced Fujita Scale for tornadoes and the Saffir-Simpson Scale for hurricanes in that the rating increases with severity from 1 to 5. Essentially, the RSI takes three factors into account: (1) The size of

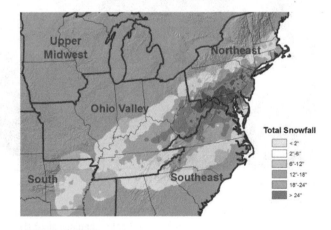

FIGURE 16.12 Snowfall during the period January 22–24, 2016. Portions of West Virginia, northern Virginia, Maryland, and Pennsylvania received in excess of two feet of snow from a nor'easter (courtesy of NOAA).

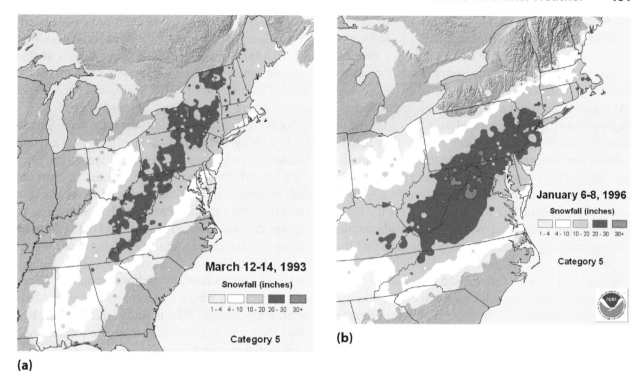

March 12-14, 1993
Snowfall (inches)
1 - 4 4 - 10 10 - 20 20 - 30 30+
Category 5

January 6-8, 1996
Snowfall (inches)
1 - 4 4 - 10 10 - 20 20 - 30 30+
Category 5

(a)

(b)

FIGURE 16.13   Snowfall from two of the snowstorms with the largest RSI in the Northeast region: (a) March 12–15, 1993 Superstorm; (b) January 6–8, 1996 Blizzard (courtesy of NOAA).

TABLE 16.1   The Regional Snowfall Index, or RSI, rates winter storms on a scale from 1 to 5 based on factors such as the size of the area affected by snow, the amount of snow, and the population affected.

Category	Description
1	Notable
2	Significant
3	Major
4	Crippling
5	Extreme

the area affected by snow; (2) the amount of snow that falls (scaled to thresholds based on climatological averages for locations in each region); and (3) the number of people affected by the snowstorm. In other words, the RSI attempts to quantify the impacts on society.

These impacts depend on the specific region(s) a winter storm strikes (in addition to those labeled in Figure 16.12, there's also a "Northern Plains and Rockies" region). For example, the winter storm of January 22–24, 2016, was rated a 4 ("Crippling") in the Southeast region, even though noticeably less snow fell there than the Northeast region. However, some of the heaviest snow in the Southeast fell in highly populated areas in Virginia, many of which are not accustomed to that much snow. So the aim of the RSI is to give each winter storm some historical context for the specific regions it affects; in the context of historic snowstorms in the Southeast, the January 22–24, 2016 storm was disruptive enough to deserve its rating of 4 on the RSI.

And the storm's "Extreme" rating in the Northeast region was also well-deserved. On the RSI, it ranked fourth out of approximately 200 snowstorms that were analyzed in this region since 1900. All in all, the storm impacted approximately 103 million people, about one-third the population of the United States, as it moved eastward from Texas and then northward along the Atlantic Seaboard. Even though the area of heavy snow in the Northeast was relatively small by historic standards, the snow that fell was concentrated on very highly populated areas, heightening the societal impacts. For the record, the only three storms to rank higher on the RSI in the Northeast region were the "100-hour Snowstorm" that began on February 22, 1969, the Superstorm of March 12–14, 1993, and the Blizzard of January 6–8, 1996 (see Figure 16.13).

We will now investigate how winter storms develop along the Atlantic Seaboard and why the East Coast of North America is a fertile breeding ground for **nor'easters**, northbound, coast-hugging low-pressure systems that typically bring significant precipitation and wind—the name derives from the strong northeast winds that blow ahead of these storms. Although these winds are the hallmark of nor'easters, high-pressure systems centered over New England or eastern Canada are accomplices because they help to sustain the cold air and strengthen the pressure gradient north of the advancing low. Revisit Figure 16.11b, the 12 UTC surface analysis on January 23, 2016 (during the historic blizzard of 2016). Note the high-pressure system centered over eastern Canada and the strong pressure gradient between that high and the coastal storm. Indeed, northeasterly winds between the high and the low gusted to 65 kt (74 mph) along the southern New England Coast, producing blizzard conditions. In addition to snow and wind, surging ocean water produced near-record water levels along the New Jersey and Delaware Coasts, causing flooding, beach erosion, and millions of dollars in damage.

Climatologically, nor'easters forge up the East Coast most frequently from December to March. As we just mentioned, their legions of gale-force winds can inflict significant beach erosion and coastal flooding. These same winds also can circulate relatively warm air inland from the Atlantic Ocean, often causing a changeover from snow to sleet, freezing rain, or even rain, especially right at the coast and just inland. At times, a nor'easter can pack the wind punch of a minimal hurricane: In fact, the March 1993 Superstorm (revisit Figure 16.13a) was called a "snow-icane" by a few meteorologists as winds gusted to 85 kt (about 100 mph) along the East Coast.

Nor'easters are some of the deepest mid-latitude low-pressure systems to affect the United States. The seed for such storms is related to geography. During winter, the natural temperature contrast between cold air over the continent and relatively mild air over the Atlantic Ocean helps to magnify the horizontal temperature gradients associated with fronts that sometimes set up along the East Coast (the presence of the Gulf Stream enhances these temperature gradients). These **coastal fronts** serve as very favorable breeding grounds for low-pressure systems to develop.

The historic snowstorm of January 2016 began as a weak low-pressure system along a stationary front in Texas on January 21. The storm moved east-northeastward, and about a day later approached the foothills of the southern Appalachians. By 00 UTC

on January 23, the low was analyzed over eastern Tennessee and was gradually weakening (see Figure 16.14a). However, the surface analysis at that time also shows another low-pressure system off the South Carolina Coast—this low would become the central figure in the historic blizzard of 2016.

A couple interesting questions arise from Figure 16.14a. Why didn't the surface low over eastern Tennessee just cross the Appalachians? And why did a second storm form along the coast? It turns out that, in general, mid-latitude surface lows tend not to cross mountain ranges. Probably the easiest way to explain this observation in the context of the Appalachians is that such a mountain-crossing storm would have to lift colder low-level air that typically gathers along the eastern foothills of those mountains (more on this process, which meteorologists call cold-air damming, shortly). Why is lifting this cold air bad news for lows? Remember that the air column over a low is the warmest (on average), lightest (weightwise) air column around. So lifting relatively heavy cold air (which would further cool on ascent) will cause any low to weaken. As a result, lows approaching the Appalachians from the west typically don't cross the mountains. Rather, they take a track similar to the Texas track in Figure 16.9.

So how did the historic blizzard of January 22–24, 2016 come about? As the original surface low gradually weakened to the west of the mountains, the supporting 500-mb trough continued to advance eastward—Figure 16.14b shows the 500-mb heights at 00 UTC on January 23, 2016. In addition to the very prominent long-wave trough, we have also identified a more subtle short-wave trough. The short-wave trough provided upper-level divergence that was critical to the development of a second low along the coast.

Also important was the large temperature contrast between land and ocean that exists along the Atlantic Seaboard during the cold season. This natural coastal front provides fertile breeding grounds for cyclogenesis. Indeed, Figure 16.14c shows surface isotherms (labeled in °C) at 00 UTC on January 23. Note the large temperature gradient along the coast, a perfect setup for cyclogenesis. Indeed, upper-level divergence ahead of the approaching 500-mb short-wave trough induced a new surface low to form (sometimes called the "secondary" low), which, in turn, strengthened in the favorable frontal zone along the coast. The rapidly-deepening low went on to produce the blizzard. In this way, the original low is

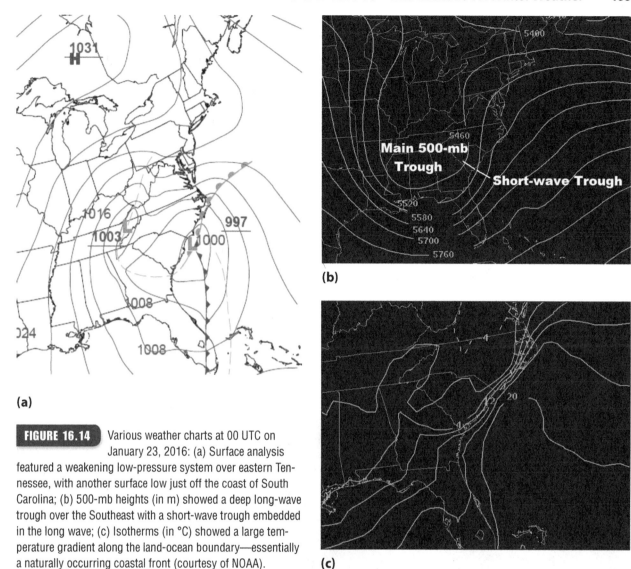

(a)

(b)

(c)

FIGURE 16.14 Various weather charts at 00 UTC on January 23, 2016: (a) Surface analysis featured a weakening low-pressure system over eastern Tennessee, with another surface low just off the coast of South Carolina; (b) 500-mb heights (in m) showed a deep long-wave trough over the Southeast with a short-wave trough embedded in the long wave; (c) Isotherms (in °C) showed a large temperature gradient along the land-ocean boundary—essentially a naturally occurring coastal front (courtesy of NOAA).

sometimes said to "transfer its energy" to the coast where a secondary low "takes over" and intensifies as it moves northward along the Atlantic Seaboard.

Let's finish our story about this storm. By 12 UTC on January 23, the intensifying offspring low had entered the early stages of occlusion along the Mid-Atlantic Seaboard as the main 500-mb low began to close off and move closer to the surface low (see Figure 16.15a and 16.15b). Meanwhile, at jet-stream level, a powerful jet streak was rounding the base of the trough, and the surface low feasted off the upper-level divergence in its left-exit region (see Figure 16.15c). Indeed, at its strongest, the surface low attained a central pressure of about 983 mb. A key takeaway here is this: many components go into a major winter storm such as this historic nor'easter.

In fact, we want to reemphasize the pivotal role of one of these components that probably does not get enough fanfare when it comes to big snowstorms in the East: a high-pressure system over eastern Canada (or thereabouts). Not only does a strong high accentuate the pressure gradient and, thus, increase the winds associated with a nor'easter, but it also supplies the low-level cold air. In fact, a common mantra for weather forecasters in winter is "Predict the high, predict the storm." Though this may fly in the face of strongly-held themes that low-pressure systems produce snow, keep in mind that a changeover from snow to ice or even rain will occur fairly quickly unless there is an ample supply of fresh, cold air to overrun. To the extent that Canadian highs provide this low-level chill, the forecasters' code of conduct makes sense.

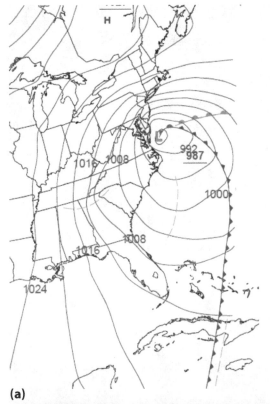

(a)

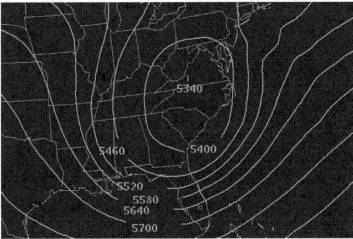

(b)

(c)

Recall that one of the keys to the historic blizzard of 2016 was that the low originating over Texas did not cross the Appalachians. Earlier we hinted that a phenomenon known as cold-air damming was, in part, responsible. In this case, the lay of the land conspires to pave the way for wintry precipitation.

Cold-Air Damming: Meteorology Meets Topography

The connection between wintry precipitation and high-pressure systems that maintain fresh supplies of low-level cold air can also be observed in parts of the southeastern United States. Georgia and the Carolinas are, by no means, pillars in the community of wintry places, but, on occasion, snowstorms and ice storms (primarily freezing rain) can cripple areas situated on the Piedmont Plateau and the eastern foothills of the southern Appalachians. Recalling that overrunning is the primary way to produce freezing rain and sleet, you might wonder why low-level cold air would not retreat quickly from such southern regions. The mechanism that provides a persistent supply of low-level cold air is a mesoscale winter phenomenon known as **cold-air damming**.

The process begins with a fairly strong high-pressure system building over the Northeast States or eastern Canada. The clockwise circulation around this high-pressure system tries to circulate cold air westward, but the Appalachians act like a dam to block its advance (see Figure 16.16a). In response, cold air spills south-southwestward along the eastern foothills of the Appalachians. The footprint of cold-air damming here is a pronounced, south-southwestward bulge in sea-level pressure isobars over Virginia, the Carolinas, and northeastern Georgia, creating a pronounced ridge of high pressure

FIGURE 16.15 Various weather charts at 12 UTC on January 23, 2016: (a) Surface analysis showed the strengthening coastal low (nor'easter) that produced an historic blizzard over parts of the Northeast and Mid-Atlantic States; (b) 500-mb heights (in m) showed a closed low just to the west of the surface low; (c) 300-mb isotachs (in kt) and streamlines show a powerful jet streak with peak winds above 120 kt (138 mph). The surface low was positioned in the left-exit region of this jet streak (courtesy of NOAA).

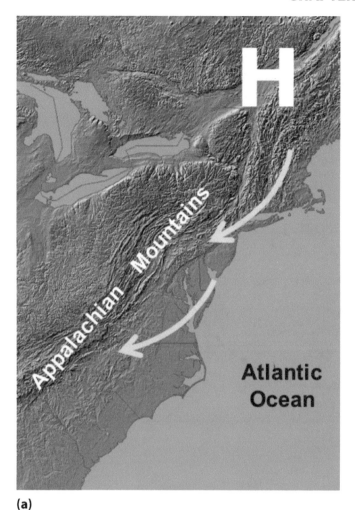

(a)

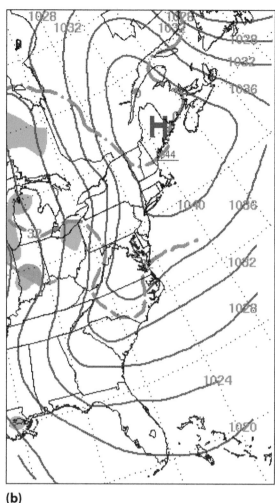

(b)

FIGURE 16.16 (a) The Appalachians play a pivotal role in cold-air damming in the East, helping to inhibit the westward progress of cold air circulating to the south of a cold high-pressure system over New England or eastern Canada (courtesy of Ray Sterner); (b) The fingerprint of cold-air damming in the pressure pattern is a wedge ridge, here seen elongating south-southwestward from the parent high in Maine. Also note the southward dip in the 32°F isotherm (dashed and blue) east of the Appalachians (courtesy of NOAA).

that meteorologists call a **wedge ridge**. Figure 16.16b shows an example of such a wedge ridge; in this case, the parent high is centered over Maine and the elongation of the isobars is very evident to the east of the crest of the Appalachians. The dashed blue contour is the 32°F isotherm—note how it dips southward, signaling the temperature footprint of the cold-air damming.

Patterns of cold-air damming increase the chance of snow and ice over the eastern foothills of the Appalachians, especially if low-pressure systems approaching from the west or south circulate relatively warm and moist air inland from the Gulf of Mexico or Atlantic Ocean, overrunning the low-level cold air dammed against the Appalachians. It's fair to say that thick clouds and precipitation actually help to maintain the cold-air damming, given that clouds limit solar heating and precipitation promotes

evaporational cooling. In fact, cold-air damming can persist several days in cloudy, damp weather patterns.

Figure 16.17a shows a classic set-up for an ice storm at 15 UTC on December 15, 2005. Prior to the ice storm, a strong high-pressure system centered over eastern Canada (not shown) promoted cold-air damming—note how the isobars bulge south-southwestward over Virginia and the Carolinas. Meanwhile, a low-pressure system approached from Alabama, setting the stage for clouds and precipitation which fell over the eastern foothills as freezing rain. The "?" on some of the station models indicates that the automated observing site could not identify the type of precipitation—to help, we've placed an asterisk at stations reporting freezing rain or "unknown precipitation." We suspect that many of

FIGURE 16.17 (a) The 15 UTC surface analysis on December 15, 2005. At the time, a wedge ridge associated with a high-pressure system centered over eastern Canada marked the footprint of cold-air damming. Overrunning clouds and precipitation associated with the approach of a low-pressure system from the south helped to maintain the cold-air damming. Asterisks mark locations reporting freezing rain or unknown precipitation (likely freezing rain); (b) The total freezing rain accumulation over the western Carolinas and northeastern Georgia from the December 15, 2005, ice storm (courtesy of the National Weather Service, Greenville-Spartanburg, SC).

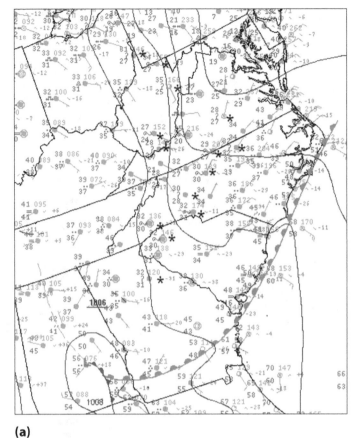

(a)

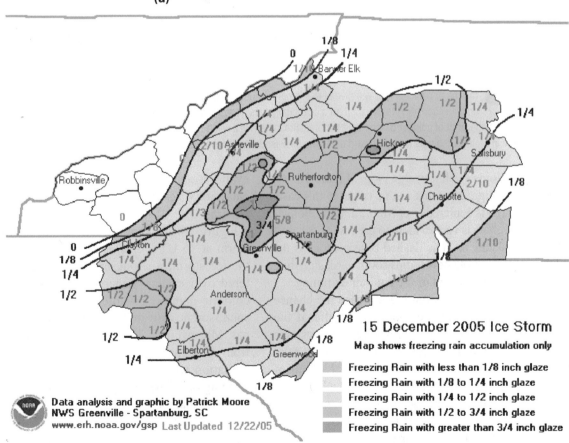

15 December 2005 Ice Storm
Map shows freezing rain accumulation only

Freezing Rain with less than 1/8 inch glaze
Freezing Rain with 1/8 to 1/4 inch glaze
Freezing Rain with 1/4 to 1/2 inch glaze
Freezing Rain with 1/2 to 3/4 inch glaze
Freezing Rain with greater than 3/4 inch glaze

Data analysis and graphic by Patrick Moore
NWS Greenville - Spartanburg, SC
www.erh.noaa.gov/gsp Last Updated 12/22/05

(b)

FIGURE 16.18 In March 2017, the roofs of three-story cabins stick out of the snow at Donner Summit, at an elevation of about 2100 m (7000 ft) in the Central Sierra Nevada Mountains. More than 400 inches of snow had fallen during the winter by this time (courtesy of Bryan Allegretto).

these reports were freezing rain, especially in light of the chart of ice accumulation shown in Figure 16.17b which focuses on the western Carolinas. Areas in blue and red experienced ice buildups of a half-inch or more. Nearly 1.5 million people from Virginia to northeastern Georgia lost power, and in some areas it took a week for the lights to come back on.

The 15 UTC surface analysis (Figure 16.17a) shows another hallmark of cold-air damming: the front draped along the Southeast Coast. Meteorologists call this feature a coastal front because it owes its existence, in part, to the natural temperature contrasts between the cold land and relatively warm sea. Also note, in Figure 16.17a, the large number of stations east of the Appalachians that reported northeast winds—the telltale sign of cold air spilling southwestward in concert with persistent cold-air damming.

There's no doubt that topography plays a leading role in the setup for cold-air damming and its associated wintry precipitation. In addition, topography is critical in Western States where a healthy winter snowpack in the Sierra Nevada Mountains is essential for providing water needs year-round (see Figure 16.18). Back in the East, topography also takes the lead in a mesoscale snowstorm that frequently occurs in late fall and early winter *after* a synoptic-scale winter storm has departed. Yes, with a cold air mass arriving and pressures on the rise, residents downwind of the Great Lakes often must brace for yet another snowstorm that sometimes exceeds the totals produced by the exiting low-pressure system. In effect, Mother Nature flips the switch that turns on the lake-effect snow machine.

Lake-Effect Snow: Your Very Own Private Snowstorm

During the period November 17-19, 2014, over five feet of snow fell just east of Buffalo, NY, but only a couple inches of snow accumulated a few miles to the north of the city (more on this historic snowstorm in just a bit). This snowstorm occurred in the wake of a cold front that was moving away from the region. But didn't we suggest earlier that the weather generally improves after the passage of a typical cold front? Indeed we did! The resolution to this apparent inconsistency lies in the infamous lake-effect snow, which reaches a crescendo from mid-November to early February as cold air charges across the relatively warm waters of the Great Lakes.

Obviously, for five feet of snow to fall near Buffalo and only a few inches to accumulate a short distance away, lake-effect snow must form in narrow bands. Sometimes, a single band develops, while at other times, multiple strips of snow form. The width of multiple bands generally ranges from 5-20 km (3-12 mi), and their length averages 20-50 km (12-30 mi). Single bands tend to have more girth, ranging from 20-50 km (12-30 mi) in width and 50-200 km (30-120 mi) in length. In light of this banded structure, it should come as no surprise that predicting the details of lake-effect snow poses a great challenge.

When multiple bands form, relatively clear spaces usually separate the bands (see Figure 16.19). As you might imagine with such an arrangement, it can snow hard under a given band, yet, only a short distance away, it might not be snowing at all! The structure

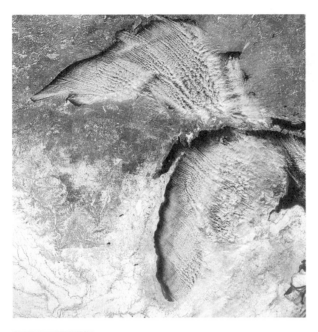

FIGURE 16.20 On a May afternoon, heating by the sun promoted widespread cumuliform clouds over land, while cooler water stabilized the lower troposphere over the lakes, and cool lake breezes suppressed cumulus clouds over surrounding lakeshore communities. Specifically note the clear areas that "shadow" Lake Huron (courtesy of NOAA).

FIGURE 16.19 A visible satellite image of Lakes Superior and Michigan on December 18, 2016, shows multiple bands of lake-effect snow that formed as cold air from the northwest flows across the relatively warm lake water (courtesy of NASA).

of these bands, which are formally known as "horizontal rolls," helps to explain the large variations in snowfall over short distances downwind of the Great Lakes in association with lake-effect snow.

The Great Lakes truly have a Jekyll-and-Hyde personality. In spring and summer, they are "islands" of tranquility because lake waters are cooler than the surrounding land. As a result, overlying low-level air is more stable, limiting convection and suppressing cumulus clouds (see Figure 16.20).

During late fall and early winter, however, the Great Lakes do an about-face. With waters much warmer than the cP or cA air masses that often flow over them, the Lakes become factories of instability that help to manufacture snowstorms with extraordinary credentials. Just ask residents of some of the areas south and east of the Lakes that average more than 250 cm (100 in) of snow per year (see Figure 16.21), most of which can be directly attributed to the lake-effect "snow machine." During a ten-day siege from February 3–12, 2007, 358 cm (141 in) of snow fell at the town of Redfield on the Tug Hill Plateau. That's almost twelve feet! Indeed, the Tug Hill, which abruptly rises about 600 m (2000 ft) east of Lake Ontario, has average annual snowfalls as high as 450 cm (180 in). For another example, travel to Upper Michigan's Keweenaw Peninsula, which juts out into Lake Superior. Near the tip of the peninsula,

FIGURE 16.21 Typical lake-effect snowbelts, shown in white, lie to the south and east of the Great Lakes.

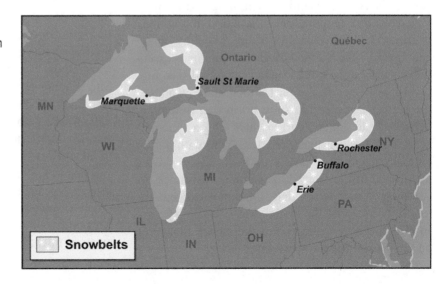

the County Road Commission maintains a "snow thermometer" (see Figure 16.22) to keep tabs on the usually heavy seasonal snowfall, most of which falls as lake-effect snow.

The fuel for the lake-effect snow machine is the large difference in temperature between the lake water and the overlying air. Cold, dry air advected from the Upper Midwest or Canada warms and moistens as it crosses the relatively tepid waters of the Lakes in late autumn and early winter, with high evaporation rates driven by large gradients in vapor pressure at the water-air interface. With invading continental polar or continental Arctic air retaining much of its chill a kilometer or two above the surface, the lower troposphere becomes unstable, setting the stage for convection and bands of cumuliform clouds that produce lake-effect snow. As a general rule, forecasters look for a temperature difference of at least 13°C (23°F) between the lake surface and the 850-mb level to signal the potential for significant lake-effect snow. Such a vertical temperature contrast represents a lapse rate close to dry adiabatic, guaranteeing the vertical overturning of air that generates roiling lake-effect clouds and squalls capable of sometimes producing 5–10 cm (2–4 in) of snow per hour. Snowfalls greater than 15 cm (about 6 in) per hour, some accompanied by lightning and thunder, have been observed in the heaviest squalls associated with lake-effect snow. Yes indeed, lake-effect snow is convective in nature.

Figure 16.23 demonstrates that the movement of air from water to land provides additional lift right at the lakeshore, where winds converge in response to the difference in friction between the underlying surfaces. Essentially, air starts to pile up along the shoreline because winds over the lake tend to blow faster than winds over the land. In other words, there's a traffic jam (of sorts) at the lakeshore, and the air rises in an attempt to lessen the congestion. The uplifting effect of this **frictional convergence**, however, generally pales in comparison to the influence of sloping terrain as air moves farther inland. For further evidence of this orographic enhancement, consider northwestern Pennsylvania, where the heaviest snowfalls usually occur inland from the lake. Erie, PA, lies along the lakeshore and averages 233 cm (91.7 in) of snow per year. About 40 km (25 mi) inland at Corry, PA, which is approximately 215 m (705 ft) higher in elevation, the annual average is closer to 330 cm (130 in), qualifying Corry as one of the snowiest communities in Pennsylvania.

FIGURE 16.22 The famous "Snow Thermometer" on U.S. Highway 41 near Copper Harbor at the tip of the Keweenaw Peninsula measures the current winter snowfall against the all-time record of 390.4 inches (992 cm) set during the winter of 1978–79. The Keweenaw Peninsula on the upper peninsula of Michigan sticks out into south-central Lake Superior. The landmark is maintained by the Keweenaw County Road Commission (courtesy of Dr. Ching-Kuang Shene, Michigan Technological University).

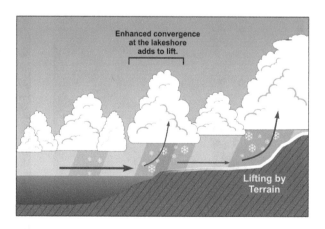

FIGURE 16.23 Lake-effect snows are typically heaviest on the upslope sides of hills and mountains, although snowfall rates can also be intense near the lakeshore, where the convergence of air is enhanced by large differences in friction between land and water.

How do forecasters predict lake-effect snow? Initially, they try to anticipate whether the developing event will feature multiple bands or a single band. Multiple bands tend to occur when low-level winds blow roughly *perpendicular* to the lake's longer axis (see Figure 16.19 and Figure 16.24). As a general rule, the bands tend to align with the average wind direction in the lower troposphere. In the case of multiple bands, the trajectory of the cold air crossing the Lake is relatively short, limiting the warming and the moistening of the air during transit (meteorologists refer to this trajectory as the **fetch**). As a result, snowfall rates are not as prodigious compared to the potential from single bands. In multiple-band events, predicting lake-effect snowfalls for specific towns is often fraught with uncertainty, given that it's next to impossible to forecast the precise location and evolution of the narrow bands.

Single bands tend to form when prevailing low-level winds blow more nearly *parallel* to the lake's longer axis. In Buffalo, NY, for example, local forecasters apply specific rules for predicting lake-effect snowfall associated with a single band. Experience dictates that a west-southwest wind (blowing from approximately 240–250°) is ideal for a single lake-effect band off Lake Erie to affect Buffalo (greater fetch means greater moistening).

During the period November 17–19, 2014, an epic lake-effect snowstorm dumped over five feet of snow in a narrow band roughly 15–20 mi (24–32 km) wide just south and east of Buffalo. Figure 16.25a is a composite radar reflectivity image at 18 UTC on November 18 that shows the single convective band that was responsible for one of the most historic snowstorms near a city known for infamous snowstorms.

Conditions were more than favorable for a notable lake-effect snow. Surface water temperatures in Lake Erie were approximately 9°C (48°F) while an unseasonably cold air mass, with 850-mb temperatures below −15°C (5°F), arrived over Lake Erie (see Figure 16.25b). With dramatically cold air moving

FIGURE 16.24 Discrete bands of lake-effect snow. Precipitating cloud bands form on the uplift side of rolls and non-precipitating dividers form on the downswing side of rolls.

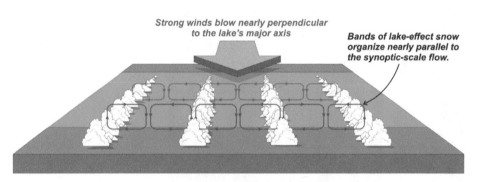

Strong winds blow nearly perpendicular to the lake's major axis

Bands of lake-effect snow organize nearly parallel to the synoptic-scale flow.

(a)

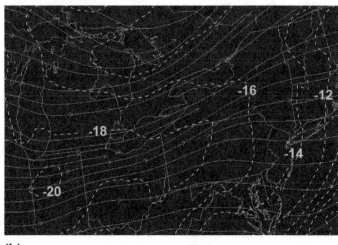

(b)

FIGURE 16.25 (a) Composite reflectivity at 18 UTC on November 18, 2014, from the radar at Buffalo, NY, showed a single band of lake-effect snow; (b) 850-mb temperatures (in °C) and streamlines at 18 UTC on November 18, 2014. Note the southwesterly wind direction, which favors heavy lake-effect show at Buffalo (courtesy of NOAA).

over relatively warm lake waters, the lower troposphere became very unstable, paving the way for convection that penetrated above 500 mb to approximately 20,000 ft (about 6000 m, high altitudes by most lake-effect standards). This extraordinarily deep lake-effect convection was accompanied by lightning, thunder, and snow squalls with snowfall rates as high as 6 inches (15 cm) per hour.

The southwesterly 850-mb streamlines shown in Figure 16.25b were indicative of the nearly perfect fetch along the lake—low-level winds blew nearly parallel to the longer axis of Lake Erie. As a result, invading low-level air had plenty of time to be dramatically warmed and moistened, setting the stage for an unwavering single band that produced prodigious snowfall on pretty much the same small area. To add insult to injury, a second lake-effect snowstorm on November 19–20 dumped several more feet of snow over practically the same area. As a result, in a space of just about four days, some places near Buffalo received more than 80 inches (about 200 cm) of snow. The second punch of heavy lake-effect snow only served to compound rescue and recovery efforts that were already underway near the end of Lake Erie's first devastating punch.

Though single bands of lake-effect snow are most common when low-level winds blow nearly parallel to a lake's longer axis, a single band can also form when winds cross more than one lake. For example, when winds blow from the north-northwest (approximately 320–330°), a single heavy lake-effect band of snow sometimes forms over Lake Huron and then extends over Lake Erie into northwestern Pennsylvania.

The Great Lakes are not the only large water bodies responsible for lake-effect precipitation. The Great Salt Lake in northern Utah produces lake-effect snow, and cold, northeast winds create "ocean-effect" snows on Cape Cod, MA, typically a few times each winter. Internationally, the lake-effect capital of the world may very well be in northern Japan. When continental polar or continental Arctic air over eastern Russia and northeastern China pours southeastward across the Sea of Japan, sea-effect snows envelope Japan's Honshu and Hokkaido islands. Tall mountains provide major-league orographic lift, especially on Hokkaido and northern Honshu, where some locations average nearly 2000 cm (787 in) of snow per year. This mechanism for ample snow undoubtedly factored in the decisions to hold the 1972 Winter Olympics in Sapporo,

on Hokkaido, and the 1998 games in Nagano, on Honshu.

Having covered lake-effect snow over the Great Lakes and East Coast winter storms, we will now shift our attention to the West Coast, where moisture plumes from the tropical and subtropical Pacific Ocean can play a pivotal role in wintertime precipitation in California and other Western States.

Winter Storms Along the West Coast: The Atmosphere Has Rivers Too

As California's rainy season began in October 2016, a large portion of the state was in the throes of a prolonged extreme and exceptional drought (see Figure 16.26a). But almost five months later, after a series of Pacific low-pressure systems brought spates of heavy rain (and flooding) during January and February, extreme and exceptional drought no longer were present in the Golden State for the first time since August 2013 (see Figure 16.26b). Needless to say, the winter of 2016–2017 was very wet by California standards.

Each bout of substantial rain that helped to ease the drought conditions was fueled by a rich supply of low-level water vapor from the tropical and subtropical eastern Pacific Ocean. For each noteworthy rain event, low-latitude moisture was rapidly transported northward toward California by strong southerly winds (more specifically, a low-level jet stream) that prevailed ahead of the cold front associated with a mature Pacific low-pressure system.

In light of what you learned in Chapter 12 about the conveyor belt model of mid-latitude lows, you might think that we're referring to the warm conveyor belt. And you'd be partly correct—we're actually talking about a narrow band *within* the warm conveyor belt. As it turns out, low-level southerly winds ahead of a cyclone's cold front sometimes organize moisture into narrow zones rich in subtropical and tropical moisture, especially when the mid-latitude cyclone is deep and thus pressure gradients are large. Meteorologists refer to such a moist, narrow band of moisture within the warm conveyor belt as an **atmospheric river** (AR, for short).

Let's quantify some of the characteristics of an atmospheric river—we'll focus on those that affect the West Coast of North America, but the characteristics of ARs are pretty much universal. As we have already mentioned, an atmospheric river is relatively narrow (compared to the wider warm conveyor), typically

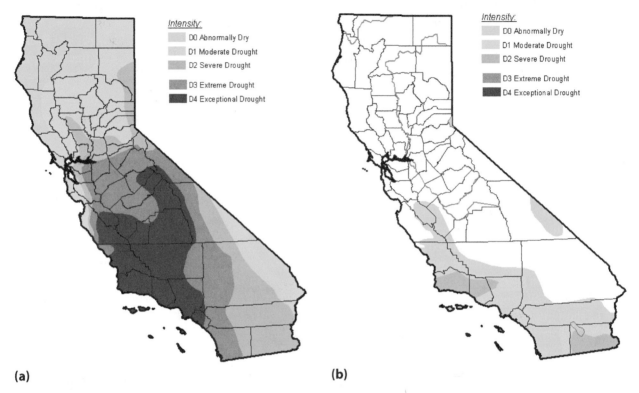

**U.S. Drought Monitor
October 4, 2016**

Intensity:
- D0 Abnormally Dry
- D1 Moderate Drought
- D2 Severe Drought
- D3 Extreme Drought
- D4 Exceptional Drought

(a)

**U.S. Drought Monitor
February 28, 2017**

Intensity:
- D0 Abnormally Dry
- D1 Moderate Drought
- D2 Severe Drought
- D3 Extreme Drought
- D4 Exceptional Drought

(b)

FIGURE 16.26 (a) Drought status in California on October 4, 2016. More than 60% of the state was in either severe, extreme, or exceptional drought; (b) Drought status in California about five months later, on February 28, 2017. By this time, only 4% of the state remained in severe drought, while extreme and exceptional drought had been wiped out (courtesy of the Drought Monitor).

about 650 km (400 mi) or less in width. As far as depth, an atmospheric river typically lies in the lowest few kilometers of the troposphere, a layer where a low-level jet stream typically would form in concert with a deep low-pressure system—the 850-mb level is a reasonable proxy for locating such a jet stream. Moreover, the layer of air in the lowest few kilometers of the troposphere over the tropical and subtropical Pacific Ocean is very rich in water vapor. That's part of the reason you shouldn't sell atmospheric rivers short just because they're not very wide. In fact, just a few ARs typically account for roughly 30–50% of the precipitation that falls along and near the West Coast of the United States during the wet season (October to March), making atmospheric rivers crucial to the water supply in the Western States.

Clearly, if atmospheric rivers are this important, it's essential to be able to locate them on weather charts. Given that ARs transport plenty of water

vapor, it may be tempting to use water vapor imagery to find them. But recall from Chapter 5 that this imagery does not routinely detect vapor in the lower troposphere, where atmospheric rivers tend to reside. Rather, informed meteorologists use analyses of a parameter known as **precipitable water** (often shortened to PWAT), which is the amount of water that would accumulate at the surface if all the water vapor in an air column were forced to condense and fall as rain. Thus, precipitable water includes all the vapor in the column, and not just what can be detected by water vapor imagery.

Figure 16.27 is an analysis of 850-mb heights and precipitable water at 12 UTC on February 20, 2017, as a potent atmospheric river was directed into central and southern California. First, note the 850-mb low just off the coast of Oregon—at this time, the surface low was very close to the 850-mb low. Next, focus your attention on the narrow band of relatively high

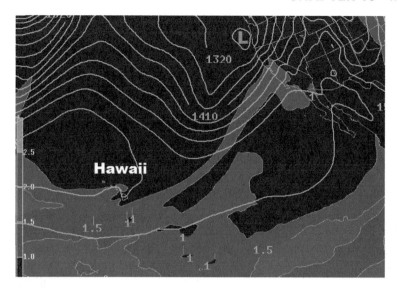

FIGURE 16.27 An analysis of 850-mb heights (in green, in m) and precipitable water (color filled, in inches) at 12 UTC on February 20, 2017. Hawaii is labeled to provide geographical reference. The narrow corridor of blue stretching from south of Hawaii toward central California was an atmospheric river.

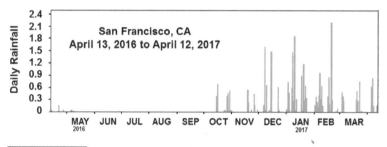

FIGURE 16.28 Daily precipitation (in inches) in San Francisco from mid-April 2016 to mid-April 2017. Each green vertical bar represents the rainfall on a particular day. Atmospheric rivers contributed to many of the high-precipitation days between October 2016 and March 2017. Note that rainfall is very unusual in San Francisco essentially from May to September (courtesy of Climate Prediction Center).

precipitable water stretching from the tropical Pacific south of Hawaii northeastward toward California (in blue, in the range 1.0–1.5 inches). This narrow band of high PWAT is an atmospheric river. To seal the deal, note the large height gradient in the vicinity of this narrow band of high precipitable water, particularly closer to the coast. This large gradient is a proxy for fast 850-mb winds—essentially, it represents a low-level jet stream. Any way you slice it, large amounts of tropical and subtropical water vapor were rapidly transported to California by this storm, and by many others during the winter of 2016–2017.

Figure 16.28 shows daily rainfall amounts in San Francisco during the year ending in mid-April 2017. By some estimates, as many 45 atmospheric rivers

affected the West Coast between October 1, 2016 and March 31, 2017, many of them contributing to precipitation in San Francisco. As a side note, Figure 16.28 also vividly illustrates how this part of California has dramatic dry and wet seasons—note the nearly complete lack of precipitation from May to September.

Although snow is rare in the city of San Francisco, it's not that unusual in some of the hills east of the city. In fact, much of the precipitation that falls in California in winter at the higher elevations falls as snow—that's particularly true in the Sierra Nevada Mountains. Even in the absence of topography, predicting snow accumulations is one of a forecaster's toughest challenges—for a variety of reasons.

Predicting Snow Accumulations: More Than Meets the Eye

Within the broad shield of stratiform (steady) snow produced by a winter storm, there are often narrow bands of heavier snow that stand out on radar imagery. These **mesoscale bands**, which are convective in nature, often produce intense snow rates of 1–2 inches (2.5–5.0 cm) per hour or greater. Figure 16.29 shows the radar reflectivity of the broad shield of snow associated with a powerful nor'easter at 09 UTC on January 27, 2015. The heaviest snow fell over eastern Long Island and southeastern Connecticut, where a slow-moving mesoscale band (indicated by yellow on the radar image) produced up to two feet of snow. Predicting where and when these bands of heavy, convective snow will set up more than a few hours in advance is a formidable forecasting challenge.

As alluded to in previous sections, the difficulties associated with forecasting snowfall can be further compounded by differences in elevation. Then there's lake-effect snow, where a few miles can sometimes mean the difference between five feet and just a few inches.

And the challenges of forecasting snowfall do not end there. Will the snow mix with, or change to, sleet, freezing rain, or just plain rain? If there is a changeover, when will it occur? Indeed, the timing of the changeover helps to determine the final tally for snowfall.

FIGURE 16.29 Radar reflectivity at 09 UTC on January 27, 2015 from the WSR-88D on Long Island, NY, showing a mesoscale snow band (indicated by yellow) within a broader area of snow. The slow-moving band produced up to two feet of snow (courtesy of National Weather Service).

Hopefully, these examples have helped you to already gain an appreciation for the challenges in predicting snowfall. That's why forecasters typically use ranges ("6 to 10 inches," for example) in order to convey the complexities and uncertainties inherent to snowfall forecasts.

Forecasters depend heavily on guidance from computer models to tailor their predictions of snowfall. The word *guidance* is important here. The models are far from perfect, but they usually give forecasters a general idea of how much liquid precipitation to expect from a storm. Forecasters then adjust the range of predicted snowfall based on topography (for example, there tends to be more snow on the windward side of mountains), and they look upstream to get an idea of how much snow the storm has already produced. When appropriate, they temper the computer forecasts based on these observations.

Experience really counts in these situations because forecasters must deal effectively with the imperfections of computer simulations that model the structure and evolution of winter storms. For example, the models can vastly underestimate the maximum snow that typically falls to the northwest of the path of the surface low (recall Figure 16.6). There, the cold conveyor belt wraps cyclonically around the low as it ascends toward the middle troposphere. An experienced forecaster takes such known computer shortcomings into consideration and adjusts the upper limit of the predicted snowfall range.

The amount of snow that accumulates during a storm also depends on temperature. You may have heard the "10 to 1" rule quoted when it comes to snowfall: In other words, ten inches of snow melts down to one inch of water. This snow-to-water ratio, however, works well only when temperatures are near 0°C (32°F). At lower temperatures, snow tends to be dry and fluffy, yielding larger ratios. For example, when surface temperatures are near –7°C (20°F), the snow-to-water ratio is often closer to 20 to 1. In such a case, forecasters expecting half an inch of liquid precipitation from a storm might call for a snow accumulation of 10 inches or so.

Then there's the issue of measuring the snow that accumulates (which, at first glance, might seem to be the simplest of all meteorological measurements). Officially, weather observers measure snowfall by using a **snowboard**. The white surface of a snowboard limits the absorption of solar radiation, which would warm the board and possibly melt some of the accumulating snow. Observers place snowboards in open areas to minimize the effects of blowing and drifting snow (around fences and houses, for example) on measurements. When windblown snow is inevitable, observers take several measurements in the vicinity of the snowboard (avoiding the largest drifts) and average them to get a representative depth. Even the average depth might be inaccurate because snow begins to compact once it settles. Also, after snowflakes reach the ground, they undergo a metamorphosis, with their building-block ice crystals shrinking inward. Both processes often lead to observers underestimating snowfall, especially when they take measurements only once a day.

Forecasters also face the challenge of managing the public's higher expectations during winter. Consider that, when meteorologists predict rain, they often don't specify how much will fall. When a snowstorm looms, however, the public expects a forecast that includes specific amounts. In reality, most people can't tell the difference between, for example, two-tenths and six-tenths of an inch of rain. But assuming a 15-to-1 snow-to-water ratio, this range converts to three to nine inches of snow. Everyone would notice this difference, and most would be upset if nine inches of snow fell when only three were predicted. An irrefutable fact of life is that the public expects specific snowfall forecasts, even though the science of predicting snow totals isn't any more accurate than predicting rain totals.

Even if forecasters accurately predict the specific snowfall at a given place, what about neighboring

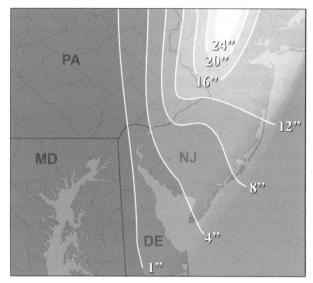

(a)

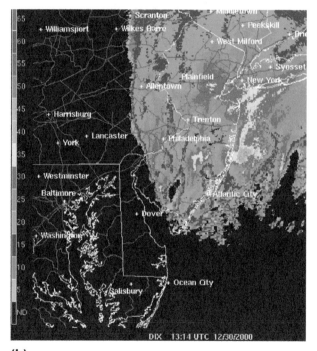

(b)

FIGURE 16.30 (a) Snowfall from the winter storm of December 30, 2000. Note the large gradient in snowfall in eastern Pennsylvania on the western side of the storm (courtesy of National Weather Service Forecast Office, Mt. Holly, NJ); (b) Doppler radar image centered on Philadelphia the morning of December 30. Predicting the sharp western edge of the snow shield was a nightmare for forecasters (courtesy of National Weather Service).

counties? Every snowstorm has an edge, and sometimes that edge is very sharp. When the snow/no-snow boundary falls in a populated area such as the Northeast urban corridor, millions of people notice. Figure 16.30a shows snowfall accumulations from the

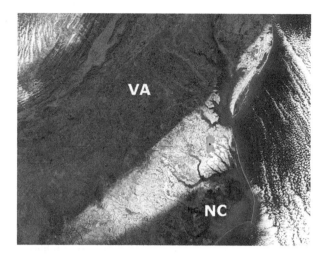

FIGURE 16.31 The narrow swath of snow left behind over the Mid-Atlantic by a storm on December 26, 2004, illustrates the difficulties forecasters face when predicting snowfall accumulations (courtesy of NOAA).

storm of December 30, 2000. Many residents of eastern Maryland and southeastern Pennsylvania awoke to a calm morning, despite predictions of a major snowstorm. As it turned out, there was a fairly sharp transition zone between heavy and light snow that ran approximately north-south through the Philadelphia metropolitan area (see Figure 16.30b): The western suburbs saw little or no snow, while parts of the eastern and northern suburbs reported more than a foot.

The sharp edge of snow cover left behind by some storms is often very evident from space the next day (when it's too late for forecasters!). Check out a classic example in Figure 16.31, which is a satellite image of the Mid-Atlantic the day after a storm on December 26, 2004, which produced a relatively narrow swath of snow from central North Carolina to the Maryland Eastern Shore. Some communities in southeastern Virginia received in excess of 30 cm (12 in) of snow, while neighboring counties to the west had much less or nothing at all.

HUMAN IMPACTS OF WINTER WEATHER: THE SEASON OF DISCONTENT

Winter weather has shaped the course of history. For example, the winter of 1941–42 was the coldest in nearly 200 years of observations over the western Soviet Union. In December 1941, the German Army had advanced toward Moscow, the Soviet capital. The city was within their binocular sights. But soon thereafter, temperatures in the field plunged to −38°C (−36°F), and blinding icy winds ripped through the inadequately dressed German ranks. A Soviet counter-offensive led by "General Winter" repelled the German offensive (in similar fashion to the 1812 retreat of Napoleon and the French from the same region). By the end of the winter, frostbite had claimed the lives of more than 100,000 German soldiers.

Today, the human impacts of winter weather are not nearly as dire as those suffered by many soldiers during World War II. Still, the need to remain aware of winter weather's threats to human safety is greater than ever.

Wind Chill: Keeping "Cool" in the Cold

During the sultry days of summer, a breeze—whether generated by nature or by a fan—brings welcome relief. That's because wind accelerates the loss of heat energy from our skin by stripping away a wafer-thin layer of warm air that normally sheaths the body. Wind also encourages the evaporation of perspiration by maintaining a relatively large vapor-pressure gradient in a thin layer of air next to our skin. As a result, evaporational cooling further enhances the cooling effects of the wind.

In winter, the cooling effect of the wind can cause discomfort that ranges from simply an annoyance to a serious health hazard. The effect of wind on comfort depends on the individual, varying with factors such as age, physical condition, and level of activity. We can estimate the wind's cooling effect on exposed human skin by the **wind-chill temperature**, more commonly called the wind-chill factor or simply the wind chill.

The first wind-chill experiments were conducted in Antarctica around 1940. Scientists exposed plastic cylinders filled with water (at various temperatures) to a variety of wind speeds and air temperatures. Then they recorded the time it took for the water

to freeze. Wind-chill values based on these experiments were used in the United States and Canada for more than half a century. But these wind-chill temperatures had no physiological basis whatsoever, and the scientific community recognized that there was plenty of room for improvement.

The National Weather Service issued a revised wind-chill index for the winter of 2001–02. The new index better incorporates a modern scientific understanding of heat transfer between skin and air in windy weather, supported by clinical trials. The updated wind-chill chart used by U.S. forecasters is shown in Figure 16.32. For most combinations of wind and cold experienced in the mid-latitudes, the new wind-chill values are higher than the old ones by anywhere from 5°F to 15°F (3°C to 8°C).

Despite the recent modification, the wind chill does not account for other meteorological factors that can affect the way you feel in cold weather, such as humidity and exposure to direct sunshine. Don't let these complicating factors persuade you to throw caution to the wind—you should still follow the general rule that frostbite can occur on exposed skin in 30 minutes or less when wind-chill values are –20°F (–29°C) or lower. As a result, forecasters routinely include low wind chills as a warning to the public to beware of the potentially dangerous combination of wind and cold.

The human body responds to changes in its environment by attempting to maintain a degree of comfort. Comfort, of course, is a qualitative feeling, but it can be roughly described in the language of accountants—in terms of surpluses and deficits. Comfort corresponds to a balance between energy production by the body (called the metabolic rate) and energy loss from the body. The former is a function of an individual's state of health, food consumption, and

FIGURE 16.32 The "revised" wind-chill values introduced by the National Weather Service in the winter of 2001–02. The various colors represent how much time is required for frostbite to develop on skin exposed to wind chills at the corresponding values (courtesy of National Weather Service).

Wind (mph)	Temperature (°F)																	
Calm	40	35	30	25	20	15	10	5	0	-5	-10	-15	-20	-25	-30	-35	-40	-45
5	36	31	25	19	13	7	1	-5	-11	-16	-22	-28	-34	-40	-46	-52	-57	-63
10	34	27	21	15	9	3	-4	-10	-16	-22	-28	-35	-41	-47	-53	-59	-66	-72
15	32	25	19	13	6	0	-7	-13	-19	-26	-32	-39	-45	-51	-58	-64	-71	-77
20	30	24	17	11	4	-2	-9	-15	-22	-29	-35	-42	-48	-55	-61	-68	-74	-81
25	29	23	16	9	3	-4	-11	-17	-24	-31	-37	-44	-51	-58	-64	-71	-78	-84
30	28	22	15	8	1	-5	-12	-19	-26	-33	-39	-46	-53	-60	-67	-73	-80	-87
35	28	21	14	7	0	-7	-14	-21	-27	-34	-41	-48	-55	-62	-69	-76	-82	-89
40	27	20	13	6	-1	-8	-15	-22	-29	-36	-43	-50	-57	-64	-71	-78	-84	-91
45	26	19	12	5	-2	-9	-16	-23	-30	-37	-44	-51	-58	-65	-72	-79	-86	-93
50	26	19	12	4	-3	-10	-17	-24	-31	-38	-45	-52	-60	-67	-74	-81	-88	-95
55	25	18	11	4	-3	-11	-18	-25	-32	-39	-46	-54	-61	-68	-75	-82	-89	-97
60	25	17	10	3	-4	-11	-19	-26	-33	-40	-48	-55	-62	-69	-76	-84	-91	-98

Frostbite Times   ☐ 30 minutes   ☐ 10 minutes   ☐ 5 minutes activity. The latter is a function of the environment, including both weather conditions (such as air temperature and wind speed) and the clothes that a person wears. When the body produces more energy than it loses, a person can become uncomfortably warm. When the body runs an energy deficit, a person can feel uncomfortably cold.

When energy loss from the body substantially exceeds energy production, internal body temperature lowers, creating a state known as hypothermia. In severe cases of hypothermia, body temperatures can fall toward a life-threatening 80°F (27°C). Hypothermia is accompanied by exhaustion and uncontrollable shivering, the latter resulting from the constriction of muscles as the body employs natural defenses to generate internal energy. Proper clothing provides resistance to the movement of air close to the body and reduces the risk of hypothermia. In contrast, contact with water increases the risk of hypothermia because water conducts energy away from the body far more efficiently than air does, so net energy loss is much more likely in wet weather.

Winter Weather Hazards: On the Road and in the Air

Perhaps the greatest menace to society in the arsenal of winter weather is freezing rain. The icy coating of glaze formed as supercooled water freezes after landing on a surface can snap power lines and trees and produce dangerously slick roads. It's no wonder that freezing rain is notorious for causing disruptions in electrical and communication services. In fact, the practice of burying communication lines grew popular after the Blizzard of 1888 (and the period of damaging freezing rain that preceded it) demolished much of the above-ground telegraph system in New York City (see Figure 16.33)

The icing of aircraft remains a hazard to aviation, though an increasingly smaller threat than in the past. In a recent report from the Federal Aviation Administration (FAA) for the period 2003–2007, icing accounted for about 4% of all weather-related aircraft accidents. Aircraft icing can occur at any time of year when an aircraft ventures into cold clouds or as rain falls on a low-flying aircraft whose fuselage temperature is lower than 0°C. When airborne, an aircraft can also accumulate ice by flying through clouds containing supercooled droplets that freeze after contact. The forward portions of the wings are one of the most susceptible parts of a plane to this kind of icing. Ice adds weight (a problem for light

aircraft), increases drag, decreases lift, and causes control problems, seriously jeopardizing the flight worthiness of the aircraft.

There are two general methods for combating aircraft icing. When icing of an aircraft occurs on the ground, de-icing solvent can be sprayed on the plane just before take-off. The solvent clings to the plane, preventing precipitation from freezing for a period of time, depending on weather conditions. In the air, many aircraft battle against icing on the wings with inflatable de-icing "boots" that are located along the front edge of the wings. When inflated or heated, these boots break up and melt ice, which is subsequently whisked off the wings (see Figure 16.34).

Driving in snow and ice is also a common source for anxiety. Until the 1960s, most areas of the United States lacked a substantial capability for clearing roads of ice and snow. Motorists relied on chains wrapped around a vehicle's tires for traction. Since then, however, a "dry pavement" policy has been in effect on interstate highways and most local roads. In addition to plows, tons of salt, ash, and chemicals have been used to keep road surfaces free of the slick hazards of winter precipitation. Salt is especially useful in snow and ice removal because it reduces the melting point of ice to around −7°C (20°F). Ash, dirt, and gravel increase traction and also contribute to melting because their albedo is lower than that of snow or ice, so they absorb more incoming solar radiation.

There are maneuvers that drivers can use to counter a skidding car, but there's nothing anyone can do once a slab of snow starts to slide down a steep mountain. During or just after a snowstorm over

FIGURE 16.33 The effects of snow and freezing rain on the campus of Princeton University in New Brunswick, NJ [about 50 km (30 mi) southwest of New York City] after the Blizzard of March 1888 (courtesy NOAA).

Weather Folklore and Commentary

Groundhog Day

Every February 2, the eyes of the nation turn to the western Pennsylvania community of Punxsutawney. Awakened from his winter slumber, a groundhog named Phil steps into the opening moments of daylight. Legend says that if Phil sees his shadow, there will be six additional weeks of winter. If he doesn't, then spring is right around the corner.

Obviously, whether Phil sees his shadow or not has no bearing on the weather in the coming weeks. Nonetheless, there is some meteorology grounded in Groundhog Day.

The tradition has its roots in the Christian calendar. February 2 falls 40 days after Christmas, and the date was set aside in the sixth century as Candlemas Day, when candles used for the rest of the year were blessed. The date came to signify the midpoint of winter by many in western Europe, and the Germans began to look to the behavior of hibernating bears to determine whether winter would continue until the spring equinox around March 20. When a group of Germans settled in Pennsylvania, they entrusted groundhogs to provide them with a forecast for the remainder of winter.

Groundhogs are true hibernants, with their core temperatures during sleep hovering a mere 0.6–1.1°C (1–2°F) above the temperature of the environment of the burrow. By contrast, bears are not true hibernants (as is popularly advertised) because they remain warm-blooded during their winter sleep (Wimps!). Though most of a groundhog's reflexes are suppressed during sleep, a groundhog retains one important reflex that can save its life—if the temperature of the burrow plummets to frigid levels capable of freezing a groundhog's tissue, the animal will awaken, aroused from slumber by some sort of natural alarm clock. In just a few hours, the awakened groundhog will regain its warm- core temperature.

Now here's the meteorology in our story. On average, surface temperatures bottom out in the latter half of January. But a groundhog's burrow is typically 1.2–1.8 meters (4–6 ft) below ground, where the temperature naturally lags behind surface air temperatures. In fact, the minimum chill of the burrow typically would not occur until sometime in February, about the time that Phil is supposed to wake up and look for his shadow.

So although you can't take Punxsutawney Phil too seriously because there's no scientific basis for his forecasts, meteorology likely does play a role in determining when Phil wakes up.

FIGURE 16.34 Ice can build up fairly quickly on an aircraft's wings during the "rest time" of the de-icing boots. This test, which simulated continuous heavy icing, was conducted in an icing tunnel. In three minutes, significant ice accumulated on the wing's leading edge. Such ice is called "intercycle ice" (courtesy of the Federal Aviation Administration).

FIGURE 16.35 A scientist from the Northwest Weather and Avalanche Center checks the structure and strength of the deep snowpack in Mount Rainier National Park in Washington on February 8, 2004 (courtesy of Tim Kirk, USDA Forest Service).

tall mountains such as the Rockies, Wasatch, or Sierra Nevada, the risk of an **avalanche**, a mass of snow cascading down a mountainside, increases dramatically (in Figure 16.35, a scientist assesses the strength of a mountain snowpack). On average, 9000 avalanches are reported in the United States each year, with an estimated 2000 avalanches in Colorado. According to the Colorado Avalanche Information Center, an average of 25 people die in avalanches annually in the United States, with two-thirds of those fatalities in Colorado, Alaska, Utah, and Montana.

Although the greatest accumulations of snow caused by orographic lift heighten the risk of avalanches on the windward side of mountains, avalanches can also form on the leeward side. Like the deposit of silt immediately downstream from a rock in a fishing stream, the air flow over mountains can pick up snow from the windward side and deposit it on the leeward side. As a result, massive accumulations of snow are sometimes found just to the lee side of mountain summits, particularly on steep slopes. An overhanging edge of snow near a summit or ridge line, called a **cornice**, is a visible sign of avalanche potential, with areas at and immediately below the edge of the cornice typically at the greatest risk.

Avalanches are generally characterized either as loose snow avalanches, which begin at a point and spread, or slab avalanches, which occur when a large, rectangular region of snow breaks free of the deeper snow and slides down the mountain. The risk of dangerous slab avalanches increases when heavy snowfall accumulates on an existing snow pack that has a solid crust of slippery ice on its surface (caused by a cycle of freezing and thawing).

POSTSCRIPT: WINTER IN THE BALANCE

Winter weather is an important cog in the machinery of the global cycles of energy and moisture. Snow and ice collectively account for approximately 80 percent of all fresh water on earth. In regional water budgets, winter weather can play a dominant role. The snow-pack over the Rocky Mountain States is a vital water resource that sustains the region's water needs for the entire year. In states such as Utah and California, water from melting snow is carefully channeled from the mountains to agricultural and urban areas. On both global and regional scales, winter precipitation is likely the source that quenches our thirst, as it moves water through the hydrologic cycle one snow crystal at a time.

Weather Folklore and Commentary

Ice Bells

On a Christmas Day several years ago, holiday-festive "ice bells" (see Figure 16.36) appeared on Spruce Creek near State College in central Pennsylvania. A very special sequence of weather conditions had to occur for these ice bells to form.

On December 23, heavy rain fell across western and central Pennsylvania as a low-pressure system passed to the west of the state, putting Pennsylvania on the "warm side" of the storm. Following on the heels of the low, an Arctic air mass arrived and temperatures started to fall noticeably on Christmas Eve.

Meanwhile, water levels on Spruce Creek crested late on December 23 after heavy rains tapered off. Thereafter, creek levels fell steadily and rather sharply on Christmas Eve night as temperatures plummeted toward −18°C (0°F). The combination of steadily falling water levels and extreme overnight cold set the stage for these incredible ice bells to form by Christmas morning (the authors believe that the ice bells formed fairly quickly overnight).

Clearly, a twig or branch needed to be partially submerged after the creek level rose in response to the heavy rains. Then, as Spruce Creek dropped, water would progressively freeze downward around the increasingly exposed branches and twigs. Interestingly, water gauges on the creek indicated that the drop in water levels slowed toward Christmas morning, allowing the fatter "bell bottoms" to take shape.

Had the precipitation been snow or light rain, the ice bells would have never formed because there would have been little or no runoff. Moreover, the heavy rains had to be followed by an Arctic air mass. This sequence of events suggests that the conditions for ice bells to form are relatively rare in central Pennsylvania.

FIGURE 16.36 Two photographs of ice bells (courtesy of Mark Wherley).

Numerical Weather Prediction

17

LEARNING OBJECTIVES

After reading this chapter, students will:

- Gain an appreciation for the fundamentals of computer modeling
- Understand the difference between the initialization time of a computer model and the time a specific forecast from the computer model is valid
- Be able to interpret each of the four panels on a standard computer prog
- Gain an appreciation for the utility of model consensus in weather forecasting
- Gain an appreciation for the utility of computer trends in weather forecasting

- Gain an appreciation for a few of the specific weather patterns that are not well predicted by computer models
- Gain an appreciation for medium-range forecasting and ensemble forecasting
- Understand that computer skill decreases dramatically several days after the model was initialized
- Gain an appreciation for the challenges of seasonal forecasting
- Understand the impact on seasonal forecasts from El Niño, La Niña and the North Atlantic Oscillation
- Be able to interpret seasonal outlooks issued by the Climate Prediction Center

We decided to publish Chapter 17 online at Penn State because we want to expose students to selected Web sites that provide real-time computer forecasts. In light of this decision, we designed some new laboratory exercises that will allow students to create their own weather forecasts for their hometown or campus. We believe that these exercises will pique student interest and make the material more meaningful and relevant. Moreover, given the continual advances in numerical weather prediction, an online chapter also affords us the opportunity to provide students and instructors with the most current computer tools and products.

You can access the online version of Chapter 17 at

http://www.e-education.psu.edu/worldofweather/

The Human Impact on Weather and Climate

18

LEARNING OBJECTIVES

After reading this chapter, students will:

- Be able to articulate the difference between the Earth's natural greenhouse effect and global warming associated with human activity
- Understand the role that the Intergovernmental Panel on Climate Change plays in compiling the science related to global warming and issuing reports that serve as the voice of scientific consensus
- Gain an appreciation for how scientists forensically reconstruct temperature trends in the distant past from ancient ice cores

- Understand how methane, nitrous oxide, and chlorofluorocarbons also contribute to an enhanced greenhouse effect
- Gain an appreciation for the complexities of global warming by recognizing the cooling effects by aerosols
- Understand the role of urbanization in the overall upward temperature trends in metropolitan areas
- Gain a basic understanding of the global climate models that predict changes in Earth's climate
- Gain an appreciation for the cycles related to the creation and depletion of stratospheric ozone
- Understand the destructive impacts of chlorofluorocarbons on concentrations of stratospheric ozone, particularly over Antarctica
- Understand the impacts of tropical deforestation on local and regional temperature and precipitation patterns

In the late nineteenth century, Charles Dudley Warner (a collaborator of Mark Twain) wrote that "Everybody talks about the weather, but nobody does anything about it." For most of human history, Warner's quip has rung true, especially as it relates to weather on short time scales and small spatial scales. For example, we cannot disarm a tornado or energize a particular cumulus cloud to produce rain on a specific plot of parched ground.

However, over the last two centuries, evidence is mounting that a burgeoning humanity, with its rapid technological developments feeding a voracious appetite for energy and materials, has become a factor in environmental change. Human population exploded in an unprecedented way during the last 200 years. It took until around 1800 for the world's population to reach one billion, but the second billion was added by 1925. Global population reached three billion by 1959, and as of late 2017 had swelled to about 7.6 billion. The world's population is expected to increase to approximately nine to ten billion by the year 2050. With this rapidly growing population comes an increasing **anthropogenic**, or human-induced, impact on our planet's weather and climate.

Figure 18.1 displays several datasets that should give you a sense for the large footprint that humanity makes on Earth. The base map is a nighttime composite visible image from space, showing lights from cities as well as from fires ablaze at the ground. These lights are powered predominantly by the burning of carbon-rich natural resources such as coal, oil, natural gas, and wood. In the process, gases and particles enter the atmosphere, changing its composition and radiative properties. The red lines indicate the routes of 87,000 daily airline flights connecting cities and cultures around the world. The blue lines mark the routes of 3500 commercial ships over the course of a year (which constitutes only 10 percent of total ocean shipping traffic). Green lines designate the world's roads, which are used by over one billion motor vehicles.

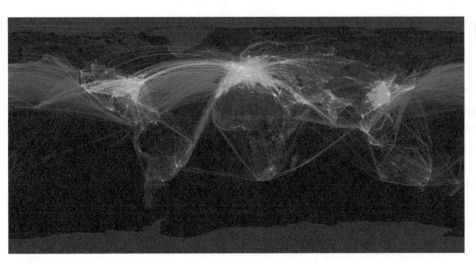

FIGURE 18.1 The footprint of humanity on Earth appears on a nighttime composite visible satellite image that captures city lights and wildfires. Superimposed on the image are the paths of 87,000 daily airline flights (red lines), the routes of 3500 commercial ships over the course of a year (blue lines), and the worlds' roads (green lines, at times masked by other lines) (courtesy of NOAA).

In total, this colorful globe shows the interconnected nature of our planet, but also hints at how much energy we use to move people and goods. At the same time, urbanization, deforestation, melting of ice at high latitudes, and other processes, have altered the planet's surface, modifying the natural exchanges of energy (and mass, such as water) between ground and air.

The discharge of a smokestack contributing to the formation of fog (see Figure 18.2a) is an unmistakable example of the human imprint on the local atmosphere. And consider the radar echo in Figure 18.2b—up to an inch of snow fell from this band in a fairly narrow swath north of Pittsburgh, PA on January 22, 2013. Not that unusual, right? Except this was manmade snow, the result of moisture and ice nuclei released from

(a)

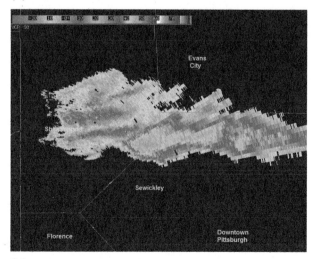

(b)

FIGURE 18.2 (a) Fog forms from the effluent of smokestacks in a central Pennsylvania valley (courtesy of Alistair Fraser); (b) Moisture and ice nuclei from the cooling towers at the Beaver Valley Nuclear Power Station in Shippingport, northwest of Pittsburgh, PA, interacting with Arctic air produced a narrow band of light snow seen here on radar at 23 UTC on January 22, 2013 (courtesy of NOAA).

cooling towers at a nearby nuclear power plant. In this chapter, we explore the extent of this human impact on weather and climate, considering global, regional, and local effects. In the process, we will travel from the tropical rain forests of the Amazon to the brutal cold of the stratosphere high above Antarctica.

THE ENHANCED GREENHOUSE EFFECT: TOO MUCH OF A GOOD THING?

In Chapter 2, we established that some atmospheric gases, primarily water vapor and carbon dioxide (CO_2), absorb much of the infrared radiation emitted by the Earth's surface. These gases represent only a small percentage of all air molecules—for example, there is so little atmospheric CO_2 that its concentration is measured in parts per million (ppm). But because of the radiative properties of these gases, Earth's average surface air temperature is about 33°C (60°F) higher than it would be if these gases were absent. This natural warming process, the greenhouse effect, has helped to keep the planet's temperature within a range conducive to life for nearly four billion years. But now, human activities are likely enhancing this natural process. This unnatural enhancement of the greenhouse effect is called **global warming**.

You could argue that it was 1988 when global warming first emerged from university seminar rooms and government reports to become a topic of public discussion and debate. In June of that year, NASA scientist James Hansen went before the Senate Committee on Energy and Natural Resources and said that he was "99 percent sure" that an enhancement of the greenhouse effect caused by human activities—that is, global warming—had been detected in the long-term temperature record. The timing was right for such a pronouncement: The thermometer read 38°C (101°F) in downtown Washington, DC, and one of the hottest and driest summers on record in the United States was in progress. Major crop areas, covering over 25 percent of the lower 48 states, had received less than half their average rainfall during the planting and growing seasons from April through June (see Figure 18.3). That summer, the Mississippi River sank to its lowest levels in over 100 years in some places, stopping barge traffic. Temperatures at cities such as Chicago, Cleveland, Pittsburgh, and Raleigh soared to all-time record levels during the summer of 1988.

Hansen's testimony received tremendous media coverage, although his statements were widely misinterpreted to mean that the record warmth and drought of 1988 could be *directly* attributed to global warming.

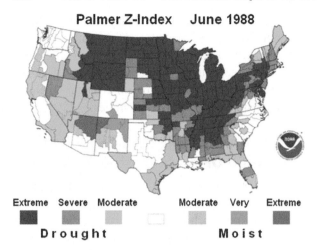

Palmer Z-Index   June 1988

Extreme Severe Moderate   Moderate Very Extreme

D r o u g h t   **M o i s t**

FIGURE 18.3 The Palmer "Z Index" for June 1988. This index measures short-term dryness (or wetness) on a monthly time scale. Based on this index, more than half the Lower 48 states, from the northern Rockies to the Deep South to the Northeast, was experiencing severe to extreme drought at this time. Wayne Palmer, a meteorologist with the U.S. Weather Bureau (now the National Weather Service), published a landmark paper about drought in 1965 in which he devised several indices that incorporated temperature and rainfall data to quantify dryness (courtesy of NOAA).

But almost overnight, the phenomenon burst into the public spotlight, and it's been there ever since.

What are the absolutes regarding global warming? What aspects of the issue are, and aren't, in dispute? Is the average surface temperature of our planet actually increasing? And if so, is the increase anthropogenic?

To help answer these questions in a scientifically rigorous way, we will frequently turn to the Intergovernmental Panel on Climate Change (IPCC, for short), a scientific body established in 1988 by the World Meteorological Organization under the auspices of the United Nations Environment Programme. The IPCC's role is to assess, on a thorough, impartial, open and transparent basis, the latest scientific and technical information produced worldwide that's relevant to anthropogenic climate change and its impacts. The IPCC does not conduct research, nor does it monitor or collect climate-related data. Rather, the IPCC bases its assessments primarily on peer-reviewed and published scientific writings. Essentially, the IPCC provides the best summary of the current science, including estimates of uncertainty. The IPCC has published five major reports, in 1990, 1995, 2001, 2007 and 2014, and also released a special report in 2018 on the impacts of a global warming of 1.5°C (2.7°F) above pre-industrial levels. Over this period, the IPCC increasingly has been acknowledged as the voice of the mainstream scientific community with regard to climate science. We will frequently reference their reports as we explore the issue of global warming in more detail.

Greenhouse Gases: A Little Goes a Long Way

The notion of an enhanced greenhouse effect is far from new. Svante Arrhenius, a Swedish scientist who won the Nobel Prize in chemistry in 1903, recognized that each year, humans were burning more wood, coal, and oil than the previous year. In 1896, he wrote: "We are evaporating our coal mines into the air." When carbon-rich fossil fuels such as coal, oil, and natural gas are burned, gaseous carbon dioxide is a by-product. Arrhenius went so far as to predict that an increase of 5–6°C (9–11°F) in earth's average temperature would accompany a doubling of atmospheric CO_2. Arrhenius's results were generally ignored—not because the greenhouse effect of CO_2 was doubted, but because most scientists believed that CO_2 was increasing at such a slow rate that it couldn't possibly lead to significant warming. In addition, most scientists at the time assumed that the excess CO_2 would dissolve in the oceans, which were known to contain 50–60 times as much carbon as the atmosphere.

More than a half-century later, however, Roger Revelle, an oceanographer and then the director of the Scripps Institution of Oceanography in La Jolla, CA, showed that the oceans actually resist taking in additional CO_2. According to his calculations, only about half the anthropogenic CO_2 would dissolve in the oceans. Revelle concluded that "human beings are now carrying out a large-scale geophysical experiment of a kind that could not have happened in the past nor be reproduced in the future."

Coincidentally, the International Geophysical Year (IGY), a period of extensive global monitoring of the atmosphere and oceans, was set to begin in 1958. Revelle, one of the IGY planners, asked Charles Keeling, a recent Ph.D. graduate from Northwestern University known for his intense interest in measuring CO_2, to organize atmospheric CO_2 measurements for the IGY. Keeling's precise measurements during 1958 at the South Pole and on Mauna Loa in Hawaii began the modern period of monitoring atmospheric CO_2. The Mauna Loa time series, shown in Figure 18.4a, is the most important documentation of recent atmospheric CO_2 trends, and has become an iconic image of the human imprint on the environment (the Mauna Loa Observatory is shown in Figure 18.4b). Atmospheric CO_2 concentrations have increased from

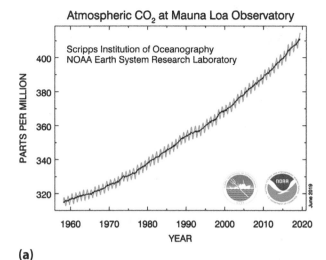

(a)

(b)

FIGURE 18.4 (a) The Keeling Curve (in red) shows average monthly atmospheric carbon dioxide concentrations, in parts per million, measured at Mauna Loa, HI, since 1958. The black curve shows the long-term trend. The wavy nature of the Keeling Curve reveals the annual cycle in CO_2 concentrations, indicating the ebb and flow of the photosynthetic activity of plants during the growing and dormant seasons. More carbon dioxide is removed from the atmosphere during Northern Hemisphere summer than during Southern Hemisphere summer because many more forests are located in the Northern Hemisphere (courtesy of NOAA); (b) The Mauna Loa Observatory (courtesy of NOAA).

about 315 ppm in 1958 to approximately 410 ppm in 2019, an increase of almost five percent per decade.

Scientists have forensically created a time series of atmospheric CO_2 concentrations that dates back hundreds of thousands of years by using less direct (and somewhat less precise) methods. For example, air comprises a small percentage of the volume of every glacier (think of the bubbles that form in ice cubes in the freezer). By removing cores of ice from the Greenland and Antarctic ice sheets (see Figure 18.5), scientists can recover "fossilized air"

(a)

(b)

FIGURE 18.5 (a) The Greenland Ice Sheet Project 2 (GISP2) drill dome obtained an ice core 3053 meters (nearly two miles) in depth, documenting over 200,000 years of Earth's climate history. The dome is connected via trenches and shafts to a subterranean network of processing and storage trenches; (b) Removing an ice core from its barrel. The surface of the ice is carefully cleaned before sawing the core into two-meter pieces. The temperature of the ice is kept at −15°C (5°F) or lower to prevent tiny cracks from developing and allowing present-day air to contaminate the fossilized air trapped in the ice (courtesy of Mark Twickler, University of New Hampshire, and NOAA).

that was trapped at the time the ice formed. These samples of bygone atmospheres show that atmospheric CO_2 levels vary naturally over time, but the beginning of the most recent increase coincides with the beginning of the Industrial Revolution about 200 years ago. Pre-industrial concentrations were stable at around 280 ppm. According to the IPCC, the current CO_2 concentration is the highest in at least the last 800,000 years.

To help explain the 30 percent increase in atmospheric CO_2 levels over the last two centuries, look no further than your oil-burning furnace or the gasoline-eater in your garage. Consider that in 1850, about 50 billion kilograms of carbon entered the atmosphere from the burning of fossil fuels: By 2014, this input had increased by a factor of nearly 200, to approximately 9800 billion kilograms, equivalent to more than one ton for each inhabitant of the planet (see Figure 18.6). The People's Republic of China overtook the United States in 2006 as the world's top producer of anthropogenic CO_2, with India, the Russian Federation, and Japan rounding out the top five countries. Together, these five nations account for slightly more than half of all global emissions. The burning of fossil fuels represents the largest human contribution to atmospheric carbon dioxide.

The amount of CO_2 input to the atmosphere from the burning of fossil fuels is known with fairly high confidence. Deforestation is another source of anthropogenic CO_2, but the magnitude of its contribution is much less certain. Wood is about 50 percent carbon, and when forests are cleared, most of their

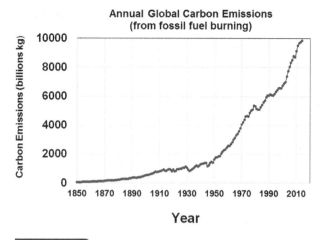

FIGURE 18.6 Annual global carbon emissions from the burning of fossil fuels since 1850, in billions of kilograms (data from http://cdiac.ornl.gov/trends/emis/meth_reg.html).

carbon is eventually returned to the atmosphere. The return rate dramatically accelerates when forests are cleared by burning, a common practice in tropical regions. Until the twentieth century, most deforestation was confined to the mid-latitudes. However, since about 1950, the focus has shifted to the lush rain forests and moist deciduous forests of tropical and subtropical South America, Southeast Asia, and Africa (see Figure 18.7). The IPCC estimates that worldwide, deforestation is responsible for as much as 30 percent of all anthropogenic emissions of carbon dioxide, though this estimate has a very large range of uncertainty.

We've seen throughout our study of meteorology that, unlike many governments, nature balances its budgets (of mass and energy). In principle, we

FIGURE 18.7 On this NASA satellite image, red dots pinpoint fires in Southeast Asia on February 28, 2017. Many of these fires are set intentionally to clear fields for spring planting (courtesy NASA/Goddard).

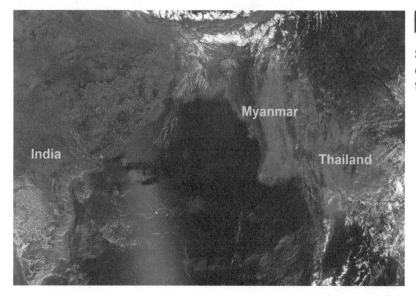

should be able to balance a global carbon budget, accounting for the inventory of our planet's carbon. The primary reservoirs are the oceans, the terrestrial surface (CO_2 is stored mainly in plants and the soil), and the geological reserves of fossil fuels. The approximate amount of carbon in these reservoirs is shown in Figure 18.8. Note that the atmosphere contains relatively little carbon when compared to the oceans and the geological reserves. But, as you might suspect, a little goes a long way.

Carbon moves naturally and continuously between the atmosphere, oceans, and terrestrial systems. For example, the atmosphere and oceans exchange CO_2 non-stop. Volcanoes belch CO_2 into the air, while plants remove it. And some rocks, in the process of weathering, chemically react with atmospheric CO_2, removing carbon from the air. The amount of carbon involved in these exchanges is also shown in Figure 18.8. We indicate the anthropogenic inputs of carbon to the atmosphere for the sake of comparison.

For the most part, the geological reservoir of carbon was out of the loop until humans began to mine and burn fossil fuels. Now, like a sleeping giant that has been stirred, this stock of coal, oil, and natural gas is active in the carbon cycle. We're burning fossil fuels a million times faster than nature created them. How is nature reacting? If we compare the estimated rate of anthropogenic CO_2 input to the air with the observed increase of atmospheric CO_2, we find that about half of humankind's contribution to atmospheric carbon has been removed from the air by the oceans and earth's terrestrial surface (plants and soil). At least for the moment, nature is undoing part of humans' CO_2 imprint on the

atmosphere. But will the oceans and the terrestrial surface continue to remove carbon at the same rate if atmospheric CO_2 concentrations continue to increase? The best available science suggests "probably not." What's clear is that the rate of input of CO_2 to the atmosphere is greater than the rate at which CO_2 is being removed, so atmospheric CO_2 concentrations are increasing.

Other Greenhouse Gases: Carbon-Copy Warming

Figure 18.9 tracks the increase in atmospheric carbon dioxide dating back to the mid-eighteenth century (the top graph, in green). However, carbon dioxide is not the only greenhouse gas that has increased during this time.

Methane (CH_4), known as marsh gas because it is a product of decay, also absorbs infrared radiation. Atmospheric levels of CH_4 have more than doubled since pre-industrial times (see the middle graph in Figure 18.9), and current concentrations far exceed the natural range of at least the last 800,000 years (as determined from ice cores). Though our quantitative understanding of the methane budget is not as comprehensive as the carbon dioxide budget, the primary sources of atmospheric CH_4 are well known. Methane is produced by the decomposition of organic matter in biological systems. For example, methane results from the decay of carbon-based garbage in landfills, and from agricultural and biological processes related to livestock digestion and rice cultivation (the production of rice worldwide has tripled in the last 50 years). In addition, methane is released during the production and distribution of fossil fuels such as coal and natural gas.

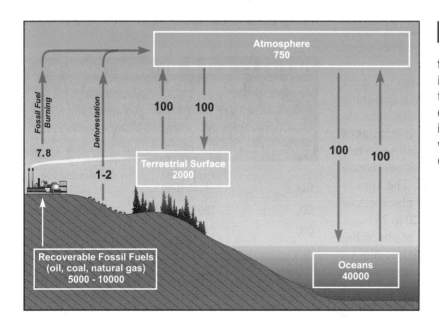

FIGURE 18.8 The principal reservoirs of carbon (in boxes) and the exchanges of carbon between reservoirs, including the anthropogenic inputs of fossil-fuel burning and deforestation. Present-day estimates of reservoir sizes are expressed in units of trillions of kilograms of carbon, while the units of the exchanges are trillions of kilograms of carbon per year.

Globally averaged greenhouse gas concentrations

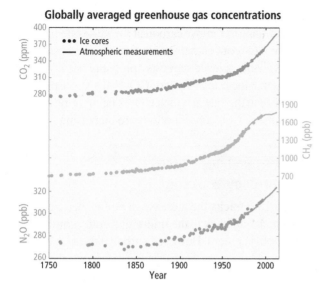

FIGURE 18.9 Observed changes in atmospheric concentrations of carbon dioxide (CO_2, green), methane (CH_4, orange), and nitrous oxide (N_2O, red) since approximately 1750. Carbon dioxide observations are in units of parts per million (ppm), while CH_4 and N_2O observations are in units of parts per billion (ppb). Data derived from ice cores are shown as dots, while direct atmospheric measurements are shown as lines (IPCC, 2014: Climate Change 2014: Synthesis Report. Contribution of Working Groups I, II and III to the Fifth Assessment Report of the Intergovernmental Panel on Climate Change [Core Writing Team, R.K. Pachauri and L.A. Meyer (eds.)]. IPCC, Geneva, Switzerland, 151 pp).

Nitrous oxide (N_2O) is another naturally occurring greenhouse gas whose concentration has increased since pre-industrial times (see bottom panel of Figure 18.9). Much of this anthropogenic increase is related to agricultural practices: Adding nitrogen to soils increases the amount of N_2O released to the atmosphere. Nitrous oxide is also released during some industrial activities and during the burning of solid waste and fossil fuels.

If you look carefully at the vertical axes in Figure 18.9, you may wonder why we're making a fuss about the potential greenhouse-enhancing ability of methane and nitrous oxide. After all, the concentrations of these gases are measured in *parts per billion*, while CO_2 concentrations are several orders of magnitude greater, measured in *parts per million*. The answer is that CH_4 and N_2O are more efficient absorbers of infrared radiation than CO_2. Methane is 20 to 25 times more effective, molecule for molecule, at absorbing infrared radiation than carbon dioxide, while nitrous oxide is 300 times more proficient. Because of this exceptional infrared-absorbing prowess, these gases are formidable players (although still minor leaguers

compared to CO_2) in enhancing the natural greenhouse effect.

So far, we've only discussed naturally-occurring gases that have increased in response to human activities. Other potent greenhouse gases, with infrared-absorbing abilities 5000 to 10000 times that of CO_2, have been created in the laboratory. For example, in 1928, researchers at General Motors were searching for a non-toxic, nonflammable refrigerant. Their work led to the development of chlorofluorocarbons (CFCs), gaseous compounds of chlorine, fluorine, and carbon that, in subsequent years, found wider use as propellants in aerosol cans and blowing agents in the production of foams such as polyurethane. Not until much later did scientists realize that CFCs were greenhouse gases (and that they threatened the ozone layer, which we will address in a later section). Fortunately, international agreements have curbed their production and use—atmospheric concentrations of two of the most common CFCs (known as CFC-11 and CFC-12) have leveled off. However, these gases (and others developed to replace them) have long atmospheric lifetimes, on the order of a few decades to a century, so their concentrations will only slowly decrease.

Table 18.1 summarizes the changes in tropospheric concentrations of the principal greenhouse gases that have increased in response to human activities. According to the IPCC, as a direct consequence of these increases, an additional 2.3±1 Watts per square meter (W/m^2) of energy now remains within the troposphere compared to pre-industrial times—this represents an increase of about one percent since pre-industrial times. With this additional energy now present in the lower atmosphere, the obvious

TABLE 18.1 Concentrations of the principal greenhouse gases that are increasing as the result of human activity. Carbon dioxide concentrations are expressed in units of parts per million (ppm), methane and nitrous oxide concentrations in units of parts per billion (ppb), and CFC concentrations in units of parts per trillion (ppt).

Gas	Pre-Industrial Concentration	Concentration (2019)
CO_2	280 ppm	410 ppm
CH_4	715 ppb	1860 ppb
N_2O	270 ppb	325 ppb
CFC-11	0	230 ppt
CFC-12	0	520 ppt

questions are: Has the average surface air temperature of the planet increased, and, if so, is the warming related to an enhancement of the greenhouse effect? Like many things in nature, the answers are not quite as simple as they appear.

Temperature Trends over the Last Century: More Ups Than Downs

Figure 18.10 shows the average annual temperature at the New York Central Park Weather Observatory since 1870, one of the longest continuous periods of record in the United States. Clearly, average temperature is highly variable from year to year, but the long-term trend in the temperature data is clearly upward, as indicated by the smoother, heavy curve: On average, temperatures are about 2.2°C (4°F) higher now in Central Park than in the late 1800s.

About 200 km (125 mi) up the Hudson River from New York City lies Albany, the state capital. Figure 18.11 shows the average annual temperature at Albany since 1900. Like New York City, average temperatures at Albany are highly variable from year to year, but, in contrast to the Big Apple, no long-term warming is evident in Albany.

We can learn several lessons from these time series. First, it's no coincidence that a large warming occurred at the weather observatory located in the heart of the most populous city in the United States (see Figure 18.12). Similar temperature increases have been observed over the last century in cities around the world. Much (if not all) of the temperature increase in Figure 18.10 can be attributed to urban-ization, the changes in land use that accompanied the build-up of the city around the weather station over the last 130 years (more on the climate of cities later in the chapter).

Second, local temperature variations are certainly important when considering the impacts of climate change on society. But imagine trying to gauge the winner of an upcoming national election by polling voters in only a few states; the degree of confidence in the results would be low, given the small sample size. Similarly, to reasonably assess global warming, a data set with enough spatial coverage to adequately sample the entire globe is necessary. But compiling

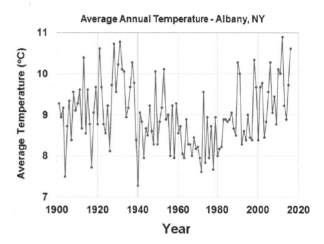

FIGURE 18.11 Average annual temperature at Albany, NY, since 1900.

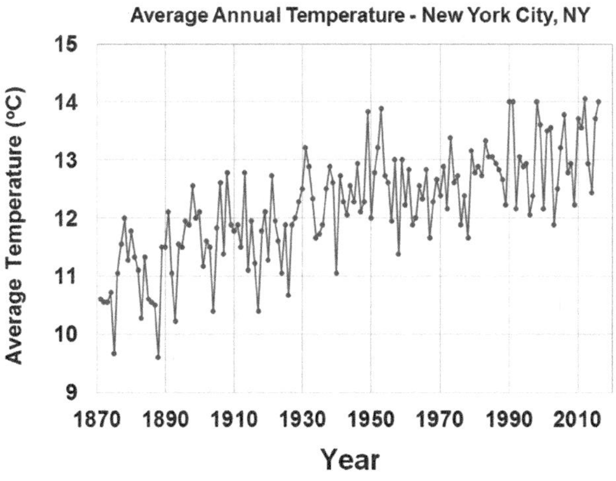

FIGURE 18.12 The urban heat island of Manhattan has grown up around Central Park, where weather observations have been taken since 1870. The increase in urbanization has contributed to a localized warming that is evident in the temperature observations (courtesy of National Weather Service, Upton, NY).

FIGURE 18.10 Average annual temperature at the weather observatory in Central Park, New York City, since 1870.

such a global data set is no small task. Reliable temperature observations are available from a large number of locations for only the last hundred years or so, and these observations are primarily in the Northern Hemisphere. Plus, 70 percent of the globe is covered by water, and relatively few observations are available over the oceans. Thus, analyses of global average surface air temperature over the last century have been culled largely from data observed on land.

Despite these difficulties, climate researchers have carefully analyzed millions of surface temperature measurements, creating time series of global average temperature back to the mid-1800s. Figure 18.13 is taken from the most recent IPCC report. The top graph shows globally averaged annual temperature anomalies (in °C) derived from observations taken over both land and oceans—the anomaly is relative to the corresponding average for the period 1986–2005. The three colors (black, blue, and orange) represent three different datasets. In the twentieth century, globally averaged temperatures generally increased to about 1940, then steadied or even

decreased slightly for several decades. Beginning in the 1970s, however, warming returned.

To get a better sense for the upturn in temperatures, check out the bottom graph in Figure 18.13 which shows the decadal averages (the gray shading indicates the extent of the uncertainty in one of the datasets). Each decade starting in the 1980s is an obvious step-up in warmth from the previous decade, and though the 2010s are not over, another step-up is very likely. Indeed, as of this writing, 2016 was the warmest year on record globally, surpassing 2015, which, in turn, had surpassed 2014. In fact, recent warming has been so dramatic that 19 of the 20 warmest years on record have occurred since 2000. Quantitatively speaking, the increase in mean global temperature from the latter half of the nineteenth century to the early years of the twenty-first century lies in the range 0.8–1.1°C (1.4–2.0°F), with most of the warming occurring since around 1980.

A couple degrees of warming in average global temperatures doesn't sound like much, especially when spread over many decades. You might be led to believe that it's no big deal. You would be incorrect. Read on.

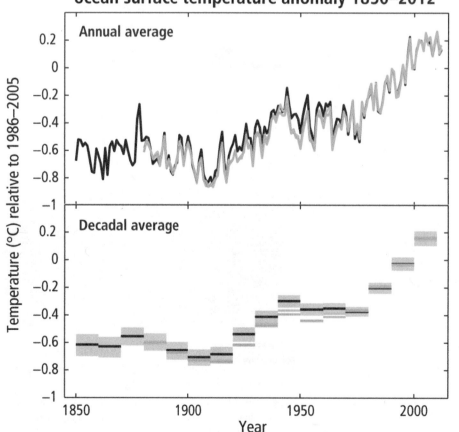

Observed globally averaged combined land and ocean surface temperature anomaly 1850–2012

FIGURE 18.13 (Top) Observed globally averaged annual surface temperature anomalies derived from observations over both land and ocean, in °C (relative to the average based on the period 1986 to 2005). The results from three different datasets are shown (black, blue and orange); (Bottom) The data from the top graph, averaged by decade (for example, the 1850s, 1860s), with the thickness of the grey shading indicating the degree of uncertainty for one of the data sets (IPCC, 2014: Climate Change 2014: Synthesis Report. Contribution of Working Groups I, II and III to the Fifth Assessment Report of the Intergovernmental Panel on Climate Change [Core Writing Team, R.K. Pachauri and L.A. Meyer (eds.)]. IPCC, Geneva, Switzerland, 151 pp).

Sea Ice, Glaciers, and Sea Level: Painting a Consistent Picture

It's important to recognize right off the bat that recent global warming has not been spread uniformly across the globe. For example, the Arctic is currently warming about twice as fast as the rest of the planet. As a result, melting of sea ice on the Arctic Ocean (as well as ice on Greenland and other high-latitude glaciers) is a footprint of the warming over the past few decades.

Figure 18.14, also taken from the most recent IPCC report, confirms a dramatic decline in the areal extent of Arctic sea ice during the months of July, August, and September, especially since around 1980. July through September is an appropriate choice for the Arctic because this period starts in the middle of meteorological summer and thus represents the prime melting season in the high latitudes of the Northern Hemisphere (we'll return to the Antarctic graph later—February is the end of meteorological summer there).

Before we launch into a discussion about melting in the Arctic, some background on freezing is in order. For starters, sea ice is simply frozen ocean water. Because of the ocean's salinity, the freezing point is lowered—for seawater, it's approximately 28–29°F (−1.6 to −2.2°C), depending on the concentration of salt. Cold water tends to sink, so a fairly deep layer of ocean water must be dramatically chilled before ice forms on the surface. Thus, most sea ice

forms over polar regions, primarily in the Arctic and Antarctic. For the record, sea ice covers, on average, roughly 25 million km² (about 9,700,000 mi²) of the Earth's surface, more than three times the area of the lower 48 states.

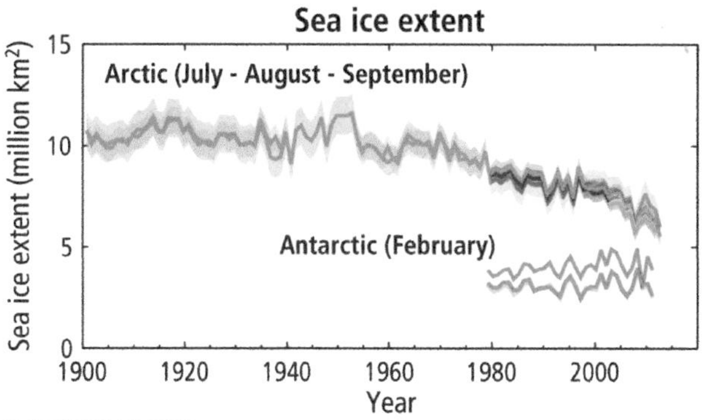

Although Figure 18.14 paints a general, visual picture, the specific statistics underlying this graph are striking. According to the IPCC, from 1979 (when satellite observations commenced) to 2012, the rate of decrease of Arctic sea ice extent was very likely about 3–4% per decade, while the decrease for the summer sea-ice minimum was very likely 9–14% per decade. And the news since 2012 is no better. Figure 18.15 shows Arctic sea ice coverage on September 20, 2016, when the sea-ice area reached its minimum that year. Note how much smaller the ice extent was on this date relative to the median for the date (which is shown in orange). This coverage tied 2007 for the second lowest minimum sea-ice extent in the observational record (September 2012 holds the record). Even more recently, Arctic sea ice extent in April 2019 (near the beginning of the melting season) was the smallest April sea ice extent since satellite records began in 1979.

Speaking of 2007, August that year marked the first time in the satellite record that the fabled **Northwest Passage** (see Figure 18.16), the short-cut navigation

Sea ice extent

FIGURE 18.14 A plot of average sea-ice extent in the Arctic (in July, August, and September) from 1900 to 2012 and in the Antarctic (in February) from 1979 to 2012, in millions of square kilometers. Note that the period of record for Antarctic sea-ice extent is much shorter than in the Arctic. Several datasets are used, which accounts for the spread during any given year (IPCC, 2014: Climate Change 2014: Synthesis Report. Contribution of Working Groups I, II and III to the Fifth Assessment Report of the Intergovernmental Panel on Climate Change [Core Writing Team, R.K. Pachauri and L.A. Meyer (eds.)]. IPCC, Geneva, Switzerland, 151 pp).

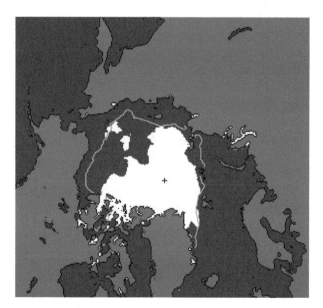

FIGURE 18.15 Extent of Arctic sea ice on September 10, 2016, when sea ice reached its seasonal minimum that year. The median sea-ice extent for this date, based on the period 1981–2010, is shown in orange. This relatively small amount of Arctic sea-ice coverage tied 2007 for the second lowest on record (courtesy of National Snow and Ice Data Center).

FIGURE 18.16 Various routes for the Northwest Passage. Once a fabled shortcut between the Northwest Atlantic Ocean and the Pacific Ocean, decreasing sea ice in the Arctic in August and September is transforming the Northwest Passage into more of a viable, late-summer route (courtesy of NOAA).

route from the Atlantic to the Pacific through the Canadian Arctic, was ice-free. This route, if navigable, would dramatically cut time and distance for ships traveling between the Atlantic and Pacific Oceans. Ordinarily, ships must use the Panama Canal or circumnavigate the southern tip of South America in order to cross from one ocean basin to the other.

Meanwhile, the ice sheet on Greenland is melting as well. By definition, an **ice sheet** is a mass of glacial land ice that covers at least 50,000 km^2 (about 20,000 mi^2). In short, ice sheets are simply very large glaciers. Not surprisingly, there are only two ice sheets on our planet: Greenland and Antarctica. Ice sheets formed on these cold land masses as winter snow didn't completely melt over the following summer. Over the course of thousands of years, layers of snow piled up into increasingly thick masses of ice. The mass of Greenland ice has steadily declined since at least 2002 (when satellite measurements began). Figure 18.17 is a plot of the mass of ice on Greenland for each month starting in April 2002. Although the melting is not uniform each year, and a seasonal cycle is clearly visible, the long-term trend is obvious.

Finally, the pronounced warming of the Arctic has also taken its toll on high-latitude **glaciers** (formally,

a glacier is an old mass of ice over land that's smaller than an ice sheet). Figure 18.18 shows Alaska's Muir Glacier in 1941 (left) and 2004 (right)—over this time, the glacier receded about 12 km (7 mi). Located at 59°N latitude in Glacier Bay National Park near Alaska's south-central coast, Muir Glacier was once classified as a tidewater glacier because it emptied into the Gulf of Alaska. Now it's labeled as a valley glacier because its terminus lies far inland from the sea (though its meltwater still drains into the Gulf of Alaska).

The melting of the Greenland ice sheet and high-latitude glaciers such as Muir (and glaciers at lower latitudes as well) have contributed to a rise in sea level. Before we delve into the topic of sea-level rise, we must address the ice on and around the continent of Antarctica.

As mentioned earlier, the other large ice sheet on Earth lies on Antarctica. The Antarctic Ice Sheet extends almost 14 million km^2 (5.4 million mi^2), roughly the combined area of the contiguous United States and Mexico. By comparison, the Greenland Ice Sheet extends about 1.7 million km^2 (656,000 mi^2).

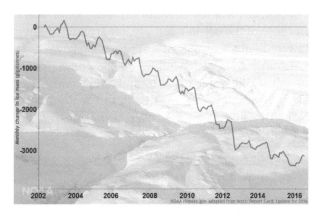

FIGURE 18.17 A plot of the mass of the Greenland ice sheet since April 2002 (when satellite measurements began) shows a steady decline. In this chart, "0" represents the ice mass as of April 2001. The unit of mass is billions of metric tons, where one metric ton equals 1000 kg (courtesy of climate.gov, data provided by Marco Tedesco/Lamont-Doherty).

FIGURE 18.18 Muir Glacier. (Left) In 1941; (Right) In 2004 (courtesy of National Snow and Ice Data Center Glacier Photo Collection).

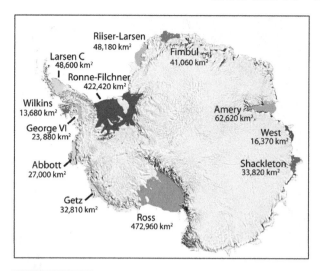

FIGURE 18.19 The ice shelves of Antarctica and their size, in square kilometers (courtesy of National Snow and Ice Data Center).

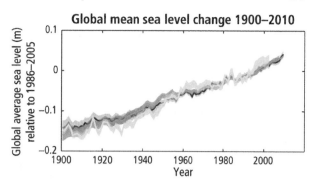

FIGURE 18.20 A compilation of various sea-level datasets shows that global sea level has, on average, increased about 0.2 m (200 mm = almost 8 inches) since around 1900 (IPCC, 2014: Climate Change 2014: Synthesis Report. Contribution of Working Groups I, II and III to the Fifth Assessment Report of the Intergovernmental Panel on Climate Change [Core Writing Team, R.K. Pachauri and L.A. Meyer (eds.)]. IPCC, Geneva, Switzerland, 151 pp).

And the Antarctic Ice Sheet is a real heavyweight, holding 30 million km³ (7.2 million mi³) of ice. Situated around Antarctica are **ice shelves**, expansive, floating slabs of ice that connect to a land mass (in this case, Antarctica—see Figure 18.19). They formed when glacial ice on Antarctica oozed into the ocean and floated, while glacial ice flowing slowly toward the sea filled in the gaps behind them.

Take a peek back to Figure 18.14 and focus your attention on the plot of Antarctic ice extent (in this case, in February, the last month of meteorological summer in the Southern Hemisphere). First, note that observations of Antarctic ice extent have not fluctuated very much since observations began around 1980—arguably, there's even been a slight increase since around 2000, in stark contrast to the trend in Arctic sea ice. Why the difference?

Let's start with the fact that that the Antarctic is a continent surrounded by ocean, while the Arctic Ocean is surrounded by land. Thus, we should not expect Antarctic ice to behave exactly the same as sea ice over the Arctic Ocean (in case you're wondering, sea ice forms in the Antarctic, but it pretty much comes and goes with the seasons). The lack of a significant long-term trend in Antarctic ice extent in Figure 18.14 has everything to do with the ice shelves. When persistent winds blow off Antarctica, ice shelves can grow in response (though it's a relatively slow process). In fact, the slight increase in Antarctic ice extent since the turn of the century is likely attributable to growth in the Ross Ice Shelf (revisit Figure 18.19) which experienced unusually

persistent offshore winds during the period starting in 2000. In recent years, September 2014 actually set a record for maximum Antarctic ice extent, though keep in mind that the data record since 1980 indicates that short-term changes to Antarctic ice extent are not very large (in the grand scheme of things).

The story of Antarctic's ice has a twist, however. In the last few decades, scientists have observed a series of ice shelf collapses in Antarctica that have occurred much more rapidly than would generally be expected. For example, since 1995, the Larsen Ice Shelf (again, revisit Figure 18.19) has lost more than 75% of its former area. The fear is that the loss of ice shelves in Antarctic will set the stage for glacial ice over the continent to start flowing unabated to the ocean where it will drift away and eventually melt in warmer ocean waters. Although the melting of ice shelves will not raise sea levels (because the ice was already floating on water), a substantial rise in sea level will occur if a lot of glacial ice over Antarctica has an unobstructed route to the sea.

We close this section by exploring global trends in sea level. Following the end of the last ice age about 20,000 years ago, global sea level rose by approximately 120 m (about 400 ft) over the next 15,000 years or so. By all indicators, global sea level then changed little until the late nineteenth century when evidence from the instrumental record suggests the start of sea-level rise. Figure 18.20 is a compilation of several sea-level datasets in different colors—shading, where shown, represents uncertainties. Observations of sea level from satellites, which began in 1993 (red line), have nearly global coverage and provide the most accurate assessment—sea level has been rising at a rate of about 3 mm per year since then, a combination of ocean thermal expansion (water expands when warmed) and the loss of ice from glaciers and the Greenland ice sheet.

Although the uncertainties associated with the details of decreasing sea-ice extent and rising sea levels might give you pause, few climate scientists argue with the trends. But do these changes result from an enhancement of the greenhouse effect caused by human activities? In other words, are we seeing the effects of global warming? Looking at the question another way: If it's not global warming, then what's responsible?

Global Warming: A Result of the Enhanced Greenhouse?

Some of the weather-observing stations used to create the time series in Figure 18.13 have, like New York City's Central Park, been affected by urbanization. To climatologists searching for evidence of warming caused by an enhanced greenhouse effect, such urban-induced temperature increases just confuse the issue. By avoiding stations that may have been contaminated by urbanization, and using idealized mathematical schemes to artificially remove the footprint of urban-induced warming from the data, we're still left with the upward trend in temperature shown in Figure 18.13.

Certainly, the warming capability of an enhanced greenhouse effect is not in question. Venus, Earth's nearest planetary neighbor, provides indisputable evidence. Because Venus is closer to the Sun, about twice as much solar radiation (per unit area) arrives at the top of Venus' atmosphere compared to Earth's. Is this the reason that Venus' average surface temperature is a toasty 455°C (850°F)? Not even close. Venus is shrouded in dense clouds (see Figure 18.21) that back-scatter 80 percent of incoming solar radiation, so Venus' atmosphere and surface actually absorb *less* solar energy (per unit area) than Earth's atmosphere and surface. In fact, based solely on the amount of solar energy Venus absorbs, its average surface temperature should be –46°C (–50°F). But surface air pressures on Venus are about 90 times greater than those on Earth, and most of Venus' atmosphere is carbon dioxide. Venus is hotter than an oven not because of its proximity to the Sun, but because of a major-league greenhouse effect that delivers about 500°C (900°F) worth of warming.

Earth's past also provides clues to the potential warming associated with an enhanced greenhouse

FIGURE 18.21 Venus is shrouded in clouds and a dense carbon-dioxide atmosphere that supports a "runaway" greenhouse effect (courtesy of NASA).

effect. Over time scales of tens of thousands of years, variations in atmospheric concentrations of CO_2 and variations in global temperature have pretty much mirrored one another. How do we know this? In addition to providing samples of ancient atmospheres, ice cores also carry information about the air temperature at the time the ice formed. Figure 18.22 shows atmospheric CO_2 concentrations (top) and surface air temperature (bottom) over the last 160,000 years (estimated from an ice core extracted at Vostok, Antarctica). Given that the fluctuations in CO_2 concentrations roughly mirrored the variations in temperature, it almost seems like an open-and-shut case. But because centuries of information are crammed into thin slices of the ice core, it is difficult to say whether the variations in CO_2 caused the temperature changes, or if natural forces caused CO_2 variations in response to temperature changes. Regardless of which comes first, the two seem to go hand-in-hand.

These ice cores clearly show that temperature is highly variable over long time scales. This natural variability makes detecting global warming a tricky business: Global temperatures have always seesawed for natural reasons, long before our species ever roamed the planet. However, since the end of the last Ice Age about 10,000 years ago, average global surface temperatures have likely varied only about 1.7°C (3.0°F)

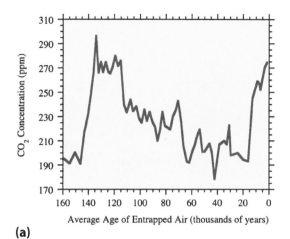

(a)

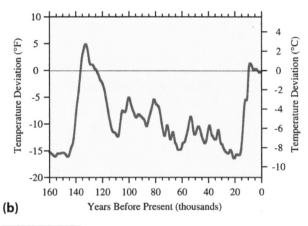

(b)

FIGURE 18.22 (a) Atmospheric CO_2 concentrations (in parts per million) over the past 160,000 years, estimated from an ice core extracted at Vostok, Antarctica. The CO_2 concentrations are derived from air samples trapped in the ice (adapted from J.M. Barnola, D. Raynaud, C. Lorius, and Y.S. Korotkevich, 1994. Historical CO_2 record from the Vostok ice core, p. 9. In T.A. Boden, D.P. Kaiser, R.J. Sepanski, and F.W. Stoss (eds.), *Trends '93: A Compendium of Data on Global Change*.); (b) Atmospheric temperatures over the past 160,000 years, estimated from an ice core extracted at Vostok, Antarctica. The temperature values are deviations from the present-day mean temperature at Vostok. Thus, negative values indicate times colder than present-day (adapted from J. Jouzel, C. Lorius, J.R. Petit, N.I. Barkov, and V.M. Kotlyakov, 1994. Vostok isotopic temperature record, pp. 593–601. In T.A. Boden, D.P. Kaiser, R.J. Sepanski, and F.W. Stoss (eds.), *Trends '93: A Compendium of Data on Global Change*).

from highest to lowest. So, relative to the past few millennia, the much more rapid temperature increase of 0.8–1.1°C since the late 19th century really stands out: It will acquire even greater significance if the rate of warming during the last few decades continues.

The flip side of the natural variability argument is that Mother Nature could be in a cooling pattern right now, offsetting some (or all) of the warming from an enhanced greenhouse effect. Ironically, human activities may actually be responsible for some unnatural cooling. How can this be? In addition to producing CO_2, many industrial processes also produce gases containing sulfur, particularly sulfur dioxide (SO_2), the principal component in acid rain. On average, sulfur pollution in the troposphere has increased in step with greenhouse gas emissions. In the air, SO_2 is converted into tiny solid particles or liquid droplets called **aerosols**, which can serve as cloud condensation nuclei. As such, they help to increase the amount of clouds and thus the amount of solar radiation back-scattered to space. There is great uncertainty in the magnitude of this aerosol effect. The IPCC estimates that since 1951, aerosols likely cooled the globe by a few tenths of a degree Celsius. Thus, ironically, it's likely that some of the warming caused by human activities is being offset by cooling caused by human activities.

Mount Pinatubo's eruption in the Philippines in June 1991 gave scientists a first-hand look at the effect of sulfate aerosols on climate. Whereas sulfur emissions from industrial activities remain mostly in the troposphere (where the sulfur is removed quickly by settling or precipitation), powerful volcanoes often eject sulfur gases into the stable stratosphere, where they can remain for months or longer. Pinatubo's contribution to the atmosphere was about 20 billion kilograms of sulfur dioxide, some of which combined with water vapor in the stratosphere to form a haze of sulfuric acid droplets that slowly dispersed around the globe during the months after the eruption (see Figure 18.23).

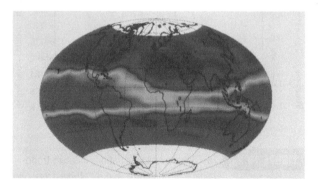

FIGURE 18.23 A NASA satellite captured a veil of stratospheric aerosols ringing the tropics during the six weeks after Mount Pinatubo's major eruption on June 15, 1991. Yellow, orange, and red mark the regions where the backscattering of solar radiation from aerosols increased by a factor of approximately 50 to 100 (courtesy of Pat McCormick, NASA Langley Research Center).

This veil of aerosols increased the back-scattering of solar radiation, reducing the amount of solar energy that reached the earth's surface. This increased scattering at least partially explains why global average surface temperatures decreased about 0.6°C (1.1°F) following the eruption (see Figure 18.24a). At the same time, the sulfuric acid haze increased the absorption of solar radiation in the stratosphere, so stratospheric temperatures increased (see Figure 18.24b).

In contrast to Pinatubo's plume, the canopy of debris produced by the eruption of Mount St. Helens in Washington State in May 1980 was many times smaller and confined mainly to the troposphere. This is true of most volcanic eruptions, including that of

FIGURE 18.25 The ash plume spewed by Mount Cleveland in the central Aleutian Islands of Alaska, on May 23, 2006, as viewed from the International Space Station. The plume moved west-southwestward from the volcano's summit. The Alaskan Volcano Observatory reported that, like most volcanic ash plumes, this debris cloud remained in the troposphere, penetrating only to about 6000 meters (20,000 ft) (courtesy of NASA).

Mount Cleveland on Alaska's Aleutian Islands in May 2006 (see Figure 18.25). In these cases, the cloud of ash and dust settles out quickly, with only short-term regional effects on weather. Similarly, the smoke clouds from the Kuwaiti oil fires following the Persian Gulf War in 1991 also remained mostly in the lower atmosphere, quelling fears that the soot would find a longer-term home in the stratosphere and consequently impact global climate.

Volcanic eruptions serve as a reminder that nature is wrapped in many layers of complexity. Almost without exception, a complicated phenomenon cannot be explained by invoking just one process. And so it is with the changes in global average surface temperatures over the last century: A number of factors likely produced these variations, including (but not limited to) enhanced greenhouse warming and aerosol cooling. But the fact remains: Humans have substantially altered the composition of the atmosphere, and continue to change its make-up at a rate much greater than natural variability. Although we cannot point our fingers at the enhancement of the greenhouse effect as the sole cause of the global temperature increase of the last century, it, at the very least, must be a significant part of the explanation. We can trace the evolution of the IPCC's take on this issue using excerpts from their five major reports:

1990 (First Assessment Report): "Our judgment is that . . . the size of this warming is broadly consistent with predictions of climate models, but it is also of the same magnitude as natural climate variability. Thus,

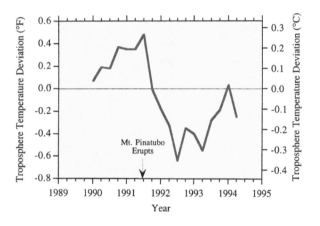

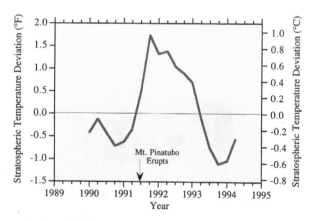

FIGURE 18.24 (a) Global average variations in lower tropospheric temperature, and (b) global average variations in lower stratospheric temperature, in the time surrounding the eruption of Mount Pinatubo in June 1991. The values plotted are deviations from seasonal averages computed from the period 1982–91 (adapted from R.W. Spencer and J.R. Christy, 1994. Global and hemispheric tropospheric and stratospheric temperature anomalies from satellite records, p. 633. In T.A. Boden, D.P. Kaiser, R.J. Sepanski, and F.W. Stoss (eds.), Trends '93: A Compendium of Data on Global Change).

the observed increase could be largely due to this natural variability."

1995 (Second Assessment Report): "The balance of evidence suggests a discernible human influence on global climate."

2001 (Third Assessment Report): "There is new and stronger evidence that most of the warming observed over the last 50 years is attributable to human activities."

2007 (Fourth Assessment Report): "Warming of the climate system is unequivocal, as is now evident from observations of increases in global average air and ocean temperatures, widespread melting of snow and ice, and rising global sea level . . . Most of the observed increase in globally averaged temperatures since the mid-20th century is very likely (at least 90 percent certain) due to the observed increase in anthropogenic greenhouse gas concentrations."

2014 (Fifth Assessment Report): "It is extremely likely that more than half of the observed increase in global average surface temperature from 1951 to 2010 was caused by the anthropogenic increase in greenhouse gas concentrations and other anthropogenic forcings together."

Clearly, over time, the IPCC has expressed increasing confidence that human activity has already contributed appreciably to global climate change. One reason for the increased confidence is that researchers attempting to simulate global and regional climate over the last century cannot reproduce the observed temperature changes without including anthropogenic influences (for example, increasing carbon dioxide concentrations). These computer simulations rely on sophisticated physical and mathematical models that virtually represent the behavior of the atmosphere. How do these climate models work? Are they reliable, and what do they project for the future? Read on.

Future Trends: Modeling into the Unknown

Projections of future climate change are sensitive to the atmospheric concentrations of the principal greenhouse gases. Will carbon dioxide concentrations (and those of other greenhouse gases) stabilize at early 21st-century levels? Or will emissions progress at a "business as usual" pace that continues the rate of increase during the last few decades (see Figure 18.6)?

In its 2014 assessment, the IPCC considered four different 21st-century scenarios for changes in atmospheric greenhouse gas concentrations, air pollution emissions, and land use (for example, from deforestation)—the IPCC called these scenarios RCPs, for "Representative Concentration Pathways." At one end of the RCP spectrum is an unlikely scenario

(called RCP 2.6) in which carbon dioxide emissions return to pre-1950 levels by the end of the century. At the other end is a high greenhouse gas emission scenario (called RCP 8.5) in which CO_2 levels approach or exceed 1000 ppm by the year 2100 (for perspective, the current CO_2 level is about 405 ppm). Two intermediate scenarios (RCP 4.5 and RCP 6.0) fall between these extremes. For the record, the numbers following "RCP" correspond to the expected changes in radiative forcing (relative to pre-industrial values, in W/m^2) resulting from the higher CO_2 concentrations associated with that scenario.

For the most optimistic scenario (RCP 2.6), global mean surface temperatures likely increase by an additional 0.3–1.7°C (0.5–3.1°F) by the year 2100. Intermediate scenarios produce a warming of 1.1–3.1°C (2.0–5.6°F), while RCP 8.5 (which features the fastest growth in greenhouse gas emissions) yields a warming of 2.6–4.8°C (4.8–8.6°F) by century's end. The accompanying sea-level rise for RCP 2.6 is estimated at 0.26–0.55 m (10–22 inches), for the intermediate scenarios 0.32–0.63 m (13–25 inches), and for the high CO_2 emission scenario 0.45–0.82 (18–32 inches).

Are these predictions far into the future reliable? How are these climate predictions produced? And why is there uncertainty in the magnitude of the predicted global warming?

Without the benefit of a twin Earth on which to experiment, climate researchers turn to the next best alternative: computer models. Climate modelers call their simulations **general circulation models** (GCMs). The rules of the climate game are the same fundamental mathematical equations used to predict short-term weather changes, but the equations are modified considerably to focus on the processes that govern longer-term climate variations.

At this point, we'd like to address a common misconception about weather and climate models that's often stated this way: If computer weather forecasts start to lose skill after just a few days into the future, how can we expect to accurately predict the climate in 100 years? Great question! The answer begins by realizing that predicting weather and predicting climate are very different problems. The ability of computer guidance to accurate predict the weather depends, in large part, on getting the initial state of the atmosphere correct. Tiny errors in the initial state – errors so small that we really can't measure them – lead to steadily declining skill (with time) in computer weather forecasts. But when we predict climate, we're not trying to forecast details of the weather in 100 years (for example, daily high temperatures and daily precipitation). Rather, we're trying to extract the statistics of climate (such as long-term averages and variability) as the Earth-atmosphere system responds to changes in CO_2 concentrations, ocean temperatures, and other factors. The bottom line is that those who cast doubt on climate models because numerical weather forecasts lose skill after several days don't understand the difference between predicting weather and predicting climate.

There are other significant differences between GCMs and the models used for day-to-day weather forecasting. By necessity, GCMs must be global in scale. In addition, a realistic climate model must include representations of the underlying land, water, and ice-covered portions of the globe, since the atmosphere actively exchanges mass (such as water and CO_2) and energy with the earth's surface. Both exchanges of mass and energy require individual model packages that have their own set of specialized mathematical rules.

The most complex climate models include three-dimensional representations of the oceans and mathematical schemes that track carbon and other atmospheric constituents as they cycle between the mass and energy model packages. These full-blown coupled atmosphere-ocean GCMs (AOGCMs) are so complicated that only the world's fastest supercomputers can handle them, yet even the most complex AOGCM is still a highly simplified version of the real-life climate system. The evolution of the complexity of GCMs from the 1970s to the time of the IPCC's Fourth Assessment Report (2007) is represented schematically in Figure 18.26. The primary

The World in Global Climate Models

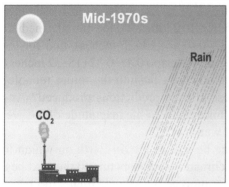
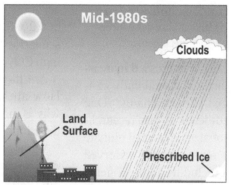

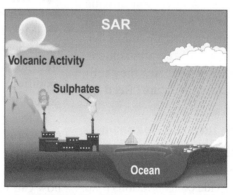
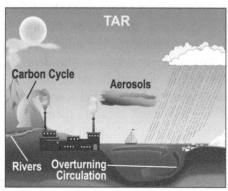
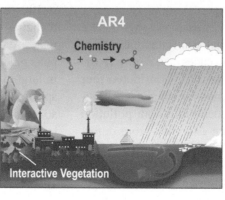

FIGURE 18.26 The complexity of global climate models has increased dramatically over the last few decades. The additional physics incorporated in the models are shown pictorially by the different features of the modeled world. FAR (First Assessment Report), SAR (Second Assessment Report), TAR (Third Assessment Report), and AR4 (Fourth Assessment Report) refer to the major IPCC reports issued in 1990, 1995, 2001, and 2007, respectively (Climate Change 2007: The Physical Science Basis, Contribution of Working Group I to the Fourth Assessment Report of the Intergovernmental Panel on Climate Change [Solomon, S., D. Qin, M. Manning, Z. Chen, M. Marquis, K.B. Averyt, M. Tignor and H.L. Miller]).

improvements in GCMs since 2007 have come in the ability to simulate ocean thermal expansion, glaciers, and ice sheets. Significant challenges still remain in simulating the behavior of the Greenland and Antarctic ice sheets, but projections of sea levels have improved nonetheless as climate science and technology advance.

Two components of the climate system—oceans and clouds—are critical pacemakers of global warming, both in the real atmosphere and in the twin earths created in AOGCMs. The oceans are critical because they are huge reservoirs of energy and carbon dioxide; Plus, compared to air, water has a large heat capacity. If the energy required to warm the global atmosphere by an average of 0.8–1.1°C (1.4–2.0°F) were mixed uniformly throughout the oceans, the increase in water temperature would be undetectable. Thus, the oceans, by virtue of their large capacity to store energy, are likely mitigating enhanced greenhouse warming in the atmosphere. The fifth IPCC report stated that, with at least 99% certainty, the heat content of the upper 700 meters (about 2300 ft) of the ocean increased during the period 1971–2010, when the oceans were more closely monitored (compared to previous times).

The oceans contain more than 50 times as much carbon as the atmosphere (recall Figure 18.8). Only about half the anthropogenic carbon dioxide released in recent decades is still in the atmosphere. Researchers believe that about half of the missing CO_2 dissolved in the surface waters of the oceans (the biosphere took in the rest). But, as CO_2 concentrations in surface waters increase, the ocean's ability to take in more CO_2 decreases. There is still a way, however, to get more CO_2 into the oceans. Surface water mixes with deeper water, which contains less anthropogenic CO_2. As a result, CO_2 concentrations decrease near the sea surface. However, the mixing process is very slow, occurring over time scales of decades or centuries. Thus, because of their huge inventory of CO_2 and their large capacity to store energy, the oceans will likely play a key role in defining the speed at which greenhouse-enhanced warming manifests itself in the atmosphere.

Clouds are the other major wild card in a greenhouse-enhanced world. Radiatively, clouds have two faces. They back-scatter incoming solar radiation, while simultaneously absorbing outbound terrestrial radiation (and emitting some of their own). Thus, clouds help to both cool and warm the planet. The process that dominates depends largely on the type of cloud. Thin ice clouds, such as high, wispy cirrus, are more efficient absorbers of infrared radiation than back-scatterers of visible radiation. But thicker, low stratus and puffy cumulus clouds, which are mostly composed of liquid water drops, are more efficient at back-scattering incoming solar radiation than absorbing outgoing infrared.

At the present time, averaged over the globe, the amount of solar radiation back-scattered by clouds is larger than the amount of infrared radiation they absorb, so clouds exert a net cooling on the planet. That is, without clouds, the average global surface temperature would be higher. But if global air temperatures increase, evaporation from the warmer oceans is also expected to increase, leading to more atmospheric water vapor and thus probably more clouds. Will an increase in water vapor, the most potent greenhouse gas of all, nullify or amplify the enhanced greenhouse? And will a warmer, cloudier planet have more ice clouds or more water clouds? We're just not sure.

With these uncertainties in mind, the IPCC has categorized potential impacts of future climate change in terms of probabilities as a way to express its confidence in a possible outcome. For example, an increase in high temperature extremes over most land areas on daily and seasonal timescales (and a decrease in low temperature extremes) is in the "virtually certain" category (at least 99% certainty), while heat waves occurring with higher frequency and longer duration is "very likely" (at least 90% certainty). An increase in the frequency and intensity of heavy precipitation events over most mid-latitude land masses and over wet tropical regions is also considered "very likely." However, changes in precipitation will not be uniform in a warming world. In fact, it's even possible that some areas could experience more frequent droughts.

This last statement might have you scratching your head. How could both the number of heavy precipitation events and the number of droughts increase? To explain this apparent contradiction, consider that higher average temperatures mean more evaporation, and thus more water vapor available for storms. As a result, the probability increases that a given storm will produce extreme amounts of precipitation. On the other hand, increased evaporation rates dry the land more quickly, increasing the probability of droughts, especially in summer over the interior of continents. So, as illogical as it may sound, the odds are that both droughts and floods could both increase on a warmer Earth.

Other potential climate change impacts, including some highly publicized ones involving storms, are more difficult to categorize. For example, the jury is still out regarding changes to the intensity and frequency of mid-latitude cyclones. The same can be said for the number and frequency of tornadoes. And climate researchers aren't sure how global warming will impact the frequency and intensity of El Niños and La Niñas.

Given the relatively high-profile nature of hurricanes (Katrina, Sandy, and Maria come to mind immediately) and the devastating impact such storms can have if they come inland, it's not surprising that the effect of global warming on tropical cyclones is front and center in any climate change discussion and remains an area of active research. Here's the current thinking:

- In some tropical basins, there are observed decadal variations in tropical cyclones whose causes, whether natural, anthropogenic, or a combination, are still being debated. Detecting trends is difficult, in part, because methods of monitoring tropical cyclones have changed over time, and, even today, differ from region to region (for example, most tropical cyclones are not measured directly by instrumented aircraft).
- The recent increase in societal impacts from tropical cyclones is mostly the result of increasing population and infrastructure along coasts.
- Most climate model simulations indicate that, if warming continues, tropical cyclones globally will be more intense, on average, but the total number of storms worldwide may actually decrease.
- Although no individual tropical cyclone can be attributed to climate change, it is sometimes possible (using an "extreme event attribution study") to estimate how climate change may have added to the severity of the event.

Thus, it's fair to say that despite the intense public fascination and (rightful) concern about the potential impact of climate change on tropical cyclones, there's still considerable uncertainty about the relationship.

How should individuals (and nations) address the potential impacts of an enhanced greenhouse effect when a number of uncertainties remain? In recent decades, several attempts have been made to forge multi-national agreements to reduce greenhouse gas emissions. For example, the Kyoto Protocol, an agreement originally drawn up in Kyoto, Japan in 1997, was eventually signed by more than 190 nations (the United States never ratified the agreement). The successor to the Kyoto Protocol is the Paris Agreement,

negotiated in 2015, which calls for voluntary and nationally determined targets for reducing greenhouse gas emissions beginning in 2020.

These agreements aside—taking immediate and dramatic action based on worst-case predictions of global warming would likely create great economic hardship, requiring a large reduction in fossil-fuel usage. Taking a "just wait and see" attitude causes no immediate economic difficulty, but the warming may be more severe if worst-case predictions of its magnitude come to pass. Between these two extremes lies a strategy to, in essence, "buy some insurance." Reducing pollution emissions, improving energy efficiency, and exploring alternative energy sources (for example, wind and solar) make sense even in the absence of an issue such as global warming. After all, most people who own a house purchase homeowner's insurance, even though the probability of their house burning down is exceedingly low.

OZONE DEPLETION: NOT ENOUGH OF A GOOD THING?

In the early 1980s, Joseph Farman, leader of a group of scientists of the British Antarctic Survey, found himself doubting some of his own observations. Since the late 1950s, British scientists had been measuring the concentration of ozone in the atmosphere above Halley Bay, on the Antarctic Coast (latitude 76°S). Ozone (O_3), a variant of oxygen, is a noxious and corrosive gas. But most of Earth's ozone is in the stratosphere, where it performs a vital function by absorbing much of the sun's harmful ultraviolet radiation. Farman's group had noticed a general decline in ozone concentrations from one October to the next above Halley Bay (see Figure 18.27), particularly since around 1975.

Farman's doubts arose because no one else had reported these dramatic ozone losses. American scientists were simultaneously measuring ozone concentrations from a NASA satellite, but reported nothing unusual. In 1984, when a second British monitoring station on the tip of Argentina detected the same ozone declines seen above Halley Bay, Farman and his team decided to report their findings. They had discovered the **ozone hole**, a region in the Antarctic stratosphere where ozone levels dramatically decrease during early spring each year (the term "hole" is catchy, but misleading—there is no "hole" in the atmosphere, merely a dramatic thinning of a single gas).

After the British article was published, the Americans reexamined their satellite data. They found that

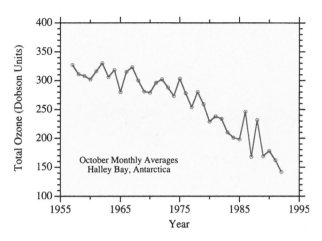

FIGURE 18.27 A time trace of average ozone concentration during October above Halley Bay, Antarctica (latitude 76°S). The graph shows the inexorable decline in October ozone concentrations from the 1950s to the early 1990s. Dobson units quantify the amount of ozone in a column of air. If all the ozone in the atmosphere could be condensed to the surface and spread uniformly, it would form a covering about three millimeters thick. This amount of ozone would correspond to 300 Dobson units; that is, one millimeter of condensed ozone is equivalent to 100 Dobson units (adapted from *Scientific Assessment of Ozone Depletion: 1994*, Executive Summary, World Meteorological Organization Global Ozone Research and Monitoring Project—Report No. 37, 1994, p. 25).

the depletion of ozone over Antarctica was in the numbers all along, but the computers trusted with interpreting the data were programmed only to recognize ozone levels within a certain range thought to encompass the "normal" variability. No one had expected that stratospheric ozone concentrations as low as Farman's group had measured would ever occur in nature, so numbers outside this range were discarded as "bad data." Figure 18.28 shows the worsening of the ozone hole from 1979 to 1995 as documented by NASA satellites.

In a way, the data were "bad:" bad news. That's because, in 1974, chemists F. Sherwood Rowland and Mario J. Molina of the University of California at Irvine discovered that chlorofluorocarbons (CFCs), man-made chemicals used in such diverse applications as refrigeration and aerosol propellants, could work their way into the stratosphere, and, through a series of chemical reactions, destroy ozone (for their work, Rowland and Molina were awarded the 1995 Nobel Prize in Chemistry). Because stratospheric ozone absorbs some of the sun's harmful ultraviolet radiation, this threat to ozone was a threat to the well-being of the planet. Years later, Rowland described his feelings about his discovery this way: "There was

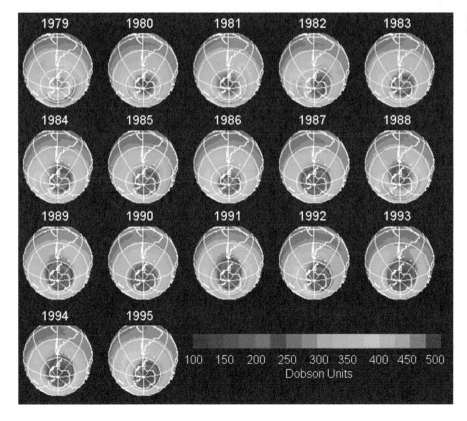

FIGURE 18.28 Average stratospheric ozone levels in October, from 1979 to 1995, as viewed from a NASA satellite. The ozone hole is the circular patch centered over Antarctica that generally increases in size and decreases in corresponding Dobson units (note the color scale) over time (courtesy of NASA).

no moment when I yelled 'Eureka!' I just came home one night and told my wife, 'The work is going very well, but it looks like the end of the world.'"

Were the dramatic decreases in ozone over the Antarctic related to Molina and Rowland's discovery of an anthropogenic process for ozone depletion? Was it possible that human-made CFCs, with relatively puny atmospheric concentrations measured in parts per trillion, were destroying a natural shield for our planet? And why was this ozone depletion observed only over the Antarctic–could such declines happen elsewhere?

Ozone: A Primer

Compared to other gases such as oxygen and nitrogen (and even carbon dioxide), ozone is a minor atmospheric constituent: On average, a sample of a million air molecules would contain only a few molecules of ozone. In the troposphere, ozone is a pollutant and a health hazard that forms on sunny days in warm weather, particularly in urban areas. Ozone's harmful effects on forest growth and crop production have been well documented. The gas is so irritating to living things that it has been bubbled through sewage to kill viruses and bacteria.

Fortunately, about 90 percent of all ozone lies in the stratosphere, approximately 15–50 km (9–30 mi) above the earth's surface (see Figure 18.29). This region of maximum ozone concentration is called the **ozone layer** (another misnomer, of sorts, because this layer of the atmosphere contains other gases as well). In the stratosphere, ozone absorbs ultraviolet radiation from the sun of wavelengths 0.28–0.32 microns (known as UVB), preventing this radiation from reaching the earth's surface. Overexposure to UVB has been linked to increases in the incidence of skin cancer, cataracts, and immune deficiencies in humans, and to adverse effects on the growth and reproduction of marine phytoplankton at the base of the ocean's food chain.

If you synthesize the previous two paragraphs, you might get the sense that ozone is a "double agent" in the atmosphere, depending on where the gas resides. And you'd be correct. In the troposphere, ozone has deleterious effects, but stratospheric ozone plays a fundamental protective role for life on Earth. The Environmental Protection Agency (EPA) has even developed a slogan to characterize ozone's dual nature: "Ozone is good up high, but bad nearby."

Ironically, stratospheric ozone owes both its existence and its destruction to ultraviolet radiation. Molecular oxygen (O_2), which we use for respiration,

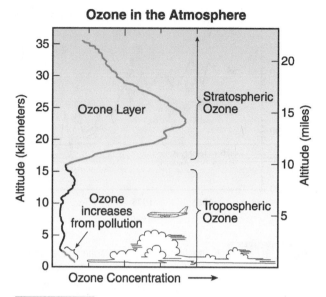

FIGURE 18.29 The average vertical distribution of ozone in the troposphere and stratosphere. Most ozone resides in the stratospheric "ozone layer." Increases in ozone near the surface occur as a result of pollution from human activities (from *Twenty Questions and Answers About the Ozone Layer: 2006 Update,* World Meteorological Organization Global Ozone Research and Monitoring Project, Report No. 50).

also absorbs ultraviolet radiation: In the process, the two oxygen atoms split. Some of the single atoms reunite to reform O_2, but some also bond with O_2 molecules forming ozone, O_3. When ozone absorbs ultraviolet radiation, the ozone molecule is destroyed, re-creating an O_2 molecule and a single oxygen atom. Ozone can also be destroyed by combining with a single oxygen atom, or by reacting with other compounds, particularly those made of nitrogen and oxygen. A balance of sorts has existed in the stratosphere between these natural processes of ozone-making and ozone-taking. But in the latter half of the 20th century and the start of the 21st century, CFCs and other human-made gases have tipped the scales in favor of destruction.

CFCs: Revisited

We encountered chlorofluorocarbons (CFCs) earlier in this chapter. They are secondary greenhouse gases, minor leaguers compared to water vapor and carbon dioxide. But CFCs (and other gases created as substitutes for CFCs) were the center of attention when it comes to ozone depletion.

Based on atmospheric concentration, the two most common CFCs were trichlorofluoromethane and dichlorodifluoromethane, more commonly known as CFC-11 and CFC-12. These gases were used as coolants in refrigeration units and air conditioners, as propellants in aerosol sprays, and as the blowing agents in the manufacture of foams such as polyurethane and Styrofoam. The estimated amounts of CFC-11 and CFC-12 released into the atmosphere during the 20th and early 21st centuries are shown in Figure 18.30.

CFC releases increased dramatically beginning around 1960, then declined in the mid-1970s after Rowland and Molina's announcement and the United States' ban on the use of CFCs as propellants in aerosol sprays in 1978. However, this decline reversed in the early 1980s because of growth in aerosol production in other countries and increased usage of CFCs in other applications. The decrease in CFC emissions beginning in the late 1980s resulted from international cooperation to limit emissions. In September 1987, 43 nations signed the Montreal Protocol, requiring a 50 percent cut in CFC production and use by 1999. Several revisions later, a ban on the production of the most damaging CFCs went into effect in 1996. As a result of these international efforts, CFC levels in the troposphere actually have begun to decrease since around 2000.

We should mention that other human-made chemicals also harm ozone, though their combined contribution to ozone depletion does not compare to the extinction of ozone by CFCs. Some of these gases, called hydrochlorofluorocarbons (HCFCs) and hydrofluorocarbons (HFCs), were created as CFC substitutes—their ozone-depleting ability is believed to be only about a tenth that of CFCs. Then there are gases that contain the element bromine. Like chlorine, bromine can react with and destroy ozone.

These other ozone-depleting gases are now monitored along with CFCs to ensure that their growth does not offset the progress made by eliminating CFCs. As of 2015, the total combined concentration of anthropogenic ozone-depleting gases in the troposphere had decreased by about ten percent from the peak concentrations observed in the late 1990s.

Despite this progress, total atmospheric concentrations of ozone-depleting gases will decrease only slowly. That's because, in the troposphere, CFCs are unsociable molecules. They do not readily react chemically with other gases. In addition, CFCs do not dissolve in water, so they are not removed easily by precipitation. As a result, much of the CFC gas that has leaked into the atmosphere is still there (CFC substitutes, such as HCFCs and HFCs, have the advantage that they are more easily removed in the troposphere by natural processes). The tossing and turning air motions characteristic of the troposphere inevitably mix some CFCs up into the stratosphere.

Once in the stratosphere, CFCs become more chemically active. At these rarified altitudes, CFCs are subject to the same exposure to ultraviolet radiation as stratospheric ozone. Absorbing ultraviolet radiation separates a chlorine atom from a CFC molecule; the chlorine is ozone's nemesis. Chlorine atoms react with ozone to form a new compound, chlorine monoxide (ClO). If the process stopped there, ozone would hardly notice: There's just not enough chlorine in the stratosphere to fight ozone one-on-one. Chlorine monoxide, however, is a very reactive substance. Some of the ClO reacts with single oxygen atoms, forming molecular oxygen (O_2) and freeing the chlorine atom to destroy more ozone. This sequence of chemical events is summarized in Figure 18.31. In their original work, Molina and Rowland estimated that one chlorine atom might go through this cycle 100,000 times, so just a little chlorine can go a long way in destroying ozone.

There is more to the story of chlorine in the stratosphere. Chlorine atoms and chlorine monoxide also react naturally with compounds of nitrogen and hydrogen, forming more exotic (but less reactive) substances such as hydrochloric acid (HCl) and chlorine nitrate $(ClONO_2)$. These compounds seem like the atmosphere's immune system against ozone depletion by chlorine: When locked in these "reservoirs" of chlorine, chlorine poses no danger to ozone. And early computer simulations of the effect of CFCs on the ozone layer suggested that ozone destruction should proceed at a leisurely pace because these

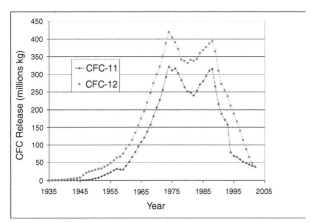

FIGURE 18.30 Annual releases of CFC-11 and CFC-12, in millions of kilograms (data from Alternative Fluorocarbons Environmental Acceptability Study, afeas.org/prodsales_download.html).

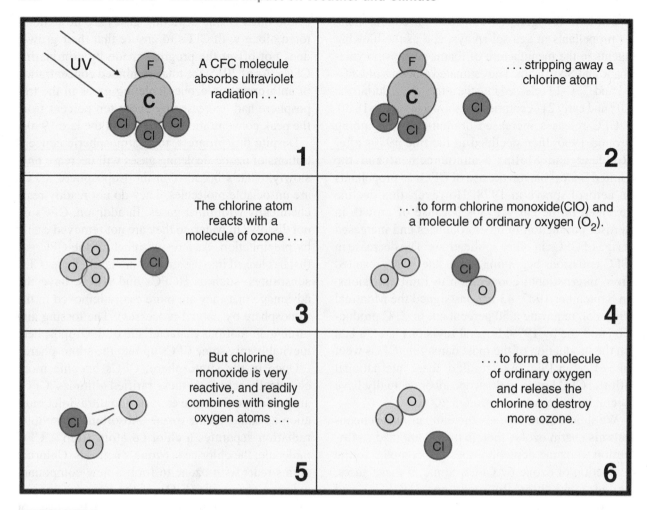

FIGURE 18.31   After ultraviolet radiation strips away a chlorine atom from a CFC molecule, the stage is set for a chain reaction that allows the same chlorine atom to repeatedly destroy ozone molecules.

reservoirs would sequester most of the chlorine. But somebody forgot to tell the atmosphere.

The Antarctic Ozone Hole: Cold and Alone

Not even Molina and Rowland expected ozone depletion to happen quickly. Their original prognosis called for a 7 to 13 percent decline in ozone levels in about a century if CFC production were not curtailed. The revelation in 1985 of the sharp drop in ozone concentrations above Antarctica suggested that something unusual was happening in the stratosphere above the South Pole. Scientists and government agencies reacted with uncharacteristic urgency to organize the National Ozone Experiment (NOZE) to Antarctica in mid-1986. The information gathered on this and subsequent expeditions clarified our understanding of ozone depletion above the continent of ice.

The Antarctic stratosphere is a very frigid place during winter. The sun never rises above the horizon during the South Polar winter (June to August), and temperatures in the Antarctic stratosphere sometimes drop below −90°C (−130°F). In response to the large temperature gradient that exists between the Antarctic stratosphere and the stratosphere over slightly lower latitudes, a belt of stratospheric winds forms during winter and encircles the continent. This ring of high-altitude circulating air, called the **polar vortex**, essentially isolates the stratosphere over the Antarctic, preventing any substantial exchange of high-altitude air between the polar regions and the mid-latitudes. As it turns out, the dramatic depletion of ozone occurs inside the polar vortex (note that this is the real definition of the "polar vortex"—the use of the term in the media in recent years to explain cold air outbreaks in winter is catchy, but not scientifically accurate).

Clouds are uncommon in the stratosphere because the air is bone-dry, with average relative humidities of one to two percent (low-level air over the summertime Sahara Desert is humid by comparison). However, air in the wintertime Antarctic stratosphere is so cold that extensive regions reach temperatures at which the meager quantities of water vapor (and even some compounds of nitrogen) deposit into ice. The resulting **polar stratospheric clouds**, or PSCs, are made of both ice and frozen compounds of nitrogen (see Figure 18.32). These clouds help to explain the dramatic ozone depletion in the Antarctic stratosphere. Although some gaseous nitrogen compounds react with chlorine, rendering it unable to destroy ozone, solid nitrogen compounds perform no such guard duty. Thus, the more nitrogen locked up in solid form in PSCs, the less gaseous nitrogen available to react with and sequester chlorine. Second, the solid surfaces of PSCs actually promote chemical reactions that free chlorine from the reservoirs of hydrochloric acid and chlorine nitrate. And chlorine on its own is the primary threat to ozone.

The end result is that when the long winter's night ends in August, more chlorine than usual is free of its nitrogen and hydrogen shackles, and the stratosphere is primed for ozone destruction. Even weak doses of ultraviolet radiation can free individual chlorine atoms, so when the sun reappears in the Antarctic stratosphere in spring, even more chlorine becomes available to deplete ozone. And with many nitrogen compounds tied up in PSCs, less gaseous nitrogen is available to react with chlorine atoms and ClO. Relatively speaking, chlorine atoms have the run of the stratosphere, and ozone pays the price. With the stratospheric air isolated by the polar vortex, ozone-richer air from lower latitudes cannot mix in to replenish the ozone. Measurements taken from aircraft flying through the edge of the polar vortex during NOZE and subsequent research missions clearly showed that, away from the polar vortex, ozone concentrations increased as levels of reactive chlorine decreased (see Figure 18.33).

Ozone levels in the Antarctic stratosphere usually bottom out in early October, shortly after the beginning of Southern Hemisphere spring. As the season progresses, temperatures over Antarctica gradually warm and the polar vortex typically begins to dissipate in late October or early November. Clumps of ozone-depleted air from the Antarctic stratosphere then mix northward, decreasing ozone levels at lower latitudes. Higher temperatures also lead to the evaporation of some PSCs, returning nitrogen compounds to the gaseous form, where they can once again detain chlorine. But "higher temperatures" are in the eye of the beholder. Indeed, stratospheric temperatures are still sufficiently low that some PSC particles keep growing until they simply fall out of the stratosphere. Thus, even in their demise, PSCs can prime the air for future ozone destruction by robbing the stratosphere of nitrogen compounds that could, in gaseous form, serve as chlorine reservoirs.

Figure 18.34 shows stratospheric ozone in the Southern Hemisphere as measured from a NASA satellite on September 24, 2006 (the South Pole lies just below the center of the image). The ozone hole is the region of

FIGURE 18.32 Iridescent polar stratospheric clouds near McMurdo Station along the coast of Ross Island in Antarctica during a past September. Found at an altitude near 24,000 m (about 79,000 ft), polar stratospheric clouds are the highest clouds in the atmosphere (courtesy of Dr. Bill Servais and the United States Antarctic Program).

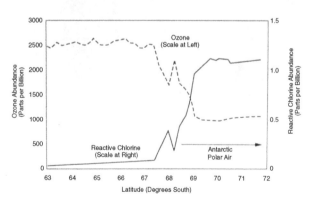

FIGURE 18.33 Measurements from high-flying aircraft on September 16, 1987, as the airplane left the ozone hole, passed through the polar vortex at about 69°S latitude, and then continued to fly towards lower latitudes (to about 63°S). The horizontal axis represents latitude, the left vertical axis indicates ozone concentrations (in ppb), and the right vertical axis shows concentrations of reactive chlorine (in ppb). The solid graph displays the variation of reactive chlorine with latitude, and the dashed plot marks the variation of ozone with latitude. As levels of reactive chlorine decreased outside the ozone hole, ozone levels increased (from *Scientific Assessment of Ozone Depletion:* 1994, Executive Summary, p.22).

lowest ozone concentrations bounded (approximately) by the dark blue color. Its areal extent on this day equaled the previous record for largest ozone hole of 29.5 million km² (11.4 million mi²), set in September 2000, an area more than three times that of the United States. The 2006 ozone hole was also a record-setter in terms of its vertical extent. For example, by early October, nearly all of the ozone in the layer between 13–21 km (8–13 mi) had been destroyed. Satellite and balloon observations showed that the lower stratosphere in late September was approximately 5°C (9°F) colder than average, helping to explain the severity of the ozone depletion in that layer (recall that lower temperatures are more favorable for the formation of PSCs).

If such significant ozone depletion can occur over the Antarctic, it's logical to ask whether the same thing could happen in the Arctic stratosphere, a little closer to home for most of the world's population (80 percent of which lives in the Northern Hemisphere)? The chlorine is there—chlorine monoxide

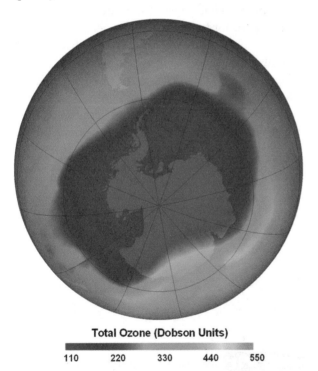

Total Ozone (Dobson Units)

110   220   330   440   550

FIGURE 18.34   During the period September 21–30, 2006, the average area of the Antarctic ozone hole was the largest ever observed, at 27.5 million km² (10.6 million mi²). This image, from September 24, 2006, shows the ozone hole at its largest extent, 29.5 million km² (11.4 million mi²), equaling a record previously set on September 9, 2000. The South Pole lies near the center of the image, while southern South America lies near the top. The area of maximum ozone depletion is indicated in purple, corresponding to ozone concentrations less than about 120 Dobson units (courtesy of NASA).

levels 100 times higher than normal have also been observed in the Arctic stratosphere.

Compared to the Antarctic, the Arctic is missing several key ingredients necessary for extensive ozone depletion. Temperatures in the Arctic stratosphere average about 10°C (18°F) higher than in the Antarctic, so fewer polar stratospheric clouds form. In addition, the underlying surface surrounding high northern latitudes is less favorable for a polar vortex to form. For example, a traveler following the Arctic Circle (latitude 66.5°N) would encounter more variable terrain and more land-water boundaries than a traveler circling the Antarctic along the Antarctic Circle. Less uniform surface characteristics lead to more storms, whose vertical reach extends to the lower stratosphere, and whose cyclonic circulations import and export stratospheric air to and from the Arctic, hampering the formation of a stable, Arctic-encircling polar vortex. Thus, air in the Arctic stratosphere isn't nearly as isolated from lower-latitude stratospheric air as in the Antarctic.

The bottom line here is that some thinning of Arctic stratospheric ozone does occur. But the extent of Arctic ozone depletion is very closely tied to how cold the stratosphere gets during any given winter, and how long that extreme chill lasts. Both factors directly impact the formation of polar stratospheric clouds. In general, ozone concentrations in the Arctic stratosphere during late winter and early spring are 10 to 20 percent lower now than they were in the mid-1970s. But an annual, recurring Arctic equivalent to the Antarctic ozone hole has not been observed.

Having said that, at the end of winter in 2011, a dramatic loss of ozone did occur in the Arctic stratosphere, prompting scientists to call the depletion an "ozone hole." Indeed, more than 80% of the ozone at altitudes of 18-20 km (11-12 mi) was destroyed, the result of an unusually persistent polar vortex in the Arctic stratosphere. Unusual events like what occurred in 2011 prompt atmospheric scientists to keep a close watch on the skies high above the Arctic for signs of more extensive ozone depletion.

A Wider View of Ozone: Global Trends

High quality observations of stratospheric ozone over low and middle latitudes began around 1979 from both ground-based instrumentation and satellites. From this time until the period 2005–2009, stratospheric ozone concentrations decreased from 3 to 10% over the mid-latitudes, with the highest losses in the Southern Hemisphere (see Figure 18.35). This ozone loss translates into roughly a 5 to 10 percent increase in the

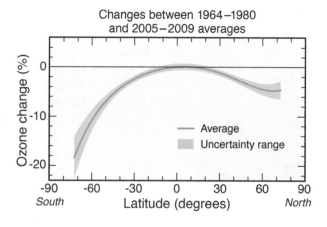

Changes between 1964–1980
and 2005–2009 averages

FIGURE 18.35 The percentage change in the total amount of stratospheric ozone, as a function of latitude, between the period 1964 and 1980 and the period 2005 and 2009. Little change has occurred over the tropics, but decreases of 3–10% have been noted over mid-latitudes (courtesy of NOAA).

amount of ultraviolet radiation reaching the surface, particularly in the Southern Hemisphere. There's been little or no decrease in ozone over tropical latitudes.

Because we've only been measuring global ozone concentrations for about 40 years (a relatively short period in the grand scheme of things), the possibility that the observed trends in ozone are solely of natural origin must be considered. For example, ozone levels are known to vary one to two percent in association with the 11-year sunspot cycle.

In addition, there are natural sources of atmospheric chlorine, including some volcanoes. And what about chlorine in the water that evaporates each day from swimming pools? While chlorine injected directly into the stratosphere by the most explosive volcanic eruptions poses a threat to ozone, chlorine that enters the troposphere from swimming pools

(or less powerful volcanic eruptions) is a different matter altogether. Although CFCs don't dissolve in water, pure chlorine does. Thus, precipitation usually removes chlorine that enters the troposphere before it ever drifts into the stratosphere. And the theory that volcanic eruptions provide the chlorine for the ozone hole is not supported by observations: Volcanoes just don't belch enough chlorine into the stratosphere to explain the amounts present there.

In the end, the five billion kilograms or so of CFCs and other ozone-depleting substances that have been released into the air since 1930 have produced current concentrations of stratospheric chlorine that are about six times the natural level. And measurements of chlorine monoxide in the lower stratosphere by NASA's Upper Atmosphere Research Satellite confirm that, on a global scale, ozone concentrations decrease as chlorine monoxide levels increase. Without CFCs, there would be no ozone hole, and anthropogenic chlorine from CFCs (and similar ozone-depleting gases) is at least partly responsible for the more modest decreases in ozone concentration over other parts of the globe in recent decades.

Although global warming and ozone depletion are often treated separately, the two issues are not independent. Consider that most anthropogenic greenhouse gases stay in the troposphere, so the warming caused by an enhanced greenhouse effect tends to be confined there. With less infrared radiation leaving the troposphere in a greenhouse-enhanced world, the stratosphere should cool. In addition, ozone absorbs ultraviolet radiation (and also some wavelengths of infrared radiation, so it's a minor greenhouse gas). Less stratospheric ozone means less absorption and thus lower stratospheric temperatures. Indeed, global radiosonde observations since the late 1950s, shown in Figure 18.36,

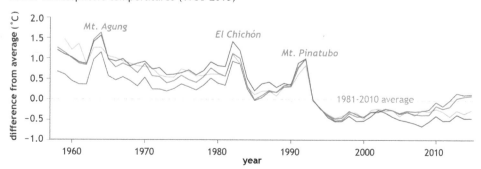

FIGURE 18.36 Globally averaged temperature in the lower stratosphere since 1958, plotted as a departure (in °C) from an average calculated using the period 1981–2010. Four different radiosonde time series are represented, but all graphs show the same general cooling trend over time, punctuated by brief warmings related to volcanic activity (NOAA Climate.gov graph, adapted from Figure 2.4(a) in BAMS State of the Climate in 2015).

Ultraviolet (UV) Index

Low	1–2	Low danger for the average person. But if you burn easily, cover up and use sunscreen.
Moderate	3–5	Take precautions, such as covering up, if you will be outside. Stay in shade near midday when the sun is strongest.
High	6–7	Apply a sunscreen of at least SPF 15. Reduce time in the sun between 11 a.m. and 4 p.m. Wear a hat and sunglasses.
Very High	8–10	Minimize exposure during midday. Liberally apply sunscreen. Unprotected skin will be damaged and can burn quickly.
Extreme	11+	Unprotected skin can burn in minutes. Reapply sunscreen every 2 hours. Best to avoid sun between 11 a.m. and 4 p.m.

FIGURE 18.37 The Ultraviolet (UV) Index quantifies the risk from exposure to ultraviolet radiation. Each category of the index includes guidelines for appropriate precautions for that level of UV exposure.

indicate a general decrease in lower stratospheric temperatures over the last five to six decades, in step with observed losses in stratospheric ozone. Various short-term warmings related to volcanic eruptions are marked, while the relatively small upward trend since the mid-1990s is likely related, at least in part, to the slow recovery of the ozone layer.

What about the future of the ozone layer? The dominant factor in the long-term outlook for stratospheric ozone is the slow rate of decrease in concentrations of ozone-depleting chemicals. Under the assumption that these chemical concentrations continue their fairly steady decrease, the current best estimate for recovery of stratospheric ozone outside of the polar regions to pre-1980 levels is the middle of the 21st century. The Antarctic ozone hole is expected to be a fixture of late winter and early spring in the Southern Hemisphere for many decades to come. The best estimate for when stratospheric ozone in the Antarctic will return to normal levels is sometime late in the 21st century.

Over the Arctic, meteorology will largely dictate ozone losses in the next few decades, with colder stratospheric winters generally leading to greater ozone depletion. Based on the best model projections, Arctic ozone concentrations are expected, on average, to return to pre-1980 levels around 2050. A caveat: These estimates of the future of stratospheric ozone also depend on other climate changes, including global warming, which (as documented earlier in the chapter) comes with its own set of uncertainties.

The issue of ozone depletion has greatly increased public and governmental awareness of the dangers of overexposure to ultraviolet radiation, which include an increase in the risk of skin cancer, cataracts, and premature aging of the skin. In 1994, the National Weather Service and the Environmental Protection Agency unveiled the Ultraviolet (UV) Index to help people plan their outdoor activities while considering the risk of adverse health effects from exposure to UV radiation.

The UV index, which was revised in 2004 in order to bring worldwide consistency to UV reporting, is shown in Figure 18.37. The color-coded index quantifies the risk of harm from unprotected sun exposure on a 1 to 11 (or higher) scale, with values of 1 or 2 representing minimal risk, and a UV index of 11 or higher indicating extreme risk. Though the actual UV level rises and falls as the day progresses, the index is reported as a prediction of the ultraviolet level at noon local time, when the sun is highest in the sky. Forecasts of the UV Index are available routinely from many sources, including the EPA Web site and many smartphone apps.

LAND USE CHANGE: IMPACT ON WEATHER AND CLIMATE

Unquestionably, the population growth and industrial development of the last 200 years has led to unprecedented changes in land use. Much of North America that was once covered by forest is now used for crops or grazing or paved with cement or asphalt. As an example, consider simply the roads we build: There are about 6,590,000 km (about 4,091,000 mi) of paved and unpaved roads in the United States. Using a reasonable estimate for average road width, we find that an area of land about the size of the state of South Carolina is now being used just for cars and trucks.

Altering the Earth's surface reorganizes the patterns of exchange of energy, water, and momentum between ground and air, sometimes with unintended consequences. The famous Dust Bowl of the 1930s is an extreme example. In the early 20th century, farming and grazing increased dramatically over parts

FIGURE 18.38 A wall of dust approaching a Kansas town during the Dust Bowl years in 1935 (courtesy of NOAA).

of New Mexico, Colorado, Texas, Oklahoma, and Kansas. Much of the natural prairie grass that covered these areas was removed. The grass helped to anchor the soil while simultaneously conserving precious moisture— these areas average only 38–51 cm (15–20 in) of rain per year. When drought conditions developed in the early 1930s, dry winds carried away much of the exposed soil (see Figure 18.38).

The changes in the atmosphere that result from changes at the surface are sometimes implied in weather forecasts. For example, meteorologists often give a range for overnight low temperatures, especially in winter: "Lows tonight will range from near 25°F in the countryside to 35°F in the city." This forecast highlights the impact of a specific and localized change in land use: the build-up of an urban area.

Deforestation affects weather and climate on a larger spatial scale, but this effect is more difficult to quantify. We first consider the general question of how the clearing of forests impacts the atmosphere. We then specifically address how the weather and climate of cities is different from that of the surrounding countryside.

Deforestation: Cutting across Natural Cycles

Deforestation continues to be a worldwide concern. Tropical rain forests, which stretch over the land surfaces near the equator, are steadily disappearing as people harvest timber and clear the landscape for farms, pastures, and roads. As we mentioned earlier, the clearing and burning of forests likely enhances the greenhouse effect. The environmental problems don't end there. Biomass burning can impact local and regional atmospheric visibility. Figure 18.39 shows hazy skies over Bolivia and northern Paraguay on a late August day in 2010 as the annual burning season was in full swing (locations of active fires are marked with red dots). The gray-white smoke spread south on the east side of the Andes Mountains (courtesy of NASA).

Deforestation outside the tropics has also raised public awareness. In the 1990s, logging in the Pacific Northwest infringed upon the habitat of the spotted owl, a threatened species. But the simple fact that forests have been cleared and the land used for other purposes has important consequences for the local and regional weather and climate.

One major difference between forested and deforested areas involves the cycling of water. Forests are water hogs. They retain water. Leaves and branches also increase the amount of surface area available for

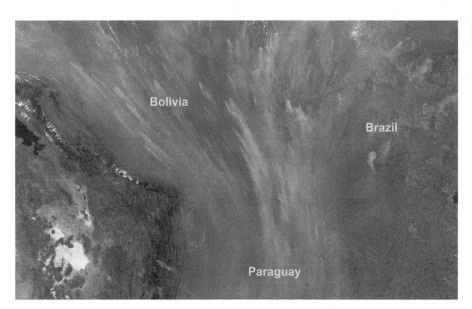

FIGURE 18.39 Red dots mark the locations of thousands of fires that were ablaze in the heart of South America on August 23, 2010, in the heart of the annual biomass burning season. Many of the fires were in Bolivia, where some fires occur naturally, but most were probably set deliberately to clear land. The smoke was so thick that 28 of Bolivia's airports had to be closed days earlier (courtesy of NASA).

evaporation, thus minimizing the loss of water into nearby rivers and streams. In contrast, more rain runs off in deforested areas, leaving less water to evaporate from the ground into the lower troposphere. Less evaporation means that more of the sun's energy is available to warm the surface, and consequently the air above. Thus, deforestation results in a general warming and dehumidification of the local lower troposphere.

The disruption of the hydrologic cycle by deforestation is most evident when tropical rain forests are slashed and burned. Girdling the equatorial regions, these forests are aptly named: They are typically very humid and rainy. As a result, they return huge quantities of water to the air by way of transpiration and evaporation, which promotes the formation of additional clouds and precipitation. Research in the Amazon Basin suggests that about half of the rain there develops from locally produced clouds (transient weather systems produce the other half). Stated more succinctly, the trees are responsible for about half the rain that falls there.

The ability of rain forests to contribute to making their own precipitation was noted more than 500 years ago by Christopher Columbus. Of his father's observations of Jamaica, Columbus's son Ferdinand wrote: "The sky, air, and climate were just the same as in other places; every afternoon there was a rain squall that lasted for about an hour. The admiral . . . attributes this to the great forests of that land; he knew from experience that formerly this also occurred in the Canary, Madeira, and Azores Islands, but since the removal of the forests that once covered those islands they do not have so much mist and rain as before."

A secondary effect of deforestation on the local atmosphere results from changes in the albedo of the surface. These changes alter the amount of radiation absorbed and back-scattered by the ground, which, in turn, alters the amount of energy transferred to the lower troposphere. In snowy climates, the difference in albedo between forested and nonforested areas is particularly noticeable during winter. Although snow-covered fields can reflect 80 percent or more of the sunlight striking them, trees usually do not retain their white covering for long after a storm: You can easily see the difference between forested and non-forested areas in Figure 18.40a, which is a visible satellite image of the northeastern United States in mid-February 2015 after a deep snowcover had been established. Much of New England had snow depths in excess of 50 cm (about 20 inches) at this time (see Figure 18.36b), but many areas there are noticeably darker—the heavily

forested White Mountains of central New Hampshire and the evergreen forests of the Adirondack Mountains in northern New York are easy to find.

Figure 18.41 also demonstrates how changes in surface covering can alter albedo. This pair of images of the same area of western Brazil vividly shows the impact of deforestation from July 2000 to July 2012. The area shown is approximately 22,500 km^2 (about 8660 mi^2, roughly the size of New Jersey). Forests are deep green while cleared areas are tan (bare ground) or light green (crops and pasture). Over the span of a dozen years, deforestation in this region followed a typical pattern as roads first penetrated the forest, followed by small farmers who initially cleared land for crops, then converted the degraded land to cattle pasture before moving on to clear additional forest for crops. Using sophisticated computer models, NASA scientists estimate that the transformation from forest to bare ground has the greatest impact on local climate, raising local temperatures as much as 1.7°C (3°F).

Deforestation in the tropics transcends weather and climate by virtue of its impact on ecosystems and biodiversity. Tropical rain forests are storehouses of plant and animal life. A few square miles of Amazon rain forest can house more species of birds than all of North America. The Malaysian Peninsula, with a total land area about the size of South Dakota, is covered by tropical forests that contain more species of trees than all of North America. And many rain forest plants have medicinal value: By some estimates, there's a one in four chance that your next prescription drug will be derived from a plant that grows in a tropical rain forest. Thus, rain forest destruction has a significant human impact beyond its potential effect on climate.

The transformation of both forests and fields into urban areas has significantly altered the exchanges of energy and water between the ground and the atmosphere. We now explore in greater detail this urban heat island effect.

The Climate of Cities: Islands unto Themselves

We expect higher global average surface temperatures in a greenhouse-enhanced world. Such predictions of global warming force us to ponder what a climate several degrees warmer than today's would be like. We tend to forget that such warmed climates already exist in localized areas: our very own cities! Over the last century, urbanization has produced temperature increases

(a)

FIGURE 18.40 (a) A visible satellite image of the northeastern United States on February 16, 2015. Skies were mainly clear over this area so, in most places, the satellite could see the ground. Despite a fairly uniform snowcover, note the differences in surface albedo. Heavily forested areas are darker, including the evergreen forests of the Adirondack Mountains which appear as the large, relatively circular darker area east of Lake Ontario (courtesy of NASA); (b) Depth of snow on the ground across the Northeast on February 16, 2015. Areas in purple had in excess of 50 cm (about 20 inches) of snow on the ground (courtesy of NOAA).

(b)

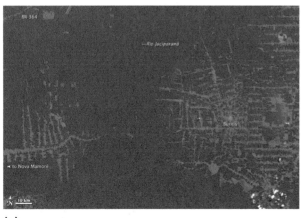

(a)

(b)

FIGURE 18.41 Changes in the Amazon rain forest in the state of Rondonia in western Brazil from July 2000 to July 2012, captured by a NASA satellite. Intact forest is deep green, while cleared areas are tan (bare ground) or light green (mainly pasture or crops) (courtesy of NASA).

in some cities that rival even the most extreme forecast of a global temperature increase on an enhanced greenhouse-warmed planet. Indeed, as cities have grown, humans have been in the business of climate modification, and not just of temperature: Cities differ from rural areas in nearly every aspect of climate.

What characteristics of a city set its climate apart from that of the neighboring countryside? Two differences stand out. First, cities are built from brick, concrete, and asphalt, materials that (on average) have higher heat capacities than the natural surface coverings of the countryside. Thus, cities absorb and store more energy in less time than rural areas. In addition, because most cities are a maze of shapes and orientations, the surface area available for absorbing energy is larger in urban areas compared to rural settings. The densely constructed urban landscape increases the absorption of radiation that is scattered (and emitted) by streets and buildings. Given that the troposphere acquires its energy primarily from the surface, cities become efficient systems for heating the air.

A second primary difference between city and rural environments is that most cities have an extensive network of sewers and drains to remove precipitation. Combined with the warmer, rapidly drying nature of city surfaces, this drainage capability keeps city surfaces drier, on average, than rural surfaces, reducing the opportunities for evaporation in the city. With less solar energy being used to evaporate water, more is available for heating the surface (and the air above).

The warmth of the urban climate over the rural environment, formally known as the **urban heat island effect**, has been detected in towns of less than a thousand residents to major metropolitan areas with populations in the millions (recall the temperature trace from New York City in Figure 18.10). In general, the larger a city's population, the larger the urban heat island effect.

Although the extra absorption of solar energy within a city keeps daytime temperatures higher, the heat island is most noticeable at night. After sunset, a city's stone and brick materials (which more efficiently store energy during the day) impart more energy to the air than the surrounding countryside, so the city air cools more slowly. Clear skies and dry air maximize the temperature difference between city and rural environments because the greenhouse effect of water vapor is weakest on such nights, paving the way for dramatic cooling in rural areas.

The urban heat island effect is most noticeable in winter, when nights are longer and the air is drier. On a clear, calm winter night, centers of major cities may be 10°C (18°F) or more warmer than surrounding snow-covered rural areas. Figure 18.42 vividly

FIGURE 18.42   Visible (left) and infrared (right) satellite views of the Upper Midwest in late March 2001, demonstrating the warmth of the metropolitan area of Minneapolis-St. Paul, MN. In the visible image, the darker circular region in southeastern Minnesota indicates the lack of snowcover in and around the urban area, while the relatively warm surfaces of the city show up darker in the infrared image (courtesy of NOAA).

illustrates the urban heat island of the Minneapolis-St. Paul, MN area. In this satellite view of the Upper Midwest, the visible imagery (on the left) captures the lack of snow cover over the Twin Cities in late March 2001. At the same time, infrared imagery (on the right) detects the warmer metropolitan area compared to the colder surfaces of the surroundings. Even within a large urban area, we can distinguish relatively warm and cooler surfaces (note the contrasts in Figure 18.43, an infrared image of Atlanta, GA).

Although the urban heat island effect is noticeable even in small cities, changes in precipitation patterns related to urbanization are generally associated with only the largest metropolitan areas. Several climatological studies have established that the presence of a major urban area leads to small increases in precipitation amounts, particularly downwind of the city, and especially during summer, when larger-scale flow patterns tend to be weak (which gives any urban-induced convective circulations a greater chance to stand out). The urban effect on precipitation appears to be strongest when the atmosphere is already primed to produce heavy, convective precipitation (showers and thundershowers). The relatively warm air over cities promotes enhanced localized instability, increasing the tendency for upward motions.

One of the first and most comprehensive attempts to measure an urban area's effect on precipitation was METROMEX, the METROpolitan Meteorological EXperiment, a multi-year project launched in June 1971 in St. Louis, MO. METROMEX found as much as a 25 percent increase in average summertime precipitation and the frequency of thunderstorms and hail in a broad region around St. Louis

and extending as far as 65 km (40 mi) east of the city. Less extensive studies around Chicago and Washington, DC, suggest a similar urban enhancement of cloud and precipitation development.

More recently, researchers using rainfall data gathered from satellites and dense observation networks have detected higher rainfall rates downwind of large cities such as Houston and Atlanta during the summer. The Atlanta study incorporated data gathered during the 1996 Summer Olympics. On three separate days, the urban heat island enhanced upward motions over the city, creating a low-level convergence zone that initiated thunderstorms. An earlier study of New York City found a similar effect on days when the larger-scale wind pattern was relatively weak. However, that same study suggested that groups of thunderstorms moving with a larger-scale weather system (such as a cold front) sometimes split and moved around the city, producing rainfall maxima on both sides and downwind of the city, and a minimum over the city itself. This same effect was noted in a similar study of summertime convective storms moving over Phoenix.

The National Lightning Detection Network (NLDN) has also helped to establish the urban effect on weather and climate. The NLDN consists of more than 100 ground-based sensors that pinpoint the location of a cloud-to-ground lightning bolt within seconds of a strike. A study of 16 midwestern cities during June, July, and August from 1989 to 1992 found an enhancement in the frequency of cloud-to-ground lightning by 40 percent or more over and downwind of the urban areas as compared to rural areas upwind. A longer-term study of Houston,

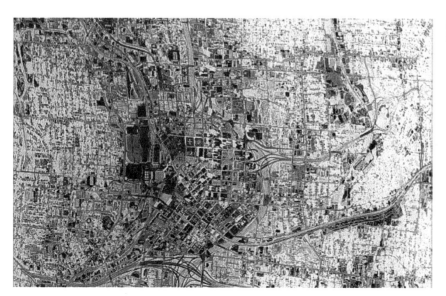

FIGURE 18.43 A close-up infrared satellite image (with resolution about 10 meters) of a section of Atlanta, GA, taken around noontime. Surfaces that are "hot" or "warm" appear darker, while somewhat cooler surfaces are shown in varying shades of white to light gray (courtesy of NASA).

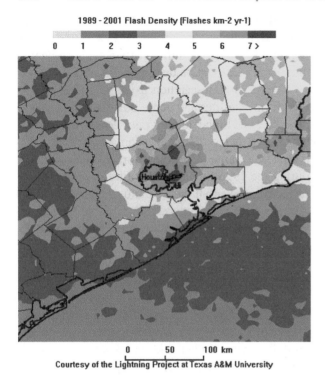

1989 - 2001 Flash Density (Flashes km-2 yr-1)

0　1　2　3　4　5　6　7 >

0　50　100 km

Courtesy of the Lightning Project at Texas A&M University

FIGURE 18.44 The distribution of cloud-to-ground lightning strikes in and around Houston, TX, from 1989 to 2001. Houston is located approximately in the center of the image. Here, flash density measures the average yearly number of strikes per square kilometer. Note how the maximum values are found over and generally downwind (east) of the city (courtesy of the Lightning Project at Texas A&M University).

from 1989 to 2001, found a 45 percent increase in the density of flashes compared to surrounding rural areas (see Figure 18.44), with the greatest increases in summer between 9 A.M. and 6 P.M.

Complicating the urban effect on precipitation is the increased number of tiny solid and liquid particles added to city air by human activity. These particles represent an increased supply of potential condensation nuclei, which, at first glance, would seem to have a precipitation-enhancing effect. However, extra condensation nuclei mean more competition for the available water vapor. This infighting can actually slow down the production of rain because individual precipitation particles can't grow as quickly. In the end, the likely effect of this delay is to shift the urban-induced precipitation maximum farther downwind away from the city.

The difference between the climates of cities and rural areas goes beyond temperature and precipitation. We list some of these differences in Table 18.2. Although the urban environment is warmer, less direct sunshine actually strikes the ground in cities than in

TABLE 18.2 A comparison of the climates of city and rural environments. These are average conditions and not necessarily representative of any specific day.

Climate Characteristic	City	Rural
Average amount of cloudiness	Higher	Lower
Average sunshine at the ground	Lower	Higher
Average wind speed	Lower	Higher
Average relative humidity	Lower	Higher
Average visibility	Lower	Higher

the countryside because the buildings intercept some of the radiation. Higher temperatures generally mean a lower average relative humidity in cities, but don't rush downtown for relief on a humid summer day—the decrease in relative humidity is primarily attributed to higher temperatures, not reduced water vapor (dew points can remain relatively high in cities during summer). The additional load of particles in city air from industry, cars, and heating systems tends to lower city visibilities. In terms of wind, the buildings that define an urban area create added friction, slowing surface winds (on average) compared to nearby rural areas. However, anyone who has ever walked through the urban canyons of a city knows that channeling effects between buildings can create localized wind gusts that far exceed any found outside the city.

In smaller cities, many of these differences are imperceptible, or may be noticed only occasionally. But in larger cities, every day is different weatherwise compared to what it would be like if the city were not there.

CLOSING THOUGHTS: THE SPEED OF CHANGE

It took from the dawn of civilization until around the year 1800 for the world's population to reach one billion, but only two centuries to add an additional five billion. There is no doubt that the combined population and industrial explosion of the last 200 years has subjected the atmosphere and surface to change at a rate unprecedented since at least the end of the last major Ice Age some 10,000 years ago. Some changes span the planet: Consider that atmospheric concentrations of carbon dioxide are increasing globally. Other changes are more regional, such as the disappearance of more than half the stratospheric ozone over the Antarctic each spring. And on a local scale, city dwellers of today live in a completely different climate than did their relatives of just a few generations ago.

In some cases, such as the climate changes induced by cities, the anthropogenic impact is unmistakable. But on a larger scale, differentiating between natural and anthropogenic changes is not as easy as it might seem because our planet is constantly changing anyway. We're just not sure how the planet would be changing if we weren't here. But, undoubtedly, it wouldn't be changing nearly as fast.

Weather Folklore and Commentary

Global Warming: Science and Policy

As you learned in this chapter, the issue of global warming gets pretty complicated. Your experience probably tells you that broaching the topic stimulates many opinions. When discussing global warming, we believe that it's important to always try to separate science from policy.

With regard to the science, there's considerable common ground in the scientific community. Burning fossil fuels puts carbon dioxide into the atmosphere; carbon dioxide is a selective absorber of infrared radiation; humans burn a lot of fossil fuels; atmospheric concentrations of CO_2 are increasing; the recent increase in atmospheric concentrations of carbon dioxide is almost entirely caused by human activity; atmospheric carbon dioxide has a warming effect (even the most skeptical scientists agree on this point).

Attributing the recent warming to human activity is a more difficult hurdle to clear. However, nearly all climate scientists who seriously research the topic and publish their results in refereed journals agree that warming has occurred over the last century (particularly since about 1980), much of it is attributable to human activities, and it's reasonable to expect that additional increases in anthropogenic greenhouse gases will result in additional warming. Most scientists also agree that, in a "burn all of earth's fossil fuel" future, the expected changes to our atmosphere will be much more dramatic than those we've already observed.

With regard to issues about policy, there will be winners and losers that accompany any climate change, whether it's natural or anthropogenic. Almost everyone agrees that poor people will be more vulnerable to climate change than rich people. Do we direct our efforts to head off climate change? Or do we make poor people richer? Or do we develop a plan that combines both strategies? Clearly, serious people can disagree on the corrective course of action we should take (or whether any action is even warranted).

In our opinion, burning all of earth's fossil fuels will cause our planet to warm and the world's climate to dramatically change. Hopefully, we can end the current planetary experiment by prudently implementing corrective measures. At this point in time, however, there is not yet a widespread consensus on how to politically and economically balance these measures.

Focus on Optics

Red Skies and Volcanoes: Twilight Tides of Erupting Crimson

For many months following the eruption of Mount Pinatubo in May 1991, twilight worldwide was sometimes accented by splashes of red that emblazoned the sky. Similar spectacular sunrises and sunsets followed the eruption of the El Chichón volcano in Mexico in 1982 (see Figure 18.45). To understand these tides of twilight crimson in the aftermath of large volcanic eruptions, we must first look at generic red sunsets (and sunrises).

First, realize that the lower atmosphere does not have to be filthy with dust, haze or pollution to turn the

(continued)

(Continued)

horizon a brilliant red at dusk or dawn. In fact, atmospheric pollutants tend to wash out and dull the luster of sunsets and sunrises. Periodic glances at the horizon near dusk or dawn in big cities such as New York City or Los Angeles would probably convince you that sunsets and sunrises are typically not all that photogenic there. So then, what is the real cause of brilliant red sunsets?

Recall from Chapter 2 that scattering of visible radiation by air molecules is not uniform across the spectrum of visible wavelengths. Because air molecules are so small, they scatter shorter wavelengths more efficiently than longer wavelengths. Also, remember that when the sun hangs low in the sky around dusk or dawn, beams of sunlight must travel a greater distance through the atmosphere to reach your eyes than they would for any other position of the sun in the sky (look back to Figure 2.15). This longer path length that solar radiation must traverse through the atmosphere when the sun is near the horizon guarantees that shorter wavelengths of visible radiation will be almost entirely scattered out of sunlight beams by the time sunlight reaches the surface, as Figure 18.46 suggests. Make no mistake—some of the longer wavelengths are scattered as well—but just not as many as the shorter wavelengths. In this way, sunlight is "reddened."

Scarlets around dusk are enhanced when cirrus and altostratus (or altocumulus) clouds reflect the fleeting, brilliant rays of the setting sun. These clouds are high enough that reddened beams of sunlight do not have to pass through the lower troposphere and the muck of dust, haze and pollution that often resides there, a path that would rob the reds of their brilliant splendor.

Tropospheric clouds do not have a monopoly on showcasing the brilliant scarlets of twilight. Invisibly thin clouds composed of dust particles and tiny droplets of sulfuric acid, which were injected into the stable, lower stratosphere by volcanoes such as Mount Pinatubo and El Chichón, also reflect reddened beams after sunset. In other words, light that has been reddened by passing through the lower atmosphere is backscattered by the bottom of these stratospheric clouds, adding to the redness of sunset (or sunrise). Indeed, about a half-hour after sunset (and about a half-hour before sunrise), the stratosphere alone is illuminated by the sun, allowing invisibly thin volcanic clouds to become brilliantly visible.

It is a true irony of life on this planet that something as catastrophic and destructive as volcanic eruptions could leave such a legacy of awesome beauty.

FIGURE 18.45 A fiery red twilight over State College, PA, in the aftermath of the volcanic eruption of El Chichón in Mexico in April 1982 (courtesy of Craig Bohren).

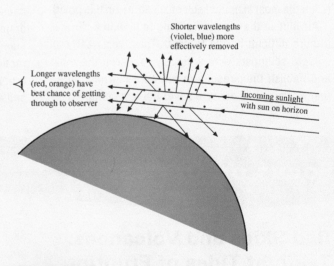

FIGURE 18.46 Around dusk or dawn, light from the sun, which is very low on the horizon, takes a longer path through the atmosphere. As a result, shorter wavelengths will be almost entirely scattered out of sunlight beams by the time light reaches the observer, leaving mostly longer wavelengths of visible light. In this way, sunlight is "reddened."

A

Absolute vorticity The sum of earth vorticity and relative vorticity.

Absolute zero Theoretically, the temperature at which all molecular motion ceases. Numerically, absolute zero is equal to $-273.15°C$, $-459.67°F$, or 0 K.

Absorptivity A measure of the amount of radiation absorbed by a substance.

Acid Rain Precipitation having a pH less than 5, resulting from raindrops or cloud droplets combining with oxides of sulfur or nitrogen or other industrial gaseous pollutants.

Adiabatic process A process that occurs without a given system (such as a parcel of air) exchanging energy with its surroundings. In an adiabatic process, expansion results in cooling and compression results in warming.

Advection fog A fog that forms when warm, relatively moist air is advected over a relatively cool surface, such as chilly ocean waters or snow-covered ground.

Advection The transport of an atmospheric property (such as temperature or moisture) by the movement of air.

Aerosols Tiny, solid particles (for example, smoke or dust) or liquid droplets suspended in the air.

African easterly wave A tropical disturbance that has a cyclonic circulation in the lower part of the middle troposphere, and that originates north of the equator over Africa and moves westward over the tropical North Atlantic Ocean. African easterly waves serve as the catalyst for the seedling low-level circulation of a lion's share of the tropical cyclones that develop over the North Atlantic Basin.

Ageostrophic wind The (typically small) component of the wind that is not geostrophic.

Aggregation The adherence of colliding snow crystals, resulting in a snowflake.

Air mass A large volume of air that is characterized by similar temperature and moisture properties at any given altitude.

Air parcel See parcel.

Air pressure The force per unit area exerted by air molecules. Alternatively, the weight of the air above a given area.

Albedo The fraction of the radiation impinging upon a surface that is back-scattered.

Alberta Clipper A fast-moving low-pressure system that originates in the vicinity of the Alberta province of western Canada and which typically brings a few inches of snow to parts of the central and northeastern United States.

Aleutian low A semi-permanent low-pressure feature of the winter pressure pattern over high northern latitudes. On charts of average pressure, it would be found just west of Alaska's Aleutian Islands.

Altocumulus clouds Middle clouds that form when ripples and rolls develop in layers or patches of stratiform clouds.

Altostratus clouds Layered clouds, typically comprised of water drops, which form at middle levels of the troposphere.

Anafront A front at which air rises on the cold side of the front, often producing a swath of stratiform clouds and precipitation there.

Analog forecasting A technique that looks to the past to predict the future, used particularly in monthly and seasonal forecasting. Forecasters determine the average jet-stream pattern in the weeks and months leading up to the present, and then look for previous years when a similar pattern occurred during the same season. They then follow the evolution of the upper-air flow during those "analog" years, note the general temperature and precipitation patterns that developed, and incorporate those patterns into their forecasts.

Anemometer An instrument that measures wind speed.

Aneroid barometer A pressure-measuring device consisting of a sealed coil of hollow tubing that has been partially evacuated of air. The tubing is mechanically connected to a needle that moves over a scale of pressure readings visible on the outside of the device. Changes in air pressure cause the tubing to compress and expand, moving the needle.

Angular momentum A measure of a spinning object's tendency to continue spinning.

Anthropogenic Resulting from the action of humans.

Anticyclone High-pressure system.

Anticyclonic vorticity Negative relative vorticity in the Northern Hemisphere.

Anvil The name given to the cloud shape formed as cirrus clouds spread out horizontally when thunderstorm updrafts reach the stable tropopause.

Apparent temperature See heat index.

Archimedes' principle The principle which states that the total upward force on a body which is completely or partially immersed in a fluid is equal to the weight of the fluid displaced by the body. In short, less dense substances "float" in more dense substances.

Atmosphere The collection of gases, particles, and clouds that comprise the envelope of air that surrounds the earth.

Atmospheric optics See optics.

Atmospheric window The band of wavelengths between approximately 10 and 12 microns that is poorly absorbed by earth's atmosphere.

Autumnal equinox The time of year at which the sun passes directly above the equator, marking the beginning of autumn. The autumnal equinox occurs on or about September 22 in the Northern Hemisphere and on or about March 21 in the Southern Hemisphere.

Average environmental lapse rate The average rate of temperature decrease with altitude. In the mid-latitudes, the average environmental lapse rate in the troposphere is approximately 6.5°C/km (3.6°F per 1000 feet).

Azores high The name given to the North Atlantic subtropical high-pressure system during the cold season. Its mean position during winter is close to the Azores, a group of islands about 1400 km (870 mi) west of Portugal.

B

Backdoor cold front A cold front that moves west or southwest in the Northeast or Mid-Atlantic States (typically in spring), allowing air over chilly Atlantic waters to advance inland.

Back-scattering The scattering of incident radiation back in the direction of the energy source.

Barograph An aneroid barometer that drives a pen, tracing a continuous time history of air pressure.

Barometer An instrument for measuring air pressure.

Bergeron-Findeisen process The process by which ice crystals suspended in a cold cloud (in which ice crystals and water droplets co-exist) grow at the expense of surrounding, supercooled water droplets.

Bermuda high The name given to the semi-permanent subtropical high-pressure area that can be found on mean pressure charts of the North Atlantic Ocean during the summer.

Bimetallic thermometer A temperature-measuring instrument that consists of two dissimilar metals that contract and expand at different rates as temperature changes.

Blizzard Conditions with sustained winds or frequent gusts of 35 mph (30 kt) or higher and visibility less than 0.25 mi (400 m) for at least three consecutive hours, caused by falling or blowing snow.

Blocking high A high-pressure system in the middle troposphere (representing a pool of relatively warm air aloft) that acts as an obstacle to the mid-latitude westerlies, forcing the flow to split into two branches.

Bow echo A crescent-shaped radar echo of a cluster of thunderstorms associated with damaging straight-line winds. A bow echo gets its name because it resembles an archery bow.

Buoyancy force The upward force exerted on an object placed or submerged in a fluid that tends to make that object float or rise.

Buys-Ballot's Law A simple guide for determining the location of low pressure relative to your location. If you stand with the wind at your back, lower pressure will lie on your left (in the Northern Hemisphere). Dutch meteorologist Buys Ballot formulated this law in 1857.

C

California current The cold ocean current that flows southward along the West Coast of the United States.

Calorie A unit of energy defined as the amount of energy required to raise the temperature of one gram of water by 1°C.

Cap cloud A stationary cloud that forms directly above a mountain peak in a stable environment.

Cabo Verde storms Tropical cyclones that form within about 1000 km (620 mi) of the Cabo Verde Islands off the west coast of Africa.

Celsius scale A temperature scale on which 0 degrees in the temperature at which ice melts and 100 degrees is the temperature at which water boils (at sea level).

Centrifugal force The apparent force in a rotating system that deflects a mass outward from its axis of rotation.

Charge separation An imbalance of electrical charge in the atmosphere, typically accomplished by a transfer of electrons. Charge separation is a prerequisite for lightning.

Chinook A warm, dry wind that forms as air sinks on the eastern side of the Rocky Mountains.

Chlorofluorocarbons (CFCs) A group of anthropogenic compounds of various combinations of chlorine, fluorine, and carbon. Chlorine from these gases is primarily responsible for the ozone hole over the Antarctic.

Cirrocumulus clouds High clouds that appear in thin, white patches. Cirrocumulus often develop when ripples or rolls form in a layer of cirrus or cirrostratus.

Cirrostratus clouds Layered, high, thin clouds made of ice crystals that often cover the entire sky and sometimes produce a halo.

Cirrus clouds Thin, wispy, high clouds, usually above 5500 m (18,000 ft), made of ice crystals.

Climate The collection of all statistical weather information (including means and extremes) at a given location for a specified interval of time.

Climatology The scientific study of climate and climate phenomena. Also, the term forecasters use to describe a prediction based on averages.

Closed-cell convection An array of cumulus cells (clouds) over the subtropical oceans characterized by rising motion at each center and sinking motion around the periphery of each cell.

Cloud A dense population of tiny water droplets and/or ice crystals that have a diameter on the order of ten micrometers.

Coastal front A boundary that owes its existence, in large part, to the natural temperature contrast during winter between cold land and relatively warm ocean. A coastal front often forms along the East Coast of the United States in concert with cyclogenesis during winter.

Cold conveyor belt A westward-flowing stream of cold air on the cold side of the surface warm front of a mature mid-latitude low that is moistened by precipitation from the warm conveyor belt and eventually rises as it circulates cyclonically around the low.

Cold front A boundary between dissimilar air masses at which colder and/or drier air (denser air) is advancing on warmer and/or moister air (less dense air).

Cold air advection The transport of colder air into a region by the wind.

Cold-air damming A mesoscale process that sometimes occurs along the eastern slopes of the Appalachians in which cold, dense, low-level air, typically transported southward on the eastern flank of high pressure over northern New England or southeastern Canada, dams up against the mountains, which halt its westward progress.

Collision-coalescence The process in which falling raindrops grow as bigger drops overtake and collide with smaller drops, and all or part of the smaller drops' water sticks, or coalesces, to the larger drops.

Compressional warming The warming incurred by a sinking parcel of air resulting from compression.

Computer model A set of mathematical instructions which, when processed by a computer, produces a virtual weather forecast.

Condensation nuclei Microscopic smoke, dust, salt, or other particles in the air that serve as surfaces onto which water vapor molecules condense.

Condensation The process of water vapor changing into water.

Conditional instability An atmospheric condition in which the stability of a layer of air depends on whether or not net condensation is occurring in parcels of vertically moving air.

Conduction The transfer of energy from molecule to molecule via molecular collisions.

Confluence A pattern of wind marked by the flowing together of two streams of air. At high altitudes just below the tropopause, the confluence of two branches of the jet stream is typically marked by a jet streak.

Conservation of Energy The principle that the total energy of an isolated system remains constant, though energy can be converted into different forms. For example, a tennis ball thrown upward has large kinetic energy initially and rather small potential energy. But, as the ball goes higher, it slows and loses kinetic energy, but its greater height translates to an increase in potential energy. The total energy of the tennis ball does not change even though its kinetic and potential energy continually change during flight.

Constant pressure surface A surface in the atmosphere on which the pressure is the same at all points on the surface.

Continental polar (cP) air mass A huge volume of relatively cold, dry air.

Continental tropical (cT) air mass A huge volume of relatively warm, dry air.

Contour An isopleth of elevation on a topographic map; more generally, a line or curve representing a constant value of a specified parameter.

Contrail A streamer of cloud formed behind an aircraft as hot, moist air exhaled by jet engines mixes with cold, dry air prevalent at high altitudes.

Convection The vertical transport of heat energy by eddies. Meteorologists also use the term to refer to the clouds that form from this process.

Convection cell A horizontal and vertical circulation consisting of a rising branch, a sinking branch, and two horizontal branches connecting the vertical currents. The circulation is driven by horizontal temperature contrasts.

Convective available potential energy (CAPE) A measure of the maximum energy available to a positively buoyant parcel of air above the level of free convection. On a plot of temperature with altitude, CAPE is qualitatively gauged as the area between the temperature sounding and the adiabat followed by the parcel, above the level of free convection.

Convective parameterization The mathematical scheme used to simulate the process of convection in a computer model.

Convergence Net horizontal movement of air into a vertical column.

Cooperative observing network A national network of about 10,000 volunteer observers in the U.S. who record daily temperature and/or precipitation in greatly diverse environmental settings, including farms, urban and suburban areas, National Parks, seashores, and mountaintops. Cooperative observers form the backbone of the long-term U.S. climate observing network.

Coriolis force The apparent force, resulting from the earth's rotation, that deflects moving objects to the right of their intended path (in the Northern Hemisphere) or to the left of their intended path (in the Southern Hemisphere).

Cornice An overhanging edge of snow near a summit or ridge line, often a visible sign of avalanche potential. Areas at and immediately below the edge of the cornice are typically at the greatest risk.

Cross-section view A view of an object which displays one horizontal dimension and the vertical dimension.

Cumuliform clouds A general term for clouds whose primary development occurs vertically, and whose presence typically signals some degree of instability in the atmosphere.

Cumulonimbus cloud A thunderstorm cloud.

Cumulus cloud A cloud with a flat base and a heaping top that exists as an individual, detached tower.

Cumulus congestus A vigorously upward-building cumulus cloud that typically has sharp outlines and often great vertical development. These clouds may produce abundant showers and sometime develop further into cumulonimbus clouds.

Cumulus stage The initial stage in the life cycle of a single-cell thunderstorm in which buoyant parcels contribute to an updraft that successively broadens the vertical extent of the fledgling thunderstorm.

Cut-off low A pool of relatively cold air in the middle to upper troposphere that has been separated, or cut off, from the steering winds of the mid-latitude westerlies.

Cyclogenesis The process by which cyclones, or lows, develop.

Cyclone An area of low pressure around which winds blow in a counterclockwise sense in the Northern Hemisphere and a clockwise sense in the Southern Hemisphere. Also, the term used for a hurricane in the Indian Ocean and near Australia.

Cyclone model A conceptual model of the evolution and common characteristics of mid-latitude low-pressure systems.

Cyclonic vorticity Positive relative vorticity in the Northern Hemisphere.

D

Density The ratio of the mass of a substance to the volume that it occupies.

Deposition The process of water vapor changing directly into ice.

Derecho A widespread, long-lived convective wind-storm produced by severe thunderstorms organized into one or more bow echoes. Typically, to qualify as a derecho, wind damage must extend over an area with major axis (measured along the wind-storm's forward path for its entire lifetime) at least 400 km (250 mi) long.

Dew point The temperature to which air must be cooled (at constant pressure) for saturation to occur.

Dew-point depression The temperature minus the dew point.

Dew-point front A front that separates air masses which differ predominantly in the moisture content of the air. An example is the dry line.

Diabatic process A process in which heating or cooling occurs as energy is added to or subtracted from the air.

Difluence (or diffluence) A pattern of wind flow in which air fans out away from a central axis that is oriented parallel to the general direction of the flow. Difluence is the opposite of confluence.

Dispersion The separation of visible light into its component wavelengths.

Dissipating stage The final stage in the life of a single-cell thunderstorm during which downdrafts eventually overwhelm updrafts, leading to the dissipation of precipitation.

Dissociate To break apart molecules into constituent atoms. For example, intense solar radiation can cause water molecules at very high altitudes to dissociate into hydrogen and oxygen atoms.

Diurnal range The difference between the daily maximum temperature and the daily minimum temperature.

Divergence Net horizontal movement of air out of a vertical column.

Doldrums The region associated with the Intertropical Convergence Zone that is characterized by light winds and abundant rainfall.

Doppler effect The detectable differences in the frequencies of radiation or acoustic waves emitted or reflected by an object that is moving toward or away from an observer.

Doppler radar A type of weather radar distinguished from conventional systems because it operates based on the Doppler effect. Doppler radar has the capability to determine the velocity of targets (precipitation and other atmospheric particles) moving toward or away from the radar based on the difference in frequency between the outgoing and returning radar signal.

Downburst The generic term for a strong downdraft of air which, upon reaching the surface, causes an outrush of damaging, straight-line winds at or near the ground.

Downdraft A sinking current; the term is typically used to describe the sinking air inside a convective cloud, such as a thunderstorm.

Downwelling long-wave radiation That part of the radiation emitted by the atmosphere that is directed toward the earth's surface.

Dropsonde A canister of weather instruments tethered to a parachute that is dropped from an airplane. Essentially, a dropsonde is a radiosonde that measures weather conditions as it descends toward the earth's surface.

Dry adiabat On a graph of altitude versus temperature, a line with a slope of $-10°C/km$ (about 5.5°F/1000 ft).

Dry adiabatic lapse rate The rate of cooling of a rising parcel of unsaturated air. Its numerical value is $10°C/km$ (about 5.5°F/1000 ft).

Dry air advection The transport of relatively dry air by the wind.

Dry conveyor belt A stream of dry air originating in the lower stratosphere or upper troposphere west of a mature mid-latitude low-pressure system that sinks as it flows cyclonically into the circulation of the low in the immediate wake of the low's cold front.

Dry line A dew-point front that often forms in the Southern Plains of the United States during the spring and early summer, marking the boundary between continental tropical air and maritime tropical air.

Dry slot The relatively cloud-free area that often forms between the rounded comma head and the clouds associated with the cold front of a mature mid-latitude cyclone. This region is produced by the cyclone's dry conveyor belt.

Dust devil A small, typically short-lived whirl of dust, sand, or other debris that spins up when strong heating by the sun promotes fast, rising columns of air near the ground.

E

Earth vorticity Vorticity, or spin, around a local vertical axis, imparted to air by the rotation of the earth.

Easterly wave A tropical disturbance, often originating over Africa, that travels from east to west in the tropical troposphere.

Eddy A turbulent swirl of air. Eddies are routinely caused by solar heating or the wind blowing across the earth's surface.

Ekman spiral A spiraling current formed as ocean surface waters, set into motion by the wind, are eventually deflected (to the right in the Northern Hemisphere) by the Coriolis force. This surface water transfers some of its momentum to a layer of water just below it, setting this sub-surface water into motion. It too is eventually deflected by the Coriolis force. This domino effect continues to spread deeper, eventually forming the Ekman spiral.

Ekman transport The bulk-water transport accomplished by the Ekman spiral. The transport is about 90° to the right of the surface wind in the Northern Hemisphere and 90° to the left of the surface wind in the Southern Hemisphere.

El Niño An anomalous warming of the topmost layers of the central and eastern tropical Pacific Ocean accompanied by relatively weak trade winds, generally occurring at irregular intervals of three to seven years.

Electrical thermometer A thermometer that relies on the relationship between temperature and electrical resistance or electrical current flow.

Electromagnetic radiation Radiation originating in a varying electromagnetic field, such as visible radiation, x-rays, radio waves, and gamma rays.

Electromagnetic spectrum The complete ensemble of all wavelengths of radiation.

Elevated convection Convection that originates from a relatively warm layer of air above the ground (as opposed to convection that develops in response to the heating of air by the ground).

Enhanced Fujita Scale (EF-Scale) A scale for classifying tornadoes and estimating their maximum wind speed based on the damage they cause, implemented in the United States on February 1, 2007, to replace the Fujita Scale (F-Scale). The EF-Scale has the same basic design as the F-Scale: six categories from zero to five representing increasing degrees of damage.

Ensemble forecasting A forecasting technique that tests the sensitivity of a computer model to changes in the initial conditions. The computer model is run multiple times with slightly different initialized states. If the model produces basically the same forecast each time for a certain day, meteorologists have a higher degree of confidence for that day's forecast.

Entrainment The intrusion of cooler, drier environmental air into a developing cumuliform cloud.

Environmental lapse rate The rate of temperature decrease with altitude, typically measured with a radiosonde. The average value of the environmental lapse rate is about 6.5°C/km (3.6°F/1000 ft).

Equator The line of latitude designated 0° that divides the earth into Northern and Southern Hemispheres.

Equatorial trough The quasi-continuous belt of low pressure that lies between the subtropical high pressure belts of the Northern and Southern Hemispheres.

Equilibrium level (EL) The level above the level of free convection at which the decreasing temperature of a buoyant parcel of air reaches the temperature of the environment. At the equilibrium level, thunderstorm updrafts no longer accelerate upward.

Equilibrium vapor pressure The vapor pressure at equilibrium (saturation), when the condensation and evaporation rates are equal. Alternatively, the vapor pressure required to saturate air of a given temperature.

Equinox A day on which the sun is directly overhead at local noon at the equator. For residents of the Northern Hemisphere, the autumnal equinox occurs on September 22 or 23 and the vernal, or spring, equinox, occurs on March 20 or 21.

Evaporation The process by which a liquid changes into a gas.

Evaporational cooling In the context of the atmosphere, the decrease of air temperature as the result of the evaporation of precipitation.

Extratropical cyclone A cyclone, or low, that derives most of its energy from horizontal temperature gradients.

Extratropical low-pressure system See extratropical cyclone. In the context of advisories issued by the National Hurricane Center, a tropical cyclone becomes extratropical (or post-tropical) when it loses its tropical characteristics. In other words, the low has moved out of the tropics, and its primary energy source is no longer the release of latent heat of condensation.

Eye The distinct, nearly circular cylinder of mostly light winds and clear or partly cloudy skies in the center of a hurricane.

Eye wall A ring of fierce thunderstorms that surrounds the eye of a hurricane, and contains the storm's heaviest rains and strongest winds.

Eye-wall replacement cycle A natural process that occurs in some major hurricanes (Category 3 or greater), characterized by an outer ring of thunderstorms wrapping around the inner eye wall and robbing it of moisture and angular momentum. The hurricane typically weakens during this phase. Once the inner eye wall collapses, a new eye wall with a larger diameter than the original eye wall takes over, and the hurricane can become as strong (or stronger) as it was before the replacement cycle began.

F

Fahrenheit scale A temperature scale on which 32 degrees is the temperature at which ice melts and 212 degrees is the temperature at which water boils (at sea level).

Fermat's Principle A law of optics which states that the path taken by light in traveling from one point to another is such that the time of travel is a minimum when compared with nearby paths.

First Law of Thermodynamics A physical law which governs how the temperature of a gas changes in response to changing its pressure and adding or extracting energy.

Flare echo A spurious radar echo in the form of an extension, or tail, beyond the edge of a severe thunderstorm. Typically such a false radar echo is associated with the presence of large hydrometeors, especially hail, in the thunderstorm.

Flash flood A flood caused by excessive rainfall in a short period of time, generally less than six hours. Also, at times, an ice jam or dam failure can cause a flash flood.

Fog A cloud with its base at the surface of the earth.

Forward-flank downdraft The downdraft produced by falling raindrops, typically found in the northeast quadrant of a supercell thunderstorm.

Fossil fuels Carbon-rich substances such as coal, oil, and natural gas that form from the decomposed remains of organic material.

Freezing rain Rain that freezes after contact with a surface having a temperature below 0°C (32°F).

Frequency The number of waves passing a given point during a specified time period (commonly taken to be one second).

Friction In the context of low-level winds, the force that acts to reduce the speed of the wind by transferring some of its momentum downward toward the ground (where the momentum dissipates). In the lower troposphere, friction also acts to alter the wind direction, forcing the wind to cross local isobars inward toward lower pressure and outward away from higher pressure.

Frictional convergence The piling up of air near the surface as a result of friction. For example, frictional convergence can occur near the shoreline on the leeward side of a large lake because winds over smoother water tend to blow faster than winds over rougher land, creating a "traffic jam" of air at the lakeshore.

Front A boundary between air masses of different densities. The density differences are caused by differences in temperature and/or moisture characteristics.

Frost A covering of ice on exposed surfaces, produced by deposition.

Frost point The temperature to which the air must be cooled (at constant pressure) for frost to form. Most meteorologists still refer to the frost point as the dew point, even though air temperatures are below 0°C (32°F).

Fujita Scale (F-Scale) A scale for classifying tornadoes and estimating their maximum wind speed based on the damage they cause, introduced in 1971 by Dr. T. Theodore Fujita. On February 1, 2007, the F-Scale was replaced by the Enhanced Fujita Scale (EF-Scale) in the United States.

Funnel cloud A whirling, cone-shaped column of air that lowers toward the ground from the base of a severe thunderstorm.

G

Gas law The law that relates the pressure of a gas to its density and its absolute temperature.

General circulation model A computer model designed to be global in scope and used to model and/or predict large-scale and/or long-term changes in global climate.

General circulation The large-scale, time-averaged circulation of air in both the horizontal and vertical, across the globe.

Geostationary satellite A satellite over a fixed point at the equator that orbits the earth at the same angular rate as the earth rotates, and thus remains stationary with respect to that point.

Geostrophic wind The idealized wind that results from a balance between the pressure gradient force and the Coriolis force.

Geostrophy An idealized state of the wind resulting from a balance between the pressure gradient force and the Coriolis force. The resulting wind is called the geostrophic wind.

Glaciation The change of cloud particles from water drops to ice crystals. The term is often specifically used to describe this process in the upper portion of a cumulonimbus cloud.

Global warming An enhancement of the natural greenhouse effect by human activities, resulting in an increase in the average air temperature near the surface.

GOES Geostationary Operational Environmental Satellite.

Gradient The change in the value of a quantity over a given distance.

Graupel Snow crystals completely coated with rime so that they fall as white, opaque, approximately round pellets. Also known as snow pellets.

Greenhouse effect The natural process by which an atmosphere is warmed by absorbing and emitting infrared radiation but allowing shorter-wavelength radiation to pass through.

Greenhouse gas A gas that contributes to the greenhouse effect. In earth's atmosphere, the primary greenhouse gases are water vapor and carbon dioxide.

Grid-point model A computer model in which the model's initialization and forecasts are computed at points on a pre-determined grid that covers the forecast area.

Ground blizzard Conditions with sustained winds or frequent gusts of 35 mph (30 kt) or higher and visibility less than 0.25 mi (400 m) for at least three consecutive hours, caused by snow already on the ground whipped up by the wind.

Ground fog Fog that forms (typically at night) in air that chills while in contact with the cooling ground. Also known as radiation fog.

Gulf Stream A well-defined current of relatively warm water that flows northward along the East Coast of the United States.

Gust front The leading edge of cooler downdraft air spreading out near the surface away from a thunderstorm. Also known as an outflow boundary.

Gustnado A small, typically weak whirlwind that forms near or along a gust front as the result of horizontal wind shear.

H

Hadley Cells The large-scale convection cells, christened by English scientist George Hadley in the 1700s, bounded by rising air near the equator and sinking air in subtropical latitudes.

Hail Onion-like balls or irregularly shaped lumps of ice produced by thunderstorms.

Hail spike See flare echo.

Halo A ring or arc that encircles the sun or moon, caused by the refraction of visible light by ice crystal clouds. More generally, any of a large class of atmospheric optical phenomena that appear as colored or whitish rings, arcs, pillars or bright spots near the sun or moon when seen through ice crystal clouds.

Harmattan winds Hot, dry northeast winds that blow over Africa's Sahara Desert.

Haze Tiny dry or wet suspended particles in the atmosphere that give the sky a milky-white appearance and reduce visibility.

Heat capacity A measure of the ratio of the amount of energy absorbed (or released) by a substance to the corresponding temperature increase (or decrease).

Heat equator For each longitude, there is a location (point) that has the highest mean annual temperature. The heat equator is the line that connects these points. Also known as the thermal equator.

Heat exhaustion An affliction that often results in the lowering of internal body temperature when excessive evaporation of perspiration dissipates body energy.

Heat flux The rate of flow of heat energy per unit time from a given surface such as water or land.

Heat index A modified temperature which attempts to quantify what the air feels like based on the actual air temperature and the water vapor content of the air. Also known as the apparent temperature.

Heat low A shallow area of low pressure created by intense heating (and thus expansion) of low-level air.

Heat stroke A potentially life-threatening condition that can occur on hot and humid days when evaporation of perspiration is suppressed, body cooling systems run at reduced efficiency, and body temperatures soar to dangerous levels.

High-amplitude pattern On an upper-level chart, a height pattern that has large meridional (north-south) undulations associated with a long-wave trough and ridge.

Hook echo A curlicue shape that sometimes appears on radar reflectivity images as precipitation coils around the mesocyclone of a severe thunderstorm.

Horizontal wind shear The change in wind speed and/or wind direction over a specified horizontal distance.

Hot towers Tall towers of cumulonimbus clouds that extend high into the tropical troposphere.

Humidity A general term used to describe the amount of moisture in the air at any given time.

Hurricane warning An advisory issued by the National Hurricane Center when hurricane conditions are expected to affect a region within 36 hours.

Hurricane watch An advisory issued by the National Hurricane Center when hurricane conditions are possible in a region within 48 hours.

Hurricane A low-pressure system of tropical origin that produces sustained winds of at least 74 mph (65 kt).

Hydrologic cycle A model which describes the alternate paths water takes as it wanders from the earth's surface into the atmosphere and back, changing phase along the way.

Hydrometeors A general term used to describe the various types of precipitation particles. Fog droplets, cloud droplets, and ice crystals are also considered hydrometeors.

Hydrostatic equilibrium Atmospheric condition in which the net vertical force on parcels of air is close to zero, with the upward pressure gradient force approximately balanced by the downward pull of gravity.

Hygrometer An instrument that measures the water vapor content of the air. One type is the hair hygrometer, which depends upon the sensitivity of the length of human hair to the relative humidity of the air.

Hygroscopic A term used to describe condensation nuclei onto which water vapor readily condenses.

Hypothermia A condition accompanied by exhaustion and uncontrollable shivering, resulting from a lowering of internal body temperature produced when energy loss from the body substantially exceeds energy production.

I

Ice-crystal process See Bergeron-Findeisen process.

Ice fog A type of fog that forms at very low temperatures when atmospheric water vapor turns directly into tiny ice crystals.

Ice nuclei Tiny particles in the air having a molecular geometry resembling the crystalline structure of ice, that act as nuclei for the formation of ice crystals in the atmosphere.

Ice pellets Tiny, transparent or translucent spherical balls of ice, typically formed by the freezing of partially melted snowflakes or raindrops during their descent through a layer of sub-freezing air. Also known as sleet.

Icelandic low A semi-permanent low-pressure feature of the North Atlantic winter, centered near Iceland.

Ideal Gas Law See Gas Law.

Inferior mirage A refraction phenomenon that makes an object appear to be below its true position. These mirages typically form when a thin layer of air right next to the ground is intensely heated.

Initialization The mathematical scheme used to represent the state of the atmosphere at the time a computer simulation begins.

In-situ observation A measurement by an instrument in contact with the medium that it is sensing.

Intertropical Convergence Zone (ITCZ) A belt of low pressure and low-level convergence of the trade winds from both hemispheres that typically lies over some tropical oceans, often marked by a band of clouds, showers, and thunderstorms.

Intertropical front A term used by African meteorologists to describe the relatively large temperature gradient that develops north of the equator between late spring and early fall. The zone of temperature contrast is produced when cool, moist southerly winds crossing the equator (in concert with the summer monsoon) meet hot, dry winds blowing over the Sahara.

Inversion A layer in the atmosphere in which temperature increases with altitude.

Isallobar A line of equal pressure tendency.

Isobar A line of equal pressure.

Isodrosotherm A line of equal dew point.

Isohyet A line of equal precipitation amount.

Isopleth General name for a line of equal or constant value of a given quantity.

Isoplething The process of drawing isopleths.

Isotach A line of equal wind speed.

Isotherm A line of equal temperature.

Isothermal Of constant temperature. The term is sometimes used to describe a layer of the atmosphere in which temperature remains constant with altitude.

Isovort A line of equal vorticity.

J

Jet streak A core of relatively fast winds embedded within the jet stream.

Jet stream A three-dimensional river of relatively fast-moving air above the ground. Two examples of jet streams are the subtropical jet stream and the mid-latitude jet stream.

K

Katafront A cold front in which air sinks on the cold side of the front. Skies often clear and air pressure rises after the front passes.

Kelvin Scale A temperature scale on which zero corresponds to the temperature at which all molecular motion ceases. A change of 1 Kelvin (1 K) corresponds to a change of 1°C.

Kinetic energy Energy resulting from molecular motion. In general, energy of motion.

Kirchoff's Law A law which states that a body with a high emission of radiation at a particular wavelength also efficiently absorbs radiation at that wavelength.

Knot A unit of speed equal to 1 nautical mile per hour, or 1.15 mph.

L

La Niña Essentially the "flip" side of El Niño, indicating an episode of relatively strong trade winds and anomalous cooling of the topmost layers of the central and eastern tropical Pacific Ocean.

Lake-effect snow Localized snows that fall on the leeward side of a lake or large body of water. These snows typically form in late fall or winter as continental polar air is warmed and moistened as it passes over the relatively warm waters of the Great Lakes.

Land breeze An offshore breeze that usually develops at night in response to the uneven cooling of land and water.

Landfall The point where a tropical cyclone's low-level center of circulation intersects a coastline. Also, the time at which this intersection occurs.

Landspout An informal term describing a typically small and weak tornado not produced by a supercell thunderstorm. Landspouts are continental cousins to waterspouts.

Lapse rate See environmental lapse rate.

Latent heat of condensation The energy released when water vapor condenses; numerically, its value is approximately 600 calories per gram.

Latent heat of evaporation The energy absorbed by water vapor during the evaporation process; numerically, its value is approximately 600 calories per gram.

Latent heat of fusion The energy released when ice freezes; numerically, its value is 80 calories per gram.

Latitude The angular distance (from 0 to 90 degrees) north or south from the equator of a point on the earth's surface, measured on the longitude line passing through the point. For locations north of the equator, "N" is appended to the degree designation; for locations south of the equator, "S" is appended. Alternatively, latitude is considered positive north of the equator and negative south of the equator.

Leeward The downwind (downstream) side.

Lenticular clouds A type of mountain wave cloud that is typically smooth and elliptically shaped, often seen singly or stacked in groups near or in the lee of a mountain ridge.

Level of free convection (LFC) The altitude at which a parcel of air, after being lifted dry adiabatically until saturation and then moist adiabatically thereafter, becomes warmer than its environment.

Lid A stable layer of warm, dry air that caps off convective cloud development in the lower troposphere.

Lifting Condensation Level (LCL) The altitude at which net condensation begins inside a rising parcel of air.

Lightning A large, visible, electrical discharge in the air produced by a thunderstorm.

Line echo wave pattern (LEWP) A connected series of bow echoes that produces a wave-like appearance on radar.

Linear momentum The product of the mass and velocity of an object moving along a straight path.

Liquid-in-glass thermometer A thermometer that depends on the expansion and contraction of a liquid, typically mercury or red-colored alcohol, in a thin opening inside a glass enclosure.

Long wave An atmospheric wave that has a wavelength of many thousands of kilometers. Such waves can be seen by plotting, for example, height contours at the 500-mb or 300-mb level. Also known as a planetary wave.

Longitude The angular distance (from 0 to 180 degrees) east or west of the Prime Meridian of a point on the earth's surface, measured on the latitude line passing through the point. For locations within 180 degrees to the east of the Prime Meridian, "E" is appended to the degree designation; for locations within 180 degrees to the west of the Prime Meridian, "W" is appended.

Low-amplitude pattern On an upper-level chart, a height pattern that is generally zonal, with limited north-south undulations.

Low-level jet stream A fast current of relatively warm, moist air, usually from a southerly direction and found between 500 and 1000 m (1600 and 3200 ft) above ground.

Low-level Somali Jet A channel of low-altitude, speedy southwest winds that typically passes over the offshore waters of Somalia. This low-level jet stream helps to fuel Southeast Asia's summer monsoon by rapidly importing moist air from the Arabian Sea.

M

Macroburst A downburst whose outrush of winds exceeds four kilometers in horizontal scale.

Major hurricane A hurricane that is Category 3 or higher on the Saffir-Simpson scale.

Mammatus clouds Pouch-shaped cloud undulations that sometimes appear at the base of a cumulonimbus cloud.

Mandatory pressure levels The pressure levels at which radiosonde data is always reported. The mandatory levels that forecasters routinely check are 1000 mb, 925 mb, 850 mb, 700 mb, 500 mb, 400 mb, 300 mb, 250 mb, and 200 mb.

Map projection A technique designed to represent the curved surface of the earth (or a portion of the earth) on a flat map. Map projections are fraught with distortions, so meteorologists must choose the one best suited for their particular application.

Maritime polar (mP) air mass A huge volume of relatively cold, moist air.

Maritime tropical (mT) air mass A huge volume of relatively warm, moist air.

Mature stage The second of three stages in the life cycle of a single-cell thunderstorm. In this stage, precipitation has reached the surface, and both strong updrafts and strong downdrafts can exist within the cumulonimbus.

Mechanical convection The production of turbulent eddies by the wind blowing over the earth's surface.

Mercator projection A map projection in which latitude and longitude lines appear as straight lines crossing at right angles. On a Mercator projection, area is increasingly distorted as distance from the equator increases. Mercator projections are often used to plot weather data in the tropics.

Mercury barometer A device for measuring air pressure that works on the principle of balancing the weight of the atmosphere against the weight of a free-standing column of mercury in an evacuated glass tube. The height of this column is a measure of the air pressure. On average, at sea level, a column of mercury approximately 29.9 inches high can be supported.

Meridians Lines of longitude.

Meridional flow An atmospheric circulation pattern in which the north-south component of the wind is pronounced.

Mesocyclone A vertical column of air, typically 5–10 km (3–6 mi) wide, that rotates cyclonically within a supercell thunderstorm.

Mesohigh A small-scale region of shallow surface high pressure created by denser, rain-cooled air from a group of thunderstorms.

Mesoscale A spatial scale of meteorological phenomena ranging from about 2 km to about 1000 km.

Mesoscale convective complex (MCC) A concentric bundle of many individual thunderstorms that are unified by a regional-scale circulation. A typical MCC persists more than six hours and has an area of at least 100,000 km^2.

Mesoscale convective system (MCS) A relatively organized group of thunderstorms that produces a generally solid area of precipitation spanning at least 100 km (about 60 mi) in at least one horizontal direction.

Mesoscale convective vortex (MCV) A mid-level cyclonic circulation that remains after a mesoscale convective complex dissipates. Occasionally, this leftover circulation is discernible on satellite imagery.

Mesovortex In the context of a hurricane, a relatively small, cyclonic swirl of low clouds that sometimes forms in the eye of a mature hurricane.

Meteogram An array of plots of weather observations (or forecasts) of temperature, dew point, pressure, wind, and other variables, over time at a given location. A meteogram typically includes 24 hours of data.

Meteorology The study of atmospheric science.

Microburst Straight-line bursts of concentrated wind produced by a strong localized thunderstorm downdraft.

Microclimate A localized region where weather conditions often differ significantly from those in nearby surrounding areas (caused, for example, by varying terrain or varying surface covering).

Micrometer A unit of length equal to one-millionth of a meter. Also known as a micron.

Microscale The spatial scale of weather phenomena that range in size from a few meters to about two kilometers. A dust devil is an example of a microscale feature.

Mid-latitudes (also middle latitudes)—The general term for the regions between the tropics and the polar regions.

Mid-latitude jet stream A three-dimensional river of relatively fast-moving air, typically several hundred kilometers wide and about a kilometer thick, that blows in a general westerly direction over the mid-latitudes just below the tropopause, at pressure levels near 300-250 mb.

Mid-level African Easterly Jet A relatively narrow band of fast easterly winds blowing across Africa (north of the equator) near the 600-mb level (about 4500 m) between late spring and early fall.

Millibar A unit of pressure. Average sea-level pressure is approximately 1011 millibars (mb).

Mixed layer A layer of the atmosphere near the ground (perhaps 1000 m or so in thickness) in which up-and-down motions have blended faster winds from aloft with slower near-surface winds.

Mixing cloud A cloud that forms when air of differing temperature and humidity properties mixes together. The condensation trail (contrail) left in the wake of jet aircraft is an example of a mixing cloud.

Mock sun See sundog.

Modified persistence A method of temperature forecasting that can be used if a location stays within the same air mass from one day to the next, but sky conditions change. The previous day's high temperature is used as a starting point, and then modified based on the change in cloud cover.

Moist adiabat On a graph of altitude versus temperature, a line that represents the rate of change of temperature of a parcel of saturated air moving vertically.

Moist adiabatic lapse rate The rate at which the temperature of a saturated parcel of air changes as it moves vertically. A representative value is 6°C/km (3.3°F/1000 ft).

Moist air advection The transport of relatively moist air by the wind.

Momentum The mass of an object multiplied by its speed.

Monsoon depression A low-pressure system that forms over the Bay of Bengal and moves northwestward toward India, bringing heavy rains during the summer monsoon.

Monsoon trough A trough of low pressure that sets up over northern India during the region's summer monsoon. More generally, a zone of low-level convergence over tropical latitudes that can provide the seedlings for tropical cyclones.

Monsoon A wind system that reverses its direction seasonally. Typically, one persistent wind regime brings protracted rains (the summer monsoon), while the other brings a period of relative dryness (the winter monsoon).

Mountain wave clouds Lines of clouds that form parallel to a mountain range as air that is close to saturation oscillates upward and downward after it crosses the mountain range.

Multicell thunderstorm A family (group) of thunderstorms in which individual thunderstorms are in various stages of development.

N

Negative buoyancy The tendency to sink as a result of the force that a fluid exerts on an object that is more dense than the fluid.

Net condensation Condition in which condensation exceeds evaporation in a system (such as a parcel of air).

Net evaporation Condition in which evaporation exceeds condensation in a system (such as a parcel of air).

Neutral equilibrium The state that describes an air parcel's tendency to quickly come to a new resting position after being gently nudged upward or downward from its original resting position. For all practical purposes, the air density of the parcel and its immediate environment are equal in this situation.

Nimbostratus Low-based, dark gray, precipitating clouds that typically produce steady precipitation over large areas.

Nocturnal inversion A ground-based atmospheric layer that forms at night in which temperature increases with altitude. Typically, such an inversion forms with clear skies, light winds, and relatively low dew points.

Nocturnal low-level jet A narrow ribbon of fast winds at altitudes generally from 500–1500 meters (approximately 1600–4800 ft) that develops at night, and often serves as a moisture feed for thunderstorm systems.

Nor'easter A strong low-pressure system that intensifies as it moves up the East Coast, typically bringing heavy precipitation. The name comes from the strong northeast winds that develop ahead of these storms.

North-Atlantic Oscillation (NAO) A variation in the latitudinal height gradient over the North Atlantic Ocean. The NAO is typically tracked by comparing the 500-mb heights at a location near Iceland with those at a point to the south near the Azores Islands.

Numerical weather prediction A branch of meteorology that deals with forecasting the weather based on solving mathematical equations using high-speed supercomputers.

O

Occluded front A front that forms when a surface low-pressure area shifts back into the colder air, allowing the cold front to overtake the warm front. An occluded front separates the advancing cold air behind the cold front from the retreating cold air ahead of the warm front.

Occlusion The process of the formation of an occluded front, associated with the life cycle of a mid-latitude low-pressure system according to the Norwegian cyclone model.

Omega block A blocking high-pressure area which, if identified by means of contours on a 500-mb height chart, resembles the Greek letter omega (Ω).

Open-cell convection An array of cumulus cells (clouds) over subtropical oceans characterized by sinking motion at each center and rising motion around the periphery of each cell.

Optics The study of atmospheric phenomena (such as halos or rainbows) produced by the interaction of visible radiation with atmospheric gases, particles, or precipitation.

Orders of magnitude Powers of ten.

Orographic lifting The lifting of air by mountains and elevated terrain.

Outflow boundary See gust front.

Overrunning The process by which relatively warm, moist (less dense) air is forced to glide up and over a wedge of colder, heavier (more dense) air, typically producing layered clouds that uniformly cover a large area.

Overshooting tops Bubbly thunderstorm cloud tops that dent the inversion in the stratosphere.

Ozone A gas whose molecules consist of three oxygen atoms. Most atmospheric ozone is in the stratosphere, where it absorbs the sun's ultraviolet radiation, preventing this harmful radiation from reaching the surface.

Ozone hole A region in the stratosphere over Antarctica and surrounding waters in which the amount of ozone decreases markedly during Southern Hemisphere spring.

Ozone layer The atmospheric layer of maximum ozone concentration, found at altitudes of approximately 15–50 km (9–30 mi).

P

Parcel (of air) An individual volume of air, perhaps hundreds or thousands of cubic meters in volume, that is assumed to act independently of neighboring volumes.

Parhelia See sundog.

Pattern recognition A forecasting technique that hinges on meteorologists recognizing familiar weather sequences from the past. Forecasters use the observed weather that subsequently evolved from a familiar pattern as the basis for their forecast.

Persistence forecasting A method of forecasting that uses weather or climate conditions (for example, sky condition and/or high temperature) from one time period as a forecast for the next time period.

Phasing The merging of troughs in separate branches of the jet stream to form a higher amplitude trough.

Photon A bundle or discrete quantity of electromagnetic energy that travels at the speed of light.

Plan view A top-down view showing two horizontal dimensions.

Planetary wave See long wave.

Polar easterlies Predominant northeast (in the Arctic) or southeast (in the Antarctic) low-level winds that blow at high latitudes, equatorward of Arctic and Antarctic highs.

Polar front A transitory, quasi-continuous boundary found in mid-latitudes that separates warm air over low latitudes and cold air over high latitudes.

Polar regions The region between the Arctic Circle and the North Pole, and between the Antarctic Circle and the South Pole.

Polar stereographic projection A map projection often used by meteorologists to plot weather data over the middle and high latitudes. This map projection sometimes appears as though the observer is suspended above the North (South) Pole looking down on a flat Northern (Southern) Hemisphere. In this case, latitude lines appear as circles.

Polar stratospheric clouds (PSCs) Clouds of ice and frozen compounds of nitrogen that form in the stratosphere over polar regions (primarily the Antarctic).

Polar vortex A belt of stratospheric winds that forms during the Antarctic winter, encircling the continent.

Polar-orbiting satellites (POES) Weather satellites at altitudes near 800 km (500 mi) whose orbits pass close to both Poles.

Positive buoyancy The tendency to rise as a result of the force that a fluid exerts on an object that is less dense than the fluid.

Positive flash A cloud-to-ground lightning stroke that delivers positive charge to the ground.

Potential evaporation An idealized proxy for the rate of evaporation in a given region. It assumes that an area of pure water at the earth's surface has the same temperature as the overlying air.

Precipitable water The amount of liquid water that would fall at a given point if all the moisture in the column of air above the point could be wrung out and forced to fall as rain (it is assumed that this column of air has unit cross sectional area).

Precipitation theory The theory that cloud electrification results from contact between different sized precipitation particles, typically ice crystals and graupel. After collision, graupel acquires a negative charge and the ice crystals acquire a positive charge.

Pre-frontal trough A surface trough in the warm sector ahead of a cold front associated with a mid-latitude low-pressure system. Squall lines sometime develop along pre-frontal troughs.

Pressure See air pressure.

Pressure gradient The rate of change of air pressure per unit distance.

Pressure gradient force The force that arises from differences in air pressure, directed from higher toward lower pressure.

Pressure tendency The rate of change of air pressure over a specified period of time, often chosen to be three hours.

Prevailing wind The wind direction that is dominant, or most frequently observed, at a particular location over a given time period.

Prime meridian The line of longitude designated 0 degrees, passing through the British Royal Observatory in Greenwich, England.

R

Radar Instrumentation used to detect objects and determine how far away they are, using the objects' ability to scatter radiation (commonly microwave or radio wavelengths) back to a receiver.

Radiation Electromagnetic energy that propagates at the speed of light, 300,000,000 m/s (186,000 mi/s).

Radiation fog See ground fog.

Radiational cooling The process by which an object loses energy by emitting radiation. The object actually "cools" only if the amount of energy emitted exceeds the amount absorbed. Thus, the ground undergoes radiational cooling during the day, but because the amount of solar radiation absorbed usually exceeds the amount of infrared radiation emitted, warming usually results.

Radiometer An instrument which detects the intensity of radiation emitted or scattered by an object at specific wavelengths.

Radiosonde A package of weather instrumentation carried aloft by a weather balloon and used to measure weather conditions (typically pressure, temperature, and humidity) above the earth's surface.

Rain shadow A region on the leeward side of a mountain where the amount of precipitation is significantly less than on the windward side.

Rear-flank downdraft A downdraft that forms in the rear (often the southwest side) of a mesocyclone in a supercell thunderstorm.

Rear-inflow jet A speedy ribbon of mid-tropospheric air flowing into the rear of a squall line. Sometimes these fast winds descend to the surface, producing damage.

Reflection The process whereby a surface returns a portion of the radiation incident upon it.

Reflectivity In radar meteorology, a measure of the amount of microwave energy back-scattered by a radar target. Reflectivity depends, in part, on the number, size, and composition of the targets.

Refraction The bending of radiation as it passes from a medium of one density into a medium of another density.

Relative humidity The ratio of the amount of water vapor in a parcel of air to the amount of water vapor that would be needed to saturate the parcel, times 100%. Alternatively, the ratio of the condensation rate to the evaporation rate, times 100%, or the ratio of vapor pressure to equilibrium vapor pressure, times 100%.

Relative vorticity Spin around a local vertical axis, relative to the earth, resulting from horizontal wind shear or curvature.

Remotely sensed observation A measurement by an instrument not in contact with the medium that it is sensing.

Retrogression The westward drift of planetary, or long, waves.

Ridge An elongated area of high pressure (on a constant height surface), or an elongated area of high height (on a constant pressure surface).

Ridge line The axis of a ridge, often indicated by a serrated line on a weather map.

Right-front quadrant The upper-right quadrant of a hurricane, looking down on the eye in the direction of its motion. In the Northern Hemisphere, this quadrant usually contains the strongest winds and produces the largest storm surge at landfall.

Riming The process by which snow crystals become coated with soft shells of ice during descent through an environment of supercooled water droplets.

Rope cloud A strung-out, stretched version of a tornado that often foretells the tornado's demise.

S

Saffir-Simpson scale A hurricane damage potential scale which relates a hurricane's wind speed to the damage it is capable of inflicting.

Santa Ana wind In southern California, a warm, dry east or northeast wind that descends from the elevated desert plateau of the Interior West.

Saturation An equilibrium at which the rate of evaporation equals the rate of condensation.

Saturation vapor pressure See equilibrium vapor pressure.

Scattering The deflection of radiation from its path by small particles in the atmosphere.

Scientific notation A numerical shorthand that uses powers of ten to write numbers, particularly useful for very small or very large numbers.

Sea breeze An onshore breeze that typically develops during the day in response to the uneven heating of land and water.

Sea-breeze front The leading edge of the sea breeze which often behaves as a "mini" cold front, lifting hot, humid air over land, generating lines of clouds and sometimes showers and thunderstorms.

Sea-level pressure Air pressure adjusted to what the pressure would be if the observing station were located at sea level.

Seasonality A general term that refers to the annual range of temperature at a location. One way to quantify seasonality is to take the difference between the highest average monthly temperature and the lowest average monthly temperature.

Self-development A feedback process by which extratropical low-pressure systems intensify. As upper-level heights decrease west of a surface low, upper-level divergence increases. In turn, the surface low intensifies, increasing the pressure gradient. As a result, surface winds accelerate, increasing cold advection, which further decreases upper-level heights. And the process continues.

Severe thunderstorm warning An advisory issued by a local National Weather Service office indicating that a severe thunderstorm is in progress or imminent. The warning may be prompted by a report from an observer or suggested by Doppler radar.

Severe thunderstorm watch An advisory issued by the Storm Prediction Center in Norman, OK, indicating that conditions are favorable for severe thunderstorms. Typically, a severe thunderstorm watch is in effect for six to eight hours and covers an area several tens of thousands of square kilometers in size.

Severe thunderstorm A thunderstorm that produces one or more of the following: (1) Wind gusts at the surface of greater than 50 knots (57.5 mph); (2) Hail of diameter 1.0 inch (2.5 cm) or larger; (3) A tornado.

Severe tropical cyclone The name given to a tropical cyclone in the Southwest Pacific Ocean and the Southeast Indian Ocean with maximum sustained winds of 74 mph (65 kt) or greater.

Shear vorticity Vorticity that results from horizontal wind shear (horizontal differences in wind speed or direction).

Shelf cloud A low, wedge-shaped cloud associated with a thunderstorm's gust front.

Short wave An atmospheric wave that has a wavelength of several hundred to a few thousand kilometers. Such waves can be seen by plotting, for example, height contours at the 500-mb level.

Siberian high A winter high-pressure feature of the Northern Hemisphere general circulation that forms over Siberia.

Sleet See ice pellets.

Smog Air pollution formed in the presence of sunlight, originally coined by combining the words "smoke" and "fog." Ozone is the primary constituent of smog.

Snowboard A flat white board on which weather observers measure snowfall, typically placed in an open area to minimize the effect of blowing and drifting.

Snowflake A general term that includes a single ice crystal or a mass of several ice crystals stuck together.

Snow crystal A single crystal of ice.

Snow flurries A very light and brief shower of snow.

Snow pellets See graupel.

Snow shower A moderate but brief period of snow which can result in a spotty, light accumulation.

Snow squall A heavy, brief burst of snow, commonly occurring in association with lake-effect snow or Arctic cold fronts.

Solar constant The amount of solar radiation received per unit area on a plane above the earth's atmosphere, perpendicular to the direct rays of the sun. Its value is approximately 1361 Watts per square meter.

Solar radiation The total spectrum of electromagnetic radiation emitted by the sun.

Solstice A day on which the sun is directly overhead at local noon at some point above either latitude 23.5°N or 23.5°S. For residents of the Northern Hemisphere, the summer solstice occurs on June 21 or 22 when the sun is directly overhead at local noon at 23.5°N, and the winter solstice occurs on December 21 or 22 when the sun is directly overhead at local noon at 23.5°S.

Sounding A set of data, often displayed in graphical form, describing the vertical structure of atmospheric variables (typically temperature and humidity) at a given time at a given location.

Source regions A region where large volumes of air (air masses) can acquire generally uniform horizontal temperature and moisture characteristics.

Southern Oscillation The reversal in air pressure across the Pacific Ocean that accompanies the waxing and waning of an El Niño.

Spectral model A computer model that describes the present and future states of the atmosphere using mathematical equations whose graphical solutions resemble waves.

Spectrophotometer A sophisticated instrument that measures the intensity of visible light back-scattered at each visible wavelength.

Spiral bands Tentacles of clouds and thunderstorms that pinwheel cyclonically around and toward the center of a hurricane.

Squall line A narrow, often linear band of thunderstorms that develops along the cold front of a mid-latitude low-pressure system or ahead of the cold front in the warm sector.

Stable equilibrium Atmospheric condition in which, if a parcel of air is given a push upward or downward, the parcel tends to return to its original position.

Static electricity An electrical discharge that results from an imbalance of electrical charge.

Station model On a weather map, a compact representation of an observing station's weather data.

Station pressure The air pressure observed at a weather observing station. Compare with sea-level pressure.

Stationary front A front that does not move or that advances at a speed of less than five knots.

Steam fog The mixing cloud that typically forms above a body of relatively warm water on a clear, cold and relatively calm night.

Stefan-Boltzmann constant The constant of proportionality in the Stefan-Boltzmann Law. Its numerical value is 5.6704×10^{-8} Watts per square meter per Kelvin to the fourth power (W/m^2K^4).

Stefan-Boltzmann Law A radiation law which states that the amount of energy per unit area emitted by an object (ideally a perfect absorber and emitter, or "black body") is proportional to the temperature of the object (in Kelvins), raised to the fourth power.

Stepped leader A discharge of electrons that emerges from the base of a thunderstorm and propagates toward the ground in a series of segments that are typically 50–100 m (165–330 ft) in length, initiating a lightning stroke.

Stevenson screen An enclosed shelter, usually painted white, used to house weather instrumentation outdoors.

Storm surge A rise in ocean levels that accompanies the landfall of a hurricane (or other strong storm).

Stratiform clouds A general term for clouds whose primary development is horizontal, as layers or sheets, and whose presence suggest some degree of stability in the atmosphere.

Stratocumulus An extensive sheet of relatively shallow, low clouds whose base has organized patterns of cumuliform (rounded) shapes.

Stratosphere The layer of the atmosphere that spans from the top of the troposphere (the tropopause) to the bottom of the mesosphere (near an altitude of 50 km [about 30 mi]). The vertical temperature profile in the stratosphere is largely isothermal or marked by a temperature inversion.

Stratus clouds Low, gray, layered clouds with rather uniform bases.

Streamlines On a weather map, lines or arrows that are everywhere parallel to the wind.

Subcooled water See supercooled water.

Sublimation The process by which a solid changes directly into a gas.

Subsidence The slow sinking of air, typically associated with areas of high pressure.

Subsidence inversion An inversion formed by the adiabatic warming of a sinking layer of air.

Subsun A bright spot of light that appears directly below the sun, produced by the reflection of visible light off the faces of large hexagonal plate ice crystals. In order to see a subsun, an observer must be above the ice crystals (for example, atop a mountain or in an airplane).

Subtropical cyclone A low-pressure system that has both tropical and extratropical characteristics.

Subtropical highs Semi-permanent high-pressure features of the general circulation in subtropical latitudes.

Subtropical jet stream Relatively fast high-level winds that blow from a general westerly direction at latitudes typically between 20° and 30° in both hemispheres, at or near the 200-mb level.

Subtropics A general term used to refer to the band of latitudes separating the tropics and the mid-latitudes, roughly from 23.5° to 30° latitude in both hemispheres.

Suction vortices Small but violent tornadoes that sometimes develop as whirling offspring inside a strong parent tornado.

Summer monsoon A season marked by prevailing onshore winds and, typically, higher rainfall. The most famous summer monsoon occurs over India and the surrounding Asian subcontinent. With moist air streaming inland from the Indian Ocean, the summer monsoon in that region is characterized by recurrent heavy rains from late spring to early autumn.

Summer solstice In a given hemisphere (Northern or Southern), the time of year when the direct rays of the sun reach the farthest poleward in that hemisphere. The summer solstice is on or about June 21 in the Northern Hemisphere and on or about December 22 in the Southern Hemisphere.

Sun pillar A long streak of light extending above or below the sun, seen only when the sun is near or below the horizon. It is produced by the reflection of visible light off the mirror-like sides of ice crystals shaped like large hexagonal columns.

Sundog A colorful spot that appears on either side of the sun, produced by the refraction of visible light through ice crystals.

Superadiabatic Characterized by a temperature that decreases at a rate greater than the dry adiabatic lapse rate (in other words, greater than 10°C/km).

Supercell A large discrete thunderstorm with a rotating updraft. A supercell is distinguished by its ability to persist for several hours owing to the separation of its updraft and downdraft. Supercells often produce large hail and sometimes tornadoes.

Supercooled water Water at temperatures below 0°C (32°F). Sometimes referred to as subcooled water.

Superior mirage A refraction phenomenon that makes an object appear to be above its true position. These mirages typically form when a layer of very cold air right next to the ground is topped by a warmer layer.

Supertyphoon A typhoon in the Northwest Pacific Basin that has a maximum sustained wind speed of at least 150 mph (130 kt).

Surface observing network A network of weather stations across the United States at which hourly surface observations are routinely recorded. Most of these weather stations are automated and at airports.

Surge region The eastward-arching apex of a bow echo where the strongest winds are usually observed.

Synoptic scale A spatial scale of weather phenomena that range from approximately 1000 kilometers to several thousand kilometers. The high- and low-pressure systems that routinely parade from west to east across the middle latitudes are examples of synoptic-scale features.

T

Teleconnection A linkage (often a statistical correlation) between weather patterns occurring in widely separated regions of the globe.

Temperature inversion See inversion.

Temperature A measure of the average kinetic energy of the molecules of a substance.

Terminal velocity The maximum velocity that a falling object attains.

Terrestrial radiation Electromagnetic radiation emitted by terrestrial, or earth-based, objects.

Thermal A small, rising current of warm air produced by uneven heating of the earth's surface.

Thermal equator See heat equator.

Thermal ridge An elongated area of higher temperatures that often protrudes poleward along or slightly ahead of a cold front associated with a mid-latitude low-pressure system.

Thermistor An electrical thermometer composed of a solid material whose resistance to the flow of electricity changes at a known rate as temperature changes.

Thermograph A thermometer that produces a continuous record of temperature.

Thermometer An instrument for measuring temperature. The most common types are liquid-in-glass, bimetallic, and electrical.

Thickness The vertical separation between two pressure surfaces. Thickness is proportional to the average temperature in the layer. A common thickness that meteorologists use to assess temperature in the lower troposphere is the 1000 to 500 mb thickness.

Thunderstorm A convective shower accompanied by lightning and thunder.

Tornado A column of air that rotates violently beneath a cumuliform cloud and whose circulation touches the ground.

Tornado Alley An informal term for the most tornado-prone region of the world, stretching roughly from eastern and northern Texas across Oklahoma and Kansas into Nebraska.

Tornado vortex signature (TVS) A pattern on Doppler velocity imagery that indicates the likely presence of a tornado. The signature is a large, abrupt shift in wind direction occurring over a small horizontal distance, from high speeds away from the radar to high speeds toward the radar.

Tornado warning An advisory issued by a local National Weather Service office indicating that a tornado has been visually spotted or indicated by radar.

Tornado watch An advisory issued by the Storm Prediction Center in Norman, OK, indicating that conditions are favorable for tornadoes to form within a specific area.

Trace (of precipitation) An amount of precipitation measuring less than 0.01 inches (0.025 cm) of liquid precipitation or less than 0.1 inches (0.25 cm) of solid precipitation.

Trade winds The winds that dominate tropical oceans, blowing away from the subtropical highs toward the Intertropical Convergence Zone. The trade winds are predominantly northeast in the Northern Hemisphere and southeast in the Southern Hemisphere.

Training The process of a series of individual thunderstorms passing over the same location, sometimes leading to flash flooding.

Transmission The process whereby radiation propagates through a substance.

Transpiration The process by which water in plants is released into the atmosphere as water vapor.

Triple point The point at which the occluded front of a mid-latitude cyclone meets the low's warm front and cold front. A new low-pressure system sometimes develops at the triple point.

Tropics That part of the earth's surface between the Tropic of Cancer (approximately 23.5°N) and the Tropic of Capricorn (approximately 23.5°S).

Tropic of Cancer The northernmost latitude line at which the sun is ever directly overhead (approximately 23.5°N).

Tropic of Capricorn The southernmost latitude line at which the sun is ever directly overhead (approximately 23.5°S).

Tropical cyclone The generic, universal term used to describe low-pressure systems that form over warm tropical seas and have an observable cyclonic circulation.

Tropical depression An organized mass of tropical clouds and thunderstorms with a cyclonic circulation and maximum sustained winds of less than 39 mph (34 kt).

Tropical disturbance A loosely organized system of showers and thunderstorms that originates over the tropics or subtropics, has a spatial extent on the order of a few hundred kilometers, and lasts for 24 hours or more.

Tropical storm An organized mass of tropical clouds and thunderstorms with a cyclonic circulation and maximum sustained winds of at least 39 mph (35 kt) but less than 74 mph (65 kt).

Tropical wave See easterly wave.

Tropopause The boundary between the troposphere and the stratosphere, characterized by an abrupt change in lapse rate.

Troposphere The layer of the atmosphere extending from the surface of the earth to the tropopause in which temperature, on average, decreases with altitude.

Trough An elongated area of low pressure (on a constant height surface), or an elongated area of low height (on a constant pressure surface).

Trough line The axis of a trough, often indicated on a weather map with a dashed line.

Typhoon A hurricane that forms in the Northwest Pacific Basin.

U

Ultraviolet radiation Electromagnetic radiation having a wavelength shorter than visible radiation but longer than x-ray radiation.

Unstable equilibrium Atmospheric condition in which, if a parcel of air is given a push upward or downward, the parcel tends to move farther away (vertically) from its original position.

Updraft A rising current of air; the term is typically used to describe the rising air inside a convective cloud, such as a thunderstorm.

Upper-air convergence Net horizontal movement of air into a vertical column at high levels of the atmosphere.

Upper-air divergence Net horizontal movement of air out of a vertical column at high levels of the atmosphere.

Upper-air observing network The network of locations at which radiosondes are routinely launched twice a day, at 00 UTC and 12 UTC. There are about 100 such locations in North America, with several hundred others worldwide.

Upwelling The rising of water (usually cold and often nutrient-rich) toward the surface from the depths of a large body of water.

Urban heat island effect The anthropogenic warming of urban areas.

V

Valley fog A ground fog that forms in a valley.

Vapor pressure That part of the total air pressure exerted by water vapor.

Vapor pressure gradient The difference in vapor pressure measured over a given distance. Qualitatively, the vapor pressure gradient across the thin interface between a warm body of water and overlying dry air is typically large.

Vault The precipitation-free chamber of a supercell thunderstorm that often serves as a depository for tornadoes.

Vector A quantity that has both a magnitude and a direction. For example, wind velocity is a vector.

Velocity couplet A juxtaposition of relatively fast inbound and outbound Doppler velocities that is the Doppler radar signature of a mesocyclone. A velocity couplet sometimes indicates the presence of a tornado.

Vernal equinox The time of year at which the sun passes directly above the equator, marking the beginning of spring. The vernal equinox occurs on or about March 21 in the Northern Hemisphere and on or about September 22 in the Southern Hemisphere.

Vertical wind shear A change in wind speed and /or wind direction with increasing altitude.

Virga Streaks or wisps of precipitation that fall from clouds but evaporate or sublimate before reaching the ground.

Visible radiation Electromagnetic radiation having a wavelength between approximately 0.4 and 0.7 microns.

Vorticity A measure of the spin around a local axis.

W

Wall cloud An apron of rotating cloud that hangs below the base of a cumulonimbus cloud, signaling that a tornado may soon develop.

Warm air advection The transport of warmer air into a region by the wind.

Warm conveyor belt A stream of relatively warm, humid air flowing poleward in the warm sector of a mature mid-latitude low-pressure system that eventually overruns the cold air mass poleward of the low's warm front.

Warm front A boundary between dissimilar air masses at which colder and/or drier air (denser air) is retreating.

Warm seclusion A rare case of the process of occlusion in which a pocket of relatively warm air forms near the low's center, completely surrounded by colder air.

Warm sector The region of relatively warm and/or moist air (less dense air) that lies between the warm front and the cold front of a mid-latitude low-pressure system.

Water vapor The gaseous form of water.

Waterspout A tornado over water.

Wave (or wave cyclone) An area of extratropical low pressure that forms and moves along an existing front.

Wavelength The distance between two crests (or two troughs) of a wave.

Wavenumber The number of waves of a given wavelength needed to encircle the earth at a given latitude.

Wedge ridge In the context of cold-air damming in the eastern United States, a pronounced southward bulge of the isobars on a surface weather map to the east of the Appalachian Mountains, extending away from a high-pressure area whose center is over eastern Canada or the Northeast United States.

Westerlies The prevailing upper-level winds that blow from a general westerly direction over the mid-latitudes.

Wet-bulb zero An adjusted altitude for the melting point of ice (0°C) which takes into account the effect of evaporational cooling on the air. Computing the altitude of the wet-bulb zero is important for hail forecasting.

Wien's Law A radiation law which states that the wavelength at which an object emits its maximum intensity of radiation is inversely proportional to the object's temperature (in Kelvins).

Wind profiler A vertically pointing radar that measures wind speed and direction to a height of approximately 15 km (9 mi) by following turbulent eddies moving with the wind above the radar site.

Wind rose A diagram used to schematically represent the frequency of wind direction (and sometimes wind speed) over some period of time, typically a month, season, or year, at a given location.

Wind shear The rate of change of wind speed or direction over a given distance in the horizontal or vertical. For example, "speed vertical wind shear" refers to the rate of change of wind speed with altitude.

Wind vane An instrument that measures the direction of the wind.

Wind-chill temperature A temperature which attempts to quantify the cooling effect of wind on exposed skin.

Wind-Induced Surface Heat Exchange (WISHE) A positive feedback loop proposed to explain the intensification of tropical cyclones. As wind speed increases, the exchange of water vapor and heat energy from the warm ocean to the atmosphere also increases, boosting the available fuel for a developing tropical cyclone.

Windward The upwind (upstream) side.

Winter monsoon A season marked by prevailing offshore winds and, typically, lower rainfall. Probably the most famous example is the winter monsoon over India and the surrounding Asian subcontinent, which is known for a noticeable lack of rain, especially from late autumn to early spring.

Winter solstice In a given hemisphere (Northern or Southern), the time of year when the direct rays of the sun reach the farthest poleward in the opposite hemisphere. The winter solstice is on or about December 22 in the Northern Hemisphere, and on or about June 21 in the Southern Hemisphere.

Z

Zonal flow An atmospheric circulation pattern that is dominated by winds blowing parallel to latitude circles. In the mid-latitudes, zonal flow refers to winds that blow in a general west-to-east pattern.